Lecture Notes in Networks and Systems 1120

The series "Lecture Notes in Networks and Systems" publishes the latest developments in Networks and Systems—quickly, informally and with high quality. Original research reported in proceedings and post-proceedings represents the core of LNNS.

Volumes published in LNNS embrace all aspects and subfields of, as well as new challenges in, Networks and Systems.

The series contains proceedings and edited volumes in systems and networks, spanning the areas of Cyber-Physical Systems, Autonomous Systems, Sensor Networks, Control Systems, Energy Systems, Automotive Systems, Biological Systems, Vehicular Networking and Connected Vehicles, Aerospace Systems, Automation, Manufacturing, Smart Grids, Nonlinear Systems, Power Systems, Robotics, Social Systems, Economic Systems and other. Of particular value to both the contributors and the readership are the short publication timeframe and the world-wide distribution and exposure which enable both a wide and rapid dissemination of research output.

The series covers the theory, applications, and perspectives on the state of the art and future developments relevant to systems and networks, decision making, control, complex processes and related areas, as embedded in the fields of interdisciplinary and applied sciences, engineering, computer science, physics, economics, social, and life sciences, as well as the paradigms and methodologies behind them.

Indexed by SCOPUS, EI Compendex, INSPEC, WTI Frankfurt eG, zbMATH, SCImago.

All books published in the series are submitted for consideration in Web of Science.

For proposals from Asia please contact Aninda Bose (aninda.bose@springer.com).

Radek Silhavy · Petr Silhavy
Editors

Artificial Intelligence Algorithm Design for Systems

Proceedings of 13th Computer Science Online Conference 2024, Volume 3

Springer

Editors
Radek Silhavy
Faculty of Applied Informatics
Tomas Bata University in Zlín
Zlín, Czech Republic

Petr Silhavy
Faculty of Applied Informatics
Tomas Bata University in Zlín
Zlín, Czech Republic

ISSN 2367-3370 ISSN 2367-3389 (electronic)
Lecture Notes in Networks and Systems
ISBN 978-3-031-70517-5 ISBN 978-3-031-70518-2 (eBook)
https://doi.org/10.1007/978-3-031-70518-2

This Springer imprint is published by the registered company Springer Nature Switzerland AG
The registered company address is: Gewerbestrasse 11, 6330 Cham, Switzerland

Preface

We are pleased to present the Computer Science Online Conference (CSOC) proceedings. This volume, titled "Artificial Intelligence Applications in Networks and Systems," brings together a collection of pioneering research and innovative advancements from experts, scholars, and professionals worldwide.

In a time when digital technologies impact every aspect of our lives, the importance of artificial intelligence (AI) in shaping the future of networks and systems cannot be overstated. The papers and articles in this volume offer deep explorations of cutting-edge methods, practices, and tools that drive progress in this dynamic field. The wide range of topics covered here highlights the vibrant and evolving nature of AI applications in networks and systems, from novel AI techniques to optimising network systems.

The CSOC conference has long been a hub of intellectual exchange, fostering collaboration and discussion that challenge the boundaries of computational methods. This year's conference has been no exception, providing a platform for a remarkable gathering of thought leaders dedicated to advancing AI for networks and system infrastructures.

We thank all the authors for their valuable contributions, the reviewers for their careful evaluations, and the organising committee members for their unwavering dedication. Their combined efforts have resulted in a collection of high-quality papers and laid a foundation for future research and development in the field.

To our readers, we hope this volume serves as both an inspiration and a resource. Whether you are an experienced professional, an emerging researcher, or a curious enthusiast, the insights and knowledge in these pages will enrich your understanding and spark your passion for AI applications in networks and systems.

Thank you for joining us on this journey. We look forward to the continued growth and evolution of the CSOC community and to witnessing future innovations that will undoubtedly arise from our collective efforts.

April 2024

Radek Silhavy
Petr Silhavy

Organisation

Program Committee

Program Committee Chairs

Petr Silhavy	Tomas Bata University in Zlin, Faculty of Applied Informatics
Radek Silhavy	Tomas Bata University in Zlin, Faculty of Applied Informatics
Zdenka Prokopova	Tomas Bata University in Zlin, Faculty of Applied Informatics
Roman Senkerik	Tomas Bata University in Zlin, Faculty of Applied Informatics
Roman Prokop	Tomas Bata University in Zlin, Faculty of Applied Informatics
Viacheslav Zelentsov	Doctor of Engineering Sciences, Chief Researcher of St.Petersburg Institute for Informatics and Automation of Russian Academy of Sciences (SPIIRAS)
Roman Tsarev	Department of Information Technology, International Academy of Science and Technologies, Moscow, Russia
Stefano Cirillo	Department of Computer Science, University of Salerno, Fisciano (SA), Italy

Program Committee Members

Juraj Dudak	Faculty of Materials Science and Technology in Trnava, Slovak University of Technology, Bratislava, Slovak Republic
Gabriel Gaspar	Research Centre, University of Zilina, Zilina, Slovak Republic
Boguslaw Cyganek	Department of Computer Science, University of Science and Technology, Krakow, Poland
Krzysztof Okarma	Faculty of Electrical Engineering, West Pomeranian University of Technology, Szczecin, Poland

Maciej Majewski	Department of Engineering of Technical and Informatic Systems, Koszalin University of Technology, Koszalin, Poland
Monika Bakosova	Institute of Information Engineering, Automation and Mathematics, Slovak University of Technology, Bratislava, Slovak Republic
Pavel Vaclavek	Faculty of Electrical Engineering and Communication, Brno University of Technology, Brno, Czech Republic
Miroslaw Ochodek	Faculty of Computing, Poznan University of Technology, Poznan, Poland
Olga Brovkina	Global Change Research Centre Academy of Science of the Czech Republic, Brno, Czech Republic & Mendel University of Brno, Czech Republic
Elarbi Badid	College of Information Technology, United Arab Emirates University, Al Ain, United Arab Emirates
Luis Alberto Morales Rosales	Head of the Master Program in Computer Science, Superior Technological Institute of Misantla, Mexico
Mariana Lobato Baes	Research-Professor, Superior Technological of Libres, Mexico
Abdessattar Chaâri	Laboratory of Sciences and Techniques of Automatic control & Computer Engineering, University of Sfax, Tunisian Republic
Gopal Sakarkar	Shri. Ramdeobaba College of Engineering and Management, Republic of India
V. V. Krishna Maddinala	GD Rungta College of Engineering & Technology, Republic of India
Anand N Khobragade	Maharashtra Remote Sensing Applications Centre, Republic of India
Abdallah Handoura	Computer and Communication Laboratory, Telecom Bretagne – France
Almaz Mobil Mehdiyeva	Department of Electronics and Automation, Azerbaijan State Oil and Industry University, Azerbaijan

Technical Program Committee Members

Ivo Bukovsky, Czech Republic
Maciej Majewski, Poland
Miroslaw Ochodek, Poland

Organising Committee Chair

Radek Silhavy, Tomas Bata University in Zlin, Faculty of Applied Informatics, email: csoc@openpublish.eu

Conference Organiser (Production)

Silhavy s.r.o.
Web: https://csoc.openpublish.eu
Email: csoc@openpublish.eu

Conference website, Call for Papers

https://csoc.openpublish.eu

Contents

Predicting Traffic Patterns in Cloud Computing Systems to Optimize Resource Usage in Organization

Sello Prince Sekwatlakwatla(✉) and Vusumuzi Malele

Department of Computer Science and Information Systems, North-West University, Potchefstroom, South Africa
Sek.prince@gmail.com, Vusi.Malele@nwu.ac.za

Abstract. One of the most vital instruments for the company to control traffic and distribute system resources effectively is the short-term traffic flow forecast. To incorporate breakthroughs like artificial intelligence and machine learning and monitor traffic, cloud computing requires focused processing resources and prediction tools. This paper proposes three traffic flow prediction tools: Gaussian process regression, linear support vector machine (SVM), and stepwise linear regression. The findings demonstrate that stepwise linear regression outperforms the models in traffic flow prediction. In the future, we propose to explore the use of a combination of machine learning models with hourly monitoring and resource allocation.

Keywords: Stepwise Linear regression · Linear Support Vector Machine (SVM) · Gaussian Process Regression

1 Introduction

Infrastructure providers are pursuing dynamic provisioning and resource restructuring depending on anticipated user demand in order to maximize both the user experience and operating costs [1]. Conventional systems may find it challenging to manage significant increases in traffic brought on by increased data needs. Data traffic has surged as a result of increased Internet and mobile application use, and cloud computing services are being adopted quickly [1–3]. Computing resources are made available as services over the Internet on demand through cloud computing [2].

Businesses are no longer required to acquire, set up, or oversee resources. They just pay for the actual amount used. Our methods for managing, accessing, and storing data and applications have evolved as a result of cloud computing. Cloud computing offers a financially viable and adaptable substitute for conventional IT infrastructure, freeing up individuals and enterprises to concentrate on their primary activities [2, 3]. For example, the automotive sector can concentrate on producing and marketing automobiles, while cloud providers handle the administration of computing resources.

R. Silhavy and P. Silhavy (Eds.): CSOC 2024, LNNS 1120, pp. 1–13, 2024.
https://doi.org/10.1007/978-3-031-70518-2_1

With the development of advanced technologies such as machine learning, artificial intelligence, and the Internet of Things, cloud computing will continue to play a key role in shaping the future of the IT industry.

There are a lot of benefits to cloud services. Examples of these benefits are.

(a) Cloud services are faster to market; consumers may launch their apps more quickly thanks to cloud development [5].
(b) Data security: Networked backups prevent data loss in the event of hardware breakdowns [3].
(c) Savings on equipment: By utilizing distant resources, cloud computing helps businesses cut costs associated with purchasing servers and other hardware. Owing to its benefits, the majority of businesses now use cloud services. Consequently, there is a severe issue with resource management and traffic flow within businesses that cannot be disregarded. To collect data on traffic that may be utilized for later decision-making and resource management, cloud network administrators need continuously monitor and analyze these networks [4]. Absence of these resource allocation and prediction tools results in low productivity, unhappy users, and inaccessible applications. Cloud computing requires focused computing resources and predictive tools to monitor traffic [1]. The existence of such problems has meant that the search for an accurate and stable model for predicting traffic and resource allocation has attracted the attention of many researchers and poses a challenge [4, 5].

The significant contributions made by this paper are as follows:

- The benefits and drawbacks of literary traffic flow forecast tools
- Three prediction techniques are proposed, namely stepwise Linear regression, linear support vector machine (SVM), and Gaussian process regression, as prediction tools for network traffic flow.
- To assess each tool's accuracy, evaluation procedures were carried out for every method.
- Comparison of the results of each technique.

In summary, this study is organized as follows: an introduction, related work, methodology, evaluation, and conclusions for future work.

2 Related Work

A critical part of the Information and Communication Technology (ICT) sector is cloud computing traffic prediction. Organizations can meet their service level agreements (SLAs) and operations level agreements (OLAs) by accurately predicting current and future traffic levels [7]. The ability to predict system traffic in cloud computing is the result of the service provider estimating a day's worth of requests and the total amount of data that will be moved [7, 8].

Since an application will need to be moved repeatedly in the future, most of the information and applications should be moved daily [8]. In numerous studies [3–5], Activity Stream's predictions were validated and met 85% of expectations. However, there has been no progress in implementing a system that predicts traffic using an application framework [5]. With cloud computing being a continuous process, it is very difficult to make a 100% accurate prediction [9].

There is a current issue in cloud computing frameworks in which a lot of information moves to the cloud and can cause system bottlenecks rather than CPU bottlenecks or storage bottlenecks [9] an example of these exceptional frameworks is Hadoop and Apache Spark. Advanced administration or scientific computation, such as PC programs, runs on many machines and resides in both open clouds and business or extended data centers [10]. In this regard, Pros and cons of traffic flow prediction tools are investigate in the existing literature to identify the techniques and technologies used to predict traffic flow was conducted. Table 1 indicate the pros and cons of traffic prediction tools.

Table 1. Pros and cons of traffic flow prediction tools

Method	Pros	Cons
Spatio-temporal Shared GRU (STSGRU) [1]	Perform better than expected in terms of prediction	The approach ignores all other aspects and just concentrates on the traffic flow input. Inadequate data
multivariate linear regression variable parameter spatio temporal (MLR-VPST) zoning model [5]	more accurate and improve production quality	The forecast error is within the process requirements' permitted range
Physics-based multiple linear regression (PB-MLR) model [6]	The model is easily scalable in real-time and has a great deal of potential for improved performance	processing with insufficient training samples
Auto Regressive Integrated Moving Average (ARIMA) models [9]	Enhances the accuracy of forecasting and boosts computational proficiency	The greatest overlap occurs between successive subseries, model orders, and time-dependent weights
support vector machines (SVM) and artificial neural networks (ANN) [8]	Accurate prediction compared to other models	Unpredictable nature makes it difficult to predict
Multiple Linear Regression Unit-Multi Region Correlation (MRC-MLRU) model [4]	Improve the accuracy of traffic flow predictions	More useful for predicting traffic flow in situations when there is little traffic data
hybrid particle swarm optimization support vector machine algorithm (PSOSVM) [7]	Improved the accuracy of SVM up to 98.13%	It is difficult to fully explore and explcit correlation mechanisms between traffic parameters

(continued)

Table 1. *(continued)*

Method	Pros	Cons
Bidirectional Gated-Recurrent Unit (BiGRU), Discrete Wavelet Transformation (DWT) [2]	Prediction accuracy is increased from 3% to 56%	Nonlinear data with random oscillations cannot be learned using basic procedures
Artificial Neural Network (ANN) [3]	Accurate prediction	Low variance can result from very little mistakes and tiny neurons, whereas overfitting can be caused by an excessive number of neurons

3 Methodology

This paper describes the technique of predicting traffic with regression models using various libraries such as stepwise linear regression, linear support vector machine (SVM), and Gaussian process regression as prediction tools.

A. Stepwise Linear Regression

Stepwise regression is a subset of hierarchical regression in which statistical methods select which predictors are included in the model [9]. Stepwise regression, forward selection, and backward elimination are the three fundamental variations of this strategy. Forward selection involves starting the model with no predictors and adding significant predictors piecemeal until a statistical stopping condition is met. When using backward elimination, the model first includes every potential predictor and then eliminates nonsignificant predictors one after another until the stopping condition is satisfied [9]. These two strategies are used in the stepwise regression method, which adds and subtracts predictors as the model is constructed.

$$Cj.std = bj\left(\frac{fx_j}{fy}\right) \quad (1)$$

where fy and fx_j are the dependent variable's standard deviations as well as the associated Cj independent variable.

Pros of the Stepwise Linear regression.

i. The capacity to manage vast lists of prospective predictor variables and fine-tune the model to identify the best predictor variables from among those available [9]. ii. Compared to previous automatic model-selection techniques, it is quicker [9]. iii. Observing the sequence in which variables are deleted or introduced might offer useful information about the predictor variables' quality [9].

Cons of the Stepwise Linear Regression.

I. data overfitting and misleading estimates [10]. II. If two predictor variables are highly associated in the model, only one of them can be included [10]. III. The confidence intervals and predicted values are too narrow [10].

B. Gaussian Process Regression

A Gaussian process is a collection of random variables that all have the same Gaussian distribution. A graphical representation of the Gaussian process is seen in Fig. 1.

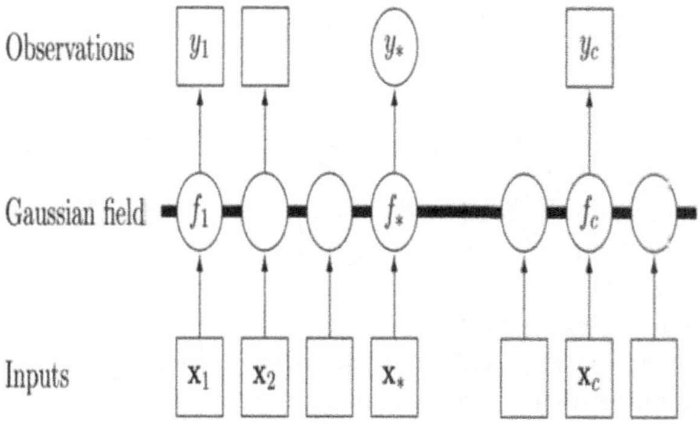

Fig. 1. Gaussian process graphical model [1].

Figure 1 the inputs are represented by $y1$ in the preceding figure, while the observations are represented by xi. The Gaussian field contains the functions fi. Every prediction $y*$ depends on every other input and observation when posterior inference is performed since f1 behave as random variables and are integrated out [11].

$$K(x) = \sum_{i=1}^{n} \beta_i \mathfrak{f}_i(x) + T(x) \tag{2}$$

When K is the answer that we are looking for, T (x) represents a Gaussian process, β is an unknown, and $\mathfrak{f}$irepresents known functions. The equation may be used to estimate β if there are n sample locations (x1, x2… xm) and corresponding sample results (y1, y2… ym). We may thus make advantage of it.

$$K(X) = \sum_{i=1}^{n} \beta_i f_i(X) \tag{3}$$

Then predict response at sites x.

Pros of the Gaussian Process Regression.

I. the Gaussian process (GP) incorporates model uncertainty direcly [12]. II. Forecast value distribution rather than a single value as the forecast [12]. III. By selecting alternative kernel functions, the model can contribute previous information and parameters about the model's structure [12].

Cons of the Gaussian Process Regression.

I. It is difficult to estimate aleatory uncertainty properly when data are scarce, such as in the middle of a data set [13]. II. The sliding window technique is one way to estimation, but it is difficult to set the window size for each dimension of the input [13].

B. Linear Support Vector Machine (SVM)

Linear Support Vector Machines (SVM) are a type of supervised learning algorithm that is used for classification, regression, and outlier identification. Is memory efficient because it makes use of support vectors, a subset of training points in the decision function [11].

$$H(c_i) = \text{sign}\left(\sum_{j=1}^{s} \propto_j \, p_j k(c_j, x_i) + b\right) \tag{4}$$

$$H\left(v, v^1\right) = \exp\left(\frac{\|v - v^1\|^2}{2y^2}\right) \tag{5}$$

The vector of values to be predicted is denoted by ci. So-called support vectors, or c_ (j), are a subset of the training set of data. The class (−1 or +1) for each piece of data p_j. There is a constant $c_{j,}$, for every each p_j. The constant b consists of a single number. Support vector count is represented by the letter s. A measure of similarity between two vectors, with 1.0 denoting identical and 0.0 denoting maximum difference, is returned by the kernel function H, which takes two vectors as inputs.

Pros of the Linear Support Vector Machine.

i. SVM classifiers thrive in high-dimensional space and have high accuracy. Because SVM classifiers only use a fraction of the training data, they use less memory [7]. ii. When there is a wide gap between classes, SVM works pretty well [7]. iii. SVM works better in high-dimensional spaces [7].

Cons of the Linear Support Vector Machine.

i. SVM does not perform well when classes in the data are not adequately separated, implying that overlapping classes exist [14]. ii. SVM on big data sets requires more time to train [14]. iii. Because SVM is not a probabilistic model, we cannot explain the classification in terms of probability [14].

4 Traffic Flow Prediction System Flow

Figure 2 shows a generic framework for traffic flow prediction, which comprises mainly of five stages: data collection, pre-processing, model structure for processing, and traffic flow predictive analysis, followed by validation and error evaluation.

5 Data Set

The volume of traffic in cloud computing is growing daily; uncoordinated traffic signal control, a lack of real-time data, and the system's continuous availability are among the crucial components that cannot be disregarded. These days, traffic congestion has a major influence. In this study, data were gathered from the online application system between 2017 and 2022 [15], where assessments of the daily data for page loads, unique visits, first-time visits, and repeat visits were performed in order to evaluate the regression models utilizing various libraries with a comprehensive set of data. MATLAB was used to carry out the experiment and display the traffic forecast findings (Fig. 3).

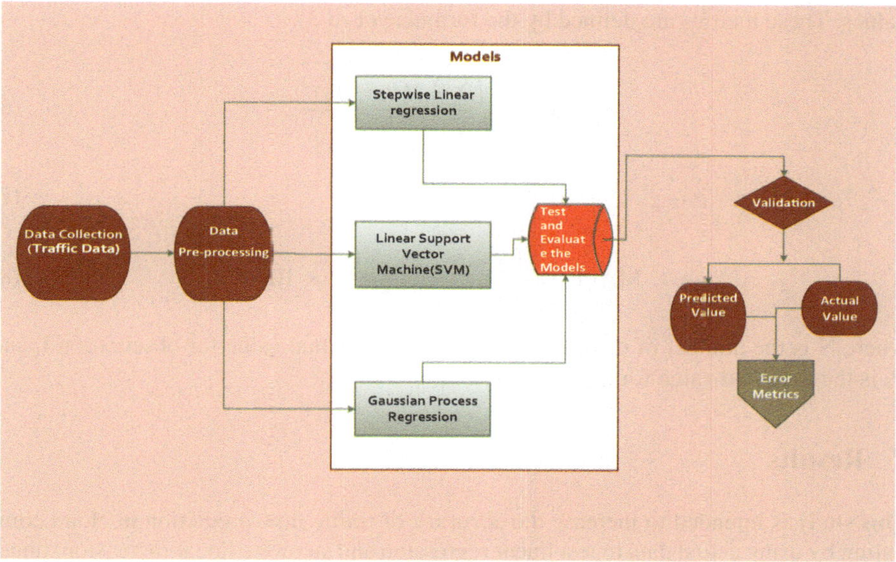

Fig. 2. Scheme for the prediction of traffic flow.

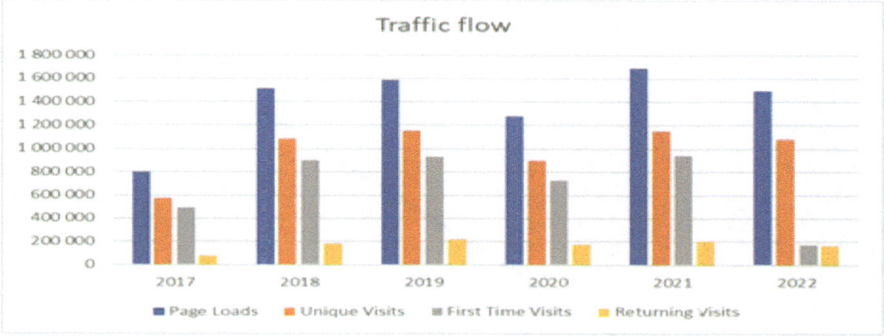

Fig. 3. Yearly Page loads and Users data analysis.

6 Evaluation Measures

Errors between actual and predicted attributes are expressed numerically using evaluation measures. The model's performance is demonstrated by the difference between the predicted and true attributes. In order to assess the matched network's performance, a test data set is fed into the system [16]. The correlation coefficient (R2), mean average error (MAE), and root mean square error (RMSE) between the actual and forecast flow series are the performance metrics used to compare the anticipated values with the actual values. These metrics are defined by the formulae (1 to 3).

$$RMSE = \sqrt{\frac{\sum_{i=1}^{N}(P_i - V_i)^2}{N}} \tag{6}$$

$$\mathrm{MAE} = \frac{\sum_{i=1}^{N}(\mathrm{P_I} - \mathrm{V_i})}{\mathrm{N}} \tag{7}$$

$$\mathrm{MAPE} = \frac{1}{\mathrm{N}}\sum_{i=1}^{N}\frac{|\mathrm{P_i} - \mathrm{V_I}|}{\mathrm{V_I}} \times 100 \tag{8}$$

where N is the number of observations, Pi, V_i is the actual value for observation I, and P_i is the expected value for observation.

7 Results

This study is intended to increase the accuracy of traffic flow prediction in cloud computing by using actual data to test linear regression and stepwise linear regression, linear support vector machines (SVM), and Gaussian process regression model.

A). Linear regression and stepwise linear regression

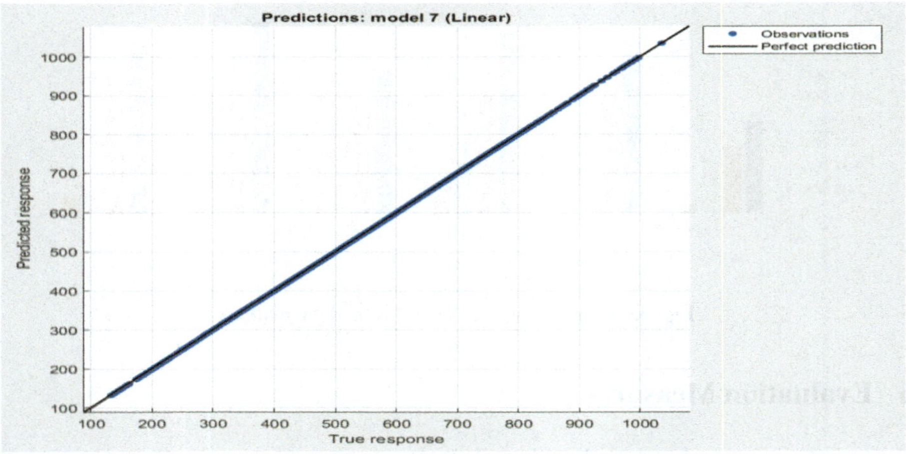

Fig. 4. Linear regression and stepwise linear regression

Figure 4 shows the results of linear regression and stepwise linear regression. The method's training time was 19.38 s, and the model's overall performance outperformed other strategies. Enhanced prediction accuracy with 3.79 MAE and 2.61 improvements in MSE validation (see Table 2).

Table 2. Linear regression and stepwise linear regression

Linear Regression and Stepwise Linear Regression	Trained Results
R-Squared (Validation)	1.00
RMSE(Validation)	5.11
MAE (Validation)	3.79
MSE(Validation)	2.61
Prediction Speed	~7400 obs/s
Training time	19.38 s
Model size	~27 kb

B). Linear support vector machine (SVM)

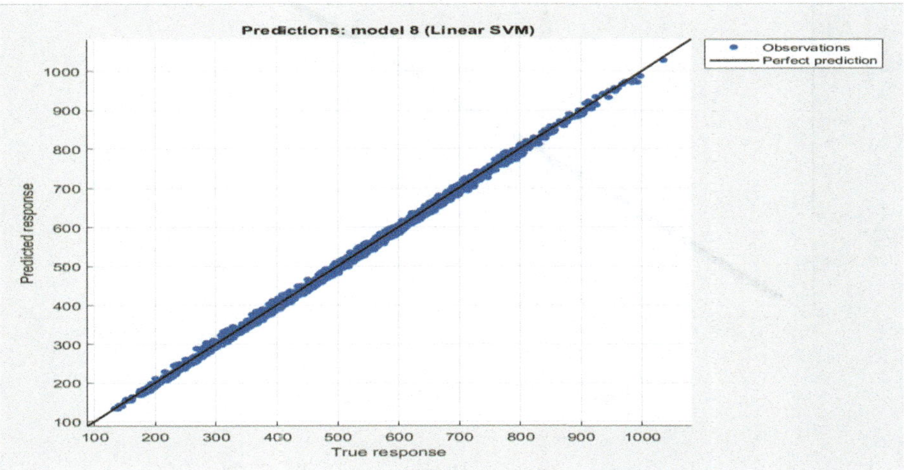

Fig. 5. Linear Support Vector Machine (SVM)

The findings of the linear support vector machine (SVM) are shown in Fig. 5. The approach required more than 6.027 s to train than the prior model. The MSE for error evaluation increased to 48.4, whereas MAE validation decreased to 5.56 (refer to Table 3). In comparison to stepwise linear regression and linear regression, there is also a 74% increase in RMSE validation.

Table 3. Linear support vector machine (SVM)

Linear Regression and Stepwise Linear Regression	Trained Results
R-Squared (Validation)	1.00
MSE(Validation)	48.4
MAE (Validation)	5.56
Prediction Speed	~21000 obs/s
Training time	6.027 s
RMSE(Validation)	6.96
Model size	~5 kb

C). Gaussian process regression

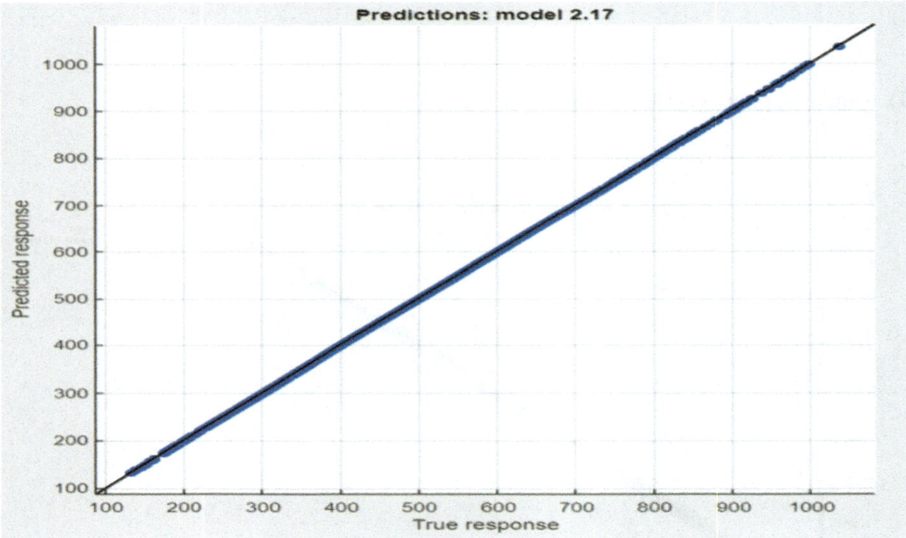

Fig. 6. Gaussian Process Regression

Figure 6 shows the Gaussian Process Regression findings. The technique required more training time more than 309.43 s than the previous two models, but the prediction accuracy increased with the minimum in error evaluation, where a 0.05 error for MAE was reported, and the validation of MSE also decreased to 1.37 (see Table 4). In comparison to linear regression, stepwise linear regression, and linear support vector machines (SVM), there is also a drop in RMSE validation to 0.01.

Table 4. Gaussian process regression

Linear Regression and Stepwise Linear Regression	Trained Results
R-Squared (Validation)	1.00
MAE (Validation)	0.05
RMSE(Validation)	0.01
Prediction Speed	309.43 s
Training time	~14000 obs/s
MSE(Validation)	1.37
Model size	~186 kb

8 Conclusions

In this paper, we contend that the ability to predict future traffic parameter values improves traffic management system performance. Stepwise linear regression, linear support vector machines (SVM), and Gaussian process regression were the techniques that were suggested. For MSE, the evaluation error of the linear support vector machine (SVM) was 48.4, and the validation of MAE likewise decreased to 5.56. Prediction accuracy for Gaussian process regression rises with the lowest error evaluation, where a reported 0.05 error was found. The prediction accuracy increased to 3.79 MAE using linear regression and stepwise linear regression, while the MSE validation increased to 2.61. Despite MAE and MSE validation, Gaussian Process Regression outperformed the other models and improved prediction accuracy with the least amount of error evaluation.

Acknowledgment. This paper is a component of the Computer and Information Science with Information Technology PhD program, registered at Northwest University's Vanderbijlpark Park Campus in South Africa's Department of Computer Science and Information Systems.

Author Contributions. Following the conceptualization of the approach, data collection, experimental platform preparation, and Matlab simulations, Prince Sello Sekwatlakwatla created an original draft. Oversight, editing, and reviewing are done by Vusumuzi Malele. Every one of the contributing authors approved the finished version.

Conflict of Interest. None of the writers disclosed that they were involved in any conflicts of interest.

References

1. Song, C., Zhou, S., Chang, W., Xiao, Y., Fu, Y., Yang, L.: A short-term demand of bike-sharing forecasting model based on spatio-temporal graph data. In: 28th International Conference on Automation and Computing (ICAC), Birmingham, United Kingdom, 2023, pp. 1–5. https://doi.org/10.1109/ICAC57885.2023.10275167

2. Mohiuddin, E., Matin, A.: Multilevel Categorization of Bengali News Headlines using Bidirectional Gated Recurrent Unit. In International Conference on Automation, Control and Mechatronics for Industry 4.0 (ACMI), Rajshahi, Bangladesh, 2021, pp. 1–6. https://doi.org/10.1109/ACMI53878.2021.9528006
3. Afandi, A., Lusi, N., Catrawedarma, I., Subono, B.R.: Prediction of temperature in 2 meters temperature probe survey in Blawan geothermal field using artificial neural network (ANN) method. Case Stud. Thermal Eng. **38** (2022). https://doi.org/10.1016/j.csite.2022.102309
4. Ivanovski, T., Zhang, G., Jemrić, T., Gulić, M., Matetić, M.: Fruit firmness prediction using multiple linear regression. In: 43rd International Convention on Information, Communication and Electronic Technology (MIPRO), Opatija, Croatia, 2020, pp. 1306–1311. https://doi.org/10.23919/MIPRO48935.2020.9245213
5. Chakraborty, A., Kumar, M., Chaurasia, N.: Secure framework for IoT applications using deep learning in fog computing. J. Inform. Secur. Appl. (2023). https://doi.org/10.1016/j.jisa.2023.103569
6. Qasaimeh, Q., Liu, J., Qasaimeh, A., Evans, J., Hamasha, S.: Predicting the life of the solder joints in electronic assemblies using physics-informed data-driven methodology. In: 22nd IEEE Intersociety Conference on Thermal and Thermomechanical Phe-nomena in Electronic Systems (ITherm), Orlando, FL, USA, 2023, pp. 1–7. https://doi.org/10.1109/ITherm55368.2023.10177645
7. Qingfeng, B., Zhang, S., Guo, J., Zhiqiang, L., Zhang, Z.: Multivariate linear-regression variable parameter spatio-temporal zoning model for temperature prediction in steel rolling reheating furnace. J. Process Control. **123**, 108–122 (2023). https://doi.org/10.1016/j.jprocont.2023.01.013
8. Chen, S., Xiaoqing, Z., Zhou, G., Fan, C., Ding, P., Chen, Q.: An online physical-based multiple linear regression model for building's hourly cooling load prediction. Energy Build. **254** (2022). https://doi.org/10.1016/j.enbuild.2021.111574
9. Gupta, A., Kumar, A.: Mid term daily load forecasting using ARIMA, Wavelet-ARIMA and machine learning. In: IEEE International Conference on Environment and Electrical Engineering and 2020 IEEE Industrial and Commercial Power Systems Europe (EEEIC / I&CPS Europe), Madrid, Spain, 2020, pp. 1–5. https://doi.org/10.1109/EEEIC/ICPSEurope49358.2020.9160563
10. Guoji, X., Huan, W., Jinsheng, W., Xuebin, C., Bing, Z.: A local weighted linear regression (LWLR) ensemble of surrogate models based on stacking strategy: application to hydrodynamic response prediction for submerged floating tunnel (SFT). Appl. Ocean Res. **125** (2022). https://doi.org/10.1016/j.apor.2022.103228
11. Liao, Z., Dai, S., Kuosmanen, T.: Convex support vector regression. Eur. J. Oper. Res. (2023). https://doi.org/10.1016/j.ejor.2023.05.009
12. Al Shedivat, M., Wilson, A.G., Saatchi, Y., Zhiting, H., Xing, E.P.: Learning scalable deep kernels with recurrent structure. J. Mach. Learn. Res. **18**, 1–37 (2017). https://doi.org/10.48550/arXiv.1610.08936
13. Kuczmarski RJ, Ogden CL, Grummer-Strawn LM, Flegal KM, Guo SS, Wei R, et al. CDC growth charts: United States. Advance data from vital and health statistics; 314. Hyattsville, Maryland: National Center for Health Statistics; 2000
14. Wang, H., Genghui, L., Zhenkun, W.: Fast SVM classifier for large-scale classification problems. Inform. Sci. **642** (2023). https://doi.org/10.1016/j.ins.2023.119136
15. Krishna, S.: Forbes Innovation, why data management needs an aggregator model. https://www.forbes.com/sites/forbestechcouncil/?sh=2786406a649b,last accessed 2023/04/21
16. Santhusitha, D., Karunasingha, K.: Root mean square error or mean absolute error? Use their ratio as well. Inform. Sci. **585**, 609–629 (2022). https://doi.org/10.1016/j.ins.2021.11.036

17. Xu, L., Liu, H., Song, J., Li, R., Hu, Y., Zhou., X, Patras, P.: TransMUSE: transferable traffic prediction in multiservice edge networks. Comput. Netw. **221** (2023). https://doi.org/10.1016/j.comnet.2022.109518
18. Mohamed, A., Hamdan, M., Suleman, K., Abdelaziz, A., Sharief, F.B., Muhammad, I., Marsono, M.N.: Software-defined networks for resource allocation in cloud computing: a survey. Comput. Netw. **195** (2021). https://doi.org/10.1016/j.comnet.2021.108151
19. Luo, J., Gong, J.: Air pollutant prediction based on ARIMA-WOA-LSTM model. J. Atmos. Pollut. Res. (2023). https //doi.org/10.1016/j.apr.2023.101761

Perceived Risk and Behavioral Intention of Customers Using Bayesian Technique

Dam Tri Cuong(✉)

Industrial University of Ho Chi Minh City, Ho Chi Minh City, Vietnam
damtricuong@iuh.edu.vn

Abstract. In recent decades, the Internet has seen significant development, transforming into an expansive worldwide marketplace facilitating the exchange of various commodities and services. Nevertheless, an increasing number of individuals engaging in online transactions has resulted in a multitude of concerns and challenges. Online customers face a wider range of risks in comparison to those who participate in traditional shop purchasing experiences. The problem of customer perceptions of risk in the context of internet purchasing has emerged as a significant current concern. Thus, the current research used the Bayesian linear regression methodology to investigate the influence of perceived risk on consumers' behavioral intention within the realm of online purchasing. The findings of the research indicate the existence of four distinct categories of perceived risk, namely product risk, financial risk, convenience risk, and non-delivery risk. These risks are evaluated negatively by customers and have a significant impact on their behavioral intention while participating in online purchase activities.

Keywords: Perceived Risk · Behavioral Intention · Bayesian Technique

1 Introduction

Over the past few decades, the Internet has evolved into a vast global marketplace for the exchange of goods and services. It has offered an alternative to the traditional in-person shopping experience, where consumers had to visit physical stores, browse for products, compare options, seek seller recommendations, and make payments at the counter. Today, consumers can comfortably shop from their homes, workplaces, or offices, browsing, ordering, paying, and receiving their products online. Many businesses have embraced the Internet as a platform for selling their products, leading to reduced marketing costs and subsequently lower product prices. This innovation has reduced the need for physical interactions, and the convenience of online shopping has been warmly received by consumers. Businesses are increasingly relying on Internet marketing to keep a competitive advantage in highly competitive marketplaces [1]. However, the growing number of consumers who shop online has given rise to a variety of new worries and roadblocks, including issues related to the safety of payments, private information protection, and the quality of goods. Online consumers encounter a greater array of risks compared to those who are engaged in conventional offline retail transactions. As

R. Silhavy and P. Silhavy (Eds.): CSOC 2024, LNNS 1120, pp. 14–22, 2024.
https://doi.org/10.1007/978-3-031-70518-2_2

a result, customer perceptions regarding risk in the context of online purchasing have become a critical contemporary concern, since they may quickly shift consumer feelings about online purchases, with significant consequences for their web-based purchasing behavior [2]. Likewise, in the field of marketing research, it has been established that risk perception has a direct influence on buying behavior and purchase intentions. To clarify, when consumers perceive a heightened level of risk, the likelihood of them making online purchases or engaging in repeat online buying diminishes. The presence of real risks can significantly influence shoppers' buying behavioral intention [3, 4]. Consumer assessments and purchase intentions are heavily influenced by risk perception. Risk perception is an important aspect of transactions via the Internet, and it's important to comprehend how the perceived risk impacts customers' Internet buying intention [1]. Therefore, a continuous assessment of consumers' perceived risk associated with online shopping intention is essential. This monitoring is necessary to gauge the extent to which these risks negatively affect consumers' online shopping intentions. This enables e-retailers to be proactive in management and reduction, ultimately fostering the growth of online shopping [5].

Numerous studies have been conducted on the effects of perceived risk on customers' behavioral intentions in online shopping (e.g. [1, 2, 5–7]). However, no research has been conducted using the Bayesian linear regression technique to examine how perceived risk influences customers' behavioral intentions in e-shopping. To address this research gap, the present study used the Bayesian linear regression approach to examine the impact of perceived risk on customers' behavioral intention in the context of online buying.

2 Literature Review

2.1 Perceived Risk

Perceived risk is the term used to describe consumers' apprehension about potential losses. It quantifies the level of risks consumers are willing to accept in their purchase decisions [8]. Consequently, this study indicates that consumers' perceived risks are anxious about the likelihood of encountering adverse consequences or situations when participating in online shopping [5]. In the Internet environment, consumers encounter limited opportunities for hands-on product evaluation, leading to an elevation in information disparity and an increased level of consumer anxiety [9]. On the other side, the discomfort experienced by consumers stems from the limited frequency of their encounters with service providers and the complicated processes associated with returning defective products via online channels. Consequently, this phenomenon leads to a reduction in the perceived level of risk associated with online transactions, resulting in a gradual decline in the sense of control individuals have over their actions and immediate environment. As a result, this harms the use of Internet technology to engage in purchasing activities [10] Moreover, consumers navigate various psychological steps during the purchase of products. One aspect frequently discussed in academic literature is the perceived risk linked to the intention to make a purchase [6]. Likewise, earlier research has indicated that perceived risk serves as a precursor to customers' behavioral intentions [2, 7].

Within the domain of e-commerce, individuals often confront various kinds of perceived risks while engaging in online buying activities. The examination of risks in the e-commerce setting is a subject of interest, as noted by Ariff et al. [5] and Ezeanya et al. [1]. These scholars have identified four primary categories that are often emphasized in the analysis of risks in the e-shopping environment: financial risk, product risk, convenience risk, and non-delivery risk. The subsequent parts of this study will analyze the four distinct categories of risks and their influence on consumers' behavioral intentions.

2.2 Financial Risk

Financial risk may be characterized as the anxiety experienced by consumers over potential losses in money incurred during online buying activities. This pertains to the gaps in prices offered between online purchasing and conventional purchasing, as well as the additional fees associated with shipping and Internet transactions [11]. The risk also is related to the possibility that consumers might realize that they have paid a greater price for the same goods and services [12]. The existing body of research has extensively examined the impact of financial risk on consumers' inclination to engage in online purchasing [4, 12].

2.3 Product Risk

Product risk refers to the potential for a bought product to deviate from its intended functionality. When people shop online instead of in a store, they are limited depending on the information, models, visuals, images, and pictures that are shown on their computer screens. This makes it hard for them to judge the level of the items they are buying. Because of this, the goods that were bought might not work the way they were supposed to [13]. Past research investigations have discovered that product risk is a factor that contributes to an individual's online behavioral intention and that this factor's effect favorably on an individual's online behavioral intention [1, 5, 7].

2.4 Convenience Risk

Convenience risk pertains to how consumers perceive the potential dangers associated with the need for a great deal of time to fix and adapt a bought product before it becomes usable [14]. Convenience risk is also associated with individuals' perceptions of risk in situations when time or effort is lost due to the necessity for product replacement or subsequent repairs [15]. In previous research, it was shown that convenience risk is an antecedent of online behavioral intention, and its influence on online behavioral intention was found to be positive [1, 16].

2.5 Non-delivery Risk

Non-delivery risk refers to the possible failure of delivery resulting from different reasons such as products being lost, items getting damaged, or items being delivered to an incorrect location after being notified of an online transaction [5]. Customers express worries over the shipping procedure, such as the possibility of harm to the goods during transit, delivery to an incorrect location, or unexpected difficulties [11]. Furthermore, consumers express concerns over potential delivery delays, which may arise from many causes, such as the failure of the delivery business to adhere to the agreed-upon timeframe for delivering bought items to clients. The apprehension over potential harm to items during their handling and transportation to customers is one aspect of the non-delivery risk [17]. Previous researchers have asserted that the risk of non-delivery serves as an antecedent to the formation of online behavioral intention [1, 5].

3 Methodology

3.1 Bayesian Linear Regression

Bayesian linear regression involves the use of Bayesian inference methodologies to analyze the statistical characteristics of models of linear regression. By employing this approach, it is feasible to offer predictions for the value of the explained variable by giving values for the explanatory variables [18]. According to Javaid et al. [19], the Bayesian linear regression model offers a comprehensive framework for scenarios in which the explained variable conforms to a normal distribution. The content may be simply described in the following manner:

$$y_i \sim N\left(\mu_i, \delta^2\right), \tag{1}$$

$$\mu_i = \beta_0 + \sum_{k=1}^{p} \beta_k x_{ki} \tag{2}$$

Within the scope of this framework, the explained variable y_i, which is also frequently referred to as the dependent variable, is constructed from i different points of data. The explanatory variable x_{ki}, which is sometimes referred to as the independent variable, is comprised of k different multivariate quantities of interest. The parameter μ_i represents the outcome of linear regression when it comes to the mean connection between the variables. The β_0 and β_k which stand for the intercept and gradients of the k values, respectively, while δ^2 represents the variability or dispersion around the average regression line. It is assumed that the variance is consistent across all the data.

Because the parameters of the Bayesian regression model are assumed to follow a distribution, it is vital to take into account the conditional link that exists between the training inputs and outputs when calculating the posterior probability for these parameters [20].

$$\text{p}(\beta|y, x) = \frac{P(y|\beta, x) * P(\beta|x)}{P(y|x)} \tag{3}$$

The posterior probability distribution of the model parameters, $\text{p}(\beta|y, x)$, is determined based on the inputs and outputs. It is obtained by multiplying the likelihood of the data – $P(y|\beta, x)$, with the prior probability of the parameters – $P(\beta|x)$ and subsequently dividing by a normalization constant [21].

$$\text{Posterior} = \frac{\text{Likelihood} * \text{Prior}}{\text{Normalization}} \tag{4}$$

The current research used the Bayesian linear regression methodology to examine the impact of perceived risk on customers' behavioral intentions in the context of online buying. The present research employs four independent variables and one dependent variable. Therefore, we provide the below equation as the Bayesian linear regression model to examine the impact of perceived risk on customers' behavioral intentions in the context of online shopping:

$$Y = BI = \beta_0 + \beta_1 FR + \beta_2 PR + \beta_3 CR + \beta_4 NR \quad (5)$$

where:

Y = BI: Behavioral intention, FR: Financial risk, PR: Product risk,
CR: Convenience risk, NR: Non-delivery risk.

3.2 Data and Sample

An online survey was utilized to gather data for the research, and 275 customers from Ho Chi Minh City, Vietnam, took part in it. The survey was conducted using a Google form. A method that did not rely on probability was used to choose the sample from the population. The criteria were evaluated using a Likert scale with five points, with one representing total disagreement and five representing total concurrence. The following items from [5, 6, 10] were utilized to assess the variables of the study via the use of a questionnaire: four items related to financial risk, four items related to product risk, four items related to convenience risk, three indicators related to non-delivery risk, and four items related to behavioral intention.

4 Results and Discussion

4.1 Results

Descriptive statistics

Table 1 displays the demographic characteristics of the sample.

Table 1. Demographics of the sample

Traits	Classification	Frequency	Percent
Gender	Male	150	54.5
	Female	125	45.5
	Total	275	100.0
Age	<18	26	9.5
	18–25	225	81.8
	26–35	13	4.7
	>35	11	4.0
	Total	275	100.0

Based on the data shown in Table 1, it can be seen that the sample consists of 150 male customers, accounting for 54.5% of the total, and 125 female consumers, accounting for 45.5% of the total. In terms of age distribution, it is seen that those below the age of 18 constitute 9.5% of the consumers, while those between the ages of 18 and 25 account for 81.8%. Additionally, individuals aged between 26 and 35 comprise 4.7% of the consumer population, and those beyond the age of 35 represent 4.0%.

4.2 Bayesian Technique

A summary of the findings that were generated by using the Bayesian linear regression technique on the posterior distribution is shown in Table 2, which can be found below.

Table 2. An overview of the posterior distribution's results

	Mean	SD	Naive SE	Quantile (2.5%)	Quantile (97.5%)
(Intercept)	8.1989	0.32082	0.0032082	7.5627	8.82803
FR	–0.4485	0.04726	0.0004726	–0.5417	–0.35532
PR	–0.4489	0.05510	0.0005510	–0.5560	–0.34060
CR	–0.1435	0.04173	0.0004173	–0.2237	–0.06276
NR	–0.1120	0.05045	0.0005045	–0.2116	–0.01451

The results shown in Table 2 indicate that the parameter estimates for each explanation variable fell within their respective 95% confidence intervals (CI), which were calculated as the range between the 2.5% and 97.5% quantiles. Furthermore, it is important to highlight that none of the weights inside the confidence interval (CI) had the value 0. This finding provides strong proof that all parameter values in the proposed model were statistically significant at a 5% significance level.

The linear regression analysis model used the Bayesian technique, which may be summarized as:

$$Y = 8.1989 - 0.4485\ FR - 0.4489\ PR - 0.1435\ CR - 0.1120\ NR$$

Furthermore, the model has an R-squared value of 0.5277, which implies that the four explanatory variables (FR, PR, CR, and NR) can explain 52.77% of the variation in customers' behavioral intention in e-shopping. This is shown by the fact that the model has this value. Similarly, the R-squared value of the model is 0.5277, and the R-squared value may vary from 0.3900 to 0.6322 (see Table 3 and Fig. 1 for further information).

Table 3. The outcome of R-squared

Min	Mean	Max
0.3900	0.5277	0.6322

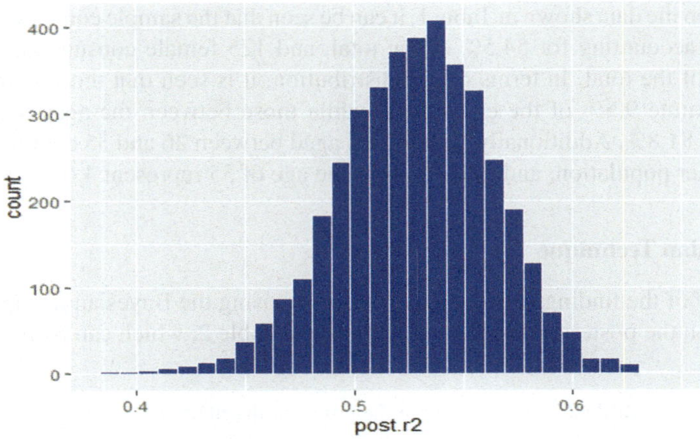

Fig. 1. The results of the R-squared

5 Conclusions and Limitations

In this study, a Bayesian technique of linear regression was employed to examine how consumers' intentions in online shopping are affected by their perception of risk. The findings of the analysis indicated that four categories of perceived risk (financial risk, product risk, convenience risk, and non-delivery risk) significantly negatively impact consumers' behavioral intention in the context of online shopping. Moreover, product risk (PR) was found to be the most negative influential risk (P = –0.4489), followed by financial risk (FR) (P = –0.4485), convenience risk (CR) (P = –0.1435), and non-delivery risk (NR) (P = –0.1120). Therefore, managers need to consider these risks to adopt a proactive approach to management and minimize them when customers engage in online product searches and intend to buy items. The reduction in consumers' perceived risk in online transactions is expected to positively impact customers' behavioral intentions in the realm of online purchasing, resulting in a potential increase in online buying activities. The present results are consistent with the research conducted by Ezeanya et al. [1], who also argued that the four types of perceived risk, namely PR, FR, CR, and NR, have a substantial adverse influence on consumers' behavioral intention within the context of online buying. The current study validates the coherence of these results.

While this research has made significant contributions to both academic knowledge and practical applications in the realm of online shopping, it does have its share of limitations. To begin with, the study's scope is restricted to utilizing the Bayesian linear regression method to investigate the influence of perceived risk on consumers' behavioral intentions in online shopping. As a result, future research should explore alternative approaches to understand how consumers' behavioral intentions in online shopping are impacted by their risk perceptions. Secondly, the Bayesian linear regression model applied in this study produced an R-squared value of 0.5277, signifying that it accounts for 52.77% of the variability in consumers' behavioral intentions in online shopping through the four kinds of perceived risk (PR, FR, CR, and NR). However, it is conceivable that there may exist other risks that influence customers' intentions to engage in online purchasing, which have not been taken into account in the present model. Therefore, it is recommended that future research efforts take into account the inclusion of supplementary risk factors to provide a more comprehensive understanding of the variations in customers' behavioral intentions while engaging in online purchasing.

References

1. Ezeanya, I.D., Obiorah, O.V., Nnaebue, C.I., Anike, R.U.: Relationship between risk perception dimensions and online purchase among undergraduates of Nnamdi Azikiwe University Awka. Nigerian Assoc. Social Psychologists. **6**, 35–43 (2023)
2. Andrian, A., Selamat, F.: The influence of consumer's perceived risk on consumer's online purchase intention in Indonesia. Adv. Social Sci. Educ. Hum. Res. **655**, 669–676 (2022). https://doi.org/10.2991/assehr.k.220404.105
3. Kovacs, M., Farias, S., Moura, F., Souza, A.: Relations between consumer relations between consumer effort, risk reduction strategies, and satisfaction with the e-commerce buying process: the development of a conceptual framework. Int. J. Manag. **28**, 316–329 (2011)
4. Alrawad, M., et al.: Assessing customers perception of online shopping risks: a structural equation modeling–based multigroup analysis. J. Retail. Consum. Serv. **71**, 103188 (2023). https://doi.org/10.1016/j.jretconser.2022.103188
5. Ariff, M.S.M., Sylvester, M., Zakuan, N., Ismail, K., Ali, K.M.: Consumer perceived risk, attitude and online shopping behaviour; Empirical evidence from Malaysia. IIOP Conf. Ser. Mater. Sci. Eng. **58**, (2014). https://doi.org/10.1088/1757-899X/58/1/012007
6. Bhukya, R., Singh, S.: The effect of perceived risk dimensions on purchase intention: an empirical evidence from Indian private labels market. Am. J. Bus. **30**, 218–230 (2015). https://doi.org/10.1108/AJB-10-2014-0055
7. Mahbob, N.N., Hasim, M.A., Ishak, M.F.: The influence of perceived risk factors on online purchase intention among millennials in Malaysia. Inf. Manag. and Bus. Rev. **15**, 50–58 (2023)
8. Zheng, L., Favier, M., Huang, P., Coat, F.: Chinese consumer perceived risk and risk relievers in e-shopping. In: 17th Symposium of the Association Information and Management 2012, AIM (2012)
9. Park, S., Nicolau, J.L.: Asymmetric effects of online consumer reviews. Ann. Tour. Res. **50**, 67–83 (2015). https://doi.org/10.1016/j.annals.2014.10.007
10. Mayo, J.A., Layante, J.P., De Lara, R.P., Macasojot, R.C., See, A.L.C., Melo, M.C.F.: Perceived risk and online purchase intention of selected public school teachers: a study in Calapan City, Oriental Mindoro, Philippines. World J. Adv. Res. Rev. **16**, 908–925 (2022). https://doi.org/10.30574/wjarr.2022.16.2.1264
11. Naiyi, Y.E.: Dimensions of consumer's perceived risk in online shopping. J. Electron. Sci. Technol. China. **2**, 177–182 (2004)
12. Dai, B., Forsythe, S., Kwon, W.S.: The impact of online shopping experience on risk perceptions and online purchase intentions: does product category matter? J. Electron. Commer. Res. **15**, 13–24 (2014)
13. Kim, I.S.: Consumers' rankings of risk-reduction strategies in E-shopping. Int. J. Bus. Res. **10**, 143–148 (2010)
14. Chang, H.H., Chen, S.W.: The impact of online store environment cues on purchase intention: trust and perceived risk as a mediator. Online Inf. Rev. **32**, 818–841 (2008). https://doi.org/10.1108/14684520810923953
15. Ko, H., Jung, J., Kim, J., Shim, S.W.: Cross-cultural differences in perceived risk of online shopping. J. Interact. Advert. **4**, 20–29 (2004). https://doi.org/10.1080/15252019.2004.10722084

16. Bhatti, A., Saad, S., Gbadebo, S.M.: Convenience risk, product risk, and perceived risk influence on online shopping: moderating effect of attitude. Sci. Arena Publ. Int. J. Bus. Manag. **3**, 1–11 (2018)
17. Claudia, I.: Perceived risk when buying online: evidence from a semi-structured interview. Econ. Ser. **22**, 63–73 (2012)
18. Ahmad, W.M.A.W., Akbar, N.A.N., Ghazali, F.M.M., Noor, N.F.M., Aleng, N.A.: Fuzzy regression model with Bayesian approach and its application to public health data. Bull. Appl. Math. Math. Educ. **2**, 19–32 (2022). https://doi.org/10.12928/bamme.v2i1.3953
19. Javaid, M.A., Harrison, J.P., Mas Ivars, D., Kasani, H.A.: A Bayesian regression analysis of in situ stress using overcoring data. IOP Conf. Ser. Earth Environ. Sci. **1124** (2023). https://doi.org/10.1088/1755-1315/1124/1/012075
20. Ali, A., Inglis, A. n., Prado, E., Wundervald, B.: Bayesian linear regression. Technical Report. 39–69 (2019). https://doi.org/10.1201/9781351165761-3
21. Zhao, X.: Application of Bayesian regression model in financial stock market forecasting. Adv. Econ. Bus. Manage. Res. **189**, 140–144 (2021). https://doi.org/10.2991/aebmr.k.210909.019

Voltage Selection Methodology for Power Lines HVDC

Ricardo Arias(✉)

Universidad Tecnológica del Perú, Natalio Sanchez street No 125,
Santa Beatriz, Lima, Perú
ricardoariasvelasquez@hotmail.com, rariasv@utp.edu.pe
https://www.utp.edu.pe/web/

Abstract. Any electric power system should maintain an energy equilibrium at every moment. This is the case for existing AC systems and will be the case for future High Voltage Direct Current (HVDC) grids based online-commutated converters (LCC) substations. In this paper, a new proposal to define the voltage in the HVDC is proposed with an analysis of the input data, the challenges, and limitations regarding the level range of the DC Voltage for interconnections between countries with special power lines parameters, power converters substations and cost balance. The case study evaluates the Interconnection Colombia - Panamá, the power line has considered the aspects related to the corona phenomenon in HVDC transmission lines according to the cable selection, substation back-to-back, submarine cable and losses; with 400–500 kV for 1700 km for 280 MW per pole or the same voltage for 950 km for 1500 MW per pole.

Keywords: HVDC · power line · methodology · voltage selection

1 Introduction

The voltage level optimization methodology and the relevant input data to establish the transmission voltage in HVDC using substations based on switching converters (LCC) is a challenges for four aspects: Grid integration, due to the integration of HVDC into existing AC grids could be challenging, requiring synchronization and control mechanisms, the second aspect is the voltage selection, the importance of the selection of the appropriate voltage level depends on factors like distance, power capacity, and cost-effectiveness, therefore, the voltage selection is the first step in the CAPEX structure [1]. On the other hand, maintenance and reliability allows us to maintain HVDC systems and ensuring their long term reliability is a complex task and the interoperability between different HVDC technologies and standards [2].

The voltage selection thought the last decades tried to define the criteria for selecting the voltage level and best range that allows balancing the design for

Thanks to Universidad Tecnológica del Perú.

R. Silhavy and P. Silhavy (Eds.): CSOC 2024, LNNS 1120, pp. 23–40, 2024.
https://doi.org/10.1007/978-3-031-70518-2_3

direct current electrical interconnection, seeking to optimize the planning and expansion of the HVDC system according to the Ref. [3]. The optimal selection of the voltage level is of vital importance and does not depend only on a technical solution, due to the long distances and geographical and environmental requirements (submarine cables, overhead lines, underground, among others) [19,20], this research implements a case study, like the cases studied by author Henderson M. et al. (2010).

According to the author Henderson M. et al. (2006) and the IEEE WG15.05.13 group, the planning of HVDC transmission lines was evaluated, concluding that the design is an economic choice for interconnection of asynchronous systems [4], the studies require the evaluation of the operation and maintenance costs, losses of substations and lines [5].

The case study will evaluate the alternatives for the Colombia - Panama inter-connection, with the perspective as follows:

- Optimal section of the driver [6], considering current research such as the authors Mazzanti G. & Marzinotto M. (2013).
- Aspects linked to the level of corona effect are considered for the determination of the clearance distances.
- Voltage selection and levels comes from the optimization of link costs updated to the reference year.
- Life cycle equal or higher to 30 years, a period for which the transmission lines are built under calls for electrical concessions considering the life cycle of the asset [7].

Finally, the paper is composed of the following sections: Sect. 2, methodology, with the mathematical model. Besides, the Sect. 3 has the description of the case study for the HVDC between Colombia and Panamá. As a complement, Sect. 4 the results, and Sect. 5 the discussion of the benefits and limits used to select the voltage limit with the power per pole and length according to the voltage selection. Finally, the last section has the conclusions.

2 Methodology

The optimization of the voltage level is carried out mainly on a technical and economic analysis based on the following guidelines regarding the voltage level. Usually, the voltage level is from 250, 300, 400 and 500 kV, based on the data (parameters, costs, etc.) currently found in the market and on experiences of installations in current operation, the costs are calculated. of capital, which is composed of four partial costs: Various standard cable sections, conversion stations (LCC), submarine cable (or submarine section) and the cost derived from the operation of the link and updated to the reference year.

Besides, from the sensitivity analysis, the cost includes: i) losses in conversion station, ii) airline losses, iii) losses in submarine cable, and iv) operation and maintenance costs. Therefore, the sum of all these costs calculated for the various values of conductor section leads to defining, by interpolating such points with an analytical function, the point of minimum cost and consequently the choice of the optimal value of conductor section within the same level of strain.

The final choice of the optimal voltage level results from the absolute minimum cost between the individual optimal solutions for the analyzed voltage levels.

2.1 Input for the HVDC

The transmission voltage, in the case of an overhead cable line, is determined by the specifications of the HVDC cable. According to the author Falahi G. & Huang A. (2015) in the case of back-to-back connections, the voltage is determined by the required power and the maximum nominal current of the thyristor [8]. However, in transmission over long distances via overhead lines, the voltage can be freely chosen. Therefore, it can be optimized, in order to reduce total transmission costs.

Considered independently, the total costs of overhead transmission lines are composed of construction costs (generally expressed as annual costs using an amortization factor) and losses per year. Therefore, the "load profile" of the transmission must be known in order to determine losses. (The commonly specified annual hours of use are not sufficient for the calculation of losses because most losses depend on I2).

In addition, the loss factor is needed, specifically the value of the losses for losses under load and the considerations of monopole and bipole energy transport with the design considerations of the grounding system according to the authors Ma J. et al. (2005) [9].

The construction costs of a bipolar HVDC overhead line (including survey, but not including land costs) can be estimated using the Eq. 1.

$$KE = A + B + Ud + C \times q \text{ [DM/km]} \tag{1}$$

where:
Ud : Voltage (KV).
Q : Cable section (mm^2).

In fact, the cost factors A, B and C were determined by the HVDC working group, [9] investigated by authors Kassakian J. and Naunin D. (1986). They have been adjusted approximately 5% per year, which are detailed in Table 1, with the analysis with quotations and engineering for the last projects. It is interesting the quantity of cables per package, it hasn't influence on construction costs.

According to the Eq. (1), it is applied for bipolar lines, and for monopolar lines the Eq. (2) with Eq. (1). Besides, the configuration of a bipolar HVDC with two monopolar lines results in an increase in the cost of the line close to 25%, [11] which is consistent with the author Westernmann D. et al. to the. (2010), for a double circuit line consisting of two bipolar systems in a tower, in Eq. (3).

$$Km = 0.63 \times \text{ KB} \tag{2}$$

Table 1. Cost factors.

Factor	1960	1990	2023	Dimension
A	6000	26000	30000	DM/km
B	115	500	750	DM/kV $\times$ km
C	30	130	200	DM/mm^2 $\times$ km

$$KD = 1.65\times \text{ KB} \tag{3}$$

On the other hand, the double-circuit line is about 17.5% less expensive than two bipolar lines. On the other hand, dividing a ±500 kV / 2,000 A bipole into two systems (each conductor with 1/2 cross section) means additional costs of approximately 15%. Furthermore, with the Eq. (1), the transmission voltage is considered, instead, the cable cross section q is used to validate the current density j, in the Eq. (4).

$$q = \frac{Idn}{j} \tag{4}$$

On the whole, the overhead lines are operated below their thermal limit for economic reasons, j is an optimization parameter, due to the optimal cross section, therefore, it could be determined by calculating the total costs using the specified current, voltage, load profile and loss value as constants and q as the variable parameter. This methodology also provides the current density j for the cable. When this Eq. (5) is applied for various HVDC lines, similar current densities could be found as long as the loss evaluation remains the same, in the Eq. (6).

$$j = 0.8 \text{ A/mm}^2 \tag{5}$$

$$\Delta u = \rho \times j \tag{6}$$

At the beginning of HVDC technology, values close to j = 1 A/mm^2 (aluminum cross section). In 2023, an economical current density between 0.7 and 0.9 $\frac{\text{A}}{\text{mm}^2}$ could be expected for the recommendations of the HVDC manufacturer in Hitachi/ABB and SIEMENS, since the value of losses has increased more than construction costs. Therefore, it is recommended to use it for quick calculations. With this value, the voltage drop along the line is a fixed amount per kilometer of line length, in the Eqs. (5) and (6). Besides, a value of 29 mm^2/km should be used for the specific resistance of the aluminum cable. By using the recommended value of Eq. (5), in this case, the voltage drop per km of line length results in Eq. (7).

$$\Delta u = 23 \text{ V/km} \tag{7}$$

These equations allow a quick assessment of efficiency in the transmission of an HVDC overhead line in Eq. (8). This equation depends on the above values and also on the line length L in km and the nominal HVDC voltage (UdN is measured at the beginning of the line on the rectifier side). The cost in the Eq. (1) and the power resistance per pole in the Eq. (9). It provides all the elements needed to determine the construction costs and loss costs for an HVDC overhead line of a given transmission power for various combinations of Ud, Id. The value of Ud, from which the minimum total costs are derived, is could be designated as the optimal transmission voltage. However, we need to consider two additional parameters:

- The maximum allowable value of the field at the surface (surface gradient).
- The degree of dependence on the cost of the voltage station.

$$\eta L = 1 - \frac{\Delta u \times L}{UdN} \tag{8}$$

$$RL = \frac{\rho \times L}{q} \tag{9}$$

Meanwhile, the line radio interference, corona losses, and ionic current depend on the surface gradient (the electric field strength at the surface of the conductor), or in the case of bundles of conductors, at the surface of the substitute equivalent conductor. The surface gradient can be reduced by dividing the conductor into a bundle of multiple conductors. There is a tension for each cross section of the conductor, for which the limit of the field strength is achieved. This limit is in accordance with the current range of transmission voltages. A further increase in stress is only possible by increasing the cross section beyond the surface gradient (qopt). The increase in construction costs will be partially offset by reduced loss costs. The costs of a station, as will be demonstrated, also depend on the voltage. This should be considered to determine the optimal transmission voltage. The relationship can be expressed in words:

- Overhead lines require the highest possible transmission voltage.
- The maximum permitted surface gradient places an upper limit on the tension.
- Station costs decrease the optimal voltage, especially in the case of relatively short lines.

In the case of long lines, station costs are less important. Figure 1 reflects the relationships described above. Therefore, curve 1 shows the voltage dependence of station costs. In curve 2, when showing the total line costs, the influence of the maximum surface gradient is considered.

To clarify these relationships, the total costs: Losses and construction are shown in Fig. 1 of an HVDC overhead line. Construction costs increase linearly with cross section. Loss costs are proportional to line resistance. This means that the losses are inversely proportional to the cross section of the conductor. The resulting curve shows a minimum cost at qopt, [12] which corresponds to the optimal cross section of the conductor Burges K. et al. (2008); with the 1) Construction cost, 2) losses cost and 3) total cost.

This feature could be observed in the upward slope of line 2 costs beyond the optimal line voltage (compare with curve 3 in Fig. 1). Curve 3 shows the total costs of an HVDC system that is at its lowest line voltage, below the optimal line voltage. The relationship between curves 2 and 1 depends largely on the length of the line. In the case of line lengths of many thousands of kilometers, the influence of the station cost becomes trivial. On the other hand, with lengths less than 1,000 km, the influence is important as shown in Fig. 1.

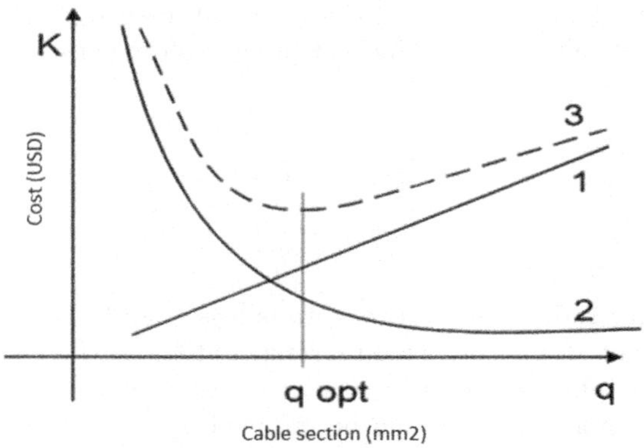

Fig. 1. Unitary components in the design of the HVDC.

2.2 Converter Stations

According to the authors Schettler F. et al. (2000) and Barker C. et al. (1998), for the "typical" HVDC station with a nominal power of 1,600 MW at ±400 kV, 2,000 A, specific costs of 160 MM/kW [13]. The cost included two converter stations for a point-to-point HVDC connection [18], including AC switches, buildings, and construction costs, [14] as indicated by the authors Van Hertem D. & Ghandhari M. (2009) and ABB (2009). It is important to indicate that this does not include property costs (purchase of land), since these vary from country to country and therefore land costs were not included [15].

In 2023, it was not necessary to apply an estimated annual rate of change to station costs as in the case of airline costs. The overall increase in costs of "conventional" HVDC station components has been slowed by falling costs, thanks to innovation in thyristor valves.

However, since 2000 HVDC station costs have been subject to a constant increase, but this increase is well below the increase for overhead lines. For quick calculations you can use 2.5% per year. This results in the specific price of a "typical" HVDC station based on 2000 costs, in the Eq. (10), as Kst2023.

$$Kst2023 = 205 \tag{10}$$

In spite of the factors with influence the cost of an HVDC station, and cause a change in specific costs, the three most important should be listed as follows:

- Station P Power.
- Transmission Voltage Ud.
- Transmission Current Id

Station power P has the strongest influence on specific costs but is complex. Whereas, for several important components, such as transformers and buffer reactors, the growth law is generally expressed in Eq. (11).

$$K = K0(\frac{(P^{(\frac{3}{4})})}{P0}) \tag{11}$$

There are totally different growth laws for valves, filters, AC switches, etc. These efforts by international committees to find a growth law for HVDC stations have concluded that the "natural" law of growth applies, with adequate accuracy, also for complete HVDC stations. The dependence of power on the specific costs of a station can therefore be calculated with the following Eq. (12).

$$Kst = Kst0 \times \sqrt{(\frac{(P0)}{P})} \tag{12}$$

where: Ksto is indicated in Eq. (10) and Po = 1600 MW.

The dependence of voltage on station costs was discussed in the previous item, with reference to its importance for the calculation of optimal transmission voltage. By determining the optimal voltage is as complex as the power dependence, in spite of the general rule, there is only a slight dependence of costs on voltage, in the range close to 500 kV, but from then on costs increase steadily. Valve building costs contribute significantly to this increase. The following Eq. (13), for the dependence of station-specific costs on voltage.

$$Fu = 0.985 + 0.015 \times (\frac{UdN}{400})^4 \tag{13}$$

Where UdN = Nominal transmission voltage in kV.

In which the dependence of the voltage, therefore on station costs is presented graphically. The pronounced increase in costs in the range above 500 kV could be clearly observed. In comparison, on the relatively minor influence of the thyristor size on the cost of the valves. The smaller the drive current, the larger the specific costs of the valve, in the Fig. 2. The influence of the current on the specific costs of the complete station cannot be neglected. For the dependence of the current, in the specific costs Eq. (14).

$$Fu = 0.96 + 0.04 \times (\frac{2000}{IdN0})^2 \tag{14}$$

where IdN = Nominal transmission current in A.

It is graphically represented in Fig. 3, the dependence of current on station costs. The increase in costs is evident below 1,500 A. Above 2,500 A, the

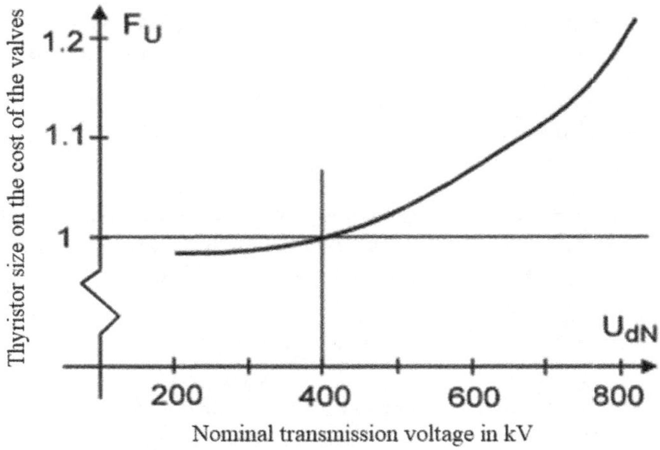

Fig. 2. Voltage dependence for the station cost.

dependence of the current on the station costs is lower. However, transmission currents greater than 2,000 A, therefore, it only occurs in long distance transmissions with a high nominal power. When the influences of the three most important parameters are combined, Eq. (15) obtained for calculating station costs.

$$Kst = P \times 205 \times \sqrt{(\frac{(1600)}{P})} \times (0.985 + 0.015(\frac{UdN}{400})^4)) \times (0.96 + 0.04(\frac{2000}{IdN})^2)) \tag{15}$$

Loss costs are also a factor in the total costs of a station. As has been observed, the losses per station are in the range of 0.6% to 0.8% of the nominal power, regardless of the operating range of the station. This can be attributed to the fact that the relatively low efficiency of smaller transformers is compensated by the high efficiency of large thyristor valves [16]. Therefore, the author Arrillaga J. (1983) recommends that a station efficiency covering both stations with this value be used. We have already presented all the elements necessary to determine the optimal transmission voltage for an HVDC system for a given power and transmission distance (i.e., the voltage at which total costs are minimized), and to allow comparison with an AC transmission system. The results allow estimating the possibilities of use for an HVDC system, [17] which has also been recommended by the IEEE Std 1240-2000 standard. For more exact tasks such as determining a budget, the cost equations found above are sufficient, however, important factors of reactive compensation must be considered, according to the analyzed topology.

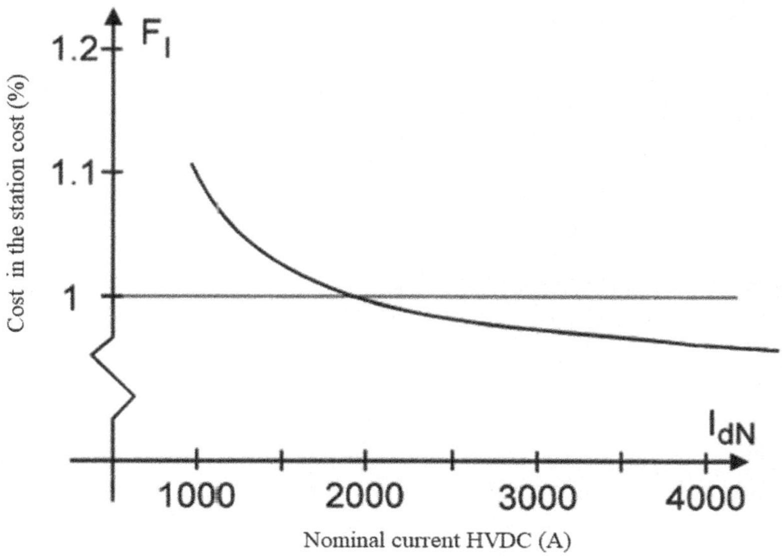

Fig. 3. Current dependence in the station cost.

3 Case study, HVDC from Colombia to Panamá

It is considered that the voltage could be freely chosen according to numeral Sect. 2, therefore, it should be optimized based on costs. Therefore, it is divided into two different categories: investment costs and operational costs. The following paragraphs indicate the main input data in the cost evaluation of the HVDC link, considering DC 250, 300, 400 and 500 kV as possible voltage levels, with Ud values and manufacturers' q ratings. The costs are discounted to the first year of construction (2016), applying the discount rate of 12%, by using Eqs. (1) and (4). As we mentioned in the previous section, an economical current density can be selected between 0.7 and 0.9 $\frac{\mathrm{A}}{\mathrm{mm}^2}$. It is recommended to use this value for quick calculations. The voltage drop along the line is a fixed amount per kilometer of the line. line length, Eq. (5) applies, with the Eq. (6), it obtains a voltage drop per kilometer in the Eq. (18) with information in the Eqs. (16), (17), and (18).

$$KB = 26000 + 500 + Ud + 130 \times Idn/j[\mathrm{DM/km}] \quad (16)$$

$$\Delta u = 23\,\mathrm{V/km} \quad (17)$$

$$\rho = 28.75 \quad (18)$$

For which the efficiency of the towers will be evaluated for a length of 600 km, according to Eq. (8) in Table 2:

Table 2. Efficiency according to the voltage.

Efficiency	Voltage (kV)
0.9540	300
0.9655	400
0.9724	500

The resistance per pole applying Eq. (9) is shown in Table 3:

Table 3. Cable section and resistivity.

Cable section	Resistivity
600	0.02875
550	0.03136364
500	0.0345
400	0.043125
350	0.04928571
300	0.0575
250	0.069
202	0.08539604

3.1 Converter Stations

There are specific costs according to the previous paragraph for 2 point-to-point HVDC converter substations, including AC switches, building and construction costs, not including the property acquisition costs in Table 4, until 2016 the cost is in millions of dollars in the Table 5.

By applying Eq. (11), the transmitted power in DC is calculated equal to 300 MW from 2017 to 2019 (monopolar configuration), and 600 MW from 2018 to 2047 (bipolar configuration), according to an expected lifetime of 30 years and considering a manufacturing period of 3 years per pole. In the case of calculation in a single-pole configuration only, a power of 300 MW has been considered for the entire useful life period, with the Eqs. (11) and (12) with the Tables 4 and 5.

Table 4. Million dollars with adjustment.

Year	Rate	Cost
2023	1	160
2024	1.025	164
2025	1.025	168
2026	1.025	172
2027	1.025	177
2028	1.025	181
2029	1.025	186
2030	1.025	190
2031	1.025	195
2032	1.025	200
2033	1.025	205
2034	1.025	210
2035	1.025	215
2036	1.025	221
2037	1.025	226
2038	1.025	232
2039	1.025	238

Table 5. Dependence of the voltage with cost.

UdN	IdN	Fu
300	2,319	0.9897
400	2,000	1.0000
500	1,611	1.0216
600	1,259	1.0609

3.2 HVDC overhead line

The analysis started with two main aspects according to the Eqs. (1) to (12), as follows:

- Laying and protection costs.
- Material costs.

The kilometer costs of the submarine cable relevant to laying and mechanical protections are not influenced by voltage. It accounts around 50% of the total cable cost per km. Material costs consist of copper and insulation costs. In the study case of insulated cable, the calculation of the section considers the balance between heat produced by the Joule effect in the conductor and heat dispersed by conduction through the insulator and the surrounding terrain. It added considerations regarding the maximum temperatures and the electrical gradients admitted by the insulator. In relation to the material used for marine cables, the copper conductor is the one with the greatest diffusion on the market and therefore allows a broader participation of suppliers in the competition for the supply of the cable. Complementary, the application with aluminum conductors is less diffuse. Currently, there are companies in Latin America installing marine cables with aluminum conductors. Their justification lies in the need to limit the weight of the installation at the high depths reached ~1,640 m.

These two different costs will be considered the aspects as follows:

- Construction costs take into consideration the assembly of towers and conductors, the construction of foundations, including grounding and rights of way.
- Material costs consist of the towers, and the set of conductors and insulation. The costs of the OHL "overhead lines" or airlines are extrapolated from the corresponding costs of HVAC lines.

3.3 Operational Cost

The Operation and Maintenance (O&M) Cost, for conversion stations, airlines are considered as a percentage of their investment costs. While for the submarine cable it is considered a fixed cost. The loss costs have the following:

- Conversion Station Losses. As mentioned, in this preliminary analysis the conversion station losses are calculated as a fixed percentage of nominal power.
- Losses of overhead lines and cables. To evaluate the losses of the overhead line and cable, resistance values are required. These values are estimated considering the simplified preliminary design of the DC connection (OHL) and cable.

4 Discussion

With the case study the alternatives one and two are considered in order to validate the cost evaluation and reliability.

4.1 Alternative 1

Initial investment for a monopolar scheme (300 MW): in this initial Stage 1, even in a one-pole contingency, operation with metallic return in permanent mode.

- Fully implemented overhead line (*) (2 poles and return), made up of: A bundle of HV-insulated conductors per pole 559 km long; a bundle of insulated MV conductors with the metallic return function and guard cables, 559 km long.
- The submarine cable (1 Pole and return) (**), made up of: An HV insulated cable 55 km long per pole.

An insulated cable for MT, 55 km long for metallic return. (*): This criterion of complete construction of the airline by the year 2012 increases the reliability of the interconnection and could also avoid an interruption of the service that can be quite onerous in the phase of expansion from the monopolar to the bipolar modality (year horizon 2016). (**): The unavailability of the submarine cable is very low (around 1 failure in the useful life with a repair time of 40 between 60 days) to consider the convenience of making the initial investment of the second submarine pole (third cable) to guarantee maximum continuity of service even in the event of cable failure.

In Fig. 4 investments to move to the bipolar scheme (600 MW): in this Stage 2, even in the event of a one-pole contingency, operation with metallic return in permanent mode is ensured for the transmission system. The laying of another submarine cable (second Pole): A 55 km long HV insulated cable.

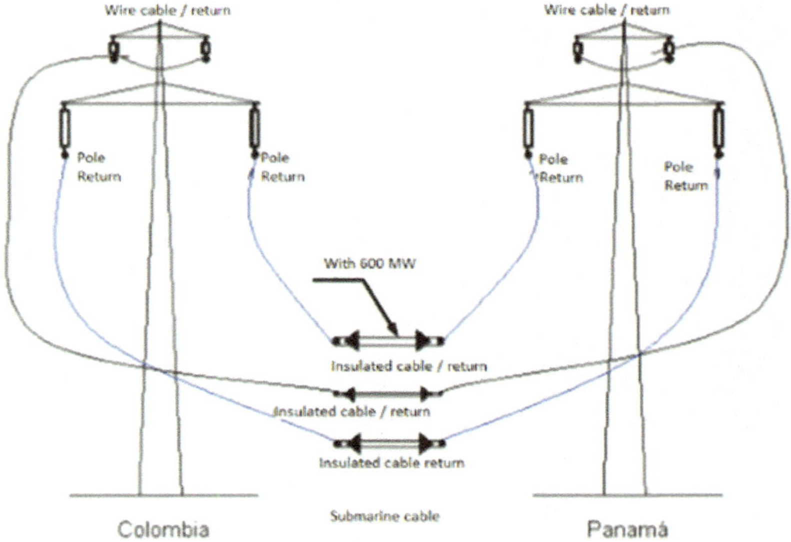

Fig. 4. Alternative 1.

4.2 Alternative 2

Initial investment for a monopolar scheme (300 MW): in this initial phase the contingency of one pole leads to the circulation of the nominal current through the ground electrode:

- Fully implemented overhead line (2 poles and guard cables), made up of: A bundle of insulated conductors per pole, each pole 559 km long. A bundle of conductors with only the function of guard cables, each 520 km long.

- The submarine cable (2 Poles), made up of: Two HV insulated cables of 55 km in length per pole.
- Two ground electrodes, it is necessary to move the ground electrodes away from the conversion stations by around 20 km to avoid interference on the components of the station itself.

In the Fig. 5, two insulated conductors for MV with the metallic return function and guard cables (it is assumed that the same inter-connection line is used to distance the ground electrode from the conversion station), of: o 20 × 2 km in length each.

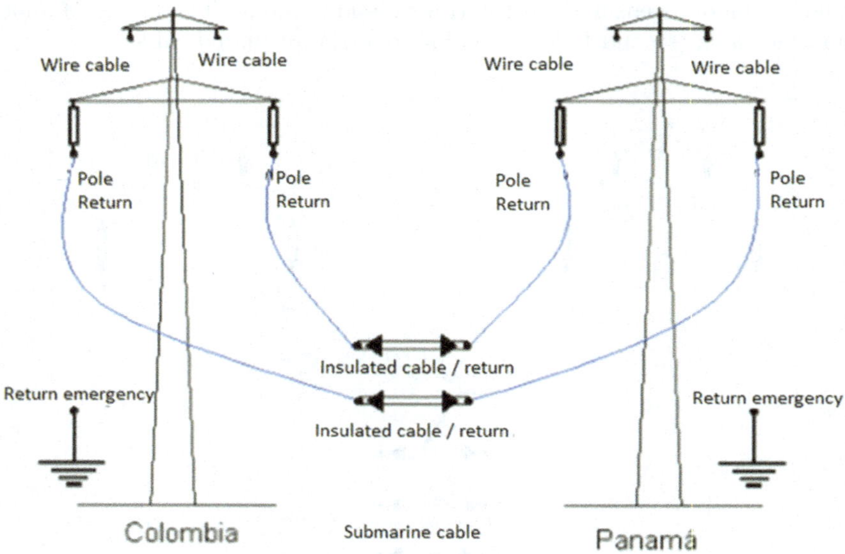

Fig. 5. Alternative 2.

To avoid exposure to fulmination in Alternative 1, so that it is comparable from this point of view with Alternative 2 protected by the wire cables, provision will have to be made in any case for the installation of the average conductors in the section. overhead line, although under normal operating conditions they will fulfill the sole function of guard cable. It is important to note that Alternative 1, faced with a contingency of one pole with few re closing operations, ensures a permanent metallic return, while Alternative 2 uses the emergency ground electrode for the maximum time granted for this type of operation. If the repair exceeds this time, the service should be completely interrupted. On the other hand, in a bipolar configuration, the contingency of the submarine cable (Pole) in Alternative 1 leads to half power operation and metallic return is guaranteed, while in Alternative 2 the operation, always at half power, is through the terrestrial electrode, and it must be taken into account that the repair times of this contingency are around 40 between 60 days and therefore the aforementioned is valid.

4.3 Discussion

The configurations adopted for the selection of the optimal voltage level; in the Table 6 proposed configurations for the execution of the line design with the comparison in the Table 7.

Table 6. Scheme and benefits.

Stage	With electrode (1)	With electrode (2)	With metallic return (1)	With metallic return (2)
Capacity (MW)	300	600	300	600
Pole number (Overhead line + submarine cable)	2	2	1	2
Wire cable	Normal	Normal	Nominal current and insulated	Nominal current and insulated
Steps for construction	2Cp + 2Cs + SW + E		2Cp + Cs + CsMV + SWi	Cs

Abbreviations and considerations:
Cp = power conductor (Pole) for overhead line.
Cs = power conductor (Pole) for submarine cable.
CsMV = MV insulated submarine cable for metallic return.
SW = guard wire, normal type SWi = guard wire, MV insulated, to function as metallic return.
E = ground electrode + connection to the substation.
Long contingencies: few days.

Table 7. Comparison of the solutions.

Information	With electrode	Metallic return	Delta capex	Benefits capex
Capex Stage 1	2Cp + 2Cs + SW + E	2Cp + Cs + CsMV + SWi (Cs + E) – (CsMV + 0,5 SW)	Proposal with metallic solutions	
Capex Stage 2	–	Cs	–Cs	Proposal with electrodes
Total Capex	2Cp + 2Cs + SW + E	2Cp + 2Cs + SW + E	E – (CsMV + 0,5 SW)	Proposal with electrodes

In the geographical area, Latin America, the only link at ±600 kV is from the Itaipú project, with a total nominal power of 6300 MW. This HVDC transmission system consists of two bipolar DC transmission lines of 785 km and 805 km, and of 3150 MW each. The 600 kV voltage has been adopted mainly because of the high voltage. The other link with DC voltage higher than 500 kV is the Cahora Bassa HVDC, with a length of 1420 km and a power of 1920 MW (bipolar). It follows that a DC voltage level higher than 500 kV is used very rarely, and for links with power much higher than 600 MW (bipolar). Consequently, the higher voltage range to be considered for technical-economic optimization is 500 kV. It turns out that excluding links characterized by short length (≤200 km) or low power (≤200 MW per pole), the voltage levels are in the range of 250 to 500 kV. Thus, this range will be considered in the optimization of the voltage, and a parametric analysis will be executed considering the levels of 250, 300, 400 and 500 kV, it can be seen in Fig. 6.

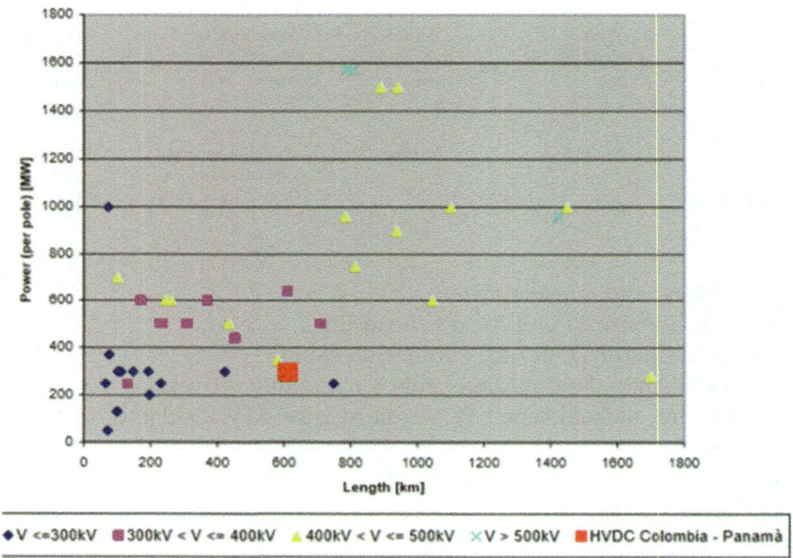

Fig. 6. Voltage selection for HVDC in the case study length in function of the power per pole in MW.

5 Conclusion

This methodology allows optimizing the selection of the voltage level for HVDC transmission lines based on converter substations (LCC), regarding the Fig. 6. In this paper, the voltage level optimization methodology has been presented considering the criteria in the transmission line and conversion units to select the representative input data in the optimal selection of the transmission voltage in HVDC, the capital costs. which is composed of the following partial costs: i) Transmission line for the various standard conductor sections, ii) conversion stations, iii) submarine cable. This cost includes: i) Losses in conversion station, overhead line, submarine cable and operation and maintenance of the link.

The sum of all these costs calculated for the various values of conductor section leads to defining, by interpolating such points with an analytical function, the point of minimum cost and consequently the choice of the optimal value of conductor section within the same level of strain.

Acknowledgments. Thank to the Universidad Tecnológica del Perú.

References

1. Ahmed, F., et al.: Dynamic grid stability in low carbon power systems with minimum inertia. Renew. Energy **210**, 486–506 (2023)
2. Arruda, C.K.C., Lima, A.C.S.: Corona modeling in HVDC transmission lines based on a modified particle-in-cell approach. Electr. Power Syst. Res. **125**(8), 91–99 (2015)
3. Taranto, G.N., et al.: Power flow control for an embedded HVDC link to integrate renewable energy in Brazil. Electr. Power Syst. Res. **211**, 108504 (2022)
4. Gagnon, J., Bertagnolli, D.: Planning issues for HVDC. In: 2006 IEEE Power Systems Conference and Exposition, pp. 28–32 (2006)
5. IEEE WG15.05.13.: A special publication for system planners. In: Transmission System Application Planning Requirements for FACTS Controllers (2006)
6. Mazzanti, G., Marzinctto, M.: Main Principles of HVDC Extruded Cable Design, p. 384. Wiley-IEEE Press (2013)
7. Cigré Working group 37.27. Ageing of the system impact on planning. CIGRE (2000)
8. Falahi, G., Huang, A.: Design consideration of an MMC-HVDC system base don 4500V/4000A emitter turn-off (ETO) thyristor. In: 2015 IEEE Energy Conversion Congress and Exposition (ECCE), pp. 3462–3467 (2015)
9. Ma, J., Dawalibi, F., Ruan, W.: Design considerations of HVDC grounding electrodes. In: 2005 IEEE/PES Transmission & Distribution Conference & Exposition: Asia and Pacific, pp. 1–6 (2005)
10. Kassakian, J., Naunin, D.: Basic Principles of Power Electronics, p. 1. Springer-Verlag Berlin Heidelberg, New York (1986)
11. Westernmann, D., et al.: Voltage source converter (VSC) HVDC for power transmission - economic aspects and comparison with other AC and DC technologies. In: CI-GRE Technical Report WG B4-46 (2010)
12. Burges, K., et al.: Study on the comparative merits of overhead electricity transmission lines versus underground cables. In: Ecofys, Technical Report (2008)
13. Schettler, F., Huang, H., Christ, l.: HVDC transmission systems using voltage sourced converters design and applications. In: Power Engineering Society Summer Meeting, 2000, vol. 2, pp. 715–720. IEEE (2000)
14. Van Hertem, D., Ghandhari, M.: Multi-terminal VSC HVDC for the European super-grid: obstacles. In: Renewable & Sustainable Energy Reviews. Elsevier (2010)
15. ABB: It's time to connect. Technical Brochure (2009)
16. Arrillaga, J.: High Voltage Direct Current Transmission. Peter Peregrinus (1983)
17. IEEE Std 1240-2000: IEEE Guide for the evaluation of the Reliability of HVDC Converter Stations (2001)

18. Barker, C., Sykes, A.: Designing HVDC transmission schemes for defined availability. In: Power Electronics Reliability - Promise and Practice. Does it Deliver? (Digest No. 1998/202, IEE Colloquium on. (1998) 4/1-411
19. Henderson, M., et al.: Planning HVDC and FACTS in New England. IEEE PES T & D 2010, pp. 1–3 (2010)
20. Liu, H., et al.: Effect of thermo-oxidative aging on thermal elongation performance of XLPE insulation for high-voltage cables. Polym. Degrad. Stab. **210**, 110291 (2023)

Varimax Rotation-Based Exploratory Factor Analysis of Crime Dataset

Natarajan Meghanathan(✉)

Department of Electrical and Computer Engineering and Computer Science, Jackson State University, 1400 Lynch Street, Jackson, MS 39217, USA
natarajan.meghanathan@jsums.edu

Abstract. We investigate the use of Exploratory Factor Analysis (EFA) on a crime dataset featuring the number of occurrences of the crime types (like Burglary, Robbery, Theft, etc.) in different location types (like Apartment, Restaurant, Street, etc.). We conduct Varimax rotation of the Eigenvectors of the retained principal components to maximize the communality score for any crime type and show EFA to be helpful in identifying the minimum number of hidden factors to dominantly represent the variations in the number of occurrences of the crime types at the different locations. We extract a mapping of any crime type to a dominating factor (the factor incurring the largest of the factor loadings under Varimax rotation). We observe the crime types mapped to the same dominant factor (different dominating factors) to mostly exhibit stronger positive or strong negative correlation (weak-moderate correlation) amongst themselves with respect to their number of occurrences at different locations.

Keywords: Exploratory Factor Analysis · Varimax Rotation · Crime Dataset · Correlation · Eigenvectors · Principal Components

1 Introduction

Exploratory Factor Analysis (EFA [1, 2]) is a technique in the machine learning literature to identify the minimum number of hidden factors that could statistically represent the values incurred for the features in a multi-feature dataset and extract a mapping of each feature to a factor (among the minimum number of factors identified) that could *dominantly* represent them. The first phase of EFA is Principal Component Analysis (PCA [3]), a widely used technique in the machine learning community for dimensionality reduction: i.e., transform a multi-feature (typically referred to as multi-dimensional) dataset to a dataset with relatively fewer dimensions but still retain the almost all the information (i.e., variations in the values incurred for the original features) in the dataset. The minimum number of hidden factors to conduct EFA for a dataset correspond to the reduced number of dimensions obtained by conducting PCA on the dataset. The second phase of EFA involves the use of a rotation technique to maximize the representation of the values incurred for each of the features in the factors. Among the few rotation

R. Silhavy and P. Silhavy (Eds.): CSOC 2024, LNNS 1120, pp. 41–50, 2024.
https://doi.org/10.1007/978-3-031-70518-2_4

techniques available in the literature [4], the Varimax rotation technique focuses on minimizing the number of factors that are needed to maximally represent any feature so that we could come up with a mapping of each feature with a factor that *dominantly* represents them.

Crime data analysis using machine learning algorithms (e.g., [5]) is a widely conducted research topic in the Data Science community. The machine learning algorithms used are typically classification algorithms ([6] such as Logistic Regression, Support Vector Machines, Linear Discriminant Analyzer, etc.) that could be used to build a model to predict the value for a feature of interest (for example, would the perpetrator of a crime be arrested or not?) in the crime dataset based on the values incurred for one or more of the other features in the dataset. Several works (e.g., [7, 8]) in the machine learning literature report that classification models built based on PCA-transformed datasets (i.e., datasets with a reduced number of dimensions and their corresponding values for the records) rather than the original datasets perform (prediction) without statistically any significant loss of accuracy, but at the same time more efficiently (due to the reduced number of dimensions in the underlying dataset used for model building).

Though certain works (e.g., [9, 10]) in the literature have recently studied the use of EFA on crime datasets, we approach the analysis from a perspective that has been hitherto not conducted. We hypothesize that the characteristics of the location type (such as Apartment, Restaurant, Street, Sidewalk, etc.) influences the number of occurrences of the crime types (such as Robbery, Burglary, Motor Vehicle Theft, Offense involving Children, etc.) as well as the correlation between them. We show the use of EFA to extract non-overlapping clusters (corresponding to the minimum number of factors) of the crime types on the basis of their correlation to their number of occurrences in the location types. We observe crime types mapped to the same factor (cluster) tend to exhibit strong positive (or negative) correlation amongst themselves with respect to their number of occurrences across the different location types, whereas crime types mapped to different factors exhibit weak-moderate correlation. We use the crime dataset (featuring the crime types and the location types where the crimes occurred) for the City of Chicago [11] for the year 2022 to showcase our analysis and the inferences that could be made from the EFA results.

The rest of the paper is organized as follows: Sect. 2 illustrates the procedure to conduct EFA using a small synthetic crime dataset. Section 3 presents the results of conducting EFA on the 2022 crime dataset for the City of Chicago. Section 4 reviews related work in the literature and highlights our contributions. Section 5 concludes the paper and presents plans for future work. Throughout the paper, the terms 'feature' and 'crime type' as well as 'factor' and 'cluster' are used interchangeably.

2 Procedure for Exploratory Factor Analysis (EFA)

In this section, we use a synthetic dataset (shown in Fig. 1) to explain the procedure to conduct EFA. The dataset comprises of 6 location types and 5 crime types, and the entries indicate the number of occurrences of a crime type in a location type over a period of time. EFA is conducted in two phases: Principal Component Analysis (PCA) of the crime dataset and Varimax Rotation of the Eigenvectors corresponding to the retained/dominating principal components. We explain each of these phases in detail below using the synthetic dataset of Fig. 1 as a running example.

Phase 1 (Principal Component Analysis): We first conduct Principal Component Analysis (PCA) of the synthetic dataset to identify the minimum number of factors (also referred to as dominating principal components) that would capture 90% or more of the variations among the features (crime types, in our case) in the dataset. Figure 1 displays the raw dataset and its standardized version (based on the mean and standard deviation of the number of occurrences for the crime types) as well as the covariance matrix (Pearson's correlation coefficient [3]) between any two of the five crime types of the dataset.

Raw Dataset

	C1	C2	C3	C4	C5
L1	80	16	70	2	99
L2	3	90	11	65	8
L3	17	50	15	29	12
L4	36	19	68	17	54
L5	80	22	82	6	93
L6	12	25	10	83	23
Mean	37.86	37.06	42.60	33.84	48.11
SD	34.27	28.47	34.05	33.06	40.47

Standardized Dataset

	C1	C2	C3	C4	C5
L1	1.23	-0.72	0.79	-0.96	1.25
L2	-1.02	1.85	-0.92	0.93	-0.99
L3	-0.60	0.47	-0.82	-0.15	-0.89
L4	-0.07	-0.62	0.75	-0.50	0.13
L5	1.22	-0.53	1.17	-0.83	1.12
L6	-0.77	-0.44	-0.97	1.50	-0.61

Covariance Matrix

	C1	C2	C3	C4	C5
C1	1.00	-0.67	0.91	-0.84	0.98
C2	-0.67	1.00	-0.67	0.50	-0.71
C2	0.91	-0.67	1.00	-0.86	0.93
C4	-0.84	0.50	-0.86	1.00	-0.79
C5	0.98	-0.71	0.93	-0.79	1.00

	Variance
PC # 1	4.1759
PC # 2	0.5295
PC # 3	0.2014
PC # 4	0.0884
PC # 5	0.0049

	EV # 1	EV # 2
C1	0.4747	0.1070
C2	-0.3748	0.8571
C3	0.4713	0.1365
C4	-0.4303	-0.4850
C5	0.4763	-0.0053

	PC # 1	PC # 2
L1	2.2453	0.0704
L2	-2.4968	0.8940
L3	-1.2097	0.2849
L4	0.8443	-0.1992
L5	2.2107	0.2401
L6	-1.5938	-1.2902

Fig. 1. Synthetic Crime Dataset and its Principal Component Analysis

PCA would determine five Eigenvectors (corresponding to the number of features/crime types in the dataset). We multiply the standardized dataset (treated as a matrix of 6 rows and 5 columns) with each of the five Eigenvectors (5 rows and 1 column for each Eigenvector) to generate the principal components (PCs, whose entries are the location types). We decide to retain the first two principal components in the sorted order of the variances of the entries (see Fig. 1) whose sum of the variances ($4.1759 + 0.5295 = 4.7054$) is at least 90% of the number of features (5) in the dataset. The number of retained/dominating principal components (two, in this case) also corresponds to the minimum number of hidden factors to be identified by EFA and the corresponding Eigenvectors are passed on to Phase 2 for Varimax rotation.

Phase 2 (Varimax Rotation): We use the entries (referred to as initial factor loadings) for the features (crime types) in the Eigenvectors corresponding to the two dominating principal components of PCA (factors in the context of EFA) as the basis to extract a mapping of the dominating factors to the crime types. A factor is a dominating factor for a crime type if it could alone capture a significant percent of the variations in the feature values. We use the communality score (in a scale of 0.0 to 1.0) [4] as the basis to quantify the percent of variations in a feature captured by the factors, computed as the sum of the squares of the factor loadings for the feature. The communality scores for the features computed based on the initial factor loadings (entries in the two Eigenvectors passed on from PCA) are in the vicinity of 0.20–0.40 for four of the five crime types and 0.88 for crime type C2. We target the communality scores for the features to be at least 0.90 in order to identify the dominating factor (among the factors) for each feature.

Varimax rotation [4] is a coordinate rotation technique to scale the factor loadings for the features so that the communality scores for each feature is maximized (in our case, to at least 0.90). For the synthetic dataset, we would need a two-dimensional coordinate system to perform Varimax rotation, as we identified two factors through Phase 1: PCA. We thus first plot (see Fig. 2) the initial factor loadings of the features in a two-dimensional coordinate system (referred to as the EV-1 and EV-2 axes). We then conduct repeated orthogonal synchronous rotations of the two Eigenvector coordinate-axes until the sum of the squares of the coordinates is maximized for each feature. We use the relevant libraries in Python (Pandas [12]) to perform Varimax rotation.

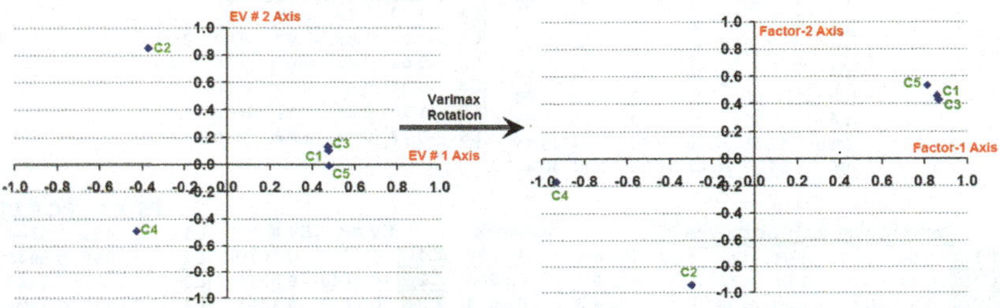

Initial Factor Loadings	EV # 1	EV # 2	Communality Score
C1	0.4747	0.1070	0.237
C2	-0.3748	0.8571	0.875
C3	0.4713	0.1365	0.241
C4	-0.4303	-0.4850	0.420
C5	0.4763	-0.0053	0.227

Final Factor Loadings	Factor-1	Factor-2	Communality Score	Dominating Factor	DFC
C1	0.858	0.459	0.947	Factor-1	0.78
C2	-0.3	-0.94	0.974	Factor-2	0.91
C3	0.865	0.432	0.935	Factor-1	0.80
C4	-0.934	-0.167	0.900	Factor-1	0.97
C5	0.812	0.536	0.947	Factor-2	0.70

Fig. 2. Varimax Rotation of the Eigenvector-Axes to the Factor-Axes

Post Varimax rotation, we refer to the coordinate-axes as the factor-axes (in the case of the synthetic dataset: Factor-1 and Factor-2 axes) and the coordinates (see Fig. 2) for the features are referred to as the final factor loadings; the communality scores (computed based on the final factor loadings) for each feature are at least 0.90. The dominating factor for a feature is then simply the factor whose absolute value for the factor loading is the largest among the factors identified. The contribution of the dominating factors to the feature values is quantified in the form of a dominating factor contribution (DFC) score for a feature, computed as the ratio of the square of the corresponding final factor loading for the dominating factor identified for the feature and the communality score for the feature. The larger the DFC (maximum could be 1.0), the larger the contribution (dominance) of the dominating factor to the feature values. In the case of the synthetic dataset, we identify crime types C1, C3, C4 and C5 to be dominated by Factor-1 and crime type C2 to be dominated by Factor-2, all of which incur a DFC of at least 0.70. Figure 2 presents the two coordinate systems and the factor loadings (initial and final) for the features as well as their corresponding communality scores, dominating factors and the DFC values.

3 EFA for the City of Chicago Crime Dataset

We now present the results of Exploratory Factor Analysis (EFA) conducted on the crime dataset [11] for the City of Chicago for the year 2022. The dataset originally had 21 features and 239,043 records. The features that are of interest to us for EFA performed in this paper are the *Crime type* (like Arson, Assault, Robbery, Burglary, etc.) and the *Location type* (Street, Restaurant, Apartment, etc.). Our hypothesis is that each location type have particular characteristic(s) that would influence the occurrence of the crime and its frequency (number of occurrences). After the pre-processing steps (including removal of records with any missing values for crime type or location type as well as removal of records whose count is less than 100 for any particular crime type or less than 10 for any particular location type), we retained 234,368 records with 15 crime types and 96 location types.

The 15 crime types and the # records for each of them are as follows: C0 – Assault, 20801; C1 – Battery, 40911; C2 – Burglary, 7593; C3 – Criminal Damage, 27237; C4 – Sexual Assault, 1562; C5 – Criminal Trespass, 4219; C6 – Deceptive Practice, 15652; C7 – Motor Vehicle Theft, 21448; C8 – Narcotics, 4731; C9 – Offense involving Children, 1863; C10 – Other Offenses, 14572; C11 – Robbery, 8963; C12 – Sex Offense, 1213; C13 – Theft, 54831 and C14 – Weapons Violation, 8772. We first build a 96 (number of location types, L) x 15 (number of crime types, C) matrix (referred to as the LC matrix) whose entries (i, j) comprise of the number of occurrences of the particular crime type j at location type i. We subject this matrix to PCA and observe just 2 dominating PCs were sufficient to capture $(10.05 + 4.06)/15 = 94.1\%$ of the variances in the crime type occurrence values in the LC matrix.

Figure 3 presents the two corresponding Eigenvectors (whose entries are the 15 crime types and also represent the initial factor loadings, contributing to communality scores less than 0.20) of the dominating PCs using which we will build a 2-dimensional Eigenvector coordinate system that will be subject to Varimax rotation (to maximize the communality scores as well as to showcase a clear alignment of the crime types to their respective dominating factors). In Fig. 3, we also show a tentative grouping (clustering) of the crime types that are likely to be influenced by the same factor. While all the 15 crime types incurred positive entries in EV # 1, crime types C1, C2, C4, C5, C6, C9, C10 and C12 incurred negative entries for EV # 2 and crime types C0, C3, C7, C8, C11, C13 and C14 incurred positive entries for EV # 2. We opine the core of these two groupings to remain the same even after Varimax rotation, with only those crime types at the periphery expected to switch clusters.

Figure 4 presents the final factor loadings for the 15 crime types and their communality scores after the Eigenvector coordinate system of Fig. 3 is subjected to Varimax rotation. In the resulting factor-coordinate system, we observe the communality scores (based on the final factor loadings) for all the 15 crime types to be above 0.90. For crime types C2, C4, C6, C9 and C12: their final factor loadings with respect to Factor-1 are at least 0.95 and less than 0.15 with respect to Factor-2. Likewise, for crime types C7, C8, C11 and C14: their final factor loadings with respect to Factor-2 are at least 0.95 and less than 0.10 with respect to Factor-1. The dominating factor contribution (DFC) scores for all these 9 crime types are 0.98 or above. Since there are only two factors, the final factor loading for any dominating factor for any feature should be above $\sqrt{0.50} = 0.7071$, leading to a DFC score of above 0.50.

Crime Type	EV # 1	EV # 2	Communality Score
C0	0.3107	0.0361	0.098
C1	0.3014	-0.0819	0.098
C2	0.2501	-0.2684	0.135
C3	0.2804	0.1999	0.119
C4	0.267	-0.2308	0.125
C5	0.2953	-0.1264	0.103
C6	0.266	-0.2365	0.127
C7	0.1781	0.3878	0.182
C8	0.1787	0.38	0.176
C9	0.2631	-0.2438	0.129
C10	0.305	-0.1087	0.105
C11	0.1897	0.3716	0.174
C12	0.2763	-0.211	0.121
C13	0.2615	0.2322	0.122
C14	0.189	0.3902	0.188

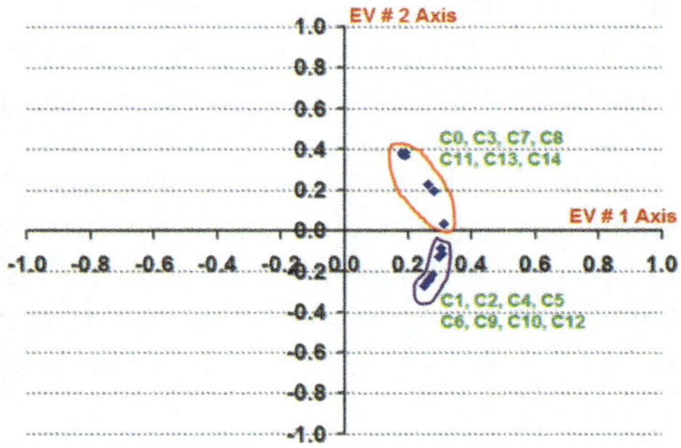

Fig. 3. Initial Factor Loadings for the Crime Types and their Distribution in the Eigenvector Coordinate System that will be subjected to Varimax Rotation

Crime Type	Factor-1	Factor-2	Communality Score	Dominating Factor	DFC
C0	0.782	0.603	0.975	Factor-1	0.63
C1	0.888	0.389	0.940	Factor-1	0.84
C2	0.96	-0.015	0.922	Factor-1	1.00
C3	0.52	0.826	0.953	Factor-2	0.72
C4	0.963	0.078	0.933	Factor-1	0.99
C5	0.922	0.303	0.942	Factor-1	0.90
C6	0.966	0.067	0.938	Factor-1	1.00
C7	0.04	0.963	0.929	Factor-2	1.00
C8	0.051	0.951	0.907	Factor-2	1.00
C9	0.967	0.05	0.938	Factor-1	1.00
C10	0.928	0.35	0.984	Factor-1	0.88
C11	0.089	0.957	0.924	Factor-2	0.99
C12	0.965	0.128	0.948	Factor-1	0.98
C13	0.434	0.847	0.906	Factor-2	0.79
C14	0.067	0.987	0.979	Factor-2	1.00

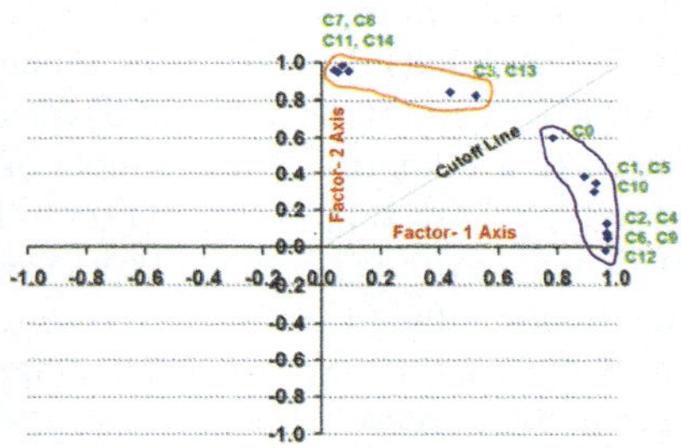

Fig. 4. Final Factor Loadings for the Crime Types and their Distribution in the Factor Coordinate System resulting from Varimax Rotation

Figure 4 presents the distribution of the final factor loadings for the 15 crime types. The relative proximity of the final factor loadings for crime types C2, C4, C6, C9 and C12 to the Factor-1 axis and for crime types C7, C8, C11 and C14 to the Factor-2 axis confirm the choice of Factor-1 and Factor-2 as respectively the dominating factors for these two sets of crime types. These final factor loadings are far away from the cutoff line (representing the DFC of 0.50) in the distribution shown in Fig. 4. The final factor loadings for the other six crime types (C0, C1, C5 and C10, dominated by Factor-1 as well as C3 and C13, dominated by Factor-2) are relatively more closer to the cutoff line, but clearly located in either of the two sides. Comparing the final clustering of the crime types in Fig. 4 to our initial prediction in Fig. 3, we notice C0 is classified as being dominated by a factor (Factor-1) that is different from our initial prediction. The other crime types are clustered as per our initial prediction.

Figure 5 presents heat map visualizations (green: strong; yellow: moderate; red: nothing or weak) of the correlations (with respect to the number of occurrences) between the crime types dominated by the same factor as well as between the crime types dominated by different factors. The correlations between crime types dominated by the same factor are moderately-strongly positive. The correlation strength decreases with increase in distance between crime types in the factor coordinate system.

		C0	C1	C2	C4	C5	C6	C9	C10	C12
Assault	C0	1	0.96	0.75	0.8	0.91	0.77	0.76	0.93	0.83
Battery	C1	0.96	1	0.85	0.91	0.93	0.84	0.85	0.95	0.91
Burglary	C2	0.75	0.85	1	0.91	0.9	0.9	0.89	0.88	0.89
Criminal Sexual Assault	C4	0.8	0.91	0.91	1	0.9	0.9	0.91	0.91	0.94
Criminal Trespass	C5	0.91	0.93	0.9	0.9	1	0.9	0.88	0.95	0.9
Deceptive Practice	C6	0.77	0.84	0.9	0.9	0.9	1	0.98	0.93	0.95
Offense involving Children	C9	0.76	0.85	0.89	0.91	0.88	0.98	1	0.93	0.97
Other Offenses	C10	0.93	0.95	0.88	0.91	0.95	0.93	0.93	1	0.93
Sex Offense	C12	0.83	0.91	0.89	0.94	0.9	0.95	0.97	0.93	1

Factor-1 Cluster: Crime Types Committed in Closed/Private Environment

		C3	C7	C8	C11	C13	C14
Criminal Damage	C3	1	0.87	0.76	0.78	0.95	0.85
Motor Vehicle Theft	C7	0.87	1	0.86	0.86	0.86	0.96
Narcotics	C8	0.76	0.86	1	0.98	0.76	0.94
Robbery	C11	0.78	0.86	0.98	1	0.81	0.94
Theft	C13	0.95	0.86	0.76	0.81	1	0.84
Weapons Violation	C14	0.85	0.96	0.94	0.94	0.84	1

Factor-2 Cluster: Crime Types Committed in Open/Public Environment

	Factor-1 Cluster								
Factor-2 Cluster	C0	C1	C2	C4	C5	C6	C9	C10	C12
C3	0.89	0.76	0.5	0.56	0.74	0.56	0.54	0.79	0.58
C7	0.59	0.37	0.03	0.11	0.34	0.11	0.09	0.4	0.14
C8	0.63	0.45	0.02	0.13	0.3	0.11	0.11	0.37	0.21
C11	0.66	0.48	0.07	0.16	0.35	0.16	0.14	0.4	0.24
C13	0.84	0.69	0.42	0.47	0.7	0.48	0.43	0.7	0.49
C14	0.64	0.43	0.04	0.15	0.35	0.14	0.13	0.41	0.2

Factor-1 Cluster vs. Factor-2 Cluster

Fig. 5. Heatmap Visualization of the Correlations between the Crime Types

Figure 6 presents a plot of the two dominating PCs whose entries correspond to the location types. The four location types Street, Residence, Apartment and Sidewalk are significantly noticeable outliers in the dataset, reporting substantial number of occurrences of the crimes compared to the rest of the location types. The orientation of the data points corresponding to these four location types correspond to the orientation of the crime types in the Eigenvector plot of Fig. 3 and to the final clustering of the factors decided through Varimax rotation. The location types Residence and Apartment are closed/private where people live; whereas, Street and Sidewalk are more open/public where people just commute. The orientation of the Residence and Apartment location types in Fig. 6 aligns with the orientation of the crime types dominated by Factor-1 in Fig. 3 and some of these crime types (especially C2: Burglary and C5: Criminal Trespass) are those that typically occur within a building. On the other hand, the orientation of the Street and Sidewalk location types in Fig. 6 aligns with orientation of the crime types dominated by Factor-2 in Fig. 3 and some of these crime types (especially C7: Motor

Vehicle Theft and C11: Robbery) are those that typically occur in the public domain. Thus, the hidden factors behind the clustering of the crime types in EFA are the crimes occurring in locations that are more closed/private and the crimes occurring in locations that are more open/public.

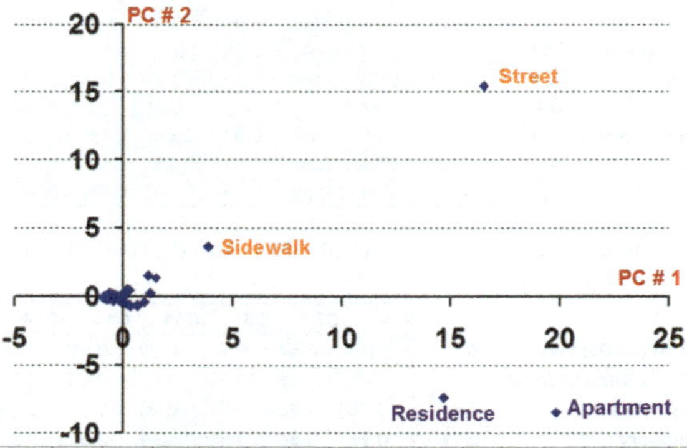

	C0	C1	C2	C3	C4	C5	C6	C7	C8	C9	C10	C11	C12	C13	C14
Street	4157	5922	72	10291	67	383	737	16168	1834	89	2676	3402	61	15571	4872
Apartment	4916	13202	2807	5024	543	870	4321	301	82	669	4929	253	305	6715	197

Fig. 6. Plot of the Principal Components and the Identification of the Closed/Private vs. Open/Public Environments as the Reason for the Clustering of the Crime Types with respect to one of the Two Factors

In Fig. 6, we also show extract of the LC matrix with respect to location types Street and Apartment and highlight the entries corresponding to crime types that incurred the largest values for the number of occurrences. The highlighted cells match to the clustering of the crime types identified through EFA. The location type ‘Street’ reported relatively larger number of occurrences for crime types {C3, C7, C8, C11, C13, C14}, whereas the location type ‘Apartment’ reported relatively larger occurrences for crime types {C0, C1, C2, C4, C5, C6, C9, C10, C12}. The above observation confirms our hypothesis that the characteristic of the location types (closed/ private vs. open/ public) could influence the number of occurrences of the crime types.

4 Related Work

The following related works have reported factor analysis of crime data. In [9], the authors conduct factor analysis on crime data in the US cities to explore correlations between *social factors* (such as province and local management costs, corruption, % population in different age groups, joblessness proportion, etc.) that could influence the occurrence of crimes. There were 14 such social factors in the crime dataset in US cities studied in [9] and four principal components were required to statistically capture at least 80% of the variances in the original dataset. After orthogonal rotation, four groupings of the social factors exhibited strong positive or negative correlation within themselves. In [10], the authors fitted an OLS regression model to predict the violent crime rate in the US (50 states and Washington DC) as a linear function of the values observed for the following four social factors: population density per square mile, unemployment rate, percentage of people living in poverty and gun ownership rate. Based on the parameters estimated for the model, [10] concluded that increases in unemployment rate, gun ownership rate and poverty rate (in this order) tend to increase the violent crime rate. Also, urban locations reported more crimes than rural locations. The work [13] reported just PCA (not factor rotation) on the crime data of the Katsina state in Nigeria and observed significant correlations between crimes such as robbery, theft and vehicle theft; moderate correlations between crimes against property and no correlation between crimes against person. In [14]: government policies and regulations, responses and management, capacity building and awareness as well as litigation were identified the main issues that could influence safety and security of the financial system in Nigeria. In a factor analysis study [15] on the generality of sex offenders vis-a-vis their criminal behavior, sexual aggressors of women were observed to be more versatile with respect to committing other crimes compared to sexual aggressors of children (the latter were observed to be more specific in the crimes committed). We also observe crime type C9 (Offense involving children) to be strongly correlated to crime types C12 (Sex offense) and C6 (Deceptive practice), but exhibit only weak-moderate correlation with a majority of the other crime types.

5 Conclusions and Future Work

The high-level contribution of this paper is the illustration of varimax-based rotation to conduct exploratory factor analysis (EFA) of a crime dataset to identify clusters of crime types (each cluster dominated by a particular factor) whose number of occurrences at different location types are moderately-strongly correlated within the cluster and weak-moderately correlated between clusters. We used the crime data for the City of Chicago for the year 2022 to showcase our analysis. With the aid of visual plots of the Eigenvectors and the corresponding dominating principal components, we observe the characteristics of the location type (especially, closed/ private vs. open/ public) influence the mutually exclusive two factors with respect to which the clusters of crime types are identified with, thus confirming our hypothesis. The procedure presented here can be applied on any crime dataset that includes similar features (such as crime types and location types) and would work even if the number of dominating factors is greater than 2. In future, we plan to apply the procedure to datasets of health records (featuring the number of occurrences of diseases among people of different regions such as counties, states, countries, etc.) to explore the hidden factors that would influence the correlated or non-correlated occurrence of the diseases within a region as well as across regions.

Acknowledgements. Research was sponsored by the Army Research Office and was accomplished under Grant Number W911NF-21-1-0264. The views and conclusions contained in this document are those of the authors and should not be interpreted as representing the official policies, either expressed or implied, of the Army Research Office or the U.S. Government. The U.S. Government is authorized to reproduce and distribute reprints for Government purposes notwithstanding any copyright notation herein.

References

1. Gorsuch, R.L.: Factor Analysis, 2nd Edition, Lawrence Erlbaum Associates (1983)
2. Loehlin, J.C.: Latent Variable Models: An Introduction to Factor, Path, and Structural Equation Analysis, 4th edn. Psychology Press (2003)
3. Strang, G.: Linear Algebra and its Applications, 4th edn. Pacific Grove (2006)
4. Darton, R.A.: Rotation in factor analysis. Statistician **29**(3), 167 (1980). https://doi.org/10.2307/2988040
5. McClendon, L., Meghanathan, N.: Using machine learning algorithms to analyze crime data. Mach. Learn. Appl.: An Int. J. **2**(1), 1–12 (2015)
6. Aggarwal, C.C. (ed.), Data Classification: Algorithms and Applications, 1st Edition, Chapman and Hall/CRC (2014)
7. Nasution, M.Z.F., Sitompul, O.S., Ramli, M.: PCA based feature reduction to improve the accuracy of decision tree c4.5 classification. J. Phys: Conf. Ser. **978**(012058), 1–6 (2018)
8. Escamilla, A.K.G., Hajjam, A., Andres, E.: Classification models for heart disease prediction using feature selection and PCA. Inform. Med. Unlocked **19**(3), 100330 (2020)
9. Doulah, M., Darwin, C., Salma, U., Hamid, M., Aktar, M.: Factor analysis of crime data. Res. Rev. J. Stat. **9**(2), 20–32 (2020)
10. Bhattacharya, A.: Analysis of the factors affecting violent crime rates in the US. Int. J. Eng. Manage. Res. **10**(5), 106–109 (2020)
11. Chicago Data Portal: https://data.cityofchicago.org/Public-Safety/Crimes-2001-to-Present/ijzp-q8t2/data
12. Pandas: https://pypi.org/project/factor-analyzer/
13. Gulumbe, S.U., Dikko, H.G., Bello, Y.: Analysis of crime data using principal component analysis: a case study of Katsina state. CBN J. Appl. Stat. **3**(2), 39–49 (2012)
14. Iwasokun, G.B., Akinyede, R.O., Fadamiro, C.D., Bello, O.A.: Factor analysis of financial crime-related issues. J. Financ. Crime **26**(1), 113–130 (2019)
15. Lussier, P., LeBlanc, M., Proulx, J.: The generality of criminal behavior: a confirmatory factor analysis of the criminal activity of sex offenders in adulthood. J. Crim. Just. **33**(2), 177–189 (2005)

Decision Support Systems and Decision Trees: An Overview

Serafeim A. Triantafyllou(✉)

Greek Ministry of Education and Religious Affairs, Athens, Greece
ser.triant@gmail.com

Abstract. Decision Support Systems (DSS) are information systems that help to solve decision problems. Specifically, they are computer-based tools designed to assist individuals or organizations in making complex decisions. These systems utilize data, models, algorithms, and user interfaces to help users analyze information and evaluate possible outcomes in decision-making processes. They usually combine: (a) knowledge databases, (b) mathematical models of the problem to be solved, (c) algorithms/techniques for solving the model, and d) a dialog system with the user (software and user interface). To better understand the various forms of decision problems, each one of them demands the introduction of proper theoretical tools. Investigating decision support systems can involve various aspects, from understanding their conceptual framework to exploring their practical applications and impact. This paper tries to contribute to a better understanding of DSS because investigating DSS is crucial for improving decision-making processes, enhancing organizational efficiency, gaining competitive advantage, understanding technological advancement, addressing complex problems, and improving user experience.

Keywords: Decision support systems (DSS) · Decision-making processes · Information Systems

1 Introduction

Decision support systems (DSS) are used in various fields such as finance, healthcare, marketing, and logistics. They assist in making both structured decisions and unstructured decisions. By employing DSS, organizations can benefit from improved decision-making processes, enhanced analysis, better insights into potential outcomes, and ultimately, more informed and effective decisions. These systems help users consider various scenarios, weigh alternatives, and understand potential risks and benefits before making a choice.

DSS are computer-based tools designed to assist individuals or organizations in making complex decisions. These systems utilize data, models, algorithms, and user interfaces to help users analyze information and evaluate possible outcomes in decision-making processes [1, 2]. There are several key components within a typical decision support system such as [2, 3]: (i) **Knowledge Database**: DSS systems often rely on a

R. Silhavy and P. Silhavy (Eds.): CSOC 2024, LNNS 1120, pp. 51–59, 2024.
https://doi.org/10.1007/978-3-031-70518-2_5

database that contains relevant data needed for decision-making. This data can include historical information, current statistics, or any other pertinent information required for analysis, (ii) **Model Base**: DSS may include models and analytical tools to help process data and predict outcomes. These models could be mathematical, statistical, or simulation-based, (iii) **User Interface**: DSS systems provide an interface for users to interact with the system. This interface can range from simple query and reporting tools to more complex visualization and decision modeling interfaces, (iv) **Decision Support Engine**: This is the core of the system that processes user input, data, and models to generate output that aids in decision-making.

DSS aims to enhance the quality of decisions, increase efficiency in decision-making processes, and provide a structured approach to handling complex problems. They do not make decisions on behalf of users but rather provide valuable information and analysis to facilitate informed decision-making.

2 The Process of Decision-Making

The following figure (see Fig. 1) illustrates the decision-making model where: S is the process which generates the result (according to the selected decision) and n is the term which represents external unpredicted factors that act to the system and are not under users' control [3–11].

Before developing DSS software, the following steps should be taken into consideration [4, 11, 12]:

- Study of the decision problem and deep understanding of the main objectives.
- Searching and finding all the alternative choices about the stated problem ("decisions").
- Problem decomposition into a mathematical model which encodes the problem structure, the uncertainty that exists when comparing decisions and results, and the preferences of the final user according to the predicted final results.
- Examining if the proposed mathematical model responds to the initially stated problem to be solved.
- Implementation and examination of the final outcomes. Conversions and repetition of the previous described stages if it is necessary.

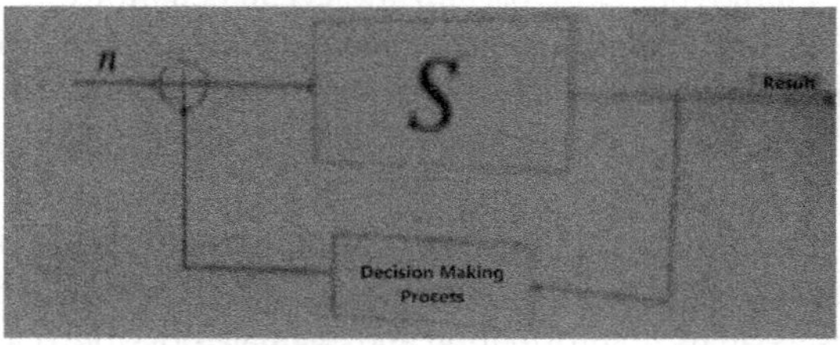

Fig. 1. Decision Making Process

3 Classification of Decision Support Systems

Decision support systems (DSS) come in various forms, each serving a specific purpose in aiding decision-making processes [2, 13]. A typical classification of decision support systems is the following:

- **Structured Decision Support Systems**
- **Database Management Systems (DBMS)**: They store, manage, and retrieve structured data efficiently.
- **Model-Driven DSS**: They use mathematical models and algorithms to analyze data and support decision-making.
- **Spreadsheet-Based Systems**: They utilize spreadsheet software for organizing and analyzing data.
- **Semi-Structured Decision Support Systems**
- **Data Warehouses**: They store and integrate large amounts of data from multiple sources for analysis and reporting.
- **Online Analytical Processing (OLAP)**: It enables multidimensional analysis of data for complex queries and insights.
- **Data Mining Systems**: Employ algorithms to discover patterns and relationships in large datasets.
- **Unstructured Decision Support Systems**
- **Knowledge-Based Systems**: Incorporate human expertise to solve complex problems by using rules and logic.
- **Natural Language Processing (NLP) Systems**: Interpret and analyze human language for decision-making purposes.
- **Intelligent Agents**: Autonomous software entities that gather information and perform tasks on behalf of users.
- **Collaborative Decision Support Systems**
- **Group Decision Support Systems (GDSS)**: Facilitate group interactions and decision-making through technology.
- **Social Decision Support Systems**: Leverage social media or collaborative platforms for decision-making in social contexts.
- **Enterprise-Wide Decision Support Systems**
- **Executive Information Systems (EIS)**: Provide top-level executives with summarized reports and key performance indicators for strategic decision-making.
- **Enterprise Resource Planning (ERP) Systems**: Integrate various aspects of a business (such as finance, HR, and supply chain) to support decision-making across departments.
- **Customer Relationship Management (CRM) Systems**: Manage interactions with customers and aid in customer-focused decision-making.

Each level or type of decision support system has its own strengths and applications, catering to different decision-making needs across various domains and organizational structures. The classification is not necessarily strict, as many systems can combine elements from multiple categories to provide more comprehensive decision support.

4 Decision Trees

Decision trees represent a flowchart for making decisions based on different conditions or attributes. In machine learning, decision trees are a powerful tool for classification and regression tasks [11, 14, 17]. They break down a dataset into smaller subsets while incrementally making decisions based on features, ultimately leading to a prediction or decision at the tree's leaves. Specifically, they recursively partition the data into subsets based on the values of input features. Each node in the tree represents a feature, and the branches represent the decision rules based on that feature. At each step, the algorithm selects the feature that best splits the data into homogeneous subsets, aiming to maximize the information gain or minimize impurity. This process continues until a stopping criterion is met, such as reaching a maximum tree depth or having subsets that are sufficiently pure [11, 14–26].

For classification tasks, decision trees assign a class label to each leaf node based on the majority class within that subset. In regression tasks, the leaf node contains the predicted value, usually the mean or median of the target variable within that subset [11, 14, 15, 17].

Decision trees are easy to interpret and visualize, allowing humans to understand the decision-making process. However, they can be prone to overfitting, especially when the tree becomes too complex. Techniques like pruning or using ensemble methods like Random Forests or Gradient Boosting help mitigate this issue [16]. Decision trees are quite versatile and find applications in various fields like finance, healthcare, and natural language processing due to their ability to handle both numerical and categorical data.

5 Solving Decision Trees Through the Calculation of the Expected Value

In decision trees, the expected value is a calculation used to determine the overall value or payoff associated with a particular decision or series of decisions. It involves considering the probabilities of different outcomes and their associated values to make an informed choice [18]. A simplified explanation is the following:

Decision Nodes: These represent points in the decision tree where a choice must be made. At these nodes, different options or paths are available.
Chance Nodes: These nodes denote uncertain events or probabilities. They represent possible outcomes and the probabilities associated with each outcome.
Terminal Nodes: Also known as end nodes or leaf nodes, these represent the final outcomes or payoffs associated with a particular path through the decision tree.

To calculate the expected value at a decision node, we can typically:

- **Assign Probabilities**: Estimate the probabilities of different outcomes resulting from the decision made at that node. For instance, if we had two choices with potential outcomes A and B, we would assign probabilities P(A) and P(B) to each.
- **Assign Values**: Associate values or payoffs with each of these outcomes. These values could represent profits, costs, utility, etc.

The expected value is then computed by multiplying each outcome's probability by its associated value and summing up these products. Mathematically, it's represented as:

$$Expected\ Value = P(A) \times Value\ of\ Outcome\ A + P(B) \times Value\ of\ Outcome\ B + \ldots$$

This calculation helps in making decisions by selecting the option with the highest expected value, as it indicates the most favorable outcome on average. This concept is extensively used in fields like finance, economics, game theory, and artificial intelligence for decision-making processes and risk analysis.

Let us consider a simple example: Supposing that we have two options: investing in Stock A or Stock B. The decision tree is constructed in the following form (see Fig. 2):

Fig. 2. Decision tree example

If someone chooses Stock A, there's a 60% chance of gaining $50, 20% chance of gaining $0, and 20% chance of losing $50. If someone chooses Stock B, there's a 50% chance of gaining $30, 30% chance of gaining $20, and 20% chance of losing $10. To calculate the expected value:

For Stock A: *Xpected value for Stock A = (Probability of outcome 1 * Value of outcome 1) + (Probability of outcome 2 * Value of outcome 2) + (Probability of outcome 3 * Value of outcome 3).*

*Expected value for Stock A = (0.6 * $50) + (0.2 * $0) + (0.2 * (−$50))*

Expected value for Stock A = $30 + $0 − $10

Expected value for Stock A = $20.

For Stock B: *Expected value for Stock B = (Probability of outcome 1 * Value of outcome 1) + (Probability of outcome 2 * Value of outcome 2) + (Probability of outcome 3 * Value of outcome 3).*

*Expected value for Stock B = (0.5 * $30) + (0.3 * $20) + (0.2 * (-$10))*

Expected value for Stock B = $15 + $6 − $2

Expected value for Stock B = $19

Therefore, based on the calculated expected values, Stock B appears to have a slightly higher expected value ($19) compared to Stock A ($20). This suggests that, on average, Stock B might yield slightly better returns compared to Stock A.

6 Using Machine Learning Algorithms and Decision Trees to Compute Expected Values

A general process of using machine learning algorithms, especially decision trees, to compute expected values is the following:

Data Preparation: Gathering of historical data that includes features (inputs) and their corresponding outcomes or values. This data should represent the decision-making scenarios we want to model.
Model Training: Use of the historical data to train a decision tree model. During training, the algorithm learns patterns and relationships between input features and outcomes.
Predicting Outcomes: Once the model is trained, it can be used to predict outcomes for new or hypothetical scenarios. For each scenario, the model provides predictions about the potential outcomes and their associated probabilities.
Calculating Expected Values: With the predicted outcomes and their probabilities, it is important to compute the expected value for each decision path. Next, we should multiply the predicted value of each outcome by its probability, and sum these products to find the expected value.
Decision-Making: We should use these expected values to guide decision-making. Opting for the decision path with the highest expected value, as it suggests the most favorable outcome on average.

For instance, in a financial context, a decision tree model trained on historical market data could predict the potential returns of different investment options. By assigning probabilities to various market conditions and the corresponding returns, the model could estimate the expected value of each investment decision.

Machine learning models, including decision trees, can assist in calculating expected values by leveraging historical patterns and relationships in data. They provide a systematic and data-driven approach to decision-making, allowing for more informed choices based on the expected outcomes of different decisions.

7 Discussion

Decision Support Systems (DSS) offer a range of usability across various industries and functions. Their primary goal is to support decision-making processes by providing valuable information, analysis, and models to aid individuals or organizations in making better choices. Here are some key aspects of their usability:

- **Enhanced Decision-Making**: DSS facilitates better decisions by providing access to structured and unstructured data, analytical tools, and models. It assists in evaluating alternatives, predicting outcomes, and exploring various scenarios, which leads to more informed and often optimal decisions.

- **Improved Efficiency**: DSS streamline decision processes by automating routine tasks, organizing data, and presenting relevant information in a user-friendly format. This efficiency can save time and resources for individuals or organizations.
- **Flexibility and Adaptability**: DSS often allows customization to suit specific needs. They can adapt to changing conditions, incorporate new data sources, and adjust models to reflect updated information, enhancing their relevance and usefulness over time.
- **Support for Complex Decisions**: In scenarios where decisions are complex, involving numerous variables or uncertain outcomes, DSS helps by breaking down complexity, providing insights, and supporting users in navigating intricate decision landscapes.
- **Cross-Functional Applications**: DSS find application across various domains such as finance, healthcare, logistics, marketing, and more. They can assist in inventory management, financial planning, strategic management, medical diagnostics, and other decision-centric fields.
- **Collaboration and Communication**: Some DSS allow collaborative decision-making, enabling multiple stakeholders to access and contribute to the decision-making process. They facilitate communication among team members, aiding consensus-building or negotiation.
- **Risk Management and Analysis**: DSS often incorporate risk assessment tools and models, allowing users to evaluate potential risks associated with decisions. This assists in developing strategies to mitigate risks or uncertainties.
- **Learning and Insights**: Over time, DSS accumulate data and learn from past decisions, providing insights into trends, patterns, and successful strategies. This learning can further refine decision-making processes.
- **Strategic Planning**: DSS can aid in long-term strategic planning by providing tools for scenario analysis, forecasting, and evaluating the implications of different strategic choices.

8 Conclusion

In summary, Decision Support Systems (DSS) offer significant usability in improving decision-making, enhancing efficiency, managing risks, and supporting various domains with their adaptability and analytical capabilities. DSS is used in various domains, including business, healthcare, finance, and logistics, among others. They assist managers, analysts, and decision-makers in assessing alternatives, understanding potential risks, and evaluating the implications of different decisions.

Thus, investigating DSS can contribute new findings to the existing literature. These findings can be used in various domains and help in improving decision-making processes. Future studies aim to investigate more this continuously evolving research domain to contribute new research findings.

References

1. Liu, S., Duffy, A.H., Whitfield, R.I., Boyle, I.M.: Integration of decision support systems to improve decision support performance. Knowl. Inf. Syst. **22**, 261–286 (2010)

2. Bonczek, R.H., Holsapple, C.W., Whinston, A.B.: Foundations of Decision Support Systems. Academic Press (2014)
3. Marakas, G.M.: Decision Support Systems in the 21st century, vol. 134. Prentice Hall, Upper Saddle River (2003)
4. Goumagias, N.D., Hristu-Varsakelis, D., Saraidaris, A.: A decision support model for tax revenue collection in Greece. Decis. Support. Syst. **53**(1), 76–96 (2012)
5. Triantafyllou, S.A.: A detailed study on the 8 queens problem based on algorithmic approaches implemented in PASCAL programming language. In: Silhavy, R., Silhavy, P. (eds.) Software Engineering Research in System Science: Proceedings of 12th Computer Science On-line Conference 2023, Volume 1, pp. 160–173. Springer International Publishing, Cham (2023). https://doi.org/10.1007/978-3-031-35311-6_18
6. Triantafyllou, S.A.: TPACK and toondoo digital storytelling tool transform teaching and learning. In: Florez, H., Gomez, H. (eds.) Applied Informatics: 5th International Conference, ICAI 2022, Arequipa, Peru, October 27–29, 2022, Proceedings, pp. 338–350. Springer International Publishing, Cham (2022). https://doi.org/10.1007/978-3-031-19647-8_24
7. Triantafyllou, S.A.: A quantitative research about MOOCs and EdTech tools for distance learning. In: Auer, M.E., El-Seoud, S.A., Karam, O.H. (eds.) Artificial Intelligence and Online Engineering: Proceedings of the 19th International Conference on Remote Engineering and Virtual Instrumentation, pp. 521–532. Springer International Publishing, Cham (2023). https://doi.org/10.1007/978-3-031-17091-1_52
8. Triantafyllou, S.A.,Sapounidis, T.: Game-based learning approach and serious games to learn while you play. In: 2023 IEEE World Engineering Education Conference (EDUNINE) (2023). https://doi.org/10.1109/edunine57531.2023.10102872
9. Triantafyllou, S.A.: Magic squares in order 4K+2. In: 2022 30th National Conference with International Participation (TELECOM) (2022). https://doi.org/10.1109/telecom56127.2022.10017312
10. Triantafyllou, S.A.: WORK IN PROGRESS: Educational technology and knowledge tracing models. In: 2022 IEEE World Engineering Education Conference (EDUNINE) (2022). https://doi.org/10.1109/edunine53672.2022.9782335
11. Hess, T.J., Rees, L.P., Rakes, T.R.: Using autonomous software agents to create the next generation of decision support systems. Decis. Sci. **31**(1), 1–31 (2000)
12. Finlay, P.N., Forghani, M.: A classification of success factors for decision support systems. J. Strateg. Inf. Syst. **7**(1), 53–70 (1998)
13. De Ville, B.: Decision trees. Wiley Interdisc. Rev.: Comput. Stat. **5**(6), 448–455 (2013)
14. Kotsiantis, S.B.: Decision trees: a recent overview. Artif. Intell. Rev. **39**, 261–283 (2013)
15. Esposito, F., Malerba, D., Semeraro, G., Kay, J.: A comparative analysis of methods for pruning decision trees. IEEE Trans. Pattern Anal. Mach. Intell. **19**(5), 476–491 (1997)
16. Barker, K., Wilson, K.J.: Decision trees with single and multiple interval-valued objectives. Decis. Anal. **9**(4), 348–358 (2012)
17. Danielson, M., Ekenberg, L., Larsson, A.: Distribution of expected utility in decision trees. Int. J. Approximate Reasoning **46**(2), 387–407 (2007)
18. Triantafyllou, S.A.: Use of business information systems to achieve competitiveness. In: 2022 13th National Conference with International Participation (ELECTRONICA) (2022). https://doi.org/10.1109/electronica55578.2022.9874433
19. Triantafyllou, S.A., Georgiadis, C.K.: Gamification design patterns for user engagement. Informat. Educ. (2022). https://doi.org/10.15388/infedu.2022.27
20. Triantafyllou, S.A.: What philosophy can teach us about games? In: Proceedings of The 7th International Conference on Social Sciences, Humanities and Education (2022). https://doi.org/10.33422/7th.icshe.2022.12.20

21. Triantafyllou, S., Georgiadis, C.: Gamification of MOOCS and security awareness in corporate training. In: Proceedings of the 14th International Conference on Computer Supported Education (2022). https://doi.org/10.5220/0011103000003182
22. Triantafyllou, S.A.: Game-Based Learning and interactive educational games for learners – an educational paradigm from Greece. In: Proceedings of the 6th International Conference on Modern Research in Social Sciences (2022). https://doi.org/10.33422/6th.icmrss.2022.10.20
23. Triantafyllou, S.A.: Understanding and designing turing machines with applications to computing. In: Farhaoui, Y., Hussain, A., Saba, T., Taherdoost, H., Verma, A. (eds.) Artificial Intelligence, Data Science and Applications: ICAISE'2023, Volume 1, pp. 142–148. Springer Nature Switzerland, Cham (2024). https://doi.org/10.1007/978-3-031-48465-0_19
24. Triantafyllou, S.A: MOOCs and EdTech Tools for distance learning. In: Proceedings of The 4th World Conference on Future of Education (2021). https://doi.org/10.33422/4th.wcfeducation.2021.12.05
25. Triantafyllou, S.A.: A chat client-server application for E-learning. In: 2023 31st National Conference with International Participation (TELECOM), Sofia, Bulgaria, pp. 1–4 (2023). https://doi.org/10.1109/TELECOM59629.2023.10409660
26. Triantafyllou, S.A.: Knowledge discovery systems: an overview. In: Silhavy, R., Silhavy, P. (eds.) Software Engineering Methods in Systems and Network Systems: Proceedings of 7th Computational Methods in Systems and Software 2023, vol. 1, pp. 150–159. Springer International Publishing, Cham (2024). https://doi.org/10.1007/978-3-031-53549-9_16

PID Control Tuning Technique for Temperature Control in Cotton Fiber Dyeing Using Neural Networks and PLC

Roguer Contreras Layme, Jorge Galindo Flores, and Guillermo Zárate Segura(✉)

Facultad de Ingeniería y Sistemas,Universidad Tecnológica del Perú, Lima, Peru
{U18102678,U17213469,E15040}@utp.edu.pe

Abstract. This research work focuses on improving temperature control in the cotton fiber treatment process by developing an advanced PID control tuning technique. The proposed methodology uses a Programmable Logic Controller (PLC) and the integration of artificial neural networks to automatically optimize the PID control parameters based on the process conditions. The study explores classical PID tuning, software autotuning, and neural network autotuning. These methods are compared in terms of reference signal tracking, response time and system stability. The results indicate that the implementation of neural networks significantly improves the responsiveness to changes in the set point, achieving a faster settling time and better damping compared to classical and software autotuning methods. Although neural networks present challenges in terms of interpretability, the benefits in control performance suggest their viability in industrial environments for the control of nonlinear or complex systems.

Keywords: Temperature control · PLC · identification techniques · neural networks · PID control

1 Introduction

Temperature is a variable widely used in manufacturing industry processes and also in the domestic sector, along with pressure and speed [1]. Being able to control the temperature is very important since it is used in a wide range of applications, among which are some processes in the furniture sector, manufacturing of clay partitions, in the asphalt industry, in food industry processes, among others; this is mentioned by authors such as [1–3]. In the textile industry, precise temperature control is essential to ensure quality and consistency in the cotton fiber dyeing process [4]. Proportional-Integral-Derivative (PID) Control is widely used to maintain temperature in control systems, offering advantages such as greater precision, resistance to adverse external conditions, and flexible temperature adjustment. This extends the life of the heating system [5]. However, manual tuning of PID parameters can be inefficient and costly due to fluctuations in the dyeing process [4].

In this context, the present research focuses on the development of an advanced PID control tuning technique using a Programmable Logic Controller (PLC) and the

R. Silhavy and P. Silhavy (Eds.): CSOC 2024, LNNS 1120, pp. 60–71, 2024.
https://doi.org/10.1007/978-3-031-70518-2_6

integration of artificial neural networks. This technique seeks to significantly improve temperature control in cotton fiber dyeing by automatically optimizing the PID control parameters based on the process conditions.

This research work dives into the theoretical and practical aspects of PID control tuning, exploring how neural networks can be implemented in an industrial environment for the determination of PID parameters that will be used in a PLC [5]. The implementation steps, from system initialization to real-time tuning, will be analyzed in detail and their effectiveness in terms of temperature control accuracy and efficiency will be evaluated. The outcome of this research has the potential to improve the textile industry by offering an automated and adaptive approach to temperature control in cotton fiber dyeing.

2 State of the Art

Several authors, such as [7–12], have investigated the tuning of PID control using techniques based on neural networks, such as NARX, ANFIS, MLP and genetic networks. -diffuse. Most of these studies employed experimental methodologies. For example, [11] developed a control model for a greenhouse and compared the performance of neural networks, such as genetic and fuzzy algorithms, in simulations with MATLAB. [12] he also used similar approaches, but this time real data was collected, and the system was modeled. [9] he described the use of artificial neural networks and the construction of a plant emulator using an MLP network, which he worked in conjunction with a neural network to adjust the parameters of the PID controller automatically

The results of these investigations were mostly positive, since they demonstrated the effectiveness of neural networks in tuning PID parameters. [13] Suggests that existing systems can be improved through the use of machine learning and neural networks. [7] he highlighted the resilience of this technique to noise and unforeseen disturbances, which represents an advance in the use of neural networks to tune PID parameters.

However, [13] he points out that, despite their potential to improve control performance, neural networks are not widely used in real-world control systems due to lack of interpretability. The PID parameters suggested by these networks cannot be easily evaluated, raising concerns about system security. This is crucial in control engineering, since interpretability is essential to ensure input-output stability. Therefore, the use of neural networks in PID parameter tuning is not yet widely accepted due to its "black box" nature [14].

3 PID Controller Development

Three methodologies were proposed to determine the parameters of the PID control, the classic method where the parameters are found theoretically, the software self-tuning method and the calculation of parameters through the use of a neural network.

3.1 PID Controller with Zieglers-Nicols (Classical Method)

The Ziegler-Nichols method is a useful and fast PID control tuning technique for simple and linear systems [15]. However, it has limitations in terms of ensuring optimal performance and may require additional adjustments. Your choice depends on the applicability and specific requirements of the system in question. To calculate the PID parameters [15] the following table is used:

Table 1. Parameter calculation using ZN in open loop [15].

Type	Kp	Ti	Td
P	T/L	Inf	0
PI	0.9(T/L)	L/0.3	0
PID	1.2(T/L)	2L	0.5L

The tuning equations for this method are:

$$K_i = K_p/T_i \tag{1}$$

$$K_d = K_p * T_d * 2 \tag{2}$$

where, K_i is the integral gain, K_p corresponds to the proportional gain, K_d is the derivative gain, T_i is the integral time and T_d is the derivative time.

The equations were obtained experimentally by carrying out multiple procedures that were carried out in controlled laboratory environments. These are based on a first-order model with delay, identified by the tangent method, for the operation of the control loop as a regulator with a ideal PID controller [16].

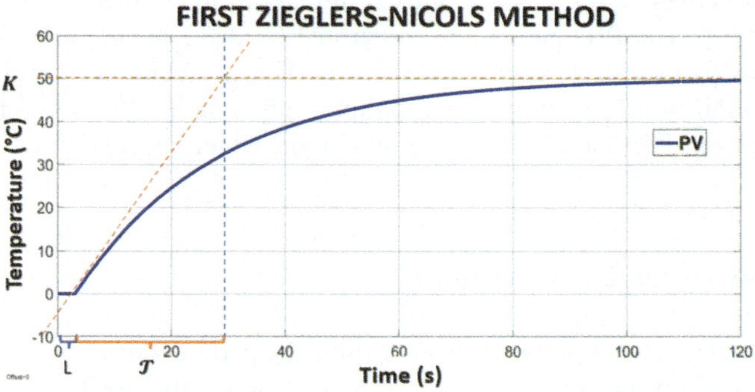

Fig. 1. Tangent plot on the curve according to the Ziegler and Nichols step response method.

From Fig. 1, it is observed that where the line intersects, the variables for the transfer function are obtained. Variables according to transfer function graph: K = 50, L = 3, T

= 23 Calculations of the variables according to the Ziegler and Nichols table (Table 1): $k_p = 0.184$, $T_i = 6$ and $T_d = 1.5$.

Calculations of PID variables: The step response method technique, which was devised by Ziegler and Nichols, was used to determine an initial set of PID controller parameters. Subsequently, these parameters were adjusted by trial and error until an adequate response was achieved [17]. We replace the obtained values of T_i and T_d in Eq. 1 and 2 respectively to obtain the final values of K(i) and K(d) for controller configuration. Being the following: K(p) = 0.184; K(i) = 0.031 and K(d) = 0.552.

The representation of the system reaction using the controller adjusted using the classical approach is presented in Fig. 2:

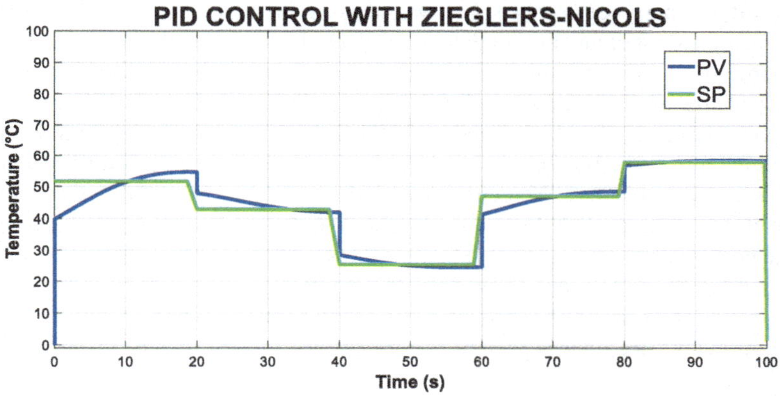

Fig. 2. System response with the controller tuned to the Ziegler and Nichols step response method.

From Fig. 2, it is observed that the PV line (process value) manages to follow the reference signal (SP), however, when there are changes in the parameters, the system fails to maintain the desired behavior by following the reference signal. Reference.

3.2 Self-tuning PID Controller (Software Method MATLAB)

MATLAB autotuning is a powerful tool for tuning PID controllers [18], especially in applications where simplicity and efficiency are critical. However, the self-tuning PID controller operates by introducing a test signal for the purpose of collecting the input and output data of the system, thus allowing the estimation of the frequency response [19]. A Simulink block with an automatic tuner has been implemented in MATLAB to calculate the plant response frequency in real time from the input and output data. The controller configuration is depicted in Fig. 3.

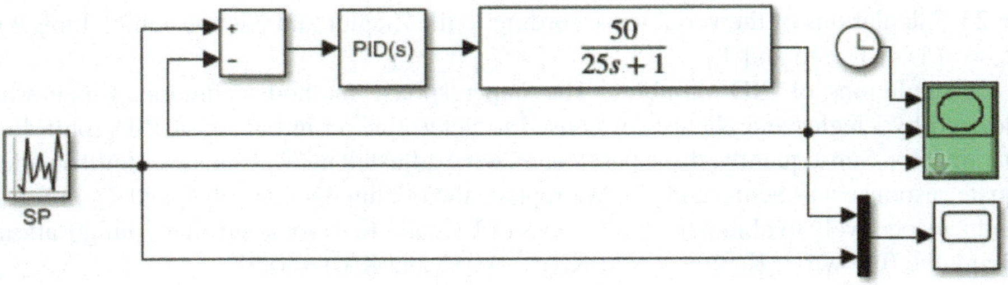

Fig. 3. PID control structure with Auto-tuning.

The autotuning process consists of two steps: in the first, a test signal is introduced into the plant and input and output data are collected to estimate the frequency response in real time. PID gains are then calculated based on this estimate. In the second step, these new gains are transferred to the PID controller, replacing the old ones, and the system becomes controlled by the updated PID, thus completing the self-tuning process. This online tuning process seeks to balance system performance and stability to achieve a specific control objective. You can determine more appropriate PID gains compared to manual calculations. The duration of the start and stop process influences obtaining accurate profits, with a longer period being necessary for complex systems. Sometimes, accuracy is sacrificed to reduce adjustment time [19].

The representation of the system reaction using the controller adjusted through Autotuning with MATLAB is presented in Fig. 4:

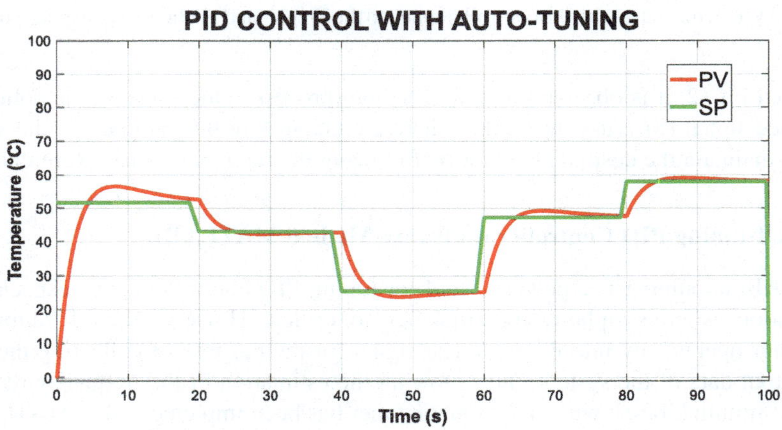

Fig. 4. System response with the controller tuned with the Autotuning method by MATLAB.

From Fig. 4, it is observed that the PV line (process value) manages to follow the reference signal (SP), however, when there are changes in the parameters, the system fails to maintain the desired behavior by following the reference signal Reference.

3.3 Self-tuning PID Controller Using a Neural Network

A self-tuning PID controller based on neural networks can be a powerful tool to improve the control of complex processes [10, 13]. However, its implementation entails challenges in terms of complexity and data requirements [7]. The choice of using a feed forward neural network to tune a PID controller is based on its learnability, adaptability, and ability to model complex systems. However, training data, network architecture, and overfitting risks must be carefully considered during implementation. The representation of the feed forward neural network is as follows (Fig. 5).

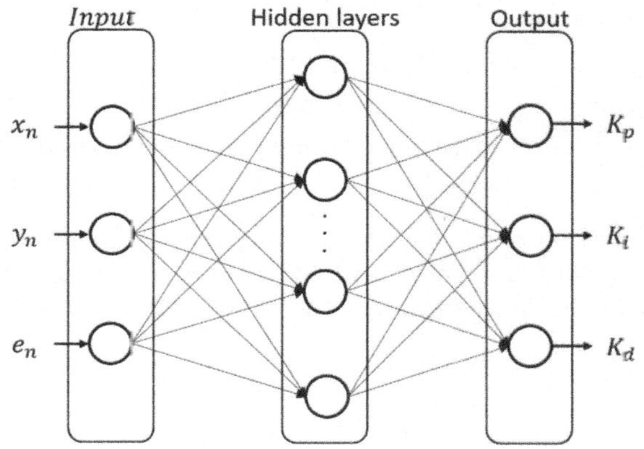

Fig. 5. Feed forward neural network block diagram.

The activation function of the neural network is the sigmoid function, it is represented in the following formula:

$$F_{(x)} = 1/(1 + e^{(-x)}) \quad (3)$$

The representation of the system reaction using the controller adjusted by neural networks with MATLAB is presented in Fig. 6:

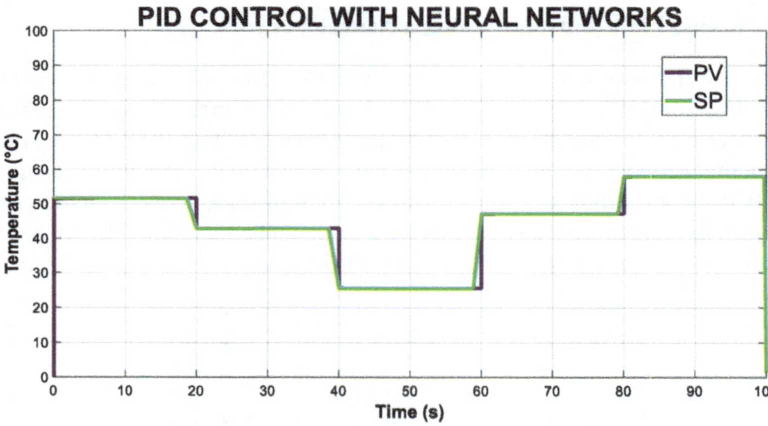

Fig. 6. System response with the controller tuned using neural networks by MATLAB.

From Fig. 6, it is observed that the PV line (process value) manages to follow the reference signal (SP) and when there are changes in the parameters, the system manages to maintain the desired behavior by following the reference signal with very little variation.

4 Results

In Fig. 7, it is observed that the classic method (yellow line) together with the autotuning method (Orange line) have an over-impulse in many areas of the graph, failing to adequately follow the reference signal (Set Point SP), as opposed to the neural network method, which manages to follow the reference signal almost perfectly. In the present work, algorithms have been developed for the control of a second-order plant using self-adjusting PID based on neural networks, obtaining satisfactory results.

In Fig. 8, it is observed that the classical method (yellow line) together with the autotuning method (orange line) the error does not tend to zero quickly, unlike the neural network method, where the error tends to zero in less time. Time in each change of SP of the temperature variable.

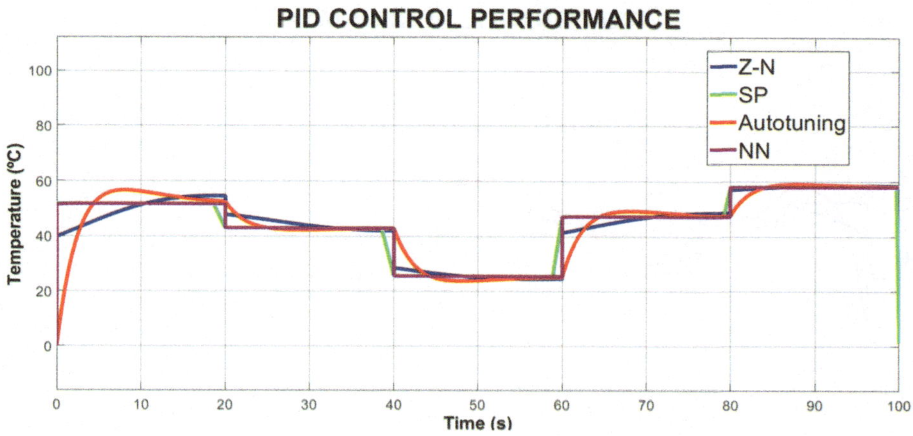

Fig. 7. System response of the 3 proposed methods by MATLAB.

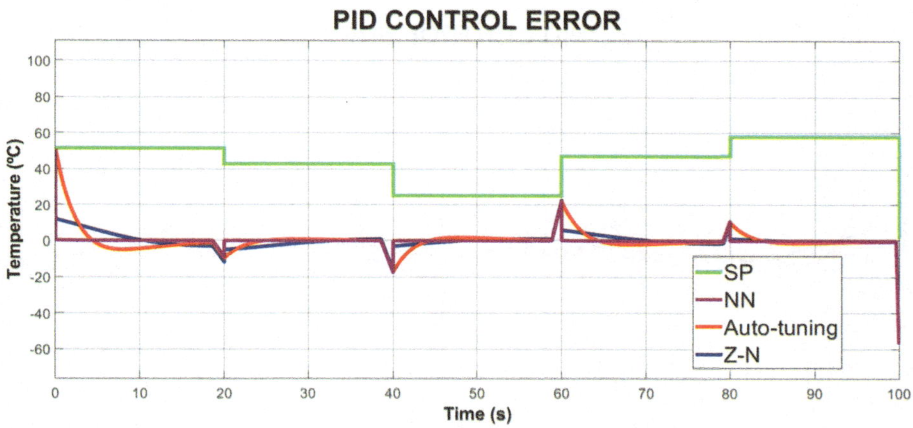

Fig. 8. Error in the system response of the 3 proposed methods by MATLAB, run 1.

In Fig. 9, it is simulated with new set point (SP) references. It is observed that the classic method (Z-N blue line) together with the autotuning method (red line) have in many areas of the graph maintaining a over impulse, failing to adequately follow the reference signal (Set Point SP), as opposed to the neural network (RN) method, which manages to follow the signal almost perfectly.

In Fig. 10, it is observed with other setpoint (SP) values that the classic method (blue line) together with the autotuning method (red line) the error does not tend to zero quickly unlike the neural network method. (RN) that the error tends to zero in less time with each change in SP of the temperature variable.

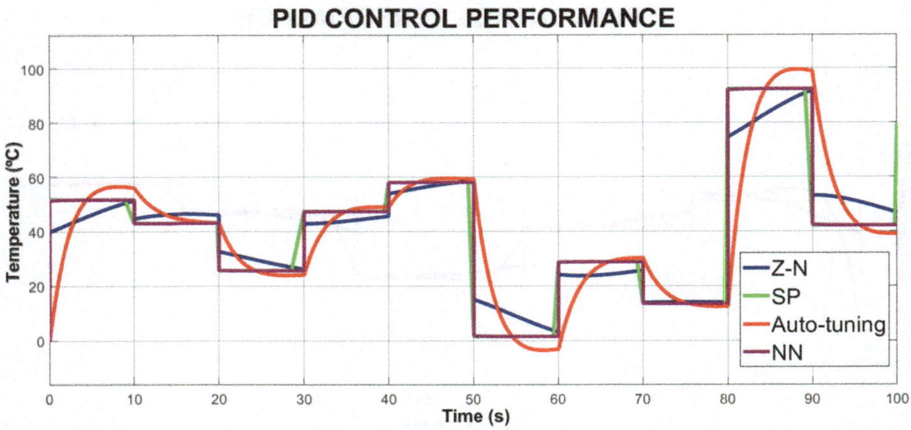

Fig. 9. System response of the 3 methods proposed in MATLAB with other set point (SP) values, run 2.

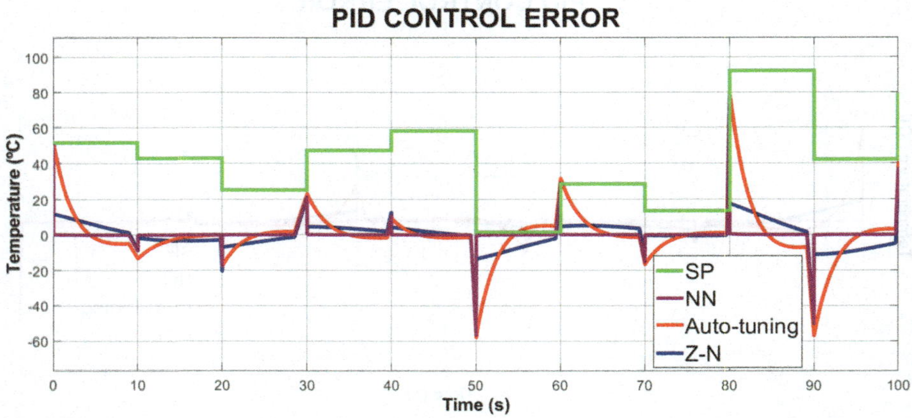

Fig. 10. Error in the system response of the 3 methods proposed in MATLAB with other set point (SP) values, run 2.

5 Conclusions

Within the field of industrial control, programmable logic controllers (PLC) are fundamental electronic devices, and the need to develop more versatile algorithms applicable in various plants or systems is of utmost importance. The application of methods based on neural networks to address adjustment problems in PID controllers offers the opportunity to control non-linear systems or systems with highly complex dynamics.

During the simulations, it was observed that this approach can confer robustness to the control of a plant. The tests revealed a good response to changes and various conditions (see Figs. 8 and 10), and despite the large number of necessary variables and the apparent significant computational load, the PID controller was able to follow the set point changes. (SP) with a response time that approaches zero (see Figs. 7 and 9).

The application of neural networks in PID controller tuning is highly superior to other techniques. Compared to traditional methods, the neural network-based approach achieved much more precise tracking of the reference signals, which significantly improved the control of the variable temperature in the dyeing of cotton fibers.

The comparison between PID tuning methods and autotuning techniques revealed that the use of neural networks outperformed these conventional techniques, highlighting their ability to address complex patterns and non-linear plants. The results obtained suggest that this technique could significantly impact the textile industry by improving the quality and uniformity of dyeing cotton fibers. The ability to control temperature more precisely and adaptively through automatic and flexible tuning of PID controller parameters could increase efficiency and reduce production costs. This approach provides more precise and efficient control by dynamically adjusting the temperature variable based on process (SP) variations.

6 Discussions

The main contribution of the implementation of the neural network method in this control system was the notable improvement in the responsiveness to changes in the set point (SP), together with an improvement in the damping factor of the output signal. (see Figs. 7, 8, 9 and 10). Furthermore, it was possible to maintain a shorter establishment time than the other two methods (classic and autotuning), which indicates that the behavior of the system is suitable for implementation in various industrial control environments.

The shared interest of several authors in improving PID controllers reflects the constant quest to increase performance in various industrial environments. The present work in this sense highlights the importance of the adaptability and speed of PID control, emphasizing the crucial role of neural networks in tuning PID controllers. Likewise, other authors, such as [17], provide perspectives on adaptability, especially in situations of system instability. On the other hand, [20] focuses on advanced PID designs, prioritizing guaranteed robustness characteristics. Together, these researches contribute to the evolution and continuous improvement of PID controllers in various industrial scenarios.

In the comparison with traditional methods, like the present work, various authors, such as [17] and [19], examine and contrast the results obtained through their approaches, highlighting the significant improvements and advantages that their proposals offer with respect to conventional methodologies.

References

1. Ruiz-Ayala, D.C., Vides-Herrera, C.A., Pardo-García, A.: Monitoreo de variables meteorológicas a través de un sistema inalámbrico de adquisición de datos. Revista de Investigación, Desarrollo e Innovación **8**(2), 333–341 (2018). https://doi.org/10.19053/20278306.v8.n2.2018.7971
2. Martínez-Mendoza, E., Sol-Sampedro, F.J., Rivas-Tovar, L.A., Toledo-Toledo, G.: Fabricación de muebles a partir de desechos de papel. Revista de Investigación, Desarrollo e Innovación **5**(2), 99 (2015). https://doi.org/10.19053/20278306.3144

3. Moreno-Anselmi, L.A., Reyes-Ortiz, Ó.J., Ruíz-Acero, J.C.: Evaluación del comportamiento mecánico de asfalto natural a partir de muestras a temperatura ambiente provenientes de Caquetá, Colombia. Revista de Investigación, Desarrollo e Innovación **6**(2), 145–154 (2016). https://doi.org/10.19053/20278306.3115
4. González Echavarría, L., Silva, M.F., Álvarez López, C., Osorio, A.R.: Generalidades de la seda y su proceso de teñido A short overview of silk and its dyeing process (2014)
5. Hernández-Arroyo, E., Díaz-Rodríguez, J.L., Pinzón-Ardila, O.: Estudio del comportamiento de un Control MPC [Control Predictivo Basado en el Modelo] comparado con un Control PID en una Planta de Temperatura. Revista Facultad de Ingeniería **23**(37), 45 (2014). https://doi.org/10.19053/01211129.2789
6. López Sotelo, J.A., Duque Marín, A., Navas, A.F.: Sintonización De Un Controlador PID En Un PLC Haciendo Uso De Inteligencia De Enjambres/Auto-tuning of a PID controller implemented in a PLC using swarm intelligence. Prospectiva **15**(1), 35–41 (2017)
7. Bari, S., Zehra Hamdani, S.S., Khan, H.U., Rehman, M.U., Khan, H.: Artificial neural network based self-tuned PID controller for flight control of quadcopter. In: 2019 International Conference on Engineering and Emerging Technologies, ICEET 2019, Institute of Electrical and Electronics Engineers Inc. (2019). https://doi.org/10.1109/CEET1.2019.8711864
8. Glushchenko, A.I.: On development of module for neural tuner to adjust D-part of PID-controller online. Procedia Comput. Sci. **150**, 208–215 (2019). https://doi.org/10.1016/j.procs.2019.02.042
9. Pirabakaran, K., Becerra, V.M.: PID autotuning using neural networks and model reference adaptive control. IFAC Proc. **35**(1), 451–456 (2002)
10. Hernández-Alvarado, R., García-Valdovinos, L., Salgado-Jiménez, T., Gómez-Espinosa, A., Fonseca-Navarro, F.: Neural network-based self-tuning PID control for underwater vehicles. Sensors **16**(9), 1429 (2016). https://doi.org/10.3390/s16091429
11. Huang, W., Xie, Y., Zhang, L., Yao, X.: Optimal design of PID controller for temperature in greenhouse system. ICIC Express Lett., Part B: Appl. **14**(5), 555–560 (2023)
12. Cera-Martínez, D., Ortiz-Sandoval, J.E., Gualdrón-Guerrero, O.E.: Sintonización de un controlador de temperatura a través de un autómata programable. Revista de Investigación, Desarrollo e Innovación **9**(1), 177–186 (2018). https://doi.org/10.19053/20278306.v9.n1.2018.8513
13. Günther, J., Reichensdörfer, E., Pilarski, P.M., Diepold, K.: Interpretable PID parameter tuning for control engineering using general dynamic neural networks: An extensive comparison. PLOS ONE **15**(12), e0243320 (2020). https://doi.org/10.1371/journal.pone.0243320
14. Yang, Z., Zhang, A., Sudjianto, A.: Enhancing explainability of neural networks through architecture constraints. IEEE Trans. Neural Netw. Learn. Syst. **32**(6), 2610–2621 (2021). https://doi.org/10.1109/TNNLS.2020.3007259
15. Adonayt, I., Ruge, R.: Control signal optimization for PID regulators based on a Wind-Up anti-reset architecture (2011)
16. Alfaro Ruíz, V.M.: Métodos de sintonización de controladores pid que operan como reguladores.
17. Lozano-Valencia, L.F., Rodríguez-García, L.F., Giraldo-Buitrago, D.: Diseño, Implementación y Validación de un Controlador PID Autosintonizado. TecnoLógicas **28**, 33 (2012)
18. Imad, R., Kairuz, V., Santos, A.I.G.: Controladores MPC y PID con autoajuste para un proceso de dinámica rápida a través de MATLAB®/Simulink® y OPC MATLAB®/Simulink® Speed control of and induction motor through MATLAB®/Simulink® and an OPC server. In: RIELAC Revista de Ingeniería Electrónica, Automática y Comunicaciones, vol. XXXVI, pp. 80–93 (2015)

19. Meng, Q., Zhang, H., Howarth, M.: Comparison between real-time auto-tuning PID and conventional PID controller for a dairy industrial evaporation process. Int. J. Model. Identif. Control **40**(1), 27 (2022)
20. Vilanova, R., Alfaro, V.M.: Control PID robusto: Una visión panorámica. Revista Iberoamericana de Automática e Informática Industrial RIAI **8**(3), 141–158 (2011). https://doi.org/10.1016/j.riai.2011.06.003

Performance Analysis of Machine Learning Techniques, Pattern Detection and Model Optimisation

Ksenia Degtyareva[1(✉)], Aleksey Borodulin[2], Vladimir Nelyub[2], Vladislav Kukartsev[1,2], and Vladimir Khramkov[1]

[1] Reshetnev Siberian State University of Science and Technology, 660037 Krasnoyarsk, Russia
sofaglu2000@mail.ru

[2] Bauman Moscow State Technical University, 105005 Moscow, Russia

Abstract. This paper focuses on the application of three data analysis techniques – decision trees, neural networks and self-organising Kohonen maps – to identify internal dependencies in a wide range of data. Based on the analysis of a dataset including patient information, the performance of these methods in classification and prediction is compared. The results emphasise the importance of method selection depending on the characteristics of the data and provide a basis for further research in the application of machine learning in various fields.

Keywords: Data Analysis · Decision Trees · Neural Networks · Self-Organising Kohonen Maps · Classification · Model Optimisation · Health · Heart Disease · Cardiovascular Disease · Cardiovascular Risk · Medical

1 Introduction

In today's era, in which information technology is evolving at a rapid pace, data is becoming a key component of informed decision-making in a variety of fields. From business and public policy to, for example, medicine, data analysis is becoming an indispensable tool for identifying patterns, predicting trends, and discovering internal dependencies.

This study focuses on the application of three powerful data analysis techniques – decision trees, neural networks, and self-organising Kohonen maps – and reveals the effectiveness of these techniques in uncovering hidden dependencies.

Data analysis today covers a wide range of tasks including classification, regression, clustering and prediction. Decision trees have been successfully applied for decision support, predicting various phenomena from a variety of data [1].

Neural networks, in turn, have been successfully applied to analyse textual data, including its interpretation, highlighting key points from the data and predicting various trends [2].

Self-organising Kohonen maps can identify common characteristics in large amounts of data, making them a valuable tool for analysing various parameters. They help to

R. Silhavy and P. Silhavy (Eds.): CSOC 2024, LNNS 1120, pp. 72–79, 2024.
https://doi.org/10.1007/978-3-031-70518-2_7

identify patterns in multivariate data and categorise them into groups, which can be useful for research and identifying new trends in different research areas [3–5].

The key step in data analysis is proper preparation of information. Data processing, normalisation, and selection of features are essential for the accuracy and validity of the results. In this context, the selected data analysis methods show their versatility, capable of processing both numerical and categorical data, as well as automatically extracting features, which simplifies the data preparation process.

The aim of the study is not only to extend the theoretical understanding of these methods, but also to provide concrete practical examples of their successful application in solving real-world analysis problems.

2 Research Methods

A decision tree is a structure that represents a tree-like model of decision making.

Tree nodes – each node in the tree represents a decision point regarding a particular attribute in the data. For example, a patient's age, cholesterol level, or ECG results can serve as attributes to divide the data into subgroups [6].

Ribs – these connect nodes and represent the conditions under which the data is split into two branches. For example, if a node separates patients based on age, an edge might represent the condition «Age < 50».

The final nodes of the tree, called leaves, represent the final classifications or predictions.

Decision trees are trained on training data, where the algorithm seeks to maximise classification performance. They provide a transparent decision structure, which simplifies the interpretation of results and ensures that key factors affecting data classification can be identified [7].

Self-Kohonen's Self-Organising Maps (SOM) is a powerful tool for analysing data in the context of our study. This method has been applied to process and visualise multivariate data.

Each node in the map corresponds to a virtual neuron representing a region in the multidimensional data space. In our case, the nodes could represent different groups of patients with certain characteristics [8].

The weighting coefficients of the nodes were adapted during training, allowing Kohonen maps to map the input data to the corresponding nodes. These coefficients are key to determining the degree of similarity of the data in different regions of the map.

The learning process of the Kohonen map was to adapt its structure to the distribution of the input data. For example, the map learning process may have highlighted areas where data on patients with certain characteristics were more likely to be found.

One of the key aspects of using Kohonen maps in our study was their ability to identify clusters of similar data. This facilitated visualisation of the structure of data, allowing researchers to highlight patterns and group patients based on similar characteristics [9].

In this paper, Kohonen maps were used to cluster groups of patients based on a variety of medical parameters. For example, they could identify groups of patients with similar age, cholesterol levels, and other clinical characteristics [10].

In this way, Kohonen's self-organising maps have made a significant contribution by providing new ways to visualise and understand complex relationships in medical data.

Neural networks are a powerful data analysis tool used to process data and identify hidden patterns.

A neural network consists of interconnected layers of neurons. A multilayer neural network including input layer, hidden layers and output layer was used in the study.

The network was trained on labelled data where each input field was associated with a specific class or characteristic. During training, the weights between neurons were adjusted to minimise prediction error and improve the overall performance of the network.

One of the strongest aspects of neural networks is their ability to automatically extract meaningful features from data. In the context of our study, this meant that the network could independently determine which characteristics were associated with the presence or absence of heart disease [11, 12].

Thus, neural networks have made a significant contribution to processing and analysis, providing a highly efficient method for classifying and identifying important patterns [13–15].

The dataset used in the study contains 1328 records. Each record represents an individual patient and includes a variety of medical attributes: Age: This attribute represents the age of the patient, Sex: This is a categorical attribute indicating the sex of the patient male (0) or female (1), Chest pain type (cp): A categorical attribute describing the type of chest pain with four possible values, resting blood pressure (trestbps): the patient's resting blood pressure value, serum cholesterol (chol): the serum cholesterol level in mg/dl, fasting glucose (fbs): a binary attribute indicating whether the fasting glucose level exceeds 120 mg/dl, resting ECG results (restecg): A categorical attribute with three possible values reflecting ECG results, maximum heart rate achieved (thalach): the patient's maximum heart rate, exercise-induced angina (exang): a binary attribute indicating whether the patient's angina occurred in response to physical activity, ST-segment depression after exercise (oldpeak): value of ST-segment depression after physical activity relative to rest, slope of ST-segment exercise peak (slope): attribute describing the slope of ST-segment exercise peak, number of large vessels (ca): number of large vessels stained on fluoroscopy, ranging from 0 to 3, thalassaemia type (thal): a categorical attribute representing the type of thalassaemia with three possible values, target attribute (target): this column serves as the dependent variable and indicates the presence (1) or absence (0) of heart disease in the patient.

3 Result and Discussion

Data were analysed using three methods: decision tree, neural networks and Kohonen self-organising maps in Deductor software. Outputs: target (presence or absence of disease), age (patient's age), sex (patient's gender). Table 1 shows the error rate after analysis.

Table 1. Error percentages

Methods	Target, %	Age, %	Sex, %
Decision tree	6,17	0	5,12
Kohonen maps	10,71	50,15	14,31
Neural Network	3,61	99,62	5,87

The decision tree showed high performance in predicting the target attribute (presence or absence of heart disease) and age. However, despite zero error in predicting age, possible overfitting of the model must be considered.

Kohonen maps encountered difficulty in predicting age and gender. More careful parameter tuning is probably required to improve performance.

The neural network showed the best overall performance, especially in predicting the presence of disease. However, the high error rate in age prediction indicates the need for additional optimisation or the use of more complex models.

The Figs. 1, 2, 3, 4, 5 and 6 depict some of the training results.

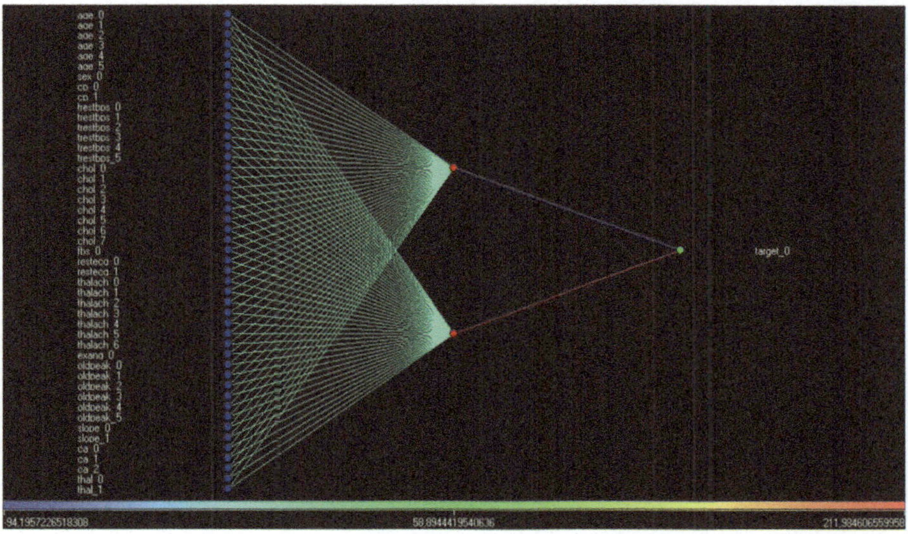

Fig. 1. Neural network graph, output: target

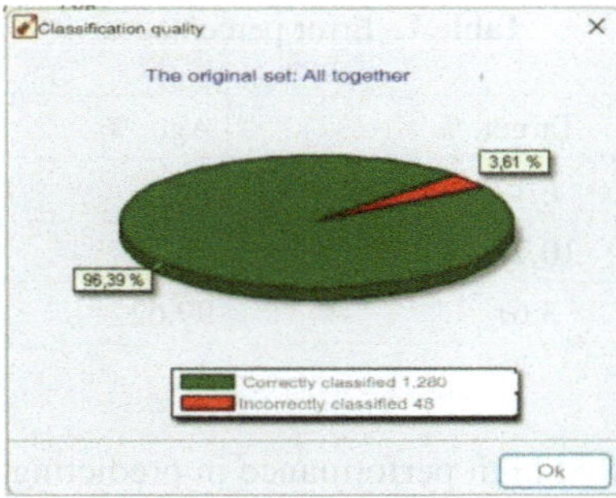

Fig. 2. Quality of classification of neural network

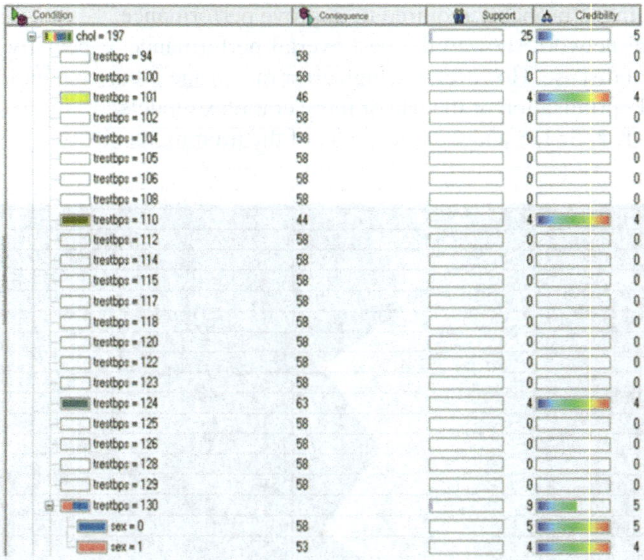

Fig. 3. One of the branches of the decision tree

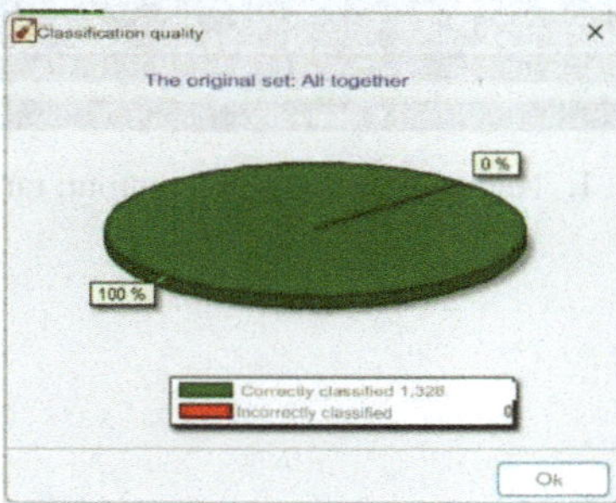

Fig. 4. Quality of classification of decision tree

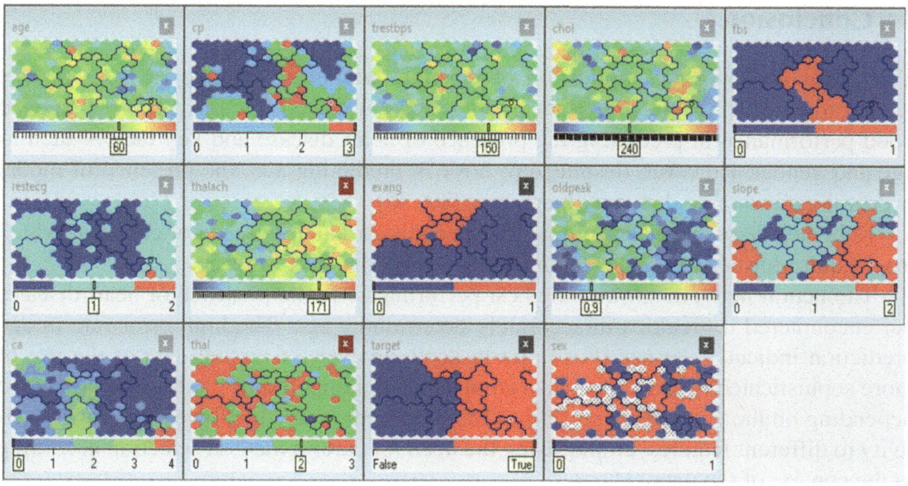

Fig. 5. Kohonen maps, output: sex

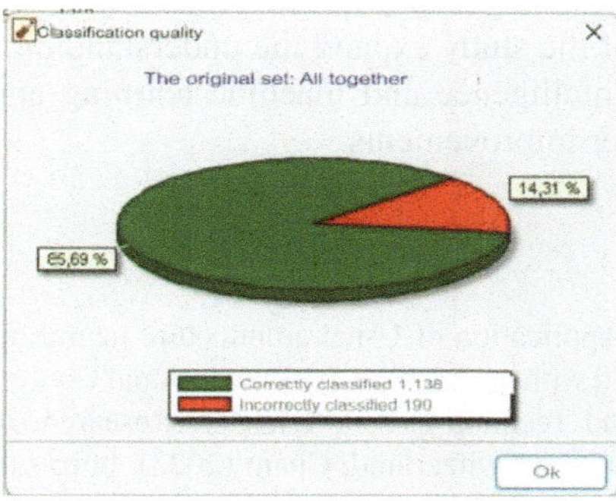

Fig. 6. Quality of classification of Kononen maps

The methods showed different sensitivity to different attributes For example, the decision tree showed high sensitivity to target, which means that its structure is good at identifying dependencies affecting the presence or absence of heart disease, and it also showed high sensitivity to age, which may be related to the fact that age is an important factor in determining heart disease risk.

All of this emphasises the importance of understanding the specificity of the data when choosing a method. This highlights the need for additional research and model optimisation, especially in the processing of textual data. Possible directions include better data preprocessing, more sophisticated neural network architectures, and more careful parameter tuning [15–18].

4 Conclusion

This paper analysed the application of machine learning techniques using a dataset related to cardiac health. Important findings were obtained: the decision tree showed good performance in predicting the presence of heart disease and key factors such as age and gender. However, despite zero error in predicting age, the presence of model overfitting on training data may be present [19, 20].

Kohonen maps were less efficient, especially in handling categorical data, which may require further parameter optimisation to improve performance.

The neural network showed the best performance in the presence of heart disease, but encountered difficulties in accurately determining age. The high error rate in age prediction indicates the need for more in-depth analysis of textual data or the use of more sophisticated models. This also emphasises the importance of choosing a method depending on the characteristics of the data. Different methods showed different sensitivity to different features, emphasising the need for careful method selection according to the context of the data [21].

Further research should include improved data processing, optimisation of method parameters and possible implementation of more sophisticated models to improve prediction accuracy [22].

Overall, the results of the study expand the understanding of the opportunities and limitations of artificial intelligence and machine learning applications, providing an important basis for further improvements.

References

1. Tynchenko, V., et al.: Application of U-net architecture neural network for segmentation of brain cell images stained with trypan blue. In: International Conference on High-Performance Computing Systems and Technologies in Scientific Research, Automation of Control and Production. Springer Nature Switzerland, Cham (2023). https://doi.org/10.1007/978-3-031-51057-1_13
2. Shutaleva, A., et al.: Sustainability of inclusive education in schools and higher education: teachers and students with special educational needs. Sustainability **15**(4), 3011 (2023)
3. Tynchenko, V., et al.: Mathematical models for the design of GRID systems to solve resource-intensive problems. Mathematics **12**(2), 276 (2024)
4. Martyushev, N., et al.: Review of methods for improving the energy efficiency of electrified ground transport by optimizing battery consumption. Energies **16**(2), 729 (2023)
5. Suprun, E., Tynchenko, V., Khramkov, V., Kovalev, G., Soloveva, T.: The use of artificial intelligence to diagnose the disease. BIO Web of Conf. **84**, 01008 (2024). https://doi.org/10.1051/bioconf/20248401008
6. Kukartsev, V., et al.: Study of the influence of the thermal capacity of the lining of acid melting furnaces on their efficiency. Metals **13**(2), 337 (2023)
7. Orlov, V., et al.: Development of a multifunctional cross-platform system for automation of energy data and resource management. E3S Web of Conferences. Vol. 460. EDP Sciences, 2023
8. Rezanov, V., et al.: Study of melting methods by electric resistance welding of rails. Metals **12**(12), 2135 (2022)
9. Kravtsov, K., et al.: Workflow automation and performance improvement based on PostgreSQL. E3S Web Conf. **458**, 09022 (2023)

10. Chernykh, N., et al.: Comparative analysis of existing measures to reduce road accidents in western Europe. In: 2023 22nd International Symposium INFOTEH-JAHORINA (INFOTEH). IEEE (2023)
11. Tynchenko, V.S., Stashkevich, A., Muzyka, P., Leontieva, A.A., Degtyareva, K.V.: Effective energy management tools: inventory management and monitoring of energy consumption by personnel. E3S Web of Conf. **458**, 01011 (2023)
12. Kurashkin, S., et al.: Mathematical modelling of waveguide paths by electron-beam welding. Procedia Comput. Sci. **200**, 83–90 (2022)
13. Semenova, E., et al.: Using UML to describe the development of software products using an object approach. In: 2022 IEEE International IOT, Electronics and Mechatronics Conference (IEMTRONICS). IEEE (2022)
14. Tynchenko, V., et al.: Software for modeling brazing process of spacecraft elements from widely used alloys. 2022 21st International Symposium INFOTEH-JAHORINA (INFOTEH). IEEE, 2022
15. Volneikina, E., et al.: Simulation-dynamic modeling of supply chains based on big data. In: 2023 22nd International Symposium INFOTEH-JAHORINA (INFOTEH). IEEE (2023)
16. Rogova, D., et al.: Software system for modeling temperature distribution during the electron beam welding. In: 2022 IEEE International Conference on Design & Test of Integrated Micro & Nano-Systems (DTS). IEEE (2022)
17. Degtyareva, K.V., et al.: Automatic monitoring system designed to detect defects in PET preforms. E3S Web Conf. **458**, 02002 (2023). https://doi.org/10.1051/e3sconf/202345802002
18. Boychuk, I., et al.: A methodological approach to the simulation of a ship's electric power system. Energies **16**(24), 8101 (2023)
19. Gantimurov, A., et al.: Investigation of the influence of geographical factors on soil suitability using a nonparametric controlled method of training and data analysis. E3S Web Conf. **431**, 03005 (2023)
20. Malashin, I., et al.: Estimation and prediction of the polymers' physical characteristics using the machine learning models. Polymers **16**(1), 115 (2023)
21. Tynchenko, V.S., Kleshko, I.I., Borodulin, A.S., Gantimurov, A.P., Nikolaev, S.V.: Using software to shape safety on the construction site. E3S Web Conf. **458**, 07003 (2023)
22. Bukhtoyarov, V., et al.: Application of machine learning algorithms for refining processes in the framework of intelligent automation. SOCAR Proc. **1**, 12–20 (2022)

The Role of Artificial Intelligence in Shaping Long-Term Career Prospects

S. A. Zamolotskii[2], V. A. Nelyub[1], V. V. Kukartsev[1,2,3], A. S. Gantimurov[1], and A. R. Glinscaya[3(✉)]

[1] Bauman Moscow State Technical University, 105005 Moscow, Russia
[2] Siberian Federal University, 660041 Krasnoyarsk, Russia
[3] Reshetnev Siberian State University of Science and Technology, 660037 Krasnoyarsk, Russia
anna_glinskaja@rambler.ru

Abstract. Students' performance in examinations plays an important role in determining academic success and future opportunities for students. There are various factors that influence academic performance. For example, there are specialized preparation courses. A study was conducted to determine how effective the preparation courses are and what other factors affect test scores for future improvement of educational policies. As a result, the factors affecting the final grade in different subjects are identified. A model predicting the possible score based on these factors is also proposed. Based on the findings, it is possible to develop and implement strategies that significantly improve student achievement.

Keywords: Access to education · education · learning opportunities · literacy

1 Introduction

Currently, artificial intelligence continues to assert its position as one of the key technological revolutions [1–4]. Artificial intelligence is some ability of a digital computer or a computer-controlled robot to learn [5], make decisions and perform actions that are characteristic of human intelligence [6, 7]. Machine learning acts as the main advantage. Artificial intelligence works with huge amounts of data, analyzes it, finds dependencies, makes predictions, and generally derives useful information for future applications [8–11].

Artificial intelligence has been applied in various areas of life from manufacturing to healthcare and finance [12]. Examples of artificial intelligence applications include: speech recognition [13], customer service, computer vision, recommendation, automated stock trading [14], fraud detection, complex calculations, big data analysis, finding optimal ways to solve problems [15] and other applications.

If we consider artificial intelligence in education, we can see its significant impact in transforming traditional learning processes and creating new opportunities for students and teachers [16, 17]. The introduction of artificial intelligence provides a more personalized, effective and accessible learning experience [18, 19]. Artificial intelligence brings the following changes to the field of education:

R. Silhavy and P. Silhavy (Eds.): CSOC 2024, LNNS 1120, pp. 80–87, 2024.
https://doi.org/10.1007/978-3-031-70518-2_8

- Adaptive learning and individualization – creating adaptive educational programs that take into account the individual needs and knowledge level of each student. Machine learning systems analyze data on students' performance and adjust materials and knowledge to their needs [20];
- Online learning and distance education – development of online platforms and distance education courses [21];
- Performance prediction – predicting the likelihood of success in examinations, which is necessary to respond more effectively to learners' needs and to warn of possible difficulties [22];
- Engagement and communication – helping to organize learning schedules and providing course information [23];
- Intelligent teacher support – application to analyzing student data, automating knowledge testing, developing instructional materials, and creating tests [24];
- Language and skills training.

Let us dwell in more detail on considering the application of artificial intelligence to analyze data in order to find patterns. Artificial intelligence is capable of finding complex relationships in data that are not obvious to humans [25]. One of the technologies used to analyze data is deep learning based on neural networks. They are capable of extracting high-level features from data, discovering complex structures and patterns. Their application is possible for any educational institution [26].

In schools and universities, students' performance in examinations plays an important role in determining academic success and future opportunities for students. Performance in examinations is closely linked to a student's grade point average, which is a major factor considered by colleges and employers when assessing applicants. In addition to college admissions, employers often use personality and aptitude exams to determine if an employee is suitable for companies or specific positions. Ultimately, high scores on exams improve both short-term and long-term career prospects.

There are various factors that affect test-taking success. For example, it could be gender, parental education level, or a special preparation course.

Test preparation courses are designed to help people prepare for standardized tests. These courses typically cover topics related to content review, test-taking strategies, and practice problems, with instructors providing guidance and feedback along the way. Test prep courses help examinees achieve higher scores through rigorous instruction and realistic practice scenarios prior to actual test dates.

It is important to determine how effective test preparation courses are and what other factors influence test scores.

2 Materials and Method

Some authors on the example of USE in physics note that elective training courses do not affect the result of the exam [27]. It is also revealed that stress negatively affects the result of exams [28]. Stress can cause psychological situation in the family, problems in relationships with friends [29]. Among other factors of stress in students are economic difficulties, combining study and work [30].

To assess the emotions of elementary and secondary school students, a special scale is proposed [31]. To reduce stress, it is proposed to include a separate course in the curriculum of universities [32].

To predict students' academic performance, it is proposed to use a model based on fuzzy sets [33]. It is also possible to predict the probability of students dropping out due to various factors: academic integration, social integration, and academic performance [34]. In e-learning, it is possible to predict academic performance based on semester work [35]. A web application has also been developed to predict academic performance [36–38].

Principal component method is used to visualize the multidimensional array.

To analyze the data, division into subsamples and investigation of mean values are planned. Artificial neural networks are also used – computational models that are able to "learn" by the example of the human brain, revealing dependencies between data.

The "Students Performance in Exams" set from the Kaggle resource, which contains the following attributes, was taken as the data under study: gender; race/ethnicity; parental level of education; free/reduced lunch; test preparation course; math score; reading score; writing score:

- gender – 2 categories: male (52%), female (48%);
- race/ethnicity – 5 categories: group A (7.9%), group B (20.5%), group C (32.3%), group D (26.2%), group E (13.1%);
- parental level of education – 6 categories: some college (22.2%), associate's degree (20.3%), high school (20.2%), some high school (19.1%), bachelor's degree (11.2%), master's degree (7%);
- lunch – 2 categories: standard (65%), free/reduced (35%);
- test preparation course – 2 categories: none (67%), completed (34%);
- math score – integer from the interval [0, 100], mathematical expectation: 66.4, standard deviation: 15.4;
- reading score – an integer from the interval [0, 100], mathematical expectation: 69.0, standard deviation: 14.7;
- writing score – an integer from the interval [0, 100], mathematical expectation: 67.7, standard deviation: 15.6.

In total, the dataset includes data on 1000 learners.

3 Results

For the study, all categorical attributes were converted into numerical attributes using LabelEncoder. Additionally, the average score for 3 tests was calculated.

The method of principal components was chosen for data visualization – displaying the combination of attributes in a two-dimensional space. The visualization is presented in Fig. 1, lighter dots indicate a higher mean score, darker dots indicate a lower mean score.

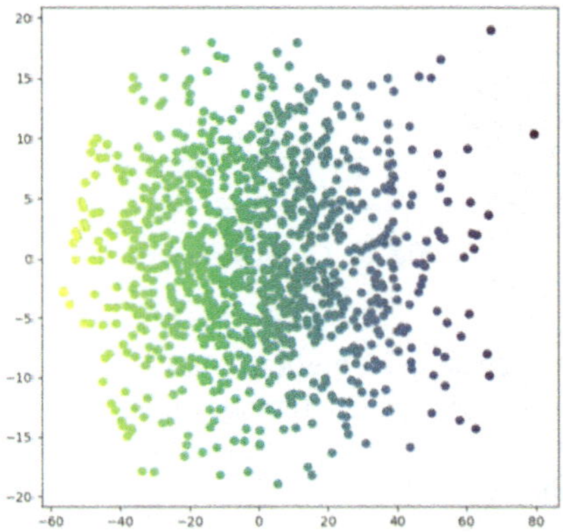

Fig. 1. Visualization by the method of principal components

In order to investigate the influence of factors on each other, correlation analysis was performed. The correlation heat map is presented in Fig. 2.

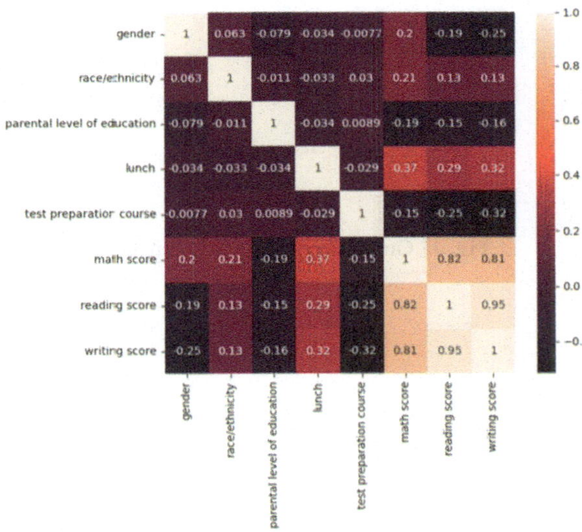

Fig. 2. Correlation heat map.

It can be seen that the grade for each subject has a strong influence on the grades for other subjects. It can also be seen that grades in each subject are affected by the availability of free/reduced lunch; the course of study has a weaker effect.

To analyze it was decided to consider dividing the initial sample into groups and compare the average score values for each subject. The results are shown in Fig. 3.

Sub-sample	*Math*	*Reading*	*Writing*
Free/favorable lunch	58.51	63.19	60.92
Regular lunch	70.61	72.10	71.37
Training course taken	69.69	74.09	74.67
No preparation course taken	64.73	66.44	64.24

Fig. 3. Group averages

It can be seen that for each subject, higher scores are found for those who have taken the training course and those who have a regular lunch.

The next data analysis method used is analysis through neural network.

It is assumed that if the evaluation depends on external factors, it is possible to build a model for the regression problem and predict the result of individual learners.

A perceptron with 256 associative elements is chosen as the neural network. Dropout, batch-normalization and early stopping are also configured. The mean absolute error (MAE) is selected as the error function.

For the study, predictions are also plotted separately for each subject for all features. Graphs with real and predicted values are shown in Figs. 4, 5, and 6.

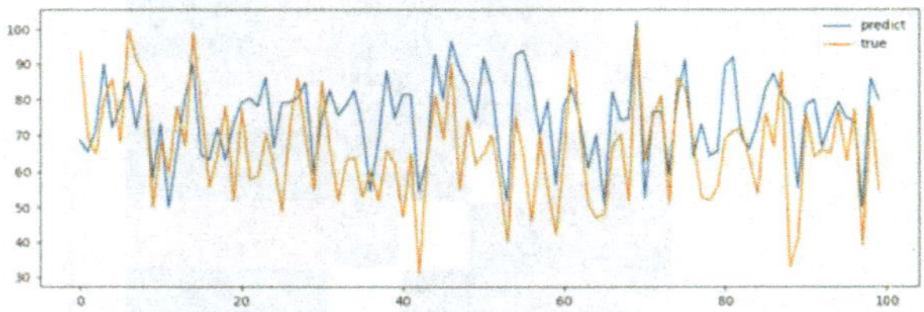

Fig. 4. Graph of results in mathematics

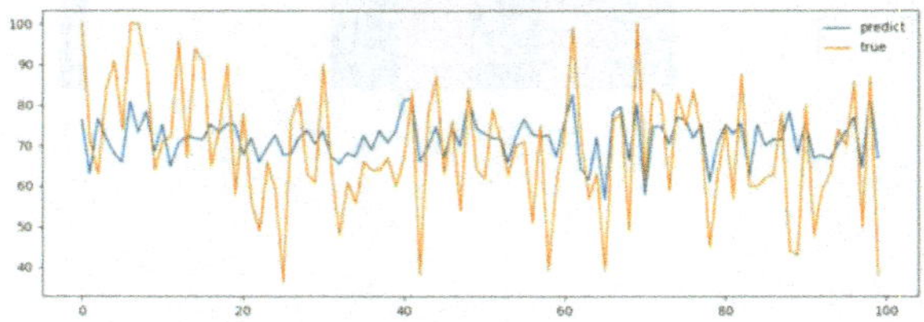

Fig. 5. Graph of results in reading

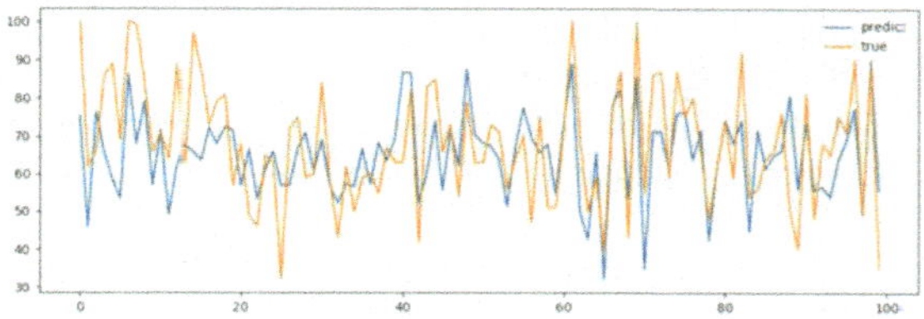

Fig. 6. Graph of results in writing

The MAE value was 13.00 for math, 10.99 for reading, and 10.53 for writing.

It can be concluded that the regressor works most accurately for writing tests – it is possible to predict the result with an average error of less than 11 points.

4 Conclusion

There are several factors that influence student achievement. These factors interact in different ways and have varying degrees of influence on different aspects of student achievement. Understanding how each factor affects performance can provide insight into what actions should be taken to optimize student achievement. While some factors are within our control, identifying those that can be influenced by schools or universities and implementing appropriate strategies can significantly improve student achievement. Overall, understanding the complexity of these factors will lead to the development of more effective educational policies.

Predicting student achievement is an important task that can help educators identify students who may need additional support and target instructional interventions to improve their academic performance. Using data from multiple sources, researchers and educators can create models that predict student achievement levels with high accuracy.

Modern technology has made it possible to conduct a study that can influence educational policy. Factors affecting academic performance were identified. A model predicting the exam grade based on these factors was also proposed.

References

1. Semenova E. et al.: Using UML to describe the development of software products using an object approach. In: 2022 IEEE International IOT, Electronics and Mechatronics Conference (IEMTRONICS). pp. 1–4. IEEE (2022)
2. Tynchenko, V.S., et al.: Energy distribution computation for induction soldered construction elements. In: AIP Conference Proceedings, vol. 2700, No. 1. AIP Publishing (2023)
3. Chernykh, N., et al.: Comparative analysis of existing measures to reduce road accidents in Western Europe. In: 2023 22nd International Symposium INFOTEH-JAHORINA (INFOTEH), pp. 1–6. IEEE (2023)
4. Malozyomov, B.V., et al.: Overview of methods for enhanced oil recovery from conventional and unconventional reservoirs. Energies **16**(13), 4907 (2023)
5. Tynchenko, V.S., et al.: Computational model of soldered details heating with considering the magnetic field distribution. J. Phys. Conf. Ser. **2373**(6), 062015 (2022)

6. Degtyareva, K.V., et al.: Automatic monitoring system designed to detect defects in PET preforms. E3S Web Conf. **458**, 02002 (2023)
7. Gantimurov, A., et al.: Investigation of the influence of geographical factors on soil suitability using a nonparametric controlled method of training and data analysis. E3S Web Conf. **431**, 03005 (2023)
8. Tynchenko, V.S., et al.: Using software to shape safety on the construction site. E3S Web Conf. **458**, 07003 (2023)
9. Tynchenko, V., et al.: Application of U-Net architecture neural network for segmentation of brain cell images stained with trypan blue. In: Jordan, V., Tarasov, I., Shurina, E., Filimonov, N., Faerman, V.A. (eds.) High-Performance Computing Systems and Technologies in Scientific Research, Automation of Control and Production: 13th International Conference, HPCST 2023, Barnaul, Russia, May 19–20, 2023, Revised Selected Papers, pp. 170–181. Springer Nature Switzerland, Cham (2024). https://doi.org/10.1007/978-3-031-51057-1_13
10. Martyushev, N.V., et al.: Production of workpieces from martensitic stainless steel using electron-beam surfacing and investigation of cutting forces when milling workpieces. Materials **16**(13), 4529 (2023)
11. Tynchenko, V., et al.: Mathematical models for the design of GRID systems to solve resource-intensive problems. Mathematics **12**(2), 276 (2024). https://doi.org/10.3390/math12020276
12. Rogova, D., et al.: Software system for modeling temperature distribution during the electron beam welding. In: 2022 IEEE International Conference on Design & Test of Integrated Micro & Nano-Systems (DTS), pp. 1–6. IEEE (2022)
13. Barantsov, I.A., et al.: Classification of acoustic influences registered with phase-sensitive OTDR using pattern recognition methods. Sensors **23**(2), 582 (2023)
14. Bukhtoyarov, V.V., et al.: A study on a probabilistic method for designing artificial neural networks for the formation of intelligent technology assemblies with high variability. Electronics **12**(1), 215 (2023)
15. Masich, I.S., et al.: Prediction of critical filling of a storage area network by machine learning methods. Electronics **11**(24), 4150 (2022)
16. Masich, I.S., et al.: Paired patterns in logical analysis of data for decision support in recognition. Computation **10**(10), 185 (2022)
17. Kurashkin, S., et al.: Mathematical modelling of waveguide paths by electron-beam welding. Procedia Comput. Sci. **200**, 83–90 (2022)
18. Martyushev, N.V., et al.: Review of methods for improving the energy efficiency of electrified ground transport by optimizing battery consumption. Energies **16**(2), 729 (2023). https://doi.org/10.3390/en16020729
19. Shutaleva, A., et al.: Sustainability of inclusive education in schools and higher education: teachers and students with special educational needs. Sustainability **15**(4), 3011 (2023). https://doi.org/10.3390/su15043011
20. Rezanov, V.A., et al.: Study of melting methods by electric resistance welding of rails. Metals **12**(12), 2135 (2022)
21. Martyushev, N.V., et al.: Provision of rational parameters for the turning mode of small-sized parts made of the 29 NK alloy and beryllium bronze for subsequent thermal pulse deburring. Materials **16**(9), 3490 (2023). https://doi.org/10.3390/ma16093490
22. Kukartsev, V.A., et al.: Study of the influence of the thermal capacity of the lining of acid melting furnaces on their efficiency. Metals **13**(2), 337 (2023)
23. Kolenchukov, O.A., et al.: The experimental research of n-butane pyrolysis using an agitator. In: SOCAR Proceedings, vol. 1, pp. 29–34 (2022)
24. Bukhtoyarov, V.V., et al.: Application of machine learning algorithms for refining processes in the framework of intelligent automation. In: SOCAR Proceedings, vol. 1, pp. 12–20 (2022)

25. Kolenchukov, O.A., et al.: Experimental study of oil non-condensable gas pyrolysis in a stirred-tank reactor for catalysis of hydrogen and hydrogen-containing mixtures production. Energies **15**(22), 8346 (2022)
26. Bashmur, K.A., et al.: Biofuel technologies and petroleum industry: synergy of sustainable development for the Eastern Siberian Arctic. Sustainability **14**(20), 13083 (2022)
27. Golik, V.I., et al.: The mechanochemical activation of leaching processes in a disintegrator. MIAB. Mining Inf. Anal. Bull. **11–1**, 175–189 (2023)
28. Panfilova, T.A., et al.: To the concept of leaching metal-containing raw materials in the dizintegrator. MIAB. Mining Inf. Anal. Bull. **11–1**, 239–251 (2023)
29. Suprun, E., et al.: The use of artificial intelligence to diagnose the disease. BIO Web Conf. **84**, 01008 (2024)
30. Orlov, V., et al.: Development of a multifunctional cross-platform system for automation of energy data and resource management. E3S Web Conf. **460**, 07002 (2023). https://doi.org/10.1051/e3sconf/202346007002
31. Kravtsov, K., et al.: Workflow automation and performance improvement based on PostgreSQL. E3S Web Conf. **458**, 09022 (2023)
32. Tynchenko, V.S., et al.: Effective energy management tools: inventory management and monitoring of energy consumption by personnel. E3S Web Conf. **458**, 01011 (2023)
33. Volneikina, E.: et al. Simulation-dynamic modeling of supply chains based on big data. In: 2023 22nd International Symposium INFOTEH-JAHORINA (INFOTEH), pp. 1–6. IEEE (2023)
34. Filina, O.A., et al.: Increasing the efficiency of diagnostics in the brush-commutator assembly of a direct current electric motor. Energies **17**(1), 17 (2023)
35. Boychuk, I.P., et al.: A methodological approach to the simulation of a ship's electric power system. Energies **16**(24), 8101 (2023)
36. Golik, V.I., et al.: Reuse and mechanochemical processing of ore dressing tailings used for extracting Pb and Zn. Materials **16**(21), 7004 (2023). https://doi.org/10.3390/ma16217004
37. Malozyomov, B.V., et al.: Designing the optimal configuration of a small power system for autonomous power supply of weather station equipment. Energies **16**(13), 5046 (2023). https://doi.org/10.3390/en16135046
38. Malashin, I.P., et al.: Estimation and prediction of the polymers' physical characteristics using the machine learning models. Polymers **16**(1), 115 (2023)

Using Rule-Based Machine Learning Method in a Hierarchical Framework Using Economic Performance of Companies

Kirill I. Kravtsov[1], Vladislav V. Kukartsev[1,2,3](✉), Vladimir A. Nelyub[3], Aleksey S. Borodulin[3], and Elena V. Suprun[1]

[1] Reshetnev Siberian State University of Science and Technology, 660037 Krasnoyarsk, Russia
vlad_saa_2000@mail.ru
[2] Siberian Federal University, Krasnoyarsk 660041, Russia
[3] Bauman Moscow State Technical University, Moscow 105005, Russia

Abstract. The article discusses important aspects of bankruptcy prediction, including early risk detection and informed financial decision making. An automated decision tree learning method can efficiently analyze multiple economic indicators, automatically identify key factors, and thus improve the accuracy and interpretability of predictions. This research approach can be a valuable tool for investors, lenders and managers, helping them to reduce financial risks and ensure the financial stability of companies. The article aims to expand the understanding of the impact of economic factors on bankruptcy and to propose innovative methods for analyzing data to make informed decisions.

Keywords: Access to Banking · Sustainable Economic · Economic Benefits · Sustainable Economic · Enterprises

1 Introduction

Forecasting company bankruptcy plays a key role in today's business and economy. The importance of this practice is due to a number of factors that affect companies, investors, creditors and economic stability [1, 2].

Investor and creditor protection: Investors and creditors play an important role in financing companies. Bankruptcy forecasting allows them to identify risks and make informed decisions about providing financial support or investment. This helps to avoid losses and minimize risks [3, 4].

R. Silhavy and P. Silhavy (Eds.): CSOC 2024, LNNS 1120, pp. 88–98, 2024.
https://doi.org/10.1007/978-3-031-70518-2_9

Effective resource management: When a company knows its financial position accurately and forecasts possible financial difficulties, it can take steps to improve its financial strength. This includes cutting costs, increasing revenues and allocating resources efficiently [5, 6].

Protecting jobs: Bankruptcy of a company often entails laying off employees and closing down the business. Predicting bankruptcy allows a company to detect problems early and take steps to save jobs and retain business. Maintaining Competitiveness: Predicting financial distress can help a company adapt to changing market conditions and remain competitive. This allows a company to be more flexible in responding to the challenges and opportunities that the market presents [7, 8].

Financial system stability: When companies cannot meet their financial obligations, it can set off a chain reaction affecting the financial system as a whole. Forecasting bankruptcy and taking timely action to prevent it helps to maintain the stability of the financial system and the economy as a whole [9, 10].

Legal requirements: In many jurisdictions, companies are required to provide information about their financial condition and bankruptcy forecasts. Failure to comply with these requirements can lead to legal consequences. Therefore, bankruptcy forecasting also involves legislation and regulation. Consumer protection: The bankruptcy of a company may affect consumers who use its products or services. Bankruptcy forecasting helps to consider the interests of customers and minimize the negative impact on them [11, 12].

Bankruptcy forecasting of a company not only serves the interests of its owners and stakeholders, but also plays an important role in ensuring the stability and sustainability of the financial system and the economy as a whole. It helps prevent financial crises, minimize losses and contribute to the healthy development of business and society [13, 14].

2 Materials and Method

The bankruptcy dataset from the Taiwan Economic Journal for 1999–2009 is used as the basis for the analysis. The definition of company bankruptcy was based on the business rules of the Taiwan Stock Exchange.

A fragment of the downloaded data is shown in Fig. 1.

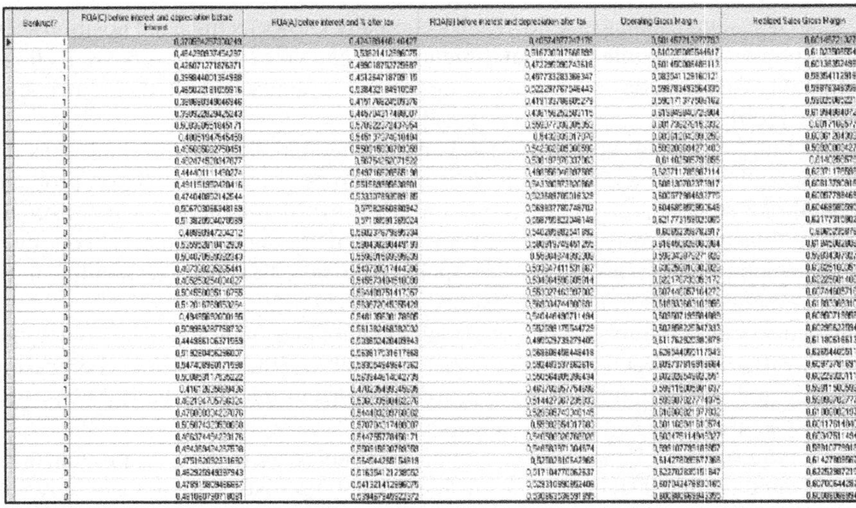

Fig. 1. Data array.

Before In order to analyze the data, you first need to process the data, for this purpose you use the tool – fill in the missing data.

Then we will remove insignificant factors using correlation analysis, the correlation threshold will be 0.1.

Correlation analysis itself is a statistical technique that is used to examine the relationship or interrelationship between two or more variables. It measures the extent to which two variables change together, either positively (as one variable increases, the other increases) or negatively (as one variable increases, the other decreases). The correlation coefficient measures the degree of this relationship, and it can take values from -1 to 1, where -1 indicates a complete negative correlation, 1 indicates a positive correlation, and 0 indicates no correlation. Correlation analysis is widely used in various fields including economics, medicine, sociology and business to identify relationships between variables and draw conclusions based on these [15, 16].

And after all the initial processing, the data set will be loaded into a decision tree to determine the factors that affect bank failures.

A decision tree is a graphical model of decision making that is used for analysis and forecasting. It is a hierarchical structure consisting of nodes (vertices) and edges, where each node represents a decision or test for a particular attribute, and each edge links the nodes to alternative decisions or outcomes. The leaves of the tree contain the final outcomes or predictions [17].

Decision trees are used to solve classification and regression problems. They are suitable for prediction for several reasons:

1. Interpretability: Decision trees are easy to interpret and explain. Therefore, they can be useful when a decision needs to be explained to stakeholders or customers [18].

2. Decision Automation: Decision trees can automatically generate rules based on training data, allowing decisions to be made without the need to manually tune the model [19, 20].
3. Handling heterogeneous data: Decision trees can handle heterogeneous data, including categorical and numeric variables, without the need for data preprocessing [21].
4. Efficiency: Decision trees can be efficient to train and use, especially on large amounts of data, and they can be extended to solve complex problems [22].
5. Prediction: Decision trees can predict the values of a target variable for new data, based on certain rules that have been learnt from training data [23].
6. Automatic feature selection: Decision trees can automatically identify the most important features for prediction, which can be useful when dealing with a large number of features [24].

The decision tree is a powerful machine learning tool that can be used to predict the bankruptcy of companies based on their economic performance [25]. However, like any method, the decision tree has its own potential limitations and problems [26]. Retrofitting is one of the main problems of the decision tree. If the tree is too deep or its growth is not sufficiently controlled, the model may adjust too accurately to the training data, which will lead to a high error on the new data [27].

Decision trees can ignore the importance of features if they do not share data efficiently at each node. This can lead to loss of information and decrease the performance of the model [28].

However, compared to other machine learning methods, this method may provide stronger arguments for superiority in predicting bankruptcy of companies.

Decision trees can deal better with non-linear dependencies between features and the target variable than linear models [29, 30]. Decision trees do not require linearity in the data and can easily handle interactions between features [31].

Decision trees are often more interpretable and easier to interpret than complex neural networks. In some cases, decision trees can learn faster and require less data to achieve good results than neural networks [32–34].

Thus, although decision trees have their limitations, their ability to work with non-linear data, easy interpretability and the ability to be included in ensembles allow them to be an effective tool for predicting bankruptcy of companies based on economic indicators.

3 Results

To begin with, a correlation analysis was performed on this dataset. A detailed view of the correlation analysis of the data is presented in Fig. 2.

86	Net Income to Total Assets	-0,315
2	ROA(A) before interest and % after tax	-0,283
3	ROA(B) before interest and depreciat...	-0,273
1	ROA(C) before interest and depreciat...	-0,261
38	Net worth/Assets	-0,250
37	Debt ratio %	0,250
19	Persistent EPS in the Last Four Seas...	-0,220
68	Retained Earnings to Total Assets	-0,218
43	Net profit before tax/Paid-in capital	-0,208
23	Per Share Net profit before tax (Yuan...	-0,201
60	Current Liability to Assets	0,194
54	Working Capital to Total Assets	-0,193
90	Net Income to Stockholder's Equity	-0,181
40	Borrowing dependency	0,177
84	Current Liability to Current Assets	0,171
91	Liability to Equity	0,167
17	Net Value Per Share (A)	-0,165
16	Net Value Per Share (B)	-0,165
18	Net Value Per Share (C)	-0,165
78	Current Liability to Equity	0,154
66	Current Liabilities/Equity	0,154
65	Working Capital/Equity	-0,147
22	Operating Profit Per Share (Yuan P'Tђ)	-0,142
42	Operating profit/Paid-in capital	-0,141
85	Liability-Assets Flag	0,139
70	Total expense/Assets	0,139
79	Equity to Long-term Liability	0,139
82	CFO to Assets	-0,115
15	Tax rate (A)	-0,110
57	Cash/Total Assets	-0,100
89	Gross Profit to Sales	-0,100
4	Operating Gross Margin	-0,100

Fig. 2. Correlation analysis of model parameters.

A correlation analysis was carried out in order to eliminate insignificant factors for the output field (bankruptcy) at this stage.

The model built using a neural network showed an error of 2.67% (see Fig. 3).

Such diagrams help us understand how accurate a forecasting model has been built by one or another method. In this case, the neural network showed an error of 2.67% in the constructed model, it is worth considering the following method – Kohonen maps.

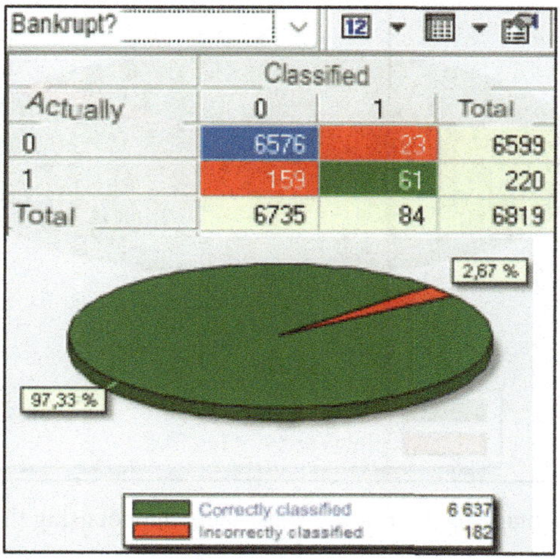

Fig. 3. Percentage of error of the constructed model using a neural network.

The cards themselves turned out to be very bad (see Fig. 4) the error was 2.98% (see Fig. 5).

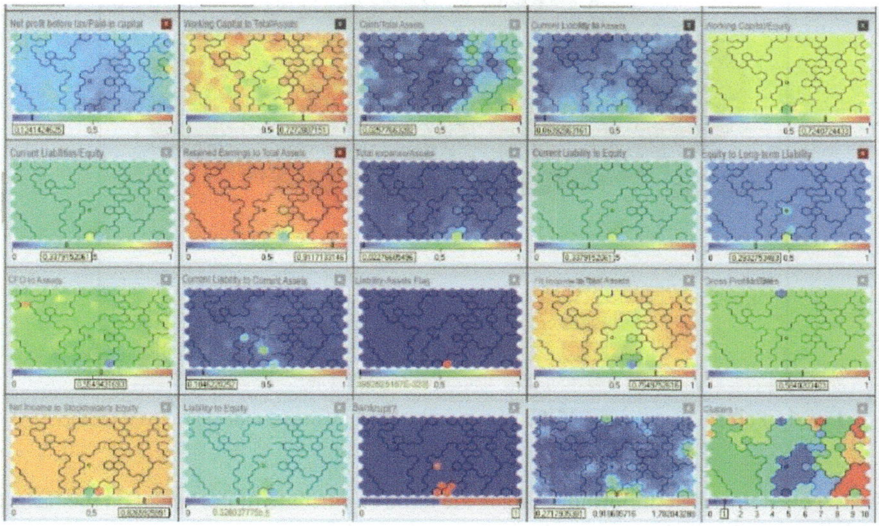

Fig. 4. Kohonen maps of the constructed model.

Since the error turned out to be higher and the maps turned out to be very bad, this method is not considered as the main one.

The study will be conducted using the "Decision Tree" method with a minimum error value.

Fig. 5. The percentage of error of the constructed model using the Kohonen map.

The model based on this method was obtained with an error of 1.6%. This is presented in detail in Fig. 6.

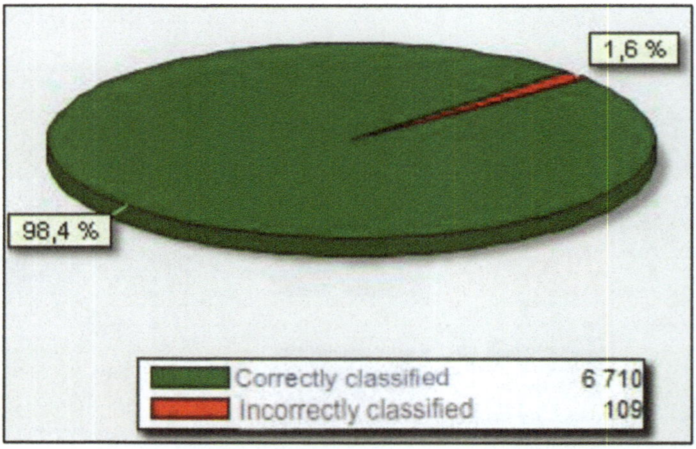

Fig. 6. Percentage of error of the constructed model using the decision tree method.

However, the most important thing is the table of attribute importance, on the basis of which it is already clear exactly which factors depend on the bankruptcy of the company. The detailed view is presented in Fig. 7.

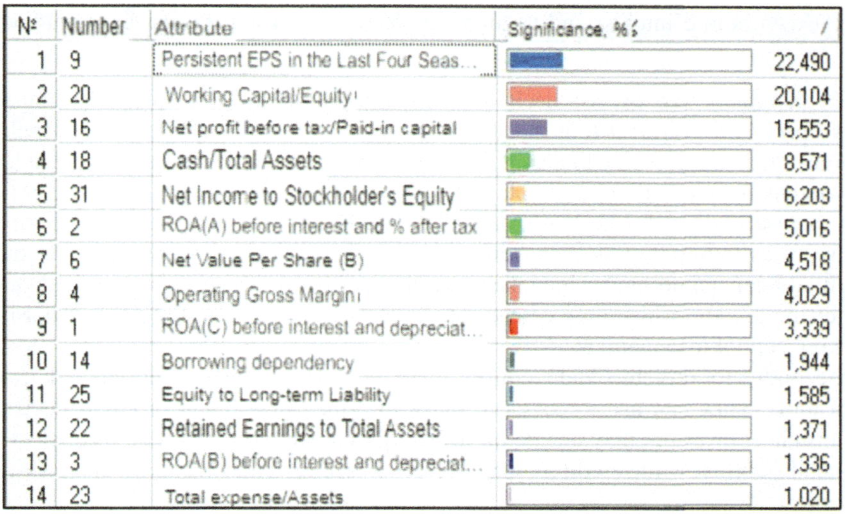

№	Number	Attribute	Significance, %
1	9	Persistent EPS in the Last Four Seas...	22,490
2	20	Working Capital/Equity	20,104
3	16	Net profit before tax/Paid-in capital	15,553
4	18	Cash/Total Assets	8,571
5	31	Net Income to Stockholder's Equity	6,203
6	2	ROA(A) before interest and % after tax	5,016
7	6	Net Value Per Share (B)	4,518
8	4	Operating Gross Margin	4,029
9	1	ROA(C) before interest and depreciat...	3,339
10	14	Borrowing dependency	1,944
11	25	Equity to Long-term Liability	1,585
12	22	Retained Earnings to Total Assets	1,371
13	3	ROA(B) before interest and depreciat...	1,336
14	23	Total expense/Assets	1,020

Fig. 7. Significance of factors from the bankrupt output field.

4 Conclusion

A decision tree allows you to identify the risks of bankruptcy of a company at an early stage, which allows you to take measures to prevent it. This is important both for the owners of the company and for investors and creditors who stand to lose money in the event of bankruptcy.

A decision tree provides a structured method of analyzing and predicting bankruptcy based on available data. This helps stakeholders to make informed decisions about investing, lending or continuing with the company.

The decision tree is easy to interpret, which helps explain to stakeholders what factors and attributes influence bankruptcy prediction. This is important for transparency and trust in decision making.

Decision trees can be automatically trained on large amounts of data, making them scalable for predicting bankruptcy of multiple companies. This is especially important in the case of portfolio asset management or investing.

Decision trees can provide high efficiency and accuracy in bankruptcy prediction, saving time and resources in decision making.

Decision trees can automatically identify the most important attributes that affect a company's bankruptcy, which helps to focus on key aspects of the analysis.

In this study, the particularly significant factors for company bankruptcy were:

- Persistent EPS in the Last Four Seasons: EPS-Net Income (significance: 22.4%).
- Working Capital/Equity (significance: 20.1%).
- Net profit before tax/Paid-in capital: Pretax Income/Capital (significance: 15.5%).
- Cash/Total Assets (significance: 8.5%).
- Net Income to Stockholder's Equity (significance: 6.2%).

- ROA(A) before interest and % after tax: Return on Total Assets(A) (significance: 4.5%).

In summary, the use of decision trees for predicting and analyzing company bankruptcy is a powerful tool that helps to improve the quality of decision-making, minimize risk and preserve financial stability for both companies and their stakeholders.

Models for predicting bankruptcy of companies provide value for both the companies themselves and for investors and creditors. Firstly, for companies, these models can serve as important risk management tools. They can use forecasts to early detect potential financial problems and take appropriate measures to prevent or mitigate them. For example, companies may consider revising their financing strategy or improving working capital management.

Secondly, for investors and creditors, these models are seen as important risk assessment tools. They can use forecasts to make informed investment or credit decisions. For instance, investors may decide on investing in a company based on forecasts of its financial stability, while creditors may decide on issuing a loan based on an assessment of bankruptcy risk.

Finally, such models can contribute to the stability of the financial market as a whole by preventing potential financial crises and minimizing risks for all market participants. By detecting and preventing company bankruptcies, models can help maintain stability and reliability in the investment environment, ultimately contributing to sustainable economic growth and development.

Research into the use of machine learning techniques such as decision trees to identify significant factors in company bankruptcy can cover various areas and expand our understanding of this problem. Future research may be aimed at a more in-depth analysis of other financial indicators that may be associated with the risk of bankruptcy of companies, such as asset turnover, debt burden, profitability indicators, and to investigate their impact on bankruptcy.

Also explore the possibility of including non-financial data such as market trends, industry characteristics, social and political factors in a bankruptcy prediction model. This can help improve the predictive power of the model.

Finally, to explore the possibility of creating more complex models, such as an ensemble of decision trees or hybrid models that combine several machine learning methods to improve the quality of forecasts.

References

1. Martyushev, N.V., et al.: Review of methods for improving the energy efficiency of electrified ground transport by optimizing battery consumption. Energies **16**(2), 729 (2023)
2. Shutaleva, A., et al.: Sustainability of inclusive education in schools and higher education: teachers and students with special educational needs. Sustainability **15**(4), 3011 (2023). https://doi.org/10.3390/su15043011
3. Rezanov, V.A., et al.: Study of melting methods by electric resistance welding of rails. Metals **12**(12), 2135 (2022)
4. Martyushev, N.V., et al.: Provision of rational parameters for the turning mode of small-sized parts made of the 29 NK alloy and beryllium bronze for subsequent thermal pulse deburring. Materials **16**(9), 3490 (2023)

5. Kolenchukov, O.A., et al.: Experimental study of oil non-condensable gas pyrolysis in a stirred-tank reactor for catalysis of hydrogen and hydrogen-containing mixtures production. Energies **15**(22), 8346 (2022)
6. Bashmur, K.A., et al.: Biofuel technologies and petroleum industry: synergy of sustainable development for the Eastern Siberian Arctic. Sustainability **14**(20), 13083 (2022)
7. Kukartsev, V.A., et al.: Study of the influence of the thermal capacity of the lining of acid melting furnaces on their efficiency. Metals **13**(2), 337 (2023)
8. Kolenchukov, O.A., et al.: The experimental research of n-butane pyrolysis using an agitator. In: SOCAR Proceedings, vol. 1, pp. 29–34
9. Bukhtoyarov, V.V., et al.: Application of machine learning algorithms for refining processes in the framework of intelligent automation. In: SOCAR Proceedings, vol. 1, pp. 12–20 (2022)
10. Golik, V.I., et al.: The mechanochemical activation of leaching processes in a disintegrator. Mining Inf. Anal. Bull. **11–1**, 175–189 (2023)
11. Panfilova, T.A., et al.: To the concept of leaching metal-containing raw materials in the dizintegrator. Mining Inf. Anal. Bull. **11–1**, 239–251 (2023)
12. Suprun, E., et al.: The use of artificial intelligence to diagnose the disease. BIO Web Conf. **84**, 01008 (2024)
13. Semenova, E., et al.: Using UML to describe the development of software products using an object approach. In: 2022 IEEE International IOT, Electronics and Mechatronics Conference (IEMTRONICS), pp. 1–4 IEEE (2022)
14. Tynchenko, V.S.. et al.: Energy distribution computation for induction soldered construction elements. In: AIP Conference Proceedings, vol. 2700, No. 1. AIP Publishing
15. Chernykh, N., et al.: Comparative analysis of existing measures to reduce road accidents in Western Europe. In: 2023 22nd International Symposium Infoteh-Jahorina (INFOTEH), pp. 1–6. IEEE (2023)
16. Orlov, V., et al.: Development of a multifunctional cross-platform system for automation of energy data and resource management. E3S Web Conf. **460**, 07002 (2023)
17. Kravtsov, K., et al.: Workflow automation and performance improvement based on PostgreSQL. E3S Web Conf. **458**, 09022 (2023)
18. Tynchenko, V.S., et al.: Effective energy management tools: inventory management and monitoring of energy consumption by personnel. E3S Web Conf. **458**, 01011 (2023)
19. Malozyomov, B.V., et al.: Designing the optimal configuration of a small power system for autonomous power supply of weather station equipment. Energies **16**(13), 5046 (2023)
20. Malashin, I.P., et al.: Estimation and prediction of the polymers' physical characteristics using the machine learning models. Polymers **16**(1), 115 (2023)
21. Malozyomov, B.V., et al.: Overview of methods for enhanced oil recovery from conventional and unconventional reservoirs. Energies **16**(13), 4907 (2023)
22. Volneikina, E., et al.: Simulation-dynamic modeling of supply chains based on big data. In: 2023 22nd International Symposium INFOTEH-JAHORINA (INFOTEH), pp. 1–6. IEEE
23. Filina, O.A., et al.: Increasing the efficiency of diagnostics in the brush-commutator assembly of a direct current electric motor. Energies **17**(1), 17 (2023)
24. Boychuk, I.P., et al.: A methodological approach to the simulation of a ship's electric power system. Energies **16**(24), 8101 (2023)
25. Golik, V.I., et al.: Reuse and mechanochemical processing of ore dressing tailings used for extracting Pb and Zn. Materials **16**(21), 7004 (2023)
26. Tynchenko, V.S., et al.: Computational model of soldered details heating with considering the magnetic field distribution. J. Phys.: Conf. Ser. **2373**(6), 062015 (2022)
27. Degtyareva, K.V., Nikolaev, S.V., Nelyub, V.A., Tynchenko, V.S., Borodulin, A.S.: Automatic monitoring system designed to detect defects in PET preforms. E3S Web Conf. **458**, 02002 (2023)

28. Gantimurov, A., et al.: Investigation of the influence of geographical factors on soil suitability using a nonparametric controlled method of training and data analysis. E3S Web Conf. **431**, 03005 (2023)
29. Tynchenko, V.S., et al.: Using software to shape safety on the construction site. E3S Web Conf. **458**, 07003 (2023)
30. Martyushev, N.V., et al.: Production of workpieces from martensitic stainless steel using electron-beam surfacing and investigation of cutting forces when milling workpieces. Materials **16**(13), 4529 (2023)
31. Tynchenko, V.V., et al.: Mathematical models for the design of GRID systems to solve resource-intensive problems. Mathematics **12**(2), 276 (2024)
32. Rogova, D., et al.: Software system for modeling temperature distribution during the electron beam welding. In: 2022 IEEE International Conference on Design and Test of Integrated Micro & Nano-Systems (DTS), pp. 1–6. IEEE (2022)
33. Tynchenko, V., et al.: Application of u-net architecture neural network for segmentation of brain cell images stained with trypan blue. In: Jordan, V., Tarasov, I., Shurina, E., Filimonov, N., Faerman, V.A. (eds.) High-Performance Computing Systems and Technologies in Scientific Research, Automation of Control and Production: 13th International Conference, HPCST 2023, Barnaul, Russia, May 19–20, 2023, Revised Selected Papers, pp. 170–181. Springer Nature Switzerland, Cham (2024). https://doi.org/10.1007/978-3-031-51057-1_13
34. Kurashkin, S., et al.: Mathematical modelling of waveguide paths by electron-beam welding. Procedia Comput. Sci. **200**, 83–90 (2022)

Features of Using Neural Network Methods to Predict the Total Electron Content of the Ionosphere in the Southern Hemisphere

Artem Kharakhashyan and Olga Maltseva(✉)

Institute for Physics, Southern Federal University, Rostov-on-Don 344090, Russia
oamaltseva@sfedu.ru

Abstract. Research of the behavior, as well as solving the problem of predicting the total electron content of the TEC of the ionosphere, is of great importance for studying and taking into account the influence of space weather on various technological systems. This influence is different in various parts of the globe, which is reflected in the dependence of the forecast accuracy on the position of the station. In the previous works of the authors, neural networks were developed using several groups of unidirectional and bidirectional, as well as recurrent and convolutional architectures, which provided high forecast accuracy based on data from stations along three meridians (European 30° E, Southeastern 110° E, and American 75° W). In the present paper, these models, with additional modifications, are used to determine the accuracy of the TEC forecast for stations in the Southern Hemisphere lying along the American meridian ~ 70° W. A comparison of the results obtained with previous estimates for the American meridian of the Northern Hemisphere showed that metrics such as Mean Absolute Error MAE, Root Mean Square Error RMSE, Mean Absolute Percentage Error MAPE follow the latitudinal structure of the ionosphere. MAE values are slightly higher in the Southern Hemisphere, and in both hemispheres are less than 0.4 TECU, with an average of 0.236 TECU in the Southern Hemisphere and 0.189 TECU in the Northern Hemisphere. RMSEs range from 0.5–1.5 TECU, with an average of 0.591 TECU in the Southern Hemisphere and 0.857 TECU in the Northern Hemisphere. MAPEs are less than 2%, with an average of 1.5% in the Southern Hemisphere and 1.9% in the Northern Hemisphere. An assessment of the forecast accuracy during three magnetic storms of different intensity (minimal Dst = −98 nT, −114 nT, −155 nT) showed the dependence of the accuracy of the forecast of ionospheric parameters on the latitude of the stations at which the measurements were carried out. The BiLSTM-FRF architecture reflects the nature of the disturbed state quite well, and the values for the BiTCN method actually coincide with the observed TEC. #CSOC1120.

Keywords: Deep Learning · Recurrent Neural Networks · Bidirectional · BiLSTM · BiGRU · Total Electron Content · Global Positioning System · Forecasting

R. Silhavy and P. Silhavy (Eds.): CSOC 2024, LNNS 1120, pp. 99–118, 2024.
https://doi.org/10.1007/978-3-031-70518-2_10

1 Introduction

Research of the ionosphere, as a part of near space, has been going on for many decades, but recently more and more works have appeared that not only generalize long-term results, but also predict the prospects for the development of space research of near space within the framework of the natural sciences. So, the paper [1] by a leading scientist in the field of solar-terrestrial physics, S. Radicella, begins with the words of the theoretical physicist M. Baranger, said in 2000, that in the 21st century, a principal tool of research will be the computer [2]. S. Radicella extends M. Baranger's words into the 21st century in the sense that the main means of research will be machine-learning tools, since natural sciences have to shift from a "linear approach" to a "complex approach", one of the characteristics of which is its nonlinearity. His paper is devoted to substantiating this statement using an example of modelling and predicting ionosphere variables as a paramount objective of research by scientists and engineers. It is noted that machine learning (ML) has been increasingly applied to ionospheric time series data analysis in the past several decades to model and forecast their evolution since the 1940s. The stages of development of these methods are given with corresponding references to papers using the example of the total electron content TEC. It is emphasized that, particularly since 2020, more sophisticated Deep Leaning tools have been applied to model and forecast the variability of ionospheric parameters like critical frequency foF2 and TEC. It is important to note, that the wide availability of TEC data in recent years made its time series the preferred parameter to be used for ionosphere research. Several papers are given as examples.

More detailed reviews of methods are given in papers that develop new methods in relation to a specific research problem. A significant number of approaches is used, from long short-term memory LSTM [3], gated recurrent unit GRU [4] to bidirectional algorithms [5], temporal convolutional networks TCN [6] and many others [7–9].

One of the tasks in space weather research is to study the behavior and forecast TEC along the meridians. In the works of the authors [10–12], the accuracy of TEC forecasts was analyzed on the European 30° E, Southeastern 110° E and American 75° W meridians of the Northern Hemisphere. The forecast methods developed by the authors have evolved along with the development of neural network approaches. In [10], a modification of the LSTM and GRU methods using bidirectional algorithms is implemented. The results are given for 2015, 2 h ahead forecast, meridian 30° E. The results were assessed using the metrics Mean Absolute Error MAE, Root Mean Square Error RMSE, Mean Absolute Percentage Error MAPE. Without bidirectional processing, the RMSE estimates changed from 2.9 TECU to 1.5 TECU with increase of latitude, MAPE changes from 8% to 16%. The use of the bidirectional architecture leads to changes in latitudinal dependence: RMSE from 1.4 TECU to 0.5 TECU, MAPE from 6% to 4%. In [11], the TCN architecture was added, along with its bidirectional modification BiTCN. For the same meridian and the same conditions, the best results were provided by the BiTCN architecture, with MAE values less than 0.3 TECU, RMSE less than 0.6 TECU, and MAPE less than 5%. Analysis of the results allowed us to leave 10 of the 20 initially considered architectures for further research: (1) LSTM-FRF, (2) LSTM-RFF, (3) GRU-FRF, (4) GRU-RFF, (5) TCN, (6) BiLSTM-FRF, (7) BiLSTM-RFF, (8) BiGRU-FRF, (9) BiGRU-RFF, and (10) BiTCN. In accordance with the metric values, three groups of architectures were identified: (1) unidirectional (indicated by (1–5)), in which MAE was in the range of 1.0–1.8 TECU, RMSE was in the range of 1.3–2.5 TECU, MAPE = 8–14%. The second group included bidirectional recurrent architectures (6–9) with MAE in the range of 0.3–1.9 TECU, RMSE in the range of 0.5–2.5 TECU, MAPE = 4–14%. The third group consisted of the BiTCN architecture, for which MAE = 0.14–0.21 TECU, RMSE = 0.3–0.6 TECU, MAPE = 1.1–2.75%.

The book paper [12] is an extension of [11], adding results for the 110° E and 75° W meridians along with reviews of methods used in the relevant areas. These reviews showed that the forecast accuracy in the considered regions was on par with the estimates for the first group. Using the methods, presented in [11] in [12], with some modifications led to the following estimates based on data from 6 stations for the Southeastern meridian 110° and the BiTCN architecture: MAE = 0.1–0.18 TECU, RMSE = 0.13–0.28 TECU, MAPE = 0.6–1% with the exception of Tunguska station (2.0%). For the American meridian of the Northern Hemisphere 75° W, these estimates based on data from 7 stations were MAE = 0.1–0.3 TECU, RMSE = 0.7–1.25 TECU, MAPE = 1.5–2.0%, with the exception of the Goose Bay station (3.5%). Thus, for all meridians, a significant improvement in forecast accuracy was obtained using the BiTCN architecture.

The purpose of this paper is to assess the accuracy of the TEC forecast for the Southern Hemisphere. The 70° W meridian was chosen because it has the largest number of stations. The study is carried out for 9 stations using the models presented in the works [10–12].

2 Materials and Methods

Section 2 includes two subsections: 2.1 Experimental data and TEC behavior and 2.2 Description of Neural Networks for TEC forecasting.

2.1 Experimental Data and TEC Behavior

The comparison of the forecasting accuracy of TEC was performed for year 2015 using recurrent and convolutional neural networks, including unidirectional and bidirectional implementations. The values of global JPL GIM-TEC maps were calculated from IONEX files with a time step of 2 h (https//cddis.nasa.gov/archive/gnss/products/ionex/) for 9 stations: Arequipa (16.4° S, 71.5° W), Bogota (4.711° N, 48.4° W), Maracaibo (10.7° N, 71.7° W), Palmer (64.7° S, 64.1° W), Ramey (18.5° N, 67.1° W), Riobamba (1.6° S, 78.7° W), Rio Grande (1.6° S, 78.7° W), Santiago (33.1° S, 70.7° W), Tucuman (26.8° S, 65.3° W). The values of the indices of solar and geomagnetic activity Interplanetary Magnetic Field IMF, proton density Np, planetary 10Kp index, disturbance storm time Dst, and solar flux F10.7 were taken from SPDF OMNIWeb Service (http://omniweb.gsfc.nasa.gov/form/dx1.html). Since most works use a forecast for 24 h ahead, this work uses exactly this period.

The behavior of the F10.7 and Dst indices is presented in Fig. 1. The year 2015 is on the descending branch of the 24th solar cycle, so a tendency for F10.7 to decrease with increasing month number can be seen. A burst in solar activity is visible on the summer solstice day of June 22, 2015. The Dst index shows strong disturbance of the ionosphere throughout the year. The most severe disturbance was observed on June 21–23 as a result of a series of three coronal mass ejections.

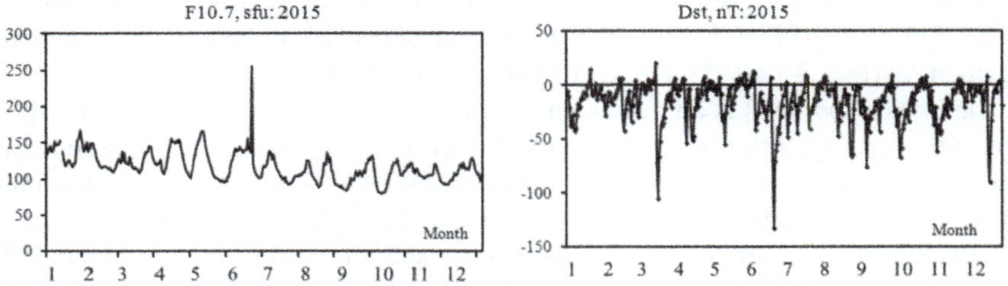

Fig. 1. Average daily values of F10.7 and Dst indices in 2015.

Figure 2 shows median values TEC(med) for several stations with different latitudes in the Southern Hemisphere (left panel) and in the Northern hemisphere (right panel).

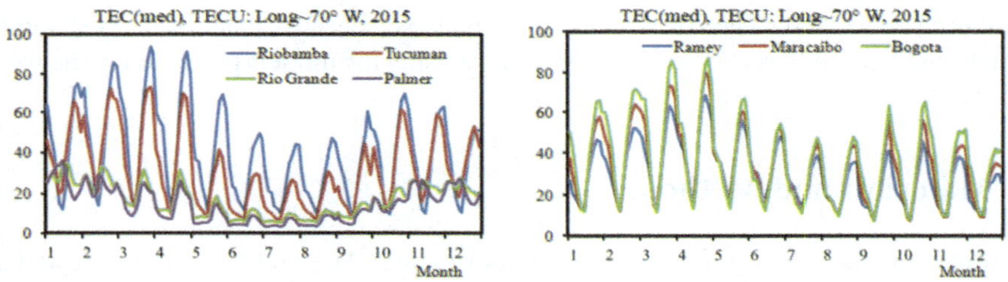

Fig. 2. TEC variations during 2015 for latitudinal networks of stations in the Southern (left panel) and Northern (right panel) Hemispheres.

The largest values and smallest values are observed in the Southern Hemisphere, i.e., here the latitudinal dependence of TEC is very strong. It's clear that TEC variations are synchronous for meridian. In addition, Fig. 2 shows an obvious semi-annual anomaly, with the high TEC values in the spring and autumn seasons, and the low values during the summer and winter seasons. The presence of two such maxima is explained by the fact that the density of the neutral atmosphere at ionospheric heights is maximal in April and October. The spring maxima are more pronounced during the descending branch of solar activity. The confirmation of semi-annual variations of ground-based and satellite measurements, which are directly related to TEC, is provided in a detailed review [13], using measurements from the CSES satellite at an altitude of ~500 km. In [1], special attention is paid to the study of such anomalies, since they reflect the complexity of the ionospheric system. It can be noted that in [12] for the ~75° W meridian in Figure B1, the TEC values for the Ramey station were the largest, but here they are the smallest. Figure 3, presenting examples of average daily variations (approximately 30 values for each month), shows the influence of geomagnetic activity.

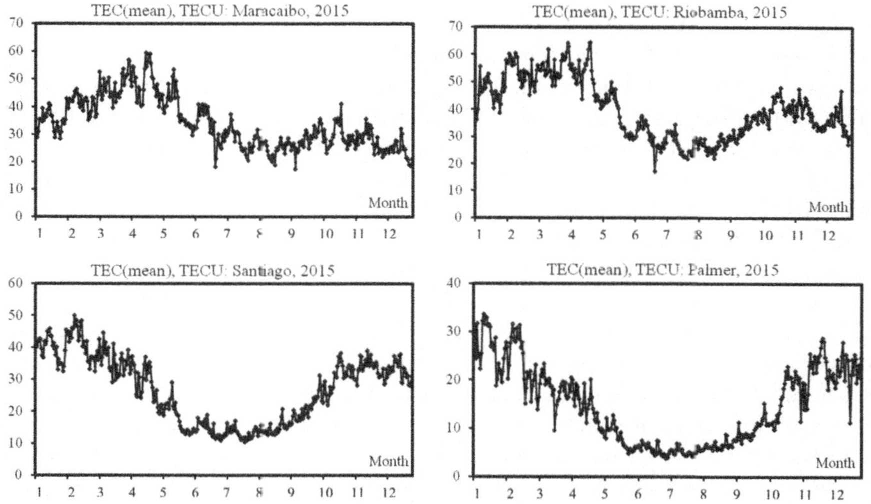

Fig. 3. Average daily TEC variations at various stations along meridian 70° W.

It can be seen that the seasonal distribution is subject to strong day-to-day variations associated with disturbed conditions. The June disturbance had the greatest impact on the TEC of the Maracaibo station in the Northern Hemisphere and the equatorial Riobamba station. The end-of-year disturbances had the greatest impact on the TEC of the highest latitudinal station Palmer in this study.

2.2 Description of Neural Networks for TEC Forecasting

This paper considers 10 models, which can be divided into 2 main groups: unidirectional and bidirectional. In each of these groups, in turn, we can distinguish recurrent and convolutional models. These models are modified versions of models proposed by the authors earlier, but in this research they are adapted to predict TEC values for 24 h in advance. Since TEC values are strictly ordered time series, the elements of which have a high degree of correlation with their previous values, and also since they are directly related to solar activity and processes occurring in the ionosphere, the solution to the problem of their forecasting can be carried out in the form of solving a regression problem, where new TEC values are determined using previous TEC values and auxiliary indices or features, which are correlated or physically related to the processes in the ionosphere which define various time scales of TEC change. Theoretical background and a detailed description of all the constituent elements of the models are given in [11]. In this section, we will describe the significant points that directly affect the structure of the models and the results obtained.

The initial dataset that was used to study forecasting accuracy included 4752 samples per station, with a time step equal to 2 h, and was compiled using data arrays for TEC, as well as solar and geomagnetic indices, described earlier: IMF, Np, 10Kp, Dst, F10.7. The models were trained independently for each station, using data for December 2014 and for the first half of 2015. The first 40% of samples from each array were used during training. The next 10% were used for validation. The remaining 50% of samples were used during testing. TEC values were passed through a sliding window function, moving in the time domain. The filter width was equal to 12 samples. The obtained sequences were used to form additional features from the previous TEC values so that the TEC_t sample corresponded to a sequence of 12 previous values TEC_{t-1}, TEC_{t-2}, … TEC_{t-12} at any given time. For the F10.7 index, which is provided on a daily basis, the daily value is used at each 2 h increment across the current data slice. Data arrays of solar and geomagnetic indices are not passed through a window function, but are organized into feature sequences. In order to correctly provide the forecast for 24 h in advance, time delay equal to 12 samples (24 h) is introduced. Every array is truncated, padded and shifted to maintain the same array sizes across the dataset, that so the TEC value at the time step t was predicted using indices at a time step $t - 1$.

The recurrent neural networks discussed in this article are based on the two most common architectures of recurrent neural networks, which have found extensive use in modeling and time series forecasting, namely Long Short-Term Memory (LSTM) and Gated Recurrent Unit (GRU). These models are integrated into multilayer deep learning neural networks. The composition of layers was changed, compared to that of [11], in such a way as to make the models suitable for providing 24 h ahead forecasts, while maintaining mostly the same structure and parameters, and achieving high accuracy and performance.

Unidirectional LSTM and GRU, as shown earlier, were inferior in efficiency compared to bidirectional architectures, however, for complete comparison, this work also proposes their modifications, similar to those made for bidirectional models. Bidirectional neural networks are capable of learning more complicated long-term dependencies between time steps compared to unidirectional architectures, and are especially useful when dealing with sequential events. For a bidirectional architecture, the cells are arranged into two layers, one of which processes the original sample in the forward direction, and the other in the backward direction. The outputs of each layer are connected either by addition or concatenation.

The structure of a generic bidirectional neural network, as well as the internal structures of LSTM and GRU cells, are shown in Fig. 4. Here i_t represents the input gate, f_t is the forget gate, o_t is the output gate, σ and *tanh* are the sigmoid and hyperbolic tangent activation functions, respectively, C_t is the internal memory state, $\tilde{C}_t$ represents the cell state candidate, C_{t-1} is the state at a timestamp $t - 1$, h_{t-1} is the output of the previous LSTM cell at a timestamp t-1, h_t is the output, x_t is the input, z_t represents the update gate, r_t is the reset gate, $\tilde{h}_t$ is the output candidate.

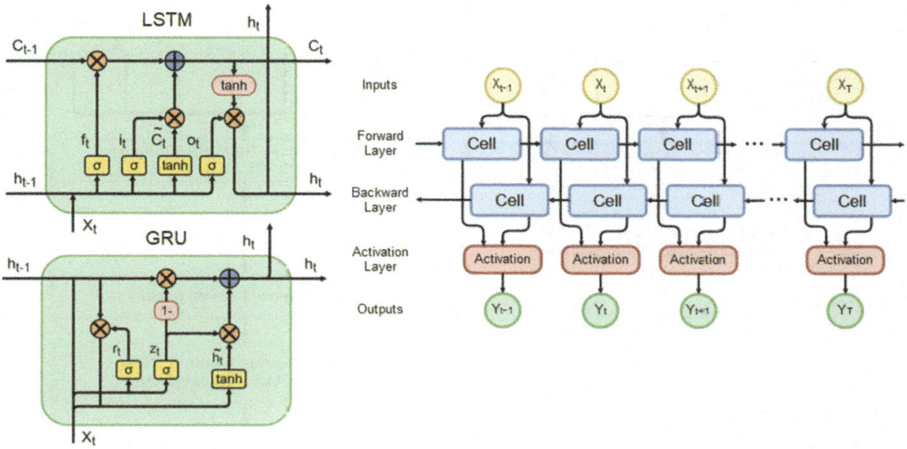

Fig. 4. Generic bidirectional neural network, and structural diagrams of LSTM and GRU cells.

Structural diagrams of unidirectional and bidirectional recurrent models are shown in Fig. 5. Here and below, the blocks designated as TEC input, Np, 10 Kp, Dst, F10.7, IMF indicate the input data of the neural network. The sliding window block denotes the procedure for applying a sliding window function, described above. Input sequences normalization layer block represents the normalization procedure. The input data normalization method has been unified compared to previous models, channel-by-channel normalization is performed in the range [0; 1], so that the maximum value in each channel was 1 and the minimum value was 0. All other blocks define specific layers of the neural network itself. The *N* parameter denotes the number of cells used in the respective layer. For fully connected layers, designated by corresponding blocks, neurons perform linear transformations of input vectors through a weights matrix, with each input neuron connected to each output neuron. Initialization of weights in fully connected layers was carried out using two different initialization procedures – Orthogonal Matrix Decomposition initialization, and Xavier Glorot's initialization [14, 15]. For the sake of readability, only the orthogonal matrix decomposition initialization is noted. Parametric rectified linear units, that were previously present, have been removed. Additional fully connected layer was added in both FRF and RFF architectures. The models of FRF series include a recurrent layer positioned between the fully connected layers, while in RFF series models a recurrent block is located right after the input layer. The recurrent block is determined by the specific model being used, as reflected in the model's name (LSTM, BiLSTM, GRU, BiGRU). The input and recurrent weights in the LSTM and GRU layers were initialized using the orthogonal matrix decomposition initialization. Batch normalization layers perform channel-by-channel normalization across the batches.

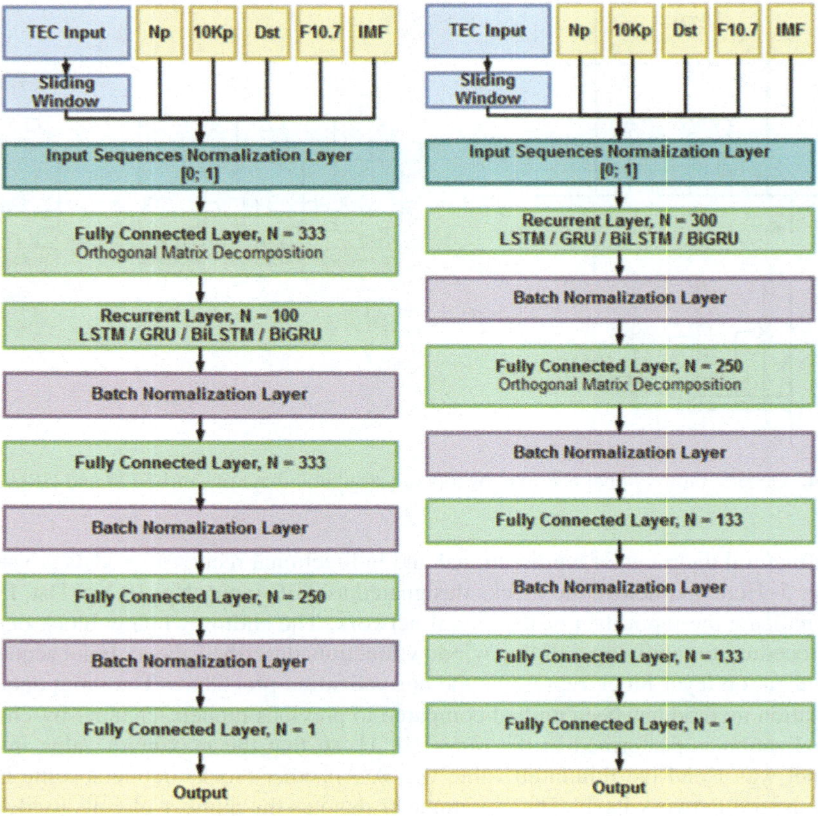

Fig. 5. Structure of the proposed modified recurrent models: FRF model (left); RFF model (right).

The convolutional neural networks proposed in this work are based on the principle of temporal convolution [16]. Temporal convolution was implemented using a series of sequential blocks of 1D convolutional layers, followed by batch normalization layers and dropout layers. The dropout layers randomly set input units to 0 with a specified probability in order to prevent overfitting. The dropout probability was set to 0.005. The unidirectional architecture included two 1D convolution layers, connected in parallel to a convolutional skip connection, that is employed to mitigate the vanishing gradient problem. In a bidirectional architecture, one branch was used for forward processing and an additional branch was used for backward processing. Causal padding was applied in every 1D convolution layer. The dilation factor in the first 1D convolution layer in each branch was equal to 1, as before, but the dilation factor in the second layer was increased to 12, compared to that in [11], that provided a significant improvement in both the accuracy and the convergence. The number of filters was set to 256, with a filter kernel size equal to 2. The outputs of the main branch, the reverse branch and the convolutional skip were connected using concatenation. Schemes for convolutional neural networks are shown in Fig. 6.

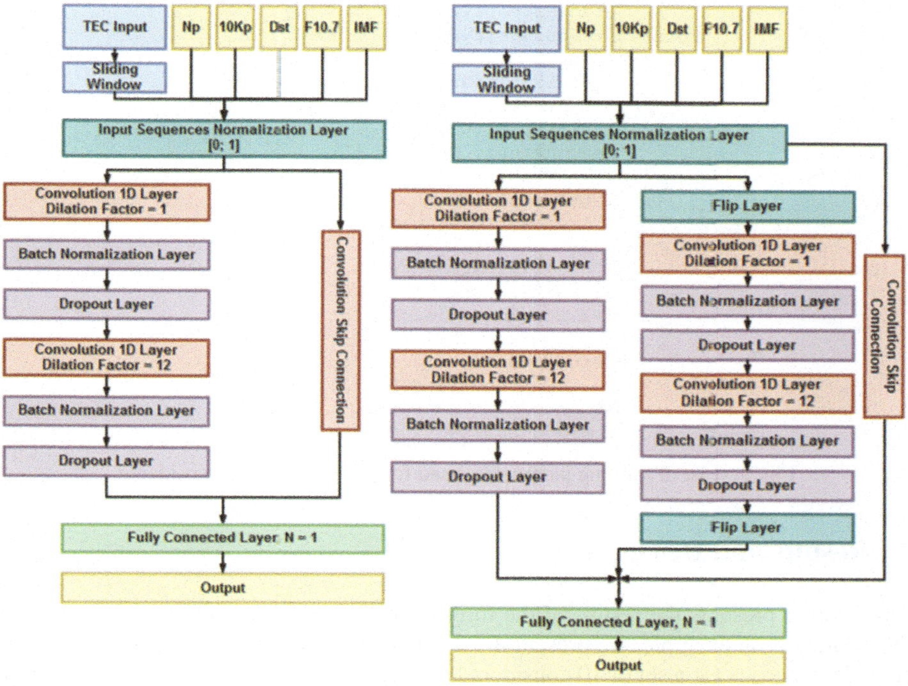

Fig. 6. Unidirectional (left) and bidirectional (right) temporal convolutional networks.

The input data normalization method has been unified compared to previous models, channel-by-channel normalization is performed in the range [0; 1], so that the maximum value in each channel was 1 and the minimum value was 0.

For uniformity, as well as easier interpretation of the results, the same metrics were used to compare accuracy as before: Mean Absolute Error (MAE), Mean Absolute Percentage Error (MAPE), Root Mean Square Error (RMSE) and Pearson's correlation coefficient (R). Optimization was carried out using the stochastic gradient descent algorithm based on adaptive estimation of first-order and second-order moments (Adam), with the mean squared error (MSE) loss function.

Development and training of neural networks were carried out on the MATLAB platform, using Deep Learning and Parallel Computing Toolboxes. The parameters used during optimization and training of neural networks are shown in Fig. 7.

```
          GradientDecayFactor: 0.9000
   SquaredGradientDecayFactor: 0.9990
                      Epsilon: 1.0000e-08
             InitialLearnRate: 0.0030
           LearnRateSchedule: 'none'
         LearnRateDropFactor: 0.1000
         LearnRateDropPeriod: 10
            L2Regularization: 1.0000e-04
     GradientThresholdMethod: 'l2norm'
           GradientThreshold: Inf
                   MaxEpochs: 150
               MiniBatchSize: 1
         ValidationFrequency: 20
          ValidationPatience: Inf
                     Shuffle: 'every-epoch'
        ExecutionEnvironment: 'gpu'
              SequenceLength: 'longest'
        SequencePaddingValue: 0
    SequencePaddingDirection: 'left'
```

Fig. 7. List of training parameters used for all architectures considered.

3 Results and Discussion

This section presents the results of the forecast accuracy study for the stations considered. First, as an example, we present the detailed results for the Tucuman station. The statistical metrics for a set of 10 considered architectures are given in Table 1. The metrics considered were MAE, MAPE, RMSE, *R*, averaged over half a year during single step 24 h ahead TEC forecasting for 2015. Such tables were calculated for 9 stations under consideration.

Table 1. Forecasting results for Tucuman station in 2015.

N	Tucuman	MAE TECU	MAPE %	RMSE TECU	*R*
1	LSTM-FRF	7.028	29.942	9.514	0.826
2	LSTM-RFF	5.276	23.007	6.976	0.910
3	GRU-FRF	5.331	22.571	7.024	0.914
4	GRU-RFF	4.951	20.931	6.560	0.917
5	TCN	4.388	19.543	5.800	0.931
6	BiLSTM-FRF	2.347	10.593	3.091	0.981
7	BiLSTM-RFF	2.247	10.113	2.919	0.983
8	BiGRU-FRF	2.380	10.666	3.095	0.981
9	BiGRU-RFF	2.206	9.783	2.899	0.983
10	BiTCN	0.188	1.012	0.262	1.000

The results are presented for the following architectures, indicated by numbers: (1) LSTM-FRF, (2) LSTM-RFF, (3) GRU-FRF, (4) GRU-RFF, (5) TCN, (6) BiLSTM-FRF, (7) BiLSTM-RFF, (8) BiGRU-FRF, (9) BiGRU-RFF, and (10) BiTCN. To better visualize the relationship between the metrics, they are shown in Fig. 8.

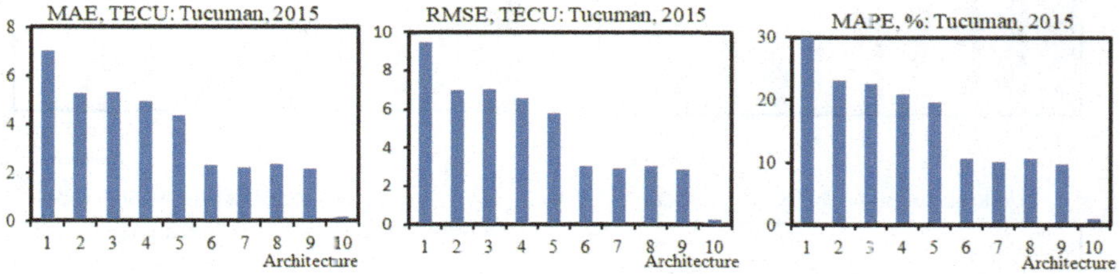

Fig. 8. Example of metrics distribution for different architectures and Tucuman station.

It can be seen that the results for the considered methods are clustered, 3 clusters can be distinguished, as in the case of other previously considered meridians. While there are differences in the results within each group, the differences between metrics are negligible, with the exception of the LSTM-FRF architecture in the first group. The average values in the first (unidirectional) group are MAE = 5.4 TECU, RMSE = 7.2 TECU, MAPE = 23.2%, for the second (bidirectional) group they are MAE = 2.3 TECU, RMSE = 3.0 TECU, MAPE = 10.3%, i.e. 2 times better than for the unidirectional group. For BiTCN, the values are given in the last row of the table and are equal to MAE = 0.19 TECU, RMSE = 0.25 TECU, MAPE = 1.0%.

The main results for all stations are presented in Fig. 9 for the architectures considered, and for 24 h ahead forecast.

The behavioral trends for all metrics are the same: in the first group, a significant latitudinal dependence of the values is visible. Only the values of the correlation coefficient are more uniform. MAE values in the first group are in the range of 2.5–9.6 TECU, in the second group – in the range from 1.2 to 3.5 TECU, in the third group – in the range of 0.1–0.33 TECU. The RMSE values in the first group range from 3.4 to 12.8 TECU, in the second group – from 1.75 to 4.4 TECU, and in the third group – from 0.15 to 1.33 TECU. MAPEs in the first group range from 15.9% to 36.7%, in the second group – from 8.1 to 13.8%, in the third group – from 0.8 to 2.0%. In the first two groups, the worst metric values refer to the Rio Grande station (latitude 53.8° S), the best values refer to the Riobamba station (latitude 1.6° S). In the third group, there is a significant reduction in errors compared to other groups, but the best results relate to different stations for different indicators.

Since the BiTCN architecture provides the best results, the latitudinal dependence in Fig. 10 is given for this model. The left panel shows the results for the Southern Hemisphere and the 70° W meridian, the right panel shows the results for the Northern Hemisphere and the 75° W meridian, considered in [12].

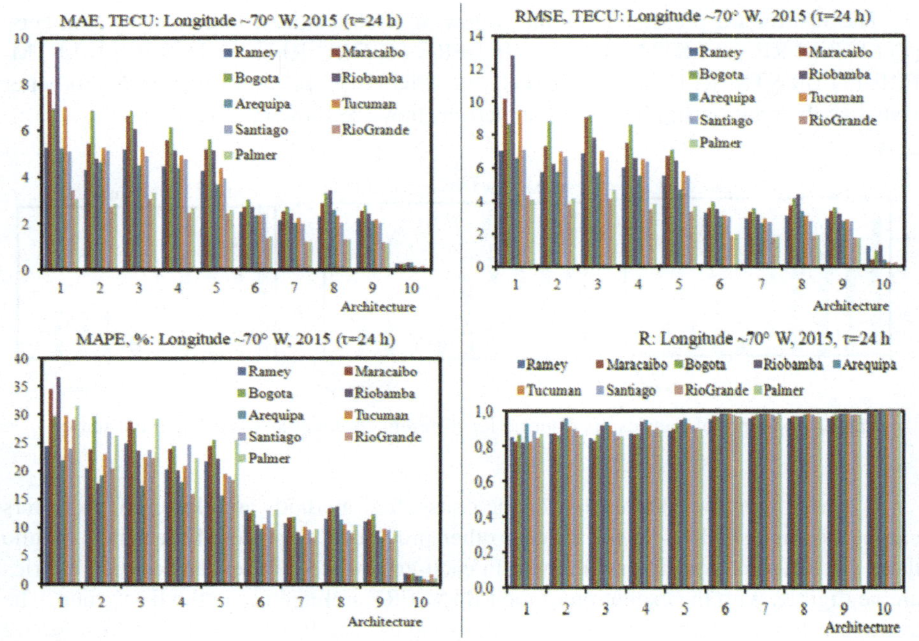

Fig. 9. Statistical characteristics of TEC forecast for different architecture and latitudes near the meridian 70° W.

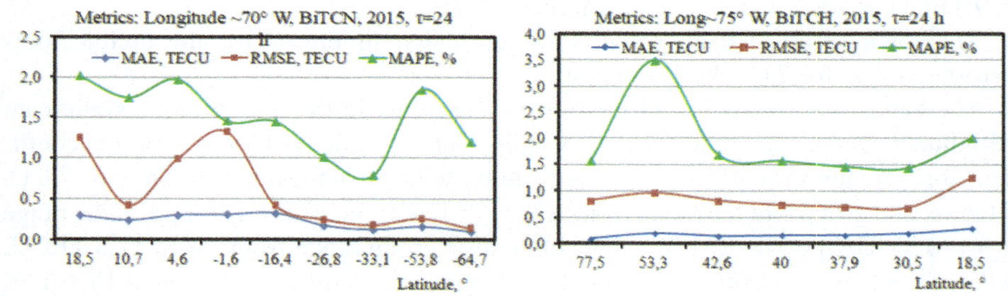

Fig. 10. Latitudinal dependence of metrics for the BiTCN architecture and two hemispheres.

It can be seen that both graphs have a common point: Ramey station (latitude 18.5° N). The MAE values on either side of this point decrease with increasing absolute value of the station latitude. The similar trend is observed for other metrics.

An increase in their values around |53°| apparently is due to the proximity to the ionospheric trough. The increase in RMSE at Riobamba station (latitude 1.6° S) is associated with proximity to the equator. MAE values are slightly higher in the Southern Hemisphere, and in both hemispheres are less than 0.4 TECU, with an average of 0.236 TECU in the Southern Hemisphere and 0.189 TECU in the Northern Hemisphere. RMSEs are in the range of 0.5–1.5 TECU with an average of 0.591 TECU in the Southern Hemisphere and 0.857 TECU in the Northern Hemisphere. MAPEs are less than 2%, except for Goose Bay station, with an average of 1.5% in the Southern Hemisphere and 1.9% in the Northern Hemisphere.

The forecast errors during the second half of year 2015 for the two stations, Tucuman and Maracaibo, on opposite sides of equator, are given in Fig. 11. Each panel for 2015 shows the values of TEC and δTEC between DoY 186 and 365 along with the Dst index, the values of which are reduced by 10 times to match scales, and the prediction accuracy is compared for two architectures: BiLSTM-FRF and BiTCN.

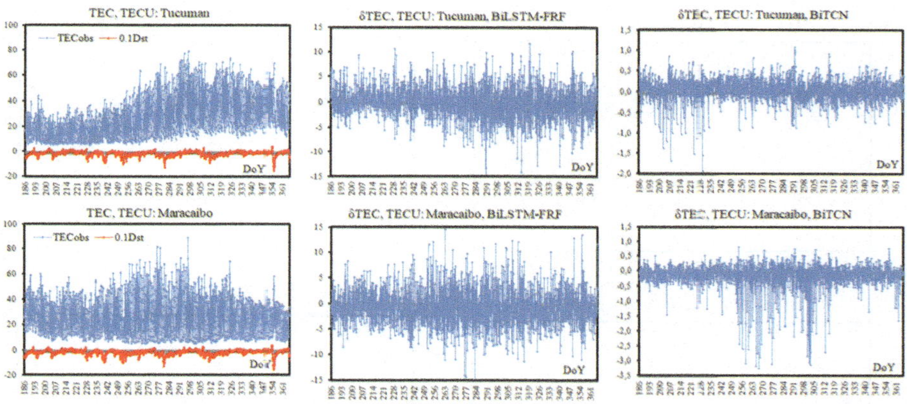

Fig. 11. Comparison of the prediction accuracy for two stations and two architectures: BiLSTM-FRF and BiTCN.

These figures show the difference in TEC behavior at the two stations. This is reflected in the δTEC values. The values for the LSTM-FRF architecture from the first group were in the range of ±40 TECU (not shown). The δTEC values for the BiLSTM-FRF architecture from the second group are in the range of ±15 TECU.

The δTEC values for the BiTCN architecture are significantly smaller and are in the range of −2.0 ÷ 1.5 TECU for the Tucuman station and −3.5 ÷ 1.5 TECU for the Maracaibo station. Significant deviations for the BiTCN architecture occur on different days for these stations, and they do not necessarily relate to periods of disturbed days. Most likely, this is the influence of a prolonged, sufficiently high level of disturbance, at which the ionosphere does not have time to return to a quiet state.

At the same time, assessing the accuracy of the forecast during periods of disturbed days is of certain interest, therefore, Fig. 12 provides examples for periods of several of the strongest disturbances: September 7–12, 2015 (DoY = 250–255) with a minimum Dst = −98 nT, October 6–10, 2015 (DoY = 279–283) with a minimum Dst = −118 nT,

19–22 December 2015 (DoY = 353–356) with a minimum Dst = −155 nT for Tucuman and Maracaibo stations. Each graph gives observational values of TEC (denoted as 'obs'), medians ('med') to see the nature of disturbances, the values of TEC for architectures BiLSTM-FRF and BiTCN are provided for comparison. Additionally, the values of the Dst index are shown, reduced by a factor of 4.

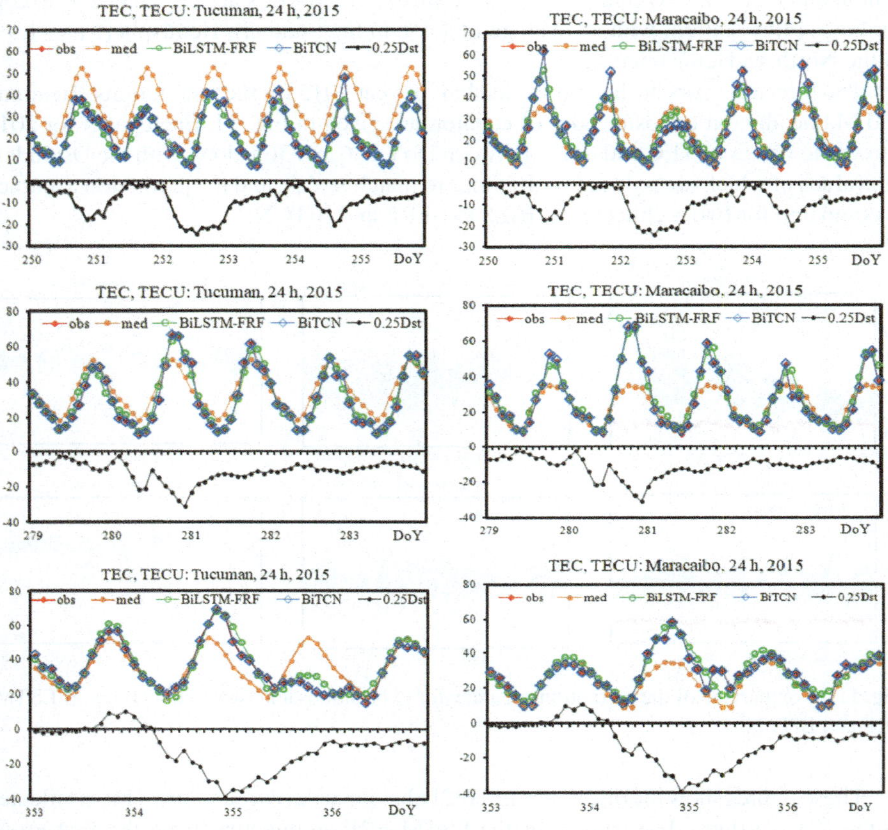

Fig. 12. Illustration of predicted performance of two architectures and two stations during three disturbed periods.

During the first magnetic storm, a constant negative response is observed at the Tucuman station. The Maracaibo station shows a mostly positive response. During the second magnetic storm, large positive deviations are visible at the Maracaibo station, even throughout the entire recovery period. During the third, strongest magnetic storm, the nature of the response is the same at both stations, but for the Tucuman station the negative response is most pronounced during the end of the recovery phase.

It should be noted that the BiLSTM-FRF architecture of the second group of models reflects the nature of the perturbed state quite well, and the values for BiTCN actually coincide with the observed TEC values. It is important to note that all the considered classes of models have their drawbacks. The disadvantages of bidirectional models, compared to unidirectional ones, include their higher requirements for the computing systems used, specifically higher RAM requirements, as well as a slower learning process, which on average is about 2 times longer in terms of computation time. Namely, for unidirectional models, the average training time for an individual station was: LSTM-RFF – 16 s, LSTM-FRF – 13 s, GRU-RFF – 12 s, GRU-FRF – 8 s, TCN – 2 s. For bidirectional models, the results are as follows: BiLSTM-RFF – 35 s, BiGRU-RFF – 23 s, BiLSTM-FRF – 16 s, BiGRU-FRF – 15 s, BiTCN – 5 s. The results are obtained using GeForce RTX 4080 and AMD Ryzen 9 7900X, but could differ greatly depending on hardware. However, the differences between recurrent and convolutional models are much more significant than between unidirectional and bidirectional. In addition, all models considered are dependent on data availability. In the absence of data for certain periods in the training samples, procedures for approximating the values of the parameters used are necessary. Solving this problem will require the development of separate approximation models or their integration into the architectures under consideration as a structural part. Also it is possible to notice, that from the point of view of experimental data in the form of global GIM maps, their permission 2 h can be not enough for revealing of effects with smaller time scales.

4 Conclusions

As noted in [1] and mentioned in the Introduction, machine learning has been applied to ionosphere time series data analysis since the 1940s. The neural network approach was used to predict such parameters as foF2, the maximum height hmF2, the maximum useable frequency MUF, TEC and others. However, the results for different parameters are scattered across different papers. A paper that simultaneously predicts three parameters (foF2, hmF2, TEC) using the same methods is [17], which gives a detailed overview of previous methods and uses machine learning regression models: decision trees (DT), random forest (RF) and support vector machine (SVM) algorithms by means of the example of the ROME ionosonde (41.9° N, 12.5° E) and the M0SE GPS receiver (41.89° N, 12.49° E). From a two-year period, 2012 data was used for training, 2013 data for testing (80% for training, 20% for testing). The inputs were sine and cosine of the day of year and hour of the day, hourly values of solar index F10.7, and geomagnetic index Ap, the present values of foF2(t), hmF2(t), TEC(t), and their values at t - 23 h. The outputs are the predicted values of foF2, hmF2, TEC at $t + 1$, namely 1 h in advance. The peculiarity of the approach was the refusal to imputing the missing values and the refusal of data scaling because the amplitudes of TEC, hmF2 and foF2 values vary non-linearly and abnormal amplitudes are not redundant due to influence of solar and geomagnetic activity. The best results for all parameters were shown by various methods: in digisonde data (foF2 and hmF2), the RF regression model has better results than other models. All models have RMSE values in the 0.86–1.05 MHz range for the foF2 parameter and the 22–28 km range for the hmF2 parameter. In TEC data, the SVM model has a lower RMSE in the range of 3.4–4.0 TECU. These results can be compared with the RMSE for the IRI-2016 model in 2013 for the Juliusruh station, the data of which were used in [10–12]. Estimates are 0.65 MHz for instantaneous foF2 values and 0.377 MHz for medians.

Among the foF2 prediction methods, the closest to the method of the given work is the method [18], in which the hybrid neural network is composed of a convolutional neural network (CNN) and bidirectional long short-term memory (BiLSTM). This method was trained and tested on 5 years (2009–2014) of foF2 data from ionosonde in Brisbane, Australia (27° 53′S, 152° 92′E). Additionally, the values of the indices F10.7, SSN, Bz, ap, and Dst were entered. 75% of data were used for training, and 12.5% for validation and testing/prediction. Four scenarios of low and high solar activity and two periods of magnetic storms were considered. Metrics included MAE, RMSE, MAPE. These metrics were compared for the LSTM-foF2, BiLSTM-foF2, IRI-2016 models with the results of the proposed CNN-BiLSTM method. Most often, the results with increasing forecast accuracy were arranged in the order: LSTM-foF2, BiLSTM-foF2, IRI-2016, CNN-BiLSTM. For all cases, the RMSE results were in the range of 0.583–1.125 TECU for 2010 and 0.778–1.816 TECU for 2014. As an example, we can give the values RMSE = 0.77, 0.878, 0.583, and 0.688 MHz for January, March, June, September 2010 and 0.837, 1.262, 0.778, 0.837 MHz for January, March, June, September 2014 for CNN-BiLSTM.

The paper [1] indicates the reasons why preference is given to the TEC parameter and provides a brief overview of publications related to the TEC forecast for two periods: before 2020 and after 2020. The first period is characterized by methods from a simple neuron network composed of several layers (an input layer, one or more hidden layers and an output layer) and recurrent networks known as Elman networks, up to radial basis function RBF. Unfortunately, most papers do not provide accuracy estimates.

The most accurate values were obtained for South Africa using a simple neuron network in the ranges RMSE 1.35–2.5 TECU, 0.95–1.25 TECU, 0.7–2.1 TECU in various works for 1 h ahead. In the paper [19], where the radial basis function (RBF) neural network improved by Gaussian mixture model method was proposed, a TEC forecast was carried out for three Chinese stations with a lead time of 0.5 h. The study period included 6 months of 2011 (the first 66.8% of days were used for training, the next 5.4% for validation, the last 27.7% for testing). The results depended on latitude and a mean relative error was about 9% and RMSE did not exceed 3 TECU. A comparison was made with the BP network (Back Propagation) method, which showed the advantage of the RBF neural network. At the conclusion of the paper, the authors stated that trained model was additionally used to predict data for January 2012, and a maximum mean relative error was 5%, and a maximum RMS error was 2 TECU.

Among all the papers mentioned in [1], one can also highlight papers [20, 21]. In the first of them, using the multilayer feed forward NN with the error back propagation algorithm, a TEC forecast with 1 day ahead was made for two stations Armi (6.062° N, 37.561° E) in Ethiopia and Ebre (40.8209° N, 0.4924° E) in Spain for 2015 using data for the period from 2011 to 2014. 80% of data were used for training, 20% for testing. The RMSE was generally larger for Armi station and it ranged from 4 TECU to 7 TECU as compared to Ebre station where it ranged from 2 TECU to 4 TECU. In the second paper, the approach of the work [20] with the same procedure for building a model and using data was extended to the case of multi-station data of the East African region. The experiment involved 13 stations, 9 of which were used to build the model, and 4 for additional testing. RMS values for 2015 were in the range of 3 TECU to 6.05 TECU at different geographical locations for both one hour and one day ahead prediction.

The given work is devoted to assessing the accuracy of the TEC forecast on the American meridian in the Southern Hemisphere in addition to the results obtained in [10–12] for three other meridians of the globe, and is using the methods developed in these papers with some additional modifications. Each paper included a detailed review of publications. Compared to the number of papers covering the meridians considered in [10–12], and presenting the results that are obtained using neural network methods, there are not so many papers for this meridian (for Bogota, Brasilia-DF stations). They show that the results are within the first or second cluster. In [22], the forecast for the next 24 h was made relying on TEC maps over the Brazilian region using data of the previous 5 days with the input indices F10.7 and Ap by means of NNs based on a multilayer perceptron approach. The database included the entire year 2014. The results are presented as MAE for different levels of solar activity: 5.44 TECU (F10.7 $\geq$ 150 sfu), 4.68 TECU (100 sfu $\leq$ F10.7 $<$ 150 sfu), 2.12 TECU (F10.7 $<$ 100 sfu). The average F10.7 for 2015 is 117.7 sfu, i.e., these results fall into the first cluster. In [23], the forecast for the next 24 h was made by the ED-ConvLSTM architecture with two modifications: ED-ConvLSTM-ND and ED-ConvLSTM-ND_mix. The methods were tested for the Brasilia-DF station, Brazil (coordinates $\varphi = -15.79°$, $\lambda = -47.89°$), and two periods of solar activity: for low activity data from 2019 and 2020 for training and validation, respectively, and for high activity data from 2014 for training and validation and 2015 for testing. For the 2014–2015 experiment, the best result was shown by the ED-ConvLSTM-ND architecture with MAE = 2.59 TECU, RMSE = 3.84 TECU. For the Bogota station, estimates were made in [24] for three days 07/27–07/31/2014 with MAE = 3.342 TECU, RMSE = 4.216 TECU. These estimates were obtained by a multivariate deep LSTM model using 10 external indices (F10.7, SSN, Dst, By, Bz, Vsw, Np, Kp, Ap, time). In [25], the LSTM model was used for this station to forecast TEC up to 5 h ahead with 30-min step in 2018 based on the 2008–2017 dataset. A feature of the method was the use of a random forest procedure to select external parameters. Of the 34 selected indices, the largest contribution was given by F10.7, Lα, ASY_D, PC, Dst. RMSE for each half hour forecast was 0.968 TECU. This value can be compared with the average RMSE value calculated by the method used in [11] for the second group and 2 h ahead, which is equal to 0.453 TECU, as well as with the values RMSE = 0.198 TECU for 2 h ahead and RMSE = 0.986 TECU for 24 h ahead for BiTCN.

In this work, results were obtained for 9 stations with a longitude near 70° W, 2015, for 24 h ahead forecast. It is shown that 10 considered architectures (1) LSTM-FRF, (2) LSTM-RFF, (3) GRU-FRF, (4) GRU-RFF, (5) TCN, (6) BiLSTM-FRF, (7) BiLSTM-RFF, (8) BiGRU-FRF, (9) BiGRU-RFF, and (10) BiTCN, including unidirectional, bidirectional, recurrent and convolutional structural groups, can be divided into three groups based on the metrics: (1) 1–5, (2) 6–9, (3) BiTCN. The first group is consisting of unidirectional models, based on LSTM, GRU, and TCN, the second group is distinguished by the inclusion of a bidirectional processing, combined with recurrent model, the third group contains the BiTCN architecture. The presented models are modified versions of the previously proposed models. The results for the Southern Hemisphere complemented previous results for the Northern Hemisphere. The comparison showed that the metrics follow the latitudinal structure of the ionosphere. MAE values are slightly higher in the Southern Hemisphere, and for both hemispheres are less than 0.4 TECU, with an average of 0.236 TECU in the Southern Hemisphere and 0.189 TECU in the Northern Hemisphere. RMSEs range from 0.5–1.5 TECU, with an average of 0.591 TECU in the Southern Hemisphere and 0.857 TECU in the Northern Hemisphere. MAPEs are less than 2%, except for Goose Bay station, with an average of 1.5% in the Southern Hemisphere and 1.9% in the Northern Hemisphere.

An assessment of the forecast accuracy during disturbances was made for three periods of varying intensity: September 7–12, 2015 (DoY = 250–255) with a minimum Dst = −98 nT, October 6–10, 2015 (DoY = 279–283) with a minimum Dst = - 118 nT, December 19–22, 2015 (DoY = 353–356) with a minimum Dst = −155 nT for Tucuman and Maracaibo stations. It is shown that the response of the ionosphere depends on the latitude of the stations, although they are situated near the same meridian. The BiLSTM-FRF architecture of the second group of models reflects the nature of the disturbed state quite well, and the values for the BiTCN method actually coincide with the observed TEC.

If we compare the obtained estimates with the values of metrics in the publications noted in the Introduction and at the beginning of this section, we can conclude that the BiTCN method provides significantly better forecast accuracy than the mentioned methods. Additionally, it can be noted that, together with the results of the papers [10–11], this conclusion covers vast regions of the globe. Moreover, the results of [11] confirmed these conclusions for the conditions of 2020 and 2022.

This indicates that the BiTCN method has the potential to provide the substantial forecast accuracy across the globe.

Acknowledgements. The TEC samples data was obtained from https://cddis.nasa.gov/archive/gnss/products/ionex/. Data on solar and geomagnetic activity was taken from SPDF OMNIWeb Service http://omniweb.gsfc.nasa.gov/form/dx1.html. The research was carried out at Southern Federal University with the financial support of the Ministry of Science and Higher Education of the Russian Federation (State contract GZ0110/23-10-IF).

References

1. Radicella, S.M.: New ways to modelling and predicting ionosphere variables. Atmosphere **14**, 1788 (2023). https://doi.org/10.3390/atmos14121788
2. Baranger, M.: Complexity, Chaos, and Entropy. New England Complex Systems Institute, Cambridge MA USA (2000)
3. Hochreiter, S., Schmidhuber, J.: Long short-term memory neural computation. Neural Comput. **9**(8), 1735–1780 (1997). https://doi.org/10.1162/neco.1997.9.8.1735
4. Cho, K., van Merrienboer, B., Gulcehre, C., Bahdanau, D., Bougares, F., Schwenk, H., Bengio, Y.: Learning Phrase Representations using RNN Encoder-Decoder for Statistical Machine Translation arXiv 2014, arXiv:1406.1078v3. [cs.CL] **3**, (2014). https://arxiv.org/pdf/1406.1078.pdf
5. Lei, D., et al.: Ionospheric TEC prediction base on attentional BiGRU. Atmosphere **13**, 1039 (2022). https://doi.org/10.3390/atmos13071039

6. Lea, C., Flynn, M.D., Vidal, R., Reiter, A., Hager, G.D.: Temporal convolutional networks for action segmentation and detection. In: Proceedings of the IEEE Conference on Computer Vision and Pattern Recognition, pp. 1003–1012. Honolulu, HI, USA, 21–26 July 2017. https://doi.org/10.1109/cvpr.2017.113
7. Natras, R., Soja, B., Schmidt, M.: Ensemble machine learning of random forest, AdaBoost and XGBoost for vertical total electron content forecasting. Remote Sens. **14**, 3547 (2022). https://doi.org/10.3390/rs14153547
8. Weng, J., Liu, Y., Wang, J.: A model-assisted combined machine learning method for ionospheric TEC prediction. Remote Sens. **15**, 2953 (2023). https://doi.org/10.3390/rs15122953
9. Murray, C., Du Bois, N., Hollywood, L., Coyle, D.: State-of-The-Art Deep Learning Models are Superior for Time Series Forecasting and are Applied Optimally with Iterative Prediction Methods, pp. 1–12 (2023). https://ssrn.com/abstract=4361707 or https://doi.org/10.2139/ssrn.4361707
10. Kharakhashyan, A., Maltseva, O.: Bidirectional recurrent neural network for total electron content forecasting. In: Silhavy, R., Silhavy, P. (eds.) Artificial Intelligence Application in Networks and Systems: Proceedings of 12th Computer Science On-line Conference 2023, Volume 3, pp. 193–207. Springer International Publishing, Cham (2023). https://doi.org/10.1007/978-3-031-35314-7_19
11. Kharakhashyan, A., Maltseva, O.: Comparison of the forecast accuracy of total electron content for bidirectional and temporal convolutional neural networks in European region. Remote Sens. **15**, 3069 (2023). https://doi.org/10.3390/rs15123069
12. Kharakhashyan, A., Maltseva O.: Comparison of the forecast accuracy of total electron content for bidirectional and temporal convolutional neural networks in European region. In: Prime Archives in Remote Sensing, pp. 1–48. Vide Leaf, Hyderabad, India (2023)
13. Zhu, K., et al.: Annual and semi-annual variations of electron density in the topside ionosphere observed by CSES. Front. Earth Sci. **11**, 1098483 (2023). https://doi.org/10.3389/feart.2023.1098483
14. Saxe, A.M., McClelland, J.L., Ganguli, S.: Exact solutions to the nonlinear dynamics of learning in deep linear neural networks. arXiv preprint arXiv:1312.6120 (2013)
15. Glorot, X., Bengio, Y.: Understanding the difficulty of training deep feedforward neural networks. In: Proceedings of the Thirteenth International Conference on Artificial Intelligence and Statistics, Sardinia, AISTATS, Italy, pp. 249–356. (2010)
16. Shaojie, B., Kolter, J.Z., Koltun, V.: An Empirical Evaluation of Generic Convolutional and Recurrent Networks for Sequence Modeling. Preprint, submitted 19 April 2018. https://arxiv.org/abs/1803.01271
17. Iban, M.C., Şentürk, E.: Machine learning regression models for prediction of multiple ionospheric parameters. Adv. Space Res. **69**(3), 1319–1334 (2022). https://doi.org/10.1016/j.asr.2021.11.026
18. Bi, C., Ren, P., Yin, T., Xiang, Z., Zhang, Y.: Modeling and forecasting ionospheric foF2 variation in the low latitude region during low and high solar activity years. Remote Sens. **14**, 5418 (2022). https://doi.org/10.3390/rs14215418
19. Huang, Z., Yuan, H.: Ionospheric single-station TEC short-term forecast using RBF neural network. Radio Sci. **49**, 283–292 (2014). https://doi.org/10.1002/2013RS005247
20. Tebabal, A., Radicella, S.M., Nigussie, M., Damtie, B., Nava, B., Yizengaw, E.: Local TEC modelling and forecasting using neural networks. J. Atmos. Solar Terr. Phys. **172**, 143–151 (2018). https://doi.org/10.1016/j.jastp.2018.03.004
21. Tebabal, A., Radicella, S.M., Damtie, B., Migoya-Orue, Y., Nigussie, M., Nava, B.: Feed forward neural network based ionospheric model for the East African region. J. Atmos. Solar Terr. Phys. **191**, 105052 (2019). https://doi.org/10.1016/j.jastp.2019.05.016

22. Silva, A., Moraes, A., Sousasantos, J., Maximo, M., Vani, B., Faria, C., Jr.: Using deep learning to map ionospheric total electron content over Brazil. Remote Sensing **15**, 412 (2023). https://doi.org/10.3390/rs15020412
23. de Paulo, M.C.M., Marques, H.A., Feitosa, R.Q., Ferreira, M.P.: New encoder–decoder convolutional LSTM neural network architectures for next-day global ionosphere maps forecast. GPS Solutions **27**(95), 1–14 (2023). https://doi.org/10.1007/s10291-023-01442-4
24. Shenvi, N., Virani, H.: Forecasting of ionospheric total electron content data using multivariate deep LSTM model for different latitudes and solar activity. J. Electr. Comput. Eng. **2023**, 1–13 (2023). https://doi.org/10.1155/2023/2855762
25. Zewdie, G.K., Valladares, C., Cohen, M.B., Lary, D.J., Ramani, D., Tsidu, G.M.: Data-driven forecasting of low-latitude ionospheric total electron content using the random forest and LSTM machine learning methods. Space Weather **19**(6), e2020SW002639 (2021). https://doi.org/10.1029/2020SW002639

Computer Science on Philosophy Perspective

Mahyuddin K. M. Nasution(✉)

Computer Science Program Study, Faculty of Computer Science and Information Technology, Universitas Sumatera Utara, Medan 20155, Sumatera Utara, Indonesia
mahyuddin@usu.ac.id

Abstract. This paper reveals the relationship between computer science and philosophy. With the aim of stating a picture of the development of computer science from a philosophical perspective, which requires concepts base on philosophy, namely language and logic in expressing the ontology of computer science into a definition. Language and logic become a forum for developing computer science through a methodology that is epistemologically appropriate to a culture or country. This relationship shows a variety of support for the development of computer science.

Keywords: Ontology · language · logic · definition · epistemology · methodology · natural law

1 Introduction

The development of computer science as a formal science not only shows that its knowledge is becoming stronger among all sciences, but this development continues to expand and spread to all other sciences and technologies [1–5], and has resulted in several scientific fields merging into computer science, while on the other hand computer science has given rise to several other scientific fields that are specifically new or an update of existing science [6–10]. However, from the existing facts, the development of computer science varies from one country to another.

No matter how high the development of computer science is in the current information era, it does not appear suddenly [12,13]. This development took place as a series of studies that produced documents as output and endlessly to provide evidence of human existence in the world, as a human effort to achieve prosperity [14–16]. This study goes hand in hand with growth and is in line with its importance in a culture [17], and started since the formation of human civilization, a human thought that was recorded in various facilities and infrastructure, such as human memory and symbols that mark the track record of that culture [18]. This thought was first organized under the name of philosophy [19]. A name for the structure of thought that emerged in the 6th century BC. In

R. Silhavy and P. Silhavy (Eds.): CSOC 2024, LNNS 1120, pp. 119–131, 2024.
https://doi.org/10.1007/978-3-031-70518-2_11

this way, previously all science—such as physics, metaphysics, psychology, or in general natural science—was part of philosophy [20,21]. Meanwhile, computer science which emerged in the 20th century AD is a derivative science in the formal field, where the emergence of computer science was centuries after the arrival of philosophy [22–24]. Therefore, the purpose of this article is to explain computer science from a philosophical perspective.

2 Review Towards Problem Definition

When computers first appeared, scientists defined computer science as a form of understanding that was easier than in later times [25]. As the definition states, computer science is "*the study of computer-related phenomena* (D1) [26,27]". A definition of an object of thought—a computer—which was the latest, final understanding at that time about an object called a computer, an object that was present in a culture with a name derived from the language of that civilization at that time, a technology was given the name computer [28]. A definition depends on the underlying philosophy, where initially science was always part of philosophy [29,30]. This definition involves terms and rules about how these terms are used so that the definition still has the potential to become a reference, just like the term phenomenon that invites the full attention of human thought at that time until now [31].

In this way, the terms involved come from already known meanings and from provisions in the rules that have been given special meanings such as the use of science and the involvement of philosophical terms [32]. Terminologically, the term science comes from the Latin *scientia*, which includes all systematic efforts to build and organize human familiarity, awareness or understanding or something in the form of facts, skills, entities and objects where explanations or predictions are possible to reveal the universe of discussion [33]. Organizing naturally within a framework of thought. Meanwhile, the term philosophy also comes from the word philosophia in Greek, namely *philein* and *sophia*, respectively meaning love and wisdom, and philosophy means love of wisdom. Thus, philosophy can be said to be a serious effort to study and express the odyssey of human thought [34,35].

Any science will consist of the components that make it up [36]. Philosophy directs language, starting from certain terms to form a basic understanding or concept about a science, then establishing certain agreements as symbols that form axioms of that definition so that it can easily gain understanding from the human mind [37]. As one of the constituent components, definition limits thinking to a certain meaning, which causes science to not only have a framework for thought, but a framework that requires the implementation of science based on its methodology and no longer following the nature of science, such as certainty or uncertainty [38]. Computer science, for example, has a short definition as "*the study of algorithms* (D2) [27,39]". The term algorithm is the name for a set of instructions for solving problems logically [40]. Logical means thinking according to the rules of logic, a branch of philosophy about thinking correctly [41]. A healthy way of thinking in which Thales initiated it into a philosophical thinking activity [42], Aristotle compiled it into a law of thinking called logic (one of the four problems of philosophy) [43], and Muhammad Ibn Musa al-Khwarizmi derived its implications in a science called algebra (*al-Jabr*) [35], a step-by-step systematic way to obtain an optimal formulation of a problem [44], and therefore scientists call its implementation—generally in computer language—as an algorithm [45].

Computer language is a language that naturally originates from the logical interpretation of the behavior of atoms in interacting with each other to form an intelligent machine, and then the machine's behavior is translated into natural language to make it easier for humans to interact with the computer machine [46,47]. In philosophy, language functions to identify objects of thought in order to produce rules and theories that apply in a science, and for scientific knowledge to arise by strictly limiting the language used, including in computer science [62]. In information structures, the recognition of human thought with language not only presents definitions of a scientific subject differently, by following grammar, but also determines punctuation to encode expressions and intonations that have different meanings [49]. Thus, it is not wrong to state that computer science is "*the study of the structure of information* (D3) [27,50]". Definitions that provide an explicit meaning of language—to obtain implications through lexical so that the computer recognizes the object of analysis, where syntax functions so that the computer does not miss-determine the process, and by involving semantics so that the computer places appropriate language symbols—to produce a synthesis where the implementation of algorithms carried out according to computer programming languages to produce information [45,51,52]. A description that is in accordance with the definition of information by ISO and ANSI which states that "*information is the meaning that a human assigns to data by means of the known conventions used in their representation* [53]".

Computer science—as a knowledge from a philosophical perspective—originates from the recognition of information, is also the meaning of data, based on the pyramid of data-information-knowledge-wisdom, and therefore data is outside the demarcation of computer science [6]. Programs and data have the same position as input to the computer and experience different treatment by the computer. Programs and data share the same place in the computer. Programs are computer languages that control computer hardware [54,55]. Data is a unit of facts encoded as a result of the collaboration of programs and hardware to produce any information, but both are associated with complexity that has the potential to reduce computer performance [56]. So, computer science is "*the study and management of complexity* (D4) [27,57]". The final definition which emphasizes that computer science involves many methods in its work serves the limitations of human computing, but requires an imitation of human computing intelligence so that it does not fall into GIGO (garbage in garbage out) [58,59]. Not only does the data potentially come from rubbish or incorrect sources, but there is also the potential for implementing inappropriate methods so that the program becomes rubbish [60]. According to philosophy, computers are a tool to achieve prosperity by increasing human intelligence through learning as the only way [30,61]. For this reason, the role of computer science is getting bigger and bigger so that it increasingly determines the daily life of every person or community. Therefore, simply expressing a definite definition of computer science is increasingly difficult. Scientists need to understand philosophy on the basis that in fact every scientific field of computer science requires ethics [62]. In addition, computer science extends to all scientific fields by introducing its methodological framework. Scientific expansion is a factor in which computer science cannot avoid the interests of other scientific fields towards it [63]. This expansion does not only include exact sciences and natural sciences, but also includes life sciences and social sciences [64].

3 Methodology Container

Language—with an understanding of terms—is to provide appropriate recognition for defining an object of human thought. Definitions, ontologically (in philosophy) [65], are to state the existence of something, just as definitions D1, D2, D3, D4, and "*computer science is the study of data, information and knowledge at different levels of complexity or abstraction, and from which other fields of scientific discipline can be developed* (D5) [66]" are to state the existence of computer science. So, language becomes a place for the existence of computer science where a definition plays a role in explaining terms that are not yet known or as updating understanding with terms that are already known, as in "*computer science is the technology that deals with the development and use of certain types of human-created artifacts* (D6) [67]". In this case, *definiendum* states the terms that need to be defined, which defining is *definiens*, so a definition takes the form $Y =_{def} abc$, where Y represents the term to be understood, the symbol $=$ (equal to) represents the word "is", while abc are terms that have been understood [68].

Words are the basic components of language together with applicable rules, grammar, and form statements such as those contained in definitions [28]. Every statement not only has meaning but also truth value, which presents logic as the ability to think or knowledge of thought, which is different from observation-based knowledge [70]. So, language embodies a branch of philosophy called logic. Logic is the key to the gate of science [71]. With logic, it has the potential to regulate methodology, which may enable the existence and essence of computer science as something that "becomes" epistemologically from ideas by abstracting from its source [72]. In this way, reasoning reveals cause and effect to provide a way to develop computer science based on logic as a procedure for computer science to work.

4 Discussion About Implication

When scientists view computer science as a whole by climbing the pyramid methodologically to the scientific peak, the choice of any working method will enter a posteriori the bottom of the pyramid in the form of induction [73,74], while a priori it causes a deduction [75]. Even though induction is an empirical way of working, its scientific knowledge cannot be separated from the role of logic and reasoning as well as their role in the way deduction works in formal science, such as computer science [76,77].

For example, nature as a source of computer science epistemology, natural law establishes a relationship between the symptoms of any natural object, phenomena that physically or chemically present the behavior of atoms or interactions between atoms in molecules in the form of mathematical geometry [78]. This interaction has become an absolute law, not an arbitrary thing [79]. In this way, natural law crowns computer science with formal theories [80–82].

The laws that govern nature appear as relationships between phenomena, allowing humans to capture them through thought and abstract them [28]. The term absolute applied to natural law is a term that has various interpretations, which encourages scientists to provide changes with newness, such as the ideas contained in definitions of computer science. Absolutes according to philosophy create complete groups of symptoms, which through classification will collect and limit these symptoms based on understanding and definition [83,84]. Natural law is not an agreement about an object, but is a complete set of objects. In this way, the logic $p \Rightarrow q$, a statement about objects that can be explained, "If p holds, then q holds" is something that forms a postulate, by filling p and q with various variety. Proof is possible using empirical facts where the data must speak in general, because it requires conclusions that may only be tested after a certain time [85]. Proving postulates with a theoretical framework requires abstraction with an understanding of mathematics (the twin brother of philosophy) as a stronger basis for evidence [86,87].

The postulates are generally supported based on the internal sources of epistemology, namely the ability to think [88]. The human mind has the power of reason in various matters that are natural, which means following natural laws. Postulate after postulate based on the basic framework of definition has accumulated into various fields of mathematics, where some of these postulates have stood behind computer science as a foundation, others have positioned themselves as supporting the development of computer science, and others are waiting for implication [89]. In general, both natural laws and human thinking abilities complement each other by involving complex and complementary developments [90]. Inductively, natural law produces the following model [91–94]:

$$f(x) = \sum_i c_i x_i \tag{1}$$

from criterion to criterion collected from an object, with that $i \in I$ is an integer, the criterion x_i is a real variable, $-\infty \leq x_i \leq \infty$, and c_i is a constant. By the rational principle, the computation must begin with an integer i and end at an integer n. Even though the n loop is linear in complexity, for $n > 2^k$, $k = 8, 16, 32, 64, \ldots$, causing the accumulation of $\sum$ over x_i will result in a computational error flag (as one of the complexity issues). From a computer science point of view, it is possible to raise i and x_i in computing when $1 \leq i \leq n = 2^k$, $0 < x_i \leq 2^k$ and $0 \leq c_i \leq \pm 2^k$ such that $|c_i x_i| \leq 2^k$ and $\sum_i c_i x_i \leq 2^k$. If one of the computational objects exceeds the computer's capabilities, scientifically determining that excess takes the form of restrictions based on arithmetic postulates such as $mod(\sum_i c_i x_i, 2^k)$ [25].

Table 1. Documents about Philosophy and Computer Science in Scopus Database (October 2023)

Number of	Philosophy (A)	Computer Science (B)
documents for all countries ($\sum x()$)	119,346	1,885,878
documents without country	16,637	89,576
countries	186	186
countries with documents ($n()$)	159	158
Average of documents ($\overline{x}()$)	751	11,936
$n_1(x(A) \geq \overline{x}(A)) \wedge n_1(x(B) \geq \overline{x}(B))$		$n_1(A \cap B) = 21$
$n_2(x(A) \geq \overline{x}(A)) \wedge n_2(x(B) < \overline{x}(B))$		$n_2(A \cap B) = 5$
$n_3(x(A) < \overline{x}(A)) \wedge n_3(x(B) \geq \overline{x}(B))$		$n_3(A \cap B) = 11$
$n_4(x(A) < \overline{x}(A)) \wedge n_4(x(B) < \overline{x}(B))$		$n_4(A \cap B) = 98$
$n_5(x(A) = 0)) \wedge n_5(x(B) < \overline{x}(B))$		$n_5(A \cap B) = 45$

Table 2. Overview of computer science development based on Table 1.

Value based on Eq. (2)	Meaning for the relationship between A and B
0.0709	strong support
0.0160	lack of support
0.0359	some support
0.4475	little support
0.1654	no support

History is one source of epistemology for computer science. History that explains not only the emergence of culture from habits of thought both from individuals and from the accumulation of communities, but also explains the emergence and development of computer science that fills that culture [95], as revealed by information from Table 1, which explains the evidence for the existence of computer science in every part of the world. However, reasoning on the data reveals that everything from ideas to innovations are born from different cultures and the emergence of science is in accordance with the ability to think and adapt to nature, and not only that it is also related to the current development of computer science, where language and logic are the the platform for philosophical development requires a *principium* of identity [96], that something is the same as itself or $a = a$, which gives rise to the principle of being recognized as similarity [97,98], such as

$$sim(A, B) = \frac{A \cap B}{A + B - (A \cap B)} \tag{2}$$

which explains that the overlap (sim) in computer science developments must be as small as possible to obtain maximum growth (of two set of characteristics A and B, for example), Table 2, or avoid imitative actions to reduce inefficiencies in the world of research and learning. Apart from that, *principium contradictioad* [99], if two opinions are contradictory it is impossible for both of them to be true at the same time. In this way, it is impossible for the development of computer science to be philosophically contradictory to each other, but it is possible to correct one another. However, if two concepts or opinions are opposite, it is impossible that both are wrong or both are right, but there is no third opinion. In principle, this principle states that naturally computer science does not have more than one alternative choice, there is only a puzzle that requires completeness [61,62].

5 Conclusion

Culture describes how thought exists in a container called language and logic, which basically has become philosophy, and then grows to give birth to science and its methodology. Today, computer science has grown and interacted with other sciences and—through documents as evidence of development—has a variety of supporting relationships between philosophy and computer science.

Acknowledgment. This paper is the result of discussion and teaching materials from the philosophy of science in the Doctoral study program at Universitas Sumatera Utara, Medan, Indonesia.

References

1. Morris, J.M.: The impact of computer science on mathematics in the engineering curriculum. Int. J. Math. Educ. Sci. Technol. **16**(3), 437–443 (1985). https://doi.org/10.1080/0020739850160313
2. Parrish, A., et al.: Computer engineering, computer science and management information systems: partners in a unified software engineering curriculum. In: Proceedings - 11th Conference on Software Engineering Education, CSEE and T 1998, pp. 67–75 (1998). https://doi.org/10.1109/CSEE.1998.658302
3. de la Iglesia, D., García-Remesal, M., de la Calle, G., Kulikowski, C., Sanz, F., Maojo, V.: The impact of computer science in molecular medicine: enabling high-throughput research. Curr. Top. Med. Chem. **13**(5), 526 (2013). https://doi.org/10.2174/1568026611313050002
4. Partovi, H.: A comprehensive effort to expand access and diversity in computer science. ACM Inroads **6**(3), 67–72 (2015). https://doi.org/10.1145/2807704

5. Gasca-Hurtado, G.P., Machuca-Villegas, L.: The impact of computer science in the real world. RISTI - Revista Iberica de Sistemas e Technologias de Informacao **2018**(29), ix–xiii (2018). https://doi.org/10.17013/risti.29.0
6. Capper, L.: Philosophy for the teaching of computer science of computer science and information technology. Comput. J. **29**(1), 83–89 (1986). https://doi.org/10.1093/comjnl/29.1.83
7. Conte, S.D., Halstead, M.H.: The technology of computer center management: a proposed course for graduate professional programs in computer science or in information systems. Commun. ACM **19**(7), 369–370 (1976). https://doi.org/10.1145/360248.360250
8. Kofahi, N.A.: The impact of computer science and information technology teaching on the growth of software industry. J. Discrete Math. Sci. Cryptogr. **7**(3), 371–386 (2004). https://doi.org/10.1080/09720529.2004.10698015
9. Nasution, M.K.M.: The birth of a science. Hist. Sci. Technol. **10**(2), 315–338 (2020). https://doi.org/10.32703/2415-7422-2020-10-2-315-338
10. Kuppler, M.: Predicting the future impact of computer science researchers: Is there a gender bias? Scientometrics **127**(11), 6695–6732 (2022). https://doi.org/10.1007/s11192-022-04337-2
11. Schelenz, L.: Diversity concepts in computer science and technology development: a critique. Sci. Technol. Hum. Values **48**(5), 1054–1079 (2023). https://doi.org/10.1177/01622439221122549
12. Corner, J.R., Wier, R.R., Nute, C.T., Rinewalt, J.R.: Emerging trends in computer science education: observations from the ACM South Central Region. In: Western Educational Computing Conference and Trade Show, pp. 75–79 (1984)
13. McGill, M.M., Thigpen, L., Mabie, A.: Emerging practices for integrating computer science into existing K-5 subjects in the United States. In: ACM International Conference Proceeding Series (2023). https://doi.org/10.1145/3605468.3609759
14. Nasution, M.K.M.: No research without publication: early mining. In: Journal of Physics: Conference Series, vol. 978 (2018). https://doi.org/10.1088/1742-6596/978/1/012063
15. Nasution, M.K.M., Munir, E.: Without research, no community service. In: Journal of Physics: Conference Series, vol. 1542 (2019). https://doi.org/10.1088/1742-6596/1542/1/012071
16. Nasution, M.K.M.: World on data perspective. World **3**(3), 736–752 (2022). https://doi.org/10.3390/world3030041
17. Wegner, P.: Three computer cultures: computer technology, computer mathematics, and computer science. Adv. Comput. **10**(C), 7–78 (1970). https://doi.org/10.1016/S0065-2458(08)60431-3
18. Griffiths, J.-M.: Track 5: social, behavioral, cultural and ethical factors - part I. Bull. Am. Soc. Inf. Sci. **26**(2), 21–22 (2000). https://doi.org/10.1002/bult.145
19. De Cock, C.: Thinking creatively about creativity: What can we learn from recent developments in the philosophy of science? Creat. Innov. Manag. **5**(3), 204–211 (1996). https://doi.org/10.1111/j.1467-8691.1996.tb00143.x
20. Hankinson, R.J.: Philosophy and science. In: The Cambridge Companion to Greek and Roman Philosophy, pp. 271–299 (2003). https://doi.org/10.1017/CCOL0521772850.011
21. Santacroce, L., Charitos, I.A., Topi, S., Bottalico, L.: The alcmaeon's school of croton: philosophy and science. Open Access Macedonian J. Med. Sci. **7**(3), 500–503 (2019). https://doi.org/10.3889/oamjms.2019.072

22. Johnsonbaugh, R., Kalin, M.: A graph generation software package. In: Proceedings of the Eastern Joint Computer Conference: Design and Application of Small Digital Computers, AIEE-IRE, pp. 151–1548 (1954). https://doi.org/10.1145/107004.107032
23. Pesante, L.H.: Integrating writing into computer science courses. In: Proceedings of the Eastern Joint Computer Conference: Design and Application of Small Digital Computers, AIEE-IRE, pp. 205–209 (1954). https://doi.org/10.1145/107004.107040
24. Vaidya, D.M., Florentin, J.S.: Preparing to live with micros. Microprocess. Microsyst. **4**(2), 53–55 (1980). https://doi.org/10.1016/0141-9331(80)90364-6
25. Nasution, M.K.M., Hidayat, R., Syah, R.: Computer science. Int. J. Adv. Sci. Eng. Inf. Technol. **12**(3), 1142–1159 (2022). https://doi.org/10.18517/ijaseit.12.3.14832
26. Newell, A., Perlis, A.J., Simon, H.A.: Computer science. Science **157**(3795), 1373–1374 (1967). https://doi.org/10.1126/science.157.3795.1373-b
27. Wegner, P.: Research paradigms in computer science. In: Proceedings - International Conference on Software Engineering, pp. 322–330 (1976)
28. Pirozelli, P., Câmara, I.: Natural language at a crossroads: formal and probabilistic approaches in philosophy and computer science. Manuscrito **45**(2), 50–81 (2022). https://doi.org/10.1590/0100-6045.2022.V45N2.PI
29. Hamming, R.W.: A philosophy for computer science or my prejudices and confusions. ACM SIGCSE Bull. **7**(4), 16–18 (1975). https://doi.org/10.1145/382218.382471
30. George, A.: Philosophy and the birth of computer science. Robot. Age **5**(1), 26–31 (1983)
31. Porta, M., Maillet, K., Mas, M.: Towards a strategy to fight the computer science (CS) declining phenomenon. Lect. Notes Electr. Eng. **103 LNEE**, 231–242 (2011). https://doi.org/10.1007/978-1-4614-0373-9_18
32. Abramson, D.: Philosophy of mind is (in part) philosophy of computer science. Minds Mach. **21**(2), 203–219 (2011). https://doi.org/10.1007/s11023-011-9236-0
33. Nasution, M.K.M., Syah, B.R., Elveny, M.: What is data science. In: Data Science with Semantic Technologies (2023). https://doi.org/10.1201/9781003310785-1
34. Kraut, R.: Plato on love. In: The Oxford Handbook of Plato (2009). https://doi.org/10.1093/oxfordhb/9780195182903.003.0012
35. Nasution, M.K.M.: Pengantar ke filsafat ilmu - Sains komputer. Filsafat Ilmu **1** (2023). https://doi.org/10.13140/RG.2.2.17498.72648
36. Gandon, F.L.: Ontologies in computer science: these new "software components" of our information systems. In: Ontology Theory, Management and Design: Advanced Tools and Models, pp. 1–26 (2010). https://doi.org/10.4018/978-1-61520-859-3.ch001
37. Ishaaq, U.M.: Computer science and philosophy: in search of a Nexus. In: IOP Conference Series: Materials Science and Engineering, vol. 662, no. 6 (2019). https://doi.org/10.1088/1757-899X/662/6/062016
38. Stoneburner, L.L.: Developing ACOG district and section computer science committees: local framework for definition and solution of information processing needs. Am. J. Perinatol. **3**(2) 139–140 (1986). https://doi.org/10.1055/s-2007-999852
39. Knuth, D.E.: Fundamental algorithm. In: The Art of Computer Programming, vol. 1. Addison-Wesley (1973)
40. Oldehoeft, R.R., Roman, R.V.: Methodology for teaching introductory computer science. ACM SIGCSE Bull. **9**(1), 123–128 (1977). https://doi.org/10.1145/382063.803373

41. De Mol, L., Primiero, G.: When logic meets engineering: introduction to logical issues in the history and philosophy of computer science. Hist. Philos. Logic **36**(3), 195–204 (2015). https://doi.org/10.1080/01445340.2015.1084183
42. Allan, D.J.: Philosophical surveys, i: a survey of work dealing with Greek philosophy from Thales to the age of cicero, 194549. Philos. Q. **1**(1), 61–72 (1950). https://doi.org/10.2307/2216501
43. Wolfson, H.A.: The amphibolous terms in aristotle, Arabic philosophy and maimonides. Harv. Theol. Rev. **31**(2), 151–173 (1938). https://doi.org/10.1017/S0017816000022276
44. Basti, G.: The philosophy of nature of the natural realism. The operator algebra from physics to logic. Philosophies **7**(6), 121 (2022). https://doi.org/10.3390/philosophies7060121
45. Turner, R., Eden, A.H.: The philosophy of computer science: introduction to the special issue. Minds Mach. **17**(2), 129–133 (2007). https://doi.org/10.1007/s11023-007-9059-1
46. Bugliarello, G., Mcnally, W.D., Gormley, J.T., Onstott, J.T.: Design philosophy, specifications, and implications of hydro, a pilot computer language for hydrology and hydraulic engineering. Water Resour. Res. **3**(2), 636–642 (1967). https://doi.org/10.1029/WR003i002p00636
47. Wang, R.-J., Xu, Y.-P., She, C., Nasution, M.: A new design for programmable logic array based on QCA-based nanotechnology. Optik **253**, 168581 (2022). https://doi.org/10.1016/j.ijleo.2022.168581
48. Vallverdú, J.: Thinking Machines and the Philosophy of Computer Science: Concepts and Principles. IGI Global (2010). https://doi.org/10.4018/978-1-61692-014-2
49. Debusmann, R., Postolache, O., Traat, M.: A modular account of information structure in extensible dependency grammar. Lect. Notes Comput. Sci. **3406**, 25–36 (2005). https://doi.org/10.1007/978-3-540-30586-6_2
50. Wegner, P.: Programming languages, Information Structures and Machine Organization. McGraw-Hill (1968)
51. Tylman, W.: Computer science and philosophy: Did Plato foresee object-oriented programming? Found. Sci. **23**(1), 159–172 (2018). https://doi.org/10.1007/s10699-016-9506-7
52. Demey, L.: The porphyrian tree and multiple inheritance: a rejoinder to tylman on computer science and philosophy. Found. Sci. **23**(1), 173–180 (2018). https://doi.org/10.1007/s10699-017-9531-1
53. Nasution, M.K.M.: Understanding data toward going to data science. Cybernet. Perspect. Syst. **LNNS 503**, 478–489 (2022). https://doi.org/10.1007/978-3-031-09073-8_42
54. Delcher, A., Kasif, S.: Some results on the complexity of exploiting data dependency in parallel logic programs. J. Logic Program. **6**(3), 229–241 (1989). https://doi.org/10.1016/0743-1066(89)90015-0
55. Xiang, C.: Study on the role of big data age in making fitness programs for individual science. Lect. Notes Data Eng. Commun. Technol. **130**, 162–166 (2022). https://doi.org/10.1007/978-3-030-99581-2_22
56. Sugiharti, E., Firmansyah, S., Devi, F.R.: Predictive evaluation of performance of computer science students of unnes using data mining based on nae Bayes classifier (NBC) algorithm. J. Theor. Appl. Inf. Technol. **95**(4), 902–911 (2017)
57. Djikstra, E.W., Dahl, O.J.: Notes on Structured Programming. Academic Press (1972)

58. Luker, P.A., Rothermel, D.: The philosophy of artificial intelligence: a general studies course with particular benefits to computer science majors. ACM SIGCSE Bull. **26**(1), 41–45 (1994). https://doi.org/10.1145/191033.191050
59. Hanson, B., Stall, S., Cutcher-Gershenfeld, J., Vrouwenvelder, K., Wirz, C., Rao, Y.: (Douglas): Garbage in, garbage out: mitigating risks and maximizing benefits of AI in research. Nature **623**(7985), 28–31 (2023). https://doi.org/10.1038/d41586-023-03316-8
60. Hyde, S.J., Bachura, E., Harrison, J.S.: Garbage in, garbage out: a theory-driven approach to improve data handling in supervised machine learning. Res. Methodol. Strat. Manag. **14**, 101–132 (2023). https://doi.org/10.1108/S1479-838720220000014006
61. Roach, S., Gate, A.Q.: Teaching software engineering in a computer science program using the affinity research group philosophy. In: Software Engineering: Effective Teaching and Learning Approaches and Practices, pp. 136–156 (2008). https://doi.org/10.4018/978-1-60566-102-5.ch008
62. Vallverdú, J.: Thinking machines and the philosophy of computer science: concepts and principles. In: Thinking Machines and the Philosophy of Computer Science: Concepts and Principles, pp. 1–435. IGI Global (2010). https://doi.org/10.4018/978-1-61692-014-2
63. Popovich, V., Vitko, A.: The concept of space in philosophy and in computer science. Adv. Geograph. Inf. Sci. **1**, 3–14 (2020). https://doi.org/10.1007/978-3-030-31608-2_1
64. Da Rosa, S., Chmiel, A., Gómez, F.: Philosophy of computer science and its effect on education: Towards the construction of an interdisciplinary group. In: Proceedings - 2015 41st Latin American Computing Conference, CLEI 2015 (2015). https://doi.org/10.1109/CLEI.2015.7360000
65. Furrer, F.J.: A short history of ontology: from philosophy to modern computer science. Informatik-Spektrum **37**(4), 308–317 (2014). https://doi.org/10.1007/s00287-012-0642-3
66. Sabin, R.E., Sabin, E.P.: Collaborative learning in anintroductory computer science course. ACM SIGCSE Bull. **26**(1), 304–308 (1994). https://doi.org/10.1145/191033.191156
67. Fouh, E., Akbar, M., Shaffer, C.A.: The role of visualization in computer science education. Comput. Sch. **29**(1–2), 95–117 (2012). https://doi.org/10.1080/07380569.2012.651422
68. Brady, R.T.: The use of definitions and their logical representation in paradox derivation. Synthese **199**, 527–546 (2021). https://doi.org/10.1007/s11229-017-1362-7
69. Tomlin, E.W.F.: Linguistic philosophy: forty years on a general statement. Camb. Q. **7**(1), 231–243 (1977). https://doi.org/10.1093/camqtly/VII.3.231
70. García, G.R.: Human rationality: explanations from cognitive science and the philosophy of logic. Topicos (Mexico) **62**, 385–402 (2021). https://doi.org/10.21555/TOP.V0I61.1168 https://doi.org/10.21555/TOP.V0I61.1168 https://doi.org/10.21555/TOP.V0I61.1168 https://doi.org/10.21555/TOP.V0I61.1168 https://doi.org/10.21555/TOP.V0I61.1168 https://doi.org/10.21555/TOP.V0I61.1168
71. Serna, M.E., Serna, A.A.: Logic in computer science. In: CISCI 2013 - Decima Segunda Conferencia Iberoamericana en Sistemas, Cibernetica e Informatica, Decimo Simposium Iberoamericano en Educacion, Cibernetica e Informatica, SIECI 2013 - Memorias, vol. 2, pp. 44–49 (2013)

72. López-Quiñones, A., et al.: Ancentral computing for sustainability: centering indigenous epistemologies in researching computer science education. TechTrends **67**(3), 435–445 (2023). https://doi.org/10.1007/s11528-022-00820-y
73. Fumerton, R.: A priori philosophy after an A Posteriori Turn. Midwest Stud. Philos **23**(1), 21–33 (1999). https://doi.org/10.1111/1475-4975.00002
74. Pananos, A.D., Lizotte, D.J.: Comparisons between Hamiltonian Monte Carlo and masimum a posteriori for a Bayesian model for apixaban induction dose & dose personalization. Proc. Mach. Learn. Res. **126**, 397–417 (2020)
75. Sokolova, T.D.: A priori in the philosophy of science: methodological aspects of research. Epistemol. Philos. Sci. **57**(3), 60–74 (2020). https://doi.org/10.5840/EPS202057342
76. Zhang, Y.: Deduction property of some logics applied to computer science. Chin. J. Comput. **22**(6), 571–576 (1999)
77. Habiballa, H., Kmet, T.: Mathematical logic and deduction in computer science education. Inform. Educ. **7**(1), 75–90 (2008)
78. Nasution, M.K.M., Sitompul, O.S.: Enhancing extraction method for aggregating strength relation between social actors. Adv. Intell. Syst. Comput. **573**, 312–321 (2017). https://doi.org/10.1007/978-3-319-57261-1_31
79. Nasution, M.K.M., Noah, S.A.: Social network extraction based on web: a comparison of superficial methods. Procedia Comput. Sci. **124**, 86–92 (2017). https://doi.org/10.1016/j.procs.2017.12.133
80. Schwarz, C.V., Meyer, J., Sharma, A.: Technology, pedagogy, and epistemology: opportunities and challenges of using computer modeling and simulation tools in elementary science methods. J. Sci. Teach. Educ. **18**(2), 243–269 (2007). https://doi.org/10.1007/s10972-007-9039-6
81. Lin, F., Chan, C.K.K.: Promoting elementary students epistemology of science through computer-supported knowledge-building discourse and epistemic reflection. Int. J. Sci. Educ. **40**(6), 668–687 (2018). https://doi.org/10.1080/09500693.2018.1435923
82. Develaki, M.: Methodology and epistemology of computer simulations and implications for science education. J. Sci. Educ. Technol. **28**(4), 353–370 (2019). https://doi.org/10.1007/s10956-019-09772-0
83. Almeida, M.B.: An unified approach to ontologies: information science, computer science and philosophy. Perspectivas em Ciencia da Informacao **19**(3), 242–258 (2014). https://doi.org/10.1590/1981-5344/1736
84. Mainzer, K.: Proof and computation: perspectives for mathematics, computer science, and philosophy. In: Proof and Computation II: From Proof Theory and Univalent Mathematics to Program Extraction and Verification, pp. 132 (2021). https://doi.org/10.1142/9789811236488_0001
85. Mainzer, K., Schuster, P., Schwichtenberg, H.: Proof and computation: digitization in mathematics, computer science, and philosophy. In: Proof and Computation: Digitization in Mathematics, Computer Science and Philosophy, pp. 1–291. World Scientific Publishing Co. (2018). https://doi.org/10.1142/9789813270947
86. Knight, J.C., Prey, J.C., Wulf, W.A.: Look back: undergraduate computer science education: a new curriculum philosophy & overview. Proc. Front. Educ. Conf. **2**, 722–727 (1997)
87. Nasution, M. K. M.: Mathematical philosophy. J. Res. Math. Trends Technol. 2(2), 45-60 (2020). https://doi.org/10.32734/jormtt.v2i2.4678
88. McDermott, R., Pirie, I., Cajander, Å., Daniels, M., Laxer, C.: Investigation into the personal epistemology of computer science students. In: Annual Conference on

Innovation and Technology in Computer Science Education, ITiCSE, pp. 231-236 (2013). https://doi.org/10.1145/2462476.2465589
89. Lande, D.R.: Development of the binary number system and the foundations of computer science. Math. Enthus. **11**(3), 513–540 (2014)
90. Hill, J., Houle, B., Merritt, S. M.: Applying abstraction to master complexity: the comparison of abstraction ability in computer science majors with students in other disciplines. In: Proceedings - International Conference on Software Engineering, pp. 15–21 (2008). https://doi.org/10.1145/1370164.1370169
91. Nasution, M.K.M.: Multi-criteria as decisions. In: IOP Conference Series: Materials Science and Engineering, vol. 1003, no. 1 (2020). https://doi.org/10.1088/1757-899X/1003/1/012118
92. Syah, R., Davarpanah, A., Nasution, M.K.M., Tanjung, F.A., Nezhad, M.M., Nesaht, M.: A comprehensive thermoeconomic evaluation and multi-criteria optimization of a combined MCFC/TEG system. Sustainability (Switzerland) **13**(23), 13187 (2021). https://doi.org/10.3390/su132313187
93. Nasution, M.K.M., Elveny, M., Syah, R.: Multiple-objective as decisions in computing. In: AIP Conference Proceedings, vol. 2714 (2023). https://doi.org/10.1063/5.0134508
94. Syah, R.B.Y., Satria, H., Elveny, M., Nasution, M.K.M.: Complexity prediction model: a model for multi-object complexity in consideration to business uncertainty problems. Bull. Electr. Eng. Inform. **12**(6), 3697–3705 (2023). https://doi.org/10.11591/eei.v12i6.5380
95. Vagelli, M.: Historical epistemology and the marriage between history and philosophy of science. In: The Past, Present, and Future of Integrated History and Philosophy of Science, pp. 96–112 (2019). https://doi.org/10.4324/9781351214827-7
96. McMullin, E.: The impact of Newton's principia on the philosophy of science. Philos. Sci. **68**(3), 279–310 (2001). https://doi.org/10.1086/392883
97. Nassution, M. K. M.: Sitompul, O. S., Elveny, M., Syah, R., Rahmat, R. F.: A similarity for new meanings. *2020 International Conference on Data Science, Artificial Intelligence, and Business Analytics, DATABIA 2020 - Proceedings*, 37-44: 2020. https://doi.org/10.1109/DATABIA50434.2020.9190316
98. Nasution, M.K.M.: A method to prove the existence of a similarity. Adv. Intell. Syst. Comput. **1295**, 243–252 (2020). https://doi.org/10.1007/978-3-030-63319-6_21
99. Porus, V.N.: Political agency of science as a theme of philosophy of science. Vopr. Filos. **2023**(9), 71–82 (2023). https://doi.org/10.21146/0042-8744-2023-9-71-82

Cyber-Physical System for Detection and Recognition of Aggression Movements with Smart Glasses

Nikolay Gospodinov(✉) and Georgi Krastev(✉)

University of Ruse, 8 Studentska Street, POB 7017, Ruse, Bulgaria
{ngospodinov,geork}@uni-ruse.bg

Abstract. This paper studies the aggression among individuals, focusing on the detection and recognition of aggressive movements by an observing person. Employing a combination of psychological and technological approaches, the paper centers on analyzing the physical and psychological indicators that precede acts of aggression. Methods for detecting and recognizing aggressive gestures, expressions, and intonations are examined through trained artificial intelligence models, in conjunction with recordings of human gestures and words within conflict scenarios. The findings underscore the significance of accurately recognizing aggression and the potential applications of such observations in real-time settings. The paper presents significant implications for the development of cyber-physical system with smart glasses aimed at supporting the recognition of aggressive behavior in society and everyday situations, aiming to mitigate conflicts and enhance interpersonal relationships.

Keywords: Cyber-physical system · Smart glasses · Artificial intelligence · Aggression · Prediction

1 Introduction

In today's world, where technological advances are gaining ground, we face challenges in protecting society form various forms of aggression. In the context of this problem, there is a need for innovative systems to detect and identify aggressive actions to prevent and manage potential threats.

Aggression is a complex and multi-layered problem that can be caused by various factors such as social, psychological, and emotional circumstances. Its causes can be related to stress, lack of communication and understanding, unresolved conflicts or even health problems. As a result of aggression, social order is disturbed, physical and emotional damage occurs, and sometimes serious consequences such as violence or conflict occur.

Physical aggression is a complex and often unpredictable type of behavior that can cause serious consequences for both those who perpetrate it and their victims. Recognizing physical aggression is essential to prevent potential conflict and protect people from violence. Interpreting physical aggression can involve different aspects of human

R. Silhavy and P. Silhavy (Eds.): CSOC 2024, LNNS 1120, pp. 132–150, 2024.
https://doi.org/10.1007/978-3-031-70518-2_12

behaviour and the environment. It can be recognized by observing certain signs and signals that include, but are not limited to:

- **Non-verbal expressions and gestures**: muscular tension, restrained or aggressive hand or body movement, threatening or outright aggressive gestures may be indicators of impending physical aggression.
- **Changes in voice tone and emotional expression**: a high or tense voice filled with obvious irritation or anger can be a warning sign of impending violence.
- **Physical signs**: visible signs of violence such as injuries, wounds or other visible signs of conflict can be direct signs of physical aggression.
- **Emotional and psychological cues**: Expressions of rage, frustration, restlessness, or internal stress may be associated with impending aggression.

In Bulgaria, the problem of aggression is extremely visible in minors. This problem can have different forms and causes. Some of the main aspects of this problem include: Social factors, Upbringing and education, Influence of media and technology, Lack of appropriate social services and support.

The following gives a summary of minors, perpetrators of crimes in the period 2013–2022 [1] (Table 1).

Table 1. Minors, perpetrators of crimes in the period 2013–2022

Minors, perpetrators of crimes	2013	2014	2015	2016	2017	2018	2019	2020	2021	2022
Murder (completed)	1	4	4	-	2	5	5	2	2	3
Murder (attempted)	2	1	2	1	1	-	1	-	-	-
Bodily damage	226	201	237	209	224	236	239	212	182	230
Fornication	56	60	47	56	56	46	57	43	46	29
Rape (complete)	7	7	8	6	7	9	2	2	3	2
Rape (attempted)	-	2	2	9	1	3	1	4	-	-
Destruction and damage to property	363	421	462	447	461	497	552	548	623	526
Robbery	223	204	157	163	192	152	162	139	113	143
Theft – general	3621	3210	3089	3019	2793	2670	2653	2222	2164	2405
Illegal confiscation of a motor vehicle	75	90	90	94	76	96	112	116	95	85
Hooliganism	337	339	326	254	259	200	235	238	269	182
Drug-related crimes	342	356	406	407	421	475	451	354	335	402
Others	754	507	532	315	478	468	473	537	497	381
Total	**6007**	**5402**	**5362**	**4980**	**4971**	**4857**	**4943**	**4417**	**4329**	**4388**

In order to reduce this problem, the use of cyber-physical systems appears as an option for prevention and detection of potential aggressive situations. Combining smart technologies such as augmented reality glasses and neural networks, these systems have the ability to analyze and recognize specific behavioral patterns and cues that may precede aggressive acts.

This innovative solution can not only provide more effective tools to prevent aggression, but also create a safer environment for society. Furthermore, these systems can help to understand and study the roots of aggression, which can point towards more targeted and comprehensive solutions to tackle this problem in the future.

Creating a cyber-physical system based on augmented reality glasses and neural networks is an innovative approach to preventing aggression and creating a safer community. Detecting early signs of potential aggressive actions can make a significant contribution to improving quality of life and protecting people from unpredictable dangers.

2 Related Work

In recent years, technology and artificial intelligence play a key role in the fight against aggression, providing innovative tools to prevent violence and protect society. Various researchers and institutions are actively working in this direction, using technology and AI to develop systems and methods that help to detect and prevent physical aggression. The importance of technology and artificial intelligence includes:

- **Data and behavior analysis**: technologies can analyze large volumes of data to predict potential aggression situations by extracting patterns from behavior or combining different parameters such as gestures, tone of voice, etc.
- **CCTV and image processing**: CCTV systems with object and gesture recognition algorithms can help in detecting potential conflict situations.
- Augmented reality and virtual training: Technologies such as augmented reality can provide training to staff on recognizing and managing aggression through simulations of real-life scenarios.
- **Warning and response systems**: Artificial intelligence can be used to send automated warnings and respond quickly when potential threats are detected.

In a study by Tomasz Sapiński (2019) [2], a body movement recognition is used for emotion recognition. The algorithm uses low level information deduced from the orientation and spatial placement of joints within the monitored skeleton to build a sequential model of emotional movement. Several deep neural networks were used in the experiment and compared to identify the emotional state of the learned motion sequences.

In another study by Shatakshi Raman (2022) [3], a machine learning models such as linear SVC, logistic regression, multinominal naïve Bayes and random forests and deep learning methods such as CNN-GRU static, TCN static, Seq2Seq are used for hate and aggression analysis in NLP.

In another study by Irfanullah (2022) [4], a two-step violence detection in surveillance video is used. In the first step violent objects are defined manually and then the system deals with the uncertainty. The second step is the availability of label dataset because manually annotation video is an expensive and labor-intensive task. In the study, a MobileNet model with convolutional neural network is used.

In another study by Enrique Bermejo (2011) [5], a Bag-of-Words framework is used for action recognition in the specific problem of fight detection, along with two of the best action descriptors currently available: STIP and MoSIFT.

3 Proposed Method

One potential answer to the issue of violence among individuals is the detection and recognition of aggressive movements. Finding the class to which an input video belongs is the aim. This can be accomplished by building and training an artificial neural network on videos. The neural network's task is to identify the class to which a given video belongs, especially considering the benefits of a method that leverages deep learning techniques in cyber-physical systems.

The process of development involves identifying and detecting potential acts of violence. An aural display of the labeled output provides the user with information about a recognized occurrence.

The current study is based on the Moverio BT-300 [6] (Fig. 1), a third generation of augmented reality glasses. Their advantages include superior image contrast, audio capabilities, and video support.

The created model is a component of a broader cyber-physical system that also consists of several other modules, such as modules for smoke detection and hazardous object identification.

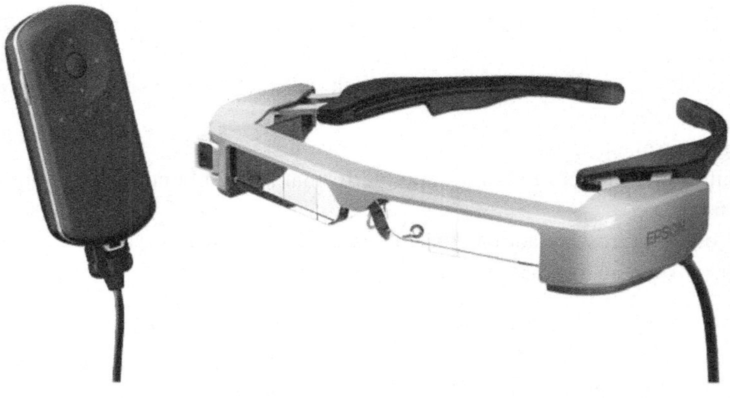

Fig. 1. Augmented reality smart glasses Moverio BT-300

BT-300 augmented smart glasses are being used to construct a prototype of a cyber-physical system for aggression detection and recognition. Moverio OS, an operating system akin to Android, powers the glasses. The following Fig. 2 gives a summary of some of the most important characteristics of the glasses (Table 2).

Table 2. Moverio BT-300 characteristics

Name	Characteristics
Driving Method	Mono Crystalline Silicon Active Matrix
Pixel Number	921,600 pixels (1280 RGB × 720)
Field of View	23 degrees (diagonal)
Camera	5 Megapixel
GPS	Yes, in Controller
CPU	Intel Atom x5 1.44GHz Quad Core
RAM	2 GB
Internal Memory	Approx. 16 GB (including Moverio OS)

4 Implementation

The cyber-physical system is based on the aggression movements detection and recognition module.

Figure 2 shows a schematic of the cyber-physical system implementation. The process consists of several steps:

- Creating a dataset containing 2000 videos. 1000 videos with aggression actions and 1000 videos without aggression actions.
- Creation of a model based on artificial neural networks for detection and recognition of aggressive actions.
- Training the created model to distinguish between aggressive actions and non-aggressive actions.
- Testing the trained model on predefined video test samples.
- Translating the model into an acceptable format for implementation in an Android-based interface.
- Detection of aggressive actions through the cyber-physical system.
- Signaling when an aggressive action is detected.

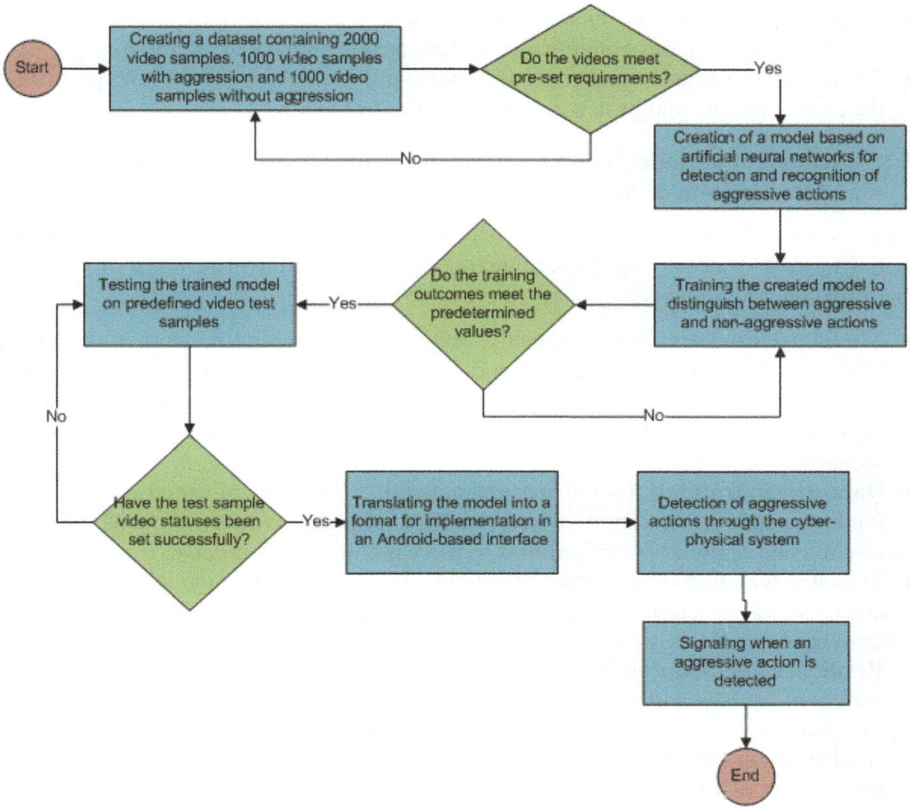

Fig. 2. Architecture of cyber-physical system implementation

4.1 Creating a Dataset Containing 2000 Videos. 1000 Videos with Aggression Actions and 1000 Videos Without Aggression Actions

Before creating the model for detecting and recognizing aggressive actions, a set of video samples is created, consisting of 1000 videos containing aggressive actions and 1000 videos containing non-aggressive actions. Video materials must meet the following requirements:

- The information in video samples should include a variety of scenarios and places where aggression may arise, such as home disagreements, workplace confrontations, sports collisions, and others.
- A range of attributes, including various ages, ethnicities, genders, and cultural contexts, should be included in the video samples.
- To assess the correctness of your model, you want data that has been divided into training and testing sets.
- High-quality video samples are favored because they provide distinct participant actions and facial expressions, good lighting, and overall good image quality.

- To prevent unintentional bias in the model, it is crucial to have a balanced data collection that contains both positive and negative examples of aggression.
- All information must be gathered and handled in compliance with the laws governing the confidentiality and protection of personal information.

4.2 Creation of a Model Based on Artificial Neural Networks for Detection and Recognition of Aggressive Actions

After creating the video sample dataset. Moving on to creating a model based on artificial neural networks for detecting and recognizing aggressive actions. The model should meet the following requirements:

- **Model architecture**: It's important to select the right neural network architecture. CNNs, or convolutional neural networks, are widely utilized and effective in image processing. Combinatorial models, such as CNNs with recurrent neural networks, or RNNs, might be useful for interpreting image sequences in more complicated scenarios.
- **Data-driven training**: To train a neural network, a vast array of varied videos featuring both aggressive and non-violent scenarios must be provided. Choosing the right training and optimization algorithms to use with this data is crucial.
- **Transfer learning or pre-trained models**: Using pre-trained models can enhance and expedite the learning process for identifying aggressive acts, particularly if they have been trained on sizable data sets.
- **Resilience and adaptability**: The network needs to be able to identify violence in a variety of settings with varying illumination, camera angles, and video quality. This calls for resilience to noise and flexibility in various environments.
- **Real-time performance**: The model needs to be quick and effective enough to process the video in real-time if real-time aggressiveness identification is needed.
- **Evaluation of accuracy and reliability**: It is important to routinely assess the neural network's capacity to recognize aggressive scenes with accuracy and reliability. This entails assessing model performance using a variety of measures.
- **Interpretability and transparency**: Despite the complexity of neural networks, it's critical to aim for greater interpretability and the capacity to comprehend how the model generates its aggressiveness predictions.

In Fig. 3, a principal scheme of the aggression actions detection and recognition model is shown.

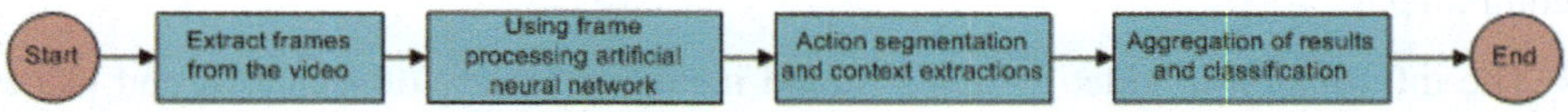

Fig. 3. Scheme of the aggression actions detection and recognition model

4.2.1 Extract Frames from Video Sample

To provide the models with input for processing, the video sample has to be divided into separate frames. This entails taking individual frames out of the video sample using video processing techniques.

4.2.2 Using Frame Processing Artificial Neural Network

Techniques like Convolutional Neural Networks (CNN), which include Depthwise Separable Convolutions (DS-CNN), is used to extract visual characteristics and features from pictures in order to assess the visual information in each frame.

Depthwise Separable Convolution (DS-CNN) is used for lowering the computational complexity of convolutional operations. Its ability to decrease the number of parameters and procedures while maintaining the key qualities of pictures has led to its widespread application in computer vision and image processing [7]. The following are some of Depthwise Separable Convolution's salient characteristics:

- **Convolution processes for channels and spaces are independent**: Two steps make up DS-CNN. To process each input channel independently, a depthwise convolution operation is carried out first. The outputs from this phase are then combined using a pointwise convolution operation.
- **Depthwise Convolution**: Each channel of the input data undergoes independent filtering at this stage. This implies that each channel receives a distinct convolution operation with a unique filter.
- **Pointwise Convolution:** The outputs from the preceding stage are combined using pointwise convolution, which comes after depthwise convolution. This process aids in combining the characteristics from the various channels by using filters with a 1×1 kernel size.
- **Decreased computational complexity:** Compared to traditional convolutional layers, DS-CNN requires fewer parameters and processing resources, which makes it appropriate for embedded and mobile systems.
- **Ability to preserve information:** DS-CNN can perform well in a variety of computer vision applications by maintaining the key visual elements of pictures even when its computational complexity is reduced.
- **Application in video analysis:** Depthwise Separable Convolution can be used to process frame sequences in video analysis, which lowers the demands on computer power.

The following Fig. 4 gives a visual presentation of the DS-CNN.

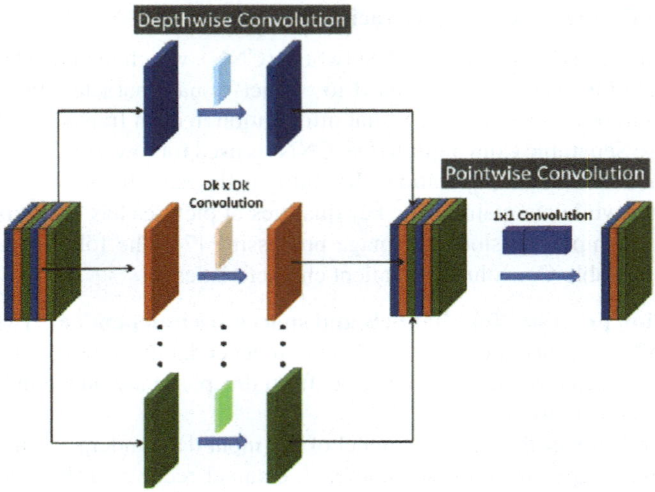

Fig. 4. Depthwise Separable Convolutions

4.2.3 Action Segmentation and Context Extraction

The sequence of frames is analyzed using the sequence processing technique: like Long Short-Term Memory (LSTM) to comprehend the context of activities and identify the presence of aggressive behavior.

Recurrent neural networks (RNNs) of the Long Short-Term Memory (LSTM) type were created to address the issues with normal RNN models while handling long-term dependencies in sequential input [9]. These are LSTM's primary characteristics:

- **Long-term memory:** Long-term dependencies in sequential data are retained by LSTM architecture. In comparison to typical RNN models, which struggle to recall information for extended periods of time because to the vanishing gradient problem, this indicates that it can remember information for a longer amount of time.
- **Information control gates**: The input, output, and forget gates are used by LSTMs to regulate the information flow within the cell. The amount of information that is remembered, lost, and transferred to the following time step is controlled by these gates.
- **Cell State:** During training, an LSTM's unique cell state preserves or filters information, enabling the model to transfer over data from one time step to the next.
- **Working with sequences:** Because LSTM can represent relationships between sequence parts, it is appropriate for handling sequential data, including text, time series, audio, and video.
- **Preventing the disappearing Gradient Problem:** Long sequence learning can cause typical RNNs to encounter the disappearing or explosive gradient problem. In contrast, LSTM has techniques that help maintain the gradient and improve learning stability.

- **Broad use:** Natural language processing (NLP), time series prediction, text production, temporal data analysis, audio and video processing, and other domains are among those that employ LSTM.

4.2.4 Aggregation of Results and Classification

One way to assess whether a sequence of actions is aggressive or not is to feed the results of the analysis of the visual information and the action context into a classification network. In this study Single Shot Detector network is used for classification.

The SSD (Single Shot Multibox Detector) is an object recognition model intended for accurate and efficient item categorization and localization in pictures[9]. When categorized, the following are some of the primary attributes of SSDs:

- **Real-Time Detection:** SSD is well-known for its quick and accurate object recognition in pictures, which makes it a good fit for applications that need to handle massive volumes of data quickly.
- **Using deep convolutional networks:** To extract visual information from pictures, SSD makes use of deep convolutional networks (CNNs), such as VGG, ResNet, or MobileNet. In the study, Depthwise Separable Convolution for extraction.
- **Several regions of interest (RoIs) in a single pass:** SSD eliminates the need for two distinct phases (as with the region-based CNN approach with regions of interest) by producing several proposals for objects (bounding boxes) and classifying the objects in them in a single pass.
- **Multiple object proposal scales:** SSD produces object proposals at several scales, making it possible to recognize things in an image of varying sizes with efficiency.
- **Default boxes**: Default boxes or predefined boxes, are used by SSD to help in object localization and classification. These boxes are preconfigured and come in various sizes and aspects for the output pictures of the convolutional layers.

The following Fig. 5 gives a visual presentation of the Single Shot Detector network.

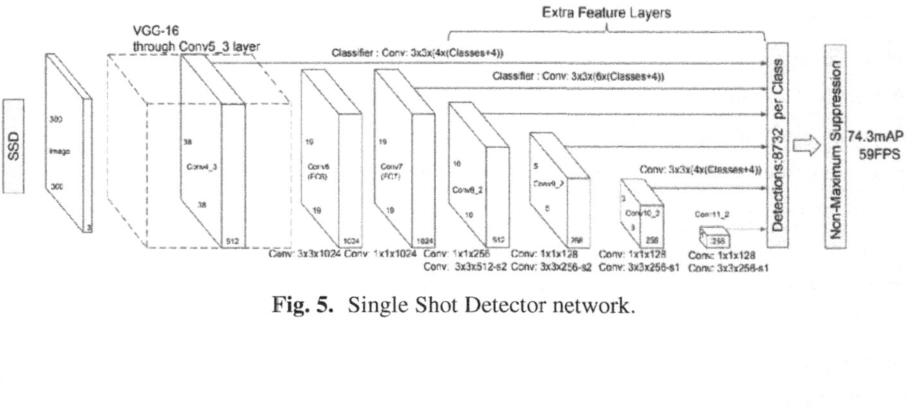

Fig. 5. Single Shot Detector network.

4.3 Training the Created Model to Distinguish Between Aggressive Actions and Non-aggressive Actions

In this part of the implementation, after creating the model for detection and recognition of aggressive actions, we proceed to train the model using the created set of video samples.

The model is trained by providing it with a total of 700 video samples (350 videos containing aggression and 350 videos that do not contain aggression). The number of video samples is chosen according to the processing power of the computer on which the model is trained. Two trainings were made:

The first training consisted of 50 epochs. Table 3 shows the main metrics when training the model.

Table 3. 50 epochs

Name	Data
Correct Predictions	3957
Wrong Predictions	237
Best Epochs	31
Accuracy on train	0.9616789817810059
Loss on train	0.11210563778877258
NonViolence (precision)	0.94
NonViolence (recall)	0.92
NonViolence (f1-score)	0.93
NonViolence (support)	1778
Violence (precision)	0.94
Violence (recall)	0.96
Violence (f1-score)	0.95
Violence (support)	2416
Accuracy (f1-score)	0.94
Accuracy (support)	4194
macro avg (precision)	0.94
macro avg (recall)	0.94
macro avg (f1-score)	0.94
macro avg (support)	4194
weighted avg (precision)	0.94
weighted avg (recall)	0.94
weighted avg (f1-score)	0.94
weighted avg (support)	4194

The following Fig. 6 gives a visual presentation of the Training validation accuracy with 50 epochs.

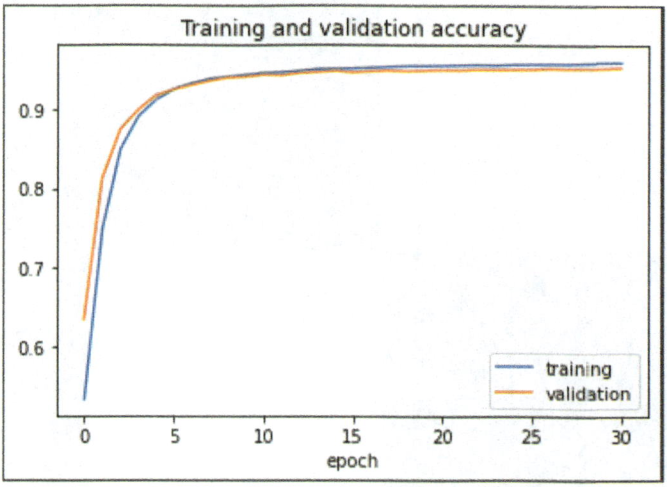

Fig. 6. Training validation accuracy with 50 epochs

The following Fig. 7 gives a visual presentation of the Training validation loss with 50 epochs.

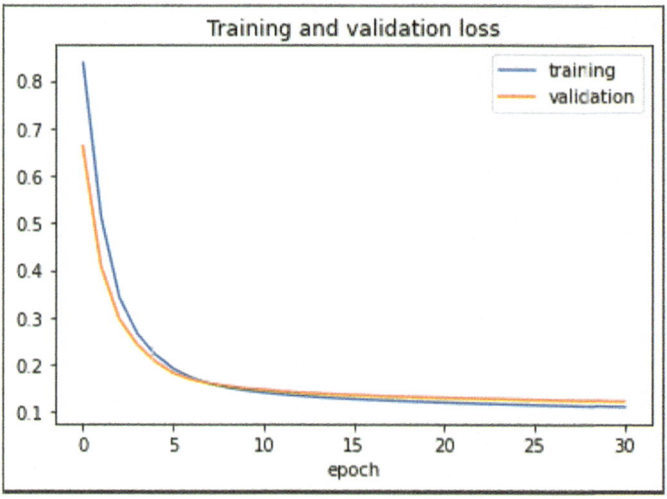

Fig. 7. Training validation loss with 50 epochs

The following Fig. 8 gives a visual presentation of the prediction with 50 epochs.

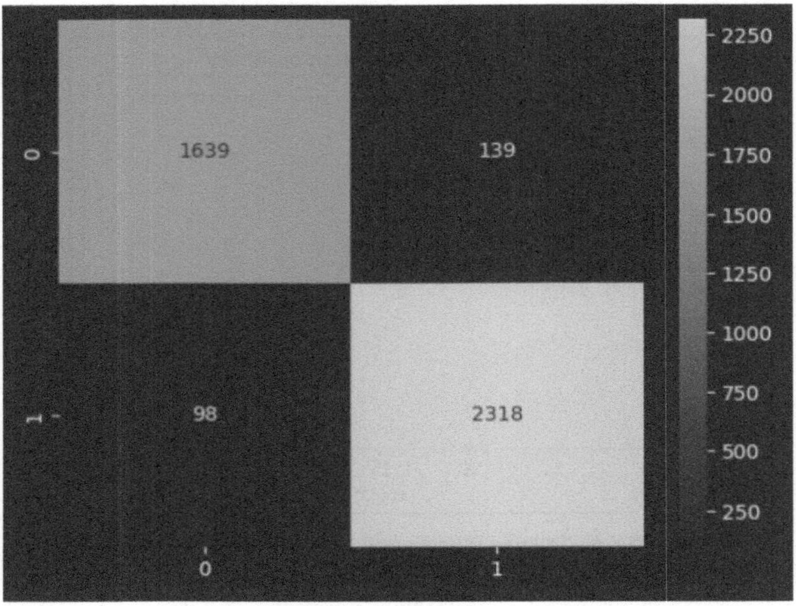

Fig. 8. Prediction with 50 epochs

The second training consisted of 100 epochs. Table 4 shows the main metrics when training the model.

Table 4. 100 epochs

Name	Data
Correct Predictions	3984
Wrong Predictions	210
Best Epochs	76
Accuracy on train	0.9622892141342163
Loss on train	0.11433202773332596
NonViolence (precision)	0.95
NonViolence (recall)	0.93
NonViolence (f1-score)	0.94
NonViolence (support)	1778
Violence (precision)	0.95
Violence (recall)	0.97

(*continued*)

Table 4. (*continued*)

Name	Data
Violence (f1-score)	0.96
Violence (support)	2416
Accuracy (f1-score)	0.95
Accuracy (support)	4194
macro avg (precision)	0.95
macro avg (recall)	0.95
macro avg (f1-score)	0.95
macro avg (support)	4194
weighted avg (precision)	0.95
weighted avg (recall)	0.95
weighted avg (f1-score)	0.95
weighted avg (support)	4194

The following Fig. 9 gives a visual presentation of the Training validation accuracy with 100 epochs.

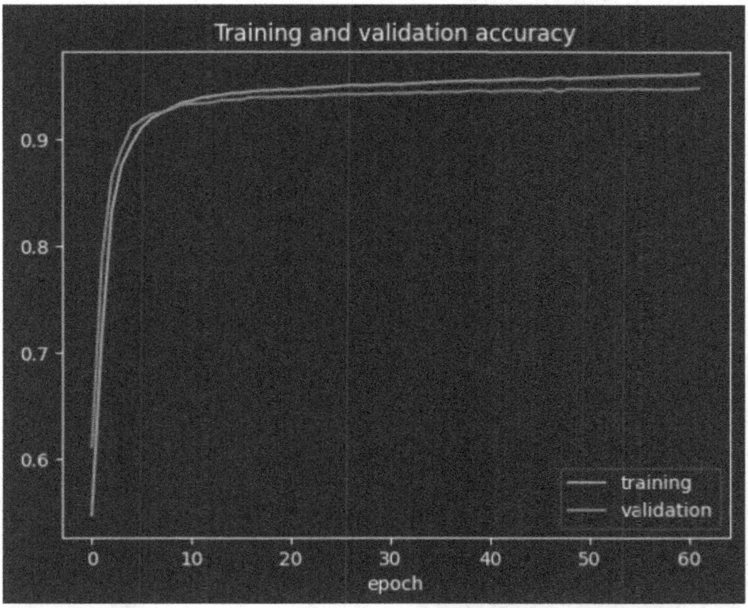

Fig. 9. Training validation accuracy with 100 epochs

The following Fig. 10 gives a visual presentation of the Training validation loss with 100 epochs.

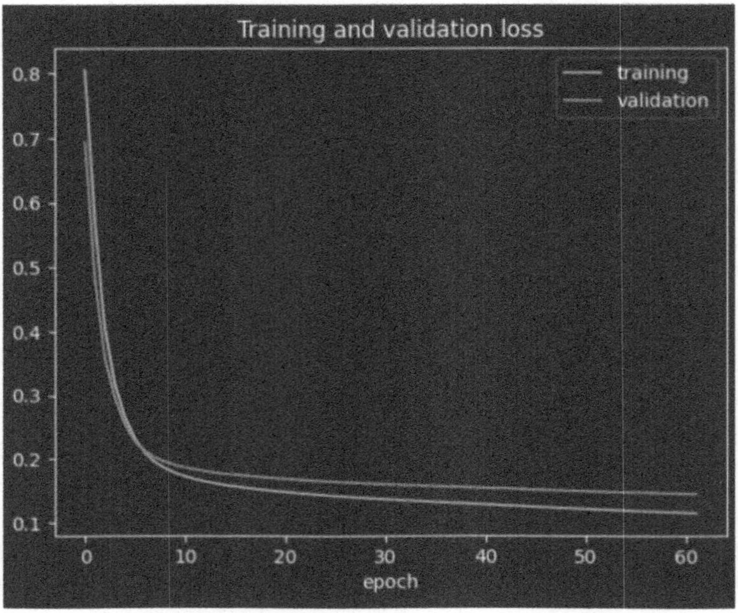

Fig. 10. Training validation loss with 50 epochs

The following Fig. 11 gives a visual presentation of the prediction with 100 epochs.

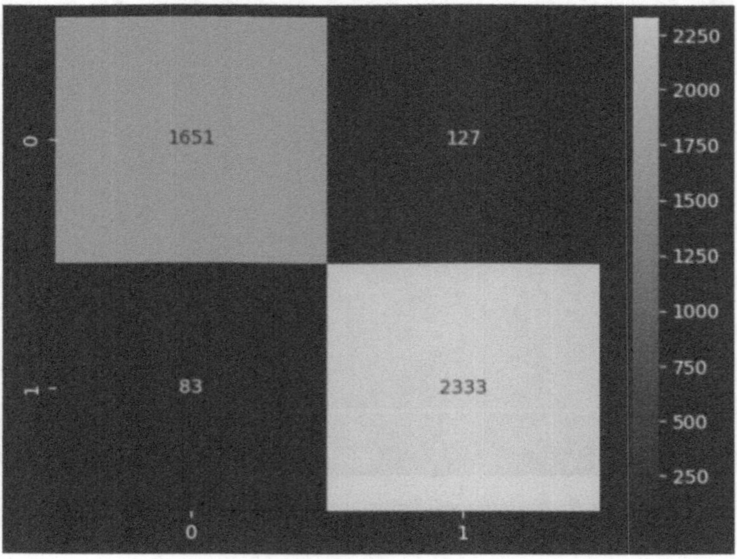

Fig. 11. Prediction with 100 epochs

4.4 Testing the Trained Model on Predefined Video Test Samples

In the next part of the implementation, the ability of the model to recognize aggressive actions is tested.

With 50 epochs, the accuracy on test is 0.9577874541282654 and loss on test: 0.116333968937397.

With 100 epochs, the accuracy on test is 0.9499284625053406 and loss on test: 0.14074040949344635.

Figure 12 shows violence detection testing result.

Fig. 12. Violence detection testing

Figure 13 shows nonviolence detection testing result.

Fig. 13. Nonviolence detection testing

4.5 Translating the Model into an Acceptable Format for Implementation in an Android-Based Interface

In the next part of the implementation, the finished model is converted into a format for implementation in the cyber-physical system that works on augmented reality glasses. In order for the model to be usable, it is refactored using the Tensorflow Lite library [10].

The conversion process uses TensorFlow Lite Converter (Fig. 14).

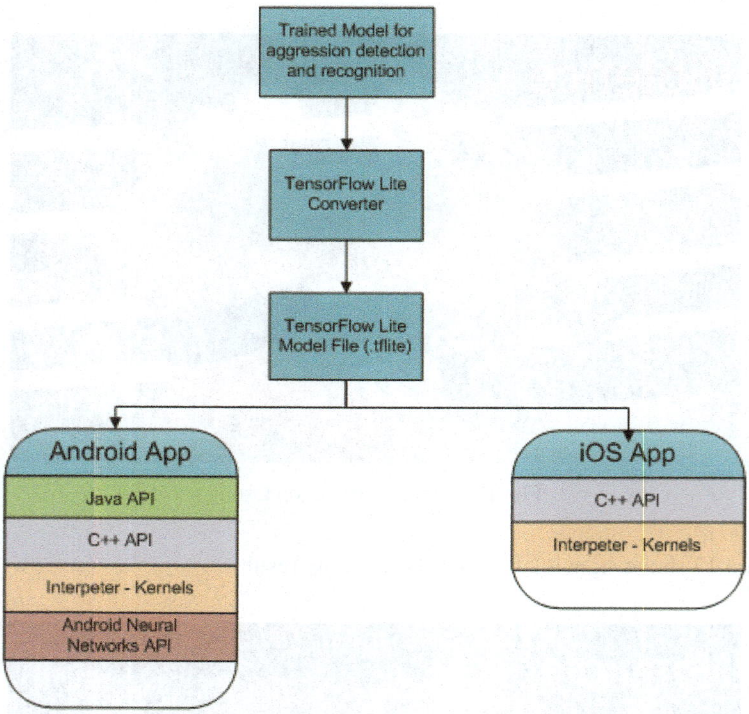

Fig. 14. TensorFlow Lite Converter

4.6 Detection of Aggressive Actions Through the Cyber-Physical System

Once the model is converted to the correct format, it is implemented in a pre-built Android-based user interface. Then there is the possibility of instant use.

4.7 Signaling When an Aggressive Action is Detected

Once operational, the cyber-physical system begins to explore its surroundings. In case of detected aggression, with the help of the built-in camera, a sound warning is sent through the headphones to the user, which signals to him about the immediate manifestation of aggression around him.

5 Conclusion

Neural network-based aggressiveness detection and recognition is a promising topic with wide-ranging applications in security, social media, video surveillance, and other domains. Using characteristics and signals linked to aggressive behavior, neural networks can assess both visual and aural data. Neural network-generated models have the ability to be multimodal, integrating data from several sources to enhance the precision of aggressiveness identification.

The quality of the training data is critical. It should be varied and reflect various locations, situations, and displays of hostility. The development and application of cyber-physical system for recognizing aggressiveness must also take privacy concerns and ethical considerations into account.

To guarantee that models are accurate, dependable, and applicable in real-world situations, transparency and model review are also essential. The sector has to be constantly evolving, and using these tools responsibly might lead to ethical and technological problems.

The results obtained during the study are more than satisfactory, as the recognition rate is very high, which ensures the security of the user of the cyber-physical system. Further testing and improvement of the model's performance to detect and recognize aggressive actions is planned in the future.

Acknowledgments. This publication is developed with the support of Project BG05M2OP001-1.001-0004 UNITe, funded by the Operational Programme "Science and Education for Smart Growth", co-funded by the European Union trough the European Structural and In-vestment Funds.

References

1. Republic of Bulgaria, National Statistical Institute: Crimes by results of proceedings and by statistical zones, statistical regions and districts. https://www.nsi.bg/en/content/3742/crimes-results-proceedings-and-statistical-zones-statistical-regions-and-districts
2. Sapiński, T., Kamińska, D., Pelikant, A., Anbarjafari, G.: Emotion recognition from skeletal movements. Entropy **21**(7), 646 (2019)
3. Raman, S., Gupta, V., Nagrath, P., Santosh, K.C.: Hate and aggression analysis in NLP with explainable AI. Int. J. Patt. Recogn. Artif. Intell. **36**(15), 22590364 (2022). https://doi.org/10.1142/S0218001422590364
4. Hussain, I.T., Iqbal, A., Yang, B., Hussain, A.: Retracted article: real time violence detection in surveillance videos using Convolutional Neural Networks. Multimed. Tools Appl. **81**(26), 38151–38173 (2022). https://doi.org/10.1007/s11042-022-13169-4
5. Nievas, E.B., Deniz Suarez, O., Bueno García, G., Sukthankar, R.: Violence detection in video using computer vision techniques. In: Real, P., Diaz-Pernil, D., Molina-Abril, H., Berciano, A., Kropatsch, W. (eds.) Computer Analysis of Images and Patterns, pp. 332–339. Springer Berlin Heidelberg, Berlin, Heidelberg (2011). https://doi.org/10.1007/978-3-642-23678-5_39
6. Epson Moverio – https://moverio.epson.com
7. Usman, M., Lee, J.A.: AMC-IoT: Automatic Modulation Classification Using Efficient Convolutional Neural Networks for Low Powered IoT Devices. In: Conference: 2020 International Conference on Information and Communication Technology Convergence (ICTC) (2020)

8. Sumon, S.A., Goni, R., Hashem, N.B., Shahria, T., Rahman, R.M.: Violence detection by pretrained modules with different deep learning approaches. Vietnam J. Comput. Sci. **7**(1), 19–40 (2020)
9. Alsalibi, A.I., Sumari, P., Alomari, S.A., Al-Betar, M.A.: Performance and reliability concern scheme for efficient garbage collection and wear leveling on flash memory-based solid state disk. Microsyst. Technol. **23**(7), 2521–2535 (2017)
10. TensorFlow Model conversion overview – https://www.tensorflow.org/lite/models/convert

Possibilities of Applying Entropy in Biomechanics

M. A. Filatov, AYu Kuhareva, N. F. Gazya(✉), T. V. Voronyuk, and I. S. Samoillenko

Surgut State University, 1 Lenina Ave, Surgut 628408, Russia
safety.ot86@gmail.com

Abstract. The Eskov-Zinchenko effect discovery in biomechanics led to proof of any samples' homogeneity loss and the loss of samples ergodicity of the of the human body functions parameters. As a result, all this was defined as type 2 uncertainty. However, in the new, third science for biosystems, type 1 uncertainty also arises. In this case, everyone is statistically the same, but the body of the subjects is objectively in different physiological, mental, environmental or physiological states. This message is intended to prove such uncertainty specifically in biomechanics (there have not been such examples yet) using the example of the use of entropy calculations in the assessment of tremorograms. #CSOC1120.

1 Introduction

Both Nobel laureates tried to detail and generalize Uncertainty and Unpredictability to biosystems [10, 11]. Nobel laureates Penrose [12] and Ginzburg [13] allegedly spoke about this. However, no one has presented real evidence of this in the last 40–50 years. Moreover, attempts to draw attention to this problem were ignored.

In 1948, W. Weaver tried to talk about this in his article [14], where he contrasted Complexity with all modern Science (he equated Complexity and Uncertainty). However, only 23 years ago, the school of Professor V.M. Eskov began to prove the special qualities of biosystems in the form of the Eskov- Zinchenko effect [1–9, 15–18]. As a result, not only the uncertainty of the 2nd type was proven, but also the type 1 uncertainty was determined in the chaos of self-organization theory. We emphasize that uncertainty of 1st type has been proven for parameters of the cardiovascular system (CVS), but there were no examples in biomechanics. This problem was generally ignored in biomechanics [1–9, 15–18].

In this article such examples are presented by us. At the same time, we study in detail the special properties of Shannon entropy using the example of TMG. All this now constitutes a new block (direction) for biomechanics and chaos of self-organization theory in general.

2 Shannon Entropy in Biomechanics

Let us immediately note that the Eskov-Zinchenko effect was initially proven based on the study of TMG and TPG samples, specifically in biomechanics. However, the presence of type 1 uncertainty for biomechanics has not been proven over the past

R. Silhavy and P. Silhavy (Eds.): CSOC 2024, LNNS 1120, pp. 151–157, 2024.
https://doi.org/10.1007/978-3-031-70518-2_13

20 years, much less over a longer period [1–9, 15–18]. It is believed that statistics works when describing any biological systems. The entire Eskov-Zinchenko effect is based on calculations of statistical distribution functions f(x), their statistical characteristics, signal spectral densities, autocorrelations A(t), etc. The calculation of Claude Shannon entropy in biomechanics is used extremely rarely and therefore we have not found an explanation.

Let us recall that entropy has always worked well in physics when some physical changes have occurred in the system. The human body has a physical basis, so the physical approach is appropriate, especially in biomechanics. However, there is very little work in this field [1–9].

It is known that the Eskov–Zinchenko effect was first discovered in biomechanics using the examples of TMG and TPG. In fact, we are talking about the ergodicity loss and the statistical repetition impossibility of any sample of motion statistical parameters.

This also applies to spectral signal density, A(t) and other biomechanical parameters within the traditional statistical approach.

It is in biomechanics that there is the lowest stability. As an example, we will present a matrix of TMG samples paired comparisons for the same subject at rest. It is obvious that in this Table 1 we have an extremely small number of pairs of TMG samples for which the Wilcoxon test is $p \geq 0.05$.

Table 1. Pairwise comparison Matrix of TMG for the same person (without load, number of repetitions $n = 15$), Wilcoxon test was used (test of differences $p < 0.05$, number of statistical matches $k = 3$).

	1	2	3	4	5	6	7	8	9	10	11	12	13	14	15
1		.00	.00	.00	.00	.00	.00	.00	.63	.00	.00	.00	.00	.00	.00
2	.00		.00	.00	.00	.00	.00	.00	.00	.00	.00	.00	.00	.00	.00
3	.00	.00		.69	.00	.00	.00	.00	.00	.00	.00	.00	.00	.00	.00
4	.00	.00	.69		.00	.00	.00	.00	.00	.00	.00	.00	.00	.00	.00
5	.00	.00	.00	.00		.00	.00	.00	.00	.00	.00	.00	.00	.00	.00
6	.00	.00	.00	.00	.00		.00	.00	.00	.00	.00	.00	.00	.00	.00
7	.00	.00	.00	.00	.00	.00		.00	.00	.00	.00	.00	.00	.00	.00
8	.00	.00	.00	.00	.00	.00	.00		.00	.00	.00	.00	.00	.00	.00
9	.63	.00	.00	.00	.00	.00	.00	.00		.00	.00	.00	.00	.00	.00
10	.00	.00	.00	.00	.00	.00	.00	.00	.00		.00	.00	.00	.00	.00
11	.00	.00	.00	.00	.00	.00	.00	.00	.00	.00		.00	.00	.00	.70
12	.00	.00	.00	.00	.00	.00	.00	.00	.00	.00	.00		.00	.00	.00
13	.00	.00	.00	.00	.00	.00	.00	.00	.00	.00	.00	.00		.00	.00
14	.00	.00	.00	.00	.00	.00	.00	.00	.00	.00	.00	.00	.00		.00
15	.00	.00	.00	.00	.00	.00	.00	.00	.00	.00	.70	.00	.00	.00	

We obtained the same pattern for signal spectral density, A(t), etc. All this proves the Esckov-Zinchenko effect, i.e. lack of samples' statistical stability in biomechanics. This is referred to in the new theory of chaos-self-organization as uncertainty of the 2nd type. This is the Esckov-Zinchenko effect, which requires the creation of a new science [1–9].

However, the reality of two uncertainties has been proven in chaos of self-organization theory. These are uncertainties of the 2nd type (in the form of the Esckov-Zinchenko effect) and type 1 uncertainty. In this case, statistics do not show differences between samples, but something actually happens to biosystems, they change [1–9]. This also proves the limitations of statistics use.

3 Type 1 Uncertainty in Biomechanics

As we noted above, the entire chaos-self-organization theory has not previously demonstrated the 1st type uncertainty for the parameters of biomechanical systems. In this case, a logical question arises: is the uncertainty of the 1st type global in nature or does it work only for parameters of CVS? There are many examples for the cardiovascular system, but no such studies have been conducted for the neuromuscular one (NMS).

We can use only two parameters in biomechanics – these are the phase coordinates x(t) (vertical position of the finger) and $dx2(t) = dx1/dt$, i.e. rate of change x(t). For these quantities we can create samples and compare them within the framework of classical statistics.

However, now we have calculated the Shannon entropy parameters H for TMG in the mode of many repetitions of tremorogram recording. For the same subject, using the TMG samples, we find the value H. In total, 225 TMG samples were recorded, for which 15 series of H samples were found. In other words, 225 samples were divided into 15 series.

There were 15 TMG samples in each such series, which provided the calculation of 15 H samples for a given subject. As a result, we received Table 2 for the subject at rest. For these samples and for the same subject, the average value of <H> was found for each such series.

By having 15 N samples in each series for all 15 series, we can build a pairwise comparison matrix of all these 225 N values for one subject (at rest). Such matrices were constructed and Wilcoxon Pij tests were found for each i-th and j-th samples of N values for several subjects.

It turned out that such matrices of paired comparisons of N sample have a very small value of numbers k. Here k is the number of samples N pairs, which, consistently exceed $p \geq 0.05$ during pairwise comparison of Pij values. This is a criterion for the of the statistical coincidence constancy of these two samples N in all statistics.

It is obvious that for samples of N entropy we have statistical stability, i.e. there is no Eskov-Zinchenko effect and no loss of ergodicity. In this regard, the possibility of using entropy N as different invariants arises.

But the situation here is much more complicated. In the absence of type 2 uncertainty for N, we have the uncertainty of the 1st type. In other words, the statistics still don't work.

As has been proven, we recorded N for the same subject, first at rest (see Table 2), and then with a static load on the limb. From a physical point of view, such a load leads to completely additional work A and this requires a change in the value of N. These are the laws of physics: doing work changes entropy. All thermodynamics is based on this.

Table 2. Statistical processing results of N-Shannon entropy behavior dynamics for tremor of the same person (number of repetitions N = 15x15).

	H1	H 2	H 3	H 4	H 5	H 6	H 7	H 8	H 9	H 10	H 11	H 12	H 13	H 14	H 15
1	3.322	3.322	3.122	3.122	3.322	3.322	3.122	2.646	3.322	3.122	3.322	3.322	3.322	3.122	2.922
2	3.322	3.322	3.122	3.122	2.922	2.846	3.122	3.322	3.322	2.722	3.322	3.322	2.922	3.122	3.322
3	3.122	3.122	3.122	3.322	3.122	3.122	3.122	3.322	3.122	2.846	3.122	3.122	3.122	3.122	3.122
4	3.122	2.922	3.122	3.122	3.322	3.122	3.122	3.122	3.122	2.722	3.322	3.322	3.122	3.322	2.922
5	3.322	2.922	3.122	2.922	3.122	3.122	2.646	3.122	3.322	2.922	2.922	3.122	3.322	3.122	3.322
6	2.922	3.322	3.322	3.122	3.322	2.922	3.122	3.122	3.122	3.122	3.122	3.322	3.122	3.122	3.122
7	3.322	2.922	3.322	3.322	3.122	3.122	3.322	2.922	3.322	3.122	3.122	3.122	3.122	3.122	3.322
8	3.122	3.322	3.322	2.522	3.322	3.122	2.846	3.122	3.322	2.846	3.122	3.122	3.122	2.922	3.122
9	3.322	3.122	3.122	3.322	3.322	3.322	2.922	3.122	3.322	3.322	3.122	3.322	3.122	2.846	3.322
10	2.922	2.922	3.322	2.846	3.122	3.122	3.322	3.122	3.322	3.322	2.922	3.322	2.922	3.322	3.322
11	3.322	3.322	3.322	3.322	3.322	3.122	3.122	3.322	2.846	3.322	3.322	2.922	3.322	3.322	3.322
12	2.922	3.322	3.122	3.322	3.322	3.122	3.322	3.322	2.922	3.122	3.322	3.122	3.122	2.922	3.322
13	2.846	3.322	3.122	3.322	3.122	2.922	3.322	3.322	2.846	2.922	2.846	3.322	3.122	3.122	3.322
14	3.322	3.322	3.122	3.322	3.322	3.322	3.122	2.922	3.122	3.322	3.322	2.922	3.122	3.322	2.922
15	2.922	2.846	3.122	3.122	3.322	3.322	3.122	3.122	3.322	3.122	3.122	3.322	3.322	2.922	3.322
< H >	3.143	3.157	3.189	3.143	3.228	3.130	3.112	3.130	3.178	3.058	3.157	3.202	3.149	3.117	3.202

We did not observe significant changes for such N samples in different series of tests for the same subject, when analyzing TMG samples before and after the statistical load. It turned out that comparisons of N samples show Wilcoxon tests, which do not differ significantly. Usually Pij is much less than 0.05 for the TMG samples themselves, but N samples did not show this. Doing work does not lead to an increase in N entropy.

Typically, these criteria are Pij $\geq$ 0.05 or more. This means that from the standpoint of stochastics, any two samples N (before and after the load) have a common population. They can statistically coincide, which is physically impossible at all (due to additional work). This contradicts thermodynamics.

The biomechanical system is in different physical states, a large amount of work A is performed, and the Shannon entropies should not coincide. However, we did not observe this in all experiments with calculations of N for the TMG of one person.

4 Discussion

TMG samples' statistical comparison for any subject at rest shows a significant difference. This proves the Eskov-Zinchenko effect and the TMG samples ergodicity loss. If any TMG sample for one subject is unique, then statistics no longer work. We need new science. Now we are creating such a new science in the form of the theory of chaos-self-organization, which is based on the Eskov-Zinchenko effect and on the second type of uncertainty.

The problem of the behavior of Shannon N entropy in the TMG parameters remained unexplored in these studies. Werther entropy N can be invariant when studying any biomechanical parameters? Multiple repetitions of TMG samples registration seemed to give a positive answer to this question. For TMG, the Eskov-Zinchenko effect was clearly observed, and we have uncertainty of the 2nd type.

However, it is known from the physics of any process that when additional work A is performed, any physical system must show changes in entropy N. In this regard, we began to measure TMG samples before and after statistical load for postural tremor. Samples N before and after the load of these measurements did not give any results. The laws of thermodynamics are violated.

It turned out that while measuring TMG samples at rest and after setting the statistical load $F = 3H$, the N values did not differ significantly.

The Wilcoxon criterion Pij in this case approaches one ($Pij \geq 0.6$). Samples N do not differ with or without load.

In this case, we are talking about the possibility of the existence of a general population (at rest and under load). This contradicts the established physical concepts of entropy N and concepts of processes with mechanical systems. If work is done entropy N must change.

The invariance (statistical) of samples N under load and at rest in the new theory of chaos – self-organization is classified as uncertainty of the 1st type. In this case, statistics cannot show differences, but the theory of chaos – self-organization methods can show real differences, especially in the mechanical system. Obviously, type 1 uncertainty also prohibits the further use of statistics in the study of biological systems. Thus, the samples N themselves do not give the Eskov-Zinchenko effect, they cannot be used as invariants, but these samples N show uncertainty of the 1st type in different physical states of the subject. As a result, we have the Eskov-Zinchenko effect for the TMG samples themselves, i.e. type 2 uncertainty, but N samples give us a different effect!

When studying the cardiovascular system (CVS) we get similar results, where type 2 uncertainty has been proven for all samples of cardiointervals, both at rest and under different physical loads. The behavior of entropy H remains unstudied.

Similar studies should be carried out not for cardiointerval samples only, but also for samples of the neurovegetative system parameters (for example, parameters of the sympathetic and parasympathetic neurovegetative system). All this allows us to obtain a more complete picture for studying the behavior of biological systems – type 3 systems, according to W. Weaver classification (1948).

5 Conclusions

The Eskov-Zinchenko effect was discovered more than 20 years ago specifically for the parameters of human biomechanical systems. The loss of ergodicity was proven on samples of TMG and TPG, that means statistical instability of parameters in biomechanics.

TMG samples are statistically unique (for one subject at rest) with a very high p probability. Typically this is $p \geq 0.95$, which is presented in statistics as discrepant results. In our work, we study the samples behavior of Shannon entropy N for TMG. This proved the absence of the Eskov-Zinchenko effect for N.

The illusion of choosing N in the form of an invariant arises. However, a comparison of N samples for one subject at rest and during physical activity showed that such samples coincided with a frequency of $p^* \geq 0.6$. This means the emergence of type 1 uncertainty for the chaos-self-organization theory and the denial of the laws of thermodynamics for physics.

From a physical point of view, this is completely impossible, because the biomechanical system (the hand and finger of the subject) is in different physical states. However, in this case, samples N have a common general population. Here another paradox arises. Our experiments deny physical concepts. We emphasize that the loss of ergodicity of the TMG is also a paradox (this rejects statistics). It is necessary to create a new science to describe biosystems.

References

1. Eskov, V.M., Bazhenova, A.E., Vochmina, U.V., Filatov, M.A., Ilyashenko, L.K.: N.A. Bernstein hypothesis in the description of chaotic dynamics of involuntary movements of person. Russian J. Biomech. **21**(1), 14–23 (2017)
2. Eskov, V.M., Filatova, O.E.: Problem of identity of functional states in neuronal networks. Biophysics. **48**(3), 497–505 (2003)
3. Eskov, V.M., Eskov, V.V., Gavrilenko, T.V., Vochmina, Y.V.: Formalization of the effect of "repetition without repetition" discovered by N.A. Bernshtein. Biophysics **62**(1), 143–150 (2017). https://doi.org/10.1134/S0006350917010067
4. Es'kov, V.M., Filatova, O.E., Ivashenko, V.P.: Computer identification of compartmental neuron circuits //. Meas. Tech. **37**(8), 967–971 (1994)
5. Eskov, V.M.: Evolution of the emergent properties of three types of societies: the basic law of human development. Emergence: Complexity Organ. **16**(2), 107–115 (2014)
6. Gavrilenko, T.V., Es'kov, V.M., Khadartsev, A.A., Khimikova, O.I., Sokolova, A.A.: The new methods in gerontology for life expectancy prediction of the indigenous population of Yugra. Adv. Gerontol. **27**(1), 30–36 (2014)
7. Garaeva, G.R., Eskov, V.M., Eskov, V.V., Gudkov, A.B., Filatova, O.E., Khimikova, O.I.: Chaotic dynamics of cardiointervals in three age groups of indigenous people of Ugra. Ekologiya cheloveka (Hum. Ecol.) **22**(9), 50–55 (2015). https://doi.org/10.17816/humeco16991
8. Eskov, V.M.: Hierarchical respiratory neuron networks. Modell. Measur. Control C **48**(1–2), 47–63 (1995)
9. Eskov, V.M.: Cyclic respiratory neuron network with subcycles. Neural Netw. World **4**(4), 403–416 (1994)
10. Prigogine, I.R.: The End of Certainty: Time, Chaos, and the New Laws of Nature, Free Press (1996)
11. Gell-Mann, M.: Fundamental sources of unpredictability. Complexity **3**(1), 13–19 (1997)
12. Penrose, R.: Newton, quantum theory and reality. In: Hawking, S.W. Israel, W. (eds.) 300 Years of Gravity. Cambridge University Press, Cambridge. (1987)
13. Ginzburg, V.L.: What problems of physics and astrophysics seem now to be especially important and interesting (thirty years later, already on the verge of XXI century)? Physics- Uspekhi **42**, 353–373 (1999). https://doi.org/10.1070/PU1999v042n04ABEH000

14. Weaver, W.: Science and complexity. Am. Sci. **36**, 536–544 (1948)
15. Eskov, V.V., Gavrilenko, T.V., Eskov, V.M., Vokhmina, Yu.V.: Phenomenon of statistical instability of the third type systems—complexity. Tech. Phys. **62**(11), 1611–1616 (2017). https://doi.org/10.1134/S106378421711007X
16. Zilov, V.G., Eskov, V.M., Khadartsev, A.A., Eskov, V.V.: Experimental verification of the bernstein effect "repetition without repetition." Bull. Exp. Biol. Med. **163**(1), 1–5 (2017)
17. Eskov, V.M., Eskov, V.V., Vochmina, J.V., Gavrilenko, T.V.: The evolution of the chaotic dynamics of collective modes as a method for the behavioral description of living systems. Moscow Univ. Phys. Bull. **71**(2), 143–154 (2016)
18. Eskov, V.M., Eskov, V.V., Braginskii, M.Y., Pashnin, A.S.: Determination of the degree of synergism of the human cardiorespiratory system under conditions of physical effort. Meas. Tech. **54**(7), 832–837 (2011). https://doi.org/10.1007/S11018-011-9812-Y

Proposal for a Model for Diabetes Detection Using Machine Learning Techniques

Benny Carrera[1], Williams Ccompi[1], Gerson Evangelista[1], and Wilfredo Ticona[1,2](✉)

[1] Universidad ESAN, Lima, Perú
{21100936,21101034,21100940}@ue.edu.pe, wmamani@esan.edu.pe
[2] Universidad Tecnológica del Perú, Lima, Perú

Abstract. Currently a very relevant problem is diabetes, this disease is the cause of thousands of deaths and tends to grow in the coming years. The objective of the present research is to detect diabetes using Machine Learning techniques. For this purpose, the Indian PIMA database (PID) was used, which was extracted from Kaggle. Preprocessing techniques such as null value treatment and feature normalization were applied. Finally, various learning models such as Multilayer Perceptron, Neural Networks, Decision Tree, Support Vector Machines, K-Nearest Neighbors, Logistic Regression, Random Forest and Gradient Boosting algorithm were developed. From the best results obtained from the evaluation of the models, it was obtained that the best of them is the one generated by the Multilayer Perceptron Neural Networks and Support Vector Machine whose metrics were superior in Accuracy with 84.62% and Recall with 76.92%; however, in Precision it was surpassed by the Random Forest algorithm with 81.81%. It follows that both the Multilayer Perceptron and Support Vector Machine models accurately predicted the onset of diabetes, thus establishing their effectiveness in this predictive task, offering invaluable assistance to the healthcare sector.

Keywords: Machine Learning · Algorithms · Diabetes · Prediction · Hyperparameter · Metrics

1 Introduction

Diabetes prediction through Machine Learning techniques has become a key area of research in the field of health and medicine due to the growth of diabetes globally. Early identification of individuals at risk of developing this chronic disease can have a significant impact on prevention and treatment. Numerous studies have explored the relationship between medical variables and the likelihood of developing diabetes, approaching the problem from a cause-and-effect perspective.

According to the World Health Organization [1] the number of people with diabetes increased from 108 million in 1980 to 422 million in 2014. In addition, between 2000 and 2019, age-standardized mortality rates from diabetes increased by 3%. In 2019, the condition was the direct cause of 1.5 million deaths and, of all diabetes deaths, 48% were under the age of 70.

R. Silhavy and P. Silhavy (Eds.): CSOC 2024, LNNS 1120, pp. 158–170, 2024.
https://doi.org/10.1007/978-3-031-70518-2_14

In the context of the Americas, according to the Pan American Health Organization (PAHO) [2] at least 62 million people live with diabetes in the Americas. If current trends continue, it is estimated that the number of people with diabetes in the region will reach 109 million by 2040. In the United States, Canada, Argentina, Chile and Uruguay, the prevalence is estimated to be between 6.1% and 8.1% of the adult population. In Brazil, Peru, Venezuela, Colombia and Cuba, the prevalence of diabetes was estimated between 5.1% and 6.0% of the adult population, while in Bolivia, Paraguay, Ecuador, Panama, Costa Rica and Guatemala it was between 4.1% and 5.0%; and in Suriname, Guyana, Nicaragua and Honduras it was between 3.1% and 4.0% of the adult population. In Peru, the National Center for Epidemiology, Prevention and Disease Control (CDC Peru) of the Ministry of Health (Minsa) reported that up to September 2022, 19,842 cases of diabetes have been registered, of which 96.5% correspond to type 2 diabetes and 1.4% to type 1 diabetes. In Peru, 3.9 cases of diabetes mellitus are registered for every 100 Peruvians over 15 years of age, according to the Demographic and Family Health Survey (Endes) 2019 [3].

The fundamental focus of this study is on the application of Machine Learning techniques to develop a highly accurate model proposal for diabetes prediction. To achieve this purpose, state-of-the-art algorithms such as Logistic Regression (LR), Random Forest (RF), Multilayer Perceptron (MLP) Neural Networks, Decision Tree (DT), support vector machines (SVM), K-Nearest Neighbors (KNN), Neural Networks (ANN) and Gradient Boosting algorithm (GBA) will be implemented. The effectiveness and performance of the model will be subjected to a thorough evaluation, using Accuracy, Precision, Recall and F1 metrics.

2 Related Works

In the work of Cuenca Andrade et al. [4] the authors used scientific databases such as ACM Digital Library, IEEE Digital Library, Science@Direct and Scopus. Quality questions were applied to evaluate the pre-selected articles. A total of 154 techniques or algorithms were identified, the most commonly used being Artificial Neural Network (ANN), Random Forest (RF) and Support Vector Machine (SVM). The most commonly used variables for diagnosis and prediction were age, body mass index and blood glucose concentration. The ANN algorithm was found to be the best for detecting type 2 diabetes due to its ability to process input variables in a large size. The SVM, ANN and RF algorithms were used where they obtained 90.1%, 88.02% and 83.59% respectively of accuracy in the PIDD dataset, perceiving in this study that the best prediction was performed by SVM.

In the study by Aftab et al. [5], the authors employed a dataset comprising diabetic patient records to train and validate the proposed model. The research utilized a split of 70:30 for training and testing datasets, optimizing classification techniques to ensure heightened accuracy in validation data. Three supervised learning algorithms were implemented: Artificial Neural Networks (ANN), Naïve Bayes (NB), and Decision Tree (DT). Furthermore, metrics such as Specificity, Sensitivity, and Accuracy were calculated to comprehensively evaluate model performance. The proposed system achieved an impressive precision rate of 95.2%. This outcome underscores the effectiveness of the amalgamation of the three algorithms, coupled with a fuzzy inference system, in accurately predicting diabetes diagnoses. These results bear significant implications for advancing our understanding of predictive modeling in the context of diabetes, contributing valuable insights to the broader scientific community.

In the research conducted by Naz and Ahuja [6], the Indian PIMA dataset (PID) provided by the National Institute of Diabetes and Digestive and Kidney Diseases (NIDDK) was used. All PID participants were female and at least 21 years old. PID was composed of a total of 768 instances, of which 268 samples were identified as diabetic and 500 as non-diabetic. The 8 most influential attributes that contributed to the prediction of diabetes are the following: times of pregnancies the patient has had, BMI, insulin level, age, blood pressure, skin thickness, glucose, DiabetesPedigreeFunction. Its methodology is structured in two stages, the first dedicated to the search for the optimal model and the second to the validation of the model. These stages involved the application of four data mining algorithms: Decision Tree (DT), Naive Bayes (NB), Artificial Neural Networks (ANN) and Deep Learning (DL). As a result, the authors conclude that the Deep Learning (DL) algorithm demonstrated the highest accuracy in diabetes detection with an Accuracy of 98.07%, which consolidates it as the most effective option for this task, according to their findings.

In another research in this case conducted by ElJerjawi and Abu-Naser [7], use was made of a dataset collected from the documentation of the Diabetic Association of the city of Urmia, consisting of 1004 samples with 9 relevant attributes. To address the task of predicting the diabetic condition of individuals, Artificial Neural Networks (ANN) were employed. The ANN was configured as a front-feed network, consisting of an input layer with 8 inputs, 3 hidden layers and an output layer with a single output. The methodology focused on the minimization of the error function during ANN training, using the Backpropagation algorithm. After the training process, a mean error function value of 0.01 was obtained, and the prediction accuracy of diabetic condition reached a respectable 87.3\%, mean error was (0.010), the number of epochs was (158,000), the number of training examples was (767), and the number of validation examples was (237). These results suggest a promising approach for the identification of diabetes in individuals from clinical data and relevant attributes.

In the work of Swapna G., Vinayakumar R. and Soman K.P [8], a method is proposed to detect diabetes using convolutional neural networks (CNN) and LSTM neural networks to analyze heart rate variability (HRV) signals, which are obtained from electrocardiogram (ECG). The features of HRV signals are extracted through these networks, which are then classified using a support vector machine (SVM). The results show an accuracy of 95.7% and an improvement in performance compared to previous methods. This system could be useful for physicians to diagnose diabetes early and accurately.

In the research conducted by Patro and Allam [9], the Indian PIMA (PID) dataset provided by the National Institute of Diabetes and Digestive and Kidney Diseases (NIDDK) was used. Several machine and deep learning algorithms were used for automatic prediction of diabetes. Among these algorithms are: Logistic Regression (LR), Decision Tree (DT), Random Forest (RF), Support Vector Machine (SVM), Neural Networks and Gradient Boosting algorithm. The following metrics were used: Sensitivity, Specifity, Precision, F-Measure. The results show that Random Forest had the best performance among the machine learning models, with an accuracy of 82.82%. In addition, the proposed

CNN algorithm achieved a classification accuracy of 96.13% with data modeling.These results suggest that the proposed approach has the potential to significantly improve the accuracy in automatic diabetes prediction compared to traditional approaches.

These studies highlight the use of machine learning techniques in addressing the diabetes prediction problem, showcasing diverse methodologies and algorithms for automating disease prediction based on the analysis of patient characteristics. This article aims to develop a robust classification model for diabetes prediction, considering the techniques along with the methodologies used.

3 Methodology

The proposed methodology was based on [6, 7], and [10]. It consists of five phases: Data Extraction, Data Preprocessing, Training, Evaluation, and Validation. Each stage includes general activities that are crucial for the development of the research. The graphical representation of the utilized methodology is shown below (Fig. 1):

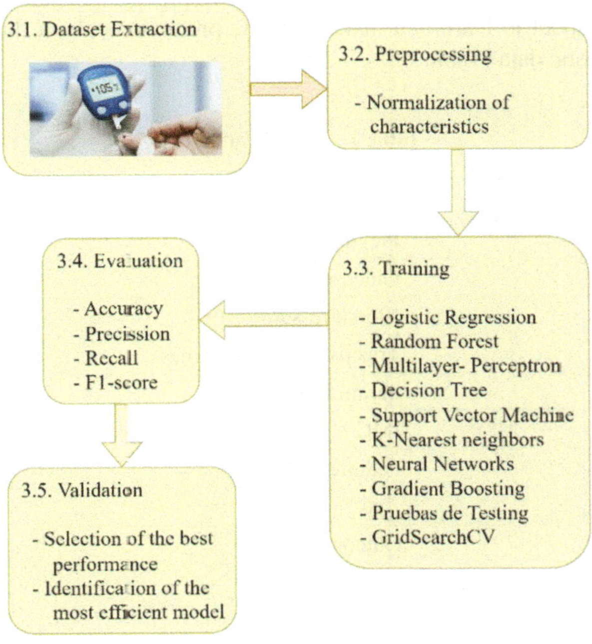

Fig. 1. Proposed methodology. Source: Own elaboration

3.1 Dataset Extraction

In the initial process, we began by identifying the most appropriate data sources. It was learned from previous research that the Pima Indian data set is robust and reliable for making diabetes-related predictions. These data contain detailed diagnostic information on patients who have been screened for diabetes.

The dataset was obtained from the URL https://www.kaggle.com/datasets/uciml/pima-indians-diabetes-database. From the data set it is known that all persons involved in the Diabetic Investigation Program were female and at least 21 years of age. The IDP consisted of a total set of 768 cases, of which 268 were classified as diabetic and 500 as non-diabetic.

In addition, the last column, called "Outcome", plays a crucial role. In it, a value of 0 is assigned to patients who have tested negative for diabetes and a value of 1 to those who have tested positive. The "Outcome" column is essential, as it will be used to train the model, being the variable to be predicted [6].

Most of the other columns, such as "Pregnancies," "Glucose," "BloodPressure," and similar ones, represent the features or attributes that will be used to make predictions about the "Outcome" label. Together, these features provide relevant information that will allow the model to learn and make accurate predictions about diabetes based on patients' diagnostic data (Table 1).

Table 1. List of Variables

N°	Variables
1	Pregnancies
2	Glucose
3	BloodPressure
4	Triceps skinfold thickness
5	Serum Insulin
6	IMC
7	Pedigree
8	Age
9	Outcome

3.2 Data Preprocessing

On many occasions, data in its original form is not directly suitable for use in a predictive model or requires certain transformations to obtain accurate results. This is especially common when working with structured data.

In the context of structured data, such as in a tabular dataset, it's common for each feature to be measured in different units. In order to a predictive model to understand and analyze this data effectively, it is advisable to standardize all features so that they are in the same unit of measurement. This makes it easier for the model to understand and compare these features appropriately [6].

$$z = \frac{x - u}{s} \tag{1}$$

where:

x: The variable or characteristic.

u: The mean of the variable.

s: The standard deviation of the variable.

The result of this process is that all standardized features have a mean equal to zero and a standard deviation equal to one. This not only helps the model to better understand the data but can also improve the convergence speed of training algorithms and make the model more robust.

3.3 Training

After all the necessary preparations have been made on the data, it is possible to train a machine learning model. This involves matching the training features (called "X train") to the training labels (called "y train"). In other words, the model will learn from the training data to understand the relationship between the features and the labels, in this case, to predict the presence or absence of the disease diabetes.

There are several training algorithms available to build predictive models. In this case, we have chosen to use six of them:

- **Logistic Regression:** Algorithm widely used for classification problems. It is based on the logarithmic relationship between the characteristics and the probability that an observation belongs to a class. Its function is to analyze the behavior of previous visitors, such as the time spent on the website and the number of items in the cart. It determines that if previous visitors spent more than five minutes on the site and added more than three items to the cart, they clicked the checkout button. With this information, the logistic regression function can predict the behavior of a new visitor on the website [11].
- **Random Forest:** An ensemble algorithm that combines multiple decision trees to make predictions. It is known for its ability to handle complex data sets and input variables. Its ease of use and flexibility have driven its adoption, as it handles both classification and regression problems. In addition, it uses both bagging and feature randomization to create a forest of uncorrelated decision trees [12].
- **Multilayer Perceptron Neural Networks:** Biologically inspired models consisting of layers of inter-connected nodes. They are known for their ability to learn complex patterns in data, but often require a larger amount of data and settings. It is a supervised learning algorithm that contains node values, activation functions, inputs, and node weights to compute the output. In addition, it works only in the forward direction. All nodes are fully connected to the network. Each node passes its value to the

next node only in the forward direction. The MLP neural network uses a somewhat back-propagation rhythm to increase the accuracy of the between-node model [13].

- **Decision Tree:** A nonparametric supervised learning algorithm, which is used for both classification and regression tasks. It has a hierarchical tree structure, consisting of a root node, branches, internal nodes, and leaf nodes. The model is a form of supervised learning, which means that the model is trained and tested on a dataset containing the desired categorization. In addition, they are known to mimic human thinking, so it is generally easy for data scientists to understand and inter-predict the results [14].
- **Support Vector Machine:** It is a robust classification and regression technique that maximizes the predictive accuracy of a model without overfitting the training data. SVM is particularly suitable for analyzing data with a large number (e.g., thousands) of predictor fields. Its goal is to find a hyperplane that best separates the data into distinct classes in a high-dimensional feature space [15].
- **K-Nearest Neighbors:** It is a nonparametric supervised learning classifier, which uses proximity to make classifications or predictions about the clustering of an individual data point. The goal of this algorithm is to identify the nearest neighbors of a given query point, so that we can assign a class label to that point. Furthermore, this algorithm belongs to the set of supervised machine learning techniques and can be used for both classification and regression problems [16].
- **Neural Networks:** Neural networks are a subset of machine learning and are at the heart of deep learning algorithms. Their name and structure are inspired by the human brain, mimicking the way that biological neurons signal to one another. Artificial neural networks (ANNs) are comprised of a node layers, containing an input layer, one or more hidden layers, and an output layer. Each node, or artificial neuron, connects to another and has an associated weight and threshold. If the output of any individual node is above the specified threshold value, that node is activated, sending data to the next layer of the network. Otherwise, no data is passed along to the next layer of the network [17].
- **Gradient Boosting algorithm:** Is a machine learning technique based on boosting in a functional space, where the target is pseudo-residuals rather than the typical residuals used in traditional boosting. It gives a prediction model in the form of an ensemble of weak prediction models, i.e., models that make very few assumptions about the data, which are typically simple decision trees. [1] and [2] When a decision tree is the weak learner, the resulting algorithm is called gradient-boosted trees; it usually outperforms random forest. [1, 2] and [3] A gradient-boosted trees model is built in a stage-wise fashion as in other boosting methods, but it generalizes the other methods by allowing optimization of an arbitrary differentiable loss function [18].

The GRID hyperparameter setting was also used with the following parameters:

1) Logistic Regression.

   - Solver = ['newton-cg','lbfgs','liblinear']
   - penalty = ['l2']
   - c_values = [100, 10, 1.0, 0.1, 0.01]
   - lr grid = dict(solver = solvers, penalty = penalty, C = c values)

2) Super Vector Machine

- kernel = ['poly','rbf','sigmoid']
- C = [50, 10, 1.0, 0.1, 0.01]
- gamma = ['scale']
- svc_grid = dict(kernel = kernel, C = C, gamma = gamma)

3) Multilayer Perceptron Neural Networks

- activation = ['identity','logistic','tanh','relu']
- solver = ['lbfgs','sgd','adam']
- learning_rate = ['constant','invscaling','adaptive']
- mlp_grid = dict(activation = activation, solver = solver, learning_rate = learning rate)

4) Decision Tree

- criterion = ['gini','entropy','log loss']
- splitter = ['best','random']
- max features = ['auto','sqrt','log2']
- clf_grid = dict(criterion = criterion,splitter = splitter, max features = max features)

5) Random Forest

- n_estimators = [10, 100, 1000,10000]
- max features = ['sqrt','log2']
- rf_grid = dict(n estimators = n estimators, max_features = max_features)

6) K-Nearest Neighbors

- n_neighbors = range (1, 21, 2)
- weights = ['uniform','distance']
- metric = ['euclidean','manhattan','minkowski']
- knn_grid = dict(n neighbors = n neighbors,weights = weights, metric = metric)

7) Neural Networks

- Grid algorithms cannot be applied.
- Gradient Boosting
- n_estimators = [1, 10, 100]
- learning_rate = [0.01, 0.1, 1]
- max_depth = [1–3]

After training these three models, the next step is to use the testing partition to evaluate their predictive ability. This partition contains data that the model has not seen during training. By comparing the performance of each of these three models in predicting diabetes disease, we will be able to determine which one is the most effective for our problem. This will help us make an informed decision about which algorithm to use in future diabetes predictions.

3.4 Evaluation

After the models have been trained using the training data, the next step is to verify how well they can make predictions on data they have not seen before, which is what we have reserved in our test data set. This is a critical part of the process of evaluating models in machine learning.

To perform this evaluation, we can use the scikit-learn library, which provides useful tools to measure the performance of our models. Visual comparison of the results is not the most effective way to do this, as the data vectors are often long and complicated. Instead, we use specialized metrics that are designed to quantify the predictive ability of a model.

In this case, the three common metrics are:

- **Accuracy:** It measures the proportion of correct predictions compared to all predictions made by the model. It is a fundamental metric for evaluating overall model performance [6].

$$Accuracy = \frac{TP - TN}{TP + FP + TN + FN} \tag{2}$$

- **Precision:** It evaluates the proportion of positive predictions made correctly in relation to all positive predictions. It is useful when we want to minimize false positives [19]

$$Precision = \frac{TP}{TP + FP} \tag{3}$$

- **Recall:** It measures the proportion of true positives relative to all true positives. It is useful when we want to minimize false negatives [19].

$$Recall = \frac{TP}{TP + FN} \tag{4}$$

- **F1 Score:** The F1 score can be interpreted as a harmonic mean of the precision and recall, where an F1 score reaches its best value at 1 and worst score at 0. The relative contribution of precision and recall to the F1 score are equal [9].

$$F1Score = \frac{precision * recall}{precision + recall} \tag{5}$$

where:

TP: True Positive.
TN: True Negative.
FP: False Positive
FN: False Negative
Table 2 shows the results of the algorithms for each metric.

Table 2. Algorithm results

Algorithms	Accuracy	Grid	Precision	Grid	Recall	Grid	F1-Score
RL	0.821	0.821	0.750	0.750	0.692	0.692	0.719
SVM	0.846	0.769	0.769	0.700	0.769	0.538	0.769
MLP	0.846	0.821	0.769	0.750	0.769	0.692	0.769
DT	0.667	0.769	0.500	0.625	0.462	0.769	0.480
RF	0.846	0.795	0.818	0.727	0.692	0.615	0.750
KNN	0.743	0.769	0.636	0.750	0.538	0.462	0.583
ANN	0.513	-	0.364	-	0.513	-	0.513
Gradient Boosting	0.795	0.769	0.727	0.7	0.615	0.538	0.666

Table 3 shows the results with the optimal hyperparameters obtained from the best models.

Table 3. Comparison of the best metrics for each model

Algorithms	Parameters	Accuracy	Precision	Recall	F1-Score
RL	random state:2023	82.1%	75%	59.2%	71.9%
SVM	kernel:" linear"	**84.6%**	76.9%	**76.9%**	**76.9%**
MLP	hidden layer sizes:(10, 10), activation:'logistic', randomstate: 2023	**84.6%**	76.9%	**76.9%**	**76.9%**
DT	random state: 2023	66.7%	50%	46.2%	48.0%
RF	random state: 2023	**84.6%**	**81.8%**	69.2%	75.0%
KNN	neighbors: 9	74.3%	63.6%	53.8%	58.3%
ANN	random state: 2023	33.3%	29%	33.3%	33.3%
Gradient Boosting	-	79.5%	72.7%	61.5%	66.6%

Multilayer perceptron and Support Vector Machine coincidentally had the same confusion matrix. See Fig. 2. Otherwise, the sensitivity of these models to identify positives was much more effective in the performance of the other models. We can perceive that the set of 8 features reached the best parameters on the side of Accuracy scores, which was 84.6% on the side of people who will have the disease of diabetes. We prioritize the best Accuracy scores since as we know it is more serious when a case of diabetes has been ignored, which can lead to the death of the victim (Fig. 3).

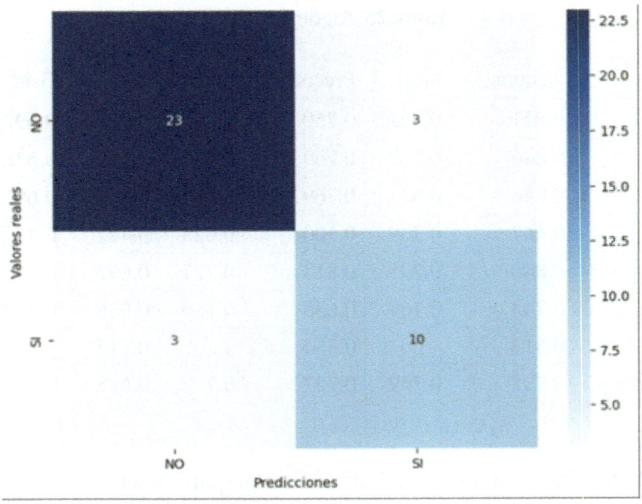

Fig. 2. Confusion Matrix of MLP and SVM. It can be observed that the majority of instances were rated correctly in the confusion matrix.

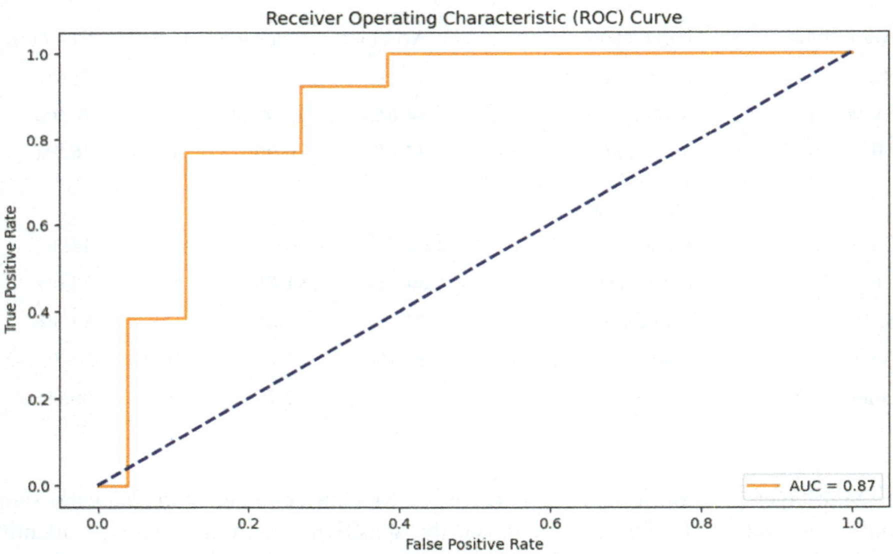

Fig. 3. Curve Roc of MLP and SVM. The AUC value is 0.87 which is very close to 1 suggesting a good performance of the model in terms of discrimination between classes.

3.5 Validation

The present work combines different methodologies and techniques from previous works to provide a solution to one of the problems that continue to impact the world today. The different experiments carried out with Machine Learning algorithms allowed to find a classification model with an accuracy of 84.6%, this model used the Multilayer Perceptron and Support Vector Machine technique.

4 Conclusions

First, the data was extracted, which was not so complicated because it was already created. Also, preprocessing was performed, where the data was normalized. With the preprocessing exposed in the work, it was enough to perform a good training of the model. Several training algorithms were used to build predictive models, among which eight were used: logistic Regression, Random Forest, Multiplayer Perceptron Neural Networks, Decision Tree, Vector Super machines, K-Nearest Neighbors, Gradient Boosting algorithm and Neural Networks. After obtaining the results of the metrics of each model, we proceed to select the one that demonstrates superior performance in the prediction of diabetes disease, which are 84.62% in Accuracy without parameters for the MLP algorithm and the same percentage for the SVM algorithm, in Precision won the Random Forest algorithm with 81.81% without parameters and in Recall won MLP and SVM without parameters with 76.92%. This validation process allows us to identify and confirm which model stands out as the most effective in the disease prediction task, which were MLP and SVM because they have 2 metrics in their favor, that is why it is guaranteed that the MLP model and the SVM model are the best choice to predict the diabetes disease. Finally, it can be concluded that a greater number of records and a greater diversity of them could improve the results evidenced. Although the model can learn to identify whether a person may or may not have diabetes, this would be more robust if trained with more data. even so, this research provides a contribution for future works that want to take it as a reference and use a similar methodology to perform diabetes disease prediction.

References

1. OMS: Diabetes (2023). https://www.who.int/es/news-room/fact-sheets/detail/diabetes
2. OPS: El número de personas con diabetes en las Américas se ha triplicado en tres décadas, según un informe de la OPS (2022). https://www.paho.org/es/noticias/11-11-2022-numero-personas-con-diabetes-americas-se-ha-triplicado-tres-decadas-segun
3. Luis, R.: Situación de la Diabetes según datos del Sistema de Vigilancia (2021). https://www.dge.gob.pe/portal/docs/tools/teleconferencia/2021/SE202021/03.pdf
4. Cuenca-Andrade, G., Guamán-Quinche, R., Álvarez-Pineda, F., Chamba-Zaragocin, W., Guamán-Quinche, J.O.: Estudio de algoritmos de inteligencia artificial más utilizados para el diagnóstico de diabetes mellitus tipo 2. CEDAMAZ **13**(1), 114–121 (2023)
5. Aftab, S., et al.: Cloud-based diabetes decision support system using machine learning fusion. Computers, Materials & Continua **68**(1), 1344–1356 (2021)
6. Naz, H., Ahuja, S.: Deep learning approach for diabetes prediction using PIMA Indian dataset. J. Diabetes Metab. Disord. **19**, 391–403 (2020)
7. El_Jerjawi, N.S., Abu-Naser, S.S.: Diabetes prediction using artificial neural network. Int. J. Adv. Sci. Technol. **121**, 55–64 (2018)
8. Swapna, G., Vinayakumar, R., Soman, K.: Diabetes detection using deep learning algorithms. ICT Express **4**(4), 243–246 (2018)

9. Patro, K.K., et al.: An effective correlation-based data modeling framework for automatic diabetes prediction using machine and deep learning techniques. BMC Bioinformatics **24**(972) (2023)
10. Pradeep Kandhasamy, J., Balamurali, S.: Performance analysis of classifier models to predict diabetes mellitus. ICGHIA 2014 **47**, 45–51 (2015)
11. AWS: ¿Qué es la regresión logística? (2023). https://www.ibm.com/es-es/topics/logistic-regression
12. DataScientest: Random Forest: Bosque aleatorio. Definición y funcionamiento (2023). https://datascientest.com/es/random-forest-bosque-aleatorio-definicion-y-funcionamiento
13. Bento, C.: Multilayer Perceptron Explained with a Real-Life Example and Python Code: Sentiment Analysis (2021). https://towardsdatascience.com/multilayer-perceptron-explained-with-a-real-life-example-and-python-code-sentiment-analysis-cb408ee93141
14. IBM: Arboles de decisión. https://www.ibm.com/es-es/topics/decision-trees
15. IBM: Funcionamiento de SVM (2021). https://www.ibm.com/docs/es/spss-modeler/saas?topic=models-how-svm-works
16. IBM: ¿Qué es KNN?. https://www.ibm.com/mx-es/topics/knn
17. IBM: What is a neural network?. https://www.ibm.com/topics/neural-networks
18. Parr, T., Howard, J.: How to explain gradient boosting (2018). https://explained.ai/gradient-boosting/index.html
19. Mercaldo, F., Nardone, V., Santone, A.: Diabetes mellitus affected patients classification and diagnosis through machine learning techniques. Procedia Comp. Sci. **112**, 2519–2528 (2017)

Leaf Condition Classification Model of Apple Tree Leaves Using Machine Learning

Mesias Picon[1], Sebastián Furuguay[1], Emerson Vilcatoma[1], and Wilfredo Ticona[1,2](✉)

[1] Universidad ESAN, Lima, Perú
{20101034,20100770,20101061}@ue.edu.pe, wmamani@esan.edu.pe
[2] Universidad Tecnológica del Perú, Lima, Perú

Abstract. Throughout time, agriculture has faced the constant challenge of determining whether plants are sick or healthy, which often leads to the application of incorrect measures. For this reason, a study was conducted to discern the health of apple plants from the analysis of their leaves. For this purpose, 5052 images were collected from websites such as GitHub and Kaggle, classifying them into healthy leaves and diseased leaves. Several image processing techniques were applied, such as grayscale, edge detection, and Salt and pepper noise filtering. Subsequently, several learning models were developed, such as Support Vector Machine, Logistic Regression, MobileNet, and Vision Transformer. The results were evaluated using metrics such as Accuracy, Precision, Recall, and F1 score. Among all models, the Vision Transformer algorithm proved to be the most effective, with superior metrics: 96.02% Accuracy and 96.02% Recall. This approach offers promising prospects for accurate identification of apple plant health status, providing a valuable tool for decision-making in agriculture.

Keywords: Plant diseases · Smart Agriculture · Machine Learning · Pest control · plant pathology · Accuracy

1 Introduction

The problematic situation is that since about 10,000 years ago, human crops have been threatened and affected by a wide range of pests that cause economic losses of approximately 200,000 to 500,000 million dollars throughout world agriculture, but the most important consequence is the food insecurity that has spread throughout history. According to FAO [1], an estimated 10–20% of the world's total apple production is lost to pests, and in Peru, 15–25%, not counting post-harvest losses. This study will focus on developing a system for classifying the leaf condition of apples, i.e., we want to classify whether the apple is healthy or diseased based on images of the leaves. For this, we will use different Machine Learning architectures.

2 Related Work

In the work of Garg [2] et al. the authors used a dataset of images taken by a UAV of plants that had the NLB disease to create a Machine Learning based model that can classify the image of a plant as diseased or healthy and quantify how dis-eased the

R. Silhavy and P. Silhavy (Eds.): CSOC 2024, LNNS 1120, pp. 171–180, 2024.
https://doi.org/10.1007/978-3-031-70518-2_15

plant was. They also used the Cascaded MRCNN technique to segment by plants and instances and then quantify how sick the plant is. They had an accuracy of 97% Dice when segmenting the diseased part of the leaves. The results showed that the Cascaded MRCNN architecture achieved a correlation of 73% and a mean squared error (MSE) of 2.0437 between predicted and actual disease severity.

Similarly, in the study by Bhugra [3] et al. they sought to solve the problem that most plant-related work required structured data generating arduous manual classification work. Also, they used three different data sets to test their model, then they opted to create a Machine Learning model capable of working with unstructured data using the Generative Inpainting Network technique which they adjust-ed by creating their Anoleaf algorithm that mainly consists of significantly improving the system's ability to locate disease areas or defects in plant leaves with greater accuracy. Accuracy. The results in the different datasets are favorable to the AnoLeaf proposal. In the dataset, regarding disease segmentation, AnoLeaf achieved an accuracy of 96.44%, in the PlantVillage dataset the same result was obtained. Finally, in the MVTec dataset, an accuracy of 93% was obtained.

According to the Jacobs [4] et al. study, early detection of plant stress is crucial to protect crops during growth. Annual yield losses in North America are as high as 11% in soybeans and 2 to 17% in corn due to plant diseases, generating significant economic losses in revenue for farmers. This work focuses on hierarchical stress classification in agricultural crops, addressing the limitations of previous CNN-based applications by using a Bayesian approach that models label probabilities in a hierarchy. Also, the study evaluated tomato, corn, and soybean datasets, revealing the effectiveness of a hierarchical classification approach. In the Tomato Dataset, with 15,892 images, a balanced accuracy of 82.1% was achieved. In the Maize Dataset, containing 8,911 images, the balanced accuracy was 68.8%, using a modified ResNet-18 network. For the Soybean Dataset, containing 6,635 images, the balanced accuracy reached 80.0\%, also with modified ResNet-18.

3 Methodology

3.1 Proposed Method

The proposed method is based on [2–4]. It consists of five phases: dataset extraction, preprocessing, feature extraction, model implementation and evaluation. Each phase contains activities essential to the development of the research. The method used is shown below (Fig. 1):

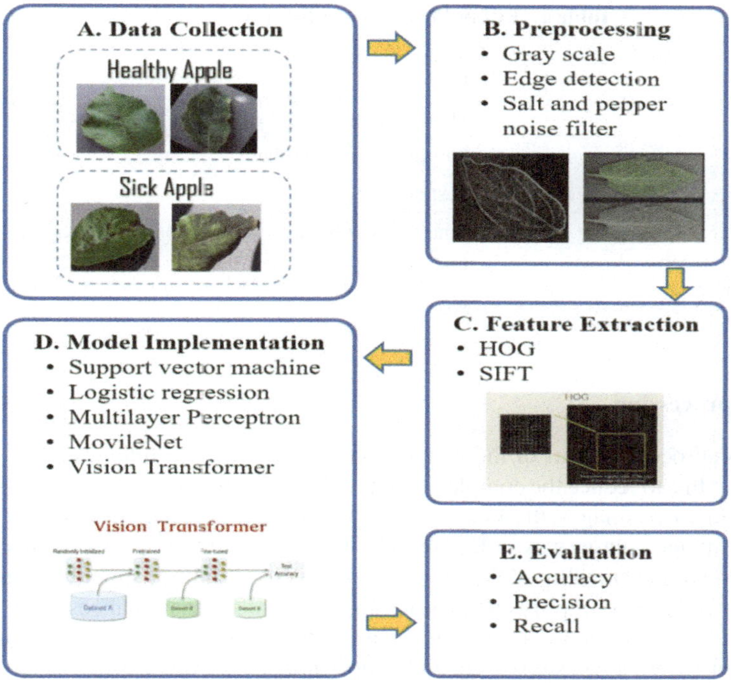

Fig. 1. Proposed Methodology. Source: Own elaboration

3.2 Data Collection

Dataset. The study was based on the use of a dataset collected from Kaggle known as the "New Plant Diseases Dataset", which includes 87,000 plant images captured in RGB format. This dataset is divided into training and test sets and encompasses 38 distinct classes representing various plants and diseases. From these classes, 5051 images were extracted specifically of apple tree leaves. Importantly, this dataset is structured such that each image is uniquely labeled as either "diseased" or "healthy". In addition, data partitioning was performed on an 80\% for train and 20\% for tests, as per the article by Jacobs [4] (Table 1).

Table 1. List of Plant Keywords (Sick or Healthy).

#	Keyword
1	Plant diseases
2	Symptoms of sick plants
3	Pest control
4	Smart Agriculture
5	Plant pathology
6	Accuracy

3.3 Preprocessing

For the preprocessing part of the image dataset, a series of techniques and steps were followed. This, to reduce the complexity at the time of analyzing the data, we also sought to highlight, in the images, the area of the plant as well as the diseased area. As a first step, since our image dataset was in RGB scale, it was converted to grayscale because this reduced the dimensionality of the image as well as its weight. It should be emphasized that this was done to simplify calculations and reduce computational costs [7]. Also, the texture and intensity of a grayscale image are sufficient to find patterns associated with plant diseases. As a last step, a salt and pepper noise filter was applied to have images with better resolution, since it was analyzed that most of the images had black and white spots [8].

3.4 Feature Extraction

Mainly, we worked with the feature extractors Histogram of oriented gradients (HOG) and Histogram of oriented gradients (HOG) [5]. HOG is an accurate choice for a feature vector for a classifier of plants as diseased or healthy. It is robust, efficient, and capable of capturing meaningful plant features. In the case of SIFT (Scale-Invariant Feature Transform) [6], this FV can automatically identify key points and generate detailed descriptors for these points, which can capture important plant features. However, the computational calculation is usually more time-consuming than that of HOG. On the other hand, using the Deep Learning techniques of Movilnet and Vision Transformer did not make use of an FV such as HOG or SIFT as these techniques do not require one.

3.5 Implementation

For the construction of the model, the "Label" column was coded, whose textual values became numerical: Sick (1) and Healthy (0). Then, the partitioning for training and testing was performed, considering 80% and 20% of the data, respectively, this new coding was joined to the records previously vectorized with 2 techniques (HOG, SIFT). Finally, 5 Machine Learning techniques were used. Table VIII shows the activities of the model-building phase (Table 2).

Table 2. Implementation Activities

Activities	Description	Tasks
Feature selection and Label Assignment	Use of Feature Extraction techniques and labeling	HOG and SIFT with label sick or healthy
Dataset partition	Partition of the dataset in training (80%) and test (20%)	Partition of the dataset in training (80%) and test (20%)
Use of Machine Learning Techniques	Training the models with the training and test data	Five different Machine learning models such as Decision Trees and Linear Regression was Used

Tuned Parameters and Values in Each Algorithm:

1) Linear Regression

- No specific parameters are mentioned for Linear Regression since it is a simple linear model and does not require hyperparameter fitting.

2) SVM

- kernel: linear, poly, rbf, sigmoid
- C: 1, 10, 100
- epsilon: 0.1, 0.01, 0.001

3) Multilayer perceptron

- Hidden layers: (10,), (10, 10), (20, 10), (20, 20)
- Activation: logsig, tanh, relu
- Alpha: 0.0001, 0.001, 0.01

4) Vision Transformer

- Epochs: 60
- Patch Size: 6
- Batch size: 256
- Resize image: 72 x 72
- Learning rate: 0.001
- Weight decay: 0.0001
- Transformer layers: 8

5) Movil net

- Image size: 224
- Batch size: 32
- Rescale: 1. /255
- Totation range: 20
- Width shift range: 0.2,
- Height shift range: 0.2,
- Shear range: 0.2,
- Zoom range: 0.2,
- Horizontal flip: True,
- Fill mode: nearest

3.6 Evaluation

Accuracy: The accuracy of a model is defined as the percentage of correct predictions with respect to the total number of predictions made, providing an overall measure of its accuracy [9].

$$Accuracy = \frac{TP + TN}{TP + FP + FN + TN} \tag{1}$$

Precision: Precision refers to the ratio of correctly identified positives to the total number of positives predicted by the classifier [9].

$$Precision = \frac{TP}{TP + FP} \tag{2}$$

Recall: Recovery is defined as the ratio of correctly identified positive results to the total number of relevant samples (i.e., all those that should have been identified as positive) [10] (Table 3).

$$Recall = \frac{TP}{TP + FN} \tag{3}$$

where:

TP: TruePositive
FP: FalsePositives
TN: TrueNegatives
FN: FalseNegatives

4 Results

The following table presents the most salient metrics of each machine learning model, along with their respective vectorizer, after optimizing the parameters (Table 4):

SVM: The following figure shows the confusion matrix of the best result with its respective feature vector (Fig. 2):

Table 3. Evaluation Results

Model	Feature E	Accuracy	Precision	Recall
SVM	HOG	92%	93%	89%
	SIFT	90%	87%	92%
MLP	HOG	89%	88%	90%
	SIFT	92%	90%	93%
Logistic Regression	HOG	89%	88%	90%
	SIFT	91%	92%	90%
Vision Transformer	None	96.02%	-	96.02%
MovileNet	None	99.40%	98.47%	99.78%

Table 4. Best Results

Model	Parameters	Accuracy	Precision	Recall
SVM	C:10 Kernel: rbf Epsilon: 0.01	92%	93%	89%
MLP	Hidden layers:(10, 10) Activation: logsig Alpha:0.001	92%	90%	93%
Logistic Regression	None	91%	92%	90%
Vision Transformer	Epochs: 60 Learning rate:0.001 Patch Size:6	96.02%	-	96.02%
MovileNet	Batch size:32 Rescale: 1./255 Totation range: 20	99.40%	98.47%	99.78%

Logistic Regression: The following figure shows the confusion matrix of the best result with its respective feature vector (Fig. 3):

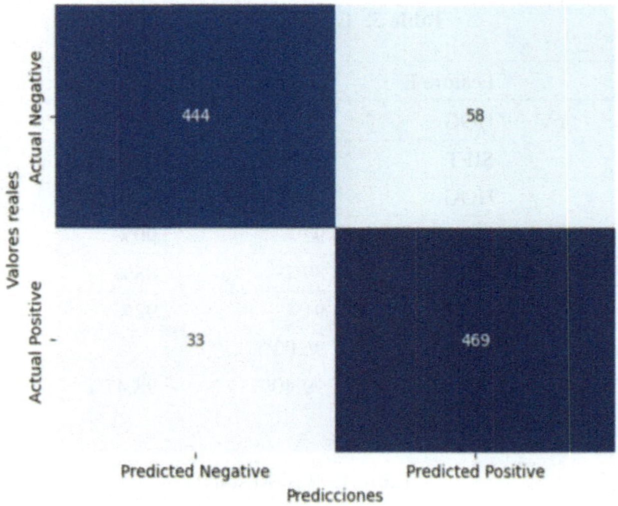

Fig. 2. SVM confusion matrix with SIFT.

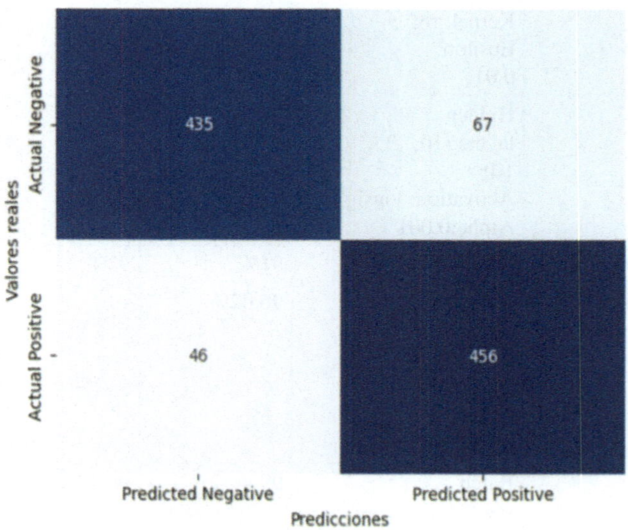

Fig. 3. RL confusion matrix with SIFT.

Multiple Layer Perceptron: The following figure shows the confusion matrix of the best result with its respective feature vector (Fig. 4):

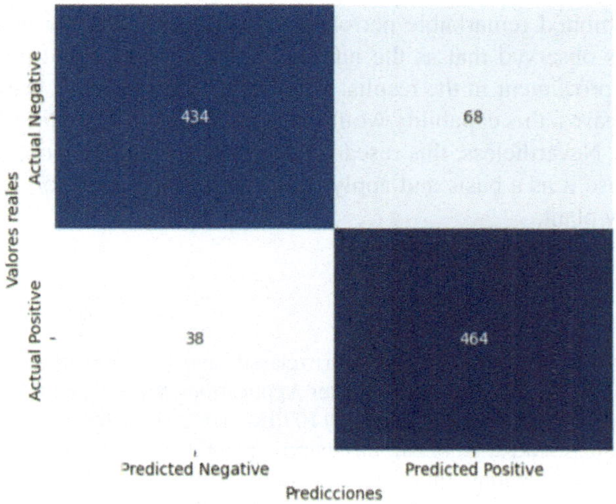

Fig. 4. MLP confusion matrix with SIFT

5 Conclusions

This paper combines various methodologies and techniques from previous work to provide a solution to one of the problems that continues to impact Peru today, the issue of food security has affected human crops for approximately 10,000 years. Pests have been a constant threat, causing substantial economic losses ranging from US200billiontoU S500 billion across global agriculture. Food insecurity, stemming largely from these losses, has been a constant throughout history. The research was based on the use of a dataset obtained from Kaggle, called the New Plant Diseases Dataset. This dataset comprises 87,000 plant images captured in RGB format and is organized into training and test sets. It includes 38 different categories representing various plants and diseases. From these categories, 5051 specific images of apple tree leaves were selected. Crucially, the structure of this dataset ensures that each image is uniquely labeled as either" diseased" or" healthy". Efficient preprocessing was carried out, which positively influenced the model results. This process was based on background information, albeit with some significant adaptations. One of the key modifications was the implementation of the Vision Transformer model. The images already had the appropriate dimensions, and techniques such as gray scaling, edge detection, and the application of a filter to reduce noise were applied. With detailed preprocessing, it was shown that effective model training was sufficient. In the field of image processing, several techniques are available that offer a significant advantage when tackling classification tasks. For this study, we chose to employ five specific techniques: Support Vector Machine, Logistic Regression, Multilayer Perceptron, MobileNet, and Vision Transformer. After conducting several training sessions, it was determined that the most effective technique was MovileNet, reaching an accuracy of 99.40%. However, the Vision Transform technique did not lag, achieving an accuracy of 96%. In addition, one of its advantages over MobileNet is the ability to use deeper and more complex architectures. It is also relevant to note that the other models also exhibited remarkable performance, all surpassing the 90% threshold. In conclusion, it is observed that as the number of images processed increases, there is a significant improvement in the results. Although our models can learn to distinguish between plant leaves, this capability would be more robust with additional training with a larger dataset. Nevertheless, this research establishes a starting point for future work that wishes to use it as a basis and apply a similar methodology for detecting the leaf condition of any plant.

References

1. Abed, S., Esmaeel, A.A.: A novel approach to classify and detect bean diseases based on image processing. IEEE Symposium on Computer Applications & Industrial Electronics (ISCAIE) **2018**, 297–302 (2018). https://doi.org/10.1109/ISCAIE.2018.8405488
2. Bhugra, A., Singh, M.K.: AnoLeaf: unsupervised leaf disease segmentation via structurally robust generative inpainting. In: Proceedings of the IEEE/CVF Winter Conference on Applications of Computer Vision (WACV), pp. 6415–6425. IEEE (2023). https://doi.org/10.1109/WACV48630.2023.00201 https://doi.org/10.1016/j.engappai.2021.104460
3. Frank, D.W., Njoroge, N., O'Brien, S.: Confidence-driven hierarchical classification of cultivated plant stresses. In: IEEE Conference on Computer Vision and Pattern Recognition, pp. 1117–1126 (2021). https://doi.org/10.1109/WACV48589.2021.00126
4. Garg, K., Bhugra, S., Lall, B.: Automatic quantification of plant disease from field image data using deep learning. In: 2021 IEEE Winter Conference on Applications of Computer Vision (WACV), pp. 1964–1971 (2021). https://doi.org/10.1109/WACV48630.2021.00201
5. Gobierno de Mexico: El cambio climatico favorecer´a el desarrollo de plagas y enfermedades de los cultivos. Gob. Mx (2023). https://acortar.link/IEXcBI
6. Kaur, S., Pandey, S., Goel, S.: Plants disease identification and classification through leaf images: a survey. Archi. Computat. Methods Eng. **26**(2), 507–530 (2019). https://doi.org/10.1007/s11831-018-9255-6
7. Khirade, S.D., Patil, A.B.: Plant disease detection using image processing. Int. Conf. Comp. Comm. Contr. Auto. **2015**, 768–771 (2015). https://doi.org/10.1109/ICCUBEA.2015.153
8. Lindeberg, T.: Scale invariant feature transform (2012)
9. Dahouda, M.K., Joe, I.: A deep-learned embedding technique for categorical feature encoding. IEEE Access **9**, 114381–114391 (2021). https://doi.org/10.1109/ACCESS.2021.3104357
10. Dahouda, M.K. Joe, I.: A deep-learned embedding technique for categorical features encoding. IEEE Access **9**, 114381–114391 (2021). https://doi.org/10.1109/ACCESS.2021.3104357
11. Sharma, D., Kapoor, N., Sood, D.: Plant disease detection techniques: a survey. In: Precision Agriculture for Sustainability, pp. 327–339. Apple Academic Press (2023). https://doi.org/10.1201/9781003435228-19
12. Toh, K.K.V., Isa, N.A.M.: Noise adaptive fuzzy switching median filter for salt-and-pepper noise reduction. IEEE Signal Process. Lett. **17**(3), 281–284 (2009)
13. Tomasi, C.: Histograms of oriented gradients. Computer Vision Sampler, 1–6 (2012)
14. Vishnoi, V.K., Kumar, K., Kumar, B.: Plant disease detection using computational intelligence and image processing. J. Plant Dis. Prot. **128**(1), 19–53 (2021). https://doi.org/10.1007/s41348-020-00368-0
15. Yim, P.J., Choyke, P.L., Summers, R.M.: Grayscale skeletonization of small vessels in magnetic resonance angiography. IEEE Trans. Med. Imaging **19**(6), 568–576 (2000)

Exploring Advantage Estimation with Higher-Order Summation of State Values in Deep Reinforcement Learning: Experimental Insights for Online Temporal Difference Approach

Martin Kaloev(✉) and Georgi Krastev

University of Ruse, Ruse, Bulgaria
{mkaloev,geork}@uni-ruse.bg

Abstract. This study investigates the potential advantages of redefining advantage calculation in actor-critic deep reinforcement learning. Departing from the conventional temporal difference approach, where the advantage is determined by the difference between the values of two consecutive states, we investigate an alternative method. Our approach involves computing the advantage as the difference between the value of the current state and the summation of multiple discounted values of subsequent states. Through a series of experiments, we explore the implications of this alternative advantage calculation on the training dynamics of actor-critic networks. Furthermore, we extend the actor's loss function by incorporating the summation of even multiple discounted rewards with appropriate gamma discounts. The experiments aim to discern whether this extended loss function enhances the learning capabilities of the actor in complex reinforcement learning tasks. We find that the advantages are contingent upon the task's reward horizon—indicating the frequency of rewards and the presence of negative rewards. Notably, the effectiveness of extended discounting is highly dependent on selecting an appropriate order of calculation, signifying the number of state values considered after initial. This order, akin to context, proves crucial in determining the efficacy of the extended discounting approach.

Keywords: Deep Reinforcement Learning · Advantage Loss Function · Temporal Difference · Experimental Insights · Computational Efficiency · Neural Network Training · Artificial Intelligence Optimization

1 Introduction

Deep Reinforcement Learning (Deep RL) is machine learning discipline. It navigating the interplay of actions and feedback loops through the computational prowess of Artificial Neural Networks (ANNs). The ANNs often referred to as tensor machines. At its core lies adaptive programming, a numerical framework shaped by Bellman, facilitating the translation of information from the action-feedback loop into a format amenable

R. Silhavy and P. Silhavy (Eds.): CSOC 2024, LNNS 1120, pp. 181–190, 2024.
https://doi.org/10.1007/978-3-031-70518-2_16

to training ANNs. The crux of Deep RL unfolds in how this processed information is harnessed. Two primary approaches surface—online and offline training of ANNs. Online training, characterized by updates after each action, contrasts with offline training, which updates at the end of an episode or after accumulating sufficient information. This dichotomy sets the stage for the exploration of various Deep RL methodologies tailored to specific task requirements. Among the myriad Deep RL types, including Q-learning, SARSA, and policy optimization, our focus gravitates towards the Actor-Critic paradigm, specifically the Actor-Critic Advantage type. In this study, we implement a modified version of Advantage Actor-Critic (A2C), building upon the foundations laid in the A3C paper. Our modification involves high-order calculations, extending the discount summation to multiple states, show at formulas 1, 2 and 3. This enhancement aims to enrich the context available for training ANNs, addressing the nuanced dynamics of learning.

$$\mathrm{LA} = \mathrm{R}_0 + \left[\sum_{t=1}^{N} \gamma^t \mathrm{V}(s_t; \theta)\right] - \mathrm{V}(s_0; \theta) \tag{1}$$

where ***LA*** is the loss calculated for the critic with a high-order calculation equal to N, also used as the advantage of action for the calculation of the actor loss.

$$\mathrm{LAR} = \left[\sum_{t=1}^{N} \gamma^{t-1} \mathrm{R}_{t-1}\right] + \left[\sum_{t=1}^{N} \gamma^t \mathrm{V}(s_t; \theta)\right] - \mathrm{V}(s_0; \theta) \tag{2}$$

where ***LAR*** is the loss calculated with discounted rewards sums to be applied to the critic with a high-order calculation equal to N, also used as the advantage of action for the calculation of the actor loss.

$$\mathrm{L}_{\mathrm{actor}} = -\log(\pi(a_t|s_t; \theta))\mathrm{A}_t; \text{ where } \mathrm{A}_t = \{\mathrm{LA}, \mathrm{LAR}\} \tag{3}$$

$\boldsymbol{V(s_t; \theta)}$ is the critic's estimate of the value of state s_t, according to the neural network with parameters θ.
$\boldsymbol{R_t}$ is the immediate reward at time step t ($s_t to s_{t+1}$).
γ is the discount factor; in our experiments, we use 0.9.
$\boldsymbol{L_{actor}}$ is the loss function used for updating the actor network.
$\boldsymbol{\pi(a_t|s_t; \theta)}$ is the probability of taking action a_t, given the state s_t, as predicted by the neural network with parameters θ.
$\boldsymbol{A_t}$ is the advantage from taking action a_t.
N is the number of orders calculated, also the size of the temporary memory buffer. When the order calculation is set to 1, it is calculated for one state after s_0; this is the standard advantage function used in A2C and A3C algorithms. When N is set to 9, this means that there are 9 summed discounted values after s_0. Furthermore, the formula (2) shows that discounting of rewards is only applied when the order N is higher than 1. From the shown formulas, it is clear that every action a_t causes the creation of one update for the parameters θ of the network. Created updates are immediately applied to the network, one by one, and not in batches. In cases of order $N = 1$, this is pure online training, while order $N > 1$, the update for action a_t is slightly delayed.

Our hypothesis is that using extended higher-order calculations will allow us to leverage the benefits of online policy-optimization algorithms, such as A2C, while incorporating the advantages of context-based Deep RL. However, we will also investigate whether the utilization of a more complex loss function, involving multiple calls to the actor-critic neural network for each action, might make the computational process of solving tasks too expensive. As we explore different Deep RL tasks, our experiments connect well with standard Atari simulations. These simulations showcase the adaptability of our modified A2C framework across tasks that share similar state and action spaces. As behavioral patterns persist and scalability remains a consistent theme, our exploration highlights the remarkable versatility of Deep RL. This adaptability extends beyond the scope of a specific task, demonstrating that insights gained in optimizing Atari gameplay can seamlessly apply to various other Deep RL challenges. Whether employed in robotics or recommendation algorithms, the agent interprets tasks as numerical data, making the machine transferable across diverse domains. If an optimization technique is discovered for an Atari game, the same approach can be applied to other domains of Deep RL tasks. This holds true as long as there is consistency in maintaining similar action and space sizes, along with a comparable reward feedback time.

2 Related Work

This paper builds upon several important concepts, and in the related work section, we discuss reference points used to construct the experimental setup. Firstly, we start with neural networks, seen as universal approximators of functions, as proposed by Hornik et al. [1]. We employ artificial neural networks (ANNs) as suggested in the paper by Hinton et al. [2]. Furthermore, we adopt the concept of viewing action tasks as sequences of mathematical operations, as proposed by Bellman in his original paper on dynamic programming [3]. Since our investigation revolves around learning from context, it is essential to reference the works of Watkins [4] and Sutton [5], where the utilization of rewards as a series of discounted values is discussed. In discussing our experiments, it is essential to acknowledge various types of RL based on data usage, including Q-learning by Watkins [6], Hindsight Experience Replay (HER) for improved sample efficiency [7], Deep Deterministic Policy Gradients (DDPG) for continuous action spaces [8], and REINFORCE: Monte Carlo Policy Gradient Methods for learning from sample trajectories [9]. Additionally, we consider deep SARSA by Riedmiller for on-policy learning [10]. As the foundation of our experiments in algorithmic mathematical programming, we rely on the formulas and code suggested in the papers by Mnih et al. [11] (A3C) and Schulman et al. [12] (PPO), where we update and extend the loss function of the critic, as described in the Introduction section.

3 Experiments and Methods

3.1 Simulation Environments

We conducted experiments across four distinct simulations from the Atari catalog [13], training on the following environments (Fig. 1):

- Boxing-ram (version 0)
- Kangaroo-ram (version 0)
- KungFuMaster-ram (version 0)
- Pooyan-ram (version 0)

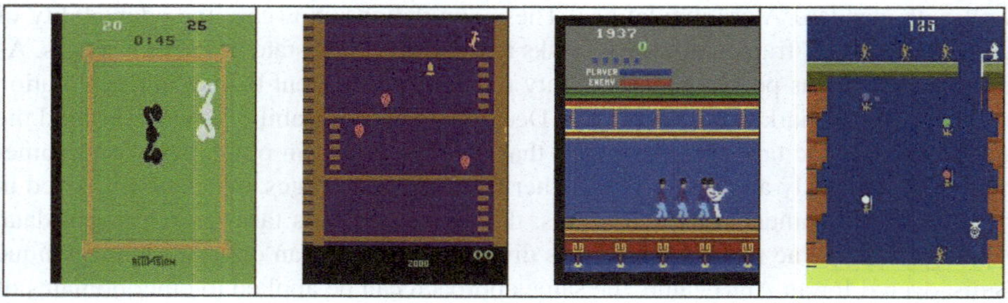

Fig. 1. Block Atari Game Simulations

3.2 Training Procedure

Customized Loss Function. We employed a customized loss function, as detailed in the introduction section, to train our reinforcement learning models. This loss function was later modified to accommodate higher-order calculations.

Training Episodes. Each simulation underwent training for multiple episodes. After completion, we restarted the simulations, initialized the artificial neural networks (ANNs), and retrained them, updating the loss function to facilitate higher-order calculations.

Actions per Episode. In order to mitigate the risk of data pollution from prolonged episodes with minimal findings, a constraint was enforced to ensure that each episode comprised less than 1500 actions. This constraint aims to maintain the relevance and informativeness of the collected data by preventing extended episodes with sparse or empty findings from influencing the overall results.

3.3 Neural Network Architecture

Input Layer. The input layer of our neural network consisted of {128} nodes, corresponding to the size of the RAM encoding representing the frame of the simulation.

Hidden Layers. Our neural network featured hidden layers with sizes 192, 96, 48, 24, and 12, respectively. The rectified linear unit (ReLU) activation function was applied to each hidden layer.

Dropout Layers. To prevent overfitting, dropout layers with a drop rate of 0.16 were strategically placed between each hidden layer.

Output Layer. The output layer comprised two parts:

- Actor Output Layer: Size {18}, utilizing the softmax activation function.
- Critic Output Layer: Size {1}, employing the linear activation function.

Our architecture implemented an actor-critic neural network with a shared body. Activation functions for the output layers included softmax for the actor output layer and linear for the critic output layer.

4 Results

Our results are presented through a total of 24 3D infographics, organized into eight families, each representing a specific simulation environment. The families are further differentiated by the loss function applied. In each 3D infographic, is found visual representations of one of three key attributes:

- Reward per Episode: Showing how rewards accumulate over episodes.
- Actions per Episode: Illustrating the number of actions taken in each episode.
- Computation Time per Episode: Providing insights into the computational efficiency of the models.

These infographics collectively offer a comprehensive overview of the experiments. In the presented infographics, episodes marked as 'Episode 0' denote instances where artificial neural networks are reset, initiating new training sessions. The variable 'N' in the higher order section represents the size of the memory buffer utilized by the agent. Specifically, when 'N = 1,' it signifies the standard order of calculating advantage, considering two states in immediate proximity, as typically applied in A2C. On the other hand, 'N = 9' indicates the calculation of advantage with an order of 9, providing insights into the impact of extended historical information on the learning process (Fig. 2).

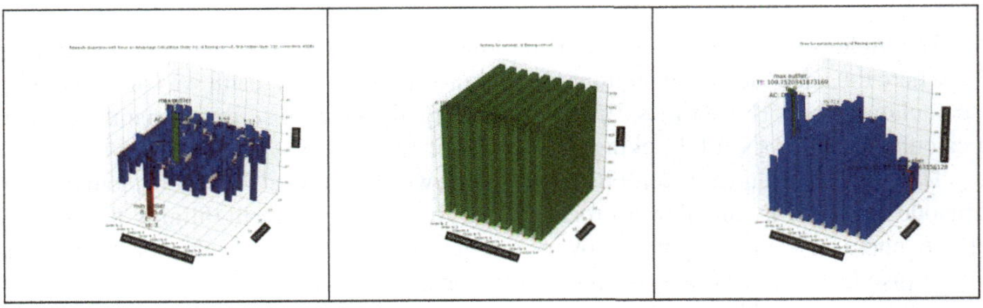

Fig. 2. Infographs for Higher-Order Advantage Calculation, Boxing-ram

The episode that exhibited the best performance is Episode 4, achieving a score of 27. This noteworthy result was attained with a higher order of calculation (N) set to 5. The most time-consuming episode was Episode 9, where the order of calculation (N) was set to 1, resulting in a computational time of 109 s (Fig. 3).

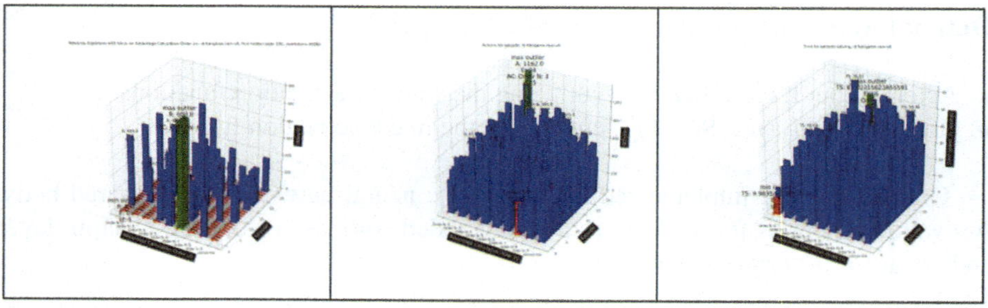

Fig. 3. Infographs for Higher-Order Advantage Calculation, Kangaroo-ram

Episodes 0 to 3 showcased the best performance for the Kangaroo-ram task, achieving high scores of 600. This result was obtained with a calculation order (N) set to 6. The most time-consuming episode was Episode 16, where the order of calculation (N) was again set to 6, with a total computational time of 62 s. Episode 24 recorded the highest number of actions per episode, reaching 1162, and this was observed with a calculation order (N) of 3 (Fig. 4).

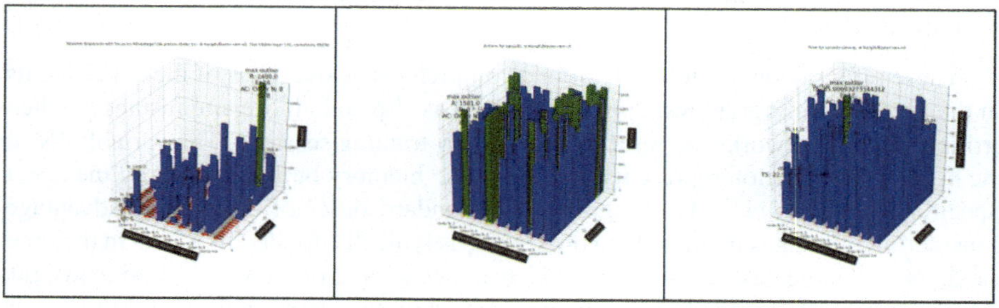

Fig. 4. Infographs for Higher-Order Advantage Calculation, KungFuMaster-ram

The highest performance for the KungFuMaster-ram task was achieved in Episode 24, where the agent achieved an impressive score of 2400. This result was obtained with a calculation order (N) set to 8. Episode 12 emerged as the most time-consuming, with a calculation order (N) of 4, taking 85 s to complete (Fig. 5).

The task Pooyan-ram demonstrated a noteworthy reward of 1780, achieved in Episode 13, with a calculation order (N) set to 4. This result highlights the agent's effectiveness in navigating the Pooyan-ram environment with a higher-order calculation. Episode 11 stands out as the most time-consuming, taking 103 s to complete, with a calculation order (N) of 1 (Fig. 6).

The task Boxing-ram, utilizing the discounted summation of rewards, achieved a notable reward of 9 in Episode 0 with a calculation order (N) of 2. On the contrary, the worst reward is observed in Episode 6, with a value of minus 28, emphasizing the challenges faced by the agent when the calculation order (N) was set to 5. Episode 2, with a calculation order (N) of 1, marked the most time-consuming episode, requiring 83 s to complete (Fig. 7).

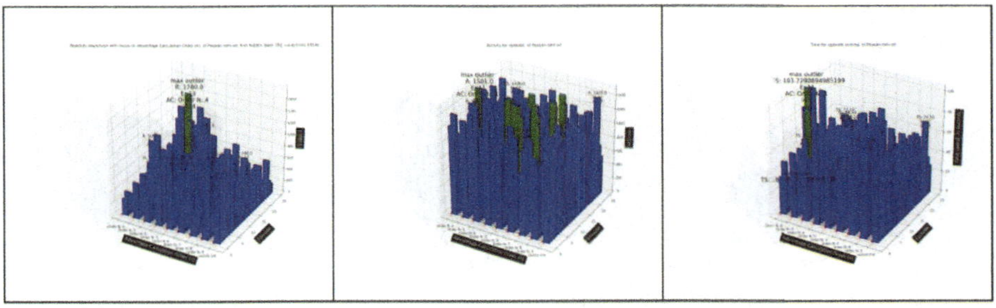

Fig. 5. Infographs for Higher-Order Advantage Calculation, Pooyan-ram

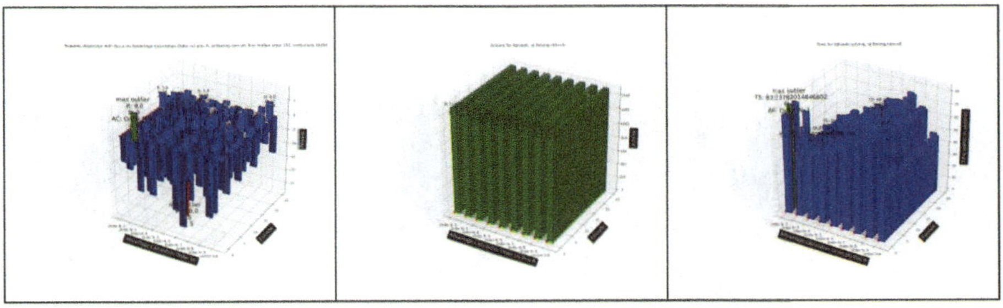

Fig. 6. Infographs for Higher-Order Advantage + Discounted Rewards, Boxing-ram

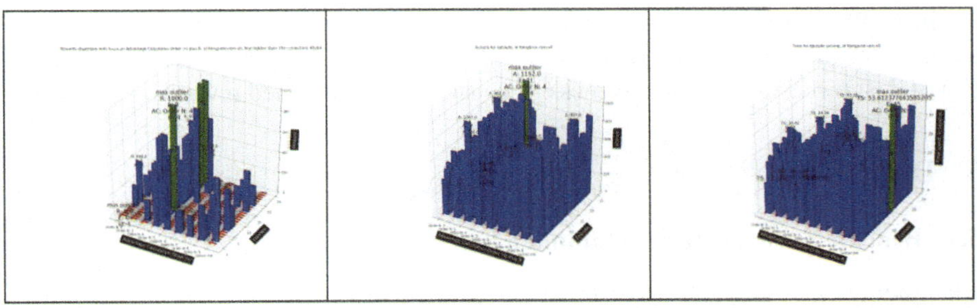

Fig. 7. Infographs for Higher-Order Advantage + Discounted Rewards, Kangaroo-ram

Kangaroo-ram task, utilizing the discounted summation of rewards, achieved a noteworthy reward of 1000 in multiple episodes, underscoring the effectiveness of this reward calculation strategy with calculation order (N) of 4. Episode 15 the most time-consuming episode at 53 s and order (N) of 9 (Fig. 8).

In the KungFuMaster-ram task, employing the discounted rewards method, the agent demonstrated exceptional performance with a substantial reward of 3600 in episode 2. Episode 12 showcased consistent excellence with an order (N) of 8, highlighting the effectiveness of this configuration. The computational time for this episode was observed to be 83 s, emphasizing the balance between computational efficiency and task performance (Fig. 9).

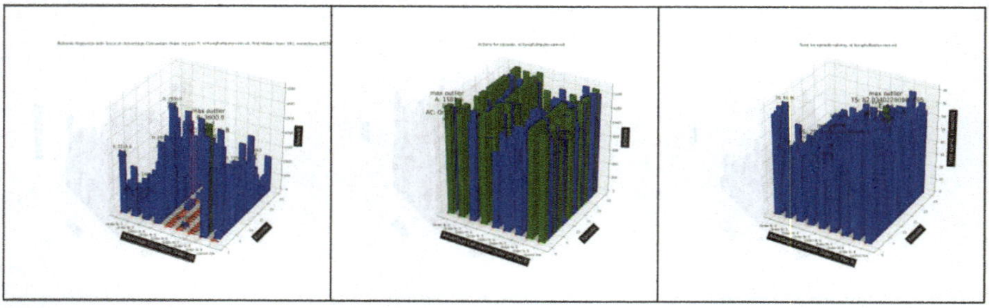

Fig. 8. Infographs for Higher-Order Advantage + Discounted Rewards, KungFuMaster-ram

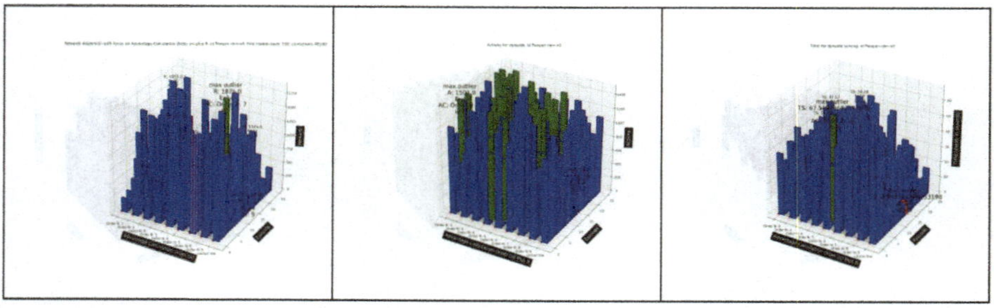

Fig. 9. Infographs for Higher-Order Advantage + Discounted Rewards, Pooyan-ram

In the Pooyan-ram task utilizing the discounted rewards approach, the agent achieved a notable reward of 1875 in episode 14. This performance was observed under an order (N) of 7, showcasing the effectiveness of considering discounted rewards.

5 Discussion

5.1 Benefits Across Different Simulations

Our findings reveal a consistent trend across diverse simulations, indicating that the application of higher-order calculations offers tangible benefits. Despite the inherent variability in simulation environments, the advantages derived from employing advanced calculation methods remain a common thread, emphasizing the robustness of our approach.

5.2 Balancing Complexity and Efficiency

While the implementation of higher-order calculations introduces increased complexity to both the agents and the programming process, our study suggests that the conventional belief in the impracticality of multiple recalls to ANNs during the training process may be unfounded. Although calculating advantages for multiple states does incur a slight increase in computation time, we emphasize that the most time-consuming aspect of the training process is the re-training of ANNs. Significantly, if training updates share similar loss values, the re-training time is further reduced, providing a practical compromise between complexity and efficiency.

5.3 Horizon of Rewards as a Crucial Factor

Perhaps the most crucial insight gleaned from our experiments is the pivotal role played by the horizon of rewards. The benefits of higher-order calculations are intricately tied to the nature of the task. Tasks demanding rapid responses with frequent positive and negative rewards after each action significantly benefit from the depth offered by higher-order calculations. Conversely, tasks with sparse rewards, yet substantial when achieved, demonstrate optimal performance when higher-order calculations account for both advantage and reward. Finally, tasks exclusively solvable through a specific trajectory or strategy exhibit enhanced efficacy with the utilization of higher-order calculations.

6 Conclusion

In summary, our study highlights the following insights in deep reinforcement learning. Contrary to expectations, higher-order calculations don't significantly increase task-solving time. This efficiency stems from the understanding that the most time-consuming aspect is the re-training of artificial neural networks, making sophisticated calculation methods practical. Tasks requiring specific trajectory of strategies or those with frequent updates of positive and negative rewards benefit significantly from a high-order advantage loss function. Tasks featuring sparse rewards but with substantial values upon receipt benefit most from a combined high-order calculation of advantage and rewards. These findings pinpoint the strategic application of high-order calculations tailored to task characteristics in deep reinforcement learning.

Acknowledgment. This publication is developed with the support of Project BG05M2OP001-1.001-0004 UNITe, funded by the Operational Programme "Science and Education for Smart Growth", co-funded by the European Union trough the European Structural and Investment Funds.

Appendix

In the following appendix, high-definition versions of all infographics presented in the paper are provided in the same order as they appear in the main text https://github.com/MartinKaloev/CSOC-2024.

References

1. Hornik, K.: Approximation capabilities of multilayer feedforward networks. Neural Netw. **4**(2), 251–257 (1991)
2. Rumelhart, D.E., Hinton, G.E., Williams, R.J.: Learning representations by back-propagating errors. Nature **323**(6088), 533–536 (1986)

3. Bellman, R.: The theory of dynamic programming. Bull. Am. Math. Soc. **60**(6), 503–515 (1954)
4. Watkins, Christopher John Cornish Hellaby. Learning from delayed rewards (1989)
5. Sutton, R.S., Andrew, G.B.: Reinforcement Learning: An Introduction. MIT press (2018)
6. Watkins, C.J.C.H., Dayan, P.: Q-learning. Machine Learn. **8**(3/4), 279–292 (1992)
7. Andrychowicz, M., et al.: Hindsight experience replay. In: Advances in Neural Information Processing Systems 30 (2017)
8. Lillicrap, T.P., et al.: Continuous control with deep reinforcement learning. arXiv preprint arXiv: 1509.02971 (2015)
9. Zhang, J., et al.: Variational policy gradient method for reinforcement learning with general utilities. Adv. Neural Inform. Process. Syst. **33**, 4572–4583 (2020)
10. Mnih, V., et al.: Playing atari with deep reinforcement learning. arXiv preprint arXiv: 1312.5602 (2013)
11. Mnih, V., et al.: Asynchronous methods for deep reinforcement learning. In: International Conference on Machine Learning, pp. 1928–1937. PMLR (2016)
12. Schulman, J., Wolski, F., Dhariwal, P., Radford, A., Klimov, O.: Proximal policy optimization algorithms. arXiv preprint arXiv:1707.06347 (2017)
13. Atari catalog: https://www.gymlibrary.dev/environments/atari/index.html

Sentiment Analysis Based on Twitter Comments Using Artificial Intelligence Techniques to Predict Peruvian Presidential Election Results

Diego Cedano[1], Mesias Picon[1], and Wilfredo Ticona[1,2](✉)

[1] Universidad ESAN, Lima, Peru
{13200107,20101034}@ue.edu.pe, wmamani@esan.edu.pe
[2] Universidad Tecnológica del Perú, Lima, Peru

Abstract. Currently, the use of social networks as the main means of communication and expression has increased. In Peru, the social network Twitter is preferred by political leaders and is key for political campaigns. The present research aims to implement a sentiment analysis model based on Twitter comments using artificial intelligence techniques to predict the results in the Peruvian presidential elections. The applied methodology was based on 5 phases: dataset construction, preprocessing, feature extractions, model implementations and Evaluation. Word2Vec and Tf-idf techniques were used for feature extraction. Finally, different machine learning models were developed, such as logistic regression (LR), Support Vector Machine (SVM), random forest (RF) and logistic regression (RL) and MLP classifier. The best model results from the combination of the SVM algorithm with Word2vec vectorization has superior performance in the metrics 0.89 Accuracy and 0.88 Precision. However, Recall is lower compared to the Tf-idf vectorization (0.87 Recall) and both vectorizations have the same result in the 0.87 F1-Score metric. Finally, the results of the second-round show PPK winning by a minimal margin over Keiko, 16.1% and 15.3% respectively. However, the percentage of undecided voters is about 34% on average.

Keywords: Sentiment analysis · Twitter · Vectorization · Comment classification · Presidential election · Twitter mining · Word2Vec

1 Introduction

One of the most important expressions of democratic alternation is the possibility for Peruvians to go to the polls to express their decision to support the candidate of their choice. Citizens seek to identify and elect the candidate who, in their opinion, best represents their feelings, aspirations and contribution to the achievement of national objectives, and who they believe will have an impact on the possibility of raising or improving the quality of life of the citizen.

According to Ramírez [1], in his article published in the methodological journal of social sciences, where he recounts the statistical origins of the opinion poll, he mentions that polls have been used since 1824, mainly in the United States, in the so-called "straw

R. Silhavy and P. Silhavy (Eds.): CSOC 2024, LNNS 1120, pp. 191–201, 2024.
https://doi.org/10.1007/978-3-031-70518-2_17

polls". This practice was common in major American newspapers to determine the voting intentions of American presidents. In this way, opinion polls managed to position themselves as a key reference for electoral processes. The latter became an X-ray of the political moment, measuring the behaviour of candidates and voters, which at the end of the elections will confirm whether they were right or wrong.

In the digital era, the use of social networks as the main means of communication and expression has increased. In the case of Peru, the reality is not alien, as studies point to Facebook as the social network preferred by the population and Twitter preferred by Peruvian politicians, according to a note published by the newspaper Gestión (2014) "Twitter: The social network preferred by political leaders" [2]. In fact, this last network was a key media for different political campaigns, where different debates were discussed, which generated a great amount of information. For example, during presidential elections, users increase the use of the platform to show their opinions and support for the candidate of their choice. In Peru, democracy is expressed through the vote, but electoral polls have often failed to predict results. According to Ramirez [1] indicates that notable cases occurred such as the Brexit and the U.S. elections with the "Straw votes" or the case of Argentina that according to Clarin [3] Alberto Fernandez won against all odds. In addition, to improve accuracy, standardized methodologies have been developed, but they are more costly and time-consuming. However, in the digital era, public opinions are massively reflected on platforms such as Twitter, which has boosted sentiment analysis as a key tool to understand political trends. This marks a shift in the way public opinion and elections are addressed through digital comments and reactions.

2 Related Work

In the work of Sazzed [4], hybrid sentiment analysis techniques were investigated to determine the polarity of comments on Twitter without relying on prior information. The study consists of two stages: the first, "Dataset Preparation," uses a self-supervised hybrid methodology that divides the dataset into five subgroups based on framework scores, performing processes such as opinion extraction and sentence- and document-level opinion calculation. In the second stage, "Classification", scores with labels "very high" and "high" are used as training data for a classifier, and Logistic Regression "LR" and Support Vector Machine "SVM" are used to classify data with labels "very low" and "zero". The SSentiA framework achieved over 87% accuracy in scoring the polarity of different datasets, with different percentages of labeled data needed for each set, such as around 32% for Clothing and between 41 and 59% for Amazon, TripAdvisor and IMBD, respectively.

Similary, Rakhmanov [5], two data sets were created and a preprocessing process with five subprocesses was carried out, including tokenization, conversion to lowercase and primitive form. CountVectorizer and TF-IDF techniques were used for vectorization, and models such as Random Forest and neural networks were evaluated. The results indicated that both Random Forest and neural networks showed similar classification performance. In addition, other studies were mentioned that focused on identifying socio-economic factors related to public sentiment on Twitter and demographic data from the

U.S. Census. Twitter data retrieval was performed with the help of the rSentiment library, followed by a feature cleaning and extraction process. Classification results revealed an accuracy of 56.18%, with a specificity of 62.32% and a sensitivity of 49.59%.

In the work carried out by Chintalapudi [6], their study was based on the analysis of 3000 medical documents provided by the International Association of Marine Electronics Companies (CIRM) during the years 2018–2020. The methodological process comprised dataset retrieval, text preprocessing, DTM matrix development, sentiment analysis and symptom visualization. One of the highlights was the application of Stemming to reduce word duplicates, as well as the use of TF-IDF to express symptoms in diagnosis. The patient feedback classification process was carried out using two techniques: Lexicon Sentiment Classifications (LSA) and ML-base sentiment classifications (Naive Bayes), choosing the second model due to previous acceptable results in the reviewed literature. In the fifth experiment, using the Naive Bayes algorithm, accuracy rates of 83.6% in negative feedback, 85% in neutral feedback and 81.7% in positive feedback were achieved, with an average model accuracy of 82%.

Finally, in the work of Atoum's [7], a software usage quality analysis framework based on the review of opinions and semantic similarities is presented. This approach does not focus on the polarity of comments, but rather on assigning them to categorize according to the standards of the software developer, addressing aspects such as efficiency, effectiveness, risk freedom, satisfaction, and context coverage. In addition, these involve a benchmark dataset with 3,013 labeled sentences and various processes for rating the software based on the categories. The experiment was conducted on web browser comments, and it is observed that traditional models perform poorly in topic prediction, while the logarithmic model (TSM) proposed by the author achieves an F1 score of 0.732. In conclusion, QinU Scoring is used to predict the in-use quality of software checks and assign categories. The resulting matrix shows browser preference from comments extracted from blogs and social networks.

3 Methodology

The proposed methodology was based on [8]. It consists of five phases: data construction, preprocessing, feature extraction, model implementation and evaluation. Then, each stage has general activities that are of paramount importance for the development of the research. The methodology used is shown graphically below (Fig. 1).

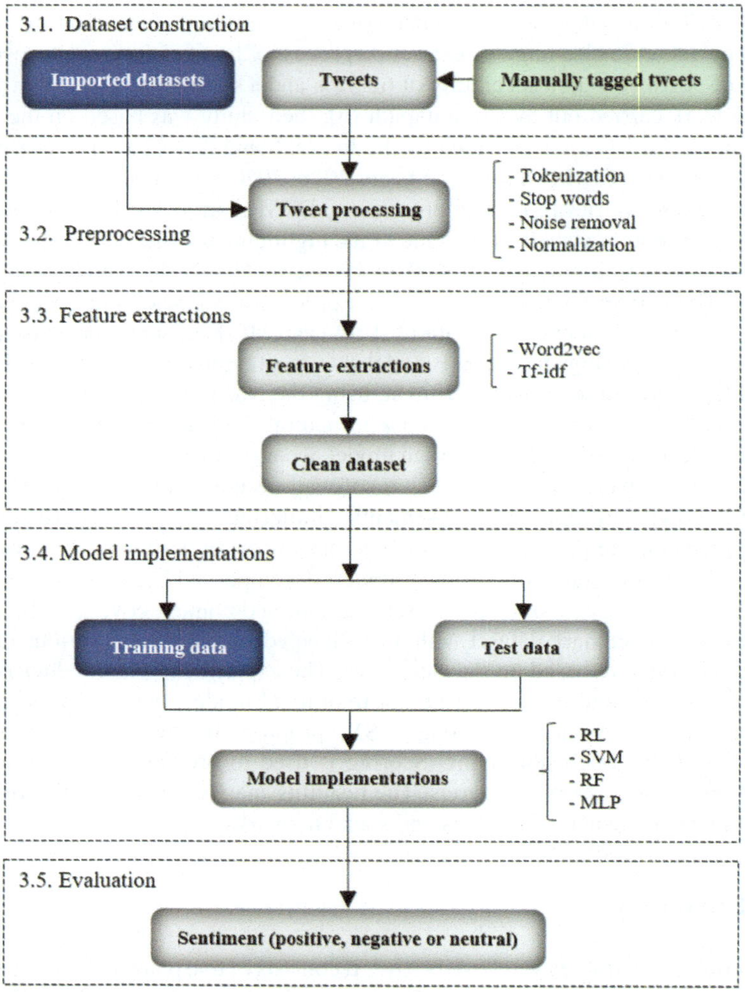

Fig. 1. Proposed methodology

3.1 Data Set Construction

This first stage is related to the detailed description to obtain the dataset. It is crucial to have a strategy because you need quality data that carries relevant information. Next, we develop the construction of the source code to obtain the tweets by scraping. For this we will use the Selenium libraries as the main library, Request, BeautifulSoup and Cycle. Table 1 shows the activity and tasks for obtaining the tweets.

Table 2 shows the tweets obtained after applying the scraping algorithms. The input configuration was applied, which is given by the starting date to extract the information, the language of the tweet, the minimum number of accounts mentioned, the number of tweets to extract, the activation of the Scroll Down and the number of days to scrape.

Table 1. Obtaining Tweets

Activity	Description	Tasks
Develop search strategy	Propose a search strategy given that the information provided by Twitter is extensive	Develop a Twitter search strategy
	Define an execution schedule, as the sweep of tweets will be carried out on a daily basis. The aim is to obtain as much data as possible on the topic	Develop an implementation schedule for each defined search
Create Twitter scraping algorithm	Develop the code to extract Twitter variables: The structure of the Twitter Search website is dynamic, so you should look closely to locate keywords and extract as many variables for each tweet as possible	Develop function to extract variables related to each tweet, such as username, text, images, date, number of retweets, number of likes, number of emoticons and URLs
	Develop code for the Twitter search: When using twitter search (https://twitter.com/search-advanced), the link has characteristics which can be parameterised to get an exact which can be parameterised to obtain an exact search	Implement a function to elaborate the query tweet, which will allow to obtain the link of the given search
	Consolidate and parameterise code: The code to be used must be parametised after the scraping and the generation of the query or tweet query	Consolidate the codes to finally parameterise the search and obtain the packaged code

Table 2. Accounts and number of tweets obtained

Accounts	Number of Tweets	%
PFuerzaPopular	10,561	5%
yonhy_lescano	8,898	4%
HDeSotoPeru	9,351	4%
Vero_Mendoza_F	7,541	3%

(*continued*)

Table 2. *(continued)*

Accounts	Number of Tweets	%
KeikoFujimori	14,571	7%
julioguzmanperu	11,453	5%
Ollanta_HumalaT	9,013	4%
BeingoleaA	8,183	4%
DanielUrresti1	9,423	4%
rlopezaliaga1	11,617	5%
George_Forsyth	7,651	4%
JuntosporPeru	8,541	4%
PartidoMorado	9,816	5%
pnacionalistap	7,378	3%
podemos_peru	8,461	4%
Renovacion_Peru	7,233	3%
Vladimir_Cerron	26,200	12%
PedroCastilloTe	8,482	4%
MartinVizcarraC	7621	3%
sigridbazan	3200	1%
ensustrece	2087	1%
martinhidalgo	7231	3%
ppkamigo	13,551	6%
Total	218,063	100%

3.2 Preprocessing

The main objective of this stage is to obtain the dataset free of noise or disturbance commonly found in social networks. Kambahampati [8] mentions that noise makes it difficult to obtain correct characteristic vectors and therefore impacts on the result. This phase is composed of two activities. The first one will allow to have the information ordered in relation to the research objective, and the second one allows to deliver clean datasets for the feature extraction stage (Table 3).

Table 4 shows the different configurations used for the initial cleaning. As a result, you get clean tweets. This cleaning is applied to the whole dataset. In addition, the tokenisation of the tweet is evident.

Figure 2 shows an example of the result after processing, focusing on cleaning up the tweet to keep the words that add the most value to the sentence.

Table 3. Preprocessing

Activity	Description	Tasks
Consolidate files by folders	Consolidate csv files and distribute them by political figures of interest to be analysed	Develop code to generate folders and distribute the scraped files by name
Cleaning up tweets	Construction and configuration of the code to perform clean-up following the steps:	

Table 4. Description of the noise cleaning function

Cleaning code	Objective
tweet = re.sub(r'\&\w*;', ", tweet)	Remove html
tweet = re.sub(r'\$\w*', ", tweet)	Remove stickers
tweet = re.sub(r'https?:\/\/.*\/\w*', ", tweet)	Delete hyperlinks
tweet = re.sub(r'[^\w\s]', ", tweet)	Delete Score
tweet = tweet.lstrip(' ')	Eliminate spaces
tweet = tweet.lower()	Convert to lowercase
tweet = ''.join(c for c in tweet if c < = '\uFFFF')	Delete unicode words

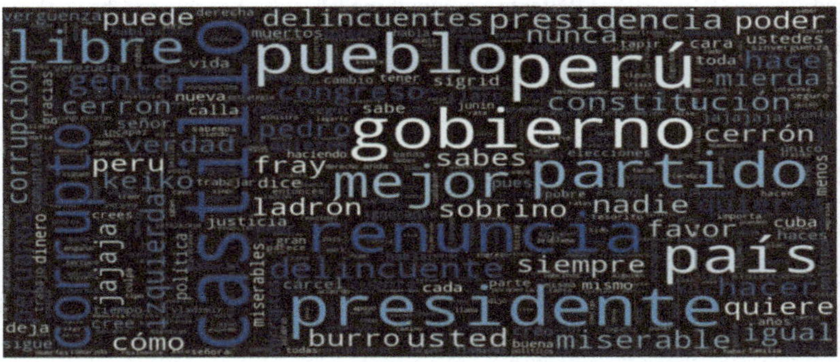

Fig. 2. Word cloud

3.3 Feature Extractions

The main purpose of this process is to display the clean sentence in a numerical representation for each comment. For this, the techniques TF-IDF and Word2vec will be used (Table 5).

Table 5. Bag of words parameters

Activity	Description	Tasks
Implement feature extraction code	Implementation of code for each of the techniques and running experiments with the characteristic vectors obtained	Implement and run code for technique Tf-idf
		Implement and run code for technique Word2vec

3.4 Model Implementations

For this stage, the RL, SVM, RF and MLP models will be used for each vectorization technique (Word2Vect and Tf-idef) (Table 6).

Table 6. Model implementations

Activity	Description	Tasks
Implement the codes to run the RL, SVM, RF and MLP models	Execution and testing using the characteristic vectors generated in the previous step for the initial models	Implement the code with the Scikit-Learn library to generate the RL, SVM, RF and MLP models, using the characteristic vector
Record results	The results are recorded to evaluate the efficiency of the modelling set-up in the next step	Store the results obtained by each model with the optimal configuration
Feedback	Better execution of models with optimal parameters	Execute feedback of results

3.5 Evaluation

To measure the performance of the models, the following metrics were used: accuracy, precision, recall and F1-Score (Table 7).

Accuracy The accuracy of a model is defined as the percentage of correct predictions with respect to the total number of predictions made, providing an overall measure of its accuracy [9].

$$Accuracy = \frac{TP + TN}{TP + FP + FN + TN} \tag{1}$$

Precision: Precision refers to the ratio of correctly identified positives to the total number of positives predicted by the classifier [9].

$$Precision = \frac{TP}{TP + FP} \tag{2}$$

Table 7. Evaluation

Activity	Description	Tasks
Run model with validation data	Run the saved model in the modelling phase and validate the results with new data (without labels)	Insertar data nueva sin etiquetar y realizar las validacicnes
Select the model with the best performance and configuration	Determine the model based on the results and comment on the results	Choose the final model that maximises the F1-score and this in relation to the information objective

Recall: Recovery is defined as the ratio of correctly identified positive results to the total number of relevant samples (i.e., all those that should have been identified as positive) [10].

$$Recall = \frac{TP}{TP + FN} \tag{3}$$

F1 score: Allow the balance between precision and recall [10].

$$F1score = \frac{Precision * Recall}{Precision + Recall} \tag{4}$$

Table 8 shows the optimal results of the techniques for each vectorisation model.

Table 8. Optimal model results

Technique	Model	Accuracy	Precision	Recall	F1-Score
RL	Tf-idf	0.86	0.86	0.83	0.85
	Word2vec	0.84	0.82	0.86	0.84
SVM	Tf-idf	**0.88**	**0.87**	**0.87**	**0.87**
	Word2vec	**0.89**	**0.88**	**0.86**	**0.87**
RF	Tf-idf	0.85	0.84	0.82	0.83
	Word2vec	0.84	0.85	0.84	0.85
MLP	Tf-idf	0.86	0.86	0.86	0.86
	Word2vec	0.87	0.87	0.84	0.85

- The RL model with the Tf-idf vectorisation has superior performance in the metrics: Accuracy, Precision and F1-Score. However, the Recall is lower compared to the Word2vec vectorisation.

- The SVM model with Word2vec vectorisation has superior performance in the metrics: Accuracy and Precision. However, Recall is lower compared to the Tf-idf vectorisation and both vectorisations have the same result in the F1-Score metric.
- The RF model with Word2vec vectorisation has superior performance in the metrics: Precision, Recall and F1-Score. However, Accuracy is lower compared to the Tf-idf vectorisation.
- The MLP model with Tf-ide vectorisation has superior performance in the metrics: Recall and F1-Score. However, with word2vect vectorisation it has superior performance in the metrics: Accuracy and Precision

4 Results

Below is a summary table of the models with the best combination of techniques and the results of the metrics. Table 9 shows the result of the ranking according to the candidates. In this case, the value 1 was taken to represent those in favour (1 = "POS"). It is evident that PPK has 56 votes in favour, while Keiko has 53 votes in favour.

Table 9. Ranking of results

Candidate	Pos	Neu	Neg	Total
Keico	53	65	64	182
PPK	56	55	54	165
Total	109	120	118	347

Table 10 shows PPK winning by a minimal margin over Keiko, however, the percentage of undecided voters is approximately 34% on average.

Table 10. Summary of results in percentage (%)

Candidate	Pos	Other (Neu + Neg)
Keico	15.3%	37.2%
PPK	16.1%	31.4%

5 Conclusions

A dataset of tweets related to politics was built. To do this, scraping techniques and Twitter APIs were used in order to collect as many tweets from political figures in Peru as possible.

Likewise, a good preprocessing was carried out, which was reflected in the results of the model. In this phase, a preprocessing function was developed to remove emoticons, user references, reference links, images, extraneous characters, etc. Using the Regex library, the clean function was designed which returns the sentence free of the above and all in lower case. Artificial intelligence techniques were used to convert sentences into vectors to be processed by machine learning models. Therefore, we worked with 2 types of word vectorization: Word2Vec and TD-IDF which allowed us to deliver the dataset with its characteristic vectors.

Several runs were carried out to obtain the optimal model with the appropriate parameters to determine the winner of the 2016 run-off election. The RL, SVM, RF and MLP Classifier techniques were used. The remaining 4 models (RL, SVM, RF and MLP Classifier) were then put into competition with their adjustment parameters and for each vectorization technique, obtaining the optimal models as a result.

Finally, the model obtained in the previous stages was applied to the tweets extracted in the date range of the second-round elections for the candidates PPK and Keiko. The result was a slight advantage for PPK over Keiko, 16.1% over 15.3%.

References

1. Ramírez, B.: Revista de Metodología de Ciencias Sociales. EMPIRIA **9**, 89–113 (2005)
2. Anonimo: Twitter: La red social preferida por los líderes políticos. Gestión. https://m.gestion.pe/politica/twitter-red-social-preferida-lideres-politicos-2130522 (2015)
3. Helfgot, M.: "Elecciones 2019: Alberto Fernández le ganó a Mauricio Macri en primera vuelta y el kirchnerismo vuelve al poder. Clarín (2019). https://www.clarin.com/politica/elecciones-2019-alberto-fernandez-gano-macri-primera-vuelta%20kirchnerismo-vuelve-poder_0_Mvhv92GS.html
4. Sazzed, S., Jayarathna, S.: SSentiA: a self-supervised sentiment analyzer for classification from unlabeled data. Mach. Learn. Appl. **4**, 100026 (2021). https://doi org/10.1016/j.mlwa.2021.100026
5. Rakhmanov, O.: A comparative study on vectorization and classification techniques in sentiment analysis to classify student-lecturer comments. Procedia Comput. Sci. **178**, 194–204 (2020). https://doi.org/10.1016/j.procs.2020.11.021
6. Chintalapudi, N., Battineni, G., di Canio, M., Sagaro, G.G., Amenta, F.: Text mining with sentiment analysis on seafarers' medical documents. Int. J. Inform. Manage. Data Insights **1**(1), 100005 (2021). https://doi.org/10.1016/j.jjimei.2020.100005
7. Atoum, I.: A novel framework for measuring software quality-in-use based on semantic similarity and sentiment analysis of software reviews. J. King Saud Univ. – Comput. Inform. Sci. **32**(1), 113–125 (2020). https://doi.org/10.1016/j.jksuci.2018.04.012
8. Kameswari, K.K., Raghaveni, J., Shankar, R.S., Someswara Rao, C.: Predicting Election Results using NLTK. Int. J. Innov. Technol. Explor. Eng. **9**, 4519–4529 (2019)
9. Dahouda, M.K., Joe, I.: A deep-learned embedding technique for categorical feature encoding. IEEE Access **9**, 114381–114391 (2021). https://doi.org/10.1109/ACCESS.2021.3104357
10. Talukder, M.A., et al.: A dependable hybrid machine learning model for network intrusion detection. J. Inform. Secur. Appl. **72**, 103405 (2023). https://doi.org/10.1016/j.jisa.2022.103405

Cryptocurrency Investments Forecasting Model Using Deep Learning Algorithms

Leonardo Enco[1], Alexander Mederos[1], Alejandro Paipay[1], Daniel Pizarro[1], Hernan Marecos[2], and Wilfredo Ticona[1,2](✉)

[1] Universidad Tecnológica del Perú, Lima, Peru
{U17302235,U18214077,U18214447,1510747}@utp.edu.pe,
wmamani@esan.edu.pe

[2] Universidad ESAN, Lima, Peru
21100534@ue.edu.pe

Abstract. The explosive growth of cryptocurrencies has attracted a considerable number of individuals willing to invest, leading to an exponential increase in their market value and trading volume. However, the cryptocurrency market is highly volatile and presents complex datasets where prediction is extremely challenging. Due to this, this article evaluates the implementation of a Prediction System for Cryptocurrency Investments for users using Deep Learning algorithms, aiming to predict the prices of six cryptocurrencies (BTC, ETH, BNB, LTC, XLM, and DOGE). For this, the use of a genetically tuned algorithm with Deep Learning and techniques based on enhanced trees are proposed to compare them. The best result obtained is that the Gated recurrent unit (GRU) model, has an average MAPE of 4%, followed by Convolutional neural networks (CNN) with an average MAPE of 7% and Direct feedforward neural networks (DFNN) of 12%.

Keywords: Artificial intelligence · Deep learning · Optimization · Forecasting · Cryptocurrencies · Investments

1 Introduction

Currently, cryptocurrencies are a digital asset that uses cryptographic encryption to guarantee its legitimacy and control the creation of new units based on the trust of the technological infrastructure, they have acquired global relevance and have attracted numerous users due to their decentralization, immutability, and security.

In recent years, the cryptocurrency market has experienced an increase in the frequency of transactions with peaks as high as a bitcoin worth 68,789.63 USD. For example, in social networks like Twitter, the volume of business and the number of participants only increases [1], reaching 320 million people with cryptocurrencies. However, the cryptocurrency market is highly volatile and complex, with heterogeneous data sets and complex interactions, making forecasts [3] or using technical analysis and deep learning techniques [4] difficult, as they may struggle to achieve optimal results.

R. Silhavy and P. Silhavy (Eds.): CSOC 2024, LNNS 1120, pp. 202–217, 2024.
https://doi.org/10.1007/978-3-031-70518-2_18

Having defined the problem, the aim is to develop a tool for prediction models to be compared. A detailed comparison of techniques based on Deep Learning (DL) and boosted trees [2] is performed using cryptocurrency datasets collected from several diverse sources. The same optimal model configuration is used on different cryptocurrencies to investigate their robustness and resilience through test and training datasets. In addition, DL techniques are selected due to their ability to discover complex structures in high-dimensional data and are compared with boosted tree-based techniques that are scalable [5] and robust to model complex data.

2 Related Work

Several quantitative research studies have been conducted in the field of financial market modeling [1] proposes a framework for predicting the cryptocurrency market using sentiment analysis of expert user opinions on Twitter. The framework consists of extracting and preprocessing the comments, performing sentiment analysis using the Vader tool, and calculating the importance coefficient and final score of each tweet. The paper also analyzes the impact of sentiment analysis methods on cryptocurrency trading. It compares the proposed model with existing methods and highlights its effectiveness in addressing challenges such as cleaning unrelated texts. The results show that existing methods tend to have a dominant neutral polarity, indicating a reluctance to trade cryptocurrencies. The proposed model aims to overcome this limitation. As a result, the paper presents a comprehensive framework combining sentiment analysis and expert opinions to predict the cryptocurrency market, which provides insight into the importance of sentiment analysis in cryptocurrency trading and offers suggestions for future work. When analyzing sentiments related to Bitcoin, the study found that approximately 57.6% of the sentiments were classified as positive. This suggests a relatively optimistic outlook towards Bitcoin in the Twitter user opinions studied.

This study has been conducted to evaluate the benefit and performance of a Deep Learning-tuned genetic algorithm and a tree-based technique [2] to predict the closing prices of various cryptocurrencies. The study consists of using convolutional neural networks, deep and recurrent units of closed models in the use of Deep Learning to evaluate the performance of tree-based models driven on six cryptocurrency data sets from multiple sources using performance metrics. In this way, it seeks to reduce the risks and challenges in the crypto market seeking a better understanding to benefit practitioners. The study collected a dataset using Datasets to train, validate, and test prediction models. The results were statistically evaluated, with the result that the models based on boosted trees exhibit unreliable predictions for cryptocurrencies when information is missing from their training set. However, there is an improvement in the predictions of most of the EVS models since the value of their predictions oscillates in a positive and stable way. A high EVS value would indicate more significant similarities between the measured and predicted values. Therefore, the average predictions of the CNN and GRU models for both scenarios in most cryptocurrencies are considered ideal since they have an acceptable EVS rate. However, the average predictions of the models based on boosted trees in the case of both scenarios can be considered virtuous. The MAPE value in the predictions return MAPE values as. For example, CNN model has the lowest average MAPE of 9%, followed by GRU with average MAPE of 11%, GBM (average MAPE of 17%), DFNN (average MAPE of 18%), XGB (average MAPE of 18%). %), and ADAB (average MAPE of 18%).

In similar studies, [3] presents a comparative analysis of three deep learning models, LSTM, GRU, and Bi-LSTM, for predicting cryptocurrency prices. The study addresses the challenges posed by the nonlinearity of the cryptocurrency market and aims to provide highly accurate price predictions. The distinct nature of time-series data and the difficulties in generating appropriate price predictions are discussed. Publicly available data from finance.yahoo.com is used, and the authors declare no conflict of interest or external funding for the research. The study emphasizes the importance of accurate predictions for investors and researchers. The Bi-LSTM model is found to provide the most accurate predictions for all three currencies, followed by the GRU model. Further investigation is suggested to enhance prediction accuracy. The implications of accurate price predictions are discussed, making the study valuable for finance and economics. Overall, the article provides a comprehensive overview of applying deep learning models to predict cryptocurrency prices, highlighting the potential implications for investors and researchers, and making a valuable contribution to the field of finance and economics. The results on the three major cryptocurrencies using both Root Mean Squared Error (RMSE) and the Mean Absolute Percentage Error (MAPE) show that the Bi-LSTM performed better in prediction than LSTM and GRU. Therefore, it can be considered the best algorithm. Bi-LSTM presented the most accurate prediction compared to GRU and LSTM, with MAPE values of 0.036, 0.041, and 0.124 for BTC, LTC, and ETH, respectively.

This study presents the circulation of cryptocurrencies as digital currencies in the computer network [4]. With the advancement of technologies this topic became popular, financial users have focused their attention on trading cryptocurrency assets, having Bitcoin as popular. These cryptocurrencies are traded like stocks, but due to its great transaction facilities, accessibility, intermediaries, costs, it offers users great opportunities to earn money involving higher risk. Therefore, the study proposes three Modified round-trip – TraderNet strategies for realizing small price fluctuations by placing modified market orders to allow the user unlimited time control of trades, managing to trade in risky states with small profit potential, overestimating unprofitable price volatility. The Smurf strategy includes lower commissions, reducing the probability of overestimating LNP profitability. And finally, the TraderNet-CR strategy integrated using the PPO algorithm and the Round-Trip strategy providing slightly higher commissions in the market. As a result, we can analyze how the modified Round-Trip strategy improves the overall performance of the agent and using the TraderNet yields PNL returns up to ten times higher in the Bitcoin market and twice as much in the other four markets (Ethereum, Cardano, Litecoin, XRP), with a cumulative PNL of 1,042.5%, cumulative returns (CR) of 10.43, Sharpe ratio (SHR) of 1.17, Sortino Ratio (SOR) of 1.68 and Maximum Drawdown (MDD) of -0.22. Allowing the broker to execute trades more efficiently and optimize his portfolio by maximizing profits and minimizing risk.

Another study [5] explores the integration of deep learning and technical analysis in the cryptocurrency market for financial forecasting. It emphasizes the significance of technical analysis as a widely adopted practice in modern financial decision-making.

The study discusses the application of deep learning techniques and the utilization of heuristics techniques in technical analysis. The article presents an extended dataset that includes popular technical analysis indicators such as simple moving average (SMA), relative strength index (RSI), moving average convergence/divergence oscillator (MACD), William's percent range (Williams %R), stochastic oscillator, and money flow index (MFI). The results compare the effectiveness of technical indicators to the model using both the "raw" dataset and the "extended" dataset. The study shows that the inclusion of these technical indicators only led to a marginal increase in precision for the test set. However, they did improve the recall of signals by more than 2% in both datasets. Overall, the article provides valuable insights into the integration of deep learning and technical analysis in the cryptocurrency market for financial forecasting. The best performing model is the GRU long-only model which has been trained on the extended candlestick data set with an accuracy of 58.36% and an average return of 0.04%. Was trained on the extended candlestick data set with an accuracy of 58.36% and an average return of 0.04%.

As demonstrated, there are various related works seeking to address the challenge of predicting cryptocurrency investments, for which it is necessary to harness cutting-edge technologies like Deep Learning algorithms and employ methodologies such as Machine Learning and Natural Language Processing. These advanced techniques play a pivotal role in automating the recognition of valuable information from extensive datasets, enabling the analysis of patterns, and identifying key characteristics that influence decision-making processes for users. By developing a robust classifier model applied to a new proprietary dataset tailored to the context of Peru, the aim is to enhance the accuracy and effectiveness of predictions in the exciting world of cryptocurrency investments.

3 Methodology

In this section of the study, we will review the data that will be used for the use of the methodology based on [1, 3] and [5]. It consists of five phases: the data set, data preprocessing, extraction of features, implementation of techniques based on Deep Learning and performance evaluation. A graph of the methodology will be shown below (Fig. 1):

3.1 Dataset

Data were collected from Yahoo Finance, UK Investing and Bitfinex as part of the study to investigate the robustness of the prediction models when considering patterns from multiple data sources. The Yahoo Finance dataset was used to both train and validate the models, while other datasets were used to evaluate the prediction models. The Yahoo Finance dataset covers six cryptocurrencies (BTC-USD, ETH-USD, BNB-USD, LTC-USD, XLM-USD, and DOGE-USD) and spans from January 1, 2018, to December 31, 2021, with a total of 1461 observations. Regarding UK Investing the investment dataset spans from July 1, 2021, to March 2, 2022 (245 observations). In addition, the Bitfinex dataset for the six cryptocurrencies is from January 1, 2021, to July 6, 2021 (187 observations). More importantly, validating models on different datasets helps to ensure that they do not match the specific characteristics of the data.

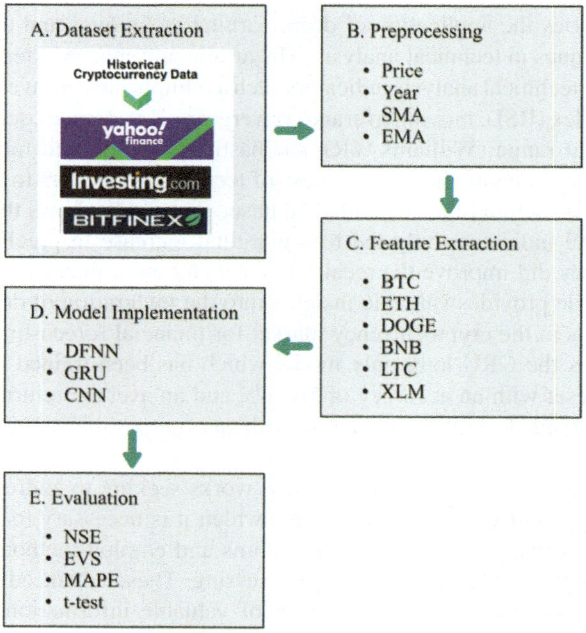

Fig. 1. Proposed Methodology

3.2 Data Preprocessing

Each dataset has five characteristics, namely, the closing price (Close), the highest price (High), the lowest price (Low), the opening price (Open) and the daily cryptocurrency volume (Volume). In addition, additional features are created, i.e., weighted average (using "Price" as values and "Year" as weights) and technical indicators that can affect prices, which include simple moving average (SMA) and exponential moving average (EMA). The reason for selecting SMA is to allow a model to recognize trends by smoothing the data more efficiently. At the same time, EMA makes it easier to dampen the effects of short-term swings. The significant difference between SMA and EMA is that EMA assigns greater weight to recent data and reacts faster to recent price movements. In addition, these technical indicators are calculated using different periods, i.e., day 3, day 10 and day 20 (Table 1).

Table 1. Description of the dataset

	BTC			**DOGE**		
	min	mean	max	min	mean	max
Open	3236.3	18408.8	67549.7	0	0.1	0.7
High	3275.4	16054	68789.6	0	0.1	0.7
Low	3191.3	15160.1	66382.1	0	0.1	0.6
Volume	2900000000.0	26000000000.0	350000000000.0	2100000.0	1000000000.0	69000000000.0
WP	8.9	50.5	185.1	0	0	0
SMA3	3244	18405.3	66511.3	0	0.1	0.6
SMA10	3397	18317.3	64698.9	0	0.1	0.6
SMA20	3524.7	18199.3	63149.3	0	0.1	0.5
EMA10	3425.1	18319.5	64411.4	0	0.1	0.5
Price	3236.8	18429.5	67566.8	0	0.1	0.7
	LTC			**ETH**		
	min	mean	max	min	mean	max
Open	23.5	2.7	387.9	84.3	933.5	4174.6
High	23.8	106.6	413	85.3	965.8	4891.7
Low	22.8	98.3	345.3	82.8	897	4718
Volume	190000000.0	2800000000.0	17000000000.0	950000000.0	13000000000.0	84000000000.0
WP	0.1	0.2	1.1	0.2	2.6	13.2
SMA3	23.6	102.6	374.4	84.7	933.2	4727.8
SMA10	24.6	102.8	347.2	89.2	926.7	4655.8
SMA20	27.4	103.1	316	97.3	917	4531.5
EMA10	25.3	102.8	342.2	90.5	926.5	4621.4
Price	23.5	6.34	386.5	84.3	935.1	4812.1

3.3 Feature Extraction

As shown in Fig. 2a the price showing non-linear dynamic features for the six cryptocurrencies versus time from January 1, 2018, to December 31, 2021. For example, in this figure also BTC indicates that the price increased significantly at the end of 2020 (about 13/ 11/2020, at index 1048 in the graph). Similarly, for the case of ETH, a significant increase in the closing price is noted around 02/02/2021, at index 1129 on the chart. A significant increase is also found in the price of BNB, which is at index 1136 on the chart, i.e., 02/09/2021.

Similarly in Fig. 2a present the summary statistics of cryptocurrencies (Yahoo Finance) from 2018 to 2021, these values have notable outlier information and as shown in the box plots containing the information to identify their distribution. For example, the closing prices of BTC in 2018 and 2020 have outliers in Fig. 2b. In addition, LTC in 2020 and 2021 in have outliers, and XLM has outliers in 2018, 2020, and 2021. However, the outliers are retained in the predictive modeling phase as they contain significant information and removing them could cause substantial loss of information.

Thus, understanding the effect of training data size on model performance is critical to advance knowledge of its generalizability, specifically in investigating the robustness of models where specific knowledge is not in the training sets. Therefore, two scenarios for training prediction models and tuning their parameters are presented. The objective is to investigate the effects of training set size dimensions on prediction quality.

The first (Scenario A) divides each Yahoo Finance cryptocurrency dataset into training (1154 days, i.e., 01/01/2018 to 02/28/2021) and test (i.e., 305 days: 03/01/2021 to 12/31/2021). Here, prominent peaks in cryptocurrency prices are missing in the training set. Scenario B divides each Yahoo Finance cryptocurrency dataset into training (1240 days, i.e., 01/01/2018 to 05/25/2021) and test set (i.e., 219 days: 05/26/2021 to 12/31/2021). Therefore, scenario B has sufficient peaks and troughs in the prices captured in the training set. For example, Fig. 2 shows the two scenarios illustrated with BNB/USD, where the highest peaks are not part of the training set (Scenario A), and some of these peaks are incorporated into the training set (Scenario B).

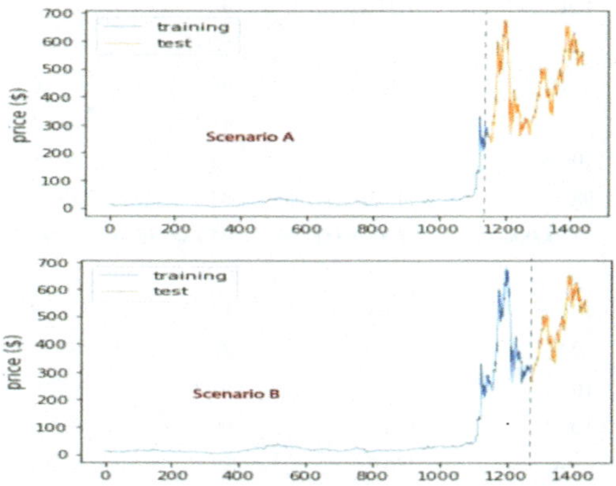

Fig. 2. Illustration of the two scenarios (A and B).

3.4 Implementation of Deep Learning

3.4.1 Direct Feed Forward Neural Networks (DFNN)

Direct feedforward neural networks, also known as feedforward neural networks or DFNNs, are a common type of artificial neural networks used in machine learning and data processing. They are called "feedforward" because information flows through the network in one direction only, from the input layer to the output layer, without the existence of cycles or feedback [9].

The basic architecture of a DFNN consists of a series of interconnected layers of neurons. The first layer is known as the input layer and is composed of nodes representing the input features or variables. Intermediate layers are called hidden layers and the last layer is called the output layer, which produces the results or predictions.

Each node in a layer is connected to all nodes in the next layer by weighted connections. These connections contain weights that are adjusted during the network training process to minimize the error between the desired outputs and the predicted outputs.

Each node also has an activation function that determines the node's output based on the weighted sum of the inputs.

DFNNs can learn from training examples using an error back propagation algorithm, which adjusts the connection weights to minimize the prediction error. This allows the network to learn pattern recognition and perform tasks such as classification, regression, and pattern recognition.

- *Step 1:* Creation of a new neural network (Feed-Forward Neural Network) with a given structure.
- *Step 2:* Hidden transfer function
- *Step 3:* TA: Training algorithm.

  o trainlm: Training algorithm based on the Levenberg-Marquardt method.
  o trainbr: Bayesian regularization-based training algorithm.

- *Step 4:* Training and validation stage of the neural network using the training data.
- *Step 5:* Neural network testing stage using separate test data.
- *Step 6:* An evaluation to determine if the neural network performance meets the predefined criteria.
- *Step 7:* Save the trained neural network parameters for later use (Fig. 3).

3.4.2 Closed Recurring Units

GRUs are a type of recurrent units used in recurrent neural networks (RNNs) that are used to capture and model long-term dependencies in data sequences. Unlike LSTMs, GRUs have a simpler structure with fewer gates, making them computationally more efficient.

The gated recurrent units in GRUs refer to the way recurrent connections and information propagation are managed over time in the network. Instead of having separate gates to control information flow and memory update, as in LSTMs, GRUs use a single gate called the "compression" or "update gate" that determines what information to pass through the recurring units [10] (Fig. 4).

- *Step 1:* Initialization of the hidden state and current input variables.
- *Step 2:* If the condition is met, the candidate state is calculated. Otherwise, skip this calculation and proceed to the next step.
- *Step 3:* Calculate the update value using the linear combination of the previ-ous hidden state and current input variables.
- *Step 4:* Calculate the forgetting value using the linear combination of the pre-vious hidden state and current input variables.
- *Step 5:* Calculate the new hidden state using a linear combination of the for-getting variables and the intermediate state (Fig. 5).

3.4.3 Convolutional Neural Networks

Convolutional neural networks (CNNs) are a type of deep neural network architecture commonly used in the processing and analysis of images and other graph-like

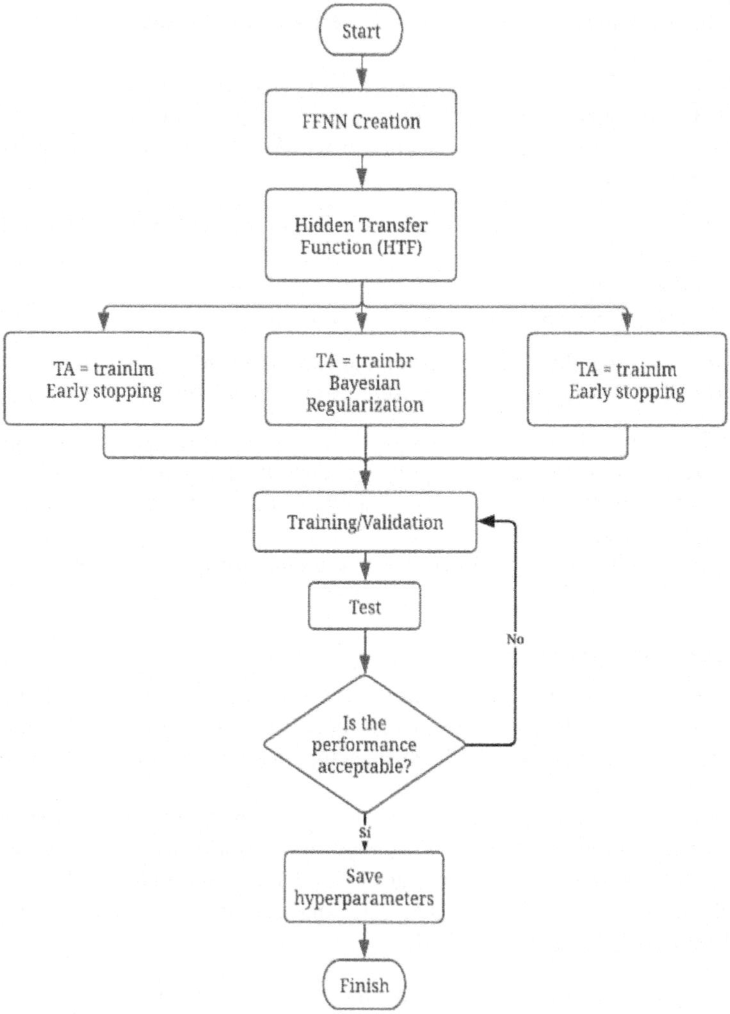

Fig. 3. Direct Feed-Forward Deep Neural Network Algorithm

data. Unlike direct feed-forward neural networks, CNNs are specially designed to take advantage of the structure and spatial features present in images.

The main feature of CNNs is the convolution layer, which applies filters or kernels to small regions of the input image. These filters extract local features, such as edges, textures, or patterns, and generate convolutional feature maps. As the network is deepened, these features are combined and refined to capture higher level information and abstraction.

In addition to convolution layers, CNNs also often include subsampling or pooling layers, which reduce the spatial dimensionality of feature maps. This helps to obtain features invariant to small translations or distortions in the images [11].

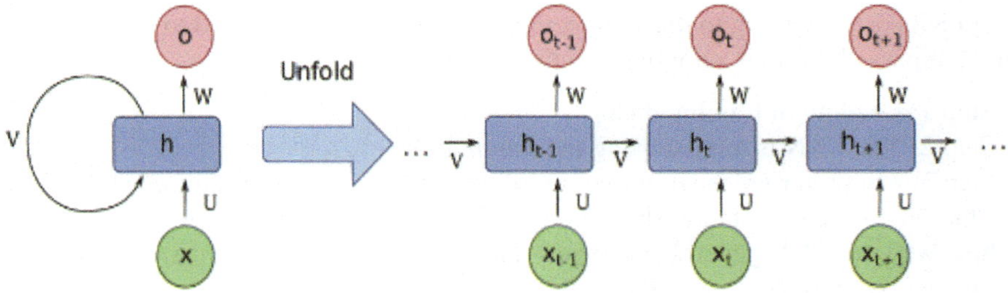

Fig. 4. Closed Recurrent Units.

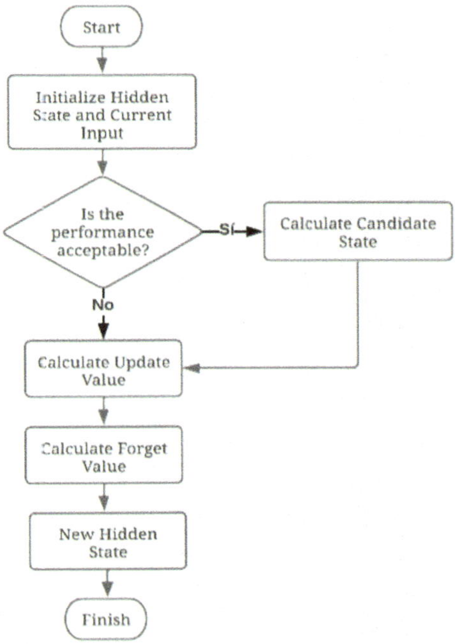

Fig. 5. Closed Recurrent Unit Algorithm.

In the feature extraction phase, the single neurons of a perceptron are replaced by array processors that perform an operation on the 2D image data passing through them, rather than a single numerical value. The output of each convolutional neuron is calculated as:

$$Y_j = g\left(b_j + \sum_i K_{ij} \cdot Y_i\right) \tag{1}$$

where the output Yj of a neuron j is a matrix that is calculated by linearly combining the outputs Yi of the neurons in the previous layer each operated on with the convolutional kernel Kij corresponding to that connection. This quantity is summed to an influence bj and then passed through a nonlinear activation function g(.).

- Step 1: Loading of training data.
- Step 2: The CNN is trained using the training set.
- Step 3: After training, the performance of the CNN is evaluated using a validation set. This validation set contains data that were not used in training and is used to evaluate how well the CNN generalizes to new data.
- Step 4: The performance of the CNN on the validation set is compared with a predefined criterion. If the condition is not met, re-sampling of the data set is performed (Fig. 6).

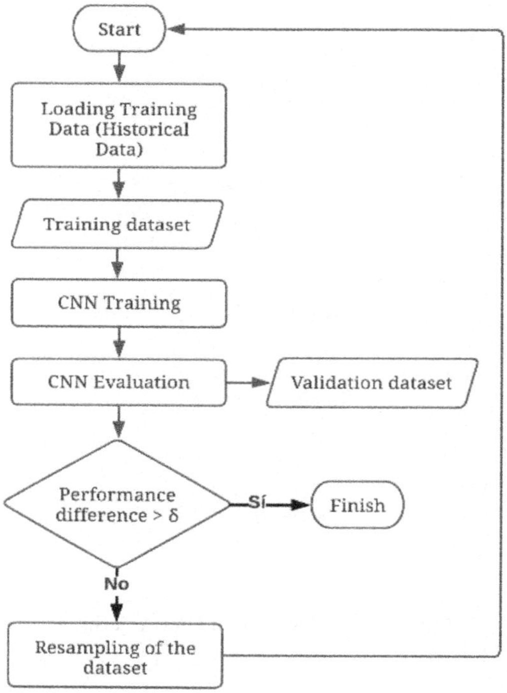

Fig. 6. Convolutional Neural Network Algorithm.

3.5 Performance Evaluation

Consequently, to finally evaluate the performance of the prediction models on test data sets (Yahoo Finance; validation sets, UK Investing and Bitfinex), a statistical analysis is performed with standard metrics to quantify how closely the predicted closing prices approximate the corresponding true prices. These metrics are:

- The Nash-Sutcliffe efficiency ratio (NSE) provides a more direct measure of the agreement between the observed closing price and the predicted values and is expressed as in the equation [12].

$$SE = 1 - \left[\frac{\sum_{j=1}^{n} (o_j - y_j)^2}{\sum_{j=1}^{n} (o_J - \overline{o})^2} \right] \tag{2}$$

Where "Yj" represents the forecasts, "Oj" represents the corresponding measured outcomes and "O" represents the mean of the measured outcome. A value of NSE closer to 1 implies that the model can successfully reproduce the observed cryptocurrency closing price. NSE = 1.0 indicates a perfect match of the model predictions with the observed values

- The explained variance score (EVS) compares the variance within the expected results to the variance in model error. This metric represents the amount of variation (dispersion) in the original data set that a model can explain and is calculated as follows.

$$EVS(o, y) = 1 - \frac{var(o - y)}{var(o)} \tag{3}$$

Where "y" is the estimated objective output or represents the corresponding objective output and "var" is the variance (i.e., the square of the standard deviation). The best possible score is 1.0 and lower values are worse for prediction models.

- The t-test illustrates the overestimation or underestimation of the data at a significance level of 95%. The t-test is thus calculated as the ratio of SS1 to SS2:

$$t = \frac{SS_1}{SS_2} \tag{4}$$

$$SS_1 = \frac{\sum_{j=1}^{n} (o_j - y_j)}{n} \tag{5}$$

Where is the average of the differences between the measured, "oj", and estimated, "yj", cryptocurrency price values, and.

$$SS_2 = \sqrt{\frac{\sum_{j-1}^{n} |(o_j - y_j) - SS_1|^2}{n - 1}} \tag{6}$$

Where a population of n > 120 and the absolute value of t <=1.96, there is no statistically significant difference between the observed and calculated data at the 95% confidence level. t-values close to zero indicate a higher precision. For a positive t-test value, the measured value is not statistically greater than the calculated value (at the 95 % confidence level). Conversely, for a negative t-test value, the calculated value is not significantly greater than the measured value at the 95 % confidence level.

- The mean absolute percent error (MAPE) measure determines the percent error per prediction and is defined in the equation [13].

$$MAPE = \frac{1}{n} \star \sum_{j=1}^{n} \left| \frac{y_j - o_j}{o_j} \right| \star 100 \tag{7}$$

The lower the MAPE, the better the performance of the model. MAPE is important in finance since gains and losses are often measured in relative values.

4 Results

A summary of the model results is presented in Table 2 below.

Table 2. Comparison of the best metrics of each model

Technique and cryptocurrency			NSE	EVS	MAPE	t-test
GRU	BTC	Data1	0.99	0.96	2%	0.79
		Data2	0.99	0.94	5%	0.93
		Data3	0.99	0.94	5%	0.97
	ETH	Data1	0.96	0.99	3%	−0.98
		Data2	0.99	0.99	2%	−0.37
		Data3	0.99	0.99	2%	−0.4
	BNB	Data1	0.99	0.99	3%	1.8
		Data2	1	1	3%	0.87
		Data3	1	1	4%	0.9
	LTC	Data1	0.99	0.99	1%	0.16
		Data2	1	1	3%	0.51
		Data3	1	1	2%	0.43
	XLM	Data1	0.96	0.96	2%	1.21
		Data2	0.99	0.99	3%	0.29
		Data3	0.99	0.99	3%	0.39
	DOGE	Data1	0.16	0.16	6%	1.78
		Data2	0.93	0.93	8%	−0.2
		Data3	0.93	0.93	8%	−0.2
	Mean		0.94	0.94	4%	0–49
DFFN	BTC	Data1	0.99	0.99	3%	2.27
		Data2	0.99	0.98	4%	1.49
		Data3	0.99	0.98	4%	1.52
	ETH	Data1	0.96	0.95	6%	1.52
		Data2	0.99	0.97	6%	1.16
		Data3	0.99	0.98	6%	1.47
	BNB	Data1	0.99	1	3%	2.09
		Data2	1	0.97	14%	−2.23
		Data3	1	0.97	13%	−2.12
	LTC	Data1	0.99	0.98	2%	1.1
		Data2	1	0.97	3%	−0.22

(continued)

Table 2. (*continued*)

Technique and cryptocurrency			NSE	EVS	MAPE	t-test
		Data3	1	0.98	2%	−0.04
	XLM	Data1	0.96	0.98	1%	0.54
		Data2	0.99	0.98	3%	−0.33
		Data3	0.99	0.98	3%	−0.24
	DOGE	Data1	0.16	0.97	16%	5.44
		Data2	0.93	0.75	67%	2.13
		Data3	0.93	0.75	67%	2.13
	Mean		0.94	0.95	12%	0.98
CNN	BTC	Data1	0.99	**0.99**	1%	0.65
		Data2	0.99	**0.99**	3%	1.08
		Data3	0.99	**0.99**	3%	1
	ETH	Data1	0.96	**0.99**	4%	2.26
		Data2	0.99	**0.97**	11%	2
		Data3	0.99	**0.97**	11%	1.87
	BNB	Data1	0.99	**0.99**	7%	3.73
		Data2	1	**0.98**	15%	2.94
		Data3	1	**0.99**	16%	3.68
	LTC	Data1	0.99	**0.99**	3%	−2.5
		Data2	1	**0.98**	4%	−1.13
		Data3	1	**0.99**	3%	−1.04
	XLM	Data1	0.96	**0.98**	5%	−3.37
		Data2	0.99	**0.98**	3%	−0.43
		Data3	0.99	**0.98**	3%	−0.49
	DOGE	Data1	0.16	**0.94**	7%	−1.23
		Data2	0.93	**0.96**	13%	1.07
		Data3	0.93	**0.96**	13%	1.07
	Mean		0.94	**0.98**	7%	0.62

The MAPE metric quantifies how close the predictions of the models are to the actual values (closing price). The predictive models yield small MAPE values but exceed 3%. For example, the GRU model has the lowest average MAPE of 4%, followed by CNN with an average MAPE of 7%, and finally DFNN with an average MAPE of 12%.

5 Conclusion

Forecasting the cryptocurrencies market is a challenging task in finance and a concern to investors due to its high volatile behavior. Therefore, this paper proposes a robust and optimal predictive models' configuration that can predict cryptocurrency closing prices with a training set having significant peaks and drops in prices not captured. Thus, the study benchmarks DL, using the models' optimal configurations to predict the daily closing prices of six different cryptocurrencies datasets collected from more than one data source. Based on the prediction results achieved in the present study, the following conclusions can be drawn, given a limited training data sample:

The DL techniques achieve a higher EVS percentage (ranging from 88% to 98%), indicating that they effectively account for the total variance in the observed data.

When comparing the robustness of the models on different test datasets (Yahoo finance, UK investing, and Bitfinex), DL models (CNN and GRU) consistently yield higher EVS values ($0.92 \leq \text{EVS} \leq 0.98$) compared to boosted tree-based models. The latter, in some cases, produce low EVS values ($0.03 \leq \text{EVS} \leq 0.50$) for ETH and DOGE, revealing the unreliability of the predicted regression for these cryptocurrencies when certain information is missing from the training set.

The residuals from the DL techniques show no defined patterns and are randomly distributed around the zero horizontal lines, satisfying the assumption of residuals having a constant variance. On the other hand, residuals from the boosted tree-based counterparts either lack symmetry to the origin or are randomly distributed due to inadequate capturing of peaks and falls of crypto prices in the training sets.

This study suggests the possibility of a single and optimal model's architecture for predicting the prices of multiple cryptocurrencies. However, predicting the financial market remains challenging due to its complex system dynamics. Nevertheless, deep learning techniques have proven to be effective modern approaches to modeling this market. Examples include their application for stock market prediction, forex market prediction, and commodity market volatility prediction. The computationally efficient deep learning models developed in this study, especially the optimized CNN model, can be adapted with minor modifications to other financial markets.

In the future, we plan to run further experiments and comparisons with statistical and time series models, including the ARIMA model, in order to improve the performance of the results and determine which model is more efficient. As well as exploring various data pre-processing and parameter fitting techniques.

References

1. Chahooki, M.A.Z., Jahanbin, K., Sutikno, T.: Cryptocurrencies investment framework using sentiment analysis of Twitter influencers. Indonesian J. Electr. Eng. Comput. Sci. **30**(2), 1068 (2023). https://doi.org/10.11591/ijeecs.v30.i2.pp1068-1079
2. Oyedele, A.A., Ajayi, A.O., Oyedele, L.O., Bello, S.A., Jimoh, K.O.: Performance evaluation of deep learning and boosted trees for cryptocurrency closing price prediction. Expert Syst. Appl. **213**, 119233 (2023). https://doi.org/10.1016/j.eswa.2022.119233
3. Seabe, P.L., Moutsinga, C.R.B., Pindza, E.: Forecasting cryptocurrency prices using LSTM, GRU, and bi-directional LSTM: a deep learning approach. Fractal Fract. **7**(2), 203 (2023). https://doi.org/10.3390/fractalfract7020203

4. Kochliaridis, V., Kouloumpris, E., Vlahavas, I.: Combining deep reinforcement learning with technical analysis and trend monitoring on cryptocurrency markets. Neural Comput. Appl. (2023). https://doi.org/10.1007/S00521-023-08516-X
5. Goutte, S., Le, H.V., Liu, F., von Mettenheim, H.J.: Deep learning and technical analysis in cryptocurrency market. Financ. Res. Lett. (2023). https://doi.org/10.1016/J.FRL.2023.103809
6. Borges, T., Neves, R.: Ensemble of machine learning algorithms for cryptocurrency investment with different data resampling methods. Appl. Soft Comput. J. (2020). https://doi.org/10.1016/j.asoc.2020.106187
7. Canh, N.P., Wongchoti, U., Thanh, S.D., Thong, N.T.: Systematic risk in cryptocurrency market: Evidence from DCC-MGARCH model. Financ. Res. Lett. **29**, 90–100 (2019). https://doi.org/10.1016/j.frl.2019.03.011
8. Chowdhury, R., Arifur Rahman, M., Sohel Rahman, M., Mahdy, M.R.C.: An approach to predict and forecast the price of constituents and index of cryptocurrency using machine learning. Physica A: Stat. Mech. Appl. **551**, 124569 (2020). https://doi.org/10.1016/j.physa.2020.124569
9. Chernodub, A.: Direct Method for Training Feed-forward Neural Networks using Batch Extended Kalman Filter for Multi-Step-Ahead Predictions. In: LNCS 8131, pp. 138–145 (2013). https://doi.org/10.48550/arXiv.1605.03764
10. Chen, W., Zhai, C., Wang, X., Li, J., Lv, P., Liu, C.: GCN- and GRU-based intelligent model for temperature prediction of local heating surfaces. IEEE Trans. Ind. Inf. **19**(4), 5517–5529 (2023). https://doi.org/10.1109/TII.2022.3193414
11. Nair, M., Marie, M.I., Abd-Elmegid, L.A.: Prediction of cryptocurrency price using time series data and deep learning algorithms. Int. J. Adv. Comput. Sci. Appl. **14**(8), 338–347 (2023). https://doi.org/10.14569/IJACSA.2023.0140837
12. McCuen, R.H., Knight, Z., Cutter, A.G.: Evaluation of the nash–sutcliffe efficiency index. J. Hydrol. Eng. **11**(6), 597–602 (2006). https://doi.org/10.1061/(ASCE)1084-0699(2006)11:6(597)
13. Kim, Sungil, Kim, Heeyoung: A new metric of absolute percentage error for intermittent demand forecasts. International Journal of Forecasting **32**(3), 669–679 (2016). https://doi.org/10.1016/j.ijforecast.2015.12.003

Proposal of a Model for the Detection of Violence Against Animals Through Convolutional Neural Networks

Juan Vargas[1], Dick Mori[1], Naydu Ramon[2], and Wilfredo Ticona[1,2](✉)

[1] Universidad Tecnológica del Perú, Lima, Peru
{U19221296,1611770}@utp.edu.pe, wmamani@esan.edu.pe
[2] Universidad ESAN, Lima, Peru
20101057@ue.edu.pe

Abstract. Animal violence is a global problem that affects millions of animals every year. It can manifest itself in various forms, such as abandonment, physical abuse, and commercial exploitation. Animal violence has a negative impact on animal welfare, but it can also have consequences for society at large. Detecting animal violence is a major challenge. Traditional methods, such as bystander reporting, may be insufficient to identify all cases of abuse. Automated detection methods, such as machine vision, may be a valuable tool to improve the detection of animal violence. This study evaluated the effectiveness of convolutional neural networks, coupled with pose estimation, in detecting animal violence in videos. A dataset of animal violence videos was created and preprocessed to improve model performance. Several detection models, such as YOLOV7, VGG-16, VGG-19, GoogleNet and ResNet, were implemented, of which, they were evaluated based on their accuracy, precision, and sensitivity. Accordingly, the results showed that YOLOV7, in combination with pose estimation, obtained the best performance metrics. The model demonstrated an accuracy of 93.49%, precision of 98% and sensitivity of 88%. suggest that YOLOV7, together with pose estimation, is an effective method for detecting animal violence. The model provides a comprehensive approach to identify violent behaviors towards animals and can contribute to the automation and improvement of the identification of such behaviors.

Keywords: Deep Learning · Convolutional Neural Networks · Violence · Animals

1 Introduction

Currently, animal protection and welfare in society reflects the global awareness of violence towards millions of living beings every year. According to Suing, G, [1], 83% have witnessed some type of animal abuse in their lifetime. In response to this worrying reality, Law 30,407, also known as the Animal Protection and Welfare Law, was enacted in Peru with the purpose of guaranteeing the safety and ethical treatment of animals in the country. Despite these legislative advances, significant challenges remain, as some people still perpetuate inhumane acts against animals.

R. Silhavy and P. Silhavy (Eds.): CSOC 2024, LNNS 1120, pp. 218–228, 2024.
https://doi.org/10.1007/978-3-031-70518-2_19

Recent years have witnessed the study of human action behavior under the spotlight of computer vision and data science. [1], violence detection is a specific task of action recognition, and it is a binary problem that consists in recognizing the presence or absence of violence. We consider "violent" a voluntary action, exerted by one subject (or more) on another (or more), to act against the will of the victims [1].

Current research allows the detection of different actions within a single image or a group of contiguous images. These approaches use techniques such as convolutional neural networks (CNN) [1], convolutional neural network bidirectional LSTM model (CNN-BiLSTM), bidirectional LSTM cells [2] and POSE estimation [3].

These solution models can adequately help in some circumstances but have poor performances individually with respect to others. Some models based on CNNs learn spatial features such as the environment, but do not solve the problem when the environment changes [1].

2 Related Work

At present, we observe remarkable advances in the technological field, and one of the most outstanding aspects focuses on the development and application of deep learning techniques, especially designed to train convolutional neural networks (CNN). These techniques have demonstrated outstanding success in detecting cases of violence.

On the one hand, Kwan-Loo, K. et al. [2] propose a new model for the detection of violent behavior, based on pose estimation (OpenPose) to differentiate between normal and violent behavior, together with CNNs called YoloV3. To carry out its training, videos are collected to build the Kranok-NV database. Subsequently, performance is evaluated using three metrics: precision, accuracy and hit rate, achieving impressive results of 98.81%, 99.80% and 98.73%, respectively.

In contrast, Accattoli, S. et al. [3] propose a new method for violence detection in videos. For feature extraction for scene classification, they use the convolutional neural network (CNN) model called C3D pre-trained, together with a support vector classifier (SVM) that served to implement an automatic violence detection system in videos. Also, to evaluate the performance, 3 datasets called Hockey Fight Dataset, Crowd Violence Dataset and Movie Violence Dataset are used. Thus, achieving an accuracy of 98.6% in the detection of violent acts in the videos.

Furthermore, Halder, R. et al. [4] proposes to employ the use of CNNs as C3D for violence detection in intelligent surveillance, in combination with recurrent neural networks, called Bi-LSTM. The proposed model integrates both spatial and temporal features to detect patterns of violence in video sequences, achieving remarkable accuracy rates on various datasets. The results obtained are particularly remarkable, with 99.27% accuracy on the "Hockey Fight" dataset, a perfect 100% on the "Movies" dataset and a solid 98.64% on "Violent-Flows". This approach proves to be the most effective and requires the least amount of time to train.

On the other hand, Irfanullah, et al. [5] employs different CNN models for violent event detection, such as AlexNet, VGG-16, GoogleNet and MobileNet on the Hockey Fight dataset. The developed methodology consists of a series of steps including Dataset acquisition, data preprocessing, model creation and training, model evaluation, optimization, and validation, all before performing the relevant tests. That said, MobileNet stands out for using few device resources, such as 10,334,030 parameters compared to the other models such as AlexNet and VGG-16 and achieves an accuracy rate of 96.6% with a loss of 13.29%.

Likewise, in their recent study, Segador R. et al. [6] have addressed violence detection in video surveillance environments using a combination of convolutional networks (3D-CNN) and the CRIMENET model, which integrates the Vision Transformer (ViT) architecture with Neural Structured Learning (NSL). These models have been applied to several databases, such as NTUCCTVFights, UBIFights, XDViolence and UCFCrime. The results reveal that CrimeNet stands out as a robust and highly effective model, achieving an impressive 99.98% on the ROCAUC metric and an AP of 99.95%, outperforming other techniques and advancing the state of the art on several datasets.

Finally, Kumar, et al. [7] succeeded in realizing a CNN model for the recognition of violent activities in videos. In his study, he performs a dataset creation called UCF-Crime, in which he includes hours of surveillance footage such as abuses, arrests, assaults and fights. Consequently, he extracted pre-trained models such as InceptionV2, InceptionV3 and InceptionV4. To achieve effective captures of these feature scenes, he combines CNNs with the Bi-LSTM technique. With this technique, special and temporal characteristics of video streams are captured, which contributes to the accurate detection of violent activities in surveillance environments. The results obtained achieve 96.40% accuracy for the InceptionV3 model with Bi-LSTM.

This project aims at creating a comprehensive dataset from various sources, followed by the extraction of key features for the effective training of a CNN, in particular, the YOLOv7 model. In conjunction, other CNN models such as VGG-16, VGG-19, GoogleNet and ResNet. In addition, a specific metric will be developed to measure the accuracy of object detection in cases of animal abuse. In conclusion, the main objective of this project is to automate and improve the identification of violent behavior towards animals, thus contributing to their protection and welfare.

3 Methodology

The proposed methodology is based on [2, 3, 5] and [6]. It will have 4 phases: Dataset creation, preprocessing, implementation models and evaluation. Then, each stage has general activities that are of utmost importance for the development of the research. The methodology used is shown graphically below Fig. 1.

3.1 Dataset Construction

First, the selection of data used in this project is described, which were chosen from the social networks Facebook and Youtube. According to IPSOS (2023), Facebook maintains its leadership with 43.3% in the use of social networks in Peru and Youtube is also considered, since although it does not have so much frequency in the country, many people expose videos of high content and quality to express their opinions, regardless of their veracity.

To obtain data from Facebook, we used the facebook-scraper library, with which we collected 58 publications from different periods and with reference to animal abuse.

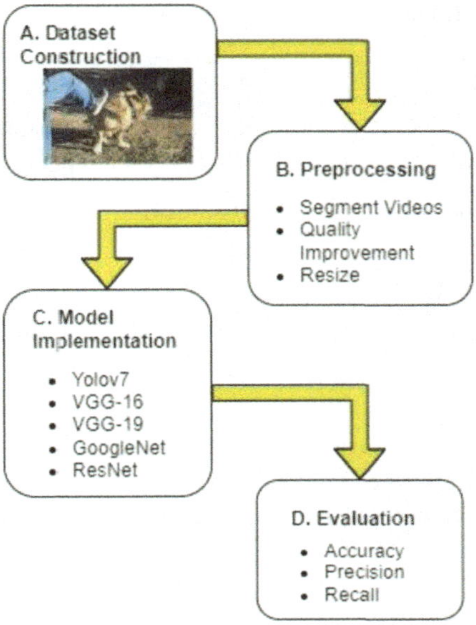

Fig. 1. Proposed Methodology

These data come from the most representative media accounts in the country, such as La República, El Comercio, Latina, Municipalities, and the most relevant influencers in the country. As for Youtube, the Youtube Data API was used to extract 65 videos. It was necessary to have defined keywords such as "mistreatment", "animal", "pet", "abuse", "cruelty". The extraction was carried out from the different departments of Peru (24) and its constitutional province of Callao. These data were stored in a public cloud.

On the other hand, to perform scraping, algorithms were developed based on the scrapy library. Since each web page had a unique structure, it was necessary to perform a proprietary analysis to create the algorithm according to its requirements. Thirty-one news items were obtained from the Municipalities website (title and link), 13 from Latina (title and link) and the rest from the influencers (title and link).

After that, the videos were renamed and those that were not useful were eliminated. Likewise, duplicate records and records not related to animal abuse were eliminated, as reflected in the keywords (Table 1). The result of this process was a dataset with 3 representative columns ("Title", "Link", "Type").

After the processes described above, the data obtained were classified into two main categories: 66 records were identified as cases of animal abuse, while 57 records were classified as situations without evidence of animal abuse.

Each step in this process, from the initial database search to the visualization of representations, is strategically designed to enrich the capability of the CNN model. The culmination of these actions provides a solid structure for analysis, validation, and iterative tuning aimed at optimizing the accuracy and sensitivity of the model in detecting cases of animal abuse in images.

Table 1 List of keywords for animal abuse

#	Key words	#	Key words
1	Abuse	8	Mistreatment
2	Negligence	9	Illegal Hunting
3	Exploitation	10	Events
4	Cruelty	11	Abandonment
5	Experimentation	12	Attacks
6	Animal	13	Animal Trafficking
7	Violence	14	Mascot

Figure 2 below illustrates the classification of the data obtained in our dataset, highlighting the distinction between cases of animal abuse and situations without evidence of abuse:

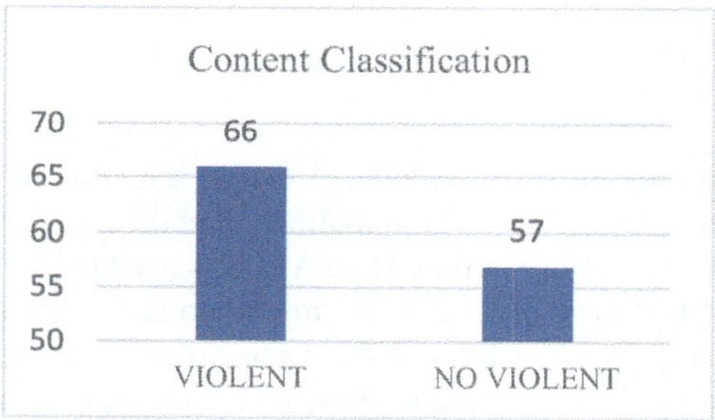

Fig. 2. General classification of videos

3.2 Preprocessing

In this phase, preprocessing of the dataset originating from Peru, which includes videos linked to cases of violence against animals, has been carried out. Below is a detailed table with the tasks and activities carried out during this unique preprocessing stage (Table 2).

In relation to the points mentioned above, such as frame segmentation, quality enhancement and video resizing. We will proceed to describe each of these in the following:

Frame sequence segmentation is a process that involved the division of each video into individual sequences. To accomplish this, thresholding was employed as a key tool. This technique was not only limited to the simple separation of frame sequences but became a perceptual tool that analyzes the distinctive visual characteristics present in

Table 2. Preprocessing activities and steps

Activities	Description	Tasks
Segment the frames of each video to focus on scenes of violence against animals	Identify and extract the relevant parts of the video, where the presence of violence towards animals can be clearly observed	Upload segmented videos showing scenes of violence against animals
Improve the quality of the frames extracted from the videos	Application of a video algorithm, which improves the quality of the videos	Dataset with higher quality videos
Resize the videos to try to maintain uniformity in the same scale	Fit all videos in the same dimension so that they can be used in the classification models	Calculate an average dimension of the videos
		Resize images

scenes of violence. Specific thresholds were explored to effectively distinguish between scenes of interest and those that are not related to acts of animal violence.

After segmenting the frame sequence, an algorithm was employed to improve the visual quality. This included contrast adjustment, sharpening, and specialized image enhancement techniques. This approach focused on optimizing the clarity of scenes of violence against animals for more accurate analysis and improved visual presentation. Contrast adjustment highlights details, sharpening defines essential boundaries, and image enhancement techniques are applied to highlight relevant aspects in each frame. This process ensures enhanced visual quality that supports detailed analysis of scenes of animal violence.

Finally, an algorithm would be used to resize the videos, ensuring uniformity in scaling. This can be crucial for effective comparison between different videos. The resizing was based on criteria such as resolution, frame size or any other relevant metric. Aiming to obtain a consistent presentation of scenes of violence against animals, facilitating their analysis and comparison.

The visual representation evidences an evolutionary process from the initial identification of violent scenes to the refined optimization of visual quality and uniformity in scale. This visual journey not only highlights the intrinsic progression of events in the video, but also offers an enhanced and comparative representation. The meticulous attention given to quality and consistency ensures a coherent presentation, facilitating a deeper and more meaningful analysis of the events captured in the sequence (Fig. 3).

3.3 Model Implementation

First, the SplitTrain is carried out, which consists of dividing the data set into two, 75% for training and 25% for testing. Consequently, a column is created in the database with certain values according to their characteristics, one called violence and the other as non-violence.

Secondly, it details how various machine learning techniques are applied, highlighting the model such as YOLOV7 as the main approach, followed by convolutional neural

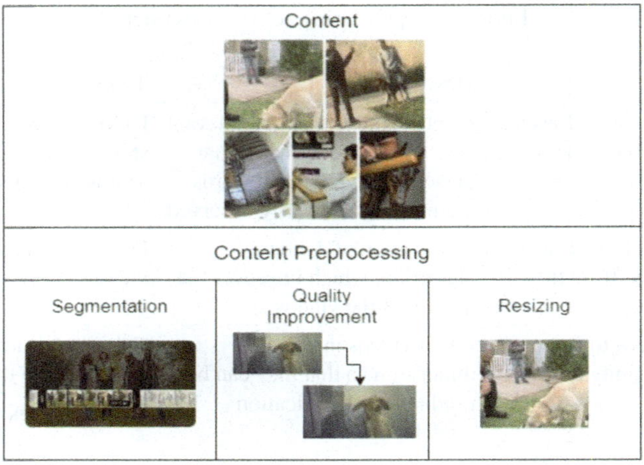

Fig. 3. Original and resulting content

network (CNN) architectures such as VGG-16, VGG-19, GoogleNet and ResNet. These architectures will be used to address specific tasks, taking advantage of their distinctive capabilities. In addition, pose estimation will be incorporated as an essential technique to understand the position and orientation of objects or body parts in images. Accordingly, certain activities presented below will be followed (Table 3).

Table 3. Model building activities

Activities	Description	Tasks
Preparation of the dataset	Variables are divided into inputs and outputs	Separation according to the values created as violence and non-violence
Training and testing of the model	Seventy-five percent is assigned to the training set and the remaining 25 percent is assigned to the test set	Determine and evaluate the relationship between the characteristics
		Evaluate and compare CNN according to their performance

It should be noted that each CNN architecture has its own specific parameters. Below, I provide a brief description (Table 4):

In addition, the pose estimation technique is added to the Convolutional Neural Network (CNN) models, which involves the use of pose models to identify key points in the videos, representing the anatomical configuration of humans and, in some cases, animals. This pose information is integrated as additional features in the main CNN analysis process that detects violence against animals. The network detects and connects key points, generating skeletons that represent the body structure of the subjects. These

Table 4. Parameters according to CNN model

Parameters	YOLOv7	VGG-16	VGG-19	GoogleNet	ResNet
Number of layers	24	16	19	21	50
Dimension Reduction	Yes	Yes	Yes	Yes	No
Model Size	Variable	Grande	Grande	Variable	Grande
Use of Layers	Conv, pooling,	Conv, pooling,	Conv, pooling	Inception, Conv, pooling modules,	Conv, pooling
Residual Connection	No	No	No	No	Yes

pose features, such as orientation and relative distance, are incorporated into the model during training, enhancing the system's ability to recognize specific patterns and contexts related to violent behaviors in human-animal interaction. This comprehensive approach enhances the accuracy of violence detection by considering both the visual appearance and spatial arrangement of body parts in the scene.

To illustrate this approach in more detail, Fig. 4 is presented, in which the individual's movement patterns can be observed with greater emphasis, highlighting especially those related to violent behaviors.

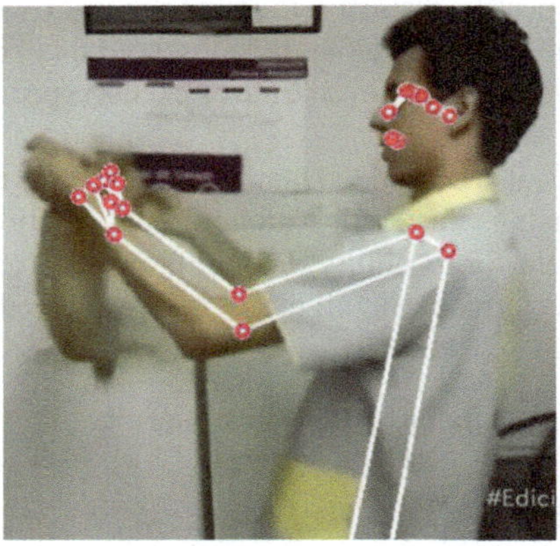

Fig. 4. Movement patterns

3.4 Evaluation

- **Accuracy:** The accuracy metric in neural networks refers to the measure of how accurate a model is. It is measured by the number of correct predictions divided by the total number of predictions [8].

$$Accuracy = \frac{TP + TN}{TP + FP + FN + TN} \quad (1)$$

- **Precision:** The precision metric in neural networks refers to the measure of fractions of positive predictions. In which, it is the number of correct positive results divided by the number of positive results predicted by the classifier [8].

$$Precision = \frac{TP}{TP + FP} \quad (2)$$

- **Sensitivity:** The sensitivity metric in neural networks measures the number of correct positive results divided by the number of all relevant samples [9].

$$Sensitivity = \frac{TP}{TP + FN} \quad (3)$$

where:

- TP: True Positives.
- FP: False Positives.
- TN: True Negatives.
- FN: False Negatives.

4 Results

After performing the respective tests, the following metrics were compiled, which are concisely presented in the following table. It is observed that both Yolov7 and VGG-19 present a high Sensitivity index. However, Yolov7 excels especially in the Accuracy and Precision metrics.

Based on these results, it is concluded that the most effective model for the detection of violence against animals is the Yolov7 model in conjunction with the Pose estimation. This model stands out for its ability to identify positive cases of violence and has demonstrated superior performance in all the metrics evaluated (Table 5).

Table 5. CNN model metrics

Models	Metrics		
	Accuracy	Accuracy (%)	Sensitivity
Yolov7	0.98	93.49	0.88
VGG-16	0.95	92.49	0.86
VGG-19	0.97	93.12	0.88
GoogleNet	0.96	91.99	0.85
ResNet	0.87	88.79	0.82

In contrast, Fig. 5 presents the confusion matrices illustrating the evaluation among the dataset of the model with the highest metrics, Yolov7, focusing specifically on true positives, false positives, and the combination of both. The results of this evaluation highlight exceptional model performance in generalizing violence detection.

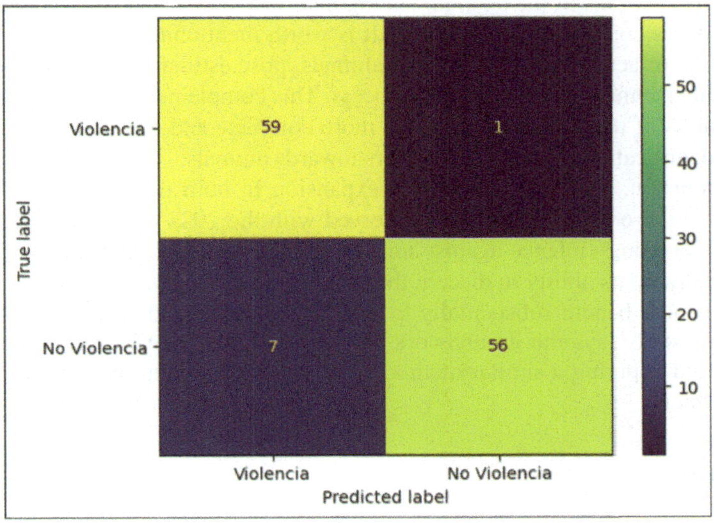

Fig. 5. Yolov7 Confusion Matrix

5 Conclusions

This document compiles visual content from various sources to address the persistent problem of violence against animals that still impacts Peruvian society. Despite the enactment of Law 30,407, a significant part of the population continues to mistreat animals in an alarming manner, which has generated indignation in the face of this abuse.

Therefore, the methodology best suited to the study was selected, which is based on 4 phases. The creation of the database was one of the most complex processes, since the content was often censored, therefore, it was necessary to perform a previous labeling and subsequent validation In the research, the Facebook scraping technique was used to extract the URLs of the videos referring to animal violence.

Likewise, the data preprocessing process in this study was based on the information collected with some specific adjustments. Initially, the inclusion of frame sequence segmentation techniques, quality improvement and resizing for the detection of animal violence by means of a CNN was considered. All this, with the aim of generating significant improvements in the results. With this adapted preprocessing, an effective training of the model was achieved, thus offering an integral approach adjusted to the particular characteristics of the data set and the study context.

For this study, different Convolutional Neural Network (CNN) architectures such as VGG16, VGG19, YOLOv7, GoogleNet and ResNet were explored. Their implementation was evaluated according to the layers provided by each one. Although the technique that stood out was Yolov7, it is essential to highlight that the CNN architectures demonstrated their effectiveness in detecting violence. Furthermore, they proved to generate robust representations of the data, indicating that, with a larger dataset, contextual capture of features could be more accurate. It is worth mentioning that, to strengthen the identification of acts of violence against animals, pose estimation was implemented as an additional technique in the analysis process. This complementary approach extended the robustness of the model, providing a more complete and detailed view for a more accurate identification of violent behaviors towards animals.

In conclusion, it is inferred that an expansion in both the number and diversity of records could enhance the results observed with the 2023 data in the classification model for detecting violence against animals using CNN. Although the current model has demonstrated its ability to discern this type of behavior in the Peruvian context, its robustness could benefit substantially from more extensive and varied training. Nevertheless, this study lays the groundwork for future research interested in using it as a reference and applying a similar methodology in the identification of acts of violence against animals.

References

1. Suing-Albito, G., Barba-Guaman, L.: Application of deep learning methods in the automatic identification and counting of wild animals. RISTI – Revista Iberica de Sistemas e Tecnologias de Informacao, 2023 (E56) (2022)
2. Kwan-Loo, K.B., Ortiz-Bayliss, J.C., Conant-Pablos, S.E., Terashima-Marin, H., Rad, P.: Detection of violent behavior using neural networks and pose estimation. IEEE Access (2022). https://doi.org/10.1109/ACCESS.2022.3198985
3. Accattoli, S., Sernani, P., Falcionelli, N., Mekuria, D.N., Dragoni, A.F.: Violence detection in videos by combining 3D convolutional neural networks and support vector machines. Appl. Artif. Intell. **34**(4), 329–344 (2020). https://doi.org/10.1080/08839514.2020.1723876
4. Halder, R., Chatterjee, R.: CNN-BiLSTM Model for violence detection in smart surveillance. SN Comput. Sci. (2020). https://doi.org/10.1007/s42979-020-00207-x. https://doi.org/10.1007/s42979-020-00207-x.
5. Irfanullah, T.H., Iqbal, A., Yang, B., Hussain, A.: RETRACTED ARTICLE: Real time violence detection in surveillance videos using Convolutional Neural Networks. Multimed. Tools Appl. **81**(26), 38151–38173 (2022). https://doi.org/10.1007/s11042-022-13169-4
6. Rendón-Segador, F.J., Álvarez-García, J.A., Salazar-González, J.L., Tommasi, T.: CrimeNet: neural structured learning using vision transformer for violence detection. Neural Netw. (2023). https://doi.org/10.1016/j.neunet.2023.01.048
7. Kumar, M., Patel, A. K., Biswas, M., Shitharth, S.: Attention-based bidirectional-long short-term memory for abnormal human activity detection. Sci. Reports **13**(1) (2023). https://doi.org/10.1038/s41598-023-41231-0. https://doi.org/10.1038/s41598-023-41231-0
8. Hernández Sampieri, R., Fernández Collado, C., Baptista Lucio, M.P.: Methodological research, 6th edn. McGraw Hill Education (2014)
9. Wang, C.-Y., Bochkovskiy, A., Liao, H.-Y.M.: Real-time violence detection in surveillance videos using Convolutional Neural Networks. arXiv. https://doi.org/10.48550/arXiv.2207.02696 (2022)

Edge Algorithm of Reconstruction 3D Model by Two of Its Images

Vladimir A. Klyachin(✉) and Elena G. Grigorieva

Computer Science and Experimental Mathematics Department, Volgograd State University, University Prospect 100, 400062 Volgograd, Russia
{klyachin.va,e_grigoreva}@volsu.ru

Abstract. The article discusses the problem of calculating the frame of a 3D model from its images obtained by two cameras. The method of such reconstruction is based on calculating the coordinates of the vertices of segments from their projections from two cameras. During reconstruction, an ambiguity arises, which is resolved when an additional condition is met, which is achieved by slightly moving the cameras. Using these assumptions and some mathematical background, we develop an algorithm for the necessary frame reconstruction.

#CSOC1120

Keywords: central projection · 3D model · segments · spatial reconstruction

1 Introduction

The task of constructing a 3D model of an object from its images finds its application in many areas of human activity [1–15]. As an example, we can cite a problem related to autonomous vehicle control. The car's autopilot is controlled according to the generated 3D traffic scene. Historical analysis of photographs leads to the need to restore architectural structures in the form of a 3D model, which can later be used both in Internet resources and to obtain a model of the building on a 3D printer. In this work, we study the problem of reconstructing a 3D model from its projections onto two planes. We use a wireframe representation of the 3D model. Therefore, the basis of the proposed approach is related to the spatial analysis of projections of individual segments. The main difficulty in solving the corresponding problem is to identify the correspondence of the projections of segments in these images to one edge of the 3D model of the object.

It should be noted that attempts to solve similar problems have been made before. Thus, it is worth noting the work [1], which presents one of the methods for estimating the position and dimensions of spatial parallelepipeds covering a car in space, based on its image in a photograph. The same techniques were used in the works [2,3]. Also, a similar problem of determining the 3D position of objects was solved in [4], which used modified convolutional neural networks

R. Silhavy and P. Silhavy (Eds.): CSOC 2024, LNNS 1120, pp. 229–238, 2024.
https://doi.org/10.1007/978-3-031-70518-2_20

and specially prepared object masks. The article [5] used a method based on frame-by-frame object tracking, and the learning model was built on the basis of recurrent neural networks from LSTM layers. Let us also note the work [6], in which, for the integral reconstruction of a three-dimensional scene, its description in the form of a grammar is proposed, and the authors construct the result of the reconstruction in the form of a special graph. It should be noted that similar reconstruction problems arise, for example, when reconstructing a three-dimensional model of a person's face and body pose from a photograph (see, for example, [7,8]). It should be noted that neural networks are most often used to solve problems of spatial image analysis. This is because convolutional neural networks can automatically detect image features during training and determine their significance based on the data set used. This approach requires a very large training data set. Therefore, we think it would be useful to be able to perform direct calculations of spatial data based solely on the characteristic features of the images of the objects themselves. Such spatial data can be useful for reconstructing their models. The possibility of obtaining such data is shown in the works of the authors. For example, in [13] it was shown that the orientation of the plane of a circle in space can be obtained from the characteristics of the elliptical region, which is the projection of the circle. In [14], an algorithm for reconstructing the profile of the visible part of the surface of rotation from its projection is obtained and described. In this paper, it is proposed the reconstruction method which is based partialy on the ideas from [9]. But here we reconstruct wireframe 3D model instead cloud of points as in the mentioned article.

2 Setting the Problem

2.1 Some Definitions

We consider some 3D model as the following geometric represantation. Let $P \subset \mathbb{R}^3$ be some finite set of points,

$$P = \{p_i = (x_i, y_i, z_i) : i = 1, 2, ..., N\},$$

where $N = |P|$ is a number of vertices. In addition, we assume that the set E of ordered pairs of integers is defined. For each $e = (n, m) \in E$ it is true

$$1 \leq n < m \leq N,$$

and there exists edge p_n, p_m. Therefore we consider a 3D model as a set of edges (segments).

Now we give a camera model in $\mathbb{R}^3$. A camera is defined by some plane $\Pi \subset \mathbb{R}^3$ which is given by equation $\langle p, \xi \rangle = d$ and a point $O \in \mathbb{R}^3$ as a center of projection. We consider an image $\pi(p)$ for any point $p \in \mathbb{R}^3$ as the intersection point of the plane Π and the ray Op. It is not difficult to calculate that

$$\pi(p) = O + t(p - O),$$

where

$$t = \frac{d - \langle O, \xi \rangle}{\langle p - O, \xi \rangle}.$$

We note that ξ be unit vector which is orthogonal to Π. Also, we consider two vectors e_1, e_2 which are parallel to the plane Π and which define orthonormal basis in Π. Therefore $\pi(p)$ can be calculated by formula

$$\pi(p) = f \cdot \frac{X}{Z} e_1 + f \cdot \frac{Y}{Z} e_2,$$

where

$$f = d = \langle O, \xi \rangle$$

is a focus length of the camera, and

$$X = \langle p, e_1 \rangle, \tag{1}$$

$$Y = \langle p, e_2 \rangle, \tag{2}$$

$$Z = \langle p, \xi \rangle. \tag{3}$$

2.2 Formulation a Problem

We consider two cameras (Π_i, O_i) $i = 1, 2$. Let $P_i \subset \Pi_i$ be two image of projections some unknown finite set $P \subset \mathbb{R}^3$ such that

$$P_i = \pi_i(P), \ i = 1, 2.$$

Morevery, we assume that P is a set of vertices of some 3D model. Therefore, we consider two projections of this 3D model and the sets of edges E_1, E_2 are known. The problem is to find P, E by P_1, E_1 and P_2, E_2. The main difficulty is to find a correspondence

$$g : \{1, 2, ..., N\} \to \{1, 2, ..., N\}$$

such that

$$\pi_1(p_i) = p^1_k,$$

$$\pi_2(p_i) = p^1_{g(k)}.$$

By $L(x, y)$ we denote a straight line joining the points $x, y \in \mathbb{R}^3$. We suppose the following

Assumption 1. *The projections $\pi_i : P_i \to \Pi_i$ are one-to-one mappings on the correspondence sets P_i, $i = 1, 2$.*

In [9] the algorithm of reconstruction of the set P was presented. But this algorithm can be reconstructs the set of points P only (Fig. 1).

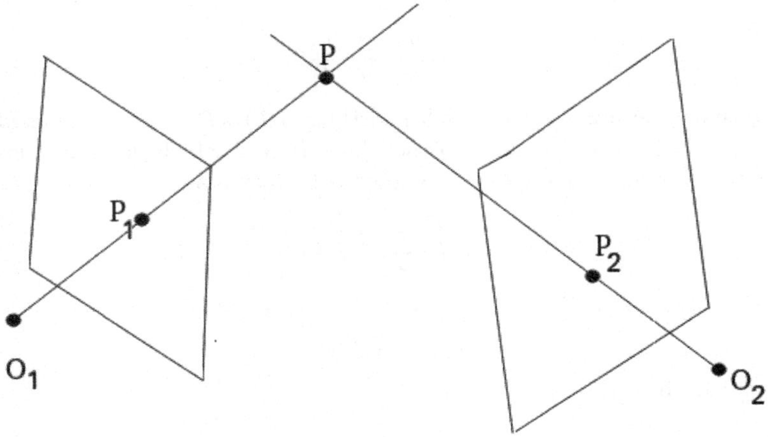

Fig. 1. 3D reconstruction of a point by its two projections

3 Reconstruction Algorithm

3.1 Mathematical Basis

We describe reconsructs algorithm for each possible edge of P in $\mathbb{R}^3$. Choose some edge $(n, m) \in E_1$. Consider points $q_n, q_m \in P_1$ and two rays $r_n = O_1 q_n, r_m = O_1 q_m$ and calculate the following projections

$$\pi_2(r_n),\ \pi_2(r_m).$$

Now we find intersections

$$P_2^n = P_2 \cap \pi_2(r_n),\ P_2^m = P_2 \cap \pi_2(r_m).$$

We denote

$$P_2^n = \{u_{i_1}, u_{i_2}, ..., u_{i_k}\},$$
$$P_2^m = \{u_{j_1}, u_{i_2}, ..., u_{i_l}\},$$

where $P_2 = \{u_i\}$ (see Fig. 2). Assuming that $P_1 = \pi_1(P),\ P_2 = \pi_2(P)$ we can find some integers $i_s, 1 \leq s \leq k,\ j_t, 1 \leq t \leq t \leq l$ such that $(i_s, j_t) \in E_2, i_s < j_t$. Further we construct two rays

$$O_2 u_{i_s},\ O_2 u_{j_t}.$$

Then the points

$$p' = O_1 p_n \cap O_2 u_{i_s},$$
$$p'' = O_1 p_m \cap O_2 u_{j_t}$$

are sought after. These intersections we find approximately. Namely, for example, we find p' as a point

$$p' = \frac{p+q}{2},$$

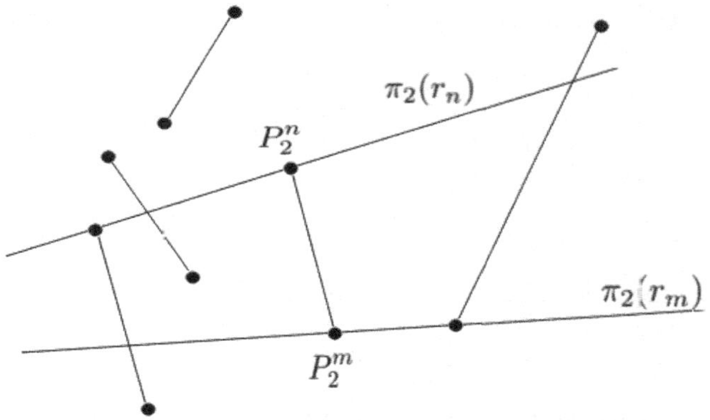

Fig. 2. 3D reconstruction of the edges.

where

$$(p, q) = \arg \min_{p \in O_1 p_n, q \in O_2 u_{i_s}} |p - q|.$$

This is due to the discrete nature of real images. Also, we note that the mentioned above integers i_s, j_t may not be found unambiguously. Therefore, even in the case of one-to-one projection mappings (Assumption 1), this can leads to ambiguity in the reconstruction of segments, as can be seen in the Fig. 3—by two projections it can be reconstructed two pairs of segments (a_1, b_1), (a_2, b_2) and $(a_3, b_3$, (a_4, b_4).

The following theorem gives some sufficient conditions for such uniqueness.

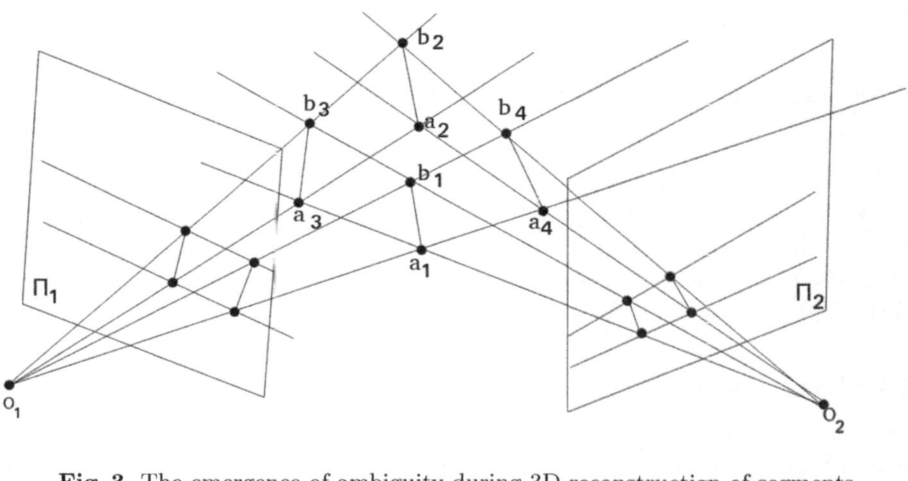

Fig. 3. The emergence of ambiguity during 3D reconstruction of segments

Theorem 1. *We consider two finite set of segments $S_i \subset \Pi_i,\ i = 1, 2$ such that*

$$S_i = \{(u, v) : u, v \in \Pi_i\},\ i = 1, 2.$$

Let $P_i \subset \Pi_i$ be a set of ends of segments. We define the following finite set $\Sigma \subset \Pi_2$ by putting

$$\Sigma = \{p : p = L(u, y) \cap L(v, z), u, v, y, z \in P_2 : (u, v), (y, z) \in S_2\}.$$

If $\pi_2(O_1) \notin \Sigma$ then there exists unique finite set S of segments $(a, b), a, b \in \mathbb{R}^3$ such that

$$\pi_i(S) = S_i, i = 1, 2.$$

Proof. We give a method of the recounstruction of each segment $(a, b) \in S$. Choose some segment $(p, q) \in S_1$. As above we construct two rays

$$O_1p, O_1q.$$

Construct the projections of these rays on the second plane

$$\pi_2(O_1p),\ \pi_2(O_1q).$$

Now we find intersections

$$A_p = \pi_2(O_1p) \cap P_2,\ A_q = \pi_2(O_1q) \cap P_2.$$

We note that

$$\pi_2(O_1) = \pi_2(O_1p) \cap \pi_2(O_1q).$$

Since $\pi_2(O_1) \notin \Sigma$ then it is found **at most one** segment $(u, v) \in S_2$ such that

$$u \in A_p,\ v \in A_q,$$

as we can see on Fig. 2. Then we construct necessary segment (a, b) as

$$a = O_1p \cap O_2u,\ b = O_1q \cap O_2v.$$

We note that the set Σ is finite, so nowhere is dense. Therefore, the condition of the Theorem 1 is almost always satisfied. Moreover, these conditions can always be met by first calculating the set $\Sigma \subset P_2$ and then choosing the location of the first camera.

3.2 Experiments

The experiments are fulfilled in Blender 3D software. Algorithm is implemented using Python programming language and NumPy, SciPy packages. We designed the classes which is demonstrated on the Fig. 4.

The matrix R in class *Camera* defines vectors e_1, e_2, ξ. The vector O is a center of projection π. The method *w2c()* calculates coordinates X, Y, Z in formulas (1)–(3). The method *reconstruct_by_edges()* fulfills directly reconstruction according with previous subsection. Reconstructed 3D model is presented by class *Model3D*. The objects of this class are described finite set of P and array E of pairs of integers—array of edges. An object of class *Model3D* can be loaded from OBJ file and it can be saved as OBJ file. Also, the method *build_edges()* build edges by the description of arrays of vertices and faces of a 3D model from OBJ.

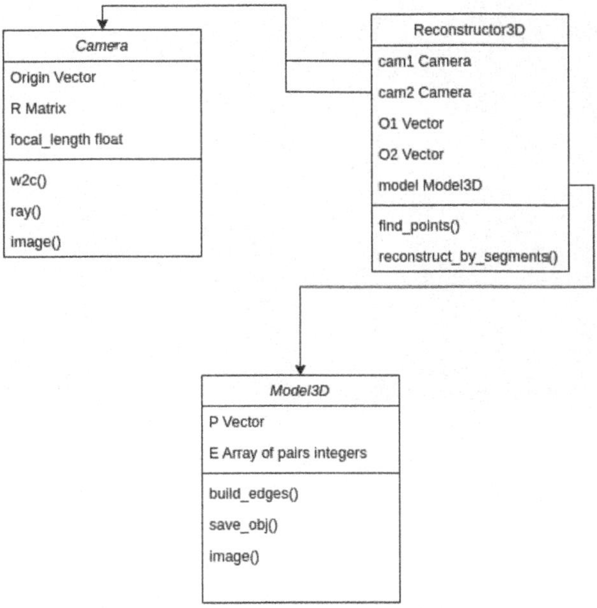

Fig. 4. Classes

Our algorithm gets two sets of edges as we can see on Fig. 5. There we can see two projections of the 3D model which are constructed from two cameras.

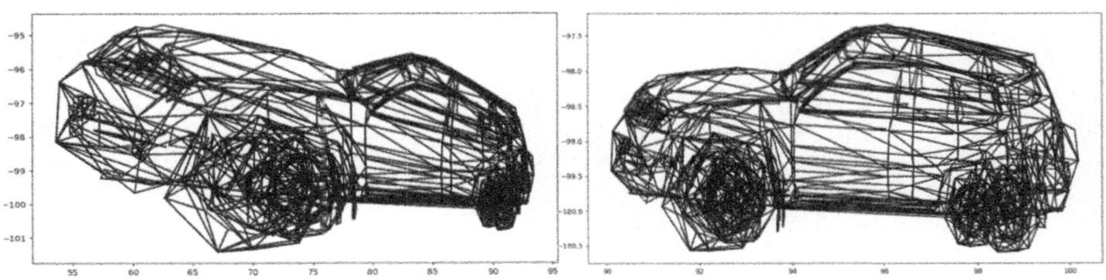

Fig. 5. The given two projections from two cameras

The result of reconstruction is saved as a OBJ file with array of points and array of edges. The reconstructed 3D model is shown below on the Fig. 6. Some other results of the experiments can be seen on the Figs. 7, 8, 9, and 10 below.

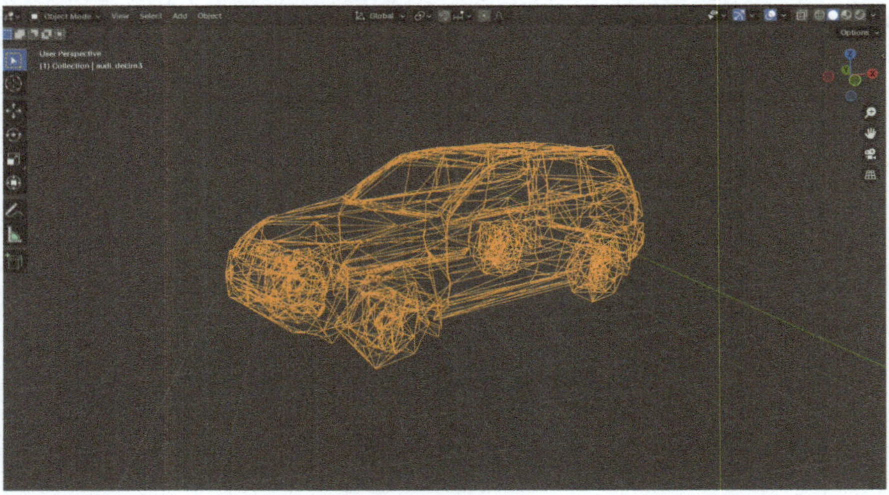

Fig. 6. The reconstructed 3D model in the Blender 3D view window

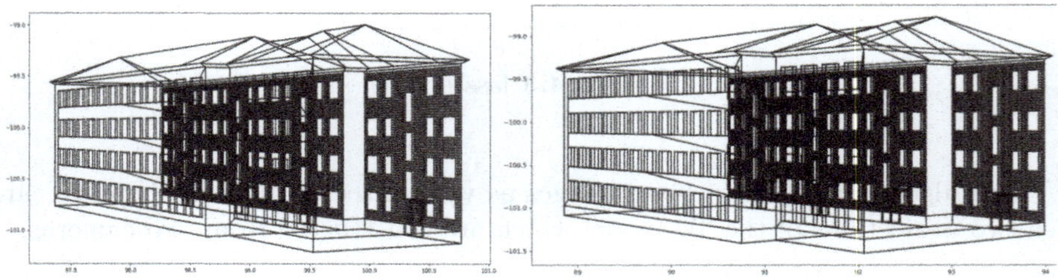

Fig. 7. The given two projections from two cameras

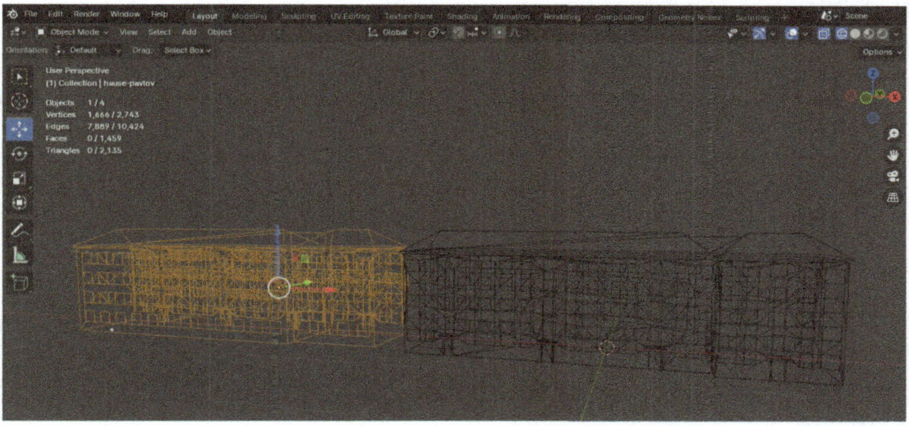

Fig. 8. Origin (on the right) and reconstructed (on the left) 3D models in Blender 3D view window

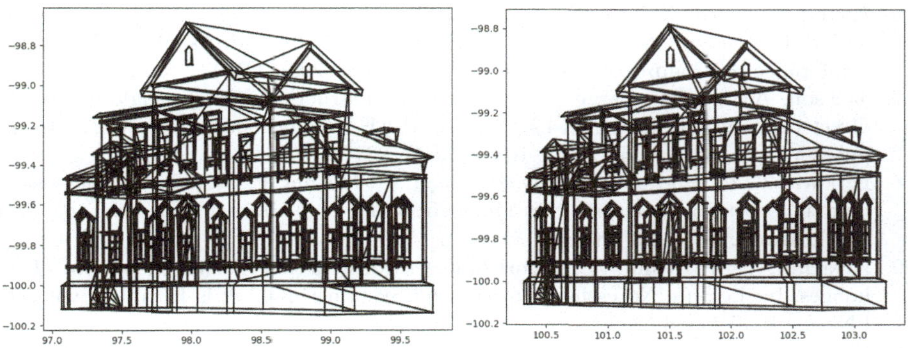

Fig. 9. The given two projections from two cameras

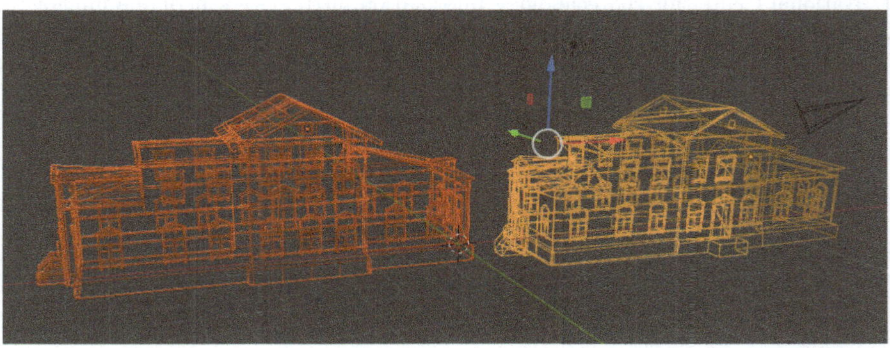

Fig. 10. Origin (on the left) and reconstructed (on the right) 3D models in Blender 3D view window

References

1. Mousavian, A., Anguelov, D., Flynn, J., Kosecka, J.: 3D bounding box estimation using deep learning and geometry. In: 2017 IEEE Conference on Computer Vision and Pattern Recognition (CVPR) (2017). https://doi.org/10.1109/cvpr.2017.597
2. Gordeev, A.Y., Klyachin, V.A.: Determination of the Spatial Position of Cars on the Road Using Data from a Camera or DVR. In: Popkova, E.G., Sergi, B.S. (eds.) "Smart Technologies" for Society, State and Economy, ISC 2020. Lecture Notes in Networks and Systems, LNNS, vol. 155, pp. 172–180. Springer, Cham (2021). https://doi.org/10.1007/978-3-030-59126-7_20
3. Gordeev, A.Y., Klyachin, V.A., Kurbanov, E.R., Driaba, A.Y.: Autonomous mobile robot with AI based on Jetson Nano. In: Arai, K., Kapoor, S., Bhatia, R. (eds.) Proceedings of the Future Technologies Conference (FTC) 2020, Volume 1. FTC 2020. Advances in Intelligent Systems and Computing, vol. 1288. Springer, Cham (2021). https://doi.org/10.1007/978-3-030-63128-4_15
4. Ren, M., Pokrovsky, A., Yang, B., Urtasun, R.: SBNet: Sparse Blocks Network for Fast Inference (2018). arXiv:1801.02108
5. Hu, H., et al.: Joint Monocular 3D Vehicle Detection and Tracking (2018). arXiv:1811.10742v3
6. Huang, S., Qi, S., Zhu, Y., Xiao, Y., Xu, Y., Zhu, S.C.: Holistic 3D scene parsing and reconstruction from a single RGB image. In: Proceedings of the European Conference on Computer Vision (ECCV), pp. 187–203 (2018)
7. Jackson, A.S., et al.: Large pose 3D face reconstruction from a single image via direct volumetric CNN regression. In: 2017 IEEE International Conference on Computer Vision (ICCV), pp. 1031–1039. IEEE (2017)
8. Ferkova, Z., Urbanova, P., Cerny, D., Zuzi, M., Matula, P.: Age and gender-based human face reconstruction from single frontal image. In: Multimedia Tools and Applications, pp. 1–26 (2018)
9. Klyachin, V., Klyachin, A.: Algorithm of reconstruction 3D model by two of its images. In: Lecture Notes in Networks and Systemsthis Link is Disabled, vol. 597 LNNS, pp. 374–383 (2023)
10. Foster, I., Kesselman, C., Nick, J., Tuecke, S.: The Physiology of the Grid: an Open Grid Services Architecture for Distributed Systems Integration. In: Technical Report. Global Grid Forum (2002)
11. National Center for Biotechnology Information. http://www.ncbi.nlm.nih.gov
12. Huang, S., Qi, S., Zhu, Y., Xiao, Y., Xu, Y., Zhu, S.C.: Holistic 3D scene parsing and reconstruction from a single RGB image. In: Proceedings of the European Conference on Computer Vision (ECCV), pp. 187–203 (2018)
13. Klyachin, V.A., Grigorieva, E.G.: Algorithm of automatic detemining of camera orientation parameters based on characteristic elements on image. In: Tendencii razvitiya nauki I obrazovaniya, no. 45–46, pp. 10–20 (2018)
14. Klyachin, V.A., Grigorieva, E.G.: A 3D reconstruction algorithm of a surface of revolution from its projection. Sib. Zh. Ind. Mat. **23**(1), 84–92 (2020)
15. Liu, P., et al.: Deep Learning to Segment Pelvic Bones: Large-scale CT Datasets and Baseline Models. arXiv:2012.08721

Identifying Factors Affecting Academic Achievement at Najran University Using Principal Component Analysis Technique (PCA)

Rasha Gaffer M. Helali[1(✉)], Nawal Ali Y. Abdelrahman[2], and Somia B. Mohammed[2]

[1] Department of Computer Science, College of Computing and Information Technology, University of Bisha, P.O Box 551, Bisha, Saudi Arabia
rashahilal@ub.edu.sa

[2] Najran University, The Applied College, Najran, Saudi Arabia
nayassein@nu.edu.sa

Abstract. The progressive expansion of educational institutions in Saudi Arabia strives to raise educational standards and better equip citizens for the future. The objective of this study is to review the factors that have a great influence on academic performance. The study conducted survey concerns opinion of faculty members in Najran University on the factors may affecting academic performance from their point of view in addition to simple survey for selected sample of student in final year in computer science and business administration departments. Course portfolio is also considered for analysis purpose in order to establish early warning indicators and improve weaknesses; principal component analysis is used to analyze collected data in addition to statistical analysis. The experimental results indicate that, with a Wight of 0.999, the course's knowledge percentage has the biggest impact on students' academic success. in addition to additional elements like the use of assistive technology and lecturer qualification level also may effect.

Keywords: Academic performance · clustering · Educational data mining · learning outcomes · student achievements

1 Introduction

The practice of passing down information through the generations is known as education in general. Reports, exams, and occasionally research are used to quantify a learner's success in meeting learning outcomes in formal education. The phrase "academic performance" is typically used in education to describe this achievement. In order to raise their performance standards and guarantee the accomplishment of current educational objectives, a large number of colleges are currently pursuing academic accreditation. It is well recognized that the degree of success in reaching students' learning objectives and academic performance is the most important focus of the ideas of quality of performance. This desire has led to the development of increasing university student academic performance as a major study area. The elements that have a major impact on

R. Silhavy and P. Silhavy (Eds.): CSOC 2024, LNNS 1120, pp. 239–250, 2024.
https://doi.org/10.1007/978-3-031-70518-2_21

pupils' performance are of interest to the researchers. Numerous studies focused on the examination of educational data to extract important insights that influence the caliber of learning. A study at the University of Hail in the Kingdom of Saudi Arabia (KSA) sought to identify the variables influencing the academic achievement of student nurses. According to the study, aspects associated to teachers, such as their relationships with students and their ability to update courses to reflect current developments in the field, have the most effects on students' academic achievement [1]. Another study by Kassu M. and Jemal A. was carried out in 2016 at Wolo University. This study's main goal is to identify the critical variables affecting students' academic performance as indicated by their CGPA. The study's data was collected through the use of a questionnaire method that covers a wide range of factors related to students' academic achievement, including the educational and sociodemographic backgrounds of both students and their parents, as well as the characteristics of lecturers, general university facilities, and parents' economic backgrounds [2]. Amel Al Shehry and Sulima Mohamed conducted a study at Najran University's Preparatory Year for Girls in 2016 to look into the elements that influence undergraduate students' academic success. The percentage of students joining the college, the length of lecturers' teaching experience, and the GPA all showed a highly significant correlation. Nonetheless, Amel Al Shehry and Sulima Mohamed (2016) found no significant correlation between staff credentials (Master-Ph.D.) and the educational background of single or dual parents and GPA [3]. In 2017, Ryan S. conducted research at an Australian institution to determine the elements that lead to academic achievement in first-year allied health and science degrees. According to their study's findings, government secondary school pupils outperformed non-government school students in terms of first-year grade point average [4]. In a study published, Yahya M. and colleagues investigated the variables influencing computing students' academic performance at the University of Tabuk's Faculty of Computers and Information Technology (FCIT). The study employed the principal component analysis technique and t-test technique to select the key factors affecting academic performance [5]. Their results reported 21 affecting factors, including the top 5 factors: gender, safari-camping, reading preferences, chronic disease, international sport fan, and availability of a WhatsApp account. Extensive literature reviews of the Extensive Data Mining (EDM) research field have been provided by Romero and Ventura [6], covering the research done in the area between 1995 and 2005, and by Nithya [7], for the period after 2005 [8]. It is remarkable that most often, the attention of researchers has been attracted and the reasons for applying data mining at higher education institutions have focused mainly on retention of students, improving institutional effectiveness, and enrolment management. The ability to predict student performance is very important in educational environments. Increasing student academic success is a long-term goal in all academic institutions [9]. In 2019, Reta R and Henok W a study investigating factors affecting students' academic achievement in science and mathematics in primary schools of Hawassa City. Their study results indicated that low studying of habits, low follow-up the progress of students by their parents, low education level of parents, and low monthly income status of parents are the major challenging for factors successful academic achievement of students in science and mathematics [10].

Getahun (2022) investigated factors affecting college student academic performance some of them are gender, age, family monthly income, study hours, stimulant use during the study and English language proficiency [11]. Their study confirmed finding from another study by helali, R, e.al (2022) [12]. According to these studies there are factors and habits may improve student success rates included parents' educational background, taking notes during lectures, having fewer absences during the semester, using extra materials to aid in understanding, finding extra materials online, and reading silently. When it comes to family support, parents' encouragement and insistence that their children study have a significant impact on the kids' academic achievement. Using a unique ergonomic appraisal framework, Gumasing Ma. J. and Frances F. have provided a method to examine the aspects affecting students' academic performance and motivation to study during online learning [13]. According to Ma. J. and Frances F. (2023), the design of the workstation, the use of a learning management system (LMS), access to technology, the way that instruction is delivered, the temperature, and the visual learning style all had a significant impact on students' learning motivation, which in turn affects their academic performance. Another study done at the University of Bosaso presents found many factors that affect students' academic achievements with a negative outcome. The major ones were student factors, such as students' learning skills, parental background, environmental causes, and peer influence [14]. The current study contributes to the existing literature by adding a new amount of data on the subject. On the other hand, the current study is an attempt to integrate a machine learning approach with a statistical approach for achieving the study purpose. The subsequence sections are organized as follows: Sect. 2 contains the following methodology. Section 3 describes the proposed model and used data set. Then the following sections highlight the analysis and results. Finally, the conclusion presented.

2 Methodology

The study goes in two directions; the first direction is to investigate the main factors that may affect student academic achievements from the instructor's point of view, this data was obtained through a questionnaire instrument. The second direction is to collect a sample of course achievements and then apply data mining techniques for analysis purposes. Figure 1 shows the research methodology.

The sample data contains 50 records collected from course portfolios and student achievement records. The attribute selection depends on the result of conducted survey and literature findings. The following sub-sections detail the following process.

A- Survey development and Data collection:

The collected data mainly depend on a simple web-based survey containing 11 questions about the main factors affecting student achievement. The targeted respondents were faculty members of the Applied College of Najran University. The survey questions were constructed to explore and find unknown features affecting students' achievements from their instructor's point of view, as presented in Table 1. Another survey was used to target students at the final level. The main purpose of the survey was to find out the satisfaction of graduate students with learning outcomes for both

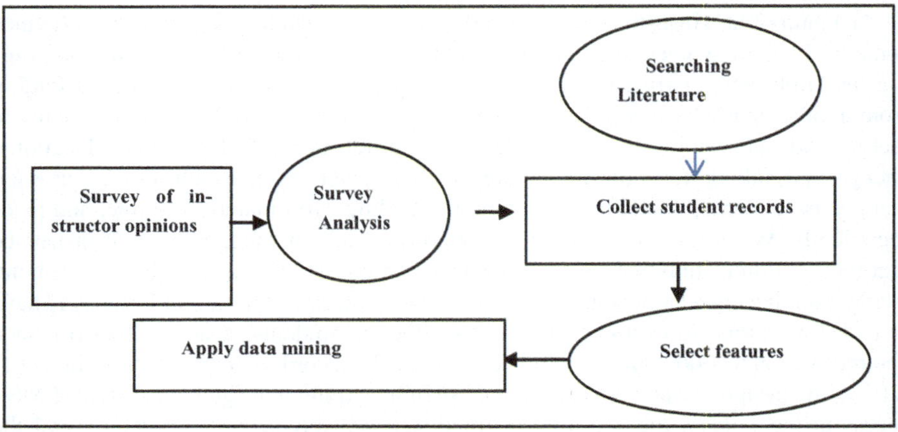

Fig. 1. Research methodology (Source: self–made)

Business Administration and Computer Science Programs. The survey consisted of four points containing nine items, as shown in Tables 2 and 3. The collected data were analyzed and the results were recorded for the next steps. Based on the survey results and literature findings, student and course achievement records were collected and certain features were selected. Finally, the data mining technique of principal component analysis (PCA) was used to analyze the collected data to identify factors that may affect academic achievement.

Surveys in Tables 2 and 3 aimed to measure the student satisfaction of learning outcomes offered by their programs in knowledge, cognitive, communication skills, and group work gained from courses. Collected data were further analyzed using a statistical analysis tool (SPSS) and results were recorded. The following sub-section shows the data analysis results.

B. Data Analysis

The surveyed sample contains a total of 30 faculty members and 50 graduate students. The analysis results of the faculty member's survey are as follows: The sample comprised 66.7% of assistant professors and 4.8 % of associate professors. More than half (66%) of the contributors were business management, 23.8% were computer science, and 9% were other specialties. 85.7% of the respondents think that the degree of the student enrolled in the college will affect his rates. For question four 42.9% said that the main factor affecting student achievement is his depends on little and simple summaries instead of main references and 38.1% think that the size of the scientific subject taught can affect it. 28.6% of faculty members select option 4 "The courses contains theoretical and practical part" as a factor affecting student performance in a certain subject. Only 14% of contributors think that the lack of references could be a factor. 42.9% of the respondent thinks the use of blackboard does not have an effect on student achievement. More than half of the contributors (71%) think that focusing on the knowledge aspect of the course in the final examination has a positive role in student achievements. Most of the respondents (95.2%) agreed with using multiple evaluation methods has a positive impact on the level

Table 1. Detailed survey questions

#	Classification	Survey Items
1	Degree	Lecturer Assistant professor Associate professor
2	Department	Medical college Computer science Human sciences Others
3	From your point of view, did the degree of the student enrolled in the college affect his rates?	Yes No Sometimes
4	From your point of view, which of the following factors can affect student achievements?	-Lake of references -The use of summaries of final exams -The size of the scientific subject taught -The courses contains theoretical and practical part
5	Do you think the use of blackboard system has an effect on improving students achievements	Yes No Sometimes
6	Focusing on the knowledge aspect has a positive role in student achievements	Yes No Some times
7	Do you think that Multiple methods of evaluation have a positive impact on the level of achievement?	Yes No Sometimes
8	Measuring the cognitive knowledge of student in written examinations without practical part does affect the achievement negatively	Yes No Sometimes
9	Do you think that the use of social media as an aid to the development of teaching methods has a positive role in academic achievement	Yes No Sometimes
10	Do you think that the financial status of student family has a positive impact on his achievement?	Yes No Sometimes
11	From your experience, what are the factors that you think may affect student academic achievement?	

of achievement. 47.6% of respondents think that measuring the cognitive knowledge of students in written examinations without practical part has a negative impact on

Table 2. Student satisfaction for learning outcomes in business administration

#		Classification	Survey Items
1	Knowledge	Knowledge of basic concepts, principles and rules	Strongly agree Agree Disagree
2		Description of modern techniques in the field	Strongly agree Agree Disagree
3	Cognitive	Application of the principles and functions of the administrative process in the activities of the organizations	Strongly agree Agree Disagree
4		Ability to analyze the problems facing the student in his work environment and to interpret the results accurately and correctly using the appropriate scientific methods	Strongly agree Agree Disagree
5		Ability to apply modern technology in analysis and decision making for practical issues	Strongly agree Agree Disagree
6	Group work	Ability to work in groups	Strongly agree Agree Disagree
7		Student can take responsibility for continuous professional development to improve student skills	Strongly agree Agree Disagree
8	Communication	The effective use of information network and modern technology in the performance of its work	Strongly agree Agree Disagree
9		Communicate effectively	

student achievement. Only 20% of the respondents think that using social media will not affect academic achievement. Question 10 asking about what are the 8 factors affecting academic achievement from the instructor's point of view, and most responses filled on family follow up and academic desire of student himself. Other issues suggested by staff members such as attending class regularly, evaluation methods used, finical issues, communication between student family and faculty members, Psychological state of the student, and reading habits. Nearly half of the respondents said that (47.6%) the financial status of a student's family has an impact on his /her achievement. The results of the satisfaction survey are as follows: For the business administration program (BA), 84% of respondents were satisfied with the knowledge aspects gained from taught courses in the program. 77% of the respondents were satisfied with cognitive issues, and 81% agreed that learning outcomes related to communication and personal responsibility is satisfied. 78% of contributors agreed about the effective use of the internet and group work aspects in their learning (Table 4).

For the computer science program (CS) overall satisfaction with learning outcomes was 66% and 67% of respondents were satisfied with knowledge aspects gained from

Table 3. Student satisfaction for learning outcomes in computer science

#		Classification	Survey Items
1	Knowledge	Knowledge of basic concepts, principles and rules	Strongly agree Agree Disagree
2		Description of modern techniques in the field	Strongly agree Agree Disagree
3		Ability to explain the various algorithms for building software	Strongly agree Agree Disagree
4		Ability to describe the components of different computer fields	Strongly agree Agree Disagree
5	Cognitive	Ability to operates and applies various computer applications	Strongly agree Agree Disagree
6		Satisfaction of practical parts	Strongly agree Agree Disagree
7		Ability to analyze and Design of algorithms to solve computer problems	Strongly agree Agree Disagree
8	Communication	Oral and written communication used effectively	Strongly agree Agree Disagree
9	Group work	Ability to work in groups	Strongly agree Agree Disagree

Table 4. List of the 4 points total results in BA program

Learning out comes	%
Satisfaction of knowledge outcomes	84%
Satisfaction of cognitive outcomes	77%
Satisfaction of communication and personal responsibility	81%
Satisfaction of communication	78%

taught courses in the program. Only 62% of the respondents were satisfied with cognitive issues, 70% agreed that learning outcome related to communication and personal responsibility is satisfied. 63% of contributors agreed about the effective use of internet and group work aspects in their learning. Table 5 List the 4 points total results.

Table 5. List of the 4 points total results in CS program

Learning outcomes	%
Satisfaction of knowledge outcomes	67%
Satisfaction of cognitive outcomes	62%
Satisfaction of communication and personal responsibility	70%
Satisfaction of communication	63%

The following section describes using of the PCA algorithm to analyze collected data.

3 The Proposed Model

3.1 Data Set Collection

The data set used in this study was obtained from a Applied college, computer science, and business administration department at Najran University. Available data contains 50 Records collected from the courses portfolio for the academic year 2019. For each course, 16 attributes were selected based on conducted surveys. Selected attributes including course code, Credit hours, percentage of knowledge aspect of the course, percentage of the cognitive aspect of the course, summary available (Yes/No), instructor use blackboard (Yes/No), course assessment methods, measuring cognitive parts through the theoretical exam, Uses of social media, availability of multiple channels for communication between the faculty member and student, student attendants, students gains financial aids, availability of course references (Yes/No), grades distribution between learning outcomes, family financial status and qualification level for the instructor. Table 6, below provides a list of attributes and their domains.

3.2 Data Mining Phase

In the following phase, we need to identify the most important features that have a larger impact on the dataset. According to previous studies, the commonly used tool in feature reduction is Principal Component Analysis (PCA) [15].

PCA is a feature reduction technique that provides weighted features in the form of main principal components. One drawback of PCA is that it ignores the outliers [15]. However, this is not a concern in our case as we are looking for common trends rather than special cases. The collected data was fed to the PCA algorithm and results were recorded. Table 7 below lists the top 5 principal components that contribute to the total variance. The features, with the largest influence on the first principal component, would have the largest impact on student achievement. The analysis reveals that the

Table 6. Course attributes.

	Attributes	Domain
1	Course id	Numeric value
2	Credit hours	Numeric value
3	percentage of knowledge aspect of the course	Numeric value
4	percentage of cognitive aspect of the course	Numeric value
5	Summary available	{yes, no}
6	Instructor use black board	{yes, no}
7	Course assessment methods	{two or less, multiple}
8	Measuring cognitive parts through theoretical exam	{yes, no}
9	Uses social media	{yes, no}
10	Multiple channel for communication between faculty member and student	{yes, no}
11	Students gains financial aids	{yes, no}
12	Course references available	{yes, no}
13	Grades well distributed between learning outcomes	{Good, Boor, Excellent}
14	Family financial status	{Good, Boor, Excellent}
15	Qualification level for instructor	{lecturer, assistant professor, associate professor}
16	Student course Achievement	{low, mid, high}

most affecting features are: Percentage of knowledge aspect of the course, using of social media and blackboard as supporting components in the courses, the qualification level for the instructor is also an affecting factor and finally measuring cognitive parts through the theoretical exam. The reached results are consistent with researched results by [13].

The first component achieves 90% over other components. Fig. 2 below shows visually how the first component has a big impact on other components. Some of the listed results confirm the findings from the study mentioned by [4] that reading preferences and using social communication may affect academic achievement. Most students prefer the knowledge parts of their courses more than the cognitive parts A study published by [16] confirmed the effect of the fourth identified factor "Qualification and experience level for an instructor" which has a great effect on student achievement.

Table 7. Lists the top principal components

#	Factors	weight
1	Percentage of knowledge aspect of the course	0.999
2	Uses social media	0.450
3	Qualification level for instructor	0.4288
4	Instructor use black board	0.382
5	Measuring cognitive parts through theoretical exam	0.150

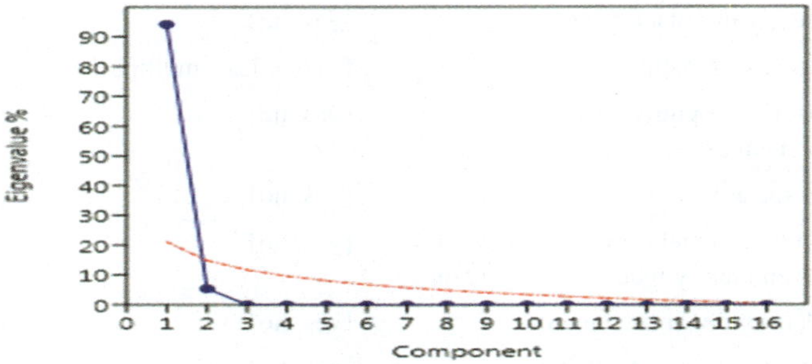

Fig. 2. Components results from PCA analysis

4 Discussion

The PCA method marks 5 factors as effective factors over 16 factors. Percentage of knowledge aspect of the course marked as the largest influence on student achievement in the course with Wight 0.999.

This factor should be considered when comparing student achievements in different courses. Students tend to gain knowledge better and more easily than cognitive aspects. This explains the high success rate in some courses in contrast to others. Somewhat unexpected, or rather this has never been considered, using an auxiliary method such as social media and blackboard to facilitate access to course scientific material in addition to, communication between students and teaching staff is also considered as affecting factor. In this point, it is important to mention that using an online platform like Blackboard in addition to normal teaching has a good impact in contrast to using such platforms instead of traditional learning may have a bad impact on student achievement as reached by [13]. In their study. One important factor that appears to have an impact on performance is the qualification level of the instructor. Students' performance improves as the degree of faculty member's increases. That means more teaching experience of faculty member affects the method of teaching and therefore, improve student learning. Finally, some courses employ theoretical exams only to measure the cognitive parts of the courses. It is better to use practical parts to measure cognitive aspects of the course or even use practical to support student learning in these aspects.

5 Conclusion

Identifying affecting factors and predicting a student's course performance can provide early warnings and timely correction roles for students and teachers. Student performance prediction is the most important direction in educational data mining [17]. The main objective of the study is the identification of highly influencing predictive variables from course portfolios and both teaching staff and students in the final year. Using PCA to identify factors affecting academic performance has shown potential and promise. It is worthwhile to further pursue more data mining techniques to enhance feature selection and therefore, improve the educational process.

Acknowledgments. The authors are thankful to the Deanship of Graduate Studies and Scientific Research at University of Bisha for supporting this work through the Fast-Track Research Support Program.

References

1. Alshammari, F., et al.: Factors affecting the academic performance of student nurses: A cross-sectional study. J. Nurs. Educ. Pract. **8**(1), 60 (2017)
2. Beyene, K.M., Jemal, A.Y.: Multilevel analysis for identifying factors influencing academic achievement of students in higher education institution: the case of Wollo University. J. Educ. Practice **7**(13), 17–23 (2016)
3. Al Shehry, A., Youssif, S.M.A.: Factors affecting academic performance of undergraduate students at Najran Preparatory Year for Girls-Najran University 2015–2016. Int. J. Asian Soc. Sci. **7**(1), 1–18 (2017)
4. Anderton, R.S.: Identifying factors that contribute to academic success in first year allied health and science degrees at an Australian university. Aust. J. Educ. **61**(2), 184–199 (2017)
5. Yahya, M., Khaled, M., Osman, A., Osman, A.: Factors influencing academic achievement of undergraduate computing students. Int. J. Comput. Appl. **146**(3), 23–28 (2016)
6. Osmanbegovic, E., Mirza, S.: Data mining approach for predicting student performance. J. Econ. Bus. **10**, 3–12 (2012)
7. Kabakchieva, D.: Student performance prediction by using data mining classification algorithms. IJCSMR **1**, 686–690 (2012)
8. Yassein, N.A., Helali, R.G.M., Mohomad, S.B.: Predicting student academic performance in KSA using data mining techniques. J. Inf. Technol. Softw. Eng. **7**(5), 1–5 (2017)
9. Agaoglu, M.: Predicting instructor performance using data mining techniques in higher education. IEEE Access **4**, 2379–2387 (2016)
10. Jetu, R.R., Wariso, H.: Identifying Factors Affecting Students' Academic Achievement in Science and Mathematics in Primary Schools of Hawassa City (2019)
11. Getahun, Kl.A.: A Bayesian approach to investigating factors influencing polytechnic college students' academic achievement. Educ. Res. Int. **2022**, 1–10 (2022)
12. Helali, R.G.M., Abdelrahman, N.A.Y., Zaroug, R.A.: Measuring the impact of study habits on student academic performance in KSA. J. Positive School Psychol. **6**, 328–335 (2022)
13. Gumasing, M.J.J., Castro, F.M.F.: Determining ergonomic appraisal factors affecting the learning motivation and academic performance of students during online classes. Sustainability **15**(3), 1970 (2023). https://doi.org/10.3390/su15031970
14. Suleiman, A.H.: Factors that affect students' academic achievement in the faculty of social science at the university of Bosaso, Garowe, Somalia. Open J. Soc. Sci. **11**(2), 446–461 (2023)

15. Li, K.F., Rusk, D., Song, F.: Predicting student academic performance. InComplex, Intelligent, and Software Intensive Systems (CISIS). In: 2013 Seventh International Conference on 2013 Jul 3, pp. 27–33. IEEE
16. Ullah, N., Almani, A.S.: Factors affecting students' academic performance: a case study of secondary schools of Makran division Balochistan, Pakistan. Webology **19**(2), 2749–2764 (2022)
17. Liu, C., Wang, H., Yuan, Z.: A method for predicting the academic performances of college students based on education system data. Mathematics **10**(20), 3737 (2022)

Model Proposal for Malware Detection Using Deep Learning on Cell Phones with Android Operating System

David Silvera[1], Pedro Molina[1], and Wilfredo Ticona[1,2](✉)

[1] Universidad Tecnológica del Perú, Lima, Perú
{U19206135,pmolina}@utp.edu.pe, wmamani@esan.edu.pe
[2] Universidad ESAN, Lima, Perú

Abstract. In recent years, with the advancement of technology, the number of malware attacks has also increased, Peru being one of the countries with more cyberattacks of this type to electronic devices. Therefore, the main objective of the proposed study is to design a Malware detection model using Deep Learning techniques in cell phones with Android operating system. For this purpose, 5 phases have been carried out in the methodology to be able to recognize Malware anomalies and present error-free data comprising the following: Imported dataset, preprocessing, feature extraction, model implementation and evaluation. In addition, different machine learning models such as DT, RF, SVM and K-NN were developed and the results were evaluated using the metrics Accuracy, Precision, Recall and F1-score. For malware detection, RF presents the highest percentage in all the indicated parameters 89.23%, 87.59% and 88.90% and in Recall it is below K-NN with 85.84%. The RF model outperforms all the algorithms applied in the model, showing that better predictive results can be obtained

Keywords: Malware detection · malware · deep learning · machine learning · android · cell phone

1 Introduction

Over the years, the use of mobile devices has become more prevalent, being used by adults, young people, and children. Today, malware is constantly being updated and uses various techniques to hide its functionality and evade anti-malware programs [1]. As a result, the use of mobile devices is increasing rapidly, and the mobile application market is also growing [2]. As such, it has led to a significant increase in threats to exploit user vulnerabilities and circumvent device security. On the other hand, malware has the potential to compromise user privacy, steal confidential information and cause personal harm [3]. This has led to new data processing models, but it has also introduced new threats and difficulties for the end user not to know when his cell phone is being tapped. The use of this information has become a lucrative target for malware designers who can exploit it for commercial purposes.

R. Silhavy and P. Silhavy (Eds.): CSOC 2024, LNNS 1120, pp. 251–268, 2024.
https://doi.org/10.1007/978-3-031-70518-2_22

According to the cybersecurity company Kaspersky Lab, [4] between 2021 and 2022 Latin American countries were victims of more than 398 million malware attacks aimed at stealing money. Furthermore, according to Fortinet [5], Peru is one of the countries with the most malware attacks only behind Mexico and Brazil, reports from Fortinet company shows that Peru suffered more than 4.7 billion malware cyber-attack attempts in the whole year 2021.

2 Related Work

Related work, Aurangzeb S. and Aleem M, [1] proposed a joint malware evaluation approach applying Deep Learning techniques, to perform such approach they had to apply their own methodology and use a dataset called KronoDroid, in order to have better robustness, furthermore, they used Random Forest (RF), K-Nearest Neighbors (KNN), XGBoost, Multilayer Perceptron (MLP) and Gradient Boosting (GB) techniques; The Random Forest (RF) technique had the best results of 95% accuracy, 95% precision, 96% recall and 95% F1-socre.

The study by Ko E. *et al.* [2] proposed a new approach in malware detection called ACAMA (API-based CNN for Android Malware Analysis) which uses API-based functions to extract a dataset using Deep Learning techniques, the DL techniques they used was, Convolutional Neural Network (CNN) composed with several layers such as: Convolution Layers, Grouping Layers, Normalization Layers and Fully Connected Layers, in addition, the method uses a ReLU (Rectified Linear Units) activation function; the result obtained was that ACAMA manages to classify 72. 456% of malware and with an accuracy of 99.8%.

In the same way, Manzil H and Manohar Naik S, [3] proposed a new DL model for Android malware category detection, for their model proposal their methodology is based on 3 phases: the first phase is data acquisition, the second phase is data preprocessing and the last phase of characterization vector generation, in addition, the techniques to be used are: RF, Decision Tree (DT), Logistic Regression (LR), Support Vector Machine (SVM), KNN, AdaBoost, MLP and CNN; the technique with the best result for this model is: RF with 98.7% accuracy, 98.7% precision, 98.7% recall and 98.7% F1-score.

Similarly Gómez A and Muñoz A, [4] proposes a DL-based attack detection and classification model, to implement the mentioned model they followed the phases of this methodology which are: static analysis, feature extraction, model training and evaluation; so, the techniques they used were: DT, RF, SVM and with Parametric Rectified Linear Unit (PReLU) activation technique, the technique with the best result is: RF with PReLU activation with a result of 99. 3% accuracy, 99.01% precision, 99.4% recovery and 99.3% F1-score.

Finally, in the paper by Rathore H. *et al.* [5] propose a strategy called GAAN (Gradient-based Adversarial Attack Network) to model different models in malware detection using DL techniques, the methodology used has 5 phases, the first phase is data collection, data mining, malware detection models, adversarial attack and adversarial defense, then the DL techniques they used are: DT, SVC, RF, Extra Trees (ET) and Gradient Boosting (GB); the technique with best result was: ET has 95.27% accuracy, 96.14% precision, 93.89% recovery and 96.34% F1-score.

With the articles mentioned above, there are several related works that seek to prevent this problem, so the Deep Learning techniques that will be used are: DT, RF, SVM AND KNN that serve to facilitate the detection of malware on mobile devices by analyzing patterns of operation and unique characteristics of malware. This paper seeks to test if the proposed method is efficient for malware detection using a dataset with the above-mentioned techniques.

3 Methodology

In this study, a five-phase methodology for malware detection was carried out [1, 3, 4]. In the first phase, the data set needed for processing was obtained. In the second phase, the dataset obtained in phase one was processed. In the third phase, it is responsible for identifying and extracting malware features. In the fourth phase, it consists of applying algorithms to create models that can classify the malware samples. Finally, in the fifth phase, the results obtained from the algorithms will be evaluated (Fig. 1).

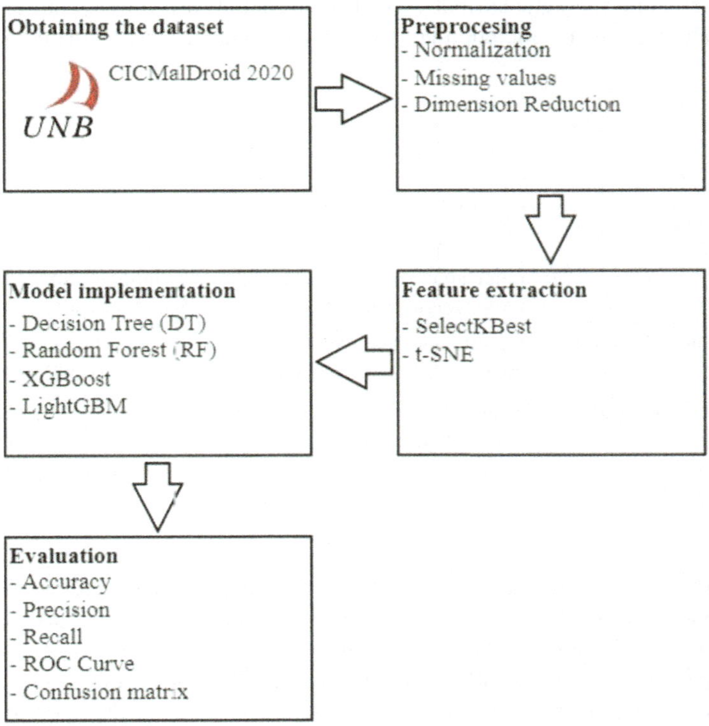

Fig. 1. Proposed methodology

3.1 Obtaining the Data Set

The dataset used in the research was acquired from the project Manzil *et al.* [3] and is available on the official website of the University of New Brunswick (https://www.unb.ca/cic/datasets/maldroid-2020.html). The dataset is referred to as "CICMalDroid 2020" and was created by the Canadian Cybersecurity Institute (CIC) with the goal of providing an up-to-date, diverse, and comprehensive database of android malware. The dataset contains 11,598 malware samples, which are classified into five distinct categories: adware, banking malware, SMS malware, mobile riskware and benign. Each malware sample has associated with it a number of attributes or characteristics, both static and dynamic, that enable its analysis and detection. In addition, the dataset includes the APK files of the malware samples, as well as the CSV and PCAP files containing the data captured during the execution of the samples in a controlled environment.

3.2 Preprocessing

In preprocessing the research dataset was imported from the "feature_vectors_static.csv" file, which contains the static attributes extracted from the APK files of Android malware samples. These attributes include the size, format, code, and behavior of each sample (Table 1).

Table 1. Preprocessing activities

Activities	Description	Tasks
Normalization	Scale characteristics to be in a range and avoid them having more weight than others	Apply normalization technique such as StandardScaler
Management of Missing Values	Identify and deal with outliers, missing data in the database	Use missing data in the database SimpleImputer
Dimension Reduction	Reduce the number of features to simplify the model	Apply data reduction, and potentially increase model performance using PCA

- *Normalization:* Normalization is essential to ensure that all features are of equal importance during ML model training. Without normalization, features with wider ranges of values could dominate the learning process, which could result in suboptimal model performance, especially in algorithms that depend on the distance between instances or those that are sensitive to the scale of the inputs, such as neural networks. Therefore, the nor-malization that has been implemented using StandardScaler from scikit-learn, which standardizes the features by removing the mean and scaling to the unit variance. This means that each feature will be transformed to have a mean of 0 and a standard deviation of 1.

The StandardScaler class of the sklearn.preprocessing library is used to standardize the data, i.e. remove the mean and scale the data to have a variance equal to 1. This improves the performance of some ML algorithms. The line 'from sklearn.preprocessing import StandardScaler' is used to import the StandardScaler class from the sklearn.preprocessing module. The next line 'scaler = StandardScaler()' creates an instance of the StandardScaler class and assigns it to the scaler variable. Finally, the code 'X_scaled = scaler.fit_transform(X)' calls the fit_transform method of the scaler instance, passing as argument the data array X, to calculate the mean and standard deviation and then transform the data to have mean zero and variance one.

- *Management of Missing Values:* Handling missing values is crucial to avoid biases and errors in the machine learning model. Algorithms typically require a complete data set, and any missing values can negatively impact their ability to learn from the data. In the dataset used, it was identified that there were no missing values, it is important to have a procedure in place to deal with them. This could be implemented using scikit-learn's SimpleImputer, which can fill in missing values with the mean, median, or mode of the corresponding columns, thus ensuring that the model is trained on a complete dataset.

The SimpleImputer class of the sklearn.impute library is used to replace missing values in a data set. The purpose of replacing missing values is to assign them a value based on an imputation strategy, such as the mean, median, or most frequent value in each column. This can improve the performance of some algorithms. The line 'from sklearn.impute import SimpleImputer' is used to import the 'SimpleImputer' class from the 'sklearn.impute' module. The following line 'imputer = SimpleImputer (strategy='mean')' creates an instance of the 'SimpleImputer' class and assigns it to the 'imputer' variable. The 'strategy='mean'' argument indicates that the mean of each column will be used to impute missing values. Finally, the code 'X_imputed = imputer.fit_transform(X)' calls the 'fit_transform' method of the 'imputer' instance, passing it the data matrix 'X' as argument. This method calculates the mean of each column of 'X', and then replaces the missing values with the corresponding mean.

- *Dimension Reduction:* Finally, applied to simplify the model, improve computational efficiency, and reduce the risk of overfitting. This is particularly important in datasets with many features, where some may be redundant or irrelevant to the prediction. Therefore, it was decided to perform dimensionality reduction using Principal Component Analysis (PCA), which transforms the dataset to a lower dimensional space while preserving as much variance as possible. This is achieved by selecting a number of principal components that account for most of the variability in the data.

This code uses the scikit-learn ML library to perform principal component analysis (PCA) on a scaled data set. PCA is a dimensionality reduction technique that projects the data into a lower dimensional space, preserving most of the variance. The line 'from sklearn.decomposition import PCA' imports the 'PCA' class from the 'decomposition' module of scikit-learn. The next line 'pca = PCA (n_components = 0.95)' takes care of creating a 'PCA' object with parameter 'n_components' equal to 0.95. This means that the PCA will select the number of components such that the percentage of variance explained is greater than 0.95. Finally, the line 'X_reduced =

pca.fit_transform(X_scaled)' applies the PCA to the scaled data set 'X_scaled' and returns the result in the variable 'X_reduced'. The 'fit_transform' method does two things: first it fits the PCA model to the data, calculating the eigenvectors and eigenvalues; and then it transforms the data, projecting it into principal component space.

3.3 Feature Extraction

After preprocessing, the feature extraction phase is an essential step in the methodology for detecting malware on Android, since it allows obtaining a numerical or symbolic representation of each malware sample to facilitate its analysis and classification. This phase consists of applying different techniques to select, extract and reduce the malware features, according to the criteria of relevance, significance, and redundancy. To this end, the following activities were carried out (Table 2):

Table 2. Feature selection activities

Activities	Description	Tasks
SelectKBest	Identify relevant features for malware detection	Use the feature importance analysis technique
		Evaluate the importance of the feature (SelectKbest)
t-SNE	Reduce the number of variables	Apply t-SNE technique to reduce dimensionality

- *SelectKBest:* In this activity, feature selection was performed, which consists of identifying the most relevant and significant features for malware detection, and discarding those that are redundant or irrelevant. For this purpose, the feature importance analysis technique was used, which allows evaluating the importance of each feature according to its contribution to the classification model. The SelectKBest function of the scikit-learn Python library was used, which selects the features with the best scores, secondly, feature extraction was performed, which consists of creating new variables that capture the patterns or relationships between the original features.

The scikit-learn machine learning library is used to perform univariate feature selection on the scaled and labeled dataset. Univariate feature selection consists of choosing the best features based on univariate statistical tests. The line 'from sklearn.feature_selection import SelectKBest, f_classif' imports the 'SelectKBest' class and the 'f_classif' function from the 'feature_selection' module of scikit-learn. 'SelectKBest' is an object that eliminates all features except the 'k' highest scoring ones. 'f_classif' is a function that calculates the ANOVA F-value between the label and each feature for classification tasks. The next line 'selector = SelectKBest(f_classif, k=10)' creates a 'selector' object with parameter 'k' equal to 10, which means that the selector will choose 10 features with highest F value. To finish, 'X_new = selector.fit_transform(X_scaled, y)' applies the selector to the scaled data set 'X_scaled' and 'y' labels, and returns the result in the variable 'X_new'. The 'fit_transform' method does two things: first it fits the selector to the data, calculating the scores and p-values for each feature; and then it transforms the data, reducing it to the selected features.

- *T-SNE:* Thirdly, dimensionality reduction was performed, which consists of reducing the number of variables used to represent each malware sample, without losing much information. For this purpose, the t-distributed Stochastic Neighbor Embedding (t-SNE) technique was applied, which allows visualizing high dimensionality data in a low dimensionality space, preserving the similarities between the points. The TSNE class of the scikit-learn Python library was used, which allows to fit and apply the t-SNE transformation to the data and obtain a representation of the data.

The scikit-learn machine learning library allows a visualization of high-dimensional data using the t-SNE algorithm, which is a technique that converts similarities between data points into joint probabilities and tries to minimize the Kullback-Leibler divergence between the joint probabilities of the low-dimensional space and the high-dimensional data. The line 'from sklearn.manifold import TSNE' takes care of importing the 'TSNE' class from the 'manifold' module of scikit-learn. The next line 'tsne = TSNE (n_components=2, random_state=42)' creates a 'tsne' object with parameter 'n_components' equal to 2 and parameter 'random_state' equal to 42. This means that the t-SNE will project the data into a two-dimensional space. Finally, the line 'X_tsne = tsne.fit_transform(X_scaled)' applies the t-SNE to the scaled data set 'X_scaled' and returns the result in the variable 'X_tsne'. The 'fit_transform' method does two things: first it fits the t-SNE model to the data, calculating joint probabilities and gradients; and then it transforms the data, reducing it to the coordinates of the low-dimensional space.

3.4 Model Implementation

For the implementation of the model, the "Label" column was coded, whose textual values became numbers: Benign (0) and Malware (1), this is linked to the past records to avoid bias (Table 3).

Then we chose the appropriate algorithms for malware detection, taking into account the type of data and evaluation criteria and these will be compared between several algorithms to evaluate the performance, also, the training of the selected algorithms is performed, applying each algorithm to the data set, adjusting its parameters with optimal values, proceeding to calculate and analyze the correlation of the characteristics of malware, then adjust the parameters to improve the predictive ability.

- *Decision tree (DT):* It is a supervised learning algorithm used for classification and regression. It consists of a hierarchical structure of nodes representing questions or conditions about data characteristics, and branches representing possible answers or outcomes. The objective is to find the optimal path from the root node to the leaf node containing the target class or value [6] (Fig. 2).

- *Random forest (RF):* It is a supervised learning algorithm used for classification and regression. It consists of a set of decision trees that are trained with different subsets of the data and different subsets of the features. The goal is to combine the predictions

Table 3. Model selection activities

Activities	Description	Tasks
Algorithm selection	Choose appropriate algorithms for malware detection	Use algorithms such as: DT, RF, XGBoost and LightGBM
Decision Tree (DT)	Split the data according to rules	Measure and contrast the performance of various methods
Random Forest (RF)	Combine several decision trees	Evaluate and compare various algorithms in terms of performance
XGBoost	Construct the optimal set of decision trees	Minimize the error between predictions and actual values
LightGBM	Build decision trees efficiently	Increase accuracy and reduce memory usage
Model training	Use the training set to train the selected models	Calculate and analyze the correlation between all the characteristics
		Adjust model parameters to improve their predictive capacity using model.fit

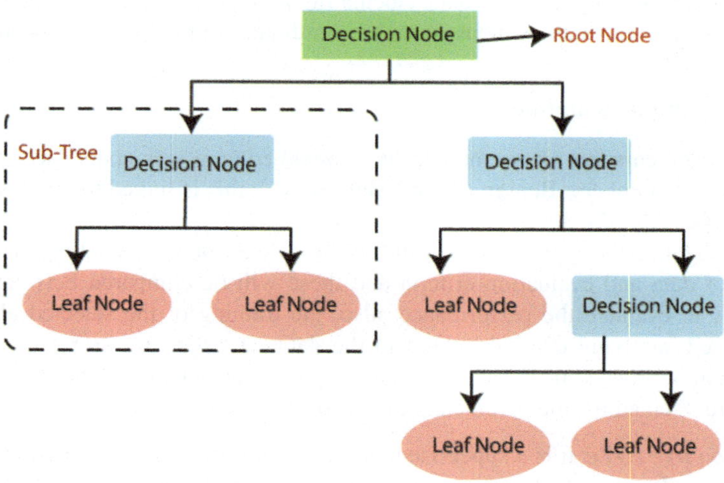

Fig. 2. Decision tree model.

of the individual trees using a voting or averaging method, to obtain a more robust and accurate prediction than that of a single tree [7] (Fig. 3).

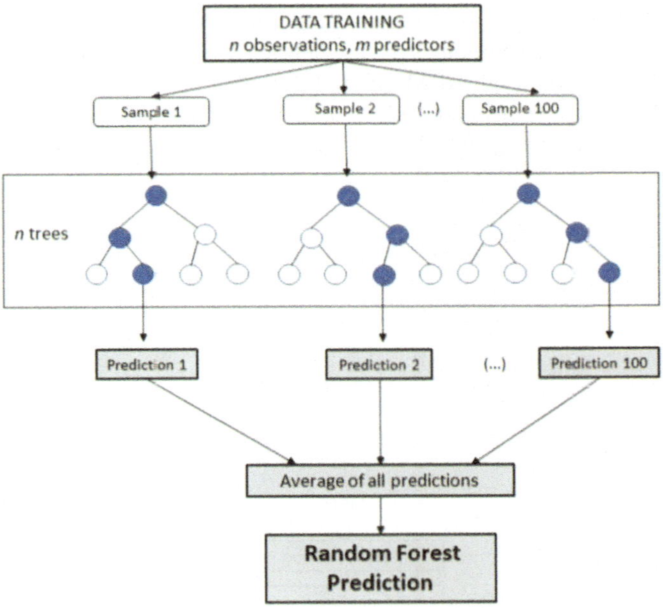

Fig. 3. Random forest model

- *XGBoost:* It is a supervised learning algorithm that uses the ensemble method called boosting, which combines the predictions of several decision trees to produce a stronger prediction. XGBoost builds decision trees sequentially, correcting the errors of previous trees, using the gradient of a loss function and regularization, pruning and sampling techniques to avoid overfitting and improve generalization [6] (Fig. 4).

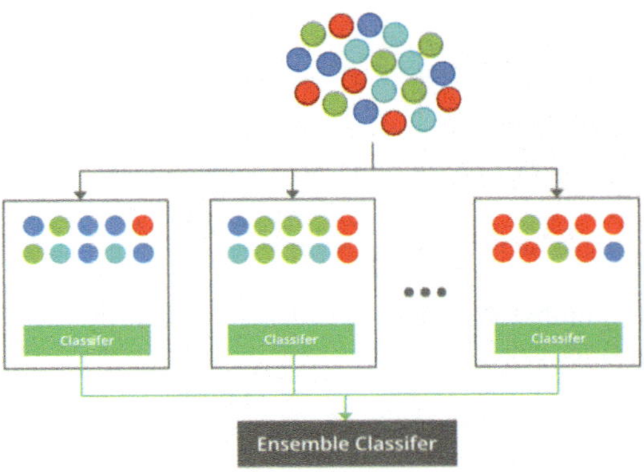

Fig. 4. XGBoost

- *LightGBM:* It is a supervised learning algorithm that uses the ensemble method called boosting, which combines the predictions of several decision trees to produce a stronger prediction. LightGBM differs from other boosting algorithms in the way it builds decision trees, using novel techniques such as gradient-based sampling and unique feature clustering, which increase efficiency and reduce memory usage [8] (Fig. 5).

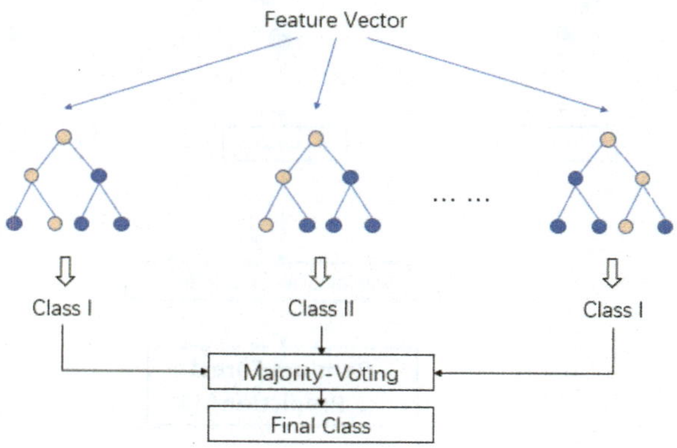

Fig. 5. LightGBM

- *Model trainnig:* It is used to train a machine learning model with the input data X_train and the output labels y_train. The model.fit(X_train, y_train) method adjusts the model parameters to best fit the data and minimize prediction error. These are the characteristics: the parameters are initialized randomly, the loss function is calculated, the model parameters are optimized, and it is repeated until a condition is met. The result is a trained model that can be used to predict new data or evaluate its performance on test data.

3.5 Evaluation

Accuracy: the overall accuracy of a model is simply the number of correct predictions divided by the total number of predictions [9].

$$Accuracy = \frac{TP + TN}{TP + FP + FN + TN} \quad (1)$$

Precision: is the number of correct positives divided by the number of positives predicted by the classifier [9].

$$Precision = \frac{TP}{TP + FP} \quad (2)$$

Recall: is the number of correct positive results divided by the number of all relevant samples (all samples must be positive) [10].

$$Recall = \frac{TP}{TP + FN} \quad (3)$$

where:

- TP = True Positives
- FP = False Positives
- FN = False Negatives
- TN = True Negatives

F1-score: allows a balance between precision and recovery [9].

$$F1 = 2x\frac{Precision\,x\,Recall}{Precision + Recall} \tag{4}$$

4 Results

After experimentation, a table is shown with a summary of the chosen models, their parameters, and metric results (Table 4).

Table 4. Comparison of the best metrics of each model

Model	Accuracy	Precision	Recall	F1-score
DT	88.91%	86.29%	86.16%	86.21%
RF	93.16%	91.67%	91.72%	91.64%
XGBoost	93.51%	92.24%	92.12%	92.13%
LightGBM	**93.82%**	**92.63%**	**92.60%**	**92.58%**

In this paper, four machine learning algorithms for malware detection are compared, but two algorithms show higher performance: the LightGBM algorithm and the XGBoost algorithm. Four evaluation metrics are used to measure the performance of the algorithms: accuracy, precision, recall, and F1-score. Accuracy is the percentage of correct predictions out of the total number of cases. Precision is the percentage of true positive cases out of the total number of predicted positive cases. Recovery is the percentage of true positive cases over the total number of true positive cases. F1-score is the harmonic mean between precision and recall.

The results obtained show that the LightGBM algorithm has a higher percentage in the parameters of accuracy, precision, recall and F1-score for malware detection, indicating that it is more effective and reliable. The data are 93.82%, 92.63%, 92.60% and 92.58% respectively. However, the XGBoost algorithm is below the LightGBM algorithm, which means that it detects fewer actual positive cases. The second-best algorithm for malware detection is the XGBoost algorithm, which has 93.51% accuracy, 92.24% precision, 92.12% recall and 92.13% F1 score. These values are lower than those of the LightGBM algorithm. It can be concluded that the LightGBM algorithm is more suitable for malware detection than the XGBoost algorithm, although the latter has a minimal difference in positive cases (Fig. 6).

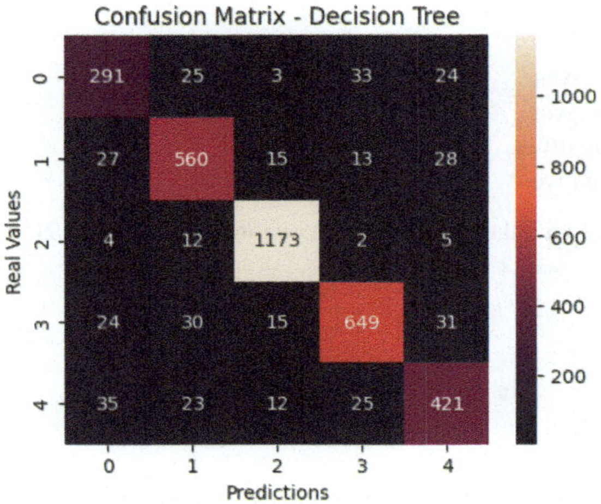

Fig. 6. DT - Confusion Matrix

The confusion matrix of the DT algorithm, using the metrics of accuracy, precision, sensitivity and specificity [11]. Accuracy is the percentage of correctly classified cases, those on the main diagonal. In this case, it is (336 + 574 + 1188 + 681 + 463) / 3241 = 0.81, i.e., the model is 81% correct. The accuracy is the percentage of true positive cases, those of the main diagonal among the total of the column. For example, the accuracy of class 0 is 336 / (336 + 27 + 4 + 24 + 35) = 0.81, 81% of the cases predicted as 0 are 0. Sensitivity is the percentage of positive cases detected, those on the main diagonal among the row total. For example, the sensitivity of class 0 is 336 / (336 + 25 + 3 + 33 + 24) = 0.79, the model detects 79% of the actual cases of 0. The specificity is the percentage of negative cases discarded, those outside the row and column of the class among the total negatives. For example, the specificity of class 0 is (560 + 1173 + 649 + 421) / (560 + 1173 + 649 + 421 + 25 + 3 + 33 + 24) = 0.97, the model discards 97% of the non-0 cases (Fig. 7).

The confusion matrix of the RF algorithm, using the metrics of accuracy, precision, sensitivity and specificity. Accuracy indicates the percentage of cases that the model classifies correctly, i.e., those on the main diagonal of the confusion matrix. In this case, the accuracy is (336 + 574 + 1188 + 681 + 463) / 3241 = 0.81, which means that the model is correct 81% of the time. The accuracy indicates the percentage of cases predicted as positive that are positive, i.e., those on the main diagonal divided by the column total. For example, the accuracy of class 0 is 336 / (336 + 27 + 4 + 24 + 35) = 0.81, which means that 81% of the cases that the model predicts as 0 are actually 0. Sensitivity indicates the percentage of positive cases that the model detects correctly, i.e., those on the main diagonal divided by the row total. For exam-ple, the sensitivity of class 0 is 336 / (336 + 25 + 3 + 33 + 24) = 0.79, which means that the model detects 79% of the real cases of 0. The specificity indicates the per-centage of negative cases that the model correctly discards, i.e., those outside the row and column of the class divided by the total negative cases. For example, the speci-ficity of class 0 is (560 + 1173 + 649

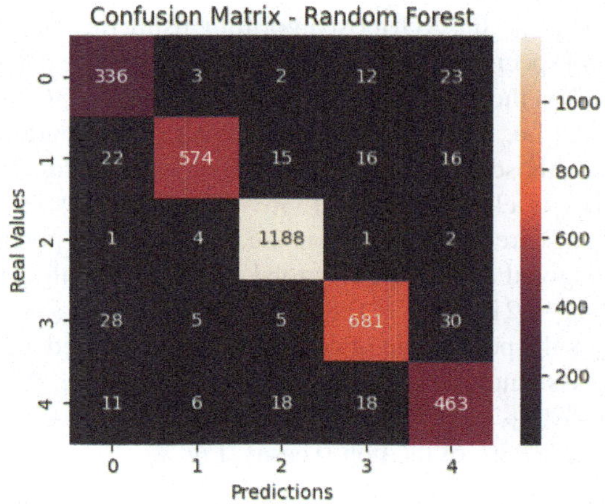

Fig. 7. RF - Confusion Matrix

+ 421) / (560 + 1173 + 649 + 421 + 25 + 3 + 33 + 24) = 0.97, which means that the model discards the negative cases (Fig. 8).

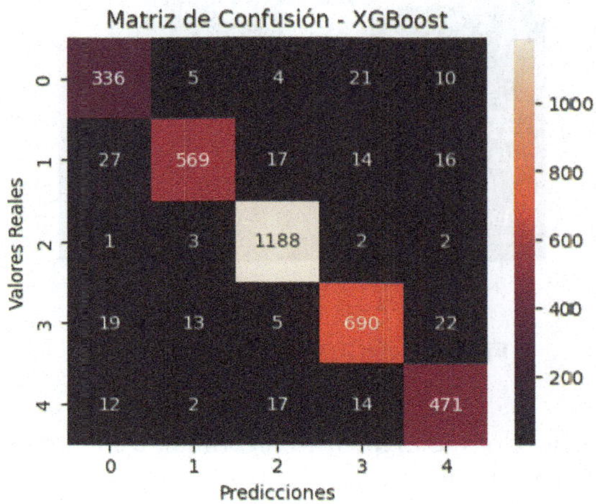

Fig. 8. XGBoost - Confusion Matrix

The confusion matrix of the XGBoost algorithm, using the metrics of accuracy, precision, sensitivity, and specificity, shows that the accuracy is the percentage of correctly classified cases, those on the main diagonal. In this case, it is (336 + 569 + 1188 + 690 + 471) / 3241 = 0.82, i.e., the model is 82% correct. The accuracy is the percentage of true positive cases, those of the main diagonal among the total of the column. For example, the accuracy of class 0 is 336 / (336 + 27 + 1 + 19 + 12) = 0.84, 84% of the cases predicted as 0 are 0. Sensitivity is the percentage of positive cases detected, those on the main diagonal among the row total. For example, the sensitivity of class 0 is 336 / (336 + 5 + 4 + 21 + 10) = 0.88, the model detects 88% of the actual cases of 0. The specificity is the percentage of negative cases discarded, those outside the row and column of the class among the total negatives. For example, the specificity of class 0 is (569 + 1188 + 690 + 471) / (569 + 1188 + 690 + 471 + 5 + 4 + 21 + 10) = 0.98, the model discards 98% of the non-0 cases (Fig. 9).

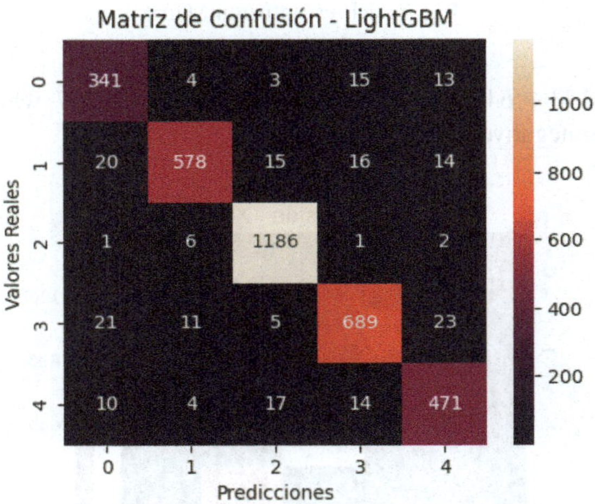

Fig. 9. LightGBM - Confusion Matrix

The confusion matrix of the LightGBM algorithm, using the metrics of accuracy, precision, sensitivity and specificity, shows that the accuracy is the percentage of correctly classified cases, those on the main diagonal. In this case, it is (341 + 578 + 1186 + 689 + 471) / 3241 = 0.82, i.e., the model is 82% correct. The accuracy is the percentage of true positive cases, those of the main diagonal among the total of the column. For example, the accuracy of class 0 is 341 / (341 + 20 + 1 + 21 + 10) = 0.86, 86% of the cases predicted as 0 are 0. Sensitivity is the percentage of positive cases detected, those on the main diagonal among the row total. For example, the sensitivity of class 0 is 341 / (341 + 4 + 3 + 3 + 3 + 15 + 13) = 0.89, the model detects 89% of the actual cases of 0. The specificity is the percentage of negative cases discarded, those outside the row and column of the class among the total negatives. For example, the specificity of class 0 is (578 + 1186 + 689 + 471) / (578 + 1186 + 689 + 471 + 4 + 3 + 15 + 13) = 0.98, the model discards 98% of the non-0 cases.

Figure 10 shows the roc curve of Decision Tree [12].

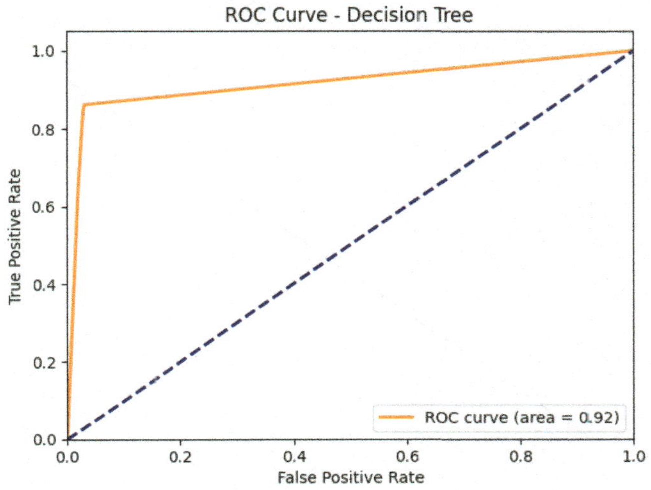

Fig. 10. DT – Curve ROC

Figure 11 shows the roc curve of Random Forest [12].

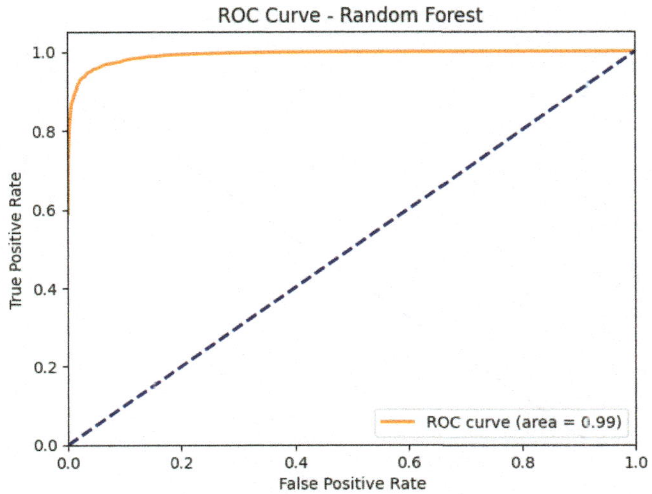

Fig. 11. RF – Curve ROC

Figure 12 shows the roc curve of XGBoost [12].

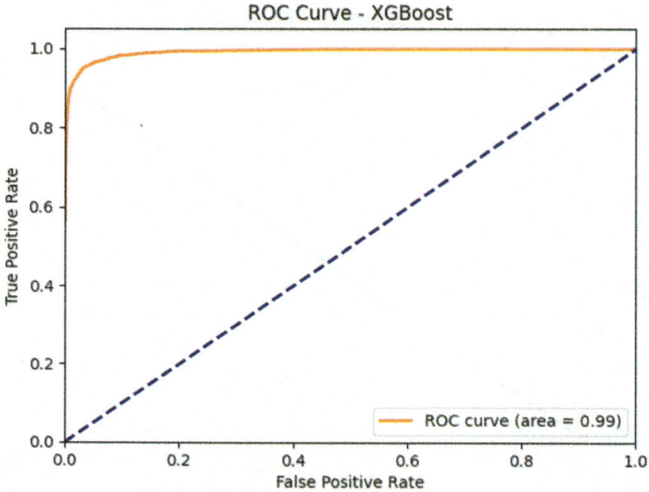

Fig. 12. XGBoost – Curve ROC

Figure 13 shows the roc curve of LightGBM [12].

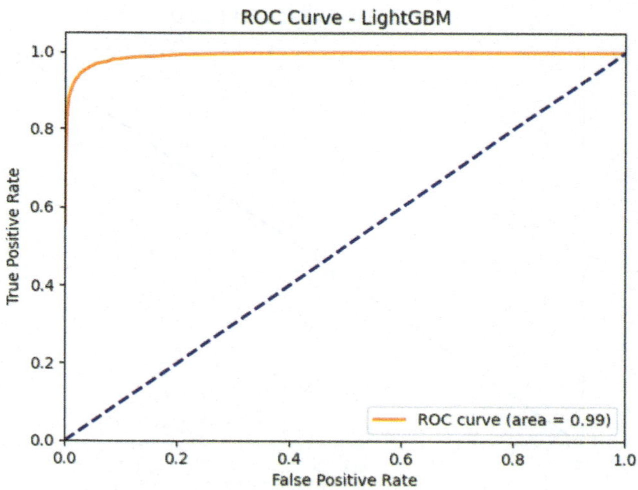

Fig. 13. LightGBM – Curve ROC

5 Conclusions

This work aims to provide a solution to one of the problems that continues to impact computer security: the entry of malware in mobile devices, especially those using the Android operating system. For this purpose, different methodologies from different sources and techniques from previous works are compiled, adapted, and combined to achieve better malware detection. The proposed methodology is defined in five main phases, which are described below:

The first consists of obtaining the dataset where the malware data is located and from where it can be obtained. The second phase is preprocessing, in which the techniques mentioned above are applied to improve the performance of the process. The third phase is featuring extraction, where a numerical or symbolic representation of each malware sample is obtained to facilitate analysis and classification. The fourth phase is the model implementation phase, where the models to be used in this project are presented and their results are trained and classified. Finally, there is the evaluation phase, where metrics from different previous works are used to classify the algorithms used, as well as the curve roc and confusion matrix.

Therefore, it is concluded that the LightGBM model has shown the best performance in the process, making it viable in the detection of malware in Android cell phones, with an accuracy of 93.82%, a precision of 92.63%, recall of 92.60% and an F1 Score of 92.58%. These values indicate that the model can correctly classify most applications, both benign and malicious, and that it has a good balance between accuracy and recall, i.e., between avoiding false positives and false negatives. Then there is the XGBoost algorithm, which had the second-best performance with an accuracy of 93.51%, precision of 92.24%, recall of 92.12% and an F1 Score of 92.13%. These values are very similar to those of the LightGBM model, but slightly lower in all metrics. Finally, it can be concluded that malware diversity is difficult to identify as it is constantly evolving, and that constant updating of models and detection techniques is required. This research provides a first scope for future works that want to take it as a reference and use a similar methodology for malware detection.

References

1. Aurangzeb, S., Aleem, M.: Evaluation and classification of obfuscated Android malware through deep learning using ensemble voting mechanism. Sci. Rep. **13**(1) (2023). https://doi.org/10.1038/S41598-023-30028-W
2. Ko, E., Kim, J., Ban, Y., Cho, H., Yi, J.H.: ACAMA: deep learning-based detection and classification of android malware using API-based features. Security and Communication Networks **2021** (2021). https://doi.org/10.1155/2021/6330828
3. Manzil, H.H.R., Manohar Naik, S.: Android malware category detection using a novel feature vector-based machine learning model. Cybersecurity **6**(1) (2023). https://doi.org/10.1186/S42400-023-00139-Y
4. Gómez, A., Muñoz, A.: Deep learning-based attack detection and classification in android devices. Electronics (Switzerland) **12**(15) (2023). https://doi.org/10.3390/ELECTRONICS12153253

5. Rathore, H., Samavedhi, A., Sahay, S.K., Sewak, M.: Robust malware detection models: learning from adversarial attacks and defenses. Forensic Science International: Digital Investigation **37** (2021). https://doi.org/10.1016/J.FSIDI.2021.301183
6. Dener, M., Ok, G., Orman, A.: Malware detection using memory analysis data in big data environment. Applied Sciences (Switzerland) **12**(17), 8604 (2022). https://doi.org/10.3390/app12178604
7. Louk, M.H.L., Tama, B.A.: Tree-based classifier ensembles for PE malware analysis: a performance revisit. Algorithms **15**(9), 332 (2022). https://doi.org/10.3390/a15090332
8. Talukder, M.A., et al.: A dependable hybrid machine learning model for network intrusion detection. J. Info. Secu. Appl. **72**, 103405 (2023). https://doi.org/10.1016/j.jisa.2022.103405
9. Alabrah, A.: A novel neural network architecture using automated correlated feature layer to detect android malware applications. Mathematics **11**(20), 4242 (2023). https://doi.org/10.3390/math11204242
10. Niveditha, V.R., Parivallal, S., Jones, M., Amandeep Singh, K., Rajasekar, P.: Malware detection in android systems using deep learning techniques. Handbook of Research on Advancements in AI and IoT Convergence Technologies, pp. 266–280 (2023). https://doi.org/10.4018/978-1-6684-6971-2.ch015
11. Naeem, H., Dong, S., Falana, O.J., Ullah, F.: Development of a deep stacked ensemble with process based volatile memory forensics for platform independent malware detection and classification. Expert Systems with Applications **223**, 119952 (2023). https://doi.org/10.1016/j.eswa.2023.119952
12. Wajahat, A., et al.: An adaptive semi-supervised deep learning-based framework for the detection of Android malware. J. Intell. Fuzzy Sys. **45**(3), 5141–5157 (2023). https://doi.org/10.3233/JIFS-231969

Comparison of Machine Learning Algorithms in the Prediction of Pisco Varieties Using Near-Infrared Spectroscopy (NIRS)

Christian Ovalle(✉) and Willian Trujillo

Technological University of Peru, Lima, Peru
dovalle@utp.edu.pe

Abstract. The objective of the research was to perform a comparison of machine learning algorithms to identify the varieties of Peruvian pisco using near-infrared spectroscopy (NIR). Twelve samples of Peruvian pisco were used, including different types of green must from which spectral data were obtained. The data were then processed, and the dataset was obtained. Classification techniques were applied using the K-Nearest Neighbors (KNN) and Naive Bayes models. The results reveal that Naive Bayes, with 55.9%, presents a slight superiority over KNN, with 50.8%, in terms of accuracy and ability to identify the Peruvian pisco variety. The Recall, F1 Score, and Precision metrics support this superiority by demonstrating Naive Bayes' ability to avoid false positives and adequately capture relevant cases. This approach integrates NIR spectroscopy and machine learning, demonstrating its effectiveness in the authentication of Peruvian pisco varieties. These findings underscore the importance of this technology in maintaining the quality and uniqueness of Peruvian pisco in international markets, consolidating its position as a vital tool in preserving the product's identity.

Keywords: Peruvian Pisco · near-infrared spectroscopy · machine learning · authentication · varieties

1 Introduction

The pisco, a distinguished Peruvian alcoholic beverage, is characterized by its unique aroma and flavor, highlighting varieties such as Puro, Mosto Verde, and Acholado [1, 2]. The authenticity and denomination of origin of these varieties are essential to preserving the quality and uniqueness of pisco, which has generated a growing international demand and a significant growth in its exports since between 2017 and 2021, they presented an average annual expansion of 6.2% [3]. Therefore, this growth is threatened by the need for an effective technological solution to guarantee the authenticity of pisco varieties, which jeopardizes the presence of Peruvian pisco in international markets and undermines consumer confidence. In this context, not only does near-infrared spectroscopy (NIRS) emerge as a promising tool for pisco authentication [4], but also the application of Machine Learning algorithms [23, 24]. Moreover, although NIRS has demonstrated its effectiveness in identifying components in liquids, its specific application in identifying pisco varieties has yet to be fully explored [5, 6].

R. Silhavy and P. Silhavy (Eds.): CSOC 2024, LNNS 1120, pp. 269–278, 2024.
https://doi.org/10.1007/978-3-031-70518-2_23

Given the complexity and subtlety of the differences between pisco strains and the growing need to preserve their authenticity, the present study focuses on predicting from the pisco varieties used, such as Moscatel, Torontel, Albilla and Italia, which belong to the Aromatic group, and those belonging to the Non-Aromatic group, which are: Negra Criolla, Quebranta, Uvina and Mollar, using NIRS spectroscopy and Machine Learning algorithms [25], to perform a comparison between them then. This is to advance towards an accurate and reliable methodology for identifying the various strains of Peruvian pisco. The choice of the varieties above is based on the diversity and representativeness of the groups, encompassing both those with distinctive aromatic profiles and those characterized by their absence of pronounced aromatic notes. Although NIRS spectroscopy and machine learning have been explored in other contexts [7–9], their specific and compelling use in the authentication of Peruvian Pisco varieties is still a developing area of research [10]. Thus, this study focuses on the convergence of two seemingly disparate disciplines: NIRS spectroscopy [11] and machine learning [12, 13], with a specific application in the authentication of Peruvian Pisco varieties [8, 14], evaluating the effectiveness of dimensional reduction algorithms to discern distinctive spectral patterns and the application of Machine Learning algorithms, such as KNN and Naive Bayes, followed by a comparison of both models to evaluate their accuracy in the classification of Pisco samples.

In turn, recent studies highlight the importance of the use of NIR spectroscopy, such as the one carried out by [15], who have highlighted the efficacy of the combination of NIR and UV-Vis spectroscopy with chemometric techniques, such as Partial Least Squares Regression (PLS-R), achieving an impressive 92% accuracy in the identification of grape varieties in wines, which highlights the relevance of integrating advanced spectroscopic technologies with analytical methods to improve product discrimination. In addition, the use of machine learning algorithms has gained ground in the classification of pisco samples, as evidenced in the work of [16], which applied algorithms such as Support Vector Machines (SVM) to achieve a promised accuracy of 88% in the classification of Peruvian piscos. On the other hand, a study focused on calibration and prediction models in wine spirits has demonstrated their effectiveness in predicting volatile phenols [17]. Although its main focus is on wines, the methodology using PLS-R calibration and near-infrared spectroscopy establishes the basis for accurately identifying pisco varieties. These models achieved coefficients of determination (R^2) higher than 0.90 for several compounds, highlighting the capability of advanced chemical analysis techniques to characterize spirits and distillates.

Previous studies have also addressed strategies to improve the identification and classification of spirit varieties. Synchronous fluorescence spectroscopy, implemented by [18] on cachaças and rums, exhibited 96% accuracy in classification, highlighting its potential for discrimination of these alcoholic beverages. Meanwhile, [19] proposed a non-invasive method to monitor ethanol and methanol levels in pisco using vibrational spectroscopy, achieving an accuracy of 98% with machine learning algorithms. However, the study by [20] evaluated the discrimination of commercial white wines, achieving an accuracy of 85% by chemometric analysis with visible and near-infrared spectroscopy. The review of trends in quality control, discrimination, and authentication of alcoholic beverages [21] highlighted the effectiveness of principal component analysis and partial least squares discriminant analysis in supporting beverage authentication. In addition, the analysis of [7], focused on detecting brandy adulteration, achieved a pre-accuracy of 94% with fluorescence spectroscopy and multivariate methods. Finally, the study that applied FTIR-ATR spectroscopy [22] to the quality control of grape distillates, although with a limited data set, demonstrated the effectiveness of this technique in chemical characterization. This analysis of a set of studies provides a comprehensive overview of techniques and approaches applied in identifying and authenticating pisco varieties, establishing a solid framework for the present study.

2 Methodology

This research analyzed 12 samples of pisco of different types (aromatic, non-aromatic, and green must) by near-infrared (NIR) spectroscopy. The NIR spectra [20] were acquired with a portable NIRvas-can spectrometer and subjected to preprocessing, standardizing the data. Then, the Naive Bayes classification model and the K-Nearest Neighbors (Knn) algorithm were applied. Validation tests were carried out to assess the classification accuracy. This comprehensive approach effectively distinguished between the pisco categories, ensuring the reliability of the results obtained [1, 16]. These steps are detailed in Fig. 1.

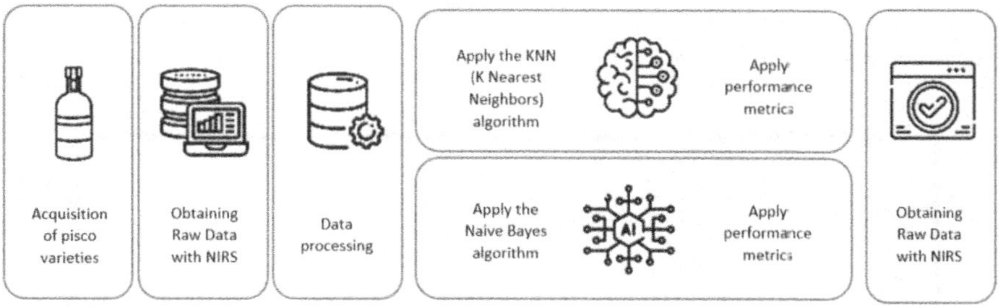

Fig. 1. Summary of the methodology.

An AvaSpec-NIR256–1.7-EVO NIR spectrometer was configured in reflectance mode, using a resolution of 200 lines/mm and a blaze of 1500 nm. This device has an operating range in the spectrum from 900 to 1750 nm, divided into 221 bands, facilitating the acquisition of data from previously collected samples. This stage is critical, as it will provide fundamental information for the following phases of the project.

Eight different types of pisco have been analyzed; however, only one measurement point has been selected for each sample, totaling 100 samples for each variant. The choice of the location of the measurement point follows a specific protocol. Initially, the pisco is placed in sample tubes and placed on a rack to ensure accurate capture of the individual spectra of each type of pisco. (See Fig. 2 (a)(b)(c)).

Following the spectra identification, a detailed spectroscopic data cleaning process, known as "data cleaning," is carried out. This process involves identifying and eliminating data that may contain noise, be incomplete, or be outliers [15–22]. In addition, a

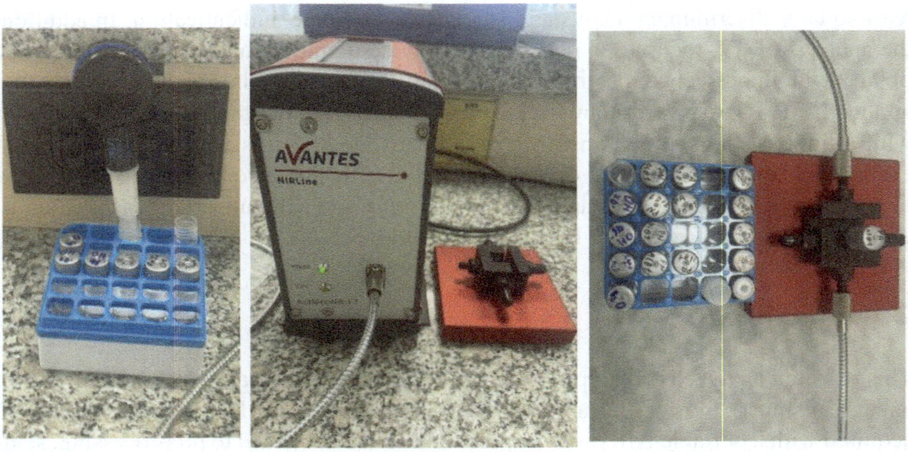

(a) (b) (c)

Fig. 2. (a) Pisco Sample Preparation, (b) NIR Data Acquisition, and (c) Sample Placement.

thorough treatment of outliers is performed to prevent them from affecting the integrity and quality of the analysis. Correction of incorrect or inconsistent data, which may arise during sampling or recording of spectral signatures, is considered a critical step in this process, as depicted in Fig. 3.

	Wavelength [nm]	Sample	Spectral Value	Classification
0	899.345263	1	0.704542	1
1	905.470353	1	0.707624	1
2	910.594255	1	0.708650	1
3	916.385616	1	0.705107	1
4	922.010113	1	0.702408	1
...	...	...	...	...
1395	1855.676502	8	0.670076	0
1396	1861.300998	8	0.665279	0
1397	1866.925494	8	0.662449	0
1398	1872.549991	8	0.651578	0
1399	1878.174487	8	0.650045	0

Fig. 3. DataFrame of the types of piscos.

A graph showing the raw spectra and the application of the Python algorithms was made where, even before, a slight difference between the Aromatic and Non-Aromatic class of Pisco can be seen. Figure 4(a) provides a visual reference to understand this process better. These columns are then standardized using the StandardScaler tool [17]. In order to facilitate the interpretation of the results, a new data frame is generated, as shown in Fig. 4(b).

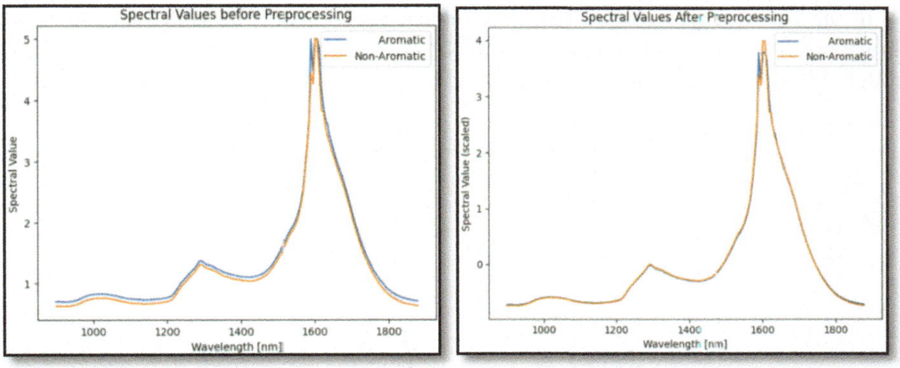

(a) (b)

Fig. 4. (a) Plot of spectra before applying StandardScaler, (b) Plot of spectra after applying StandardScaler.

Specific actions are performed to prepare the spectroscopic data for subsequent analyses. This includes importing relevant libraries such as Seaborn, pandas, numpy, Scipy.signal, StandardScaler from sci-kit-learn, and mat-plotlib.pyplot. These libraries are used for statistical functions, data manipulation, data standardization, and visualization of analysis results.

The training process, which focuses on building and fitting the KNN and Naive Bayes models, is carried out. In the case of the KNN model, the essential functions of the sklearn.Neighbors' library is imported, including "KNeighborsClassifier," which is used to build a classification model. In addition, "train_test_split" is incorporated to perform the division of the DataFrame into training sets (x_train, y_train) representing the values to be predicted (Wavelength and Spectral Value) and test sets (x_test, y_test) corresponding to the target value (Classification). (See Table 1).

Table 1. AI model used.

AI Models	Mathematical formula
Naïve Bayes	$P\left(\frac{y}{x}\right) = \frac{P(x/y)xP(y)}{P(x)} P\left(\frac{y}{x}\right) = \frac{P(x/y)xP(y)}{P(x)}$
KNN Classifier	$P\left(\frac{y}{x}\right) = \frac{P(x/y)xP(y)}{P(x)} P\left(\frac{y}{x}\right) = \frac{P(x/y)xP(y)}{P(x)}$

After creating an instance of the KNeighborsClassifier model, the training process is started using the training sets (x_train, y_train) through the "fit" method. Subsequently, the characteristics of the test set (x_test) are used to carry out predictions using the "predict" method, and the results obtained are stored in the "y_pred" variable. This procedure allows the model's performance to be evaluated by comparing the predictions generated with the actual values of the test set.

In the Naive Bayes model, the fundamental functions of the sklearn.naive_bayes library are imported, including "GaussianNB," which is used to build a Naive Bayes model. After that, the algorithms are applied the same way as KNN, as shown in Fig. 5.

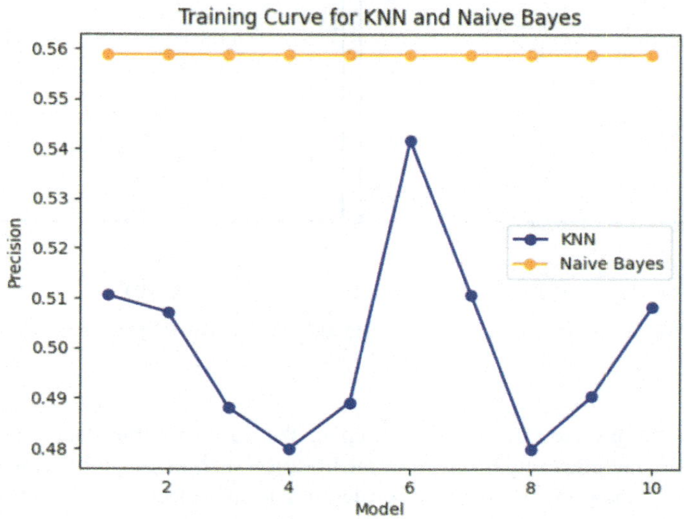

Fig. 5. Training curve.

Subsequently, evaluation criteria and performance metrics are defined for the previously selected models. These metrics play a crucial role in evaluating the effectiveness and accuracy of the models in the prediction task. Measures such as mean square error, accuracy, sensitivity, specificity, and other relevant metrics adapted to the project's specific objectives are also considered. These metrics are fundamental for evaluating and comparing the performance of artificial intelligence models in classifying grape types. Table 2 provides a visual representation of this evaluation process and metrics.

Table 2. Selected performance metrics.

Performance metrics	Formula
Precision	TP = (TP + FP)
Recall	TP = (TP + FN)
F1-Score	F1 = 2*((recall*precision)/(recall + precision))

In the last stage, whether the selected models adequately fulfill the primary purpose of accurately classifying the type of pisco according to the established business objectives is verified. This phase involves an exhaustive review using an accuracy test (Accuracy). This meticulous evaluation of the results is aligned with the previously defined success criteria to ensure that the models align with the planned objectives (see Fig. 6).

```
# Realizar predicciones
knn_predictions = knn_model.predict(X_test)
nb_predictions = nb_model.predict(X_test)

# Evaluar rendimiento
knn_accuracy = metrics.accuracy_score(y_test, knn_predictions)
nb_accuracy = metrics.accuracy_score(y_test, nb_predictions)

print(f'Accuracy KNN: {knn_accuracy}')
print(f'Accuracy Naive Bayes: {nb_accuracy}')

Accuracy KNN: 0.48928571428571427
Accuracy Naive Bayes: 0.5714285714285714
```

Fig. 6. Selected performance metrics.

3 Results

By analyzing in detail the results derived from the application of the K-Nearest Neighbors (KNN) and Naive Bayes algorithms in the classification of Pisco varieties by spectroscopy (NIRS), significant insights are revealed that provide a detailed view of the performance of both models.

In addition, a graphical representation of classification metrics such as Recall, F1 Score, and their Accuracy, is performed to amplify the comparability between KNN and Naive Bayes. This visual approach provides an intuitive understanding of the relative performance of both models, as shown in Table 3.

Table 3. Metrics results.

Metrics	KNN	Naïve Bayes
Precision	0.5083333333333333 – 50.8%	0.5588235294117647 – 55.9%
Recall	0.4357142857142857 – 43.6%	0.6785714285714286 – 67.9%
F1 Score	0.4692307692307692 3 – 46.9%	0.6129032258064516 – 61.3%

In turn, Fig. 7 adds the strategic inclusion of a graph of the classification metrics performed, as it not only adds a visual component to the presentation of results but also significantly deepens the comparative understanding of the effectiveness of both algorithms, providing an analytical tool that facilitates the intuitive interpretation of the relative performance of the models. Also, when examining the metrics graphs, it can be seen that Accuracy, which measures the proportion of correctly classified instances among the optimistic predictions, reveals that Naive Bayes slightly outperforms KNN. This difference highlights the ability of Naive Bayes to avoid false positives and provide more accurate classifications in the positive category.

Also, in terms of Recall, which evaluates the ability of a model to identify all relevant instances correctly, Naive Bayes shows a significantly higher performance than KNN.

This disparity suggests that Naive Bayes has a more remarkable ability to capture and recognize positive instances, which can be crucial when identifying all relevant instances is a priority. Moreover, the F1 Score metric, which combines Precision and Recall, supports the superiority of Naive Bayes over KNN, indicating a more effective balance between accuracy and completeness in predictions.

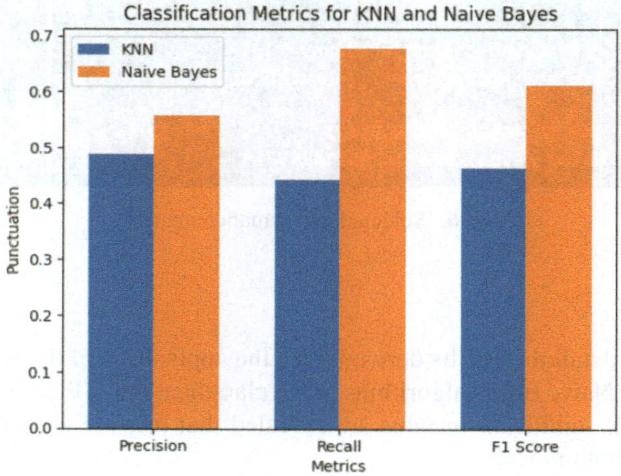

Fig. 7. Rating Metrics Chart.

The present research examines the results obtained by applying the K-Nearest Neighbors (KNN) and Naive Bayes algorithms in classifying Pisco varieties by spectroscopy (NIRS). According to the results, Naive Bayes slightly outperforms KNN in accuracy. Naive Bayes exhibits significantly superior performance, suggesting a more remarkable ability to capture and recognize positive cases. In addition, the F1 Score metric supports the overall superiority of Naive Bayes over KNN, indicating a more effective balance between Accuracy and completeness in predictions. These results have important implications for model choice in specific applications.

4 Conclusion

Deep learning techniques improve the diagnosis of autism spectrum disorder (ASD) by improving the accuracy of diagram classification. Artificial intelligence and automation are needed to improve the efficiency and fairness of this process. Although progress has been made, issues still need to be addressed, such as the requirement for diverse data and validation in specific clinical contexts to improve the model's generalizability. To refine the process, interdisciplinary collaboration between artificial intelligence experts, health professionals, and ASD specialists is essential. This study emphasizes not only a novel method for diagnosing ASD but also the ongoing importance of multidisciplinary research and the creation of cutting-edge technological tools for early identification and support of individuals with ASD and their families.

Acknowledgments. We want to thank those who contributed to this article. We are sincerely grateful for all the support and facilities the Universidad Tecnológica del Perú provided.

References

1. Huertas, L.: Cronología de la Producción del Vino y del Pisco. Universidad Ricardo Palma, Lima, Perú (2010)
2. Instituto Nacional de Defensa de la Competencia y de la Protección de la Propiedad Intelectual: Guía práctica de la denominación de origen pisco. Indecopi (2019)
3. Centro de investigación de Economía y negocios globales (CIEN): Panorama del Sector Agroindustrial: Cacao y derivados. Nota De Inteligencia Comercial (2022)
4. Yang, J., et al.: TeaNet: deep learning on near-infrared spectroscopy (NIR) data for the assurance of tea quality. Comput. Electron. Agric. **190**, 106431 (2021)
5. De-La-Cruz, C., et al.: Application of machine learning algorithms to classify Peruvian Pisco varieties using an electronic nose. Sensors **23**(13), 5864 (2023)
6. Reda, R., et al.: Optimized variable selection and machine learning models for olive oil quality assessment using portable near infrared spectroscopy. Spectrochim Acta A Mol Biomol Spectrosc. **303**, 123213 (2023)
7. Markechová, D., Májek, P., Sádecká, J.: Fluorescence spectroscopy and multivariate methods for the determination of brandy adulteration with mixed wine spirit. Food Chem. **159**, 193–199 (2014)
8. García Martin, J.F.: Potential of near-infrared spectroscopy for the determination of olive oil quality. Sensors **22**(8), 2831 (2022)
9. Ellis, D.I., Muhamadali, H., Xu, Y., Eccles, R., Goodall, I., Goodacre, R.: Rapid through-container detection of fake spirits and methanol quantification with handheld Raman spectroscopy. Analyst **144**(1), 324–330 (2019)
10. Acevedo, F.J., Jiménez, J., Maldonado, S., Domínguez, E., Narváez, A.: Classification of wines produced in specific regions by UV-visible spectroscopy combined with support vector machines. J. Agric. Food Chem. **55**(17), 6842–6849 (2007)
11. Pasquini, C.: Near infrared spectroscopy: Fundamentals, practical aspects and analytical applications. J. Braz. Chem. Soc. **14**(2), 198–219 (2003)
12. Janiesch, C., Zschech, P., Heinrich, K.: Machine learning and deep learning. Electron. Mark. **31**(3), 685–695 (2021)
13. Garcia-Rios, V., Marres-Salhuana, M., Sierra-Liñan, F., Cabanillas-Carbonell, M.: Predictive machine learning applying cross industry standard process for data mining for the diagnosis of diabetes mellitus type 2. IAES Int. J. Artif. Intel. **12**(4), 1713–1726 (2023)
14. Callejón, R.M., Amigo, J.M., Pairo, E., Garmón, S., Ocaña, J.A., Morales, M.L.: Classification of Sherry vinegars by combining multidimensional fluorescence, parafac and different classification approaches. Talanta **88**, 456–462 (2012)
15. Bower, J.A.: Statistical Methods for Food Science: Introductory procedures for the food practitioner (2009)
16. Iñón, F.A., Garrigues, S., De la Guardia, M.: Combination of mid- and near-infrared spectroscopy for the determination of the quality properties of beers. Anal. Chim. Acta **571**(2), 167–174 (2006)
17. Anjos, O., et al.: PLS-R calibration models for wine spirit volatile phenols prediction by near-infrared spectroscopy. Sensors **22**(1) (2022)
18. Silveira, A.L., Barbeira, P.J.S.: Synchronous fluorescence spectroscopy and multivariate classification for the discrimination of cachaças and rums. Spectrochim Acta A Mol Biomol Spectrosc **270** (2022)

19. Menevseoglu, A., Aykas, D.P., Hatta-Sakoda, B., Toledo-Herrera, V.H., Rodriguez-Saona, L.E.: Non-invasive monitoring of ethanol and methanol levels in grape-derived pisco distillate by vibrational spectroscopy. Sensors **21**(18) (2021)
20. Cozzolino, D., Smyth, H.E., Gishen, M.: Feasibility study on the use of visible and near-infrared spectroscopy together with chemometrics to discriminate between commercial white wines of different varietal origins. J. Agric. Food Chem. **51**(26), 7703–7708 (2003)
21. Arslan, M., et al.: Recent trends in quality control, discrimination and authentication of alcoholic beverages using nondestructive instrumental techniques. Trends Food Sci. Technol. **107**, 80–113 (2021)
22. Anjos, O., Santos, A.J.A., Estevinho, L.M., Caldeira, I.: FTIR-ATR spectroscopy applied to quality control of grape-derived spirits. Food Chem. **205**, 28–35 (2016)
23. Wójcicki, K.: Application of NIR spectroscopy for sport supplements identification **3**(52), 32–37 (2018)
24. Wójcicki, K.: Application of nir spectroscopy for whisky identification and determination the content of ethanol. Current Trends in Commodity Science **123** (2015)
25. Ander, G.: Tipos y niveles de investigación. Metodologia De La Investigación (1992)

Formal Model of the Internet of Things Network

Olga Isaeva(✉)

Institute of Computational Modelling SB RAS, Krasnoyarsk, Russia
isaeva@icm.krasn.ru

Abstract. The paper presents the formalization of model of the Internet of Things (IoT) network designed for monitoring technological premises with telecommunications equipment in our science center. The physical implementation of the IoT network includes measuring devices, telecommunications environment, servers for data collection and application software solutions. For the information interaction, the "Publisher-Subscriber" scheme and "Message Queuing Telemetry Transport" (MQTT) protocol are used. An ontology has been created which describes the network architecture. It contains the classes, representing the concepts of the subject area, relations, individuals, data properties, its ranges of change and critical values limiting the attributes of the ontology elements. The ontology describes the IoT network architecture and the properties of devices that collect, transmit, store, and process data. Objects in the ontology have their own digital representation in databases. They contain measurement results, precedents for anomalous behavior of devices, statistical and spectral characteristics of data. The formalization made it possible to identify hidden dependences between objects, relate them to the characteristics of the physical processes controlled by IoT the IoT devices and solve practical problems. An example of the problem of choosing a broker and its settings for modifying the network structure is considered, whose solution is made on the basis of the constructed ontology. #CSOC1120.

Keywords: Internet of Things · Message Queuing Telemetry Transport · Publisher · Broker · Subscriber · ontology · delay analysis · frequency analysis

1 Introduction

The innovations, defined by the concept of the Internet of Things and related with embedding of electronics into everyday physical objects and their integration within the global cyberphysical infrastructure, contribute the development of the entire Information and Communication Technologies sector. The widespread deployment of spatially distributed devices paves the way to a whole new class of services and applications which are able to leverage the interconnection of physical and virtual realms [1].

The problem of semantic fragmentation and uncertainty in terminology is aggravated by the fact that any solutions there require synergistic activities of specialists in different fields of knowledge, such as telecommunications, electronics, informatics, data analysis, artificial intelligence and other sciences [2]. At all levels of immersion in the subject area, new technology poses serious challenges to all specialists.

R. Silhavy and P. Silhavy (Eds.): CSOC 2024, LNNS 1120, pp. 279–290, 2024.
https://doi.org/10.1007/978-3-031-70518-2_24

At the component level, the IoT is based on the notion of "thing", which complements the existing telecommunications domain with such properties of entities as identifying themselves, interacting with the environment and communicating among themselves for building networks of interconnected objects. The organization of these processes is an important task for this level.

At the system level, the IoT is a dynamic and distributed networked architecture composed of a very large number of objects producing, transmitting and consuming information. Here, the functions of autonomous management and distribution of data analysis mechanisms on the cloud, fog and edge segments should be included in a set of effective solutions [3].

From the service level perspective, the main issue relates to transforming data streams generated by IoT objects into functionalities or resources provided by services. This requires the creation of a standardized digital representation of IoT networks which would allow integrating and arranging various devices and methods into convenient smart-tools for end users.

An information platform is required, providing semantic integration of different aspects of the technology and creating a uniform glossary of concepts and relations based on knowledge of the subject area. Numerous studies use semantic modeling and construct an ontology to solve this problem.

The purpose of the study is to create an ontology containing formalization of the basic concepts of the IoT-technology. The architecture deployed within the corporate network of the Science Center, which monitors technological premises with telecommunications equipment [4], was chosen for the research. Functional layers are implemented in the network which combine sensors, net-transport, services, application solutions and the protocol "Message Queue Telemetry Transport" (MQTT) is used to support the interaction of logical entities [5]. Various approaches and examples of the IoT applications for creating the ontology are considered. A review was made of the literature data describing the features of the MQTT protocol and scheme of information interaction between the sources and consumers of information.

MQTT is widely known as a lightweight and flexible messaging protocol based on the "Publish-Subscribe" communication pattern. Communications using MQTT are asynchronous, each client continues to perform its tasks without downtime during data wait cycles. Messages are exchanged through a central server, whose clients can be either senders or receivers of the required information at the same time. IoT devices have limited security control capabilities, which can be broadened by formalizing device profiles and precedents for typical and abnormal behavior [6].

In some studies, structured data of the communication sessions are added with the information from unstructured request logs [7, 8]. For attack detection and consequence mitigation, the paper [9] suggests modeling the behavior of IoT authorized customers and building scenarios of the behavior of the device network during different types of attacks. Detection of abnormal traffic is completed via machine learning methods; however, the authors in [10, 11] note that large data volumes create problems for their usage. The study in [12] presents an automated choice of the features necessary and sufficient for model training and setup. In some studies, the specialists create public sets of data on the basis of the MQTT protocol and use the indicators that they contain to search for anomalies [13]. In [14, 15] an approach was suggested to building a proactive behavior system for the operation industrial devices based on the creation of time series and it was integrated in the security system together with other methods of analysis. Ontology can not only provide specialists with the possibility to exchange data about anomalies and cyber threats within a single stable knowledge structure, but it also allows speeding up and simplifying the operation of the machine learning modes used for analyzing and building attack vectors, thereby increasing the protection level of the IoT-systems.

Taking into account the review, important components of the technology involved in solving the problems of ensuring the reliability and security of IoT were identified and adapted to the implementation of our network. They formed the basis of the ontology, complementing the standard presentation of the digital profile of the IoT devices with indicators constructed as a result of analysis and aggregation of the data collected by sensors. The ontology will allow the identification of the relations between the concepts in order to analyze abnormal behavior of the network elements and to detect failures of the equipment or changes in its operating modes.

2 Methods

The research method in this work is semantic modeling, formalization and ontology construction. The IoT-network architecture in the Science Center is built in accordance with the definition of the Internet of Things as a dynamic global infrastructure of distributed nodes equipped with built-in tools to measure, understand and change external environment [16].

As a software tool, we used the public editor of ontologies and data bases Protégé [17], supporting Web Ontology Language (OWL) as the most popular ontology language, with a web-interface for collaborative work of the research groups. The editor gives the definition of an ontology as a formal explicit description of concepts (classes), samples of classes, their properties (functions and attributes) and the restrictions on properties. This tool allows building of a categorized hierarchical structure of the elements and provides with the methods of dependency monitoring, hidden knowledge reconstruction, and creation and execution of semantically-aware data queries. To create an ontology, the following items have been defined: the characteristics describing the scheme itself and the participants of informational interaction, indicators of network activity and properties of the processes supervised by IoT sensors.

The network built according to the "Publish-Subscribe" communication scheme with the roles Publisher, Broker, Subscriber linked by a common Topic [18]:

Publisher – is a device that generates date about the state of the objects under monitoring. In our case, these are temperature, humidity, etc. sensors that send messages to brokers on certain topics.

Broker – is a server and software (in our implementation we use Mosquitto MQTT) that receives all data from the Publishers and then distributes messages to the appropriate destination Subscribers.

Subscriber – is a client (an application or a device) that receives and consumes data in accordance with the preset thematic subscriptions.

Topic – is a message subject that allows publisher and subscriber to exchange data with a defined semantic.

The publishers in our scheme are measurement devices CL-210-E (ICP DAS production), they monitor temperature, humidity, dew point and fine dust concentration. The broker function is performed through a Kubernetes cluster (K8s) with 10 nodes and a Hyper-V virtualization system. Life cycles of the containers are centrally controlled by Rancher platform. The subscribers are the software located on smartphones, iPads or other devices and using data in technological premises monitoring system. The ontology has been constructed this scheme (Fig. 1).

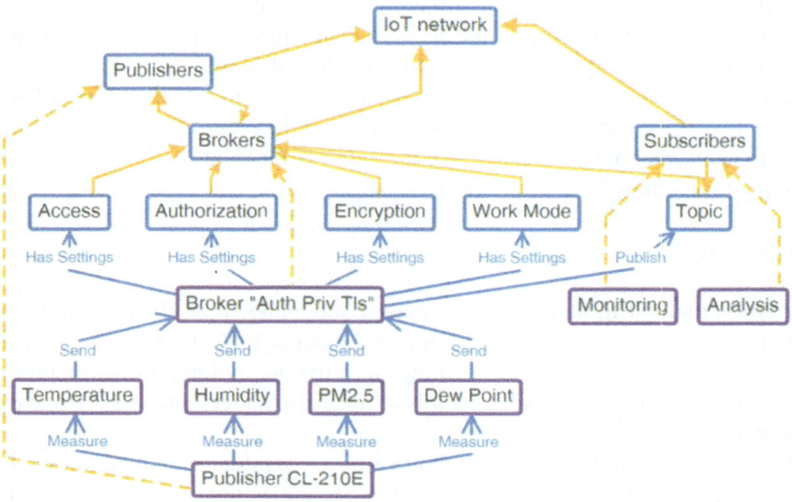

Fig. 1. Fragment of the implemented scheme "Publish-Subscribe".

Based on the graphical representation, the structures of the constructed ontology were formalized. The formal notation for the constructed classes:

$$IoT = \{P, B, S, T, R\},$$

where IoT we designated the class that describes the structure of the IoT network, P is a set of publishers, B – brokers, S – subscribers, T – topics, R – the set of transport layer device set within our network that describes the route for data transmission.

A graphical representation of the "Publisher" class (Fig. 2) has been created. Its formal description:

$$P =< N^P, M^P, DP^P, PS^P, B^P >,$$

where N^P is the unique name of device, $M^P = (V, lim(V)) = \{(v_l, lim(v_l))\}$ – the set of pairs defining the type of measurement and restrictions on measurement values, $V = \{v_l\}(l = 1, \ldots, L)$, L – number of measurement types. In our implementation, the sensors perform four types of measurements ($L = 4$): $v_1 =$ *"Temperature"*, $v_2 =$ *"Humidity"*, $v_3 =$ *"Dew Point"*, $v_4 =$ *"PM2.5"*. The $lim(v_l)$ takes values: *opt* – optimal value, *maxl* – maximum permissible value, *alc* – range of changes. $DP^P =< f(v_l), g(v_l) >$

– characteristics calculated from the collected data of each type of measurement, the function f shows the periodicity of data, g – statistical characteristics, such us: *max* – the largest value in the data sample, *min* – the least value, *med* – the center point (median), *av* – the average value, *dis* – the deviation from the mean (dispersion), *rc* – data change rate and etc., PS^P – sensor properties: *ar* – accuracy, *pp* – publication period, *pr* – the type of protocol, *nt* – network types, *lc* – location, *vf* – information about verifications performed, $B^P \subseteq B$ – brokers to which the publisher sends data.

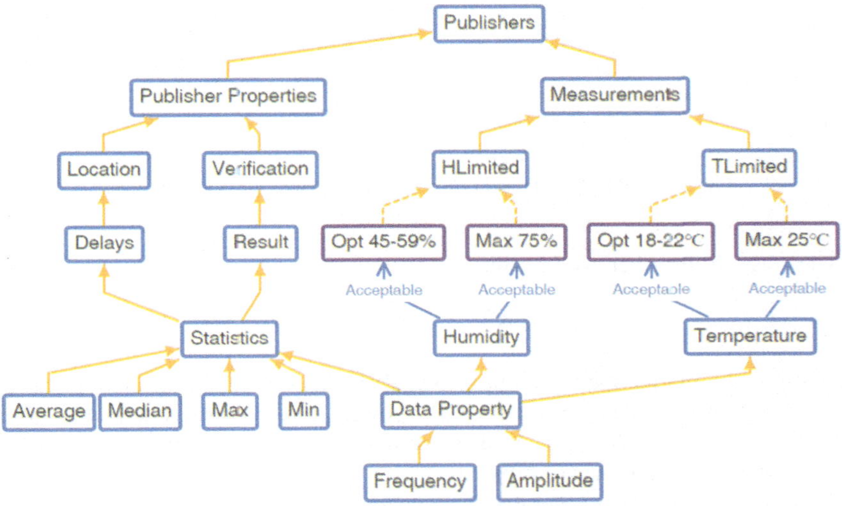

Fig. 2. Graphical representation of the "Publisher" class (fragment).

The topics $T = \{(N, V)\}$ are a set of pairs combining the names of publishers $N = \{n_i\}$, and the types of measurements they perform $V = \{v_{ij}\}$, where $p_i \in P$ – the publisher, n_i – the name of p_i, v_{ij} – the type of measurement for the publisher p_i. Since each publisher can perform several different dimensions, several topics are created for it.

The "Broker" class (Fig. 3) and other classes included in the concept of an Internet of Things network are graphically described. Formally, the "Broker" class can be described as:

$$B = < N^B, PS^B, ST^B, W^B, T^B >,$$

where N^B is the unique name of cluster node, PS^B – the properties: *dtd* – data transfer delay, *cpu* – available CPU power, *tp* – time to process data, ST^B – settings specified to the broker: *fsd* – frequency of sending data, *QoS* – service flags and other operating modes, W^B – setup options include access choices ("*Private*" - only via corporate network, "*Public*" – via Internet), authorization ("*Login/Pass*", "*Anonymous*"), encryption ("*TLS*" – Transport Layer Security protocol, "*Open*" – no encryption), $T^B \subseteq T$ – a subset of topics that such broker processes.

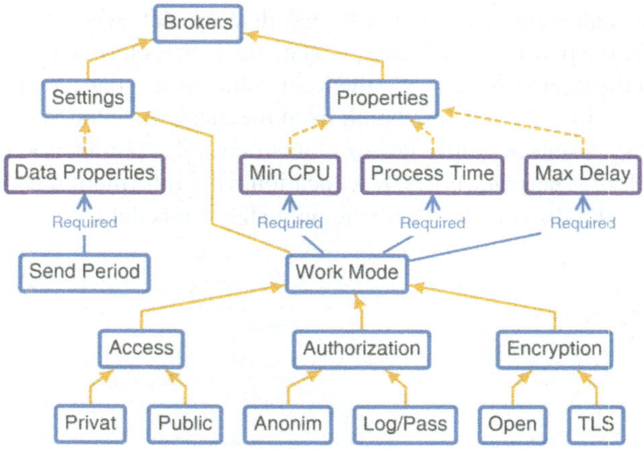

Fig. 3. Graphical representation of the "Broker" class (fragment).

The class "Subscriber" is described:

$$S = < N^S, PS^S, A^S, B^S, T^S >,$$

where N^S is the unique name of subscriber, PS^S – the properties of an application or a device: data transfer delay, available CPU power, time to process data, A^S – a set of applications (software), $B^S \subseteq B$ – brokers from whom subscribers receive data, T^S – the topics to which applications are subscribed.

All sets have their own generalized characteristics, presented in the ontology in the form of classes. Parameters, associated with a particular implementation, are represented as properties or individuals. For example, the sensor $p \in P$ is defined in the ontology (in Fig. 1) as the individual "*Publisher CL-210E*" with the types of measurements $v_l \in V$ (the classes: v_1 – "*temperature*", v_2 – "*humidity*") and constraints $lim(v_l)$ (individuals: "*Opt*" – the optimal value and "*Max*" – the maximum value, in Fig. 2).

The output mechanism that works in the editor allows you to detect hidden dependencies, control the structure and relationships between data samples. To perform logical inference, the functions of union, intersection, and selection under complex conditions have been introduced:

- $A \cup B|C$ – returns all elements from the A or B for which condition C is true. For example: $P \cup B|b \in B^P$ – returns all $p_i \in P$ which send its data to the broker b.
- $A \cap B|C$ – returns all elements that are in A and B for which condition C is true. For example: $T^B \cap T^S|b \in B^P$ – returns all topics $t_l \in T$ for selected b for which have at least one subscriber.
- Let us denote $a[Pr]$ or $Pr(a)$ – the property with the name Pr of class or individual. For example: $p[N^P] =$"*Publisher CL-210E*", where N^P – is the name of publisher p.
- Select $aA|a[Pr] = C$ – returns the individual a such that for its property Pr the condition C is true. For example: $b \in B|b[N^B] = Broker$ "*Auth Priv Tls*" – returns the individual with name "*Auth Priv Tls*".
- Calculate $|A|$ as a cardinality of the set (the number of elements).

The ontology structure is ready, it is now being finalized to become functionally complete. In the current form, the ontology allows solution of practical tasks, and one of those is choosing and configuring the broker.

3 Results

Changing in the IoT-network structure requires security setting and configuration of the data output parameters associated with the observed processes. To do this, dependencies are extracted from the ontology using the inference engine, describing classes, class instances, relationships between them, and restrictions on data. The following are some results of using the ontology to change the structure of the network. The tasks of choosing a broker when adding a new publisher or subscriber, as well as choosing the settings of a new broker, are considered. Each change does not affect other elements of the network structure.

3.1 Choosing the Broker

In order to add a new publisher or subscriber to the network, you need to select the broker with whom they will interact.

The end-to-end delay or response time must be taken into account by choosing a broker so that the path from sending data to receiving it by the end user is the best. In the paper [19], end-to-end delay is defined as the delay in response time from the publisher to the subscriber through the broker. This delay is computed as the sum of delays of publishing a message to the broker, time of the broker processing and the subscriber receiving the message:

$$R(b)_{p\to s} = R_{p\to b} + R_b + R_{b\to s},$$

where, $R(b)_{p\to s}$ is the end-to-end delay, $b \in (B^P \cap B^S)$ – the broker, $R_{p\to b}$ – the delay induced by the transmission network between b and p, R_b – the average time within the broker and $R_{b\to s}$ is the delay induced by the transmission network between b and s.

In practical cases end-to-end response time will be associated with all possible communications between publishers and subscribers. The time will change according to the they location. We can compute the worst-case related to a topic and a broker as:

$$R(b)_{p\to s} = \sum R(b,t)_{p\to b},$$

where $R(b,t)_{p\to b}$ is end-to-end delay for broker b and $\forall t \in T^B \cap T^S = \{(N, V)\}$ – all topics in $b, N = \{n_i\}$ – the name of publisher p, $V = \{v_{ij}\}, v_{ij}$ – the type of measurement.

$$R(b,t)_{p\to s} = max(R_{pi\to b}) + R_b(t) + max(R_{b\to sj}),$$

where $R_{pi\to b}$ is the delay induced by the transmission network between b and p_i, p_iP such that $p_i[N^P] = n_iN$ – the publisher whose name appears in the topic $t = \{(N, V)\}$, $R_b(t)$ – average time of working with measurement results of type $v_{ij} = t[V]$ from a publisher p_i at a broker b, $R_{b\to sj}$ is the delay between b and s_j, $ts_j[T^S]$.

The ontology made it possible to select subscribers, brokers and publishers related by common topics and, for each case, to consider query statistics. In addition, we can calculate how loaded each broker is in the existing configuration. In a simple case, this is determined by the number of topics processed:

$$S(b) = |b[T^B]|,$$

where $|b[T^B]|$ is the topics of broker b, $[*]$ – cardinality of the set.

To add a new publisher or subscriber to the scheme, it is recommended to select a broker with the shortest end-to-end response time – $R(b)_{p \to s}$, least busy – $S(b)$ and with the best capabilities (minimum processing time, high CPU – specified in the ontology and etc.). Combining conditions for multicriteria selection is the task of the next study.

Moreover, to select a broker, it is proposed to take into account such an important characteristic as access security. Formalization allows us to group network log data [19] by brokers, that differ in security settings, types of requests and their sources. We can select various settings options specified in the S^{TB} and analyze network logs of the brokers that have these settings.

For example, when choosing the broker, take into account the distribution of the number of illegitimate requests from various sources by month (Fig. 4).

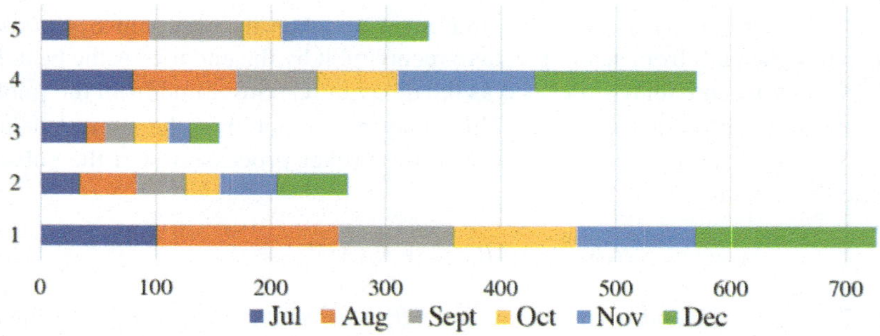

Fig. 4. Number of illegitimate requests by month for brokers 1–5

Using this information, we can choose a broker that receives fewer illegitimate requests and therefore it is less known and more secure. We can use the same statistics to select the security policy settings of the new broker.

3.2 The Broker Setup

The brokers provide the data from the publisher to the subscribers, including the mobile ones with limited resources for data processing and storage. The data output frequency should take into account its rate of changes, which will reduce the volume of mailings and the load on communication channels, while keeping adequacy of presentation of the ongoing processes at the same time.

From the ontology structure, publishers and parameters were selected that are necessary to configure each broker. For broker $b \in B^P$ which we will configure, select all topics $T_b = b[T^B]$. For each topic $t_i T_b$, $t_i = (N, v)$ we will do:

1. Select a publisher $p \in P|p[N^P] = N$;
2. Get $M = p[M^P] = (v, lim(v))$ for p, v;
3. Get restrictions $X_{min} = M[lim(v)[alc]]$, alc – allowable change;
4. Get $DP = p[DP]$, where $DP =< f(v), g(v) >$ for p, v;
5. Find the rate of the processes $S_{max} = DP[g(v)[rc]]$, rc – data change rate;
6. Calculate the period P_S during which critical changes will not have time to occur;
7. Calculate the maximum frequency F_{max} using $f(v)$.
8. Determine the minimum sampling period P_{min}.
9. Set $b[ST^B[fsd]] = \min(P_S, P_{min})$, fsd – frequency of data sending, which must be specified in the b settings.

The following is an explanation of these steps (for detail see [20]). The Steps 1–4 are intended to select from the ontology the publisher and its properties on the basis of which calculations will be made.

The Step 5 is designed to find the maximum rate of change in the indicator values per unit time and then to determine the time during which unacceptable changes will occur under the given restrictions. In general, the it can be obtained based on the collected statistics by simulating critical states: $S_{max} = \frac{\max(|x(t_{n+1}) - x(t_n)|)}{\Delta t}$, where $\Delta t = t_{n+1} - t_n$ is the time between observation, $n = [0, N - 1]$, N – the number of samples.

At the Step 6, the period is calculated during which critical changes will not occur. For each observed indicator, we set the X_{min} value of its allowable change, then the period is calculated based on the rate of the processes: $P_S = {X_{min}}/{S_{max}}$.

The Step 7 shows how to determine the rate at which the data changes. To calculate $f(v)$, the results of measurements of type v are taken. We will consider them as a function of a discrete argument that takes arbitrary positive values at time intervals: $x(t) = \{x(n\Delta t)\}$, where $x(nT)$ is the n-th result, $n = [0, N - 1]$, $\Delta t = t_{n+1} - t_n$ – the time between observation (Fig. 5).

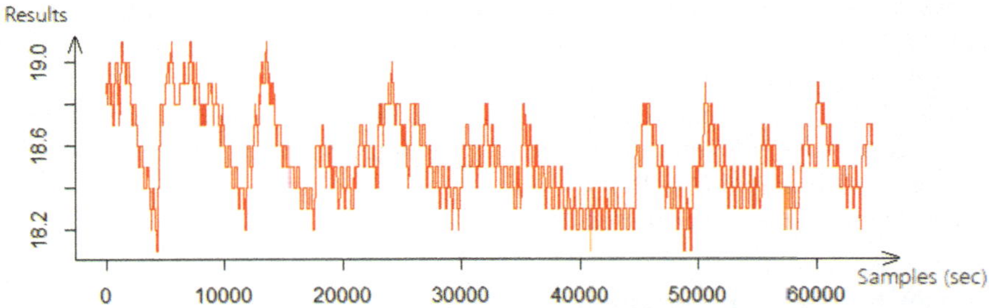

Fig. 5. Measurement results (v type).

The function $x(t)$ – is periodic and discrete. In the general case, the property of periodicity is achieved by repeating the considered data with a period $N\Delta t$. For a periodic function, a spectral decomposition in the form of a Fourier series can be performed for $k = [0, N - 1]$. The Fourier decomposition coefficients can be obtained by the discrete transform formula, which depends only on the index of the input signal $X(k) =$

$\sum_{n=0}^{N-1} x(n)e^{-i\frac{2\pi}{N}nk}$. The decomposition allows for N measurements of values $x(t)$ to obtain N samples on one repetition period of the spectrum $X(k)$ (Fig. 6).

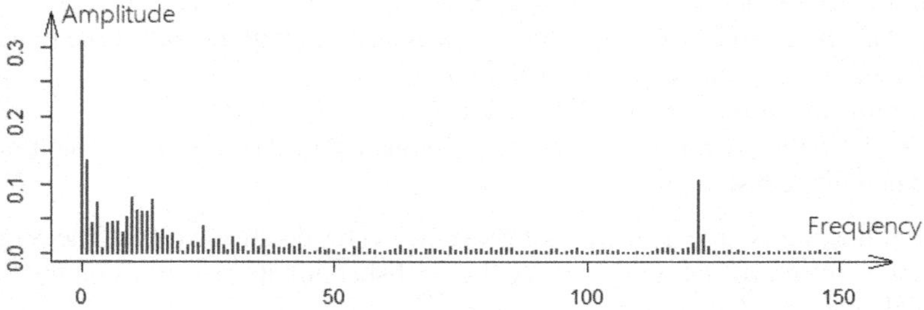

Fig. 6. Frequency spectrum.

Its first harmonic at $k = 0$ reflects the constant component of the data. Other frequencies (for $k \geq 1$) are signal harmonics. We need the limit ε for choosing the values of the amplitudes that will be taken into account in the analysis. The minimum ε can be determined from the accuracy of the IoT-sensor: $\varepsilon = p[PS^P[ar]]$, where PS^P are the properties of sensor p, ar – its accuracy. We choose decomposition coefficients $X(k)$ such that $Re(X(k)) > \varepsilon$ and the condition $X(k-1) \leq X(k) \geq X(k+1)$ is satisfied. By k we determine the oscillation period corresponding to the chosen $X(k)$, by dividing the observation interval NT the sample number: $P(k) = {}^{NT}/_{k}$. The frequency is $F(k) = {}^{1}/_{P(k)}$. For k, satisfying condition, we choose the maximum frequency of oscillations of the signal spectrum, which determines the sampling period $Fd = \max F(k)$.

According to the Nyquist-Shannon sampling theorem if a signal spectrum is limited by F_{max} frequency, it can be definitely restored by its discrete samples taken in time intervals, at the frequency that is at least twice as high as the maximum frequency of the signal. It follows that in order to observe the data, it is sufficient to perform measurements with a sampling frequency $F_{max} \geq 2Fd$.

At the Step 8, the minimum sampling period is calculated using the formula: $P_{min} = {}^{1}/_{F_{max}}$. At the last, 9th step, the resulting sampling period for the operation of IoT devices is defined as $\min(P_S, P_{min})$.

Such calculations are performed on each Publisher for new broker.

4 Discussions and Conclusion

The conducted semantic modeling has allowed to formalize the description of the Internet of Things network in our scientific organization. The ontology contains the objects describing the concepts of the subject area, relations, samples of the objects, data properties including the its change ranges and the critical values limiting the attributes of the ontology elements, settings of the IoT network architecture and description of the properties of devices that performing data collection, transmission, storage, processing. Ontology objects describe the "Publisher-Subscriber" information interaction scheme and the features of working with the MQTT protocol.

The standard presentation of the digital profile of the IoT devices usually takes into account only the network activity data: transmission frequency, network delays, packet sizes, flags "service quality", time of setting and holding the connection, etc. In new ontology it is complemented with indicators constructed as a result of analysis and aggregation of the temperature and humidity data collected by sensors, their statistical and frequency characteristics, observed anomalies and outliers.

Each of the individuals of the objects in the ontology has its own digital presentation in data bases, including the results of measurement and statistical characteristics. It combines knowledge and allows revision of the objects from different sides. For example, for Broker select and setup not only the statistics of requests from network logs is used, but also the characteristics obtained through the analysis of the processes monitored by IoT devices. This is important because IoT devices have limited capabilities in power consumption, memory and CPU usage, communication bandwidth and security control. It is shown that to solve the problem of choosing a broker with which a new device (publisher or subscriber) will interact, it is necessary to calculate the end-to-end delay or response time. This delay is computed as the sum of delays of receiving to the broker a message on a specific topic, time of the broker processing and the subscriber receiving the message. Since the ontology contains a detailed description of the structure and topics for each broker, this task is easily solved.

The collected statistics on brokers with different security settings, types and sources of illegitimate requests and the precedents for typical and abnormal behavior can be taken into account not only when choosing a broker but also to configure its security policies. We also showed how to use characteristics obtained through the analysis of the processes monitored to setup the broker. It allows a several time reductions of the amount of the transmitted data and of the IoT devices' activity time, providing monitoring of the events in accordance with the dynamics of their change. This approach is applicable for setting of the nonvolatile publishers where the reduction of frequency of data transmission and the extension of rest periods will extend the time of their autonomous function.

We regard the further development of the ontology through achievement of its functional fullness in terms of creation of a monitoring system for detection of network elements' abnormal behavior, equipment failures or changes in operation modes.

Acknowledgements. This work is supported by the Krasnoyarsk Mathematical Center and financed by the Ministry of Science and Higher Education of the Russian Federation in the framework of the establishment and development of regional Centers for Mathematics Research and Education (Agreement No. 075-02-2024-1378).

References

1. Miorandi, D., Sicari, S., Pellegrini, F., Chlamtac, I.: Internet of things: Vision, applications and research challenges. Ad Hoc Netw. **10**(7), 1497–1516 (2012)
2. Atzori, L., Iera, A., Morabito, G.: The Internet of Things: A survey. Comput. Netw. **54**, 2787–2805 (2010)
3. Javed, A., Heljanko, K., Buda A., Främling K.: CEFIoT: A fault-tolerant IoT architecture for edge and cloud. In: IEEE World forum on Internet of Things, pp. 813–818. Singapore (2018)

4. Isaeva, O.S., Kulyasov, N.V., Isaev, S.V.: Creation of a simulation stand for studying of the Internet of Things' technologies. AIP Conference Proceedings, p. 2647 (2022)
5. Patel, C., Doshi, N., Novel, A.: MQTT security framework in generic IoT model. Procedia Computer Science **171**, 1399–1408 (2020)
6. Meidan, Y., Bohadana, M., Mathov, Y., Mirsky, Y., Breitenbacher, D.: BaIoT: Network-based detection of IoT botnet attacks using deep autoencoders. IEEE Pervasive Comput. **17**(3), 12–22 (2018)
7. Isaev, S., Kononov, D.: Analysis of the dynamics of Internet threats for corporate network web services. CEUR Workshop Proceedings **3047**, 71–78 (2021)
8. Kononov, D., Isaev, S.: Development of secure automated management systems based on web technologies. IOP Conference Series: Materials Science and Engineering **537**(5), 1–13 (2019)
9. Haripriya, A., Kulothungan, K.: Secure-MQTT: An efficient fuzzy logic-based approach to detect dos attack in MQTT protocol for Internet of Things. J. Wirel. Commun. Network. **90**, 1–15 (2019)
10. Bhattacharyya, D., Kalita, J.: Network anomaly detection: A machine learning perspective. CRC Press, London (2014)
11. Nassif, A., Talib, M., Nasir, Q., Dakalbab, F.: Machine learning for anomaly detection: a systematic review. IEEE Access **9**, 78658–78700 (2021)
12. Omar, S., Ngadi, A., Jebur, H.: Machine learning techniques for anomaly detection: an overview. Int. J. Comp. Appl. **79**(2), 32–41 (2013)
13. Dissanayake, M.: Feature engineering for cyber-attack detection in Internet of Things. Int. J. wirel. Microw. Technol. **11**(6), 46–54 (2021)
14. Vaccari, I., Chiola, G., Aiello, M., Mongelli, M.: MQTTset, a new dataset for machine learning techniques on MQTT. Sensors **20**(22), 6578 (2020)
15. Cook, A., Misirli, G., Fan, Z.: Anomaly detection for IoT time-series data: a survey. IEEE Internet Things J. **7**(7), 6481–6494 (2019)
16. Rozik, A.S., Tolba, A.S., El-Dosuky, M.A.: Design and implementation of the sense Egypt platform for real-time analysis of IoT data streams. Adv. Inter. Things **6**(4), 66–91 (2016)
17. Musen, M.A.: The Protégé project: A look back and a look forward. AI Matters. Association of Computing Machinery Specific Interest Group in Artificial Intelligence, 1–4 (2015)
18. Dizdarević, J., Carpio, F., Jukan, A., Masip-Bruin, X.: A Survey of communication protocols for Internet of Things and related challenges of fog and cloud computing integration. ACM Comput. Surv. **51**(6), 1–29 (2019)
19. Hmissi, F., Ouni, S.: An MQTT brokers distribution based on mist computing for real-time IoT communications (Preprint). https://doi.org/10.21203/rs.3.rs-695717/v1
20. Isaeva, O.S., Kulyasov, N.V., Isaev, S.V.: Formation of adaptive publications from the Internet of things data broker. Info. Contr. Sys. **5**(120), 23–31 (2022)

Detection of Driver Drowsiness Using Artificial Intelligence and Machine Learning

Jannathl Firdouse Mohamed Kasim[1(✉)] and Muhammad Shameer G. Ameerudeen[2]

[1] Information Technology Department, Al Zahra College for Women, Muscat, Sultanate of Oman
jannathlfirdouse@zcw.edu.om

[2] Department of Computer Science and Engineering, Ponjesly College of Engineering, Nagercoil, India

Abstract. Drowsy driving can be caused by a variety of things, such as not getting enough sleep, long commutes, working shifts, and taking certain drugs. In addition, variables like drinking or using drugs and underlying medical illnesses like sleep disorders can make it worse when a person is driving while drowsy. It happens when a driver cannot retain complete focus and alertness while driving because of exhaustion or sleep deprivation.

According to estimates, drowsy driving may contribute to up to 40% of traffic accidents, with potentially serious effects. It leads to the design of a system for real-time hardware and software solutions to detect and prevent drowsy driving to address this problem. On the hardware side, a camera is used to gather information regarding the conduct and physical condition of the driver. This considered things like camera distance, eye movements, and facial expressions. Software algorithms are created to examine this data in real time and find patterns that point to drowsy driving. The system can take several steps to warn the driver and avoid an accident when it detects drowsy driving. These procedures include alarm sounding and warning message display. This could help to avoid an accident. The accuracy of the proposed system is 86.34% and the time taken for detecting the drowsiness is 70.67 s. The estimated overall speed is 14.29 frames per second.

Keywords: drowsy driving · camera distance · eye movements · facial expressions · alarm sounds

1 Introduction

Fatigued driving is a serious problem that can lead to crashes and fatalities. In the world, drowsy driving is believed to result in thousands of fatalities and hundreds of thousands of injuries each year. To solve this problem, numerous technologies and techniques have been created to identify driver drowsiness. Using machine learning algorithms to identify driver drowsiness is one of the most promising technologies. To find patterns and anomalies that point to drowsiness, these algorithms can examine a wide variety of data sources, including physiological signals, video footage, and speech patterns. This proposed work aims to create a project using Python and OpenCV which is a module

R. Silhavy and P. Silhavy (Eds.): CSOC 2024, LNNS 1120, pp. 291–299, 2024.
https://doi.org/10.1007/978-3-031-70518-2_25

in Python to detect the drowsiness of the driver. OpenCV is a powerful tool that can be used to analyze and process visual data from cameras and other sources. In the context of driver drowsiness detection, OpenCV can be used to extract features from camera footage that indicate signs of fatigue, such as slow eye movements or blinking. It extracts frames from the live video stream and analyzes them for changes in facial expression or head position. OpenCV provides functions for detecting faces, extracting facial landmarks, and analyzing facial expressions, which can be used to identify signs of drowsiness.

The problem statement defined driver sleepiness as a severe issue that can cause collisions and traffic fatalities. Driver drowsiness detection aims to create a system that can recognize when a driver is tired or falling asleep behind the wheel and warn them to pull over or take a break. Various sensors, cameras, and machine learning algorithms that can examine a variety of data sources, such as physiological signals, video footage, and driver behavior, can be used to achieve this. The goal of driver drowsiness detection is to create a system that can recognize when a motorist is getting sleepy or tired while driving and warn them to stop or pull over for a rest. Depending on the situation and the system's aims, the precise goals of driver drowsiness detection can change. However, a few typical goals are as follows:

In the case of improving road safety, driver drowsiness detection systems can assist reduce accidents brought on by driver fatigue by spotting and alerting drivers to indicators of drowsiness. For increasing driver awareness, driver drowsiness detection systems can assist drivers in being more aware of their level of sleepiness and taking action to prevent accidents by tracking driver behavior and physiological signals. In the case of reducing driver stress, driver drowsiness detection systems can help lower driver stress and enhance overall driving performance by alerting drivers to indicators of tiredness. Driver drowsiness detection uses real-time feedback to let drivers know how tired they are. Systems can support drivers in taking rests as needed, minimizing discomfort and tiredness. Thus, it enhances driver comfort while driving.

The motivation behind this proposed system is to help avoid accidents brought on by driver drowsiness, which is a major cause of traffic fatalities and accidents around the world. Long driving hours, lack of sleep, poor sleep quality, and circadian rhythms are just a few of the complicated issues that can contribute to driver tiredness.

2 Background History

The current technologies for detecting driver drowsiness evaluate driver behavior and physiological signs using a combination of sensors, cameras, and machine learning algorithms. Eye tracking technologies watch the driver's eyes with infrared cameras to look for indicators of tiredness like slowed blinking or diminished eye movement. Head movement systems track the driver's head movement and look for drowsy behaviors such as frequent head nodding using accelerometers or gyroscopes. Vibrations in the steering wheel can be a sign that the driver is losing control of the vehicle, and steering wheel vibration systems use sensors to detect these vibrations. A driver's speech patterns are analyzed by audio analysis systems using microphones to look for drowsiness-related indicators such as a drop in speech volume or monotone speaking. To identify drowsiness, sleep monitoring systems use sensors to track the driver's movements, breathing, and heart rate as they sleep. Driver behavior analysis systems analyze the driver's behavior, such as speed, acceleration, and braking, to detect indicators of tiredness using cameras and machine learning algorithms.

The most popular methods for spotting driver intoxication employ image-based technology. The eyes, lips, and head are just a few of the facial features that can be utilized to spot many of the visual traits that people who are tired display Visual sensors or cameras can capture these somnambulistic actions. Then, from these data, a few features can be identified, and by using computer vision algorithms, they are evaluated to visually examine the driver's physical condition to detect drowsiness in a non-invasive way. According to how the movements of the lips, eyes, and head are observed, image-based systems can be broadly divided into three groups [1]. The literature uses a variety of image-based elements.

These include the frequency of blinking, the longest period that the eyes may be closed [2], the percentage of eyelid closure [3], the eye aspect ratio [4], the curvature of the eyelids [5], the frequency of yawning [6], the MAR [7], the frequency of opening the mouth [8], the frequency of head nodding [9], and the analysis of head movement [10]. Combinations of these characteristics have also been considered [6, 7, 11]. A set of research papers referring to our proposed work are presented in this section.

As observed by A. Ebrahimi, 2018, fatigue and driver sleepiness are two of the most frequent factors in car accidents. Due to several variables, the number of deaths as well as fatalities substantially increases each year. Therefore, a system that warns the motorist is needed to decrease traffic accidents and protect lives and public property. In this study, a Driver Drowsiness Detection System (DDDS) based on eye movement was developed and put into use to safeguard drivers against collisions. The suggested DDDS employs a technique for detecting driver drowsiness based on the behaviors of the driver and their visual features employing eye movement (closed or open) by using a high-resolution camera. The main advantages are the higher accuracy in detecting driver drowsiness and the real-time detection of drowsiness. The higher complexity of implementing the deep learning algorithm requires both software and hardware requirements for the implementation. It requires a larger amount of data for the training is considered as the drawback of this work.

As explained in the research by A. S. Vidhya, 2019, smart or intelligent cameras have been developed because of advances in computer vision technology to monitor driver drowsiness. Furthermore, sounding an alarm all the time, the driver begins to get sleepy. Using this technology lowers the danger of accidents. To determine the status of the human eye while operating a vehicle, a novel framework is being suggested in this research that makes use of deep learning. We may train our model and categorize the data using built-in deep learning methods like transfer learning. Additionally, to acquire the best accuracy for our model, we will use CNN as our algorithm. Non-invasive methods of detection and continuous monitoring are the advantages of this system.

Accidents occur most frequently when a driver is sleepy, and this is true anywhere in the world. Many drivers experience fatigue and drowsiness due to lack of sleep and exhaustion, which frequently causes traffic accidents (M. A. H. Zulkefli,2021). The greatest strategy to prevent driving accidents brought on by tiredness is to warn the motorist. There are many ways to identify tiredness. In this research, we propose a deep learning-based method to identify driver drowsiness. Convolutional neural networks, a subset of deep learning, are what we employ. To find sleepiness, we employed the Face and Eye regions. The employed technique is the Yawing Detection Dataset (YawDD) and the Closed Eye in the Wild dataset (CEW) which had a mediocre level of precision of 96%. The advantages are a) drowsiness can be quickly detected because of deep learning systems' real-time data processing capabilities and b) DL algorithms are a non-invasive technique of detection because they don't require the driver to make any physical touch. The disadvantage is that deep learning implementation can be pricey in terms of both hardware and software expenditures. The efficiency of deep learning systems for identifying sleepiness in professional drivers has received little attention.

A face-monitoring drowsiness-detection system was reported by Moujahid et al., 2021 [6] that used a hand-crafted compact face texture descriptor to capture the most noticeable sleepy aspects. They started by capturing three signs of drowsiness: head nodding, frequency of yawning, and blinking rate. Then, to achieve compactness, they used pyramid multi-level face representation. Finally, they used a non-linear SVM classifier, which had an accuracy of 79.84%.

A driver drowsiness-detection architecture utilizing ResNet, AlexNet, FlowImageNet, and VGG-FaceNet was introduced by Dua et al. [12]. These models are created using behavioral features (such as head, mouth, and eye movements) and facial expressions that are retrieved from the driver's footage. The four deep learning algorithms were fed simulated driving movies. The four models' outputs were fed into a straightforward ensemble averaging technique and a SoftMax classifier, which produced an overall accuracy of 85%.

A sleepiness detection system based on the EAR measure was created by Maior et al. [13]. Inputs for machine learning algorithms including the multilayer perceptron, RF, and SVM classification models were created using the EAR values for successive frames. According to their evaluation findings, the SVM performed the best accuracy rate. The EAR measure was also applied in [14], which investigated the use of drowsiness as a binary SVM classifier input. With the highest accuracy, the model identified the driver's state of inattention.

As observed in Yaman Al Badawi et al., 2023, the technique for detecting driver tiredness in real-time utilizing non-invasive visual features is presented. These characteristics are taken from videos that were captured using a dashboard-mounted camera. The suggested system locates the regions of interest using facial landmarks and face mesh detectors, from which mouth aspect ratio, eye aspect ratio, and head pose features [15] are extracted and fed to three different classifiers: random forest, sequential neural network, and linear support vector machine. The suggested system has been tested using the National Tsing Hua University driver sleepiness detection dataset, and results indicate that it can accurately identify and alert sleepy drivers up to 99% of the time.

3 Proposed System

The current technologies for detecting driver drowsiness have several drawbacks that restrict their precision and usefulness in preventing accidents brought on by driver fatigue. Many of the current technologies for detecting driver drowsiness are expensive and necessitate large hardware, software, and upkeep expenditures Widespread adoption may be hampered by this, particularly for small enterprises and individual drivers. Some of the current technologies for detecting driver drowsiness need to be calibrated, which might take time and require special training. This may reduce their usefulness and dependability in actual applications. Accurate driver behavior data, including speed, acceleration, and braking, is crucial for many of the driver sleepiness detection systems currently in use. However, gathering and analyzing this information can be challenging, particularly in situations when stressed-out or distracted drivers may be present. Some of the systems are not very accurate, especially when it comes to spotting fatigue in drivers who are not behaving normally. False positives or false negatives may result from this, which can be harmful in practical applications. Lack of specificity and limited integration are the other drawbacks. The proposed systems have speed, usability, scalability, and cost-effectiveness as their merits. The module included in the proposed system is the Open-Source Computer Vision system (OpenCV).

It is a library that offers numerous tools and algorithms for processing images and videos, finding features, and identifying objects. It was created by Intel initially, and the OpenCV community currently looks after it. Numerous applications, such as those based on computer vision, robotics, surveillance, and autonomous cars, use OpenCV extensively. The features it offers for processing images and videos are extensive and include picture acquisition, image modification, image enhancement, object identification, object recognition, and picture segmentation. The flow diagram of the proposed work is given in Fig. 1. The initial step of the proposed work includes the capturing of video input, detecting the features of the face, and feature extraction. After that, depending on the organs like the head, eye, and mouth the classification is carried out by applying the Convolutional Neural Network, Support Vector Machine, and Hidden Markov Models. The drowsiness is checked and if it is true then the warning message is displayed, and it will alert the driver to concentrate while driving. Otherwise, the steps are repeated from the beginning.

4 Result and Discussion

The implementation of this system is carried out with a laptop configuration of 2 GB RAM, SSD of 300 MB, the processor Intel Pentium inside, and the GPU of Intel 82945G express. For the development environment, OS Windows 7 and PyCharm with Dlib in Python 3.11 are used. The sample results screen for the different statuses such as active, sleeping, and drowsiness are shown in Figs. 2, 3 and 4 respectively. For the active status, no warning signal is produced. But for the two states sleeping and drowsiness detection, the warning alarm will be generated to alert the driver.

The parameters such as accuracy, time taken for drowsiness detection, and the overall speed of the proposed system are calculated by using Eqs. (1), (2), and (3) as follows:

$$\text{Accuracy} = \frac{\text{Total Number of Frames}}{\text{Number of Correct Predictions}}\frac{\text{Total Number of Frames}}{\text{Number of Correct Predictions}} \quad (1)$$

$$\text{Time Taken for Drowsiness Detection} = \text{End Time} - \text{Start Time} \quad (2)$$

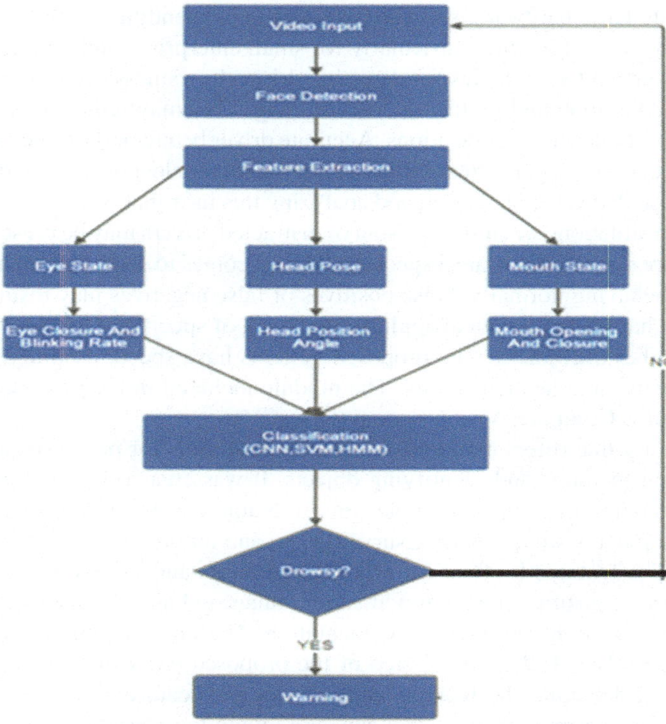

Fig. 1. Flow diagram of the proposed system.

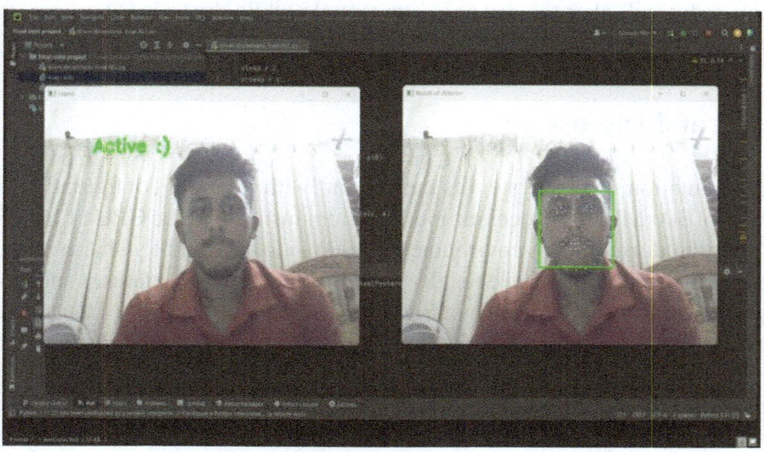

Fig. 2. Detection of Active status

$$\text{Overall Speed} = \frac{\text{Total Number of Frames}}{\text{Elapsed Time}} \frac{\text{Total Number of Frames}}{\text{Elapsed Time}} \quad (3)$$

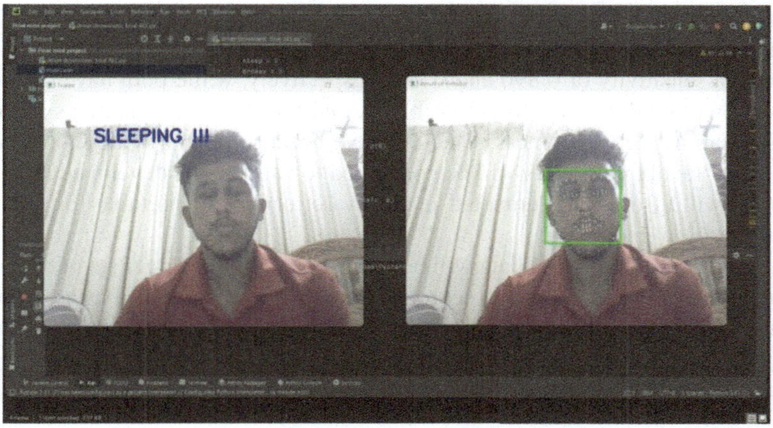

Fig. 3. Detecting the sleeping status

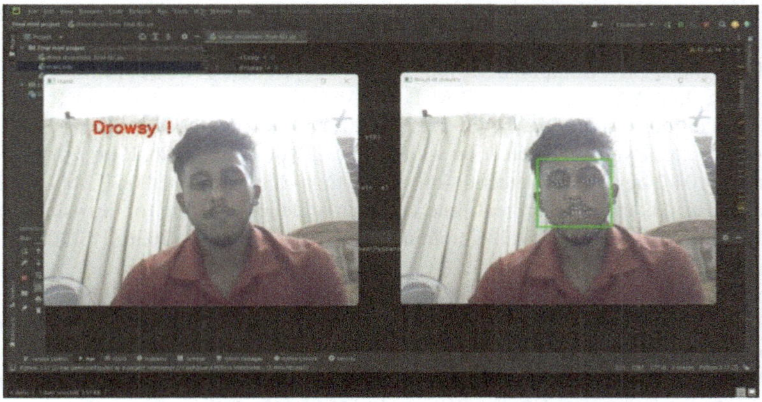

Fig. 4. Detecting the drowsy status

The accuracy of the proposed system is 86.34%, the time taken to detect drowsiness is 70.67 s, and the overall speed is estimated as 14.29 Frames Per Second.

5 Conclusion

In this research, a low-cost, real-time monitoring system for driver drowsiness based on ocular behavior and machine learning is provided. Here, webcam-captured streaming video is used to calculate aspects of visual behavior such as eye aspect ratio, mouth opening ratio, and nose length ratio. To identify driver drowsiness in real time, an adaptive thresholding method has been devised. With the generated synthetic data, the system is created correctly. Machine learning methods have been utilized for categorization after storing the feature values.

Here, we have looked at the Bayesian classifier, FLDA, and SVM. The FLDA and SVM have been found to perform better than the bayesian classifier. Both FLDA and SVM have a sensitivity of 0.896 and 0.956, respectively, while both have a specificity of 1. Work will be done to incorporate FLDA and SVM into the created system to do the classification (i.e., sleepiness detection) live because they provide more accuracy. A pilot study involving drivers will be conducted to validate the developed system, and the system will also be implemented in hardware to make it portable for automobile systems.

The testing and training of data will be carried out with the larger set of data and the performance measures such as F1- score, and precision call calculations are to be done as the future development. Also, we plan to develop a mobile application system with the proposed system for efficient and safe driving.

References

1. Albadawi, Y., Takruri, M., Awad, M.: A review of recent developments in driver drowsiness detection systems. Sensors *22*, 2069 (2022). **[Google Scholar]** **[CrossRef]** **[PubMed]**
2. Bamidele, A.A., et al.: Non-intrusive driver drowsiness detection based on face and eye tracking. Int. J. Adv. Comput. Sci. Appl. 10 (2019). **[Google Scholar]** **[CrossRef]**
3. Lin, S.T., Tan, Y.Y., Chua, P.Y., Tey, L.K., Ang, C.H.: Perclos threshold for drowsiness detection during real driving. J. Vis. **12**, 546 (2012). **[Google Scholar]** **[CrossRef]**
4. Rosebrock, A.: Eye Blink Detection with Opencv, Python, and Dlib. https://pyimagesearch.com/2017/04/24/eye-blink-detection-opencv-python-dlib/. Accessed on 7 May 2022
5. Tayab Khan, M., et al.: Smart real-time video surveillance platform for drowsiness detection based on eyelid closure. Wirel. Commun. Mob. Comput. **2019**, 2036818 (2019). [Google Scholar] [CrossRef]
6. Moujahid, A., Dornaika, F., Arganda-Carreras, I., Reta, J.: Efficient and compact face descriptor for driver drowsiness detection. Expert Syst. Appl. **168**, 114334 (2021). **[Google Scholar]** **[CrossRef]**
7. Sri Mounika, T., et al.: Driver Drowsiness Detection Using Eye Aspect Ratio (EAR), Mouth Aspect Ratio (MAR), and Driver Distraction Using Head Pose Estimation. In: ICT Systems and Sustainability, pp. 619–627. Springer, Berlin/Heidelberg, Germany (2022). **[Google Scholar]**
8. Celecia, A., Figueiredo, K., Vellasco, M., González, R.: A portable fuzzy driver drowsiness estimation system. Sensors **20**, 4093 (2020). **[Google Scholar]** **[CrossRef]** **[PubMed]**
9. Sikander, G., Anwar, S.: Driver fatigue detection systems: A review. IEEE Trans. Intell. Transp. Syst. **20**, 2339–2352 (2018). **[Google Scholar]** **[CrossRef]**
10. Coetzer, R., Hancke, G.: Driver fatigue detection: a survey. In: Proceedings of the AFRICON 2009. Nairobi, Kenya, pp. 1–6 (2009). **[Google Scholar]**
11. Liu, W., Qian, J., Yao, Z., Jiao, X., Pan, J.: Convolutional two-stream network using multi-facial feature fusion for driver fatigue detection. Future Internet **11**, 115 (2019). **[Google Scholar]** **[CrossRef]**
12. Dua, M., Singla, R., Raj, S., Jangra, A., Shakshi: Deep CNN models-based ensemble approach to driver drowsiness detection. Neural Comput. Appl. **33**, 3155–3168 (2021). **[Google Scholar]** **[CrossRef]**
13. Maior, C.B.S., das Chagas Moura, M.J., Santana, J.M.M., Lins, I.D.: Real-time classification for autonomous drowsiness detection using eye aspect ratio. Expert Syst. Appl. **158**, 113505 (2020). **[Google Scholar]** **[CrossRef]**

14. Rasna, P., Smithamol, M.: SVM-based drivers drowsiness detection using machine learning and image processing techniques. In: Progress in Advanced Computing and Intelligent Engineering, pp. 100–112. Springer, Berlin/Heidelberg, Germany (2021). **[Google Scholar]**
15. Popieul, J.C., Simon, P., Loslever, P.: Using driver's head movements evolution as a drowsiness indicator. In: Proceedings of the IEEE IV2003 Intelligent Vehicles Symposium. Proceedings (Cat. No. 03TH8683), pp. 616–621. Columbus, OH, USA (2003). **[Google Scholar]**

Modeling for Cost Minimization Applied to the Allocation of Vehicles in Road Transport Companies

Francisco Vando Carneiro Moreira[1], Plácido Rogério Pinheiro[1(✉)], and Carolina Ferreira Araujo[2]

[1] University of Fortaleza, Fortaleza, CE, Brazil
Placido@unifor.br
[2] Aberta University, Lisbon, Portugal

Abstract. This research presents an Integer Linear Programming (ILP) model for optimising the dynamic allocation of resources in European road transport companies. The main focus is improving operational efficiency and minimising various costs, such as maintenance, fuel, salaries, tolls, insurance, delays, and environmental impacts, within the European regulatory context, particularly Portuguese Regulation Reg. (EC) No. 561/2006. The model incorporates specific constraints on driving time limits, rest requirements, and interactions between departments and vehicles. The results indicate a reduction in operational costs and an improvement in logistic efficiency. Implementing the model using LINGO software demonstrated the ability to adapt to changes in demand and regulations, maintaining compliance and promoting safety. In addition, a tool was developed to facilitate sensitivity analysis and strategic decision-making, allowing for flexible adjustments to changes in demand or operational conditions. The objective is to contribute to logistic optimisation and provide a practical and robust tool for transportation sector managers. The importance of technological innovation in logistics operations management and addressing challenges in complex transportation environments is highlighted. The insights generated by the model have the potential to positively influence logistic management practices, emphasising a more strategic and sustainable approach in the road transport sector.

Keywords: Integer Linear Programming · Dynamic Allocation of Transport · European Road Transport · Transportation Regulations · Operational Efficiency · Logistic Management

1 Introduction

In European road transport's dynamic and regulated environment, resource allocation efficiency becomes critical for companies ' operational and financial success [1]. This study addresses the inherent complexity of managing truck fleets, focusing on optimising the allocation of transportation assets to maximise efficiency and minimise operational costs. Adopting an Integer Linear Programming (ILP) approach allows for incorporating multiple variables and constraints, including driving and rest time limits, various

R. Silhavy and P. Silhavy (Eds.): CSOC 2024, LNNS 1120, pp. 300–306, 2024.
https://doi.org/10.1007/978-3-031-70518-2_26

costs, and specific regulations such as Portuguese Regulation Reg. (EC) No. 561/2006. This work aimed to develop and validate a model that provides an optimised solution for the dynamic allocation of trucks, considering operational and regulatory variables and assessing its effectiveness in reducing costs and improving operational efficiency. Relevant studies in this area include the works of [7–10], which provide insights into supply chain management and business logistics. Additionally, the application of Artificial Intelligence, as discussed in [11], offers significant advancements in road transport logistics optimisation.

Furthermore, the course on logistics optimisation [12] and the work on vehicle routing with fixed heterogeneous fleets [13] provide relevant academic contexts for this study. In addition, the methods developed and presented in the survey on prioritising maintenance work orders in thermal power plants [2, 3] provide a relevant context for applying multicriteria optimisation models in complex industrial and logistical scenarios. The LINGO Software [4, 5] supports the model's implementation as the solver and a tool developed in parallel to facilitate real-time sensitivity analysis and strategic decision-making. This research aims to contribute to logistics optimisation, providing insights for companies in the road transport sector seeking to enhance their operations and resource management strategies.

2 Methodology

The model proposed in this study utilises the Integer Linear Programming (ILP) approach to optimise resource allocation in road transport companies. The methodology involved the integration of operational and regulatory variables, including maintenance costs, fuel, salaries, tolls, insurance, and environmental impacts, as well as driving and rest time constraints by Portuguese Regulation Reg. (EC) No. 561/2006. For the model's implementation, the LINGO Software with a solver was used, and concurrently, a tool was developed to assist in sensitivity analysis and adjustments according to changes in operational demands. The methodology also included the simulation of various scenarios to test the model's effectiveness under different operational conditions. Model validation was conducted by comparing the results obtained with historical data and the company's operational standards, ensuring the accuracy and practical applicability of the model. This methodological approach aims to provide a comprehensive and flexible solution for resource management in road transport, enabling companies to respond efficiently to market dynamics and current regulations.

3 Modelling

A detailed modelling approach was adopted, dividing the essential variables in road transport fleet management into Operational and Regulatory categories. Operational variables are multifaceted, including administrative costs, vehicle load capacity, financial penalties for delivery delays, maintenance costs, fuel expenses, driver remuneration, toll fees, insurance, and considerations regarding environmental impacts. This aspect is crucial for understanding road transport operations' efficiency and economic viability. On the other hand, Regulatory Variables focus on the legal constraints that affect fleet operations. In this study, we delimited the regulatory variables only to include driving time limits and mandatory rest periods for drivers, aligned with European legislation such as Portuguese Regulation Reg. (EC) No. 561/2006. These constraints are essential to ensure the safety and well-being of drivers, as emphasised by [15] in their study on the regulatory framework for cargo road transport. Both categories are critical for responsible transport operations management, requiring a balance between operational efficiency and regulatory compliance.

3.1 Mathematical Formulation

The model was formulated to represent the relationships between variables and constraints, aiming to find a solution that minimises total costs while meeting regulatory requirements and operational needs. Incorporation of Binary Constraints: Essential for modelling "yes" or "no" decisions, such as allocating a truck to a specific department on a given day. The challenge of efficiently managing vehicles over time, known as the Vehicle Dynamic Allocation Problem (VDAP), is common in various fields, including road, rail, and maritime freight transportation. Still, it is a daily challenge for traffic managers. One of the case studies in the literature focused on controlling an extensive fleet of trucks for road cargo transportation [14].

3.2 Model Formulation

The objective function is a weighted sum of the costs associated with allocating trucks to meet the demands of the departments. The encompassed costs include transportation, maintenance, fuel, labour, capacity, tolls, insurance, administration, delays, and environmental impacts. Such costs are crucial in optimising transportation logistics, as highlighted in [16] in their supply chain analysis:

$$
\begin{aligned}
\min\Bigg(\sum_{j\in J}\Bigg(\sum_{i\in I}\Bigg(\sum_{k\in K} & \mathrm{SetAssign}(j,i,k) \\
& \times[\mathrm{SetAvailabilityCost}(j,i) \\
& +(\mathrm{SetTimeToDelivery}(j,i)+\mathrm{SetLevelPerformanceService}(j,i)) \\
& +\sum_{m\in M}\mathrm{SetCostTransport}(j,i,k,m)+\sum_{m\in M}\mathrm{SetCostMaintenance}(j,k,m) \\
& +\sum_{m\in M}\mathrm{SetCostFuel}(j,k,m)+\sum_{m\in M}\mathrm{SetCostDriver}(j,k,m) \\
& +\sum_{m\in M}\mathrm{SetCostCapacity}(j,k,m)+\sum_{m\in M}\mathrm{SetCostToll}(j,k,m) \\
& +\sum_{m\in M}\mathrm{SetCostInsurance}(j,k,m)+\sum_{m\in M}\mathrm{SetCostAdmin}(j,k,m) \\
& +\sum_{m\in M}\mathrm{SetCostDelay}(j,k,m)+\sum_{m\in M}\mathrm{SetCostEnvironmental}(j,k,m)]\Bigg)\Bigg)\Bigg)
\end{aligned}
$$

The objective of the function is to minimise the weighted sum of the presented costs, subject to the model's constraints. The minimisation process will attempt to find the optimal allocation of trucks that meet the departmental demands at the lowest possible cost.

4 Implementation and Analysis

4.1 Dashboard

An interactive dashboard was developed to facilitate the visualisation of results, sensitivity analysis, and simulation of different operational scenarios, as shown below (Fig. 1):

Fig. 1. Dashboard

5 Analysis of Results

The research proposes an ILP model to optimise the dynamic allocation of resources in road transport operations. The focus is maximising efficiency and minimising costs while adhering to driving and rest time constraints and the specific relationships between departments and trucks. The main result, the optimised allocation, shows that the model achieved an optimised allocation of trucks for each department daily, balancing supply and demand, as depicted in the graph below. The graph analysis focuses on comparing the current allocation of vehicles in various departments with a reference allocation, highlighting the percentage reduction in this allocation. The visual representation, with blue bars for the current allocation and red bars for the reference, allows for a clear assessment of potential over-allocation or under-allocation of vehicles. If the blue bars surpass the red ones, it would indicate possible over-allocation, while the opposite would signify under-allocation. The percentage reduction line complements the analysis, indicating the efficiency of the current allocation compared to the ideal one. High values in this line would suggest a significant margin for optimisation, pointing to an inefficient use of vehicles. On the other hand, low values indicate that the current allocation is close to the ideal. This analysis is crucial for fleet strategic management, as it identifies areas for potential optimisation, aiding in decision-making to enhance operational efficiency and reduce costs in various departments (Fig. 2).

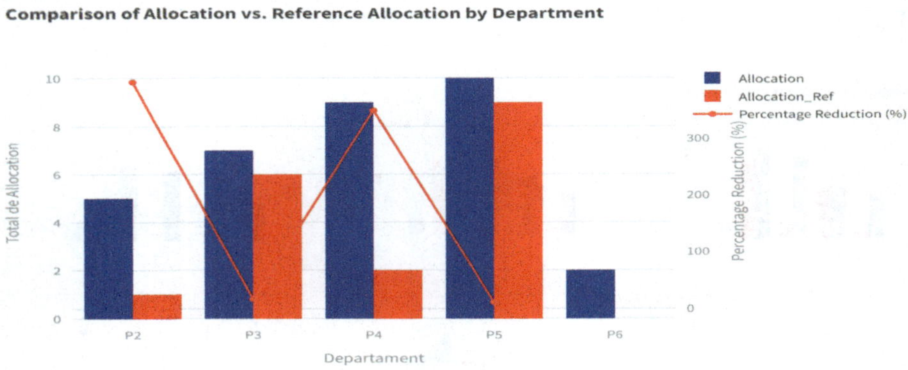

Fig. 2. Allocation

Cost Minimisation: The heat map analysis reveals the distribution of values associated with different types of trucks across various departments. Colour variations in the graph indicate the levels of these values, with more intense colours representing higher values. In this case, it allows for a quick identification of which departments exhibit higher or lower activity and efficiency concerning the types of trucks used. That said, the analysis helps understand the suitability of different types of trucks for specific tasks in each department. Anomalies or unusual patterns highlighted in the graph may signal areas requiring additional optimisation attention (Fig. 3).

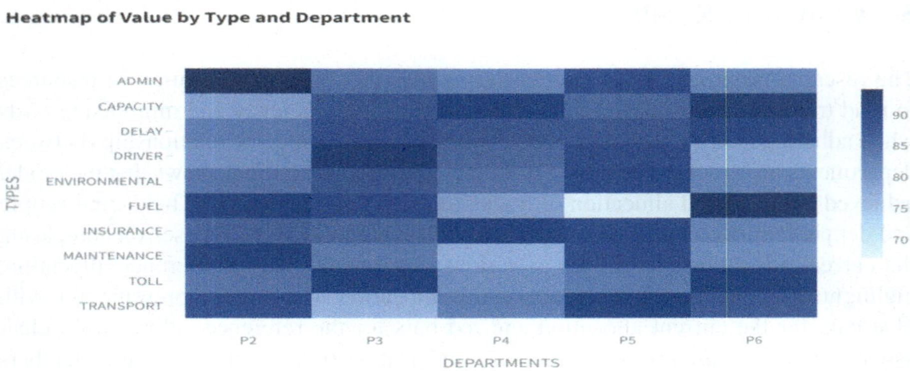

Fig. 3. Heatmap

Regulatory Compliance: The model suggests ensuring drivers comply with driving and rest time requirements, maintaining compliance with relevant regulations, and promoting safety. The analysed graphs provide insights into different departments' average rest and driving times. They reveal variations in rest periods and highlight differences in driving times between departments. The direct comparison of these two aspects in the graphs allows for an assessment of the balance between work and rest, indicating potential areas for adjustments in shift and operational management to optimise performance and ensure the well-being of drivers (Fig. 4).

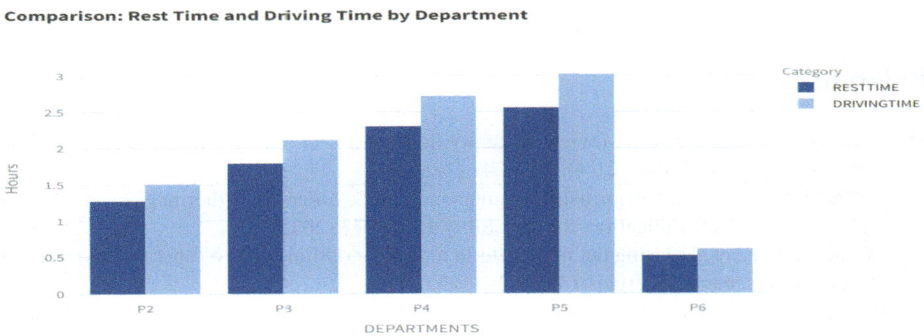

Fig. 4. Conformidade

Contributions and Practical Implications: The model presented an integrated and holistic approach, encompassing various facets of road transport operations. The comprehensive cost analysis provided a detailed view of the financial impacts of resource allocation. The results offered insights for strategic logistics and transport management decision-making, demonstrating the model's applicability.

6 Final Considerations

The conducted research reveals that the proposed model can serve as an effective tool to address the complex and multifaceted challenges in the road transport sector. It demonstrates the ability to improve regulatory compliance, a crucial aspect in increasing legal and regulatory rigour. Furthermore, the model signals that it can contribute to optimising operational efficiency, facilitating resource management, and cost reduction, essential aspects for the economic sustainability of transport companies. The flexibility and adaptability of the model are its most prominent highlights. It efficiently adapts to changes in market demands and conditions, an indispensable feature in this era of technological evolution. This adaptability ensures that the model remains relevant even in the face of changes in logistics sector dynamics. For future research, the focus will be on expanding the model to include more detailed analyses of the environmental impacts of road transport. This will involve incorporating variables such as greenhouse gas emissions, fuel efficiency, and the ecological footprint of freight transportation. Such expansion would align the model with growing environmental concerns and provide companies with tools for more sustainable and responsible operations.

Furthermore, the potential integration with Machine Learning and Big Data analysis methods is highlighted [6]. This integration could offer promising insights and real-time decision-making, enhancing the model's effectiveness in predicting and managing operational and logistical complexities. These emerging technologies would enhance the current model's functionality and open new possibilities for innovative and efficient practices in the transport and logistics sector.

Acknowledgements. Plácido Rogério Pinheiro thanks the National Council for Scientific and Technological Development (CNPq) for developing this project.

References

1. Dejax, P.J., Crainic, T.G.: A review of empty flows and fleet management models in freight transportation. Transp. Sci. **21**(4), 227–247 (1987)
2. Andrade, S.J.M., et al.: Prioritising maintenance work orders in a thermal power plant: a multicriteria model application. Sustainability **15**, 54–73 (2023)
3. Moraes, C.C.F., et al.: Using the multicriteria model for optimization of operational routes of thermal power plants. Energies **14**, 3682–3699 (2021)
4. Baptista, M.N., Campos, D.C.: Metodologias Pesquisa em Ciências - Análise Quantitativa e Qualitativa (in Portuguese), 2nd edn. Grupo GEN - LTC, Rio de Janeiro (2016)
5. LINDO Systems Inc.: LINGO User's Manual. Disponível em: <https://www.lindo.com/downloads/PDF/LINGO.pdf>. Accessed on: 08 Nov. 2023
6. Pinheiro, P.R., Tamanini, I., Pinheiro, M.C.D., Albuquerque, V.H.C.: Evaluation of the alzheimer's disease clinical stages under the optics of hybrid approaches in verbal decision analysis. Telematics Inform **35**(4), 776–789 (2018)
7. Ballou, R.H.: Gerenciamento da cadeia de suprimentos: logística empresarial. Bookman Editora, São Paulo (2009)
8. Bertaglia, P.R.: Logística e a cadeia de abastecimento. Ed. Saraiva, São Paulo (2006)
9. Bovo, D.: Estudo de simulação da reconfiguração do layout logístico para redução de custos e eliminação de gargalos. Revista Gestão Industrial (2015). D.O.I.: https://doi.org/10.3895/gi.v11n3.2822
10. Caixeta-Filho, J.V.: Pesquisa operacional. 2ª edição, Atlas, São Paulo (2004)
11. Portal NTC: A Inteligência Artificial no Transporte Rodoviário de Cargas: Avanços e Benefícios. COMJOVEM Campinas (2023). Disponível em: www.portalntc.org.br
12. Melo, R.: Curso: Capacitação em Introdução à Otimização Logística. Universidade Federal da Bahia, Salvador, Bahia, Brasil (2023)
13. Dutra, Q.F.S.: Problema de roteirização de veículos com frota heterogênea fixa, janelas de tempo, entregas fracionadas e limitações de acesso. Escola Politécnica, Universidade de São Paulo, São Paulo, Dissertação de Mestrado (2023)
14. Powell, W.B., Sheffi, Y., Thiriez, S.: The dynamic vehicle allocation problem with uncertain demands. In: International Symposium on Transportation And Traffic Theory **9**, 357–374 (1984)
15. Togan, S.: The regulatory framework in road freight transportation. The Liberalization of Transportation Services in the EU and Turkey. Oxford Academic (2016)
16. Chopra, S., Meindl, P.: Supply Chain Management: Strategy, Planning, and Operation. Pearson (2013)

Development of Selective Electrodes for Drug Detection

Zamira Abdurakhmanova[1], Ergashboy Abdurakhmanov[2(✉)], Zulfiya Murodova[2], Shokhista Salimova[2], Eldor Ismoilov[2], Shokhjakhon Suvonkulov[2], and Beknazar Shukrulloyev[2]

[1] Samarkand Medical University, Samarkand, Uzbekistan
[2] Samarkand State University, Samarkand 140104, Uzbekistan
ergash50@yandex.com

Abstract. Methods for obtaining plasticized membranes have been developed based on synthesized ionophores. The physico-chemical properties of EAC, the effect of membrane and solution composition on the Potentiometric characteristic of ISE have been determined. The main electrochemical indicators of the developed ISE are studied: such parameters as the linear range of the signal (electrode function), the slope of the electrode function, the lowest limit of ion detection, and the dynamics of the electrode signal (response time).

Keywords: Medicinal Substances · Novocaine · Phosphoromolibcate · Phosphorovolframate · Electrode Function

1 Introduction

In the Republic of Uzbekistan, great attention is paid to control the production and composition of medicinal substances. In this direction, certain fundamental and practical results were achieved. The head of our state, in the"strategy of action for the further development of the Republic of Uzbekistan", defined important phases on further reform of the health sector, above all on reducing the incidence rate of the population and increasing the life expectancy of the population [1]. In this direction, it is relevant to intensively develop methods for monitoring the composition of pharmaceutical products and the technological processes of their extraction, develop modern operational methods for analyzing the composition of products and, on their basis, create highly effective sensors that solve various problems.

Today, pharmacy departments with a wide network of Pharmaceutical Industries and control and analytical units, which are rapidly developing in the world, need analytical methods widely used in the control of drugs in ready-made forms, various medicinal substances in plant raw materials for the purpose of Drug Control [2, 3]. The ionometric method of drug detection is the most common instruments based on this method (ionomers), available in almost every laboratory, so data on ionometric methods for detecting various medicinal substances are of particular importance.

R. Silhavy and P. Silhavy (Eds.): CSOC 2024, LNNS 1120, pp. 307–314, 2024.
https://doi.org/10.1007/978-3-031-70518-2_27

There is a lot of research in the world on the creation of highly effective ion-selective materials (ISMs) for ion-selective electrodes (ISE) of medicinal substances (Drugs). Due to the speed and simplicity of the equipment under non-laboratory analysis conditions, potentiometry based on the use of an ion-selective electrode is an alternative to time-consuming and expensive methods such as high-performance liquid chromatography, capillary electrophoresis, and spectral analysis [4, 5]. For this reason, the creation and research of a new high-performance Ise based on the existing electrode active ingredients is one of the important tasks for the application of drugs. Such electrodactive substances include phosphomolybdenum, phosphovolfram, and other heteropolyacids (HPAs) and their derivatives, which are quickly found and in some cases significantly outperform natural ionophores in selectivity and sensitivity [6].

The purpose of this work is to develop plasticized membrane ion-selective electrodes based on tungsten, molybdenum heteropolycyslots for express analysis of medicinal substances and to check the effects of various factors on their conductors.

To achieve the stated goal, the synthesis of low-soluble electrode-active compounds of medicinal substances with PVA is the choice of optimal conditions, verification of the structure and physico-chemical properties of EAC obtained using modern methods of analysis. Research has been carried out in the field of developing technology for obtaining plasticized membranes with the resulting ionophores, establishing the effect of the composition of the ISE membrane, which determines drugs, on its selectivity.

2 Materials and Methods

As an electrode-active substance, compounds based on phosphovolframatic acid (PVA) and medicinal preparations were used: novocaine, lidocaine, pyridoxine and diphenhydramine.

The syntheses were carried out at room temperature (25 °C). Aqueous solutions of appropriate drugs (novocaine et al.) have been added to aqueous solutions of phosphorovolfram heteropolyacid (GPA) heated to 25 °C while continuously stirring. After a while, the white sediment sinks. Then the sediment was separated from the liquid, washed with cold distilled water and dried for a day. Given the three bases of phosphoromolibdate and phosphorovolframate HPA, as well as the possibility of protonating three molecules of Novocaine, the molar ratio of HPA to novocaine was taken to be appropriate ratio 1:3.

In the next stage, ion assosiates of PVA and drug preparations were synthesized. To obtain phosphovolframolidocaine assosiate, a solution of 100 ml of 0.01 M of phosphovolframatic acid (m = 2.8802 g) and a solution of 100 ml of 0.01 m of lidocaine are prepared. The prepared 20 ml of PVA solution and 60 ml of lidocaine solution (PVA-lidocaine ratio = 1:3) are poured into a glass with a capacity of 200 ml, and the resulting precipitate is kept for 8 h, then filtered in a shotta filter (filter pores size 40 nm). The precipitate is washed several times with distilled water until the chloride ions are completely lost, then the precipitate is air-dried at room temperature for one day. The final products are yellow powders.

The plasticized film membranes of Ise were prepared by dissolving PVA, a dioctylphthalate plasticizer, and an electrode active compound (EAC) in Tetrahydrofuran with incessant mixing. The EAC content was 5–15% compared to the weight of the membrane; the ratio of PVA-plasticizer was 1:2 by weight the membrane electrode consists of a cylindrical teflon body, to the end of which a plasticized membrane is attached. A ~ 1.5 ml internal comparative solution containing a potential determining ion ($1 \bullet 10^{-5}$ – $1 \bullet 10^{-4}$ M) and 1–2 drops of 3 M KCl was introduced into the electrode body. As an internal comparative electrode, a silver wire was obtained, covered with a layer of silver chloride.

The table shows the results of the synthesis of ion associations, which are administered on the basis of the preparations PVA and drotaverin, novocaine, lidocaine and dimedrol (Table 1).

Table 1. The results of the synthesis of eam based on the medicinal preparations of PVA and novocaine, lidocaine, drotaverin and dimedrol.

	The general formula of EAC is	Color of EAC	Initial salt	By yield $H_3[P(W_3O_{10})_4]$	Solubility of the resulting EAC
Starting component- 25 ml 0,01 M $H_3[P(Mo_3O_{10})_4]2H_2O$ solution					
1	$Nov_3[P(Mo_3O_{10})_4]2H_2O$	light green	NovHCl	96.7	less soluble
2	$Lid_3[P(Mo_3O_{10})_4]\ 2H_2O$	light green	LidHCl	98.1	weak blur
3	$Dim_3[P(Mo_3O_{10})_4]\ 2H_2O$	dark green	DimHCl	90.5	less soluble
4	$Drot_3[P(Mo_3O_{10})_4]2H_2O$	dark green	DrotHCl	92.8	weak blur
Starting component- 25 ml 0,01 M $H_3[P(W_3O_{10})_4]*2H_2O$ solution					
5	$Nov_3\ [P(W_3O_{10})_4]2H_2O$	White	NovHCl	97.6	less soluble
6	$Lid_3[P(W_3O_{10})_4]2H_2O$	light yellow	LidHCl	97.1	weak blur
7	$Dim_3\ [P(W_3O_{10})_4]2H_2O$	dark yellow	DimHCl	94.7	less soluble
8	$Drot_3[P(W_3O_{10})_4]2H_2O$	Brown	DrotHCl	95.5	weak blur

3 Results

Examination of the ISE ni couriers, which detect the developed drugs, and calibration of the electrode are carried out in the following order. Before measuring the potential, the electrodes were prepared for work in the prescribed manner. The investigations were based on drugs (novocaine, dimedrol, drotaverin hydrochloride, lidocaine hydrochloride) and ionselective electrodes made using ionophores based on fosfornovolframatic acid (Nov.- PVA, lid.- PVA, dim.- PVA, drot.- PVA). The electrode potential was 10–2 mol/L at a temperature of 25 °C according to EXAM 8.213–76 by measuring the ECP of the electrode system under investigation. Determined using solutions. It was found that the concentration for electrodes developed for the detection of various drugs depends on the concentration of the substance in solution, which determines the signal value of the electrode in the range of 10-1–10-7 mol/L. The results of this experiment are presented in Table 2.

Table 2. The dependence of the signal of ises based on phosphorus tungsten on the concentration of the detectable drug in solution.

№	Amount of the drug in solution, mol / l	Electrode signal, E mV			
		Novok-PVA	Lidok- PVA	Dimed- PVA	Drotav- PVA
1	10^{-7}	−79,1	−132,2	−49,0	−37,5
2	10^{-6}	−78,1	−135,7	−52,2	−35,5
3	10^{-5}	−56,8	−117,7	−54,2	−30,5
4	10^{-4}	−23,9	−85,3	−38,7	17,8
5	10^{-3}	0,8	−41,5	−20,8	30,1
6	10^{-2}	30,9	−7,6	12,0	59,1
7	10^{-1}	59,1	37,3	37,8	90,5

The main electrochemical indicators of how ISE include parameters such as the linear range of the signal (electrode function), the slope of the electrode function, the lowest limit of ion detection, the coefficient of selectivity and the dynamics of the electrode signal (response time). All parameters of the signal other than its dynamics are determined by the expression E - pA that the e-electrol potential depends on pA. Using the indicators of the ISE, a conclusion is made about the operation of the electrode. The main characteristic of the measuring electrode is the performance interval (Nernst region) of the electrode function – the part of the ion detection range that has a straight-line appearance of the dependence of the signal on the concentration. That is, the range of linear dependence of the electrode potential (E) on the inverse logarithm of the activity of the detectable ion A (pA).

Developed from results obtained on the determination of the execution interval of the Nernst region of the membrane electrode, developed on the basis of fosfornovolframate and the corresponding drug, the electrodes showed the possibility of detecting drugs in the 10^{-1}-10^{-5} mol/l bands of the concentration in accordance with the ionophore content.

In this case, it was observed that the highest detection range of the ISE studied corresponds to the ISE based on lidocoin PVA. The results of determining the electrode function of this electrode are presented in Fig. 1, from which we see that the electrode developed has the potential to detect lidocaine in their concentration 10^{-1} mol/L to 10^{-5} mol/L, in accordance with the ionophore composition.

Another indicator of a drug-detecting selective electrode is the steepness (K) of its electrode function, which is the value:

$$K = X / U$$

is found by the equation. Where X = ΔE and U = ΔpA.

According to the Nernst coefficient, the value of the steepness calculated theoretically in two-valence ions is 29 mV. Electrode pound steepness percentage:

$$A\% = K/29 \cdot 100.$$

It is determined by the equation and this is used to determine if the coupler is working the electrode correctly.

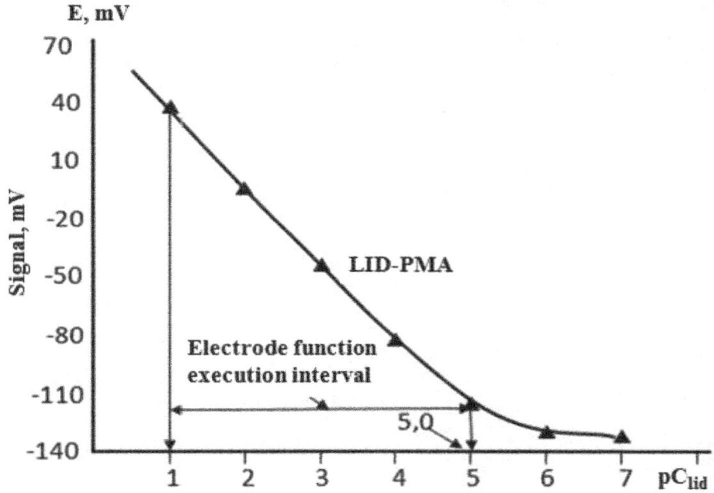

Fig. 1. Lidocoin drug detection Lidoc is an electrode function execution interval of ISE developed on the basis of PVA.

4 Discussions

From the results of determining the slope of the electrode function of the Ise (Fig. 2), developed on the basis of the EAC with Novok-PVA composition, the value of anicized K is: K = ΔE/ΔpA = 130/4, 7 = 27.7, and the slope percentage is (27.7/29)100 = 95.5%. The results of the study of the slope of the electrode function for electrodes, developed on the basis of memranes of different composition, are presented in Table 3.

From table 3 we see that the highest indicator of the electrode function equal to 97.9% of the slope magnitude value corresponds to Dim$_3$ $[P(W_3O_{10})_4]2H_2O$ and the lowest 92.8% value of that function corresponds to Lid$_3$ $[P(W_3O_{10})_4]2H_2O$.

The lower limit of Ion selective electrodes in drug detection is extrapolated to the linear sections of the E - pA dependence to determine the potential detection limit (C_{min}, p), one of the major indicators of the electrode. In this case,the ion detection limit corresponds to the value $-lgC_{min}$, p, on the abscissa axis of the intersection point obtained. The detection limit of Novocoin selective electrode (Nov-PVA) and lidocaine selective electrode (lid-PVA) detected in experiments was found to be above 10^{-5} mol/L.

Table 3. The values of the slope magnitude of the electrode function corresponding to the corresponding ions of the developed ises (n = 5, P = 0,95, Cion = 10^{-2} mol/l).

№	Composition of the ionophore	Defining component	Slope magnitude of electrode function (x ± ΔX), Mv	Slope percentage of electrode function (x ± ΔX), %
1	$Nov_3\,[P(W_3O_{10})_4]2H_2O$	Novocaine	27.7 ± 0.1	95.5 ± 0.4
2	$Lid_3[P(W_3O_{10})_4]2H_2O$	Lidocoin	26.9 ± 0.2	92.8 ± 0.7
3	$Dim_3\,[P(W_3O_{10})_4]2H_2O$	Dimedrol	28.4 ± 0.2	97.9 ± 0.5
4	$Drot_3[P(W_3O_{10})_4]2H_2O$	Drotaverin	27.3 ± 0.3	94.1 ± 0.6

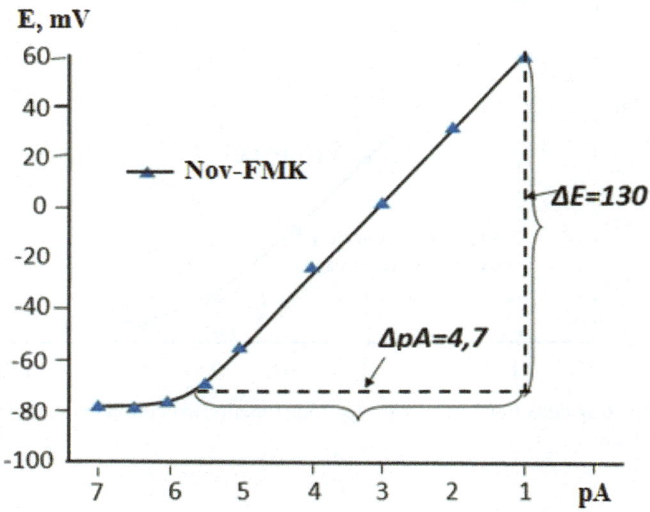

Fig. 2. Novok is the slope of the electrode function of the ISE, developed on the basis of the PVA-containing EAC.

The results of finding the lowest concentration values of the detectable ion in the determination of drotoverin using the ISE developed in Fig. 3 are given below.

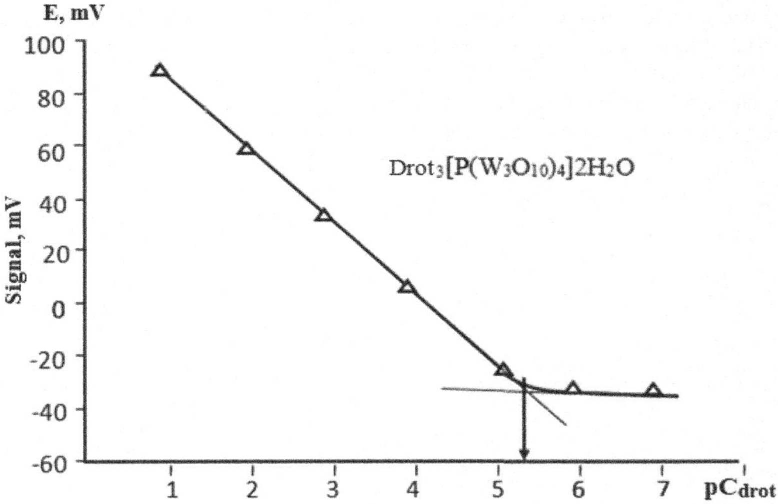

Fig. 3. $Drot_3[P(W_3O_{10})_4]2H_2O$ containing ISE lower limit for drotcvirin detection using containing ISE.

5 Conclusions

For express analysis of drugs, plasticized membrane ion-selective electrodes based on tungsten, molybdenum heteropolyacislotae have been developed and the effects of various factors on their ccnductors have been investigated.

With the PVA anion, the optimal conditions for the synthesis of low-soluble electrode-active compounds (EACs) of medicinal substances were selected and the electroanalytic properties of the obtained EACs were evaluated using modern methods of analysis, the obtained EAC structure and physicochemical properties based on drugs and PVA have been determined.

References

1. Matveychuk, Y., Rakhmanko, E.M.: Application of Zn(NCS)42– and Co(NCS)42–selective electrodes in the analysis of natural objects and food products. Journ. analyt. chem. **71**(5), 513–518 (2016). https://doi.org/10.7868/S0044450216030075
2. Matveychuk, Y., Rakhmanko, E.M., Zhilko, V.V.: New film molybdate and tungstate selective electrodes based on higher ChAS with different steric availability of the exchange center. Methods and Obj. Chem. Analy. **12**(1), 24–30 (2017). https://doi.org/10.17721/moca.2017.24-30
3. Egorov, V.V., Repin, V.A., Ovsyannikova, T.A.: Influence of ionic association on the selectivity of electrodes reversible to organic cations. Journ. Analyt. Chem. **47**(9), 1685–1692 (1992)
4. Ceresa, A., Sokalski, T, Pretsch, E.: Influence of key parameters on the lower detection limit and response function of solvent polymeric membrane ion-selective electrodes. J. Electroanalyt. Chem. **501**(N1P), 70–76 (2001)

5. Smirnova, A.L., Khitrova, V.L., Schwake, A.: Membranes for ion-selective electrodes containing a neutral carrier and ionogenic groups opposite in charge sign. Ion Exchange and Ionometry **10**, 201–214 (2000)
6. Kharitonov, S.V.: Electrochemical characteristics of ion-exchange membranes selective to anions of sulfonamide drugs. Electrochemistry **37**(12), 1490–1495 (2001)

Packet Detection and Classification by Sequential Data Decoding

Melnik Ivan, Kalygin Georgy, and Kulikov Konstantin(✉)

Vladimir State University, Gorkogo Str. 87, Vladimir 600000, Russia
kulikov@vlsu.ru

Abstract. The article considers an algorithm for detecting and classifying a packet in the absence of a priori information about the packet structure. The algorithm allows to optimize hardware costs when implementing a receiving device. The structure of the demodulator is given. The algorithm can be used to expand the coverage area of 5G/6G systems, implement multi-channel heterogeneous communication systems. The results supported by the State Assignment of the Vladimir State University (FZUN-2023–0004). #CSOC1120.

Keywords: packet data transmission · serial decoding · hardware resources

1 Introduction

Multistandardness–the usage of a set of standards for the implementation of communication, is a characteristic of 5G [1], 6G [2] networks, software-defined radio systems [3]. Heterogeneity of data transmission structures is used in multichannel systems, for example, the GMR satellite system [4], in the Internet of Things (IoT) [5].

Packet data transmission is used for information exchange. Network devices can use different standards for data transmission, while the receiving device must process all possible variants of signals (packets). In heterogeneous networks, the problem of time synchronization is acute [5, 6] under the condition of limited hardware resources. Standard packet processing is based on knowledge of the packet structure and signal parameters, in the presence of a set of packet options, the complexity of processing, as well as hardware costs, increases linearly with an increase in the number of packet options used. An urgent task is to reduce the hardware resources of the device when processing packets with a set of structures in the absence of a priori information about the packet structure or its parameters.

2 Methods

There are P types of packets that must be captured by the receiver. In general, the package contains the following fields (Fig. 1):

Guard interval (QI) (it can be used at the beginning and end of the package).

R. Silhavy and P. Silhavy (Eds.): CSOC 2024, LNNS 1120, pp. 315–320, 2024.
https://doi.org/10.1007/978-3-031-70518-2_28

- Header (HD), contains information for temporary synchronization sync sequence SS, system information SI (identifier of the source and recipient of information, packet parameters, signal parameters, etc.). Individual header elements can be protected from transmission errors or have means of detecting the absence of errors (parity bit, CRC).
- Data, the field contains the transmitted data, algorithms are always used for this field that increase the reliability of their transmission – noise-resistant codes, error detection codes. Information about the size of the data field, the codes used, is contained in the header of the package.

It is necessary to develop the structure of the receiving device for processing a packet from a given set of P packets.

2.1 Standard Algorithm

The standard algorithm for receiving a packet: temporary synchronization over SS, packet type detection, system information allocation and data processing in accordance with certain parameters.

P correlation receivers [7] or matched filters are used for time synchronization. Each correlation receiver calculates a correlation function Ri(τ) of input signal s(t) = sI(t) + i sQ(t) and function ki(t)

$$R_i(\tau) = \int_0^{T_i} s(t)k_i(t-\tau)dt, \quad (1)$$

where i = 1,…,P,

$$R_i(\tau) = \int_0^{T_i} s(t)k_i(t-\tau)dt, \quad (2)$$

where i = 1,…,P, and $k_i(t)$ – function corresponding to SSi;

Ti – duration SSi.

In digital form, the calculation of the correlation function is performed using the formula

$$\mathrm{R_i(l)} = \sum_{\mathrm{m=0}}^{\mathrm{N_i-1}} \mathrm{s(m)k_i(m-l)}, \quad (3)$$

where i = 1,…,P and Ni – number of discrete samples SSi.

The complexity of the implementation of the time synchronization algorithm is determined by the number of operations of complex multiplication and complex addition

$$A = \sum_{i=1}^{P} N_i, \quad (4)$$

It depends almost linearly on the number of package options.

Since hardware multipliers are the most critical hardware elements, their number must be minimized.

2.2 Sequential Decoding

The algorithm for processing a packet by sequential decoding of data: decoding of the data block for SI detection, packet type detection, system information extraction, data processing in accordance with certain parameters.

The received data is processed in blocks, the size of which corresponds to the maximum length of the SI + CRC field of all packets.

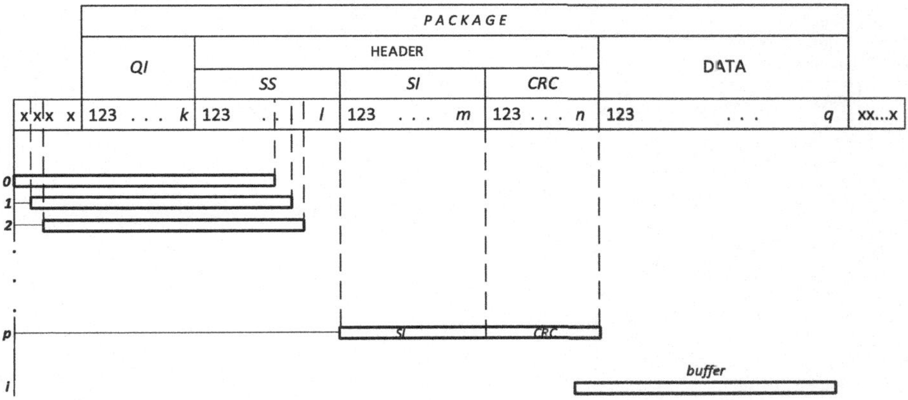

Fig. 1. Buffering data processing

The block is stored in a buffer, processing is performed iteratively. The buffer data processing process is schematically shown in Fig. 1.

Filling the buffer with input stream data (i = 0). Then, at each iteration, the data is shifted and samples of one new data character are recorded. Buffer data is demodulated/decoded, checked for CRC convergence, these actions are performed for each packet type. If the CRC does not converge for any type of packet, the next iteration of processing is performed. If the CRC matches for one of the package variants (i = p), a decision is made on the type of package, system information is extracted from the data, data corresponding to the package structure and system information are stored in memory. During SI processing, parameters for data processing (time, frequency synchronization) are determined (calculated), which are stored in memory and used when processing packet data.

2.3 Demodulator Architecture

Figure 2 shows a block diagram of the demodulator and its component blocks. The received data (quadrature samples of the signal) and a set of parameters required at different stages of processing are stored in memory. In the process of obtaining new parameters, the contents of the memory are adjusted.

The process of demodulating a data packet consists of two stages. The purpose of the first stage is the primary synchronization and determination of the packet parameters, it is implemented by the sequential decoding algorithm discussed above, the "Shift buffer" and "Synchronization, detection and classification modules" blocks are used. When defining package parameters, package data and package parameters are stored in memory.

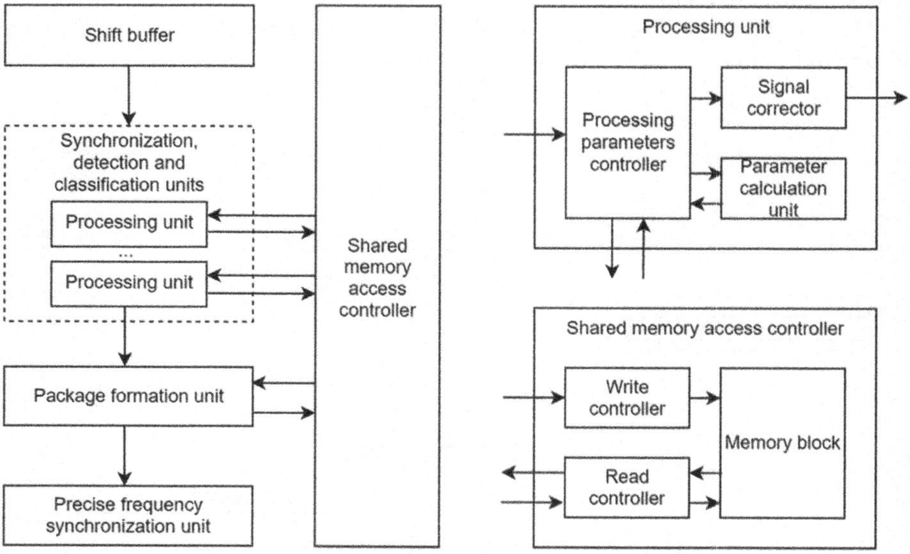

Fig. 2. Block diagram of the device

The purpose of the second stage is to form a package for further processing. The "Signal Fine Tuning Unit" evaluates the frequency and phase of the signal, which are used to adjust the signal samples stored in memory.

The shift buffer is a controller and a block of memory, the size of which is sufficient to process the packet header. Buffer data is used by "Processing Blocks".

The sequence of processing blocks solves the tasks of reducing the bit depth of data, provides character and frequency synchronization, determines the type and parameters of the packet.

Each processing unit consists of a controller, a parameter calculation unit and a corrector. When receiving new data, the controller accesses information about the current operating mode and previously calculated parameters, applies this data to calculate new parameters and adjust the signal. The updated parameters are stored in shared memory and can be used by all processing units.

The processing units can operate in two modes. In the initial setup mode, the parameters are calculated. In the packet generation mode, the data in memory is corrected, and the last saved parameters are used to correct the signal. The packet generation unit selects the operating mode and generates data packets with additional parameters obtained during processing of the packet identifier.

The sequence of processing blocks solves the tasks of reducing the bit depth of data, provides character and frequency synchronization, determines the type and parameters of the packet.

Each processing unit consists of a controller, a parameter calculation unit and a corrector. When receiving new data, the controller accesses information about the current operating mode and previously calculated parameters, applies this data to calculate new parameters and adjust the signal. The updated parameters are stored in shared memory and can be used by all processing units.

The shared memory access controller consists of read, write and data storage control units. The recording management module organizes recording operations by forming a tree structure of multiplexers that support the AXI-Stream protocol. Similarly, the task of organizing read operations is solved, and the task of transferring data from memory to further processing devices (depending on the type of data) is solved by duplicating transactions with their subsequent filtering. Figure 3 shows a diagram of the organization of access to shared memory.

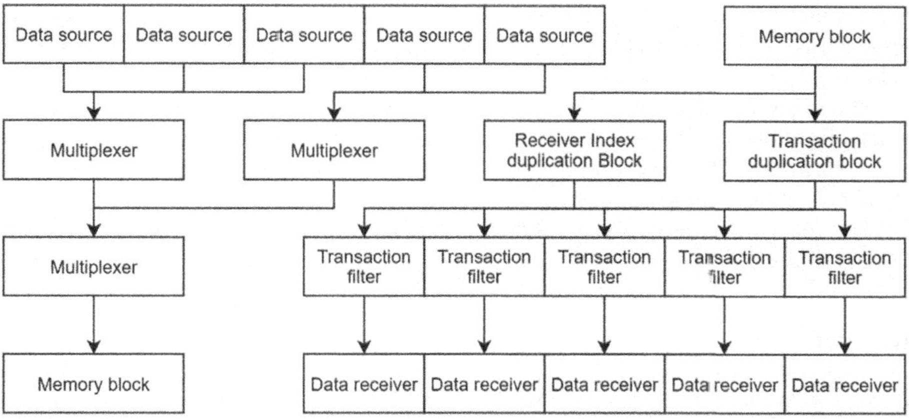

Fig. 3. The scheme of organizing access to shared memory

3 Results

The efficiency of processing algorithms is estimated by the hardware resources used and the processing speed.

The algorithms of packet detection and classification under consideration include SI and CRC data processing, in the standard version one variant of the packet structure is used, in the second one all variants are analyzed, while the hardware resources used are the same, since packet processing of various types can be performed in parallel, data processing speeds in both methods are equal.

In the second option, the use of multipliers and adders is not required, in the time of processing the data in memory is constantly analyzed, it requires the implementation of optimal, from the point of view of the equipment, procedures for working with memory.

The proposed algorithm was implemented to process data containing 9 types of packets. Table 1 shows the hardware resources of two packet processing options in FPGA XCKU115.

Table 1. Hardware resources.

Primitive type	Number of elements	
	Standard algorithm	Sequential decoding of data
REGISTER/SDR	1902	947
CLB/LUT	75	436
RAMB36 Tiles	0	3
ARITHMETIC/DSP	63	0

4 Conclusion

Sequential decoding of data makes it possible to significantly reduce the number of multipliers used in solving problems of detection, packet classification and synchronization in heterogeneous systems.

The sequential decoding algorithm is implemented in a system based on the possibility of gradual correction of packet samples stored in memory and accessible to all processing units. The correction is performed in accordance with the parameters obtained at the next stage of processing the package.

The results supported by the State Assignment of the Vladimir State University (FZUN-2023–0004).

References

1. Agiwal, M., Roy, A., Saxena, N.: Next generation 5G wireless networks: a comprehensive survey. IEEE Commun. Surveys Tuts. **18**(3), 1617–1655 (2016)
2. Tataria, H., et al.: 6G wireless systems: vision, requirements, challenges, insights, and opportunities. Proc. IEEE. **109**(7), 1166–1199 (2021)
3. Sinha, D., Verma, A.K., Kumar S.: Software defined radio: Operation, challenges and possible solutions. In: 2016 10th International Conference on Intelligent Systems and Control (ISCO). – IEEE, pp 1–5 (2016)
4. ETSI TS 101 376–1–3 V3.1.1 (2009–07) Technical Specification GEO-Mobile Radio Interface Specifications (Release 3) Third Generation Satellite Packet Radio Service; Part 1: General specifications; Sub-part 3: General System Description GMR-1 3G 41.202
5. Meehan J., et al.: Integrating real-time and batch processing in a polystore. In: 2016 IEEE High Performance Extreme Computing Conference (HPEC). pp 1–7 (2016)
6. Yiğitler, H., Badihi, B., Jäntti, R.: Overview of time synchronization for IoT deployments: clock discipline algorithms and protocols. Sensors **20**(20), 5928 (2020). https://doi.org/10.3390/s20205928
7. Hu, H., Yuan, C.: The design on FPGA-based correlator in GPS receiver using ISE. Int. J. Intell. Inorm. Technol. Appl. **3**, 92–97 (2010)

Mapping of Sakhalin Power Grid Vulnerability Based on Network Analysis

Andrey Karpachevskiy(✉) and Maria Lipovetskaya

Lomonosov Moscow State University, 1, Leninskie Gory, Moscow, Russia
karpach-am@yandex.ru

Abstract. The vulnerability assessment methodology proposed for substations focuses on enhancing the conventional SVI index by incorporating a synthesized indicator to evaluate predisposition to high vulnerability in power transmission lines. This indicator amalgamates criteria such as operational time, transmission capacity, and line length, assigning weights based on their relative importance. Utilizing the Analytic Hierarchy Process (AHP), weight coefficients for each criterion are calculated to derive the overall vulnerability of power system elements. The assessment involves calculating SVI indicators for substations based on weighted shortest distances from energy sources, considering alternative energy supply routes. Additionally, the methodology evaluates the load on power lines using centrality measures, reflecting the lines' importance in transmitting electrical energy. Through simulation, cascading failures are modeled, highlighting critical points of failure and their impact on the power system. The results demonstrate the vulnerability of substations and power lines, particularly in coastal and central regions, emphasizing the need for network expansion and backup power supply routes to enhance system resilience. Zoning analysis identifies regions at risk of power outages, underscoring the importance of strategic planning and infrastructure development for ensuring reliable energy supply in the Sakhalin region. #CSOC1120.

Keywords: Vulnerability Assessment · Power System · Cascading Failures · Network Resilience · Energy Infrastructure · GIS

1 Introduction

In the majority of studies within the realm of transport geography, nomothetic models predicated on graph theory are employed to delineate the architecture of transportation networks [1]. Concomitant methodologies are adopted in research focused on the analysis and assessment of energy system vulnerability. Electrical power infrastructures are conceptualized as graphs, which include vertices connected by edges. Herein, the edges represent electric power lines, whereas vertices correspond to generation stations and substations [2]. The intrinsic structural vulnerability of the electric grid is quantified by the ratio of operational stations and substations to their respective counts following contingencies.

R. Silhavy and P. Silhavy (Eds.): CSOC 2024, LNNS 1120, pp. 321–334, 2024.
https://doi.org/10.1007/978-3-031-70518-2_29

Various strategies exist to evaluate the power system vulnerability. Not only does this attribute take into account the topology, but it also integrates additional parameters pertinent to the system's operation and external negative influences. A salient methodology for appraising the structural vulnerability of the electric grid is the selective elimination of network vertices [3], simulating assaults on power system nodes (substations), premised on the notion that spontaneous node failures are more probable. The approach encompasses determining the centrality metric of a graph predicated on shortest paths. Subsequent to each stochastic vertex exclusion, computations are performed anew to ascertain the shortest paths linking all substations to the nearest power plants, in addition counting such paths per substation. An overloading simulation for a substation is triggered if the resultant burden surpasses the original by a factor of two. System robustness within the study is gauged by the proportion of remaining vertices relative to the initial quantity. A shortcoming of these algorithms is their omission of variable transmission line capacities and diverse generation capacities of power plants.

The aforementioned vulnerability assessment method concentrates on unanticipated disruptions and on the topological configuration of energy nets but does not accommodate ancillary operating parameters, including, for instance, the inclusion or exclusion of local supplemental energy-generating facilities. In cases where the power architecture integrates auxiliary low-capacity generation plants, such as Distributed Energy Systems (DES), decentralized production at the distribution tier can supplant a portion of the centralized output at the transmission tier. This transition mitigates the electrical transmission distance, thereby amplifying the network's overall performance and resilience. Zhang and Chen [4] delineates two modes to appraise the vulnerability of electric grids incorporating analogous local energy sources, utilizing simulated cascade attacks. In the initial scenario, the resilience assessment draws on a comprehensive vulnerability metric – the efficiency index, which assesses the efficacy of electrical conveyance via each network line [5].

The Global Efficiency Index (GEI) serves as an indicator of network performance, premised on the postulate that the efficiency of electric power transmission between two nodes is inversely proportional to the spatial separation between them. This index encapsulates aggregated capacity of the electrical grid [6]. The energy system's susceptibility is evaluated based on this index; specifically, the magnitude of decrement in the global efficiency metric following a cascade failure provides a measure of vulnerability. Nevertheless, when considering the incorporation of local power generation nodes, the GEI may not suffice as a reliable metric for assessing the power system's stability. Consequently, in the second model of cascade failures, a reformed measure, termed the Modified Global Vulnerability Index, was utilized for vulnerability assessment [7]. This index accounts for the influence of individual node's generation capacity on the overall efficiency of the system.

Despite its limitations, the GEI has been extensively applied in the analysis of energy system vulnerability. The primary drawback of this efficiency measure is the underlying assumption that electrical transmission occurs exclusively along the shortest path, a scenario which often diverges from the actual network configuration where energy disperses through various conductive routes during transit from the generator to the consumer. Additionally, the power output of generators is not factored in. The Structural

Vulnerability Index (SVI) [8] builds upon the efficiency index but incorporates both the generator node's power and the maximal permissible load on the remaining nodes. Moreover, unlike earlier approaches, this formulation considers the actual impedance instead of the shortest path distance between nodes i and j:

$$SVI = \frac{1}{n_g n_l} \sum_{i \in V_g j \in V_l} \frac{P_{gj}}{P_{li} \exp(Z_{ijeq})}, \quad (1)$$

here P_{gj} denotes the allowable generation at the source node j, P_{li} is the peak load at node i and Z_{ijeq} signifies the equivalent resistance (electrical distance) between nodes i and j. Accordingly, this index captures not only the topological framework of the networks but also additional parameters pertinent to their structure and functionality, as previously discussed.

Analysis of the power system's robustness may also involve identifying its most fragile components, such as during simulations of transitory outages affecting certain generators or transmission lines. To appraise the criticality of such spontaneous failures the Continuity Vulnerability Index (CVI) could be used [8], which builds upon the aforementioned SVI. Given that a failure of one or more units results in an extended electrical transfer distance between nodes, SVI values diminish; the extent of this reduction is captured by the CVI.

Comparable method revolves around identifying the weakest links within the grid and is predicated on computing the variation in efficiency indices before and after network element severance [6]. The distinction with this approach is its consideration of severing all connections to the impacted node, rather than focusing solely on the node itself.

A narrowly tailored approach for pinpointing the least stable segments within a power grid involves the scrutiny of vulnerable transmission lines [9], which strives to formulate a comprehensive Line Vulnerability Index (LVI). The determination of the most critical transmission lines is contingent upon an examination of the network's complexity, as well as employing a risk index predicated on probabilistic power flow analyses to gauge the propensity for overloading each individual line. By leveraging the calculated composite index, a model of cascade failure within the electrical network is constructed that incorporates the magnitude of the energy flow to preclude overloads on specific lines.

Through an investigation of aggregated details regarding the physical, geographical, and climatic variables in the Sakhalin region, a hypothesis was posited that the power grids in this locale exhibit an intrinsic susceptibility to cascading failures, instigated by natural disasters such as ice storms and strong winds. It is paramount to note that the energy system is not only externally segregated, lacking access to reserve power from adjacent territories, but is also comprised of multiple disjointed energy nodes. Consequently, such an energy framework necessitates specialized scrutiny and stability analysis to forestall potential catastrophic repercussions.

2 Materials and Methods

The vulnerability assessment methodology for substations we are developing is based on the aforementioned SVI index, modifying the measure of distance between objects. Instead of using just the electrical resistance of the shortest path between the energy source and the substation or a single specific parameter, we propose to use a synthesized indicator of predisposition to high vulnerability for each line. This indicator is comprehensive and formed from three criteria of power line vulnerability: the degree of wear (operational time at the time of the study), transmission capacity in terms of the maximum allowable current, and the line length. The proposed synthesized indicator essentially acts as the weight of each line in graph theory and is used for further analysis of the vulnerability of power system elements. To calculate the resulting weight of the lines, the vulnerability criteria were ranked according to their presumed importance: the degree of wear has the greatest weight in the evaluation, while the line length has the least. A "1–2–3" ranking scale was chosen, where "1" signifies an equal importance of the compared criteria, which have the same significance for a higher-level element. "3" indicates a moderate superiority of one criterion over another, with preliminary evaluation suggesting slightly greater importance of one criterion compared to the other. The value "2" indicates an intermediate degree of one criterion's superiority over another. Let us state that the criterion for line age is three times more important than the criterion for line length, and the criterion for transmission capacity is twice as important as the length. Please note that in reverse situations, the line length criterion is 1/3 times more important than the line age criterion and 1/2 times more important than the transmission capacity. Then, using the Analytic Hierarchy Process (AHP), a weight coefficient for each criterion was calculated. This requires pairwise determining the degrees of superiority between criteria. The sums of the obtained values in each column are then calculated, followed by the normalized values of the criteria's degrees of superiority. The obtained values are averaged row by row and are the weight coefficients. All calculations and obtained coefficient values for vulnerability criteria are presented in Table 1.

Table 1. Calculation of weight coefficients for criteria of vulnerability of power transmission lines using AHP.

	Length	Capacity	Age	Weight coefficient
Length	1	0.5	0.333333	0.247691
Capacity	2	1	0.333333	0.479509
Age	3	3	1	0.893074
Sum	6	4.5	1.666667	
Normalized coefficients				
	Length/sum	Capacity/sum	Age/sum	
	0.545455	0.15	0.047619	
	1.090909	0.3	0.047619	
	1.636364	0.9	0.142857	

The resulting weight for each line is calculated as the sum of the appropriately weighted parameters of wear, transmission capacity, and length. Thus, the greater the resulting cost value, the more important and, consequently, vulnerable the line is. The longer the service life of the wire or cable, and the longer the length of the line, the less stable it is. The greater the transmission capacity, the greater the damage that can be caused by the failure of the line itself or the associated nodes.

The vulnerability assessment of electrical substations in the network based on the modified SVI index consists of calculating the SVI indicator for each node using the formula:

$$SVI_{P_{li}} = \frac{P_{gj}}{P_{li}\left(Z_{ij}\right)}, \quad (2)$$

where P_{li} is the nominal power of substation i, P_{gj} is the nominal power of the power source j, and Z_{ij} is the shortest distance between the source and the substation, calculated based on the resulting weight of each line obtained above.

Since the same substation may not only be accessible from a single power station, i.e., in the event of a failure in the power flow path from one source to a node, there is an alternative possibility of energy supply from another source through other nodes, it is necessary to calculate for each substation the weighted shortest distance from all possible sources and sum the obtained values.

The numerical characteristics of substations obtained after calculation are interpreted as follows: the higher the value of the SVI index, the more stable it is and vice versa, i.e., a substation's vulnerability is inversely proportional to the sum of weighted distances between the substation and accessible energy sources.

In addition to calculating the vulnerability of substations, it is also necessary to determine the load on the power transmission lines, as the degree of their loading affects the resilience and consequences of failures at adjacent nodes of substations and power plants. The load on power lines is determined by the centrality measure of the graph based on the calculation of the shortest paths between power stations and substations (betweenness centrality), i.e., the number of shortest paths between the sources and consumers of energy passing through a given line is calculated for each line.

The calculated centrality measure is interpreted as follows: the higher the value, the greater the number of shortest paths that pass through the line, and therefore the more loaded it is. The higher the load on the line, the more important it is, as it is the most frequently used element in the path of transmitting electrical energy from sources to consumers. In the event of failures on the most loaded lines or adjacent nodes, the areas supplied with electricity are at risk of losing power.

In the event of failure of substations or power lines, it is necessary to assess the possibilities of redistributing transmitted electrical energy to other power lines to avoid blackouts. For this, it is necessary to simulate cascading failures in the Sakhalin region's power system.

Accident modeling is carried out according to the following scheme:

1. the most loaded power line is removed, i.e., the one with the highest centrality value;
2. for the remaining lines, the centrality measure is recalculated;
3. the SVI index for substations is recalculated (since the calculation of the shortest paths is only made for substations that are accessible from power sources, a de-energized substation will not have any values if adjacent lines that turned out to be the most loaded are removed);

4. if the recalculated centrality values are twice or more the original values on the lines, an accident is simulated on these lines;
5. if the centrality values have not changed or have decreased, the removal of lines with the highest initial centrality values continues;
6. the algorithm is repeated until the power system disintegrates into separate de-energized areas.

3 Results and Discussion

3.1 Implementation of the Vulnerability Assessment Methodology for Power System Elements

The vulnerability modeling of the Sakhalin region's power system is carried out using the NetworkX library of the Python programming language. This library allows working with an imported shapefile as a graph. To implement the algorithm for determining the stability of the nodes of the Sakhalin region's power system, three different graphs were obtained: a graph based on power transmission lines, a graph based on power substations, and a graph based on power plants. To calculate the distance from each source to each accessible substation, it is necessary to organize a loop of traversal through the elements of the power plant graph. Since each substation can be powered not just by one power plant but by several, it is necessary to add the calculated weighted shortest distances from each energy source to the accessible substations to the substation layer as separate attributes. To obtain the resulting SVI vulnerability index, it is necessary to sum up the individual indices calculated depending on the power of the power plant from which the shortest distance was calculated. To assess the load on transmission lines, it is also necessary to use the NetworkX library to calculate centrality metrics for each element.

3.2 Vulnerability Assessment of the Sakhalin Region's Power System Elements

The most vulnerable substations, whose SVI index < 0.002, are those that are farthest away from energy sources. These are mainly substations located along the coast or in the center of the island. The density of substations across the territory is uneven; it is important to note that in areas with a high concentration of substations, i.e., in places with the highest consumption, the most vulnerable are substations with high power (from 50 to 150 MVA) and correspondingly high voltage (up to 220 kV). Low-power substations (from 5 to 20 MVA) and low voltage (35/10 kV) have higher SVI values in cases where they are the only receivers of electricity in the territory. The most vulnerable substations with a voltage of 220 kV include substations in coastal major cities such as Korsakov, Nevelsk, Kholmsk, Tomari, Makarov, Shakhtersk, Uglegorsk, Smirnykh, and Tymovskoye. Substations of lower voltage but with high SVI values are those of such small settlements as Chekhov, Krasnogorsk, Krasnopole, Vakhrushev, Leonidovo, Lesogorskoye, Boshnyakov, Onor, Alexandrov-Sakhalinsky, Mgachi, and Dagi (Fig. 1).

The listed substations have adjacent power lines with high centrality values (centrality values greater than 60), i.e., those that are integral parts of the transport of electricity. Such lines are the most loaded due to their location—they connect remote parts of the island to the sources, for example, the southern part of the island (Tomari and Kholmsk)

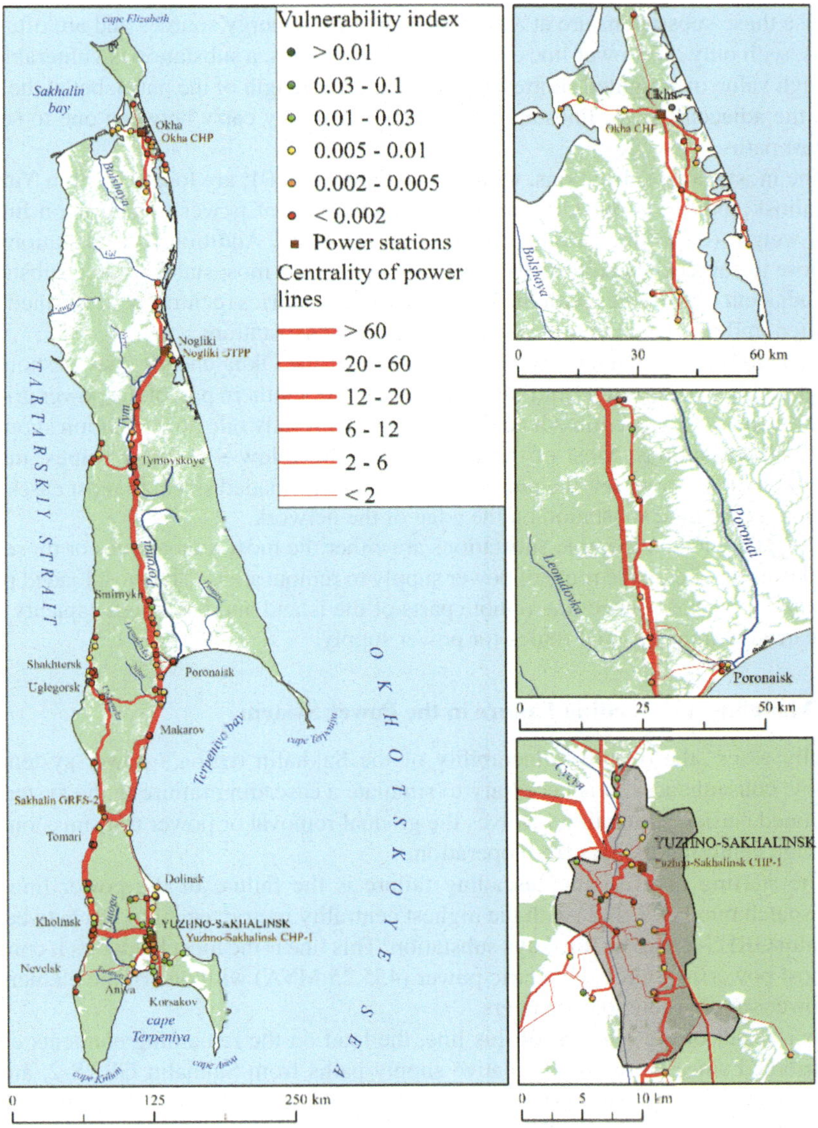

Fig. 1. Map of power grid elements vulnerability in Sakhalin

and the central part (Poronaysk, Makarov, Uglegorsk) to the southern energy sources Sakhalinskaya GRES-2 and Yuzhno-Sakhalinskaya TEC-1. Also, the most loaded lines adjacent to vulnerable high-voltage substations are lines connecting the central and northern (Tymovskoye, Nogliki) parts of the island with the northern Noglikskaya GES.

The above-mentioned vulnerable low-voltage substations have adjacent power lines with average or low centrality values, i.e., they have a relatively small load. This is because these substations are at a great distance from energy sources and are often end points, with only one power line extending to them. Thus, a substation is vulnerable due to a high value of weighted shortest distance (greater length of the path), but at the same time, the adjacent power lines are not loaded, since they carry between one to several shortest paths.

The most stable substations, with an SVI index > 0.01, are located within Yuzhno-Sakhalinsk and on its outskirts due to a large number of power transmission lines of equal weight (these lines are relatively short and new). Additionally, substations that are close to sources of electric power are also among the most stable. Stable substations have adjacent power lines with the lowest centrality metrics (centrality < 6) since there are often multiple paths to transmit energy to stable substations.

Separately, the vulnerability and congestion of the Okha energy district should be considered. This area is not centrally connected to the southern part of the power grid and exists in isolation. The entire Okha energy district has only one power station supplying individual substations. Most of these substations have low SVI index values, making them vulnerable. However, the power lines are not overloaded since, in most cases, they only supply a single substation on the edge of the network.

Thus, the most vulnerable substations are either the most remote ones or those with high power that lie on the route of power supply to remote areas. The most loaded power lines are those that connect the remote parts of the island and have high capacity since they carry the most optimal routes for power supply.

3.3 Modeling a Cascading Failure in the Power System

To fully assess the overall vulnerability of the Sakhalin region's power system and possible consequences, it is necessary to simulate a cascading failure in the system. As mentioned earlier, simulation involves the gradual removal of power transmission lines with high centrality metrics from operation.

The starting point of the cascading failure is the failure of the power line with the dispatch number T-232 (with the highest centrality metric, equal to 88) between the Sakhalin GRES-2 and the Ilinskaya substation. This line is the most loaded as it connects the most powerful source of electric power (455.25 MVA) with one of the elements on the power supply route to consumers.

As a result of the removal of this line, the load on the remaining elements evenly redistributed since there are alternative supply paths from Sakhalin GRES-2, and the overall load on the power system decreased. For the second removal, another one of the backup lines, specifically T-233, was taken out of service. After this iteration, there were no lines with low weights left on the power supply route from Sakhalin GRES-2, resulting in energy flow being redistributed through non-optimal routes, and the load on the adjacent lines to the source increased.

The load on the 220 kV power line with dispatch number D-8 increased the most. This line acts as a backup during the power supply to the Tomari district, as a result, when the main power flows are disconnected, the shortest paths to the substations are transferred to this line. After the removal of this line, the southern part of Sakhalin almost loses the possibility of power supply from Sakhalin GRES-2, as the only method of transferring electric power from this station remains the D-5 line, which is very long.

As a result, the main share of power supply falls on the Yuzhno-Sakhalinsk Central Heating Plant (CHP-1). Lines D-9 and D-12 are the main power supply routes for the western part of the island from the CHP, therefore, they are the most loaded. After their removal, the power flow is evenly redistributed to the backup 110 kV lines. Due to the removal of the lines through which the largest number of optimal routes pass, the overall load on the power system decreases, as it is evenly redistributed through the backup lines, thereby reducing the risk of the main power lines' failure. Subsequent failures are simulated on the main power supply routes from Yuzhno-Sakhalinsk CHP-1 with the highest centrality values, which are the 110 kV lines S-21, S-20, S-5, and S-14 with S-13. As these paths are removed, the transmitted electric power is redistributed to parallel lower voltage 35 kV lines, as a result of which the load on them increases insignificantly and failures are then simulated on these lines. As a result of the simulation, it is found that part of the power system in the southern part of the island disintegrates into separate de-energized districts, which can no longer receive energy from either Sakhalin GRES-2 or Yuzhno-Sakhalinsk CHP-1.

Meanwhile, the mentioned energy sources continue to supply the central and accessible northern parts of the island, however, power is only provided by the most extended and loaded 220 kV lines with dispatch numbers D-5, D-3, and D-6, which lack a backup power transmission capability.

Thus, with the removal of the main flows, the central and northern parts of the island lose power supply from the most powerful power stations. However, the northern Nogliki GTES continues to function and transmit energy. In this part of the island, the most loaded are the high-voltage lines D-13 and D-2 and the medium-voltage line S-55. If D-13 is removed, central Sakhalin completely loses power supply, as this line has no backup neighbors. If S-55 is removed, a large part of the northern settlements are de-energized.

It is also important to note that the S-53 and S-54 lines, which are the only connecting links between Nogliki GTES and consumers, are highly loaded. In case of failure on these lines, Nogliki GTES loses the rationale for power generation.

A separate part of the entire power system of the Sakhalin region is the northernmost part of the island, which has no possibility of power supply other than from the Okha CHP. Most of the settlements in this district are supplied directly from the CHP along a single power transmission line with multiple branches, and in rare cases, the power flow reaches consumers via two adjacent power lines. With such a network configuration, it is clear that if any power line fails, the consumer settlements are de-energized. When simulating a cascading failure, the power system disintegrated into separate de-energized districts upon the second removal of lines with high centrality metrics (Fig. 2).

Special attention needs to be paid to the changes in the vulnerability index SVI at the substations of the power system. As the load on the power lines adjacent to the substation nodes increased/decreased, the vulnerability index changed accordingly: it decreased, i.e., vulnerability increased due to the growth of the weighted shortest distance, and it increased, i.e., the stability of the node improved. As power lines were removed, the overall trend of substation vulnerability was characterized as increasing until the substations became unreachable and, accordingly, lost the ability to receive power supply.

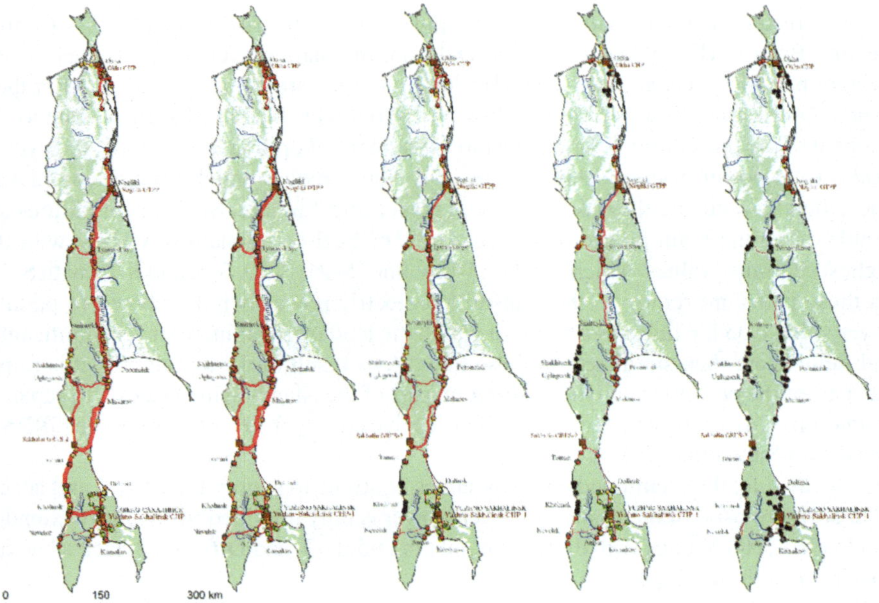

Fig. 2. Stages of cascading failure in Sakhalin power system.

3.4 Assessment of the Consequences of a Cascading Failure in the Power System for Populated Territories

As a result of the gradual failure of power transmission lines, the power system of the Sakhalin Region ceases to be intact and disintegrates into 11 parts, of which only two districts – Nogliksky and Yuzhno-Sakhalinsk – do not lose power supply. The probability of a blackout in these parts of the power system is extremely low because most of the substations are not vulnerable, and the load on the power lines is not high. The remaining districts and their substations are de-energized because the main routes of electricity transmission from the sources to the most powerful substations in the area are out of order, and the power lines within the districts were not removed during the simulation due to their low load.

3.5 Zoning of the Sakhalin Region According to the Degree of Power Supply Reliability

Based on the analysis of the obtained indicators of vulnerability and load on the elements of the power system, as well as based on the simulation of cascading failures, zoning of the populated part of Sakhalin was carried out according to the degree of vulnerability of the adjacent territories (see Fig. 3). The starting point for zoning is all populated localities of the Sakhalin Region obtained from the OSM. A Voronoi diagram was constructed based on these points: such surface division allows for determining the closest territories to each point in the context of their irregular spatial distribution.

After constructing the Voronoi diagram based on the points of populated localities, it is necessary to determine the stability of each cell of the obtained surface depending on the degree of vulnerability of the substation closest to the populated locality. The identification of the closest substations to the points was carried out taking into account the administrative-territorial division of the region. The degree of stability was determined by the average value of the SVI vulnerability index for all stages of the failures. Since a large part of the Sakhalin territory is not populated and occupied by small villages without electricity, and therefore does not require an assessment of the reliability of power supply, the obtained Voronoi polygons need to be edited, preserving the assessment of stability only in the populated areas.

It is also important to note that the most remote territories of the island, occupied by one or two small settlements, have their own local power supply like diesel power plants. Such regions do not make sense to assess the stability of power supply due to their dependence on a single source. A similar situation is typical for the islands of the Kuril chain, for this reason, this territory is also not subject to analysis.

Zoning of the Sakhalin Region by the degree of vulnerability of its elements allows identifying territories that are at risk of a blackout. The least stable territories are the coastal districts of the island and its central part. In the coastal regions, the greatest number of remote substations is concentrated, supplying electricity to no less remote localities. Often, such substations have only one possible way of supplying electricity, i.e., in case of a failure of one of the elements on the path from the source, the population of the district completely loses the possibility of supply. The central parts of Sakhalin are also most vulnerable due to the specific roles of power lines and substations – the former are links between sources and remote consumers, and the latter are transit points on the supply path. If such linking elements fail, the power system of the region disintegrates into several parts.

The most stable regions are those close to energy sources. However, it's interesting to note that, for example, areas around the Nogliki Thermal Power Plant (TPP), which are supplied directly from the source, are several times less stable than similar areas closest to the Yuzhno-Sakhalinsk Thermal Power Plant-1 (TPP-1). This is because the city of Yuzhno-Sakhalinsk and the surrounding areas have a large number of power lines that are a) less worn out, b) serve as backups for each other in terms of electricity transmission.

As a result, the power system in the Sakhalin region is quite vulnerable, with risks of power outages in remote populated areas. This is due to the following reasons:

1. The structure of the network: the network is highly extended from south to north, which necessitates the construction of extended power transmission lines,
2. Insufficient development of the power system at the time of the study: a large number of settlements can only be powered by one route, i.e., there is no backup power supply,
3. The terrain of the territory: most of the island is occupied by mountains, this factor complicates the possible expansion of the network, increasing the number of its components to reduce the load on the lines.

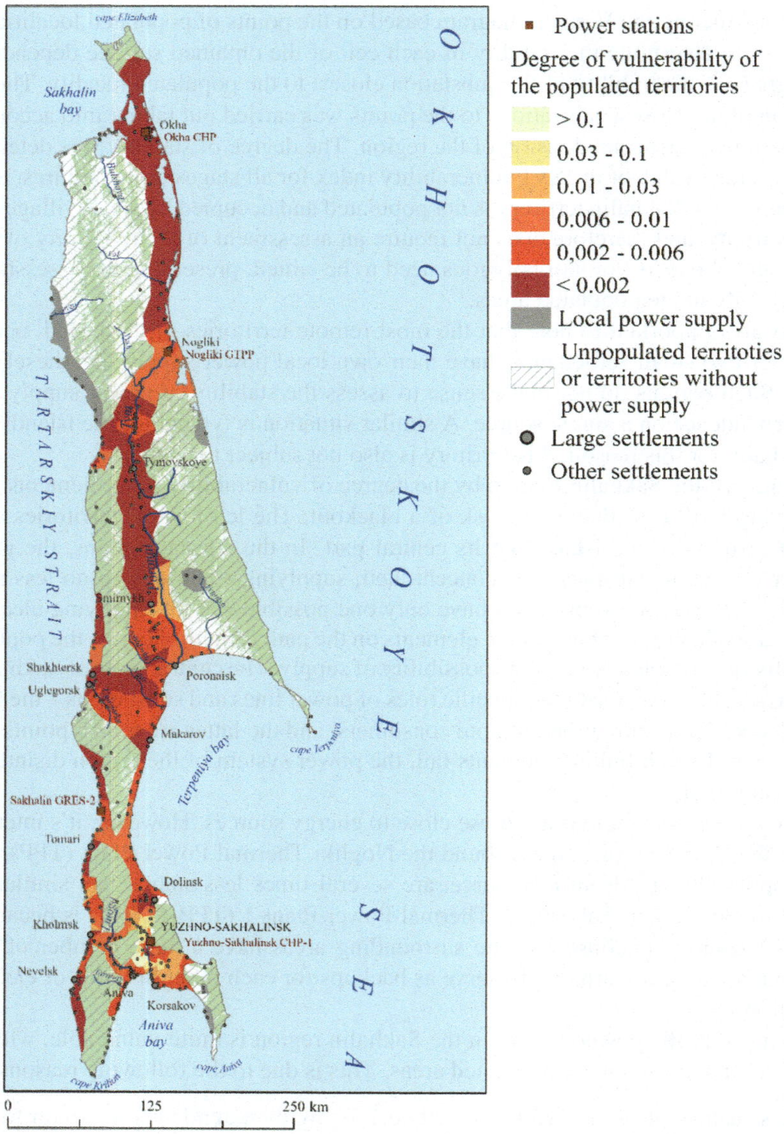

Fig. 3. Power supply stability in the Sakhalin Region

4 Conclusion

The energy system of the Sakhalin region is isolated and does not have the capability of backup power supply in case of serious accidents; moreover, it consists of separate energy districts. Both the external and internal isolation increase the risks of accidents related to the spatial configuration of the network and, consequently, contribute to the vulnerability of both individual elements and the entire energy system as a whole. This necessitates the analysis of the vulnerability of the region's energy system.

During the study, a current database of the energy system was developed, including all power transmission lines, electrical stations, and substations. Information on criteria affecting vulnerability was entered for each element. For the lines, these criteria are age, capacity, and length. These criteria were synthesized into a general weight indicator that was used in calculating the vulnerability of both the lines and the substations. For substations and power sources, the criteria are the nominal capacities.

As a result, maps of vulnerability assessment of substations and lines, stages of cascading failure, and the regions of Sakhalin that were de-energized as a result of the accident were obtained. The findings were summarized on a map of the regional zoning of Sakhalin's populated territory according to the degree of power supply resilience.

Accordingly, it was revealed that the Sakhalin region's energy system is quite vulnerable. The most vulnerable substations in the region are low-voltage substations in coastal areas, as well as high-voltage substations that act as "transition points" between sources and consumers. The most loaded and, consequently, most vulnerable, are lines adjacent to vulnerable high-voltage substations, because they are also the main elements of optimal power supply paths. Zoning the territory by the degree of resilience shows that much of the island is at risk of losing power supply in the event of accidents. This is due to the insufficient number of backup power supply routes, as well as the extended structure of the network due to the specific physical and geographical conditions.

Acknowledgements. This research was funded by Russian State Task N 121051400061–9 "Development of methods and technologies for cartography, aerospace sensing in research of nature and society".

References

1. Tarkhov, S.A.: Evolutionary morphology of transport networks. Universum, Smolensk – Moscow, Russia (2005)
2. Harary, F.: Graph Theory. Mir, Moscow, Russia (1973)
3. Faddeev, A.M.: Assessment of vulnerability of the power systems of Russia, nearby foreign countries, and Europe to cascading failures. Vestnik Moskovskogo Universiteta. Seriya 5. Geografiya **1**, 46–53 (2016)
4. Zhang, Y., Chen, S.: Complex networks theory based vulnerability analysis of power grid with distributed generation. In: Proceedings of the 2nd International Conference on Computer Science and Electronics Engineering (ICCSEE 2013). Atlantis Press, China (2013). https://doi.org/10.1049/cp.2012.2165
5. Ding, M., Han, P.: Small-world topological model based vulnerability assessment algorithm for large-scale power grid. Autom. Electr. Power Syst. **30**, 7–10 (2006)
6. Arianos, S., Bompard, E., Carbone, A., Xue, F.: Power grid vulnerability: a complex network approach. Chaos **1**, 013119 (2009). https://doi.org/10.1063/1.3077229
7. Wang, Y., Mei, S., Mao, Y., Liu, F.: Vulnerability assessment of power grid with distributed generation based on complex network theory. J. Syst. Sci. Math. Sci. **30**, 859–868 (2010). https://doi.org/10.12341/jssms09002

8. Liu, C., Xu, Q., Chen, Z., Bak, C.: Vulnerability evaluation of power system integrated with large-scale distributed generation based on complex network theory. In: Proceedings of the 47th International Universities Power Engineering Conference (UPEC), pp. 1--5. IEEE, Uxbridge, Middlesex, United Kingdom (2012). https://doi.org/10.1109/UPEC.2012.6398605
9. Li, X., Qi, Z.: Impact of cascading failure based on line vulnerability index on power grids. Energy Syst. **13**, 893–918 (2022). https://doi.org/10.1007/s12667-021-00435-x

Development of an Application for the Study of the GGH Cryptographic Scheme

N. A. Urban and E. A. Melnikova(✉)

Russian State Social University, Moscow, Russia
urbannikolai@mail.ru

Abstract. In the first part contribution presents the main approaches to the development of post-quantum algorithms and reviews progress in the field of quantum computers and post-quantum cryptographic encryption systems. Further, article considers the most promising approaches, that are based on the using of theory of lattices. Then, authors reviews in more detail the Goldreich Goldwasser Halevi encryption scheme based on the theory of lattices. In the second part, article presents a software package that allows the user to study the main functions of the GGH scheme: key generation algorithms, encryption algorithms and message decryption. The article specifies the main purposes of the modules of the software package, describes the user interface of the program. Also, the software package makes it possible to carry out an attack on the encryption scheme using the Lenstra-Lenstra-Lovasz algorithm. The article considers the details of applying this algorithm, mentioning the difficult to solve problems, which are being the base of cryptographic scheme and are being solved by LLL-algorithm. This application can be used as part of a laboratory complex when studying. #CSOC1120.

Keywords: Post-quantum encryption algorithms · Cryptosystem · Qubit · Goldreich Goldwasser Halevi scheme · Lenstra-Lenstra-Lovasz algorithm · Lattice-based cryptography

1 Introduction

Back in 2001, the growth rate of computing power of quantum computers made it possible to predict that quantum computers capable of efficiently solving cryptanalysis problems of currently used cryptosystems will be available in the range of 2028–2033.

In a report, published by National Institute of Standardization (NIST) of United States [1], experts warn about the ability of quantum computers to make the use of the RSA algorithm unsafe by 2030. However, the increase in computing power of quantum computers has already exceeded the predicted indicators, and the efficiency of the developed computing complexes increases with the scaling of such systems.

Information security in the modern digital world depends on the resistance of modern cryptographic systems to various attacks. And, if modern classical computers are already capable of processing colossal amounts of information, then quantum computers, due to the specifics of computing, are able to effectively solve problems from the category of difficult-to-calculate ones [2].

R. Silhavy and P. Silhavy (Eds.): CSOC 2024, LNNS 1120, pp. 335–343, 2024.
https://doi.org/10.1007/978-3-031-70518-2_30

In November 2022, IBM introduced the most powerful quantum processor in the world–osprey, which has 433 qubits. On its basis, the quantum system two[1] multiprocessor system is being prepared. And in April 2023, the canadian company D-Wave presented the results of using its D-wave Advantage quantum processor with 5760 qubits, which showed the highest efficiency in 3D spin glass optimization problems, problems belonging to the class of difficult-to-solve optimization problems. D-Wave is currently working on a prototype of the sixth generation of its quantum computer–advantage 2, which will have 7 thousand qubits[2].

The advent of powerful supercomputers and quantum computers has compromised the security of classical encryption. The development of science has led to the creation of new digital signature and encryption schemes, since the main systems currently used, RSA, ECDSA, are not quantum-resistant [3]. In the case of using classical computers, to achieve sufficient cryptographic strength, too long a key length is required [4].

Thus, the task of finding and developing cryptographic algorithms that are resistant to quantum computing was formed, and to date, various cryptographic systems have been proposed that can be reliable if the key size is sufficient.

Since 2016, the NIST has been holding an international competition of algorithms for forming a standard for quantum-secure cryptosystems, which is currently nearing completion. Under the technical committee of Rosstandart (TC-26), a domestic set of quantum-resistant cryptosystems is being developed, which in the near future will complement the family of national standards. A similar competition to select a post-quantum cryptographic system was also held in China by the Chinese Association for Cryptographic Research [5–7].

From this perspective my main objection was developing a software implementation of one of the post-quantum cryptographic systems that allows to study the principles of operation of the cryptosystem interactively and the possibility of its hacking in controlled environment.

2 Modern Approaches to Building a Post-quantum Cryptographic System

The following approaches are used to develop post-quantum algorithms [8]:

- use of multidimensional quadratic systems [9, 10];
- building electronic signatures based on one-way hash functions [11];
- application of the theory of algebraic coding [12];
- use of elliptic curve isogenies [13];
- application of lattice theory [14–16];
- quantum hashing [17, 18].

[1] IBM Unveils 400 Qubit-Plus Quantum Processor and Next-Gen Quantum System Two https://www.hpcwire.com/off-the-wire/ibm-unveils-400-qubit-plus-quantum-processor-and-next-gen-quantum-system-two/.

[2] D-Wave Demonstrates First-Ever Coherent Quantum Spin Glass Dynamics on More than 5,000 Qubits https://www.hpcwire.com/off-the-wire/d-wave-demonstrates-first-ever-coherent-quantum-spin-glass-dynamics-on-more-than-5000-qubits/.

In order to counteract quantum computing, the concept of learning with errors is well suited, the essence of which is that an error is intentionally introduced into simple computational problems, which makes them practically impossible to solve in a reasonable time using known methods.

The main developers of post-quantum cryptography use this concept as part of their tasks. For example, there is an approach that uses error correction codes (the McElice system and their modifications [19]), and an approach based on lattice theory (the GGH scheme, the NTRUEncrypt system and their modifications [20]).

To date, the NIST standardization competition has identified the best post – quantum encryption schemes [21], among which one scheme is suitable for implementing key encapsulation (PKE/KEM)–Crystalls-Kyber [22] and two schemes suitable for implementing digital signatures: Crystalls-Dilithium and SPHINCS +. While the SPHINCS + is based on the construction of one-way hash functions, the rest are representatives of lattice theory.

Cryptographic primitives based on lattice theory problems have very strong cryptographic stability, and they are performed quite efficiently on computers. Their cryptographic stability is based on such hard-to-compute lattice theory problems [23] as:

- finding the shortest vector (Shortest vector problem);
- finding the closest vector (Closest vector problem);
- finding the shortest independent vectors (Shortest independent vectors problem).

Goldreich-Goldwasser-Halevi (GGH) is the first signature scheme based on lattices. In this scheme, the message must be hashed to the space on which the grid was projected, and the signature for this hash in this space is the nearest node of the grid [12].

The algorithm is based on a function with a hidden input. This function, in turn, relies on the complexity of solving the problem of finding the closest vector. The function is implemented by adding error vector *e* to the ciphertext vector [24].

Thus, the private key in the scheme is the error vector e and two matrices:

- orthogonal matrix – matrix *B*, whose rows form the basis of the lattice and are represented as orthogonal vectors;
- unimodular matrix – a matrix *U* whose determinant is equal to one.

The public key is the open matrix *B1* obtained by multiplying the unimodular matrix *U* by the basis of the lattice *B*.

To encrypt a message, do the following calculation:

$$c = Bm + e \tag{1}$$

where *c* is the ciphertext vector, *m* is the source message vector, and *e* is the error vector of selected *n* dimension.

For decryption, we calculate:

$$m = c * B^{-1} * U^{-1} \tag{2}$$

In 2006, a modified version of the system was hacked by Nguyen and Regev. Nevertheless, the development of the cryptosystem is considered promising [12, 25].

3 Developed Software Application and Experiment Conducting

A program that allows you to demonstrate and use the main principles of the GGH encryption scheme, including key generation, encryption and decryption algorithms, as well as the implementation of an attack on this cryptosystem.

The software product was written in the C++/CLI programming language or otherwise C++ /Cmmon Language Infrastructure, which was convenient for implementing support.Net Framework, while maintaining the performance, memory allocation, and deallocation benefits of C ++, while supporting the implementation of the interaction interface through the Windows API.

When opening the program, the user is greeted by the key generation tab. The generation itself consists of forming a private and public key. A private key consists of a lattice basis, a unimodular matrix, and an error vector. When initializing a custom form, a function is registered that sets ready-made values for the private key, which is convenient so that you can immediately start getting acquainted with the device of this cryptographic scheme.

The user is also given the opportunity to enter the values of the private key itself: the unimodular matrix and the lattice basis. Moreover, if the entered data is not correct, the user is prompted a message, for example, that the input vectors should be linearly independent.

The error vector is set in advance, it is different for each dimension and usually takes small positive or negative values.

The program interface is shown in Fig. 1.

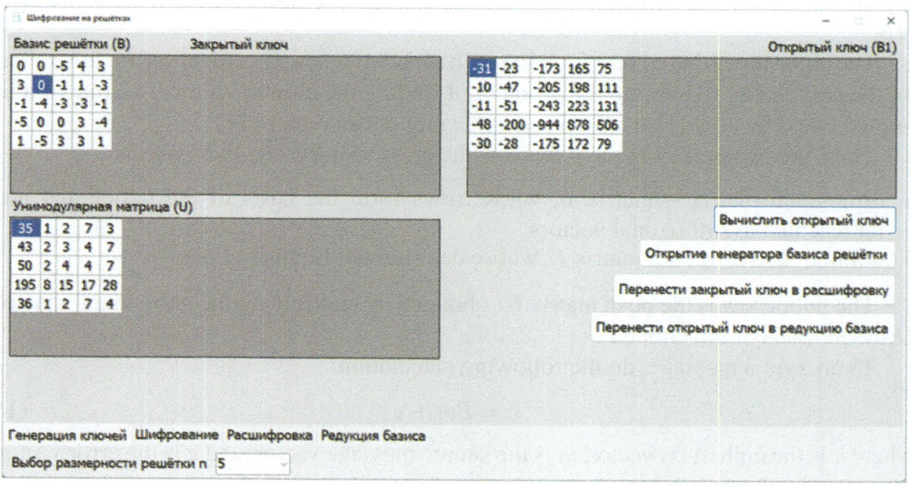

Fig. 1. Interface of the developed program

The user is given the opportunity to use the generator of basic lattice vectors. The generator is based on pseudorandom algorithms. The values obtained depend on the current moment in time, which increases the protection against collisions, that is, the probability of generating exactly the same matrices is quite low. The shape with a lattice basis generator is shown in Fig. 2.

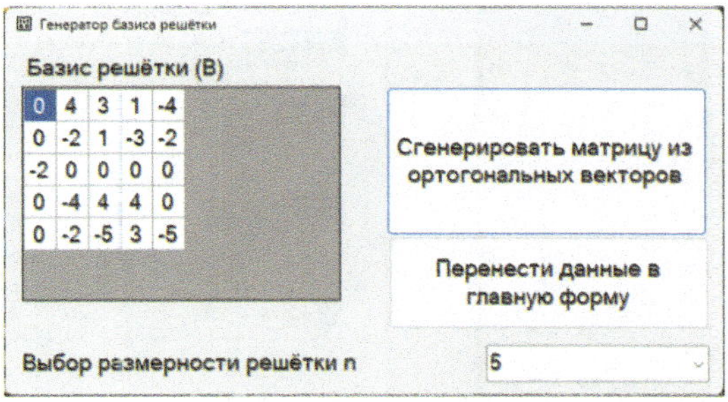

Fig. 2. Appearance of the generator form

In the encryption tab, you can encrypt a message that looks like integers or single characters using this encryption scheme, Fig. 3.

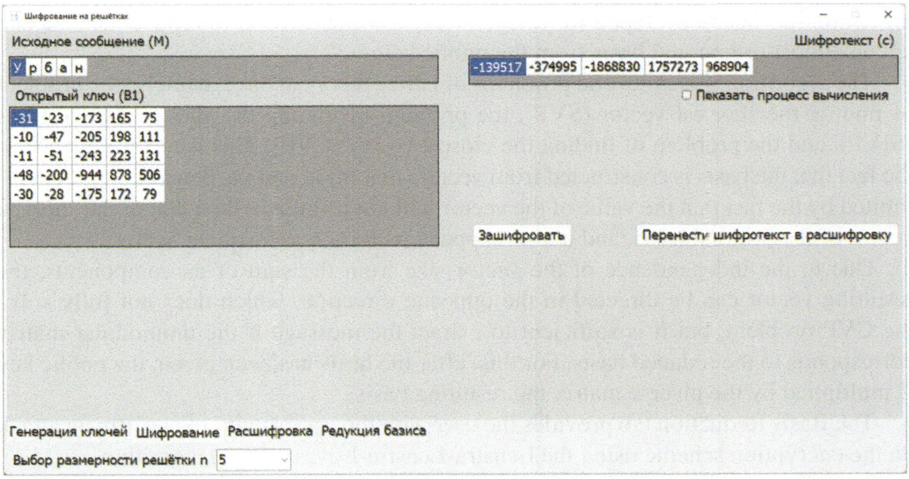

Fig. 3. Encryption tab appearance

The Decryption tab shows the process of calculating the original message from the ciphertext, Fig. 4.

4 Conducting an Attack on the Cryptosystem

Small dimension of the data used for the private key and the error vector is a vulnerable point of this scheme, and if the ciphertext and the public key were intercepted during the exchange of encrypted messages, then an attack based on the Gram-Schmidt process can

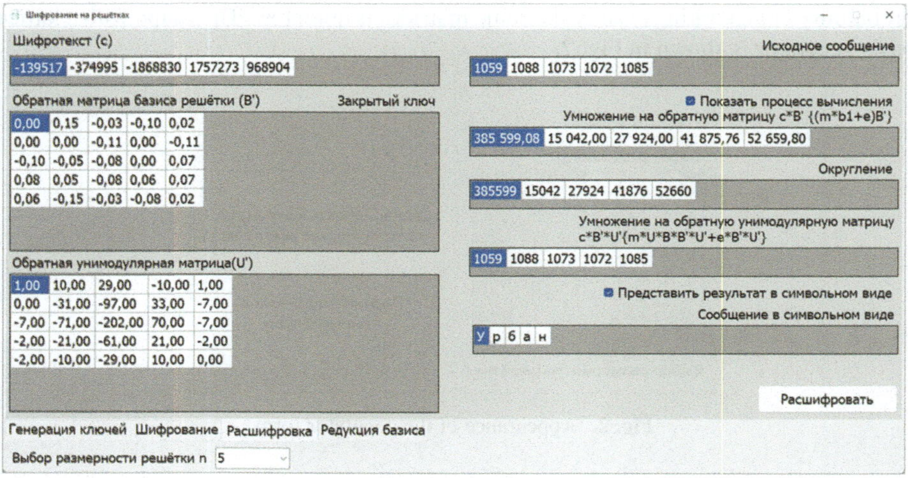

Fig. 4. Decryption tab appearance

be performed, which finds real vectors in a reasonable amount of time, but a modified algorithm has been developed for integer values, which is used to reduce the size and get an almost orthogonal basis from the public key.

The algorithm solves several problems of lattice theory at once, namely: the problem of finding the shortest vector (SVP), the problem of finding the shortest lattice basis (SIVP), and the problem of finding the closest vector (CVP). This is a consequence of the fact that the basis is constructed from vectors that fill it, and the generation process is limited by the fact that the value of the vector will not be greater than that of the shortest vector of the original basis, and it also happens with each subsequent vector.

Due to the independence of the vector size from the sign of its components, the resulting vector can be directed in the opposite direction, which does not fully solve the CVP problem, but it is sufficient to extract the message if the unimodular matrix corresponds to the reduced basis. For this, after the basis has been given, the public key is multiplied by the inverse matrix the resulting basis.

The Basis Reduction tab provides the user with the opportunity to perform an attack on the encryption scheme using the Lenstra-Lenstra-Lovasz (LLL) algorithm.

The appearance of the basis reduction tab is shown in Fig. 5.

As a result of the conducted experiments, it becomes clear that the dimensions for cryptographic stability, even for classical binary calculations, are far from optimal, and the LLL algorithm copes with reducing the basis to the original dimension, or at least as fast as generating a private key based on random values.

Also, measurements were made to carry out the attack, taking into account recommendations for improving cryptographic strength, namely: the use of the most orthogonal vectors in the lattice basis, as well as filling the basis with mutually simple elements, did not lead to misfires in the calculation of the reduced basis, even by 5 dimensions, and each time they were used to calculate the corresponding source messages. This indicates the extraordinary efficiency of the attack algorithm on small dimensions.

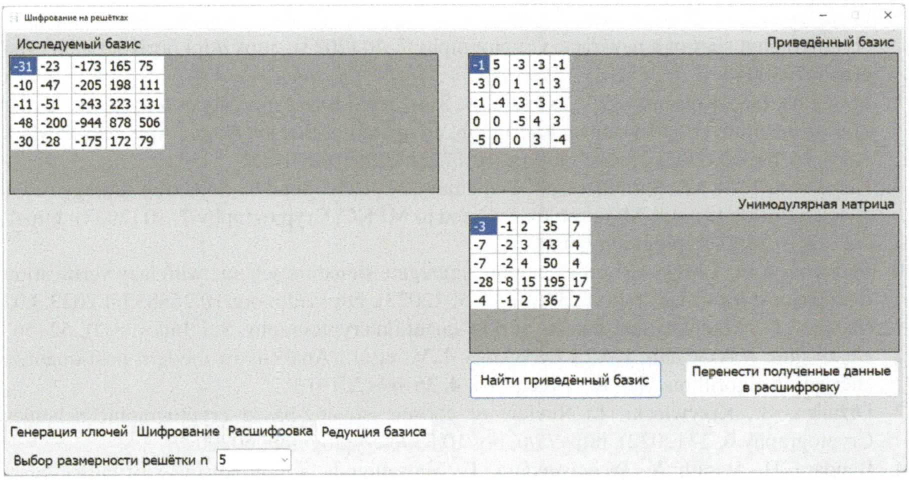

Fig. 5. Basis reduction tab appearance

5 Conclusion

We analyzed existing approaches to constructing cryptographic systems that are resistant to quantum computing, among which the most promising approach was noted–the approach using lattice theory. The developed software package allows the user to explore the main functions of the GGH cryptosystem: key generation, encryption and decryption, as well as conducting an attack using the LLL algorithm. This application can be used as part of a laboratory complex when studying cryptographic information security tools.

References

1. Report on Post-Quantum Cryptography, URL: https://nvlpubs.nist gov/nistpubs/ir/2016/NIST.IR.8105.pdf
2. Matveev, G.V.: Separation of the secret in the post-quantum period. G.V. Matveev, Information systems and technologies: materials of the international scientific congress on informatics. In 3 parts, Minsk, October 27–28, 2022/Editorial Board: S.V. Ablameyko (Chief Editor) [and others]. Volume Part 1. – Minsk: Belarusian State University, pp 95–101 (2022). https://elibrary.ru/KOIRBJ
3. Kirichenko, E.A.: Quantum superiority as a threat to cybersecurity and post-quantum methods of cryptography. IMSIT Bulletin. **1**(85), 37–39 (2021)
4. Lukashev, A.V., Shabunya, V.V., Sarafannikov, V.S., et al.: Quantum aspiration as a threat to information security. Int. J. Inform. Technol. Energy Effic. **86**(32), 97–101 (2023)
5. Nazarenko, A. P.: The current state of post–quantum cryptography in Russia and abroad. In: A.P. Nazarenko, E.V. Dmitriev, Synchronization systems, signal generation and processing. 12(6): 77–83(2021). https://elibrary.ru/FNKIVP
6. Minbaleev, A.V.: Ensuring information security of mining industry equipment in the quantum era. In: A.V. Minbaleev, M. A. Berestnev, K. S. Evsikov, Izvestiya Tula State University. Earth Sciences. 1: 567–584 (2023). https://doi.org/10.46689/2218-5194-2023-1-1-567-584

7. Dam, D.-T., Tran, T.-H., Hoang, V.-P., Pham, C.-K., Hoang, T.-T.: A survey of post-quantum cryptography: start of a new race. Cryptography **7**, 40 (2023). https://doi.org/10.3390/cryptography7030040
8. Malygina, E.S., Kutsenko, A.V., Novoselov, S.A., et al.: Basic approaches to the construction of post-quantum cryptosystems: description, comparative characteristics. Appl. Discr. Math. Appl. **16**, 58–65 (2023). https://doi.org/10.17223/2226308X/16/16
9. Kurokawa, T., Ito, T., Shinohara, N., Yamamura, A., Uchiyama, S.: Selection strategy of F4-style algorithm to solve MQ problems related to MPKC. Cryptography **7**, 10 (2023). https://doi.org/10.3390/cryptography7010010
10. Makarov, A.O.: LMQSAS- post-quantum aggregate signature scheme with lazy verification. Bezopasnost Informac. Tehnol. **30**(3), 30–50 (2023). https://doi.org/10.26583/bit.2023.3.02
11. Matveev, G.: Mathematical aspects of post-quantum cryptography. Sci. Innov. 8(2): 52–56
12. Bukovshin, V.A., Chub, P.A., Cherkesova, L.V., et al.: Analysis of modern post-quantum encryption algorithms. Sci. Rev. Tech. Sci. **4**, 36–44 (2019)
13. Drzazga, B., Krzywiecki, Ł: Review of chosen isogeny-based cryptographic schemes. Cryptography **6**, 27 (2022). https://doi.org/10.3390/cryptography6020027
14. Bandara, H., Herath, Y., Weerasundara, T., Alawatugoda, J.: On advances of lattice-based cryptographic schemes and their implementations. Cryptography. **6**(4), 56 (2022). https://doi.org/10.3390/cryptography6040056
15. Nguyen, T.-H., Pham, C.-K., Hoang, T.-T.: A high-efficiency modular multiplication digital signal processing for lattice-based post-quantum cryptography. Cryptography **7**, 46 (2023). https://doi.org/10.3390/cryptography7040046
16. Camacho-Ruiz, E., Martínez-Rodríguez, M.C., Sánchez-Solano, S., Brox, P.: Timing-attack-resistant acceleration of NTRU round 3 encryption on resource-constrained embedded systems. Cryptography **7**, 29 (2023). https://doi.org/10.3390/cryptography7020029
17. Orlov, M.A.: Evaluation of statistical properties and cryptographic stability of random sequences obtained by an IBM quantum computer. Inform. Technol. Sec. **30**(1), 14–26 (2023). https://doi.org/10.26583/bit.2023.1.01
18. Azman, A.V.: Authentication of the distribution of quantum keys using post-quantum cryptography. In: A.V. Azman, R.N. Rastamkhanov, I.R. Tsukanov. Izvestiya Tula State University. Technical sciences. 1: 29–35 (2023). https://doi.org/10.24412/2071-6168-2023-1-29-35
19. Chizhov, I.V.: Structural attack on cryptosystems of the Mcelys-Sidelnikov type, built on the basis of combining random codes with Reed-Muller codes. Int. J. Open Inform. Technol. **8**(6), 24–33 (2020)
20. Cherkesova, L.V.: Development of a modification of the post-quantum NTRUENCRYPT algorithm to counter an attack based on a selected ciphertext. L. V. Cherkesova, E. A. Revyakina, N. G. Lyashenko. Fundamental and applied aspects of computer technology and information security: Collection of articles of the All-Russian Scientific and Technical Conference, Taganrog, April 10–15, 2023. – Taganrog: Southern Federal University, 2023. – pp. 20–22. – URL: https://elibrary.ru/VACUPC
21. Karakaya,A., Ulu, A.: A Review on Latest Developments in Post-Quantum Based Secure Blockchain Systems. В материалах 11-го Международного симпозиума по цифровой криминалистике и безопасности (ISDFS) 2023 года, Чаттануга, Теннесси, США, 11–12 мая 2023 года; стр. 1–6
22. J. Bos et al., CRYSTALS - Kyber: a CCA-secure module-lattice-based KEM. In: 2018 IEEE European Symposium on Security and Privacy (EuroS&P), London, UK, pp. 353-367 (2018). https://doi.org/10.1109/EuroSP.2018.00032
23. Ortiz, J.N., de Araujo, R.R., Aranha, D.F., Costa, S.I.R., Dahab, R.: The ring-LWE problem in lattice-based cryptography: the case of twisted Embeddings. Entropy **23**, 1108 (2021). https://doi.org/10.3390/e23091108

24. Terentyev, A.D. Development of a software package for studying the basics of cryptography on lattices. In: A.D. Terentyev, M. A. Negrozov, E. A. Melnikova, Modern information technologies in education, science and industry: XX International Conference, XVIII International Competition of Scientific and Methodological works, International Competition "Neural Network Drawing": proceedings, Moscow, November 11–12, 2021. – Moscow: Limited Liability Company "Ekon-Inform Publishing House", (2021) https://elibrary.ru/YEUZHB
25. Yadav, V.K., Verma, S., Venkatesan, S.: An efficient and light weight polynomial multiplication for ideal lattice-based cryptography. Multimed Tools Appl **80**, 3089–3120 (2021). https://doi.org/10.1007/s11042-020-09706-8

Application of LSTM Neural Networks for Air Quality Index Class Forecasting

Natalia M. Lychenko[1(✉)] and Anastasija V. Sorokovaja[2]

[1] Kyrgyz-Russian Slavic University, Bishkek, Kyrgyz Republic
n.m.lychenko@krsu.kg

[2] Institute of Mechanical Engineering and Automatics, National Academy of Sciences of the Kyrgyz Republic, Bishkek, Kyrgyz Republic

Abstract. The problem of forecasting the Air Quality Index (AQI) is crucial for preventing the adverse effects of air pollution on people and improving the environmental situation. This issue is particularly relevant for Bishkek, a city significantly affected by air pollution. This work addresses the task of forecasting AQI as a classification problem and develops a classifier based on an LSTM neural network for its solution. The classification was based on meteorological parameters corresponding to AQI values measured every 3 h, which serve as the input variables for the neural network. Additionally, the daily amount of coal burned at the thermal power plant in Bishkek, a major pollution source, was used as an extra input variable. The developed classifier can predict AQI as one of four possible classes: AQI $\leq$ 50 ("Good"), 50 $<$ AQI $\leq$ 100 ("Moderate"), 100 $<$ AQI $\leq$ 150 ("Unhealthy for sensitive groups"), and AQI $>$ 150 ("Unhealthy", "Very Unhealthy", "Hazardous"). Computational experiments were conducted, varying the depth of AQI forecasting (up to 2 days ahead) and the length of input variable sequences (up to 4 days of data history) due to the inertia of air pollution processes. Incorporating the pollution factor as an additional input increased the accuracy of AQI class prediction to 83%. Further expansion of the observational history will enable the creation of a classifier for predicting AQI classes for AQI $>$ 200. #CSOC1120.

Keywords: Air Quality Index · Meteorological Parameters · Pollution Factor · Forecast · LSTM Classifier

1 Introduction

The problem of air pollution is one of the most important environmental problems of the modern industrial world. For the Kyrgyz Republic and, in particular, for the city Bishkek, the problem of air pollution is extremely important. This is related to the rapid increase of motor vehicles within the city limits, the unsatisfactory performance of filters in the fuel and energy complex, an increase in emissions from the private sector during the heating season and city's thermal power plant (TPP) running on coal - the most environmentally harmful sources of energy. Additionally, the high level of pollution is facilitated by the physiographic and climatic conditions of Bishkek, which are characterized by the occurrence of intense surface and elevated temperature inversions [1].

R. Silhavy and P. Silhavy (Eds.): CSOC 2024, LNNS 1120, pp. 344–356, 2024.
https://doi.org/10.1007/978-3-031-70518-2_31

One of the integrated indicators of atmospheric air pollution is the Air Quality Index (AQI) [2]. The AQI is a piecewise-linear function of the concentration of pollutants in the atmosphere: sulfur dioxide (SO2), nitrogen dioxide (NO2), particulate matter less than 10 μm (PM10) and less than 2.5 μm (PM2.5), carbon monoxide (CO), and ozone (O3). For each of these pollutants, the United States Environmental Protection Agency (EPA) has established national air quality standards. The calculation of AQI is based on the ratio of the measured average concentration of the pollutant to the standard (acceptable) concentration of that pollutant. An AQI value of 100 generally corresponds to the national air quality standard for the pollutant. Since the AQI for nitrogen dioxide, sulfur dioxide, and carbon monoxide is most often below 50, the most harmful air pollutants are ozone and $PM_{2.5}$ - particulate matter, which, due to its small size, can infiltrate directly into the human blood.

In general, there are 6 levels (categories) of the AQI such as good, moderate, unhealthy for sensitive groups (unhealthy-), unhealthy, very unhealthy (unhealthy +), and hazardous. The Air Quality Classication and explanations are shown as follows [2] (see Table 1).

Table 1. The AQC and explanations.

Category	Range	Explanation
Good	[0,50]	Very satisfactory air quality
Moderate	[51,100]	Satisfactory air quality
Unhealthy−	[101,150]	Unhealthy air quality for sensitive population
Unhealthy	[151,200]	Dangerous air quality for sensitive population
Unhealthy+	[201,300]	Dangerous air quality for the whole population
Hazardous	[301,500]	Very harmful air quality

An important task in ensuring environmental problems is the construction of a model for forecasting the air pollution level based on current data for a given region. If necessary, adequate forecast models will allow decisions to be made to limit emissions of the main sources of air pollution until pollution levels drop to acceptable levels.

The main factors taken into account in the task of forecasting air pollution are meteorological parameters, as they largely determine the nature of processes occurring in the atmosphere. Much research on this topic has been published in the literature, for example [3–6]. In [7], based on regression analysis, the correlations between air pollution in the city of Bishkek and the associated meteorological parameters (air temperature, humidity, dew point, atmospheric pressure, precipitation intensity, wind strength) were investigated for different (cold and warm) seasons of the year, corresponding to heating and non-heating periods. Both positive and negative correlations were identified. It is also effective to use available information on air pollution factors. We were able to identify significant correlations between the level of atmospheric air pollution in the city and the amount of coal consumed daily by the Bishkek TPP. Moreover, the impact of the pollution factor associated with TPP emissions of PM2.5 into the city air has a non-stationary character, conditioned by the heterogeneity of the type of coal used (based on the place of extraction).

As shown in [7], the degree of air pollution is influenced by meteorological conditions not only at the current moment but also by the history of changes in these conditions. This indicates the inertia of pollution processes, i.e., changes do not occur instantaneously but gradually along with the change in accompanying factors. It is reasonable to assume that predictive models that take into account the history of observations will perform better than models based solely on the assessment of current data. For analyzing historical data, recurrent neural networks are used in particular [8]. However, the degree of air pollution may respond to different factors at varying speeds. The history of each factor must be assessed differently – for some factors, only the most recent data may be important, while the influence of others may be felt over a longer period. Considering this fact, in work [9], the use of LSTM-networks for AQI classification is proposed. The architecture of these networks contains so-called filters, which during training are adjusted to selectively retain/forget information about various factors and thus can weigh the influence of each factor over time [10].

In this paper, the task of forecasting the AQI is considered as a problem of air quality classification forecasting, depending on meteorological parameters and one of the pollution factors - the number of tons of coal burned daily at the Bishkek city TPP. The classifier is developed based on an LSTM neural network, whose architectural features are most suitable for the task at hand. The article is organized as follows. In Sect. 2, we describe the dataset and the forecasting approach. After that, in Sect. 3, we present the results of computational experiments and discuss the obtained results. Finally, Sect. 4 concludes the paper.

2 Materials and Methods

2.1 Dataset

In the study, data on atmospheric air pollution in Bishkek, published on the website [11], were used. The data includes the AQI and the concentration of PM2.5 particulate matter in the air ($\mu g/m^3$), starting from 06.02.19 to the present (with a measurement interval of 1 h). Also we have access to open information about meteorological data [12]: air temperature, dew point, humidity, wind strength, precipitation intensity, and atmospheric pressure. Figure 1 shows the AQI time series for Bishkek for the period from 09.02.2019 to 14.09.2023, with intervals every 3 h: 0, 3, 6, 9, 12, 15, 18, 21 h.

The graph shows a clear seasonal component associated with the start/end of heating seasons in the city. The obtained autocorrelation function (ACF) and partial autocorrelation function (PACF) of the AQI time series (see Fig. 2) also indicate the non-stationarity of the presented time series and their 24-h periodicity. This is due to the repetition of the daily history in pollution processes: in the morning hours, the pollution level is usually lower than in the evening, which is mainly explained by the physiographic and climatic conditions of the city.

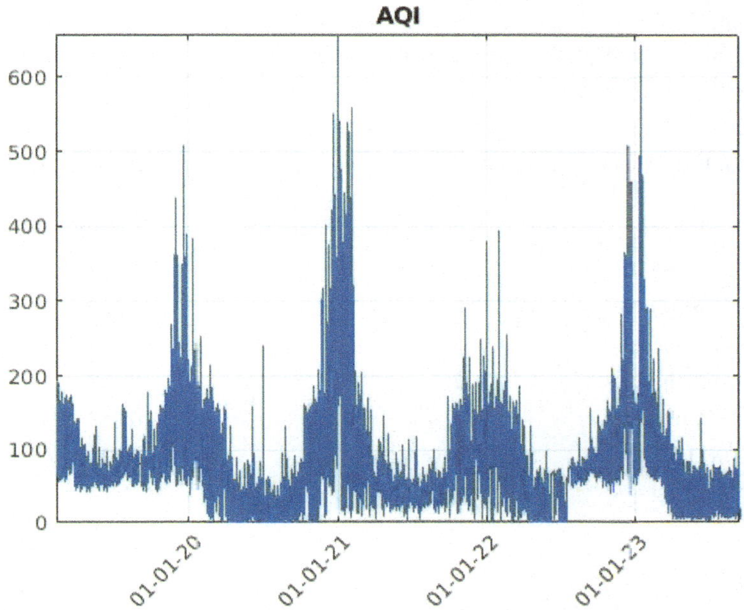

Fig. 1. Variation in dataset.

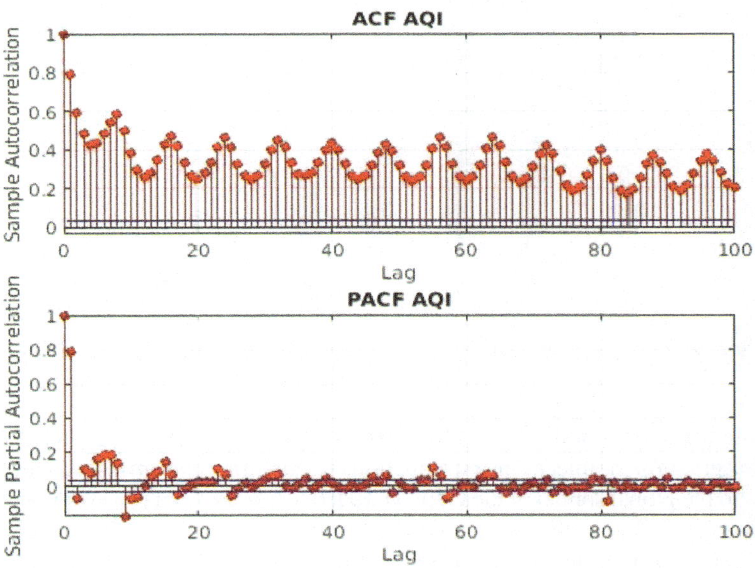

Fig. 2. AAutocorrelation and partial autocorrelation functions of AQI time series.

In analyzing AQI data from 06/02/19 to 31/03/23, the number of measurements indicating a specific AQI class (category) and their percentage ratio were calculated (see Table 2 and Fig. 3).

Table 2. Measurement AQI value distribution by class, with a set of 6 classes.

№	Category	Number of measurements	Relative share
1	Good	2458	0.218
2	Moderate	5309	0.469
3	Unhealthy⁻	1630	0.144
4	Unhealthy	1326	0.117
5	Unhealthy+	360	0.032
6	Hazardous	236	0.021

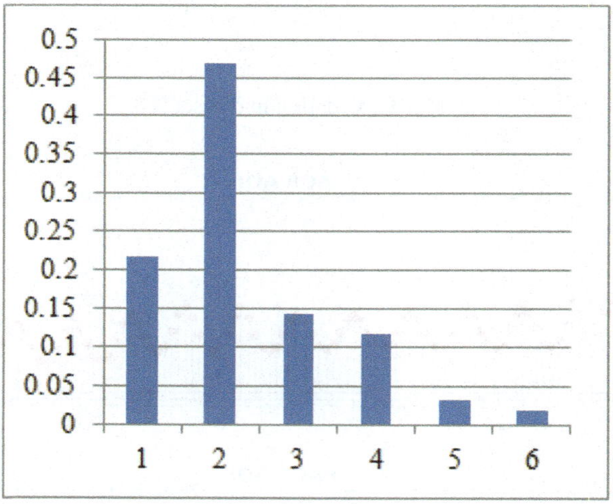

Fig. 3. Measurement entry value distribution by AQC, with a set of 6 classes.

During this analysis, the period from 01.04.20 to 31.07.20 was excluded from the dataset observation sample, as the introduction of quarantine and emergency measures in response to the global spread of the COVID-19 led to a significant (on average threefold) reduction in PM2.5 concentrations compared to similar periods of other years. Figure 4 shows graphs of curves displaying the measured values of PM2.5 in Bishkek for the time periods from April 1 to September 30, 2019 and 2020.

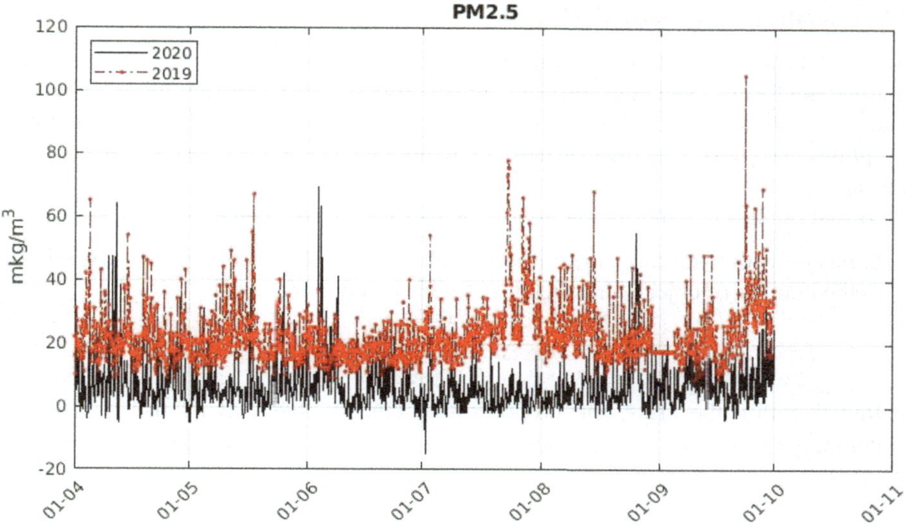

Fig. 4. PM2.5 concentrations in the air of Bishkek in April-September 2020 and in April-September 2019.

2.2 AQI Forecast as a Classification Problem

The task of forecasting the AQI is considered as a problem of classifying AQI into categories (see Table 1) depending on meteorological conditions and air pollution factors corresponding to certain measured AQI values.

Classification, in a general sense, is the task of dividing a set of objects into classes based on a training sample – a finite set of objects for which classes are predefined by an expert or by other means. The solution to the classification problem involves developing an algorithm to determine the class membership of objects outside the training sample.

Mathematically, the classification task can be formulated as follows. A set of object descriptions X and a set of class numbers Y are given. An object description $x \in X$ is represented by a feature vector $x = [x^1, x^2, \ldots, x^n]$, called the feature description of the object x. There exists an unknown mapping $y': X \rightarrow Y$, whose values are defined on the training sample. The task is to construct an algorithm $a: X \rightarrow Y$, capable of classifying any object $x \in X$.

For the task of AQI classification, the features x^j include the measured AQI value, meteorological data corresponding to this AQI value (air temperature, atmospheric pressure, relative humidity, wind speed, dew point temperature), and information about the pollution factor – the number of tons of coal burned on the day of measurement at the city's TPP; class numbers Y correspond to categories of air pollution. For the task of predicting the AQI class, Y is the class number representing the level of air pollution at a future point in time, depending on the values of the feature vector x at previous time points.

2.3 Classification Quality Evaluation

To assess the accuracy of the classification (essentially, the accuracy of prediction), error matrices are formed, and then corresponding metrics (1–3) [13] are computed. The error matrix A is an $n*n$ matrix, where n is the number of classes represented in the data sampling. The element of the matrix A_{ij} contains a value that indicates how many times the classifier identified class j as class i.

Based on the error matrix, the following metrics are calculated:

- Precision within the class: the proportion of objects truly belonging to a class relative to the number of objects that the classifier assigned to that class

$$precision_c = \frac{A_{c,c}}{\sum_{i=1}^{n} A_{c,i}} \tag{1}$$

- Recall within the class: the proportion of objects classified by the classifier as belonging to a class relative to all the documents that truly belong to that class

$$recall_c = \frac{A_{c,c}}{\sum_{i=1}^{n} A_{i,c}} \tag{2}$$

- Then, a metric that combines precision and recall is calculated, known as the F-score, which characterizes the classification accuracy (prediction accuracy)

$$F = \frac{2 * precision * recall}{precision + recall} \tag{3}$$

Therefore, the higher the F-score, the better the classifier performs.

2.4 Forecasting Approach

The input vector of the classifier, denoted as x, includes 7 elements: relative humidity, wind speed, air temperature, dew point temperature, atmospheric pressure at the weather station level, as well as the corresponding values of AQI and the number of tons of coal burned daily at the city's TPP.

The length of the sequence of the input vector (historical data vector) is denoted as S. In other words, the input vector is $x = [x(t\text{-}1), x(t\text{-}2),\ldots,x(i\text{-}s)]'$. P represents the forecast horizon (how many steps ahead AQI is predicted).

The output vector of the classifier consists of probabilities that, in P steps from time $t - 1$, AQI will belong to one of the predicted classes. The predicted class is determined by the highest probability.

To train the classifier with given values of S and P, all possible sequences of input vectors of length S are generated, and they are associated with the values of AQI at time points that are P steps away from the moments of the last values of these sequences.

Time series cross-validation is used to prevent overfitting and evaluate model performance. From the dataset, 5 non-overlapping parts are extracted, each of which serves as a test set once, while the remaining parts are used as training sets. The F-score of the classifier is determined in each experiment by taking the average of the precision F-scores (3) obtained at each cross-validation stage. The results of each experiment are then averaged.

2.5 Long Short-Term Memory (LSTM) Neural Network

In this work, it is proposed to use LSTM neural networks for AQI classification. LSTM networks are a type of recurrent neural network (RNN) that are well-suited for time series forecasting tasks. LSTM models have been successfully applied to predict air pollutant concentrations by capturing long-term dependencies and patterns in temporal data [14].

In a standard recurrent network [8], a module (building block) consists of a single layer of neurons with a typical activation function. However, an LSTM network module (LSTM cell) is not just one, but four layers that interact in a special way (see Fig. 5).

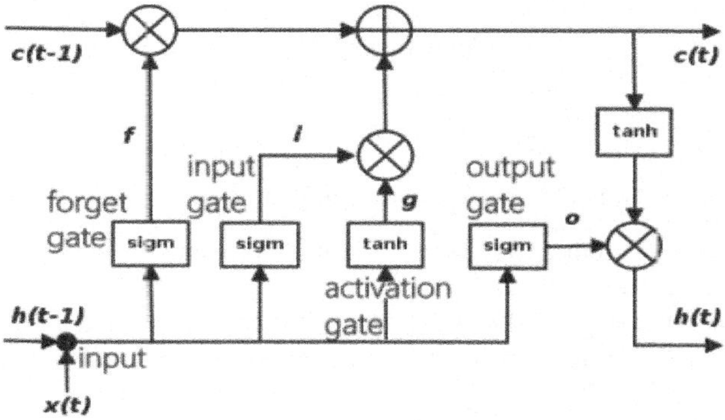

Fig. 5. A single LSTM cell: [10].

The notations here are as follows:
x_t—input vector at time t,
h_t—output vector at time t,
C_t—state vector at time t,
W_k—parameter matrix of layer k, i.e., the weights of the connections,
b_k—bias vector of the outputs of layer k,
f_t—forget gate vector at time t,
i_t—input gate vector at time t,
o_t—output gate vector at time t.

The key component of an LSTM network module is the state vector Ct. The state vector is regulated by special gate that control the deletion and updating of information within it. A sigmoid layer in the gate determines what proportion of information to overwrite in the state vector (0 – write nothing, 1 – write everything).

The LSTM functions as follows [15].

1. The forget gate layer decides what information should be removed from the state vector. For each number in the state C_{t-1}, it returns a number from 0 to 1 (where a zero value means that the information should be forgotten):

$$f_t = \sigma\left(W_f \cdot \left[h_{t-1}, x_t\right] + b_f\right)$$

2. The Sigmoid Layer of the Input Gate Determines Which Values Should Be Updated, and the Tangent Layer Constructs a Vector of New Values, C ~ t, that Can

Be Added to the State Ct.

$$i_t = \sigma\left(W_i \cdot \left[h_{t-1}, x_t\right] + b_i\right)$$

$$C \sim_t = \tanh\left(W_c \cdot \left[h_{t-1}, x_t\right] + b_C\right)$$

3. To udate the state, the old state must be multiplied by f_t and then added to i_t*$C \sim t$:

4. The output of the LSTM network is the state C_t with an applied activation function, where the output gate determines which specific elements of the state to output:

$$h_t = o_t * tanh(C_t)$$

$$o_t = \sigma\left(W_o\left[h_{t-1}, x_t\right] + b_o\right)$$

The structure of the LSTM network used in the classifier is as follows:

1. Input layer with a length of 7 (7 features in the input vector), which takes a sequence of feature vectors of length *S*;
2. The first hidden layer is a feedforward layer with 100 neurons and a hyperbolic tangent activation function. It maps input vectors to longer vectors to introduce noise into the data;
3. The second hidden layer is a regularization layer that changes some percentage of the output values of the previous layer to prevent overfitting (it introduces noise);
4. The third hidden layer is the LSTM network with 50 neuronal modules and a hyperbolic tangent activation function, serving as the primary classification layer;
5. The fourth hidden layer is a feedforward layer with 10 neurons and a hyperbolic tangent activation function;
6. The output layer is a layer with 4 neurons and a softmax activation function. This function weights the inputs and predicts the activation probabilities of each neuron. The sum of neuron outputs always equals 1.

All layers in the network are fully connected, meaning each neuron is connected to every previous neuron, and for recurrent layers (LSTM network), each input of the layer is also connected to every output of the layer.

3 Results and Discussion

3.1 Data Analysis and Normalization

Data analysis showed that the data on temperature, dew point, amount of coal, AQI have a bimodal distribution, which is characteristic of the superposition of two normal distributions, corresponding, evidently, to the cold and warm seasons of the year. Data decomposition of these data by seasons was performed, allowing for distributions that closely resemble normal distributions within one season. This enabled the application of Z-score normalization for data when forming the corresponding input vectors for the classifier.

The distribution of atmospheric pressure values at the station level is also quite close to normal. This feature is also normalized using the Z-score. Relative humidity, which ranges from 0 to 100% (or 0 to 1 in relative units), does not require normalization. Wind speed takes only a few discrete values, so normalizing them doesn't make sense.

Analysis of the number of measurements indicates that in most cases, AQI values of the "Moderate" class are recorded. The number of measurements for the "Unhealthy + " and "Hazardous" classes is insufficient to solve the AQI classification problem for all classes. Therefore, these classes are integrated into one class along with the "Unhealthy" class (see Fig. 6), and the problem of classifying AQI into 4 merged classes is then solved.

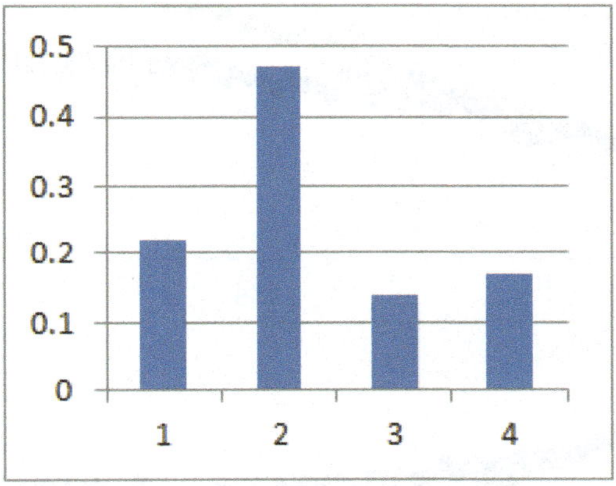

Fig. 6. Measurement AQI value distribution by class, with a set of 4 merged classes.

3.2 AQI Classification into 4 Merged Classes

Thus, the output vector of the classifier is determined by parameters that characterize the probabilities of assigning the classifier's output to different classes of the Air Quality Index (AQI) from the possible four classes: AQI50 ("Good"), 50 < AQI100 ("Moderate"), 100 < AQI150 ("Unhealthy-"), AQI > 150 ("Unhealthy", "Unhealthy + ", "Hazardous").

To solve the classification task, two data files were prepared for the observation period from 06.02.2019 to 31.03.23: one with historical AQI observations [11] and the other with historical meteorological observations [12], with a 3- hour interval.

Since the analysis shows an imbalance in the amount of data in the classes, it was necessary to balance them for classifier training. To do this, the same number of samples as the smallest class (the "Unhealthy-" class) is randomly selected from the data sampling of each class.

The software implementation of the tool for conducting computational experiments extracts data, aligns them in time, normalizes the data using a specified normalization function, and generates all possible sequences of input vectors of length S and associates them with AQI values at time points distant by P steps from the last values of these sequences.

During the computational experiments, S and P were varied.

The classifier demonstrates quite acceptable training properties. Figure 7 shows the graphs of accuracy (F-score) and loss of the network with parameters $S = 16$, $P = 8$ depending on the epoch number for the training and validation sample data. To stop the classifier's training, a threshold for changes in the loss function is used: if the loss function changes by less than 0.01 for 20 epochs, the training is stopped.

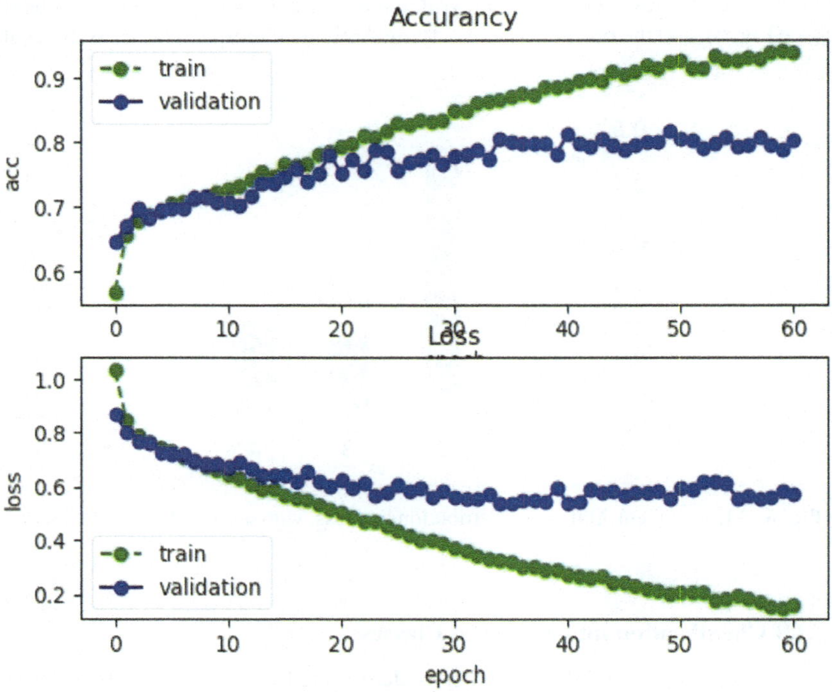

Fig. 7. The process of training the LSTM network with parameters S = 16 and P = 8.

Table 3 presents F-score of the classifier accuracy in predicting the four AQI classes depending on the length of input vector sequences (S) and the forecast depth (P) with and without considering daily amount of coal burned at the city's TPP.

As seen from the table, the classifier demonstrates quite acceptable accuracy in forecasting up to 4 days ahead ($P = 32$) both with and without considering daily amount of coal burned at the city's TPP: the forecasting accuracy $F > 0.71$ with a required measurement history of one day or more. As the forecast depth decreases, the accuracy naturally increases. The best accuracy, $F \cong 0.80$ and $F \cong 0.83$, corresponds to a forecast depth of $P = 8$ (i.e., one day ahead), with a required measurement history of $S = 16$ (two days back). Depending on the forecast depth P and the length of the samples of data history S input to the neural network classifier, the accuracy of AQI class prediction with considering daily amount of coal is 0.03 to 0.05 higher than without considering this data.

Table 3. F-score of the classifier accuracy

S	P	F- score (without TTP's coal)	F- score (with TTP's coal)
8	4	0.7641	0.7945
8	8	0.7449	0.7707
8	16	0.7286	0.7749
8	24	0.7270	0.7567
8	32	0.7139	0.7736
12	4	0.7669	0.8023
12	8	0.7728	0.8013
12	16	0.7571	0.7930
12	24	0.7556	0.7876
12	32	0.7456	0.8010
16	4	0.7895	0.8297
16	**8**	**0.8007**	**0.8287**
16	16	0.7773	0.8196
16	24	0.7772	0.8178
16	32	0.7733	0.8123

4 Conclusions

Thus, the work has demonstrated the development of an LSTM classifier for AQI with 4 classes: AQI50 ("Good"), 50 < AQI100 ("Moderate"), 100 < AQI150 ("Unhealthy-"), AQI > 150 ("Unhealthy", "Unhealthy + ", "Hazardous"). AQI classification is performed based on meteorological factors: air temperature, humidity, dew point, atmospheric pressu re, and wind strength. AQI forecasting for 4 classes is possible up to 4 days in advance with an accuracy of at least 71%.

Taking into account the factor of atmospheric pollution - the daily amount of coal burned at the TPP in Bishkek - has increased the accuracy of class prediction to 83%. This provides a basis to assert that considering this factor, as well as meteorological factors, is of great importance for solving the problem of predicting air pollution levels in the city.

Further increasing the observation history will allow the prediction of classes with AQI > 200, which is a very important task. This is because solving this task will allow government organizations. in the case of very poor air quality that can pose significant health risks to citizens, to declare an environmental emergency and take measures to limit emissions from major sources of air pollution until meteorological conditions become more favorable.

References

1. UNDP and UNEP (2022). Air quality in Bishkek: Emission source assessment and roadmap to promote air quality management. Bishkek and Nairobi https://www.undp.org/sites/g/files/zskgke326/files/2022-10/Качество_воздуха_в_Бишкеке_RU_3.pdf, last accessed 2024/02/02
2. Kanchan, K., Kumar, A., Gorai, P.: A review on air quality indexing system. Asian J. Atmosph. Environ. **9**(2), 101–113 (2015)
3. Liu, Y., Wang, P., Li, Y., Wen, L., Deng, X.: Air quality prediction models based on meteorological factors and real-time data of industrial waste gas. Sci. Rep. **12**, 9253 (2022). https://doi.org/10.1038/s41598-022-13579-2
4. Hong, W.Y.: Meteorological variability and predictive forecasting of atmospheric particulate pollution. Sci. Rep. **14**, 14 (2024). https://doi.org/10.1038/s41598-024-52617-z
5. Zhang, Y., Yang, M., Yang, F., Dong, N.: A multi-step prediction method of urban air quality index based on meteorological factors analysis. Web of Conf. **350**, 1010 (2022). https://doi.org/10.1051/e3sconf/202235001010
6. Mampitiya, L., et al.: Machine learning techniques to predict the air quality using meteorological data in two urban areas in Sri Lanka. Environments **10**, 141 (2023). https://doi.org/10.3390/environments10080141
7. Lychenko, N.: Regression analysis of meteorological factors and concentrations of PM25 particles in the atmospheric air of Bishkek. Probl. Autom. Control **2**(37), 5–15 (2019). https://doi.org/10.5281/zenodo.3594649
8. Sehovac, L., Grolinger, K.: Deep learning for load forecasting: sequence to sequence recurrent neural networks with attention. In IEEE Access **8**, 36411–36426 (2020). https://doi.org/10.1109/ACCESS.2020.2975738
9. Lychenko, N., Sorokovaya, A.: Application of LSTM neural networks for classification of air quality index in Bishkek. Probl. Autom. Control **1**(38), 70–80 (2020). https://doi.org/10.5281/zenodo.3904130
10. Stojov, V., Koteli, N., Lameski, P., Zdravevski, E.: Application of machine learning and time-series analysis for air pollution prediction. CIIT 2018 (2018). http://hdl.handle.net/20.500.12188/20987, last accessed 2024/02/02s
11. AirNow Depatment of State, https://airnow.gov/index.cfm?action=airnow.global _summary #U.S._Department_of_State $Bishkek, last accessed 2024/02/02
12. Weather schedule rp5.ru. Weather archive in Bishkek https://rp5.ru/Weather_archive_in_Bishkek, last accessed 2024/02/02
13. Galdi, P., Tagliaferri, R.: Data mining: accuracy and error measures for classification and prediction. In: Encyclopedia of Bioinformatics and Computational Biology. In: ABC of Bioinformatics, 431–436. Elsevier (2018)
14. Mitreska, E., et al.: Methods for urban air pollution measurement and forecasting: challenges, opportunities, and solutions. Atmosphere **14**, 1441 (2023). https://doi.org/10.3390/atmos14091441
15. Zhao, X., Zhang, R., Wu, J.-L., Chang, P.-C.: A deep recurrent neural network for air quality classification. J. Inform. Hiding Multimedia Signal Process. **9**, 2 (2018)

An Alternative Approach to the Study of the Conway's Universal Finite Automaton

Boris Melnikov[1(✉)] and Aleksandra Melnikova[2]

[1] Shenzhen MSU–BIT University, International University Park Road 1, Dayun New Town, Longgang District, Shenzhen 517182, Guangdong, China
bf-melnikov@yandex.ru

[2] Moscow University of Finance and Law, Vvedenskogo Str. 1a, 117342 Moscow, Russia

Abstract. This article discusses another approach to the concept and construction of the so-called universal automaton, through the definition of an equivalent COM automaton, and this definition is constructive, it is possible to build the specified automaton. Some well-known estimates of the exact upper bound of the size of a universal automaton for a regular language are given, which can be given using some nondeterministic finite automaton having k states, and a series of examples is considered that shows how fast the number of blocks can grow, i.e. the size of a universal automaton, if we are given the dimensions of both canonical automata. The property is given that in a table specifying the relation # associated with a COM automaton equivalent to a universal automaton, the number of columns cannot be significantly greater than the number of rows #CSOC1120.

Keywords: nondeterministic finite automata · regular languages · the extension of automata · universal automaton

1 Introduction

In the theory of regular languages, there are problems associated with the minimization of nondeterministic finite automata (NFA). Moreover, minimization can be considered by various criteria, primarily by the number of states, as well as the number of edges, or by mixed criteria and star-altitude minimization. In this article, we consider an approach to solving these problems using the so-called universal automaton for a given regular language. It is known that the minimization problem, for example, by the number of states, is PSPACE-complete (see [1]), but at the same time, sufficiently effective algorithms can be proposed that give sufficiently good results. Some of these algorithms are mentioned, in particular, in [2, 3].

The concept of a universal automaton was first implicitly introduced by Conway in [4], and then used, as shown by Arnold, Dicky and Nivat, in solving minimization problems, in their work [3] they pointed out that any minimal NFA for a given regular language is isomorphic to a certain sub-automaton of the universal automaton. The same automaton was used in solving the problems of minimizing NFA by the number of states

R. Silhavy and P. Silhavy (Eds.): CSOC 2024, LNNS 1120, pp. 357–362, 2024.
https://doi.org/10.1007/978-3-031-70518-2_32

of L. Polak in the article [2], where he formulates various conditions for subsets of states of the universal automaton and explores the relationships between them.

In this article we will consider an alternative approach to the construction of a universal automaton used in [5].

2 Preliminaries

We shall use the notation and preliminaries of [5, 6]. Let us repeat the main ones. We shall consider nondeterministic finite automaton $K = (Q, \Sigma, \delta, S, F)$, the language of a nondeterministic finite automaton K is denoted by L(K). The input language of the state $q \in Q$ will be denoted by $L_K^{in}(q)$, the output language of the state $q \in Q$ will be denoted by $L_K^{out}(q)$ (i.e. the languages of automatons $(Q, \Sigma, \delta, S, \{q\})$ and $(Q, \Sigma, \delta, \{q\}, F)$ accordingly).

The canonical automaton will be denoted by the entry $\tilde{L}=(Q_\pi, \Sigma, \delta_\pi, \{s_\pi\}, F_\pi)$, without the useless ("dead") state. The canonical to inverse automaton will be denoted by $\tilde{L^R} = (Q_\rho, \Sigma, \delta_\rho, \{s_\rho\}, F_\rho)$. Earlier in the article [4], the definition and algorithm for constructing a special binary relation $\# \subseteq Q_\pi \times Q_\rho$ were considered.

We will also use the following definitions of pseudo-blocks and blocks. If for some pair $P \subseteq Q_\pi$ and $R \subseteq Q_\rho$ the condition is met

$$(\forall A \in P)(\forall X \in R)(A\#X),$$

then $B = (P, R)$ will be called a pseudo-block, it is convenient to designate $P = \alpha(B)$ and $R = \beta(B)$.

If for some pseudo-block $B = (P, R)$ do not exist:

- no state $A \in Q_\pi \backslash P$ such that $((P \cup \{A\}), R)$ is also a pseudo–block;
- no state $X \in Q_\rho \backslash R$ such that $(P, (R \cup \{X\}))$ is also a pseudo–block,

then B will be called a block.

The set of blocks for a given regular language L will be denoted by the entry Q_{COM} (this designation will be clear from the following). A subset $Q = Q_{\text{COM}}$.

will be called a covering subset of blocks if for any $A \in Q_\pi$ and $X \in Q_\rho$.

such that $A\#X$, there exists a block $B \in Q$ such that $A \in \alpha(B)$ and $X \in \beta(B)$.

3 Conway's Universal Finite Automaton and Automaton COM

For a given regular language L, we will consider a universal automaton, which according to [2, Def. 2.4], we will denote by U_L.

Now let us consider the definition of the COM automaton from work [5] in the form of a five, with a set of states Q_{COM} is defined in the previous paragraph.

Definition 1. Let $\tilde{L}$ and cbe canonical automata for the regular language L, we define the automaton.

$$COM(L) = (Q_{\text{COM}}, \Sigma, \delta_{\text{COM}}, S_{\text{COM}}, F_{\text{COM}}),$$

where:

- $S_{\mathrm{COM}} = \{B \in Q_{\mathrm{COM}} | \alpha(B) \ni s_\pi\}$;
- $F_{\mathrm{COM}} = \{B \in Q_{\mathrm{COM}} | \beta(B) \ni s_\pi\}$;
- for some pair $B_1, B_2 \in Q_{\mathrm{COM}}$ (equality $B_1 = B_2$ is valid) and some letter $a \in \Sigma$, we assume

$$\delta_{\mathrm{COM}}(B_1, a) \ni B_2 (\text{i.e. } B_1 \overset{a\delta_{\mathrm{COM}}}{\rightarrow} B_2)$$

if and only if the following conditions are simultaneously met:

$$(\forall A \in \alpha(B_1))(\delta_\pi(A, a) \neq \emptyset) \& (\delta_\pi(A, a) \subseteq \alpha(B_2)); \tag{1}$$

$$(\forall Y \in \beta(B_2))\big(\delta_\rho(Y, a) \neq \emptyset\big) \& \big(\delta_\rho(Y, a) \subseteq \beta(B_1)\big). \tag{2}$$

Note that considering canonical automata having a possible "dead" state, we get the same values of the functions φ^{in} and φ^{out}, since they cannot contain "dead" states. Therefore, the definition of the COM(L) automaton does not depend on the consideration (or non-consideration) of "dead" states.

Condition (1) can also be written as follows:

$$(\forall A \in \alpha(B_1))(\delta_\pi(A, a) = \{B\}) \& (B \in \alpha(B_2)).$$

Of course, condition (2) can be written in the same way.

In the article [5] (see Theorem 1) it is proved that in this way a certain automaton coincides with the universal Conway automaton up to the re-designation of states. In that article, not only the definition is given, but also several different algorithms for constructing the COM(L) automaton.

As noted in the article [7], a universal automaton is constructed in such a way that this U_L automaton has no more than 2^n states if the size of the minimum deterministic automaton is n, and this boundary is rigid. It is interesting to have a fairly complete description of the history of the development of the concept of a universal automaton by the authors of this article. In the same article, the limits of the size of a universal automaton are estimated: since the definition of a nondeterministic automaton with n states can give at most an automaton with 2^n states, we get 2^{2^n} – the upper bound of the size of a universal automaton with respect to the minimum a non-deterministic automaton that accepts a language, and this boundary is not rigid. The authors of [7] provide proof that the hard bound is given by Dedekind numbers, and also that in the case of a unary alphabet, the worst cases can be combined, but then the definition does not lead to an increase by 2^n. But such automata are of little interest for studying the properties of regular languages.

One of the invariants of regular languages constructed using canonical automata (canonical for a given language and canonical for an inverse language) is a basic automaton, which can also be used in various problems of the theory of regular languages, in particular, in minimization problems according to various criteria. A number of articles by authors have been written about this (see [8–10], etc.).

In this section we will consider a series of examples that shows how fast the size of a universal automaton can grow (i.e. in terms of the definition 1 is the number of blocks) if we are given the sizes of both canonical automata. According to the authors, these examples complement those considered in [5].

4 Using Examples for Some Estimation

To begin with, consider the following matrix of size 12×12:

$$\begin{pmatrix} 1&0&0&0&1&1&1&1&1&1&1&1 \\ 0&1&0&0&1&1&1&1&1&1&1&1 \\ 0&0&1&0&1&1&1&1&1&1&1&1 \\ 0&0&0&1&1&1&1&1&1&1&1&1 \\ 1&1&1&1&1&0&0&0&1&1&1&1 \\ 1&1&1&1&0&1&0&0&1&1&1&1 \\ 1&1&1&1&0&0&1&0&1&1&1&1 \\ 1&1&1&1&0&0&0&1&1&1&1&1 \\ 1&1&1&1&1&1&1&1&1&0&0&0 \\ 1&1&1&1&1&1&1&1&0&1&0&0 \\ 1&1&1&1&1&1&1&1&0&0&1&0 \\ 1&1&1&1&1&1&1&1&0&0&0&1 \end{pmatrix}$$

Moreover, such matrices can be considered for any $x > 2$ (if we denote by $x \times x$ the dimensions of the "block of zeros") and any $n > 2$ such that $n \bmod x = 0$ (denoting the dimensions of the matrix by $n \times n$). In this example, $x = 4$, $n = 12$. In addition, you can take blocks of zeros of different sizes. However, we will not consider formulas for elements of such matrices.

Based on the properties proved in [11–13], it can be shown that for each such matrix we can construct an appropriate nondeterministic finite automaton that has exactly such a relation table # and that such an automaton always exists.

If we number the rows of the matrix with numbers from 1 to 12, we can select blocks as follows: select exactly one row from the set {1, 2, 3, 4}, then exactly one line from the set {5, 6, 7, 8} and also exactly one line from the set {9, 10, 11, 12}. When building blocks, the numbers of the selected columns must match the numbers of the previously selected rows. When selected in this way, there are at least $4^3 = 64$ blocks in the example.

Now let

$$n = a_1 + a_2 + ... + a_k$$

and there are blocks of zeros of sizes $a_1, a_2, \ldots, a_k$.; then the corresponding number of blocks corresponding to the states of the canonical automaton will be equal to $a_1 \cdot a_2 \cdot \ldots \cdot a_k$. We can set the task (for a fixed value of n) of finding the maximum of this value. To solve it, we need to take into account that the following representations take place

$$4 = 2 + 2, 2 \cdot 2 = 4 > 4;$$
$$5 = 2 + 3, 2 \cdot 3 = 6 > 5;$$
$$6 = 2 + 2 + 2 = 3 + 3, 3 \cdot 3 > 2 \cdot 2 \cdot 2$$

thus, each term greater than or equal to 4 can be divided into twos and threes, with such a division the product will not decrease. It is also obvious that with such a representation of terms, it is necessary to give preference to triples, as in the last inequality.

The desired maximum has one of the following types depending on the remainder when dividing n by 3:

$$3 \cdot 3 \cdot \ldots \cdot 3$$
$$3 \cdot 3 \cdot \ldots \cdot 3 \cdot 2 \cdot 2$$
$$3 \cdot 3 \cdot \ldots \cdot 3 \cdot 2.$$

Let us limit ourselves to the case of (n mod 3) = 0 for simplicity of reasoning. Then the number of blocks will be at least $3^{n/3}$.

This series of examples shows that the size of a universal automaton can grow exponentially (with a base of $3^{1/3}$) with respect to the size of canonical automata, i.e. with respect to $n = \min(|Q_\pi|, |Q_\rho|)$.

Recall that our examples do not give the exact number of blocks that can be formed if we only know about the canonical $\tilde{L}$ and canonical to mirror $\tilde{L^R}$ automata the value of $n = \min(|Q\pi|, |Q\rho|)$.It also makes no sense to combine the series of examples described here with an obvious estimate of $n = 2k$ for the number of states of the canonical automaton: the resulting function $3^{n/3} = 3^{2^k/3}$ grows much faster than D(k). This only means that languages with a relation # similar to the matrix in the example cannot have equivalent automata with a number of states much smaller than n.

In a table specifying the relation #, the number of columns cannot be significantly greater than the number of rows (see [8, 9]). Indeed, let the canonical automaton $\tilde{L}$ contain n states. Then n states also contains an inverse automaton $\tilde{L^R}$, which defines an inverse language. This means that the canonical automaton for this inverse language $\tilde{L^R}$ contains no more than 2^n states. The same number limits the number of possible columns of the relation table #, in which there are n rows. There cannot be, for example, a relation # table having 4 rows and 17 columns (since 17 > 24). Such assessments allow us to better understand the essence of the issue.

References

1. Jiang, T., Ravikumar, B.: A note on the space complexity of some decision problems for finite automata. Inf. Process. Lett. **40**(1), 25–31 (1991). https://doi.org/10.1016/S0020-0190(05)80006-7
2. Polák, L.: Minimalizations of NFA using the universal automaton. Int. J. Found. Comput. Sci. **16**(5), 999–1010 (2005)
3. Arnold, A., Dicky, A., Nivat, M.: A note about minimal non-deterministic automata. Bull. EATCS **47**, 166–169 (1992)
4. Conway, J.H.: Regular Algebra and Finite Machines. Chapman and Hall, Ltd (1971)
5. Dolgov, V.N., Melnikov, B.F.: Construction of a universal finite automaton. I. From theory to practical algorithms. Vestnik Voronezh State University. Ser. Phys. Math. **2**, 131–139 (2013)
6. Melnikov, B.: Once more on the edge-minimization of nondeterministic finite automata and the connected problems. Fund. Inform. **104**(3), 267–283 (2010)

7. Lombardy, S., Sakarovitch, J.: The universal automaton. Logic Automata Texts Logic Games **2**, 457–504 (2008)
8. Melnikov, B., Vakhitova, A.: Some more on the finite automata. Korean J. Comput. Appl. Math. (J. Appl. Math. Comput.) **5**(3), 495–506 (1998)
9. Melnikov, B., Melnikova, A.: Some properties of the basis finite automaton. Korean J. Comput. Appl. Math. (J. Appl. Math. Comput.) **9**(1), 135–150 (2002)
10. Melnikov, B., Melnikova, A.: Extended basis finite automaton. Part I. The basic definitions. Int. J. Open Inf. Technol. **7**(10), 1–8 (2018)
11. Melnikova, A.: Some properties of the basic automaton. Bulletin of the Voronezh State University. Ser. Phys. Math. **2**, 184–189 (2012)
12. Zuzanova, M.: Some Algorithms for Equivalent Transformation of Nondeterministic Finite Automata. Togliatti State University, Togliatti (2010)
13. Melnikov, B.: Extended nondeterministic finite automata. Fund. Inform. **104**(3), 255–265 (2010)

Application of Kohonen Self-organizing Maps for Balancing Test Questions in E-Learning

Hoang Phuong Nguyen[1,2], Faycal Bensalah[3], Jorge Alberto Esponda-Pérez[4], Ashot Gevorgyan[5,6](✉), Diloram Zakirova[7], and Roman Tsarev[8,9]

[1] University of Social Sciences and Humanities, Ho Chi Minh City, Vietnam
[2] Vietnam National University, Ho Chi Minh City, Vietnam
[3] Lab LERSEM, ENCG, University Chouaib Doukkali, El Jadida, Morocco
[4] UNICACH, Tuxtla Gutiérrez, México
[5] Yerevan State University, Yerevan, Armenia
ashot.gev@ysu.am
[6] National University of Architecture and Construction of Armenia, Yerevan, Armenia
[7] Tashkent State University of Economics, Tashkent, Uzbekistan
[8] MIREA - Russian Technological University (RTU MIREA), Moscow, Russia
[9] Bauman Moscow State Technical University, Moscow, Russia

Abstract. The paper concerns the issue of clustering and balance of test questions on knowledge, skills and abilities of university students. Opportunities of developing nowadays electronic learning has a great number of advantages in providing learning material and control of obtained knowledge, skills and abilities of students over traditional methods and means: flexibility, accessibility, cost-effectiveness, interactivity, multimedia, individualization and differentiation of learning, high degree of students' involvement. Formation of knowledge, skills and abilities in students is the most important process in learning. The level of knowledge, skills and abilities determines the success of university students in their future profession. That is why at the stage of designing electronic tests, it is important for the teacher to develop questions based on what the student should know and be able to do, what skills to possess. One of the main functions of the learning management system Moodle, as an e-learning tool, is the diagnostic function. LMS Moodle allows to optimize and automate the process of controlling the mastery of learning material. Thanks to this environment the teacher has the opportunity to easily develop test questions of open and closed types. However, in order for these test questions and e-testing in general to become effective, valid and reliable, it is very important for the teacher to keep the balance of test questions in terms of their categories, focus on testing knowledge, skills and abilities. Each category should contain a roughly equal number of questions so that e-testing can provide a complete view of student learning outcomes. The study involved the use of Kohonen self-organizing maps for clustering of test questions, which allowed to assess the balance of available questions in terms of knowledge, skills, and abilities and to make adjustments in the bank of test questions to achieve a high degree of efficiency and objectivity of e-testing.

Keywords: E-Learning · Moodle · Kohonen Network · Kohonen Map · SOM · Neural Network · E-Testing · Clustering

R. Silhavy and P. Silhavy (Eds.): CSOC 2024, LNNS 1120, pp. 363–372, 2024.
https://doi.org/10.1007/978-3-031-70518-2_33

1 Introduction

The ever evolving e-learning nowadays opens up more and more learning opportunities. E-learning is based on the use of information and communication technologies to transfer knowledge and control learning. It is characterized by flexibility, accessibility, interactivity, multimedia, efficiency and cost-effectiveness [1–5]. This means that thanks to e-learning tools, students can work with the learning material at any time, in any place, referring to it the necessary amount of times, receive feedback from the teacher and other students, as well as increase the speed of learning and memorization of the learning material, the level of stability of their knowledge, track their results, follow individual educational trajectories and reflect on their learning activities [6–10].

The need for highly qualified specialists is increasing with the development of the economy worldwide, the expansion of various industries, especially high-tech industries, in particular those related to nanotechnology [11–17]. The demand for high-level specialists is determined by the rapid development of information systems, big data technologies, data mining, artificial intelligence, a variety of systems for processing and analyzing information [18–22]. There is a necessity for high-level specialists in other areas, such as economics, law, public administration, social work, medicine [23–27]. Preparing specialists of such level requires the use of modern teaching methods, as well as modern electronic learning environments.

Currently, there are many e-learning tools, which include e-textbooks, online platforms, digital educational resources and mobile applications. One of the most popular e-learning tools is the electronic educational environment Moodle, which optimizes learning processes not only for students but also for teachers. Such platforms create e-learning courses for distance and blended learning, develop modules with interactive material, assignments and tests [1, 28–30]. Learning management system Moodle allows to plan, conduct, manage training sessions, virtual classes and electronic courses [31, 32].

LMS Moodle allows creating, conducting, automatically checking, evaluating and analyzing electronic tests [5, 33–35]. The function of automatic checking and evaluation of assignments saves the teacher's time and also eliminates the subjective factor. Electronic testing does not require additional software and can be conducted both in face-to-face classes in the classroom and remotely, this can be configured in LMS Moodle based on the course structure. The test itself has various settings, such as time of passing the test, number of attempts, which questions from the test bank to use and so on. It is important to note that all information about the test results is stored, which allows for multilateral analysis of the test results, as well as the test questions themselves, and the formation of individual tests based on them. Thanks to LMS Moodle it is possible to provide feedback at the end of the test: providing a report on the correctness of the student's answers, quantitative assessment of correct answers, demonstration of the correct answer to the student.

However, the degree of validity and reliability of e-testing depends, among other things, on the structure of tests, to which special attention should be paid. LMS Moodle has great functionality in this area: the teacher can create closed-type questions such as Multiple Choice, True or False questions, Fill in the blank questions, Matching questions, as well as open-type questions. Thus, students should demonstrate their knowledge, skills and abilities in various forms: choose the correct answer from the list of proposed answers, fill in the blanks, match, establish correspondence between elements or their sequence, determine true or false statements, give a detailed, brief or numerical answer, describe, insert missing elements, etc. [36].

Control and determination of the level of knowledge, skills and abilities, which characterize the level of students' preparation in general, is the main purpose of electronic testing. Thus, some types of test questions (mostly of the closed type for choosing the correct answer from a list, matching elements or determining the correct statement) can best test students' knowledge of specific facts, concepts, principles and law; others, to a greater extent, are suitable for testing skills, i.e. the application by students of the acquired knowledge in practice; and determining the level of students' abilities, i.e. the ability of students to use their skills automatically. Open-ended, fill-in-the-blanks or computable types of questions reflect skills and abilities most objectively.

In the process of learning, students acquire various universal, general professional and professional knowledge, skills, and abilities, the level of which should be controlled equally [37–39].

To create an effective e-test, a teacher needs to maintain a balance between different types of questions and categories. Each of them has its advantages and disadvantages, so it is important to use them in the right proportion. For example, too many open-ended questions can make the test very long and complicated, while an overabundance of closed-ended questions makes testing easy and non-objective, such tasks may not give a deep understanding of the formed knowledge, skills and abilities. It is impossible to test only the acquired knowledge, or only skills or abilities, it will not give an objective assessment. Electronic testing should reflect the complex of formed knowledge, skills, abilities, as these three phenomena are closely related to each other, there should be a balance between these categories. A university may have requirements to the ratio of test questions by categories of knowledge, skills, abilities. For example, it could be 10 closed-ended questions in the knowledge category, 10 closed-ended and 10 open-ended questions in the skills category, and finally, 10 closed-ended and 10 open-ended questions in the abilities category.

Neural networks, which are experiencing rapid development today, are also applied in the field of e-learning [40–47]. In our study, Kohonen network was used to create tests with a balance of test question types and categories [48]. This artificial neural network, also known as Kohonen Self-Organizing Map or simply Self-Organizing Map (SOM), is used for data clustering and consists of multiple neurons, each of which is associated with all input data. When data is fed to the input of the network, the neurons are activated depending on how closely they match the input data [49–51].

Kohonen self-organizing maps, as a further extension of neural networks, are highly visual; they are designed to visually represent multidimensional properties of objects on a two-dimensional map [52–54]. So, thanks to Kohonen's self-organizing maps, in our study the data were divided into clusters corresponding to different categories. Without the effort and involvement of a teacher, our test questions were assigned to a specific category (knowledge, skills, abilities), which allowed us to evaluate the balance of the test questions.

2 Methods

Kohonen self-organizing map (Fig. 1) was used to solve the problem of clustering test questions into knowledge, ability, and skill categories. This is a self-learning neural network, training of this neural network takes place without a teacher. This aspect is fully appropriate for the problem stated as there is no training sample to train the neural network. When solving the problem of clustering test questions, the Kohonen network is trained without a teacher and understands by itself to which category, i.e. to which cluster the question should be assigned.

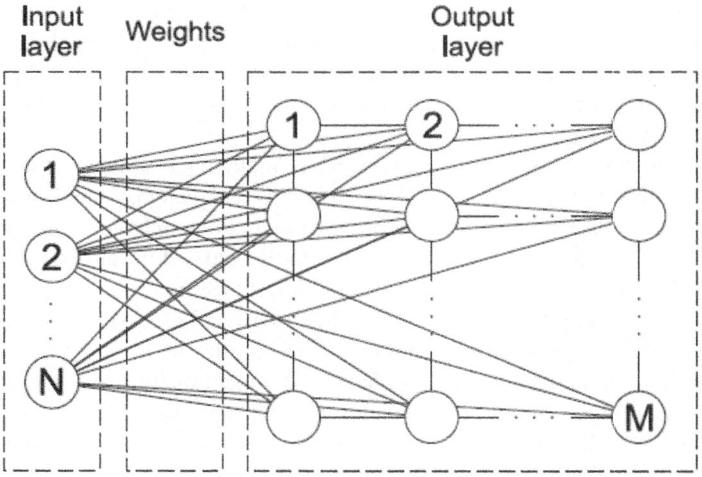

Fig. 1. Kohonen Network.

Kohonnen self-organizing maps have been used in solving a wide variety of tasks, in particular, identifying patterns in the obtained data, data compression, modeling, forecasting. This neural network in practice confirmed the effectiveness of its work and showed the ability to successfully solve the tasks. A detailed description of theoretical aspects and practical examples of the use of self-organizing Kohonen maps can be found in the literature [48–54]. In general, the application of Kohonen self-organizing maps can be described as follows.

Given: a sample $\boldsymbol{X}$ consisting of objects $X_i = \{x_{i1}, \ldots, x_{id}\}$, $i = 1, \ldots, N$, where d is the number of attributes that describe an object. The size of $\boldsymbol{X}$ is d by N. W is the matrix of neuron weights, its size is $d \times M$, where M is the number of neurons; $W_j = \{w_{j1}, \ldots, w_{jd}\}$. D is a $M \times M$ matrix of distances between neurons in the layer. In fact,

D represent the topology of neurons. α is the cooperativity; γ is the learning rate decay rate; E is the number of epochs.

Algorithm:
1. Initialize W
2. E times:
   1. For each object X_i of X:
      1. $p = \arg\min \| X_i - W_i \|$ // find the closest neuron to the object
      2. $\forall j$: $h_{p,j} = e^{D_{p,j}/\sigma^2(t)}$
      3. $\forall j$: $w_j = w_j + \mu h_{p,j}(x - w_j)$
   2. $\mu = \mu \cdot \gamma$ // μ characterizes the learning rate
   3. $\sigma = \sigma \cdot \alpha$ // σ indicates the "elasticity" of the network (map)

Once the map topology resulting from the above algorithm is generated according to the input data, a unified distance matrix (U-matrix) is required to solve the clustering problem. The U-matrix allows to visualize the Kohanen map. In this case, in addition to neurons, the distance between them is displayed. Based on the U-matrix, clusters can be easily identified.

3 Results

To create a balanced bank of test questions for intermediate and final tests in LMS Moodle, the test questions were categorized into three clusters according to the categories "knowledge", "skills", "abilities" and "skills" using Kohonen self-organizing maps. The result of applying Kohonen self-organizing map to test questions is shown in Fig. 2.

The Kohonen self-organizing map visualizes the results and allows to evaluate the balance of test questions. In case of inconsistency, the teacher can make changes to the e-test: modify, add or delete test questions in order to achieve a balance between questions of different categories. After that it is possible to re-analyze the test question bank to assess the balance of test questions taking into account the changes made.

It should be noted that the proposed approach is universal and can be applied in analyzing test questions of any course in any university. It can be useful both in terms of ongoing evaluation and improvement of the course by the teacher, as well as during the inspection and accreditation of the university by higher organizations such as departments or ministry of education.

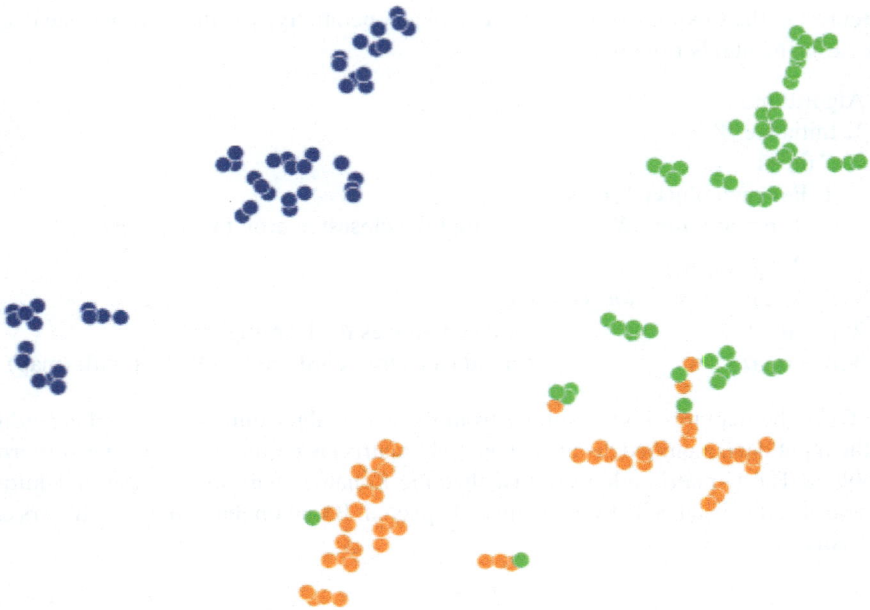

Fig. 2. Test questions categorized into three clusters.

4 Conclusion

Traditional teaching methods and ways of assessing students' knowledge, skills and abilities are significantly inferior to progressive e-learning in many aspects. Firstly, e-learning tools, in general, provide accessibility and visibility of educational material. Secondly, e-learning tools, in particular LMS Moodle, are tools for effective, economical and optimal diagnostics of students' results, they are more flexible and adaptive in comparison with traditional methods and tools. Thirdly, electronic testing, which can be created using LMS Moodle, reduces the psychological burden and anxiety of students, as well as teachers, who no longer have the need to independently check test tasks, the subjective factor is eliminated. In addition, electronic testing, motivates university students to improve their results and always causes more interest and concentration, as well as contributes to individualization and differentiation of learning.

However, the development of test questions for e-testing is not an easy methodological task for the teacher, because the teacher needs to keep a balance between the types of questions and their categories. In pedagogical practice, teachers composing tests are not able to visually assess whether the number of test questions is evenly distributed in accordance with the categories of know, be able, and possess. Failure to observe this balance leads to unobjective results and unreliability of the electronic test. Creating different types of test questions, thanks to LMS Moodle, provides the teacher with many options for controlling both knowledge separately, and skills and abilities. In this matter, it is very important to keep this proportion so that the test questions are in equal amount for each category. Kohonen self-organizing maps, designed to solve the problem of data clustering, facilitate analysis through data visualization and allow to identify imbalance of categories of test questions. Thanks to this network, in our study, the disruption of this balance was detected, which allowed us to adjust the question bank to achieve a consistent and objective picture of the acquired knowledge, skills and abilities of students. The proposed approach and the developed computer program can be applied for automatic evaluation of test questions for all disciplines taught at the university, which are implemented in the form of electronic educational courses.

References

1. De Castro, M.G.A., García-Peñalvo, F.J.: Systematic review of Erasmus+ projects labelled as good practice and related to e-learning and ICT: some case studies. Heliyon **9**(12), e22331 (2023). https://doi.org/10.1016/j.heliyon.2023.e22331
2. De Medio, C., Limongelli, C., Sciarrone, F., Temperini, V.: MoodleREC: a recommendation system for creating courses using the moodle e-learning platform. Comput. Hum. Behav. **104**, 106168 (2020). https://doi.org/10.1016/j.chb.2019.106168
3. Joy, J., Pillai, R.V.G.: Review and classification of content recommenders in E-learning environment. J. King Saud Univ. Comput. Inf. Sci. **34**(9), 7670–7685 (2022). https://doi.org/10.1016/j.jksuci.2021.06.009
4. Nagy, V., Duma, L.: Measuring efficiency and effectiveness of knowledge transfer in e-learning. Heliyon **9**(7), e17502 (2023). https://doi.org/10.1016/j.heliyon.2023.e17502
5. Tsarev, R., et al.: Improving test quality in E-learning systems. Lect. Notes Netw. Syst. **723**, 62–68 (2023). https://doi.org/10.1007/978-3-031-35317-8_6
6. Al-smadi, A.M., Abugabah, A., Al Smadi, A.: Evaluation of E-learning experience in the Light of the Covid-19 in higher education. Procedia Comput. Sci. **201**, 383–389 (2022). https://doi.org/10.1016/j.procs.2022.03.051
7. Baabdullah, A.M., Alsulaimani, A.A., Allamnakhrah, A., Alalwan, A.A., Dwivedi, Y.K., Rana, N.P.: Usage of augmented reality (AR) and development of e-learning outcomes: an empirical evaluation of students' e-learning experience. Comput. Educ. **177**, 104383 (2022). https://doi.org/10.1016/j.compedu.2021.104383
8. Deetjen-Ruiz, R., et al.: Applying ant colony optimisation when choosing an individual learning trajectory. Lect. Notes Netw. Syst. **723**, 587–594 (2023). https://doi.org/10.1007/978-3-031-35317-8_53
9. Kacetl, J., Semradova, I.: Reflection on blended learning and e-learning – case study. Procedia Comput. Sci. **176**, 1322–1327 (2020). https://doi.org/10.1016/j.procs.2020.09.141
10. Wu, I.-L., Hsieh, P.-J., Wu, S.-M.: Developing effective e-learning environments through e-learning use mediating technology affordance and constructivist learning aspects for performance impacts: moderator of learner involvement. Internet Higher Educ. **55**, 100871 (2022). https://doi.org/10.1016/j.iheduc.2022.100871
11. Ang, A.S., Sukhov, S.V., Dogariu, A., Shalin, A.S.: Scattering forces within a left-handed photonic crystal. Sci. Rep. **7**, 41014 (2017). https://doi.org/10.1038/srep41014
12. Baryshnikova, K.V., Kadochkin, A.S., Shalin, A.S.: Nanostructural antireflecting coatings: classification analysis (a review). Opt. Spectrosc. **119**(3), 343–355 (2015). https://doi.org/10.1134/S0030400X15090040
13. Hashim, M.A.M., Tlemsani, I., Mason-Jones, R., Matthews, R., Ndrecaj, V.: Higher education via the lens of industry 5.0: strategy and perspective. Soc. Sci. Hum. Open **9**, 100828 (2024). https://doi.org/10.1016/j.ssaho.2024.100828

14. Kovrov, A., Novitsky, A., Karabchevsky, A., Shalin, A.S.: A photonic nanojet as a tunable and polarization sensitive optical tweezer. Annalen Phys. **530**(9), 51800129 (2018). https://doi.org/10.1002/andp.201800129
15. Kuznetsov, A.V., Valero, A.C., Tarkhov, M., Bobrovs, V., Redka, D., Shalin, A.S.: Transparent hybrid Anapole metasurfaces with negligible electromagnetic coupling for phase engineering. NanoPhotonics **10**(17), 4385–4398 (2021). https://doi.org/10.1515/nanoph-2021-0377
16. Ngepah, N., Saba, C.S., Kajewole, D.O.: The impact of industry 4.0 on South Africa's manufacturing sector. J. Open Innov. Technol. Mark. Complex. **10**(1), 100226 (2024). https://doi.org/10.1016/j.joitmc.2024.100226
17. Novitsky, D.V., Shalin, A.S., Novitsky, A.: Nonlocal homogenization of PT-symmetric multilayered structures. Phys. Rev. A **99**, 043812 (2019). https://doi.org/10.1103/PhysRevA.99.043812
18. Boychuk, I.P., et al.: A methodological approach to the simulation of a ship's electric power system. Energies **16**, 8101 (2023). https://doi.org/10.3390/en16248101
19. Kozlova, A., Kukartsev, V., Melnikov, V., Kovalev, G., Stashkevich, A.: Finding dependencies in the corporate environment using data mining. E3S Web Conf. **431**, 05032 (2023). https://doi.org/10.1051/e3sconf/202343105032
20. Kukartsev, V.V., Kravtsov, K.I., Suprun, P.S., Gek, D.K., Pinchuk, I.A.: Advancements in network-based management systems for enhanced business services. E3S Web Conf. **460**, 07003 (2023). https://doi.org/10.1051/e3sconf/202346007003
21. Liu, Y., Qiao, H., Wang, J., Jiang, Y.: Influencing mechanism of the intellectual capability of big data analytics on the operational performance of enterprises. Heliyon **10**(3), e25032 (2024). https://doi.org/10.1016/j.heliyon.2024.e25032
22. Orlov, V., Tynchenko, V., Nizameeva, A., Shalaeva, D., Ageev, D.: Development of a multifunctional cross-platform system for automation of energy data and resource management. E3S Web Conf. **460**, 07002 (2023). https://doi.org/10.1051/e3sconf/202346007002
23. Derindag, O.F., Canakci, M., Tsarev, R.: Information and communication technologies in e-commerce and e-governance. J. Phys. Conf. Ser. **1399**(3), 033110 (2019). https://doi.org/10.1088/1742-6596/1399/3/033110
24. Kulju, E., Jarva, E., Oikarinen, A., Hammarén, M., Kanste, O., Mikkonen, K.: Educational interventions and their effects on healthcare professionals' digital competence development: a systematic review. Int. J. Med. Inform. **185**, 105396 (2024). https://doi.org/10.1016/j.ijmedinf.2024.105396
25. Pinto, R., Žilka, M., Zanoli, T., Kolesnikov, M.V., Gonçalves, G.: Enabling professionals for industry 5.0: the self-made programme. Procedia Comput. Sci. **232**, 2911–2920 (2024). https://doi.org/10.1016/j.procs.2024.02.107
26. Polovchenko, K.A.: Influence of the constitutional court on the transformation of vital national interests of Bosnia and Herzegovina. Eur. Politics Soc. **24**(3), 410–420 (2023). https://doi.org/10.1080/23745118.2022.2044600
27. Simon, R.: Appraising social work curriculum in India: learners' perspective. Horizon **29**(3), 117–141 (2021). https://doi.org/10.1108/OTH-01-2021-0029
28. Fernando, W.: Moodle quizzes and their usability for formative assessment of academic writing. Assess. Writ. **46**, 100485 (2020). https://doi.org/10.1016/j.asw.2020.100485
29. Tsarev, R., Azizam, S.H., Sablinskii, A., Potekhina, E., Gogoleva, I., Nikolaeva, I., Ikonnikov, O.: Gamification of the graph theory course finding the shortest path by a greedy algorithm. Lect. Notes Netw. Syst. **723**, 209–216 (2023). https://doi.org/10.1007/978-3-031-35317-8_18
30. García, D.S., Otcheskiy, I., Alkhafaji, M.A., Keshta, I., Gabriyelyan, S., Gevorgyan, A.: The use of digital educational resources in the educational process. In: Silhavy, R., Silhavy, P. (eds.) Networks and Systems in Cybernetics. CSOC 2023. Lecture Notes in Networks and Systems, vol. 723, pp. 478–485. Springer, Cham (2023). https://doi.org/10.1007/978-3-031-35317-8_43

31. Chen, C.-J., Tsai, H.-J., Lee, M.-Y., Chen, Y.-C., Huang, S.-M.: Effects of a Moodle-based E-learning environment on E-collaborative learning, perceived satisfaction, and study achievement among nursing students: a cross-sectional study. Nurse Educ. Today **130**, 105921 (2023). https://doi.org/10.1016/j.nedt.2023.105921
32. Hachicha, W., Ghorbel, L., Champagnat, R., Zayani, C.A., Amous, I.: Using process mining for learning resource recommendation: a moodle case study. Procedia Comput. Sci. **192**, 853–862 (2021). https://doi.org/10.1016/j.procs.2021.08.088
33. Dobashi, K., Ho, C.P., Fulford, C.P., Lin, M.-F.G., Higa, C.: Learning pattern classification using moodle logs and the visualization of browsing processes by time-series cross-section. Comput. Educ. Artif. Intell. **3**, 100105 (2022). https://doi.org/10.1016/j.caeai.2022.100105
34. Kaur, P., Kumar, H., Kaushal, S.: Affective state and learning environment based analysis of students' performance in online assessment. Int. J. Cogn. Comput. Eng. **2**, 12–20 (2021). https://doi.org/10.1016/j.ijcce.2020.12.003
35. Romero, E., García, L., Ceamanos, J.: Moodle and Socrative quizzes as formative aids on theory teaching in a chemical engineering subject. Educ. Chem. Eng. **36**, 54–64 (2021). https://doi.org/10.1016/j.ece.2021.03.001
36. Karapetyan, M.: Teaching languages in the digital age: incorporating machine translation. Transl. Stud. Theory Pract. **3**(6), 58–69 (2023). https://doi.org/10.46991/TSTP/2023.3.2.058
37. Ahmed, M.M.H., Indurkhya, B.: Investigating cognitive holding power and equity in the flipped classroom. Heliyon **6**(8), e04672 (2020). https://doi.org/10.1016/j.heliyon.2020.e04672
38. Rutskaya, K.A., et al.: Mastering software engineering with the help of the level model of competencies acquisition. J. Phys. Conf. Ser. **1515**, 022087 (2020). https://doi.org/10.1088/1742-6596/1515/2/022087
39. Akhmetjanov, M., Ruziev, R.: Fundamentals of modeling fire safety education. Inform. Econ. Manag. **1**(2), 0301–0308 (2022). https://doi.org/10.47813/2782-5280-2022-1-2-0301-0308
40. Alatrash, R., Priyadarshini, R., Ezaldeen, H., Alhinnawi, A.: Augmented language model with deep learning adaptation on sentiment analysis for E-learning recommendation. Cogn. Syst. Res. **75**, 53–69 (2022). https://doi.org/10.1016/j.cogsys.2022.07.002
41. Churaev, E., Savchenko, A.V.: A standalone software for real-time facial analysis in online conferences and e-lessons. Softw. Impacts **16**, 100507 (2023). https://doi.org/10.1016/j.simpa.2023.100507
42. Gomede, E., De Barros, R.M., De Souza Mendes, L.: Deep auto encoders to adaptive E-learning recommender system. Comput. Educ. Artif. Intell. **2**, 100009 (2021). https://doi.org/10.1016/j.caeai.2021.100009
43. Gruzenkin, D.V., et al.: Neural networks to solve modern artificial intelligence tasks. J. Phys. Conf. Ser. **1399**(3), 033058 (2019). https://doi.org/10.1088/1742-6596/1399/3/033058
44. Jardim, R., Delgado, C., Schneider, D.: Data science supporting a question classifier model. Procedia Comput. Sci. **199**, 1237–1243 (2022). https://doi.org/10.1016/j.procs.2022.01.157
45. Khakimzyanov, R., Ali, S., Kalmuratov, B., Phuong, N.H., Karnaukhov, A., Tsarev, R.: Applying a recurrent neural network to implement a self-organizing electronic educational course. Lect. Notes Netw. Syst. **733**, 30–40 (2023)
46. Lunev, D., Poletykin, S., Kudryavtsev, D.O.: Brain-computer interfaces: technology overview and modern solutions. Mod. Innov. Syst. Technol. **2**(3), 0117–0126 (2022). https://doi.org/10.47813/2782-2818-2022-2-3-0117-0126
47. Zakharenko, D.V.: Using deep learning neural networks to classify toxic comments on social media. Inform. Econ. Manag. **2**(4), 0119–0133 (2023). https://doi.org/10.47813/2782-5280-2023-2-4-0119-0133
48. Kohonen, T.: An introduction to neural computing. Neural Netw. **1**(1), 3–16 (1988). https://doi.org/10.1016/0893-6080(88)90020-2

49. Krishnan, K.J., Mitra, K.: A modified Kohonen map algorithm for clustering time series data. Expert Syst. Appl. **201**, 117249 (2022). https://doi.org/10.1016/j.eswa.2022.117249
50. Pan, Y., Zhang, L., Li, Z.: Mining event logs for knowledge discovery based on adaptive efficient fuzzy Kohonen clustering network. Knowl. Based Syst. **209**, 106482 (2020). https://doi.org/10.1016/j.knosys.2020.106482
51. Briukhanova, E.R.: Utilizing zeroning neural networks approach for resource management in distributed dynamic computing network. Mod. Innov. Syst. Technol. **3**(4), 0301–0310 (2023). https://doi.org/10.47813/2782-2818-2023-3-4-0301-0310
52. Motegi, R., Seki, Y.: SMLSOM: the shrinking maximum likelihood self-organizing map. Comput. Stat. Data Anal. **182**, 107714 (2023). https://doi.org/10.1016/j.csda.2023.107714
53. Nowak-Brzezińska, A., Horyń, C.: Self-Organizing Map algorithm as a tool for outlier detection. Procedia Comput. Sci. **207**, 2162–2171 (2022). https://doi.org/10.1016/j.procs.2022.09.276
54. Vlaović, ŽD., Stepanov, B.L., Anđelković, A.S., Rajs, V.M., Čepić, Z.M., Tomić, M.A.: Mapping energy sustainability using the Kohonen self-organizing maps - Case study. J. Clean. Prod. **412**, 137351 (2023). https://doi.org/10.1016/j.jclepro.2023.137351

Flood Modeling in the Urban Area Model

Abdikerim Kurbanaliev(✉), Burulgul Oichueva, and Myrzagul Rakhmanberdi kyzy

Osh State University, Osh, Kyrgyzstan
abdikerimkurbanaliev@oshsu.kg

Abstract. Modeling floods on tall buildings and major structures is important for flood risk assessment in urban areas. However, this is also a challenging task due to the difficulty of obtaining detailed topography and monitoring data to build and validate the model. Catastrophic floods that occur in our country and around the world cause enormous damage to the national economy and often lead to human casualties. This research presents the results of numerical modeling of a challenging hydrodynamic process based on time dependent 3D RANS equations for describing the dynamics of a air-water double phase flow with a free surface. Verification of the used model was tested on the example of the classical problem of a symmetric dambreak. Particular consideration is paid to the accuracy of calculating the water level at different reservoir gauges. #CSOC1120.

Keywords: Modelling · Floods · Risk Assessment · Topography · Monitoring · Water Level · Reservoir Gauges

1 Introduction

Flash floods in cities, which often have a dramatic impact urban area and public life, are an emerging and challenging problem. Because of living community complexity, the diversity of three-dimensional simulation modeling makes it difficult to study and implement urban storm inundation modeling [1–4].

The problem of modeling the symmetrical destruction of a dam in a rectangular channel is considered (Fig. 1). Here, the 1 m × 2 m × 0.8 m water reservoir is filled with a 0.6 m high water column. This task corresponds to the conditions of experimental work [5].

In the experiment, five sensors were used to measure the height of the water flow, the coordinates of which are given in Table 1.

Table 1. Coordinates of measuring sensors.

Reference points	P1	P2	P3	P4	P5	P6	P7	P8	P9
x, m	0.18	0.48	1.0	1.0	1.722	2.0	2.6	2.6	3.4
y, m	1.0	0.4	1.16	1.0	1.0	1.0	0.6	1.4	1.0

R. Silhavy and P. Silhavy (Eds.): CSOC 2024, LNNS 1120, pp. 373–389, 2024.
https://doi.org/10.1007/978-3-031-70518-2_34

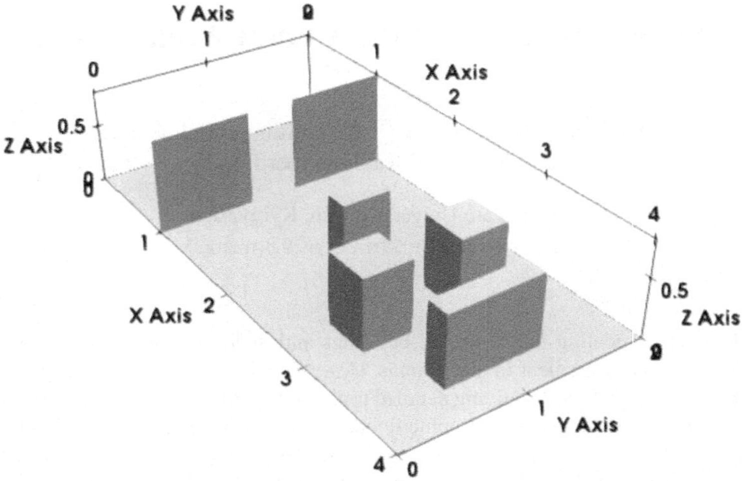

Fig. 1. Urban area model.

2 Mathematical Model

The 3D numerical model of flooding process reads [6, 7]:

$$\frac{\partial}{\partial x_i}(\rho u_i) = 0 \tag{1}$$

$$\frac{\partial}{\partial t}(\rho u_i) + \frac{\partial}{\partial x_j}\left(\rho u_i u_j + \overline{\rho u_i' u_j'}\right) = -\frac{\partial p}{\partial x_i} + \frac{\partial \tau_{ij}}{\partial x_j} \tag{2}$$

where:

u_i – average velocity components,

ρ – density,

p – average pressure,

$\tau ij = \mu(\frac{\partial u_i}{\partial x_j} + \frac{\partial u_j}{\partial x_i})$ – average viscous stress tensor,

μ – dynamic viscosity.

Averaging is performed over time, and the prime means the fluctuation part. In the presence of external forces, for example, gravity, these equations must be supplemented with the appropriate terms.

Usually, the closure of systems of Eqs. (1–2) is based k-ε - turbulence model. For general flows, Reynolds stresses can be written as [6, 7]:

$$\rho\overline{u_i u_j} = \mu_t\left(\frac{\partial u_i}{\partial x_j} + \frac{\partial u_j}{\partial x_i}\right) - \frac{2}{3}\rho\delta_{ij}k \tag{3}$$

One can calculate quantities k and ε by solving next equations [6, 7]:

$$\frac{\partial(\rho k)}{\partial t} + \frac{\partial(\rho u_j k)}{\partial x_j} = \frac{\partial}{\partial x_j}\left[\left(\mu + \frac{\mu_t}{\sigma_k}\right)\frac{\partial k}{\partial x_j}\right] + P_k - \rho k \tag{4}$$

$$\frac{\partial(\rho\varepsilon)}{\partial t}+\frac{\partial(\rho u_j\varepsilon)}{\partial x_j}=C_{\varepsilon 1}P_k\frac{\varepsilon}{k}-\rho C_{\varepsilon 2}\frac{\varepsilon^2}{k}+\frac{\partial}{\partial x_j}\left[\left(\frac{\mu_t}{\sigma_\varepsilon}\frac{\partial\varepsilon}{\partial x_j}\right)\right] \tag{5}$$

where:

$P_k = \mu_t\left(\frac{\partial u_i}{\partial x_j}+\frac{\partial u_j}{\partial x_i}\right)\frac{\partial u_i}{\partial x_j}$ – the rate production of turbulence kinetic energy by the mean flow,

$\mu_t = \rho C_\mu k^2/\varepsilon$ – eddy viscosity coefficient:

$$\sigma_\varepsilon = 1.3,\ C_\mu = 0.09,\ C_{\varepsilon 1} = 1.44,\ C_{\varepsilon 2} = 1.92,\ \sigma_k = 1.0.$$

3 Numerical Model

The simplicity of computational domain shape allow us to divide it into hexahedral finite volumes [8]. For OpenFOAM [9] blockMesh utility is used. The size of each computational cell is equal to of 0.02 × 0.03 × 0.02 m (Fig. 2).

Fig. 2. Computational cells.

The boundary conditions used in this work are given in Table 2

Tables 3 and 4 shows schemes of discretization and methods for solving of linear algebraic equations system.

Table 2. Used boundary conditions.

Border	Variables					
	alpha.water	ε	k	nuTilda	p_rgh	U
left	zeroGradient	epsilonWallFunction	kqRWallFunction	zeroGradient	fixedFluxPressure	fixedValue
out	zeroGradient	zeroGradient	zeroGradient	zeroGradient	fixedValue	zeroGradient
front	zeroGradient	epsilonWallFunction	kqRWallFunction	zeroGradient	fixedFluxPressure	fixedValue
back	zeroGradient	epsilonWallFunction	kqRWallFunction	zeroGradient	fixedFluxPressure	fixedValue
lower	zeroGradient	epsilonWallFunction	kqRWallFunction	zeroGradient	fixedFluxPressure	fixedValue
atmosphere	inletOutlet	inletOutlet	inletOutlet	inletOutlet	totalPressure	PressureInletOutletVelocity

Table 3. Used schemes for discretization.

Terms	Schemes
	Euler
ddtSchemes	
	Gauss linear
gradSchemes	
	Gauss linear upwind grad(U)
Velocity	Gauss vanLeer
alpha	Gauss linear
k, ε	Gauss upwind
divSchemes	Gauss upwind
Laplacian	Gauss linear corrected
interpolationSchemes	linear
snGradSchemes	corrected

Table 4. Solver methods.

Quantity	Method	Tolerance
alpha water	smoothSolver	10^{-8}
	symGaussSeidel	
pcorr	PCG preconditioner DIC	10^{-5}
p_rgh	PCG preconditioner DIC	10^{-7}
U, k, ε	smoothSolver symGaussSeidel	10^{-6}

4 Results and Discussion

The water volume fraction evolution at various values of the water column in the reservoir is shown in Fig. 3.

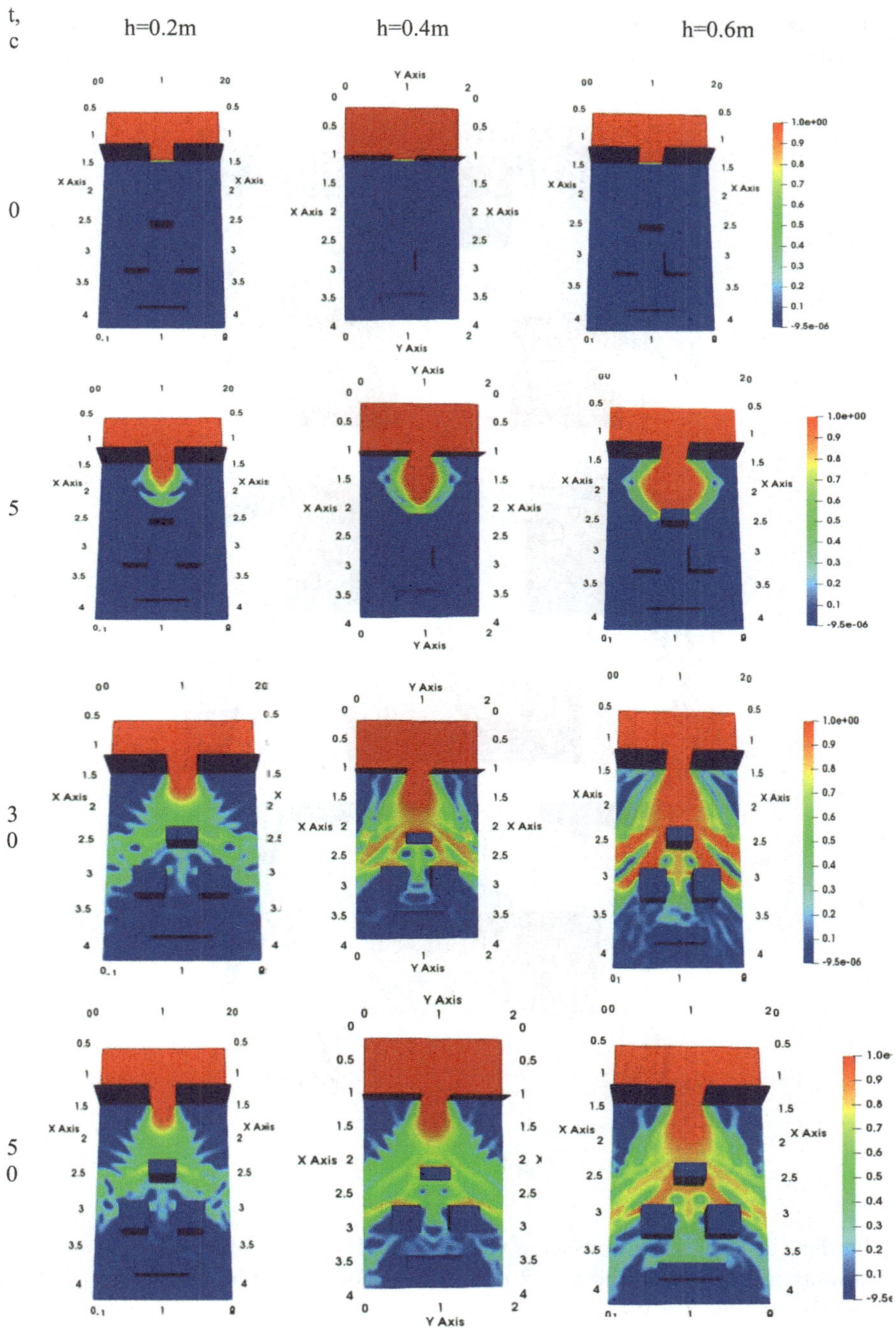

Fig. 3. Distribution of the volume fraction of water.

The following figures show the distribution field of the vector of velocity, where the recirculation zones after the first building are clearly visible (see Fig. 4).

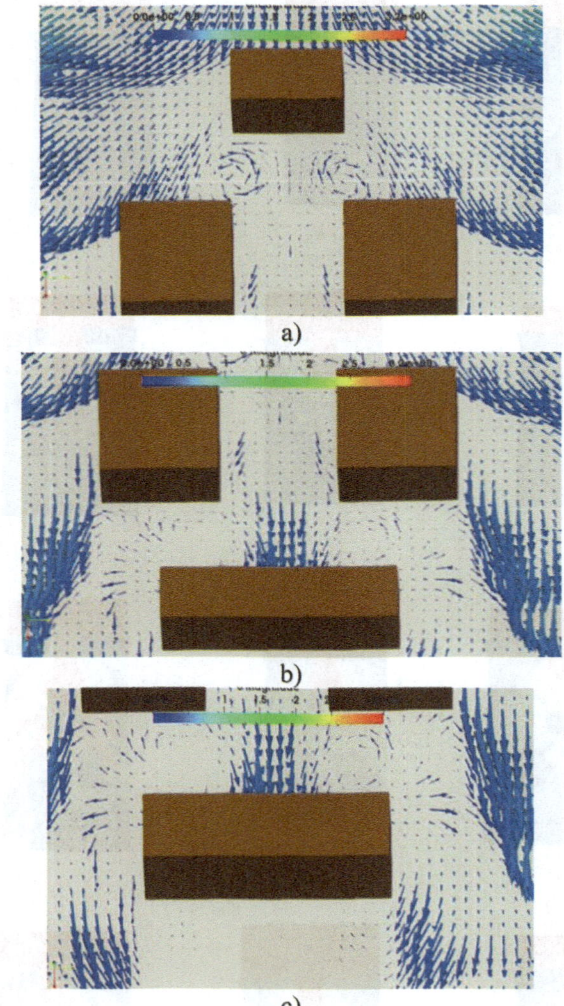

Fig. 4. Velocity field.

Further, in Fig. 5, the transverse velocity profiles U_x, volume fraction of water alpha.water along central line x = 2 m, z = 1.05 for 3 different initial water column height (H = 0.2 m, H = 0.4 m, H = 0.6 m) are shown.

Figure 6 shows a graph of the numerical results of the water flow height (solid) and the appropriate experimental data - markers [5] for the initial water column height H = 0.6 m.

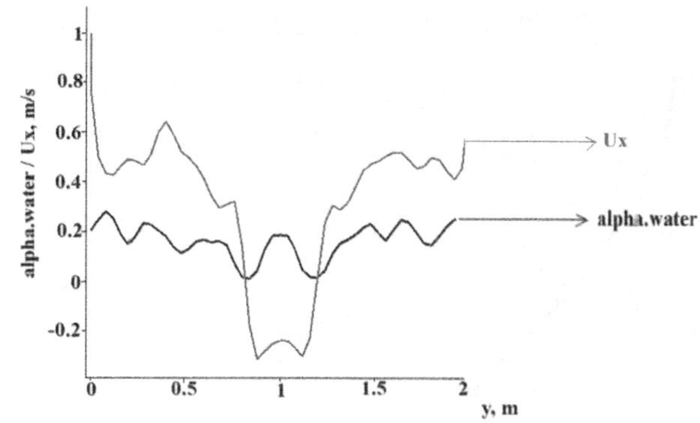

(H=0.2m)

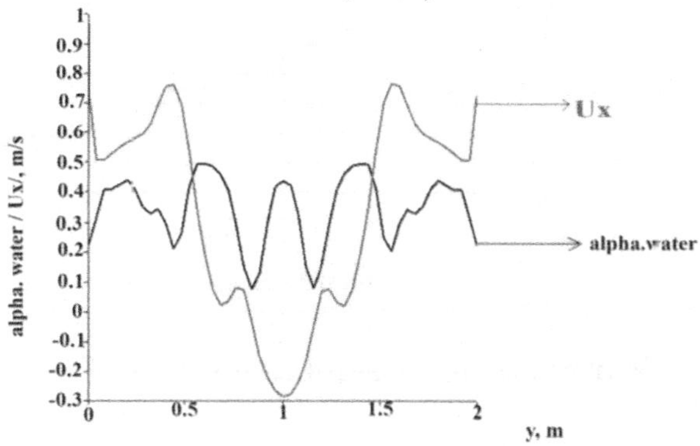

(H=0.4m)

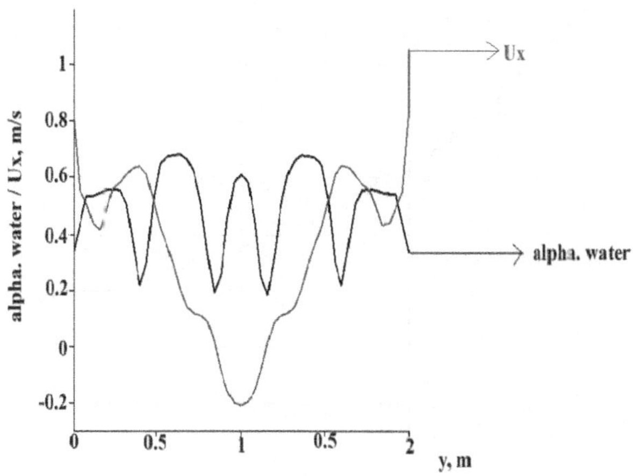

(H=0.6m)

Fig. 5. Cross profiles U_x alpha.water.

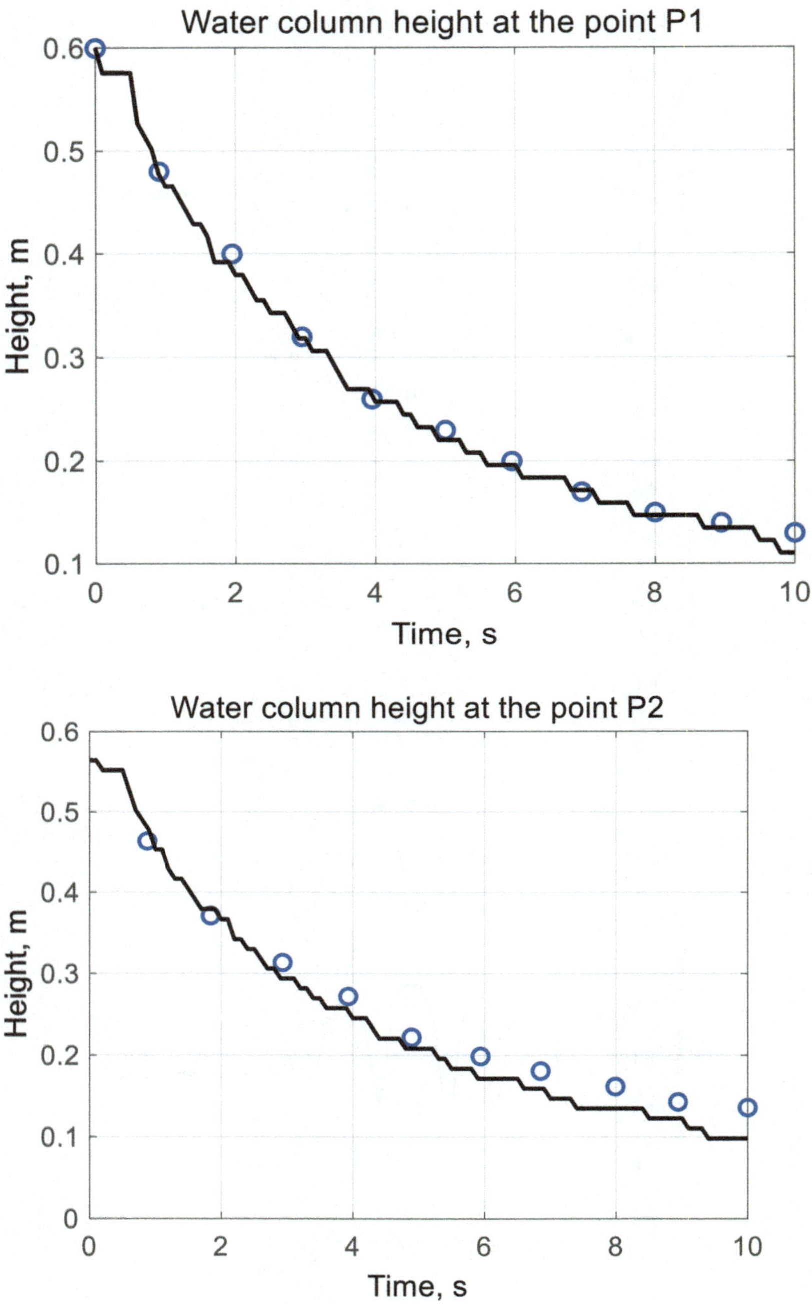

Fig. 6. Evolution of the height of the water flow at the reference points P1, P2, P3, P4 and P5.

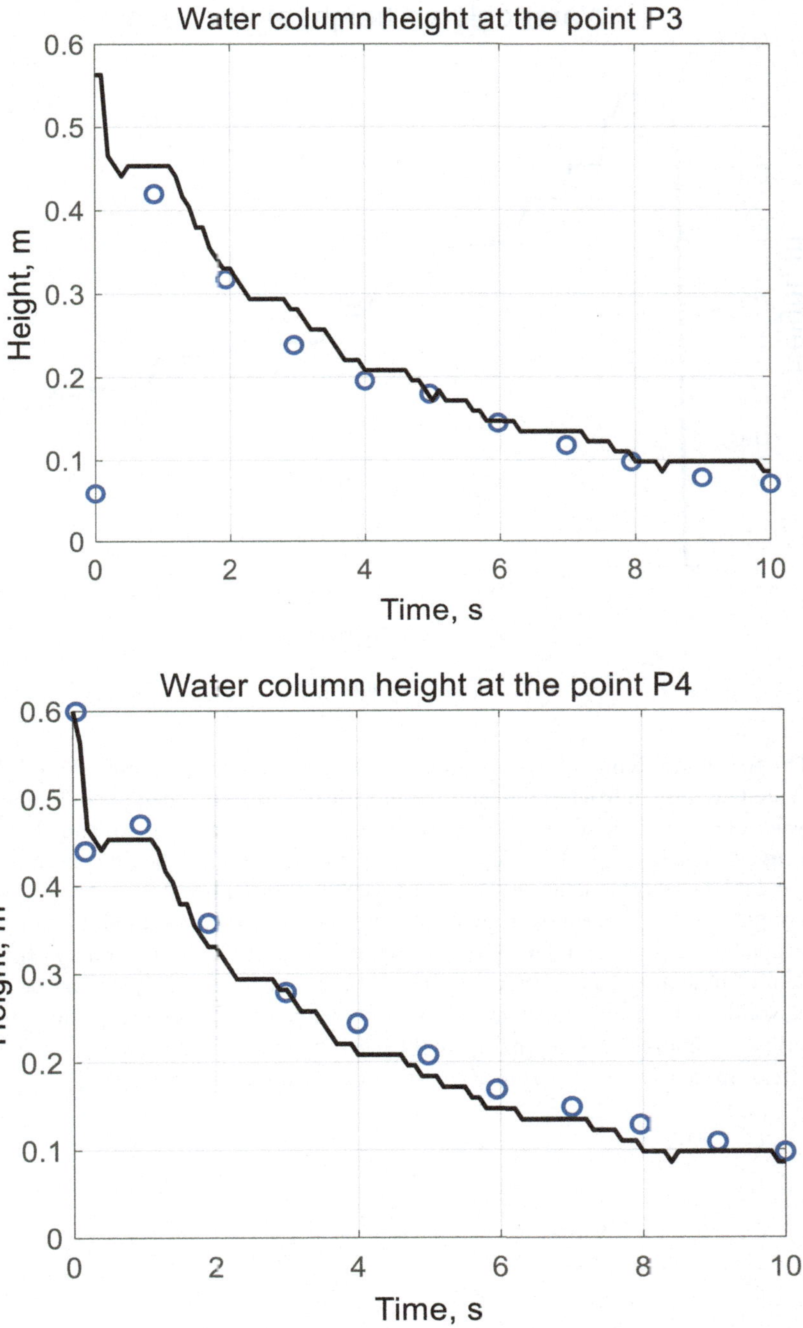

Fig. 6. (*continued*)

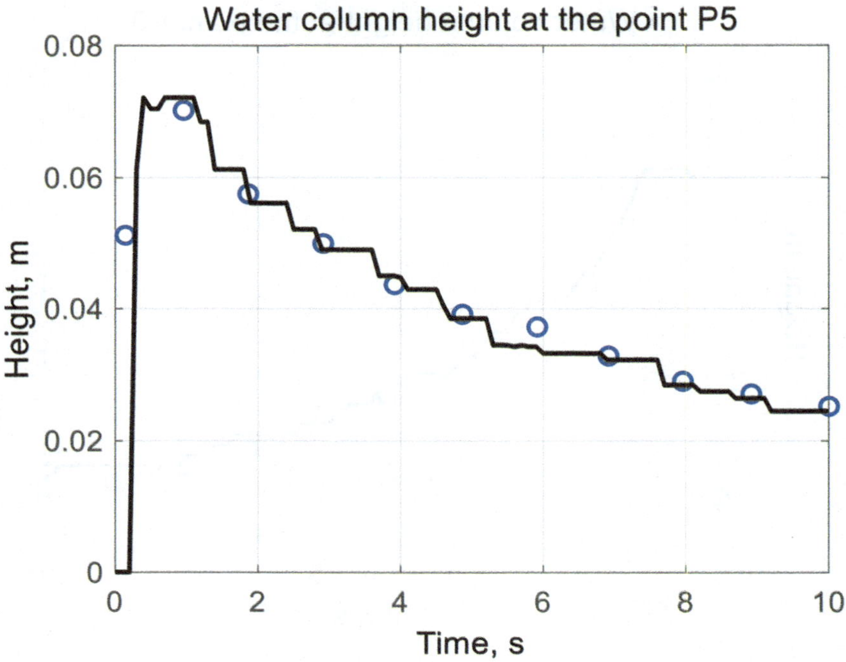

Fig. 6. (*continued*)

The following figures show the time evolution of pressure at points P6, P7, P8 and P9 located directly in front of the building models for the case when the initial height of the water column in the reservoir was equal to (H = 0.2 m, H = 0.4 m and H = 0.6 m) respectively. The leading edge of the breakthrough wave reaches the first building within 0.5 s and the pressure reaches a value of about 800 N/m^2 (Figs. 7, 8 and 9).

Comparison of experimental data [5] with the numerical simulation results show an acceptable good agreement. A slight fluctuation of the numerical data at the initial moment of time at point 0 is probably due to the influence of the boundary conditions on the solid walls of the reservoir and the finite value of the baffle retraction speed. In numerical simulation, it is assumed that the partition is removed instantly, i.e. the speed is endless. In fact, in the experiment, this speed has some finite value.

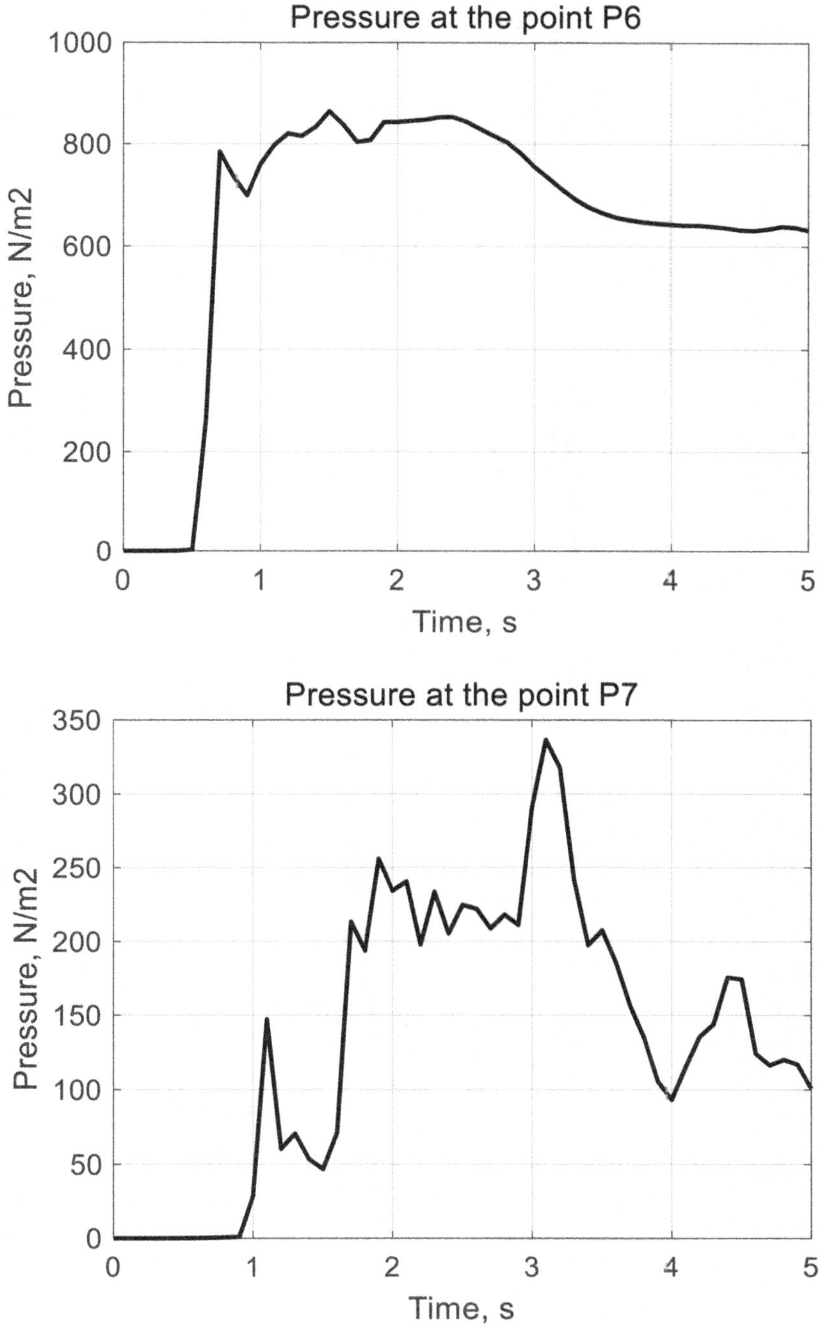

Fig. 7. Pressure change at points P6, P7, P8 and P9. H = 0.2 m.

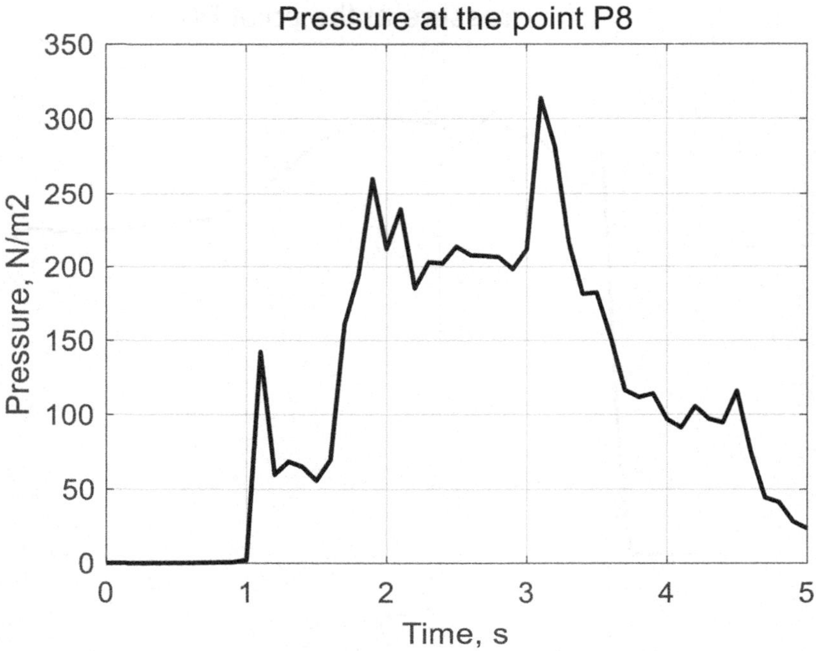

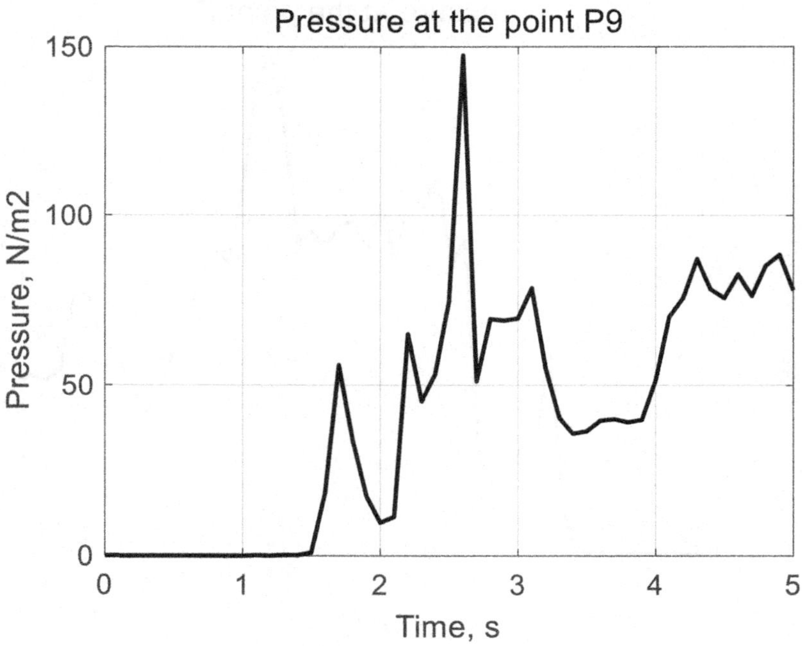

Fig. 7. (*continued*)

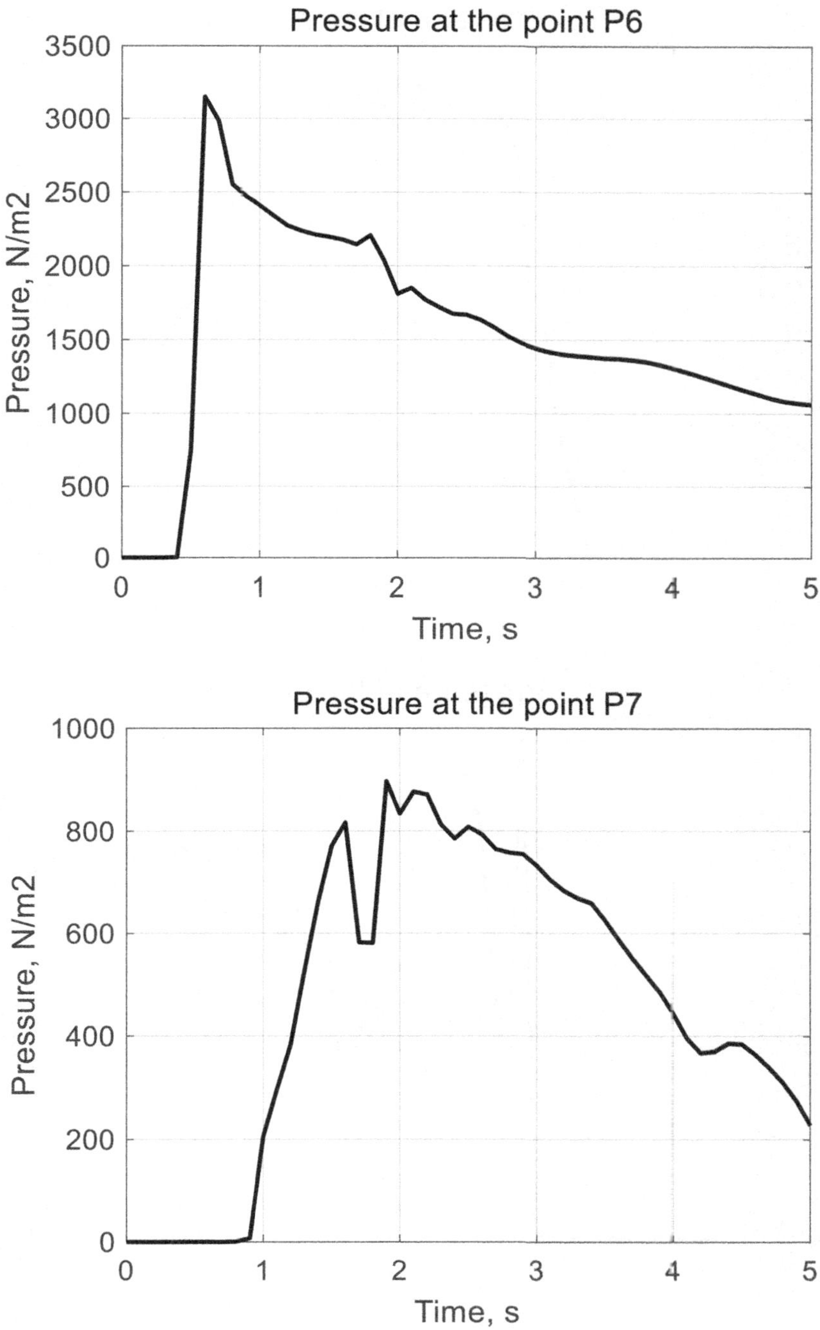

Fig. 8. Pressure change at points P6, P7, P8 and P9. H = 0.4 m.

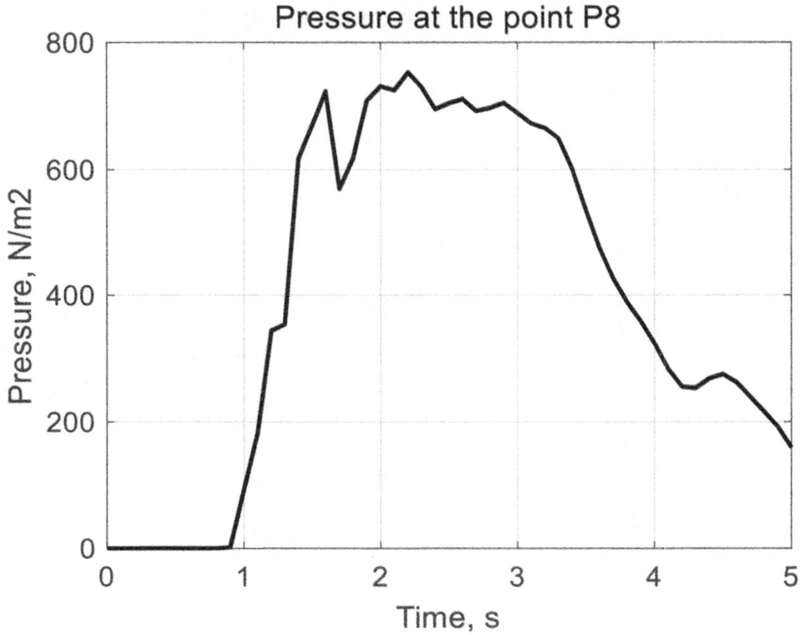

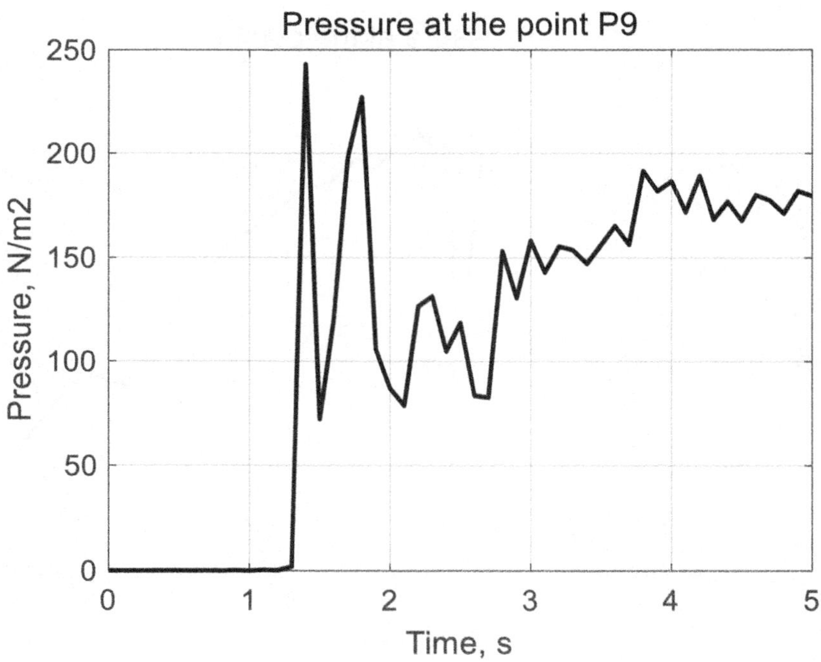

Fig. 8. (*continued*)

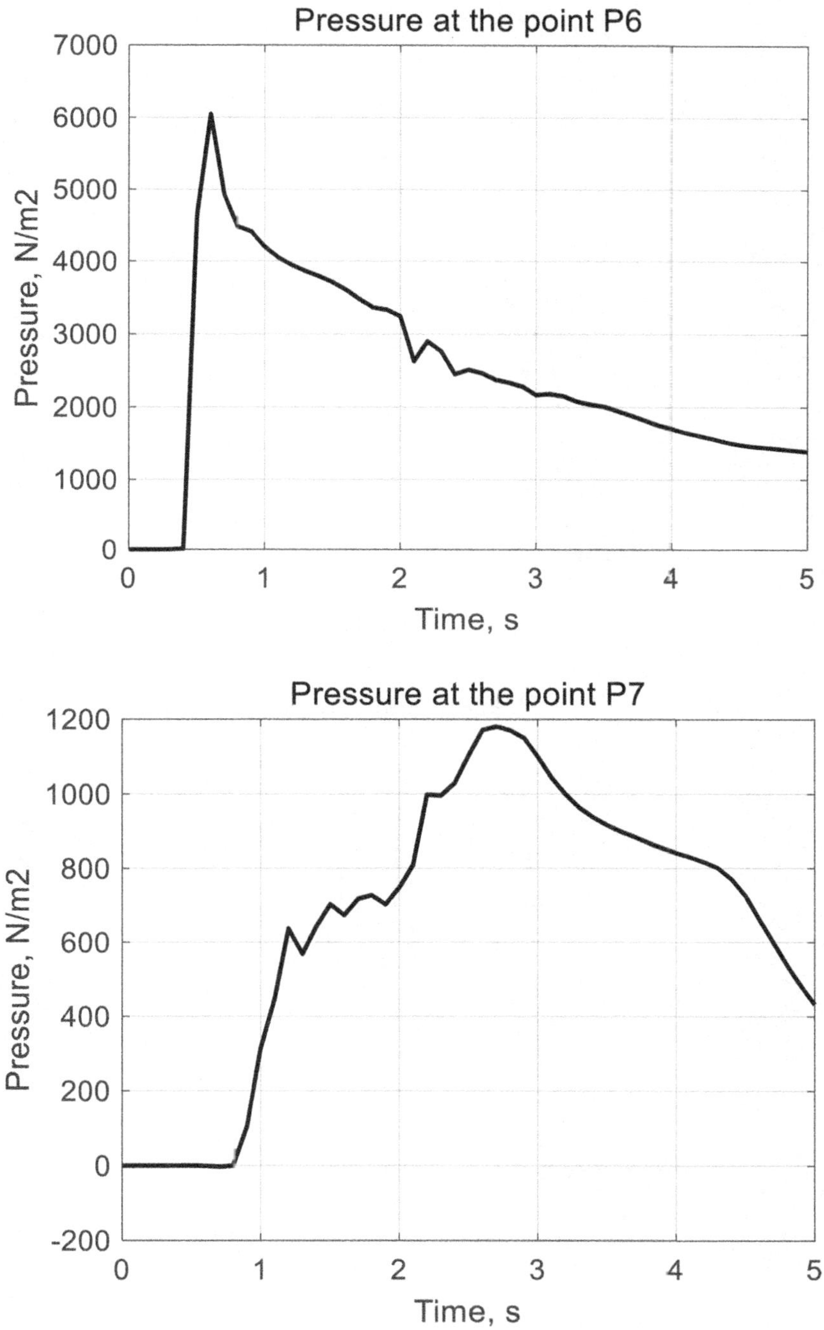

Fig. 9. Pressure change at points P6, P7, P8 and P9. H = 0.6 m.

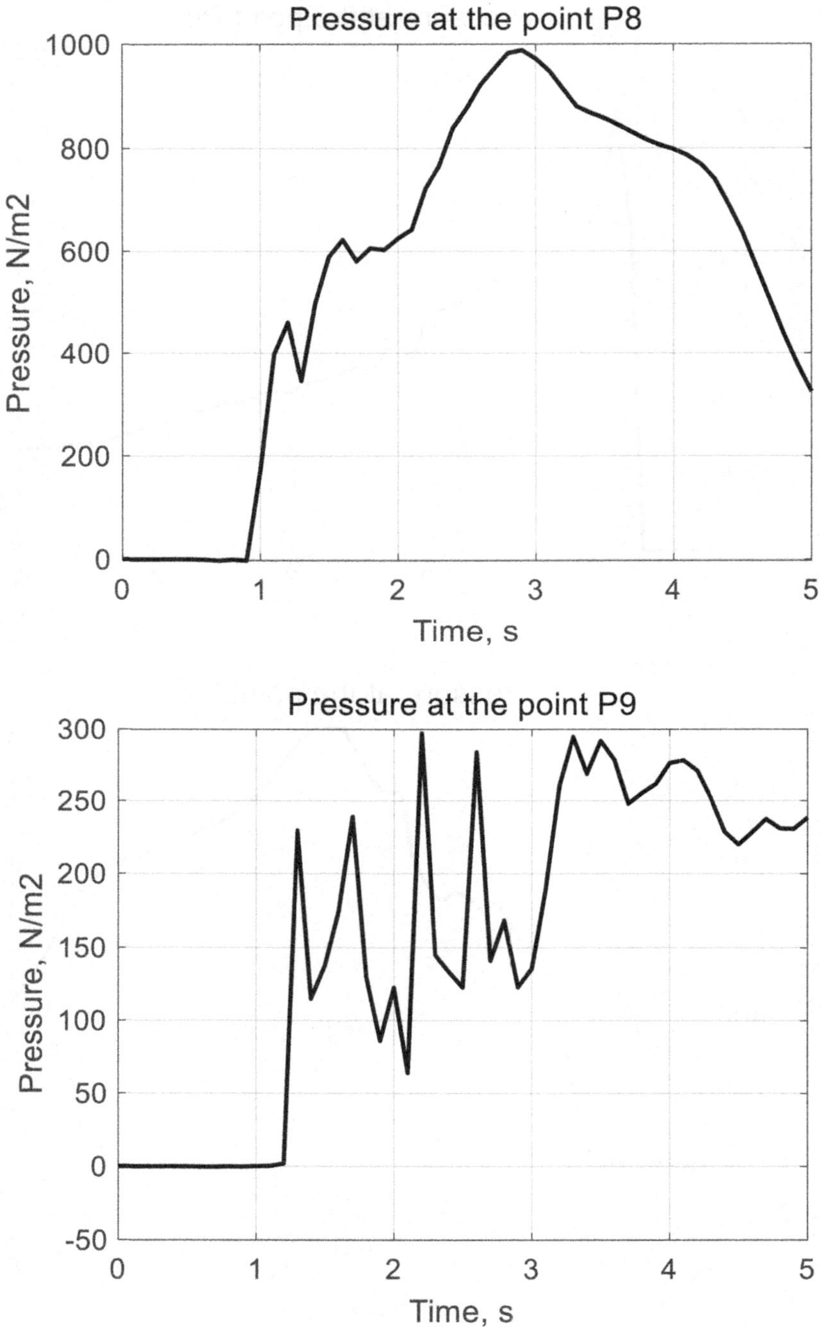

Fig. 9. (*continued*)

5 Conclusion

Modeling floods on tall buildings and major structures is important for flood risk assessment in urban areas. However, this is also a challenging task due to the difficulty of obtaining detailed topography and monitoring data to build and validate the model. Catastrophic floods that occur in our country and around the world cause enormous damage to the national economy and often lead to human casualties. This research presents the results of numerical modeling of a challenging hydrodynamic process based on time dependent 3D RANS equations for describing the dynamics of an air-water double phase flow with a free surface. Verification of the used model tested on the example of the classical problem of a symmetric dambreak. Comparison of experimental data [5] with the numerical simulation results show an acceptable good agreement. Particular consideration paid to the accuracy of calculating the water level at different reservoir gauges.

References

1. Abdolmaleki, K., Thiagarajan, K.P., Morris-Thomas, M.T.: Simulation of the dam break problem and impact flows using a Navier-Stokes solver. In: Proceedings of the 15th Australasian Fluid Mechanics Conference. The University of Sydney, Sydney, Australia (2004)
2. Kurbanaliev, A.Y., Oichueva, B.R., Alieva, C.M., Mamaev, S.S.: Mathematical modeling and visualization of large-scale flooding in the flood plain of Naryn River using OpenFOAM. In: Lecture Notes in Networks and Systems, vol. 503, pp. 357–65 (2022)
3. Kurbanaliev, A.Y., Oichueva, B.R., Maksutov, A.R., Obodoeva, G.S.: Using openFoam multiphase solver interFoam for large-scale modeling. In: Proceedings of the World Congress on Engineering and Computer Science, WCECS, pp. 366–370. San Francisco, USA (2019)
4. Kurbanaliev, A.Y., Oichueva, B.R., Kalmurzaeva, A.T.: Critical comparison of different package versions OPENFOAM on spillway modeling problem. Comput. Technol. **1**, 44–57 (2021). https://doi.org/10.25743/ict.2021.26.2.004
5. Fraccarollo, L., Capart, H.: Riemann wave description of erosional dam break flows. J. Fluid Mech. **461**, 183–228 (2002)
6. Ferziger, J.H., Peric, M.: Computational Methods for Fluid Dynamics. Springer Verlag, Berlin (2002). https://doi.org/10.1007/978-3-642-56026-2
7. Versteeg, H.K., Malalasekera, W.: An Introduction to Computational Fluid Dynamics. Pearson Education Limited, Edinburg (2007)
8. Patankar, S.V.: Numerical Heat Transfer and Fluid Flow. Hemisphere, New York (1980)
9. OpenFOAM v7 User Guide. https://doc.cfd.direct/openfoam/user-guide-v7/index. Accessed 02 Feb 2024
10. Moukalled, F., Mangani, L., Darwish, M.: The Finite Volume Method in Computational Fluid Dynamics. Springer International Publishing, Switzerland (1980). https://doi.org/10.1007/978-3-319-16874-5

Optimizing the Feature Selection Problem with Pendulum Search Algorithm: Binarization Strategies and Their Impacts

Felipe Cisternas-Caneo[1(✉)], Broderick Crawford[1], Ricardo Soto[1], José Barrera-García[1], Marcelo Becerra-Rozas[1], Giovanni Giachetti[2], Jose Manuel Gomez-Pulido[3], and Alberto Garces-Jimenez[3]

[1] Pontificia Universidad Católica de Valparaíso, Valparaíso, Chile
{felipe.cisternas.c,marcelo.becerra.r}@mail.pucv.cl,
{broderick.crawford,ricardo.soto,jose.barrera}@pucv.cl

[2] Universidad Andres Bello, Santiago, Chile
giovanni.giachetti@unab.cl

[3] Departamento Ciencias de la Computación, Universidad de Alcalá, Madrid, Spain
{jose.gomez,alberto.garces}@uah.es

Abstract. Technological advances and the digitization of information have allowed us to obtain a large amount of data from different processes such as medicine, commerce, and mining, among others. All this data has been used as input by different researchers in machine learning techniques to speed up the decision making process of professionals. Machine learning techniques are susceptible to data, so cleaning it to remove irrelevant and redundant information is necessary. This removal of information is known as the Feature Selection Problem. This work presents the Pendulum Search Algorithm (PSA) applied to solve the Feature Selection Problem. As the PSA is a metaheuristic designed for continuous optimization problems, a binarization process is performed using the Two-Step Technique. In particular, we have used four transfer functions and three binarization rules to observe their impact when binarizing PSA. Preliminary results indicate that the V-Shaped transfer function and the Standard binarization rule is the binarization scheme that obtains the best experimental results.

Keywords: Pendulum Search Algorithm · Feature Selection Problem · Binarization Schemes · Combinatorial Optimization · Two-Step Technique

1 Introduction

In recent years, the employment of various machine learning algorithms has garnered significant interest within the scientific realm. This surge in popularity stems primarily from two factors: firstly, the availability of substantial computational resources, and secondly, the abundance of extensive data sets. These

R. Silhavy and P. Silhavy (Eds.): CSOC 2024, LNNS 1120, pp. 390–402, 2024.
https://doi.org/10.1007/978-3-031-70518-2_35

algorithms have demonstrated their efficacy in several domains, including image analysis, natural language processing, and sentiment analysis.

Nevertheless, the copious amounts of data accessible for analysis often include redundant or irrelevant variables for classification, which can diminish the accuracy of outcomes and the efficiency of the predictive models. Owing to this vulnerability to the nature of input data, achieving dependable results is challenging without thorough data preprocessing.

Feature selection is a prevalent preprocessing technique, primarily focusing on eliminating irrelevant and/or redundant information. Its objective is to retain only the most representative data from the dataset, thereby enhancing the overall quality and relevance of the information used for analysis [23].

We find metaheuristics among the techniques used to solve the Feature Selection Problem. Metaheuristics are general-purpose algorithms that, with simple modifications, can solve different optimization problems [4,30].

As observed in the work [28], the vast majority of existing metaheuristics in the literature are designed to solve continuous optimization problems; therefore, to solve the feature selection problem, they must be adapted since this is a binary combinatorial optimization problem.

In the literature, there are different ways to binarize continuous metaheuristics [7], the most used being the two-step technique. This technique binarizes the metaheuristics by applying a transfer function and, subsequently, a binarization rule.

We have employed the guidelines presented in this work in [6]. We have implemented a binarized version of the Pendulum Search Algorithm (PSA), a recently developed metaheuristic designed to solve continuous optimization problems. Such binarization is performed so that PSA can solve problems such as the Feature Selection Problem. In particular, we have applied four different transfer functions and three different binarization rules to observe the impact they have on the results obtained. Also, since we need a classifier to obtain the performance metrics, this work used three different classifiers: K-Nearest Neighbors (KNN), Random Forest, and XgBoost. Preliminary results indicate that the V-Shaped transfer function and the standard binarization rule perform best.

The paper is organized as follows: Sect. 2 defines the Feature Selection Problem, Sect. 3 reviews how metaheuristics solve the Feature Selection Problem, Sect. 4 defines the Pendulum Search Algorithm, Sect. 5 defines our proposal Binary Pendulum Search Algorithm, the results obtained are shown in Sect. 6 to end with the conclusion in Sect. 7.

2 Feature Selection Problem

The Feature Selection Problem (FS) is a combinatorial optimization problem whose search space is defined as 2^n, and n represents the number of features in the datasets. Also, it is a multi-objective optimization problem, and the aim is to minimize the number of selected features and to maximize the quality of the prediction or classification results obtained.

In this way, the metaheuristic algorithm searches for different combinations and calculates its Z objective function based on the number of selected features and the precision of the results obtained by the classification or prediction algorithm that is trying to be optimized, according to the following equation:

$$min\,Z = \alpha \cdot Y_R(D) + \beta \cdot \frac{|R|}{|N|} \tag{1}$$

Where $Y_R(D)$ represents the classification or prediction error of the ML algorithm, $|R|$ is the number of selected features, and $|N|$ the total number of features, with $\alpha \in [0, 1]$ and $\beta = (1 - \alpha)$, parameters that regulate the importance of the quality of the results and the size of the subset, respectively.

As seen in Eq. 1, we need the error rate, and this is a classifier performance metric. This metric is constructed from the confusion matrix obtained after classifier validation. With this, the error rate is constructed as follows:

$$Error\,rate = \frac{FP + FN}{FP + FN + TP + TN} \tag{2}$$

3 Metaheuristics Solving Feature Selection Problem

In the literature, we find several applications of metaheuristics solving the Feature Selection Problem in different fields of study such as health [25,26,31,35], text classification [24,33], image classification [3,10,12] or intrusion detection [18,27,29].

Most of the metaheuristics applied in these problems are metaheuristics designed to solve continuous optimization problems, so it is necessary to make a modification so that they can solve binary combinatorial optimization, such as the Feature Selection Problem.

3.1 Two-Step Techinques

In the literature, there are different ways of binarizing continuous metaheuristics [7], but the most widely used is the Two-Step Technique [15]. As the name suggests, the binarization process is performed in two steps: (1) Application of a transfer function transfers the continuous value to a value within the range [0,1], and (2) Application of a binarization rule which determines the assignment of a 1 or a 0.

In the literature exists different transfer functions [7], in the Table 1 and Fig. 1 shows the transfer functions were used in this work.

On the other hand, the authors indicate five different binarization rules in [15], but we used only three binarization rules in this work. This binarization rule is shown in Table 2.

Table 1. Transfer Functions

Transfer Functions			
S-Shaped	V-Shaped	X-Shaped	Z-Shaped
$T(d_w^j) = \frac{1}{1+e^{\frac{-d_w^j}{3}}}$	$T(d_w^j) = \left\lvert \frac{2}{\pi} arctan\left(\frac{\pi}{2} d_w^j\right)\right\rvert$	$T(d_w^j) = \frac{1}{1+e^{\frac{d_w^j}{3}}}$	$T(d_w^j) = \sqrt{1-20^{d_w^j}}$

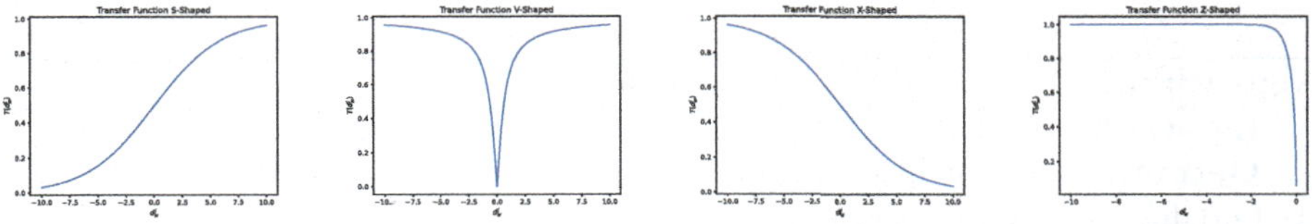

(a) S-Shaped (b) V-Shaped (c) X-Shaped (d) Z-Shaped

Fig. 1. Transfer Functions

4 Pendulum Search Algorithm

The Pendulum Search Algorithm (PSA) is a new population-based metaheuristic created in 2022 by Nor Azlina Ab. Aziz and Kamarulzaman Ab. Aziz [1]. This metaheuristic was designed to solve continuous optimization problems, and the harmonic motion of the simple pendulum inspires it. The equation of movement is very similar to the one proposed in Sine Cosine Algorithm. The authors incorporate an exponential function, which would improve the balance between exploration and exploitation [2].

The search agents are initialized randomly, and their position is updated using Eq. 3.

$$X_{i,j}^{t+1} = X_{i,j}^{t} + pend_{i,j}^{t} \cdot (Best_j - X_{i,j}^{t}) \quad (3)$$

where $X_{i,j}^{t}$ is the position of the i-th solution in the j-th dimension in t-th iteration, $Best_j$ it is the position of the best solution in the j-th dimension in the t-th iteration and $pend_{i,j}^{t}$ is a parameter which is calculated as follows:

Table 2. Binarization Rules

Name	Binarization Rule
Standard	$X_{new}^j = \begin{cases} 1\ if & rand \leq T\left(d_w^j\right) \\ 0 & otherwise \end{cases}$
Complement	$X_{new}^j = \begin{cases} Complement\left(X_w^j\right)\ if & rand \leq T\left(d_w^j\right) \\ 0 & otherwise \end{cases}$
Elitist	$X_{new}^j = \begin{cases} X_{Best}^j\ if & rand < T\left(d_w^j\right) \\ 0 & otherwise \end{cases}$

$$pend_{i,j}^{t} = 2 \cdot e^{(-t/tmax)} \cdot cos(2 \cdot \pi \cdot rand) \tag{4}$$

where t is the current iteration, $tmax$ is the maximum number of iterations and $rand$ is a uniform random number between [0,1]. The pseudo-code of PSA is shown in Algorithm 1.

Algorithm 1. Pendulum Search Algorithm

Input: The population $X = \{X_1, X_2, ..., X_i\}$
Output: The updated population $X' = \{X'_1, X'_2, ..., X'_i\}$ and $Best$

Initialize random population X
Evaluate the objective function of each individual in the population X
Identify the best individual in the population ($Best$)
for *iteration* (t) **do**
  for *solution* (i) **do**
    for *dimension* (j) **do**
      Update $pend_{i.j}^{t}$ by Eq. (4)
      Update the position of $X_{i,j}^{t}$ using Eq. (3)
    end for
  end for
  Evaluate the objective function of each individual in the population X
  Update $Best$
end for
Return the updated population X' where $Best$ is the best result

5 Binary Pendulum Search Algorithm

As mentioned in Sect. 4, PSA is a metaheuristic designed to solve continuous optimization problems, so it is necessary to convert the solutions to the binary domain to solve the Feature Selection. We applied the Two-Step technique to binarize the Pendulum Search Algorithm in the present work. In particular, we use a combination of the four transfer functions and the three binarization rules discussed in Sect. 3.1. The pseudo-code of B-PSA is shown in Algorithm 2.

6 Experimental Results

In this paper, we compare different transfer functions and binarization rules, all applied to the Binary Pendulum Search Algorithm. Thus, we have 12 different versions of the Pendulum Search Algorithm. The versions that include the letter "S" refer to the use of the S-Shaped transfer function, the versions that include the letter "V" refer to the use of the V-Shaped transfer function, the version that include the letter "X" refers to the use of the X-Shaped transfer function, the version that include the letter "Z" refers to the use of the Z-Shaped transfer function, the versions that include the word "STD" refers to the use of the standard binarization rule, the versions that include the word "COM" refers to the use of the complement binarization rule, the version that include the word "ELIT" refers to the use of the elitist binarization rule.

Algorithm 2. Binary-Pendulum Search Algorithm

Input: The population $X = \{X_1, X_2, ..., X_i\}$
Output: The updated population $X' = \{X'_1, X'_2, ..., X'_i\}$ and *Best*

Initialize binary random population X
Evaluate the objective function of each individual in the population X
Identify the best individual in the population (*Best*)
for *iteration* (t) **do**
  for *solution* (i) **do**
    for *dimension* (j) **do**
      Update $pend^t_{i,j}$ by Eq. (4)
      Update the position of $X^t_{i,j}$ using Eq. (3)
    end for
  end for
  Binarization of population X
  Evaluate the objective function of each individual in the population X
  Update *Best*
end for
return the updated population X' where *Best* is the best result

On the other hand, the datasets used in this work are those proposed by the UCI Website [5]. Table 3 shows the characteristics of the resolved instances. The first column represents the solved instance, the second column refers to the number of features that compose the datasets, the third column refers to the number of samples in the datasets, the fourth column represents the number of classes to be classified, and the fifth column refers to whether the dataset has missing data.

To calculate the objective function discussed in Sect. 2, we need to use a classifier. There are different classifiers in the literature but we use three: K-Nearest Neighbors (KNN) [14], Random Forest [9], and XGBoost [13]. The datasets were separated into two: 75% of the data was used as training data and 25% of the data was used as validation data. The population size for the three metaheuristics is 20 individuals, and 100 iterations were performed.

Tables 4, 5, and 6 show the fitness obtained for each version of the algorithm in each instance. The first column of this table indicates the versions of the Binary Pendulum Search Algorithm used and the second, third, and fourth columns are repeated for each instance. The first of them indicates the best fitness obtained, the second of them indicates the average obtained and the third of them indicates the standard deviation.

Table 3. Instances used

ID	Instance	No of Features	Sample size	No of Class	Missed Value?
DB1	Ionosphere	34	351	2	No
DB2	Connectionist Bench (Sonar, Mines vs Rocks)	60	208	2	No
DB3	Divorce	54	170	2	No
DB4	Breast Cancer Wisconsin - Diagnostic (wdbc)	30	569	2	No
DB5	Breast cancer wisconsin (Original)	9	699	2	Yes

Analyzing Table 4 we can see that the version that includes the V-Shaped transfer function and the standard binarization rule is the best in all five instances.

Table 4. Fitness with classifier KNN

BSS	DB1			DB2			DB3			DB4			DB5		
	best	avg	dev-std	best	avg	dev-std	best	avg	dev-std	best	avg	dev-std	best	avg	dev-std
S-STD	0.073	0.084	0.006	0.03	0.051	0.007	0.032	0.036	0.008	0.013	0.022	0.004	**0.019**	**0.019**	**0.0**
V-STD	**0.015**	**0.034**	**0.009**	**0.002**	**0.056**	**0.023**	**0.0**	**0.008**	**0.013**	**0.011**	**0.02**	**0.005**	**0.019**	**0.02**	**0.002**
X-STD	0.058	0.078	0.011	0.027	0.058	0.015	0.031	0.032	0.0	0.013	0.019	0.004	**0.019**	**0.019**	**0.0**
Z-STD	0.029	0.036	0.009	0.049	0.082	0.021	**0.0**	**0.006**	**0.011**	0.019	0.024	0.003	**0.019**	**0.019**	**0.001**
S-COM	0.043	0.059	0.013	0.05	0.063	0.011	0.03	0.031	0.0	0.013	0.021	0.004	**0.019**	**0.019**	**0.0**
V-COM	0.042	0.059	0.017	0.049	0.086	0.021	**0.0**	**0.027**	**0.015**	0.014	0.027	0.007	**0.019**	**0.019**	**0.0**
X-COM	0.03	0.064	0.012	0.051	0.065	0.01	0.002	0.02	0.014	0.013	0.019	0.003	**0.019**	**0.019**	**0.0**
Z-COM	0.042	0.069	0.013	0.053	0.094	0.018	0.001	0.029	0.012	**0.011**	**0.025**	**0.009**	**0.019**	**0.019**	**0.001**
S-ELIT	0.029	0.061	0.018	0.027	0.083	0.025	0.029	0.032	0.009	0.019	0.027	0.006	**0.019**	**0.022**	**0.003**
V-ELIT	0.042	0.067	0.019	0.025	0.077	0.022	**0.0**	**0.021**	**0.013**	0.012	0.027	0.009	**0.019**	**0.021**	**0.003**
X-ELIT	0.029	0.065	0.02	0.071	0.091	0.014	**0.0**	**0.024**	**0.011**	**0.011**	**0.028**	**0.009**	**0.019**	**0.02**	**0.001**
Z-ELIT	0.029	0.063	0.016	0.053	0.086	0.02	0.029	0.029	0.0	**0.011**	**0.02**	**0.007**	**0.019**	**0.019**	**0.001**

Analyzing Table 5, we can see that the version that includes the V-Shaped transfer function and the standard binarization rule, the S-Shaped transfer function, and the elitist binarization rule are the best in 4 instances.

Analyzing Table 6, we can see that the version that includes the V-shaped transfer function and the standard binarization rule is the best in all five instances.

Tables 7, 8 and 9 show the number of features selected for each classifier. The first column refers to the binarization scheme under study, and the second and third columns are repeated for each data set. The first refers to the average of

Table 5. Fitness with classifier Random Forest

BSS	DB1			DB2			DB3			DB4			DB5		
	best	avg	dev-std	best	avg	dev-std	best	avg	dev-std	best	avg	dev-std	best	avg	dev-std
S-STD	0.032	0.042	0.006	0.029	0.044	0.011	0.032	0.032	0.0	0.029	0.031	0.003	**0.019**	**0.019**	**0.0**
V-STD	**0.015**	**0.03**	**0.011**	0.049	0.077	0.019	**0.0**	**0.0**	**0.0**	**0.02**	**0.033**	**0.007**	**0.019**	**0.02**	**0.002**
X-STD	0.03	0.039	0.006	0.029	0.054	0.011	0.003	0.019	0.014	0.022	0.027	0.004	**0.019**	**0.019**	**0.0**
Z-STD	**0.015**	**0.035**	**0.011**	0.053	0.087	0.018	**0.0**	**0.0**	**0.0**	0.023	0.035	0.005	**0.019**	**0.019**	**0.0**
S-COM	0.029	0.036	0.007	0.028	0.061	0.018	0.001	0.002	0.001	**0.02**	**0.031**	**0.006**	**0.019**	**0.019**	**0.0**
V-COM	**0.015**	**0.039**	**0.01**	0.072	0.086	0.011	**0.0**	**0.016**	**0.014**	0.022	0.032	0.009	**0.019**	**0.02**	**0.001**
X-COM	0.016	0.033	0.008	0.027	0.059	0.02	0.001	0.003	0.001	0.022	0.028	0.002	**0.019**	**0.019**	**0.0**
Z-COM	0.029	0.043	0.008	0.072	0.097	0.017	**0.0**	**0.013**	**0.014**	0.024	0.034	0.006	**0.019**	**0.019**	**0.0**
S-ELIT	**0.015**	**0.044**	**0.013**	0.026	0.1	0.031	**0.0**	**0.016**	**0.014**	**0.02**	**0.035**	**0.007**	**0.019**	**0.022**	**0.003**
V-ELIT	0.042	0.048	0.009	**0.002**	**0.085**	**0.035**	**0.0**	**0.019**	**0.014**	0.027	0.036	0.006	**0.019**	**0.019**	**0.001**
X-ELIT	**0.015**	**0.043**	**0.013**	0.071	0.1	0.022	**0.0**	**0.016**	**0.014**	0.027	0.038	0.008	**0.019**	**0.02**	**0.002**
Z-ELIT	0.029	0.039	0.008	0.072	0.088	0.011	**0.0**	**0.013**	**0.014**	0.027	0.032	0.006	**0.019**	**0.019**	**0.001**

Table 6. Fitness with classifier XgBoost

BSS	DB1			DB2			DB3			DB4			DB5		
	best	avg	dev-std	best	avg	dev-std	best	avg	dev-std	best	avg	dev-std	best	avg	dev-std
S-STD	0.019	0.03	0.005	0.029	0.044	0.011	0.004	0.007	0.008	0.039	0.047	0.005	**0.025**	**0.025**	**0.0**
V-STD	**0.016**	**0.028**	**0.007**	**0.002**	**0.057**	**0.025**	**0.0**	**0.001**	**0.0**	**0.027**	**0.037**	**0.007**	**0.025**	**0.025**	**0.0**
X-STD	0.018	0.028	0.006	0.028	0.045	0.014	0.003	0.004	0.001	0.03	0.043	0.006	**0.025**	**0.025**	**0.0**
Z-STD	0.019	0.029	0.005	0.03	0.078	0.025	**0.0**	**0.006**	**0.011**	0.036	0.038	0.006	**0.025**	**0.025**	**0.0**
S-COM	0.017	0.029	0.004	0.028	0.061	0.015	0.002	0.002	0.0	0.028	0.04	0.005	**0.025**	**0.025**	**0.0**
V-COM	0.029	0.036	0.006	0.029	0.08	0.024	**0.0**	**0.016**	**0.014**	0.036	0.046	0.005	**0.025**	**0.025**	**0.0**
X-COM	**0.016**	**0.028**	**0.005**	0.028	0.057	0.014	0.002	0.002	0.0	0.028	0.04	0.007	**0.025**	**0.025**	**0.0**
Z-COM	**0.016**	**0.03**	**0.006**	0.052	0.08	0.017	**0.0**	**0.006**	**0.011**	0.036	0.045	0.007	**0.025**	**0.025**	**0.0**
S-ELIT	0.029	0.038	0.009	0.049	0.078	0.022	**0.0**	**0.016**	**0.014**	0.028	0.05	0.012	**0.025**	**0.025**	**0.0**
V-ELIT	0.029	0.035	0.009	0.049	0.085	0.025	**0.0**	**0.019**	**0.014**	0.036	0.048	0.009	**0.025**	**0.025**	**0.0**
X-ELIT	0.029	0.037	0.009	0.049	0.064	0.02	**0.0**	**0.011**	**0.014**	0.036	0.043	0.007	**0.025**	**0.026**	**0.001**
Z-ELIT	0.019	0.034	0.008	0.071	0.088	0.015	**0.0**	**0.003**	**0.008**	0.036	0.045	0.005	**0.025**	**0.025**	**0.0**

selected characteristics, and the second is the standard deviation. By reviewing Table 7 we can observe very interesting things. For example, for the DB2 data set, the best results using KNN and observable in Table 4 are with V-STD. This experiment obtained a fitness of 0.002 with an average of 14.6 selected features. On the other hand, the X-ELIT experiment was the worst, obtaining a fitness of 0.071 with an average of 7.6 selected features. With this, we can see that the V-STD experiment correctly eliminates irrelevant and redundant features. On the other hand, selecting fewer features is not an indicator that we will have better performance, and with these two experiments, it is verified. This analysis is extensible for other classifiers.

Table 7. Total Features Selected with classifier KNN

BSS	DB1		DB2		DB3		DB4		DB5	
	avg	dev-std	avg	dev-std	avg	dev-std	avg	dev-std	avg	dev-std
S-STD	12.8	2.4	33.5	2.7	24.4	3.5	15.1	1.8	4.1	0.3
V-STD	4.2	1.1	14.6	5.5	3.5	1.1	8.1	2.2	4.2	0.8
X-STD	11.2	1.9	26.5	3.1	17.5	1.9	14.6	2.2	4.3	0.4
Z-STD	3.3	1.1	17.0	8.8	3.5	1.4	8.1	4.5	4.5	0.7
S-COM	6.3	1.8	19.5	2.7	10.3	2.5	9.5	1.8	4.1	0.3
V-COM	3.5	1.2	16.7	11.8	3.1	0.9	10.6	5.4	4.1	0.3
X-COM	7.7	0.9	20.5	3.8	10.8	2.3	9.6	2.9	4.1	0.3
Z-COM	3.5	0.5	15.0	9.3	3.7	1.1	8.4	4.1	4.4	0.6
S-ELIT	3.7	1.1	10.8	4.2	3.0	1.0	7.0	2.4	4.1	0.9
V-ELIT	3.7	1.2	12.1	4.4	3.3	1.0	7.5	1.9	4.3	0.7
X-ELIT	3.7	0.9	7.6	2.6	2.3	1.1	7.2	2.5	4.6	0.9
Z-ELIT	3.2	1.3	14.3	9.5	3.3	0.7	8.2	3.3	4.4	0.6

Table 8. Total Features Selected with classifier Random Forest

BSS	DB1		DB2		DB3		DB4		DB5	
	avg	dev-std	avg	dev-std	avg	dev-std	avg	dev-std	avg	dev-std
S-STD	16.9	3.8	33.2	4.6	19.4	1.4	14.6	2.5	4.0	0.0
V-STD	5.3	1.5	14.1	8.3	2.5	1.7	5.9	2.1	3.9	0.3
X-STD	13.1	3.2	28.8	5.3	17.2	4.6	13.8	1.3	4.0	0.0
Z-STD	4.6	1.1	20.7	10.3	2.6	1.1	9.9	4.0	4.0	0.0
S-COM	8.3	2.4	21.1	3.8	13.2	4.0	9.6	1.6	4.0	0.0
V-COM	4.9	2.2	14.3	8.0	5.2	6.6	7.1	3.7	4.5	0.9
X-COM	9.4	1.8	22.2	5.5	14.3	3.8	9.8	2.8	4.0	0.0
Z-COM	5.5	3.5	19.2	10.4	3.1	1.2	8.1	5.4	4.0	0.0
S-ELIT	5.1	1.3	9.7	2.3	4.4	1.6	5.1	1.6	4.1	1.0
V-ELIT	4.4	1.2	11.3	5.1	3.5	1.6	7.9	2.4	4.2	0.6
X-ELIT	4.9	1.4	13.0	8.2	3.1	1.8	7.1	2.7	3.8	0.4
Z-ELIT	4.9	2.8	18.9	11.2	3.0	1.0	7.7	3.9	4.5	0.8

Table 9. Total Features Selected with classifier XgBoost

BSS	DB1		DB2		DB3		DB4		DB5	
	avg	dev-std	avg	dev-std	avg	dev-std	avg	dev-std	avg	dev-std
S-STD	17.3	3.0	32.5	2.9	24.5	4.3	12.9	1.8	3.1	0.3
V-STD	6.8	2.6	11.0	2.9	4.6	2.6	5.9	1.7	3.0	0.0
X-STD	15.0	3.0	27.5	2.7	22.5	3.8	12.8	1.8	3.0	0.0
Z-STD	6.0	3.5	20.4	10.2	3.3	1.5	3.9	0.9	3.0	0.0
S-COM	9.6	2.0	20.2	3.7	12.5	1.1	8.4	2.7	3.0	0.0
V-COM	6.5	5.1	18.4	9.1	5.5	3.1	6.4	4.5	3.1	0.3
X-COM	9.2	1.9	20.9	5.0	12.5	2.0	8.7	2.0	3.0	0.0
Z-COM	6.4	4.5	19.4	8.7	5.1	2.0	4.7	2.0	3.2	0.4
S-ELIT	5.1	1.2	9.2	3.8	4.4	2.0	5.4	1.8	3.2	0.4
V-ELIT	5.3	1.7	12.1	4.9	4.5	1.7	4.9	1.0	3.2	0.4
X-ELIT	6.9	5.3	13.5	3.3	4.1	1.9	4.7	1.1	3.5	0.7
Z-ELIT	6.0	3.6	15.6	7.5	5.3	2.7	7.6	4.1	3.1	0.3

7 Conclusion

In the present work, we demonstrate that the Binary Pendulum Search Algorithm is competitive in solving the Feature Selection, considering both the fitness obtained and the f-score metric. With this, we were able to demonstrate how simple it can be to adapt a continuous metaheuristic to solve binary combinatorial problems. Furthermore, we can observe that the transfer functions and the binarization rule influence the results obtained, where we can highlight the S-shaped transfer function and the standard binarization rule as the best.

However, as indicated by authors in [7], there are different ways to binarize continuous metaheuristics. Such as different transfer functions (U-shaped [8], O-Shaped [17], X-Shaped [20], Z-Shaped [32], time-varying S-Shaped [21] and time-varying V-Shaped [11]), other binarization rules [15], clustering-based binarization techniques [19] or even binarization supported by machine learning techniques such as Q-Learning [16] or SARSA [22]. The No Free Lunch Theorem [34] tells us that no algorithm is capable of reaching the global optimum of all existing optimization problems. Therefore, another binarization technique could be implemented in future work to validate whether the good performance obtained in the present work is maintained. In addition, our proposal could be validated with other datasets and even applied to another combinatorial problem, such as the Set Covering Problem or the Knapsack Problem.

Acknowledgements. Felipe Cisternas-Caneo is supported by the National Agency for Research and Development (ANID)/ Scholarship Program/DOCTORADO NACIONAL/2023-21230203.

The work of Broderick Crawford Labrin is supported by the Spanish Ministry of Science and Innovation Project PID2019-109891RB-I00, under the European Regional Development Fund (FEDER).

Jose Barrera-García is supported by National Agency for Research and Development (ANID)/ Scholarship Program/DOCTORADO NACIONAL/2024-21242516.

Marcelo Becerra-Rozas is supported by National Agency for Research and Development (ANID)/ Scholarship Program/DOCTORADO NACIONAL/2021-21210740.

References

1. Aziz, N.A.A., Aziz, K.A.: Pendulum search algorithm: an optimization algorithm based on simple harmonic motion and its application for a vaccine distribution problem. Algorithms **15**(6), 214 (2022)
2. Rahman, T.A., Ibrahim, Z., Aziz, N.A.A., Zhao, S., Aziz, N.H.A.: Single-agent finite impulse response optimizer for numerical optimization problems. IEEE Access **6**, 9358–9374 (2018)
3. Adeli, A., Broumandnia, A.: Image steganalysis using improved particle swarm optimization based feature selection. Appl. Intell. **48**, 1609–1622 (2018)
4. Agrawal, P., Abutarboush, H.F., Ganesh, T., Mohamed, A.W.: Metaheuristic algorithms on feature selection: a survey of one decade of research (2009–2019). IEEE Access **9**, 26766–26791 (2021)
5. Asuncion, A., Newman, D.: UCI Machine Learning Repository (2007)
6. Barrera-García, J., Cisternas-Caneo, F., Crawford, B., Gómez Sánchez, M., Soto, R.: Feature selection problem and metaheuristics: a systematic literature review about its formulation, evaluation and applications. Biomimetics **9**(1), (2024)
7. Becerra-Rozas, M., et al.: Continuous metaheuristics for binary optimization problems: an updated systematic literature review. Mathematics **11**(1), 129 (2022)
8. Beheshti, Z.: UTF: Upgrade transfer function for binary meta-heuristic algorithms. Appl. Soft Comput. **106**, 107346 (2021)
9. Breiman, L.: Random forests. Mach. Learn. **45**, 5–32 (2001)
10. Canayaz, M.: MH-covidnet: diagnosis of covid-19 using deep neural networks and meta-heuristic-based feature selection on x-ray images. Biomed. Signal Process. Control **64**, 102257 (2021)
11. Chantar, H., Thaher, T., Turabieh, H., Mafarja, M., Sheta, A.: BHHO-TVS: A binary Harris Hawks optimizer with time-varying scheme for solving data classification problems. Appl. Sci. **11**(14), 6516 (2021)
12. Chen, B., Chen, L., Chen, Y.: Efficient ant colony optimization for image feature selection. Signal Process. **93**(6), 1566–1576 (2013)
13. Chen, T., Guestrin, C.: Xgboost: a scalable tree boosting system. In: Proceedings of the 22nd ACM SIGKDD International Conference on Knowledge Discovery and Data Mining, pp. 785–794 (2016)
14. Cover, T., Hart, P.: Nearest neighbor pattern classification. IEEE Trans. Inf. Theory **13**(1), 21–27 (1967)
15. Crawford, B., Soto, R., Astorga, G., García, J., Castro, C., Paredes, F.: Putting continuous metaheuristics to work in binary search spaces. Complexity **2017**, (2017)

16. Crawford, B., et al.: Q-learnheuristics: towards data-driven balanced metaheuristics. Mathematics **9**(16), 1839 (2021)
17. Rafet Durgut and Mehmet Emin Aydin: Adaptive binary artificial bee colony algorithm. Appl. Soft Comput. **101**, 107054 (2021)
18. Eesa, A.S., Orman, Z., Brifcani, A.M.A.: A novel feature-selection approach based on the cuttlefish optimization algorithm for intrusion detection systems. Expert Syst. Appl. **42**(5), 2670–2679 (2015)
19. García, J., et al.: A binary machine learning cuckoo search algorithm improved by a local search operator for the set-union knapsack problem. Mathematics **9**(20), 2611 (2021)
20. Ghosh, K.K., Singh, P.K., Hong, J., Geem, Z.W., Sarkar, R.: Binary social mimic optimization algorithm with x-shaped transfer function for feature selection. IEEE Access **8**, 97890–97906 (2020)
21. Kahya, M.A., Altamir, S.A., Algamal, Z.Y.: Improving whale optimization algorithm for feature selection with a time-varying transfer function. Numer. Algebra Control Opt. **11**(1), 87 (2021)
22. Lemus-Romani, J., et al.: A novel learning-based binarization scheme selector for swarm algorithms solving combinatorial problems. Mathematics **9**(22), 2887 (2021)
23. Liu, H., Lei, Yu.: Toward integrating feature selection algorithms for classification and clustering. IEEE Trans. Knowl. Data Eng. **17**(4), 491–502 (2005)
24. Yonghe, L., Liang, M., Ye, Z., Cao, L.: Improved particle swarm optimization algorithm and its application in text feature selection. Appl. Soft Comput. **35**, 629–636 (2015)
25. Medjahed, S.A., Saadi, T.A., Benyettou, A., Ouali, M.: Kernel-based learning and feature selection analysis for cancer diagnosis. Appl. Soft Comput. **51**, 39–48 (2017)
26. Nadimi-Shahraki, M.H., Banaie-Dezfouli, M., Zamani, H., Taghian, S., Mirjalili, S.: B-MFO: a binary moth-flame optimization for feature selection from medical datasets. Computers **10**(11), 136 (2021)
27. Anjum Nazir and Rizwan Ahmed Khan: A novel combinatorial optimization based feature selection method for network intrusion detection. Comput. Secur. **102**, 102164 (2021)
28. Rajwar, K., Deep, K., Das, S.: An exhaustive review of the metaheuristic algorithms for search and optimization: taxonomy, applications, and open challenges. Artif. Intell. Rev., 1–71 (2023)
29. Gauthama Raman, M.R., Somu, N., Kirthivasan, K., Liscano, R., Shankar Sriram, V.S.: An efficient intrusion detection system based on hypergraph-genetic algorithm for parameter optimization and feature selection in support vector machine. Knowl. Based Syst. **134**, 1–12 (2017)
30. Sadeghian, Z., Akbari, E., Nematzadeh, H., Motameni, H.: A review of feature selection methods based on meta-heuristic algorithms. J. Exp. Theor. Artif. Intell., 1–51 (2023)
31. Sudha, M.N., Selvarajan, S.: Feature selection based on enhanced cuckoo search for breast cancer classification in mammogram image. Circ. Syst. **7**(4), 327–338 (2016)
32. Sun, W.-Z., Zhang, M., Wang, J.-S., Guo, S.-S., Wang, M., Hao, W.-K.: Binary particle swarm optimization algorithm based on z-shaped probability transfer function to solve 0–1 knapsack problem. IAENG Int. J. Comput. Sci. **48**(2), 294 (2021)
33. Thirumoorthy, K., Muneeswaran, K.: Feature selection using hybrid poor and rich optimization algorithm for text classification. Pattern Recogn. Lett. **147**, 63–70 (2021)

34. Wolpert, D.H., Macready, W.G.: No free lunch theorems for optimization. IEEE Trans. Evolut. Comput. **1**(1), 67–82 (1997)
35. Zivkovic, M., Jovanovic, L., Ivanovic, M., Krdzic, A., Bacanin, N., Strumberger, I.: Feature selection using modified sine cosine algorithm with covid-19 dataset. In: Evolutionary Computing and Mobile Sustainable Networks: Proceedings of ICECMSN 2021, pp. 15–31. Springer (2022)

The Software Complex for Computer Simulation of the Motion Control System

Mikhail N. Smirnov(✉) and Maria A. Smirnova

St. Petersburg State University, 7/9 Universitetskaya nab., St. Petersburg 199034, Russia
{mikhail.smirnov,mariya.smirnova}@spbu.ru

Abstract. The paper is devoted to construction the special the software complex for analysis, synthesis and modeling of the control system for marine vessels. The complex is designed in MATLAB-Simulink package. It contains mathematical and computer model of the controlled object, mathematical and computer model of controller and tools for visualization. This software complex is universal for all objects with similar mathematical model and it can be easily modified for any marine vessel. In the paper the structure of software complex is given. Its work is illustrated by an example of real vessel #CSOC1120.

Keywords: Controller · Simulation · Software complex

1 Introduction

The computer modeling allows to us significantly reduce the cost of developing models compared to non-computer modeling methods and conducting field tests. It makes possible to build symbolic computer models of the objects for which it is impossible to build physical models (e.g. models of objects studied in climatology). It serves as an effective means of modeling complex systems in engineering, economics, and other fields of activity. It is the technological basis of automated design systems.

Particularly computer modeling is used for simulation of automatic control systems for different marine vessels. Such systems allow us save time and money resources during the developing new marine vessels. Also automatic control systems help to spend resources more efficiently and make right decisions in difficult weather conditions. Various automatic controllers are discussed in [1–12]. In this paper the software complex for computer modeling is proposed and described.

2 The General Structure of the Modeling Complex

The main purpose of the simulation complex is to ensure the goals of research and development of mathematical models and control algorithms, as well as testing mathematical, algorithmic and software developed on the basis of the conducted research.

This research was supported by the Russian Science Foundation, project N 23-71-01001,https://rscf.ru/project/23-71-01001/.

R. Silhavy and P. Silhavy (Eds.): CSOC 2024, LNNS 1120, pp. 403–411, 2024.
https://doi.org/10.1007/978-3-031-70518-2_36

This section provides a detailed description of the general structure, basic and additional elements of the software, as well as complex control tools.

The application software of the modeling complex includes simulation models of control objects, external influences and technical tools. In addition, the software includes simulation models of control algorithms built in the Simulink system, which are designed to simulate dynamic processes in closed systems. Combining these elements into a single complex and managing its operation is carried out using specialized software elements.

Let's consider the general structure of the complex. The scheme of the upper hierarchical level is shown in Fig. 1.

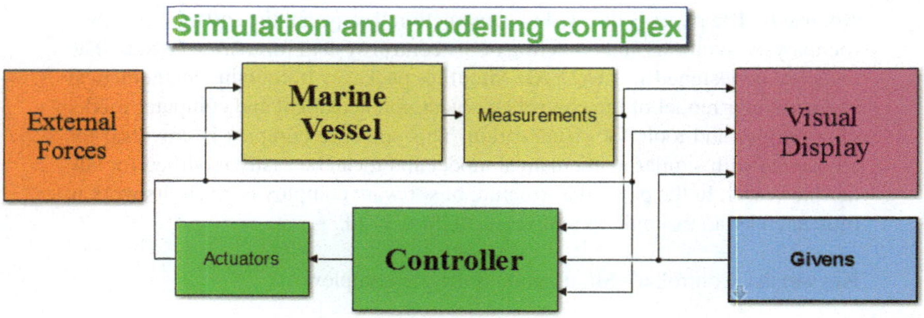

Fig. 1. The general structure of the complex.

The main modelling blocks include elements having the following functional purpose:

1. The block **Marine Vessel** represents the dynamics of the marine vessel as a control object.
2. The block **Controller** simulates regulators that provide the desired program motion of the vessel at a given course and/or depth, as well as the dynamics of the executive organs.
3. The block **Visual Display** provides a concentration of information displayed on the screen for the user of the complex. The main part of the subsystem consists of blocks that concentrate and distribute information according to the corresponding display blocks.
4. The block **Givens** is responsible for setting constant values of the command signal.
5. The block **Measurements** contains the values of measured variables.
6. The block **Actuators** simulates the dynamics of rudders.
7. The block **External Forces** represents the disturbances act on the vessel.

The schemes of the main subsystems are shown in Figs. 2, 3, 4, 5, ö and 7. The developed software complex is universal; it can be used for modelling of the motion of the surface marine vessels and of the underwater vehicles. In case of underwater objects there are two control channels – horizontal rudders, providing the depth motion, and vertical rudders, providing the motion in the horizontal plane. If necessary, the developed software complex can be easy modified for a specific movable object.

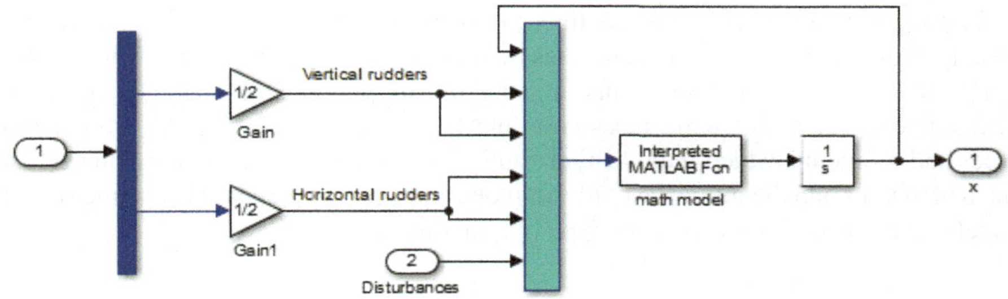

Fig. 2. The block **Marine Vessel**.

The block **Marine Vessel** represents the mathematical model of the vessel. The general nonlinear mathematical model of the marine vessel has the following form

$$m_x \frac{dV_x}{dt} + mh \frac{d\omega_z}{dt} = Q_x,$$

$$m_y \frac{dV_y}{dt} + \lambda_{26} \frac{d\omega_z}{dt} = Q_y,$$

$$m_z \frac{dV_z}{dt} + (\lambda_{34} - mh) \frac{d\omega_x}{dt} + \lambda_{35} \frac{d\omega_y}{dt} = Q_z,$$

$$J_x \frac{d\omega_x}{dt} + (\lambda_{34} - mh) \frac{dV_z}{dt} = N_x,$$

$$J_y \frac{d\omega_y}{dt} + \lambda_{35} \frac{dV_z}{dt} = N_y,$$

$$J_z \frac{d\omega_z}{dt} + \lambda_{26} \frac{dV_y}{dt} + mh \frac{dV_x}{dt} = N_z,$$

$$\begin{aligned}\dot{\xi} =& V_x \cos(-\phi) \cos\psi - V_y(\sin\theta \sin(-\phi) + \cos(-\phi) \cos\theta \sin\psi) + \\ &+ V_z(-\cos\theta \sin(-\phi) + \cos(-\phi) \sin\theta \sin\psi),\end{aligned}$$

$$\dot{\eta} = V_x \sin\psi + V_y \cos\psi \cos\theta - V_z \cos\psi \sin\theta,$$

$$\begin{aligned}&\dot{\zeta} = V_x \sin(-\phi) \cos\psi - V_y(-\sin\theta \cos(-\phi) + \sin(-\phi) \cos\theta \sin\psi) + \\ &+ V_z(\cos\theta \cos(-\phi) + \sin(-\phi) \sin\theta \sin\psi),\end{aligned}$$

$$\dot{\psi} = \omega_z \cos\theta + \omega_y \sin\theta,$$

$$\dot{\theta} = \omega_x - tg\psi\left(\omega_y \cos\theta - \omega_z \sin\theta\right),$$

$$\dot{\phi} = -\frac{1}{\cos\psi}\left(-\omega_y \cos\theta + \omega_z \sin\theta\right).$$

In these equations V_x, V_y, V_z are linear velocity projections, ω_x, ω_y, ω_z are angular velocity projections, λ_{ij} are attached masses and moments of inertia, h is metacentric height, J_x, J_y, J_z are inertia moments, m_x, m_y, m_z are generalized masses, θ, ϕ, ψ are Euler angles, ξ, η, ζ are centre mass coordinates, Q_x, Q_y, Q_z, N_x, N_y, N_z are external forces and moments, which include hydrostatic force, thrust force, lift, propeller thrust, attached masses and the influence of the aft plane and steering planes. This mathematical model can be specified for a certain type of a marine vessel.

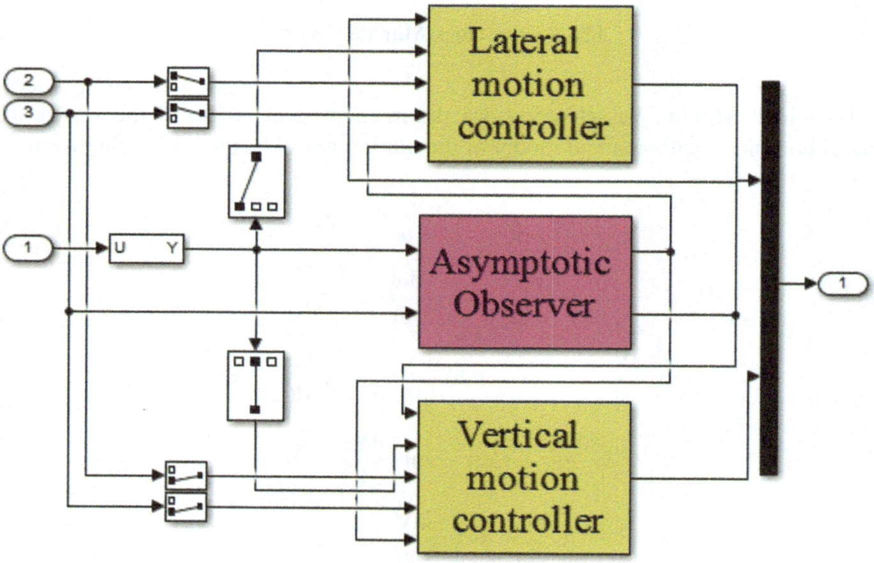

Fig. 3. The block **Controller**.

The block **Controller** is consists of three main parts: Lateral motion controller, Vertical motion controller and Asymptotic observer. The mathematical model of the asymptotic observer is

$$\dot{\mathbf{z}} = \mathbf{Az} + \mathbf{B}\delta + \mathbf{G}(\mathbf{y} - \mathbf{Cz}),$$

where **z** is the estimation of the system state vector.

In the Fig. 4 the lateral motion controller is presented. This controller is constructed for compensating external disturbances in horizontal plane. Note that it can be modified for the specific vessel and for the specific task. Similarly the vertical motion controller has the identical form that is shown in Fig. 5.

The block **Actuators** is shown in the Fig. 6. It represents the drive dynamics mathematical model

$$\dot{\delta} = \mathrm{u}.$$

Here δ is the state vector of the control system, exactly, it is the vector of vertical rudders δ_r and horizontal rudders δ_s deviations. Also this block takes into account technical

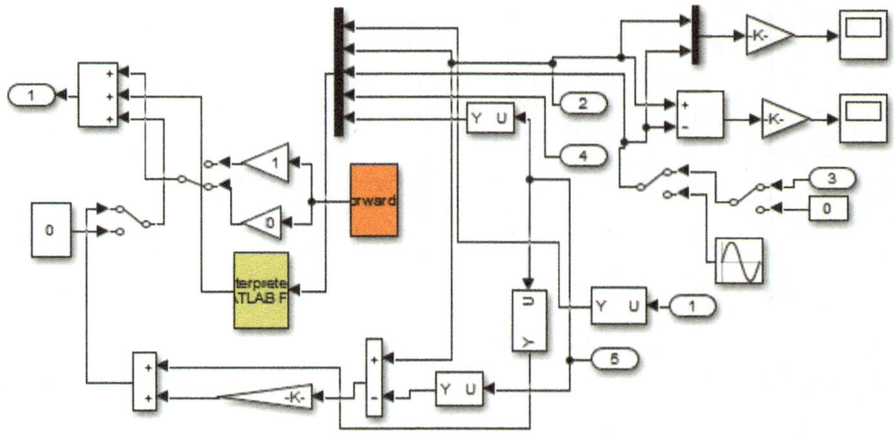

Fig. 4. The lateral motion controller.

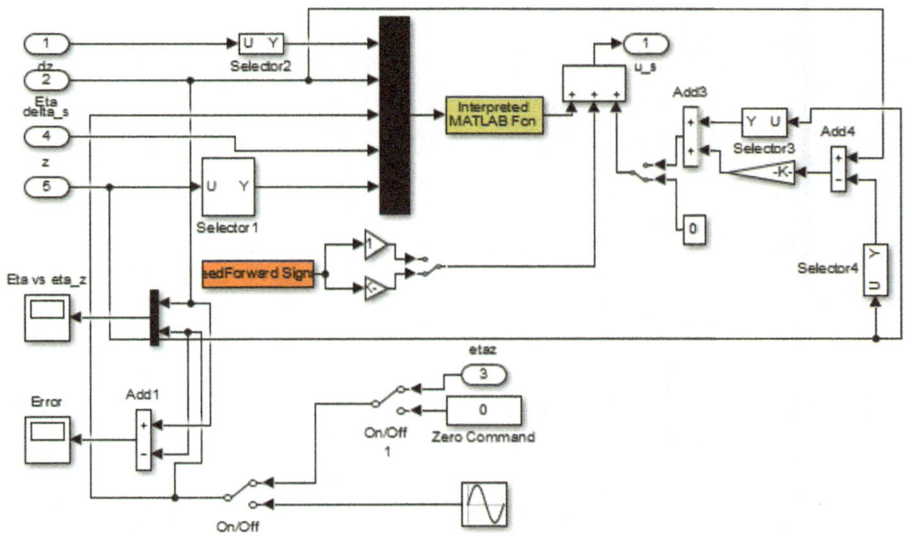

Fig. 5. The vertical motion controller.

restrictions on the rudders deviation and on their speed. Usually $|\delta| \leq 30°$, $u_{\max} = 3°/\,\text{sec}$, but these restrictions can be changed if necessary.

Figure 7 represents the block **Visual Display**. In consists of Scope blocks for display the simulation results.

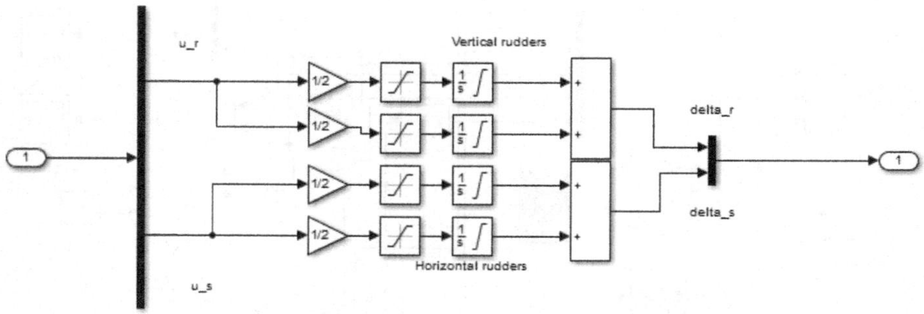

Fig. 6. The block **Actuators**.

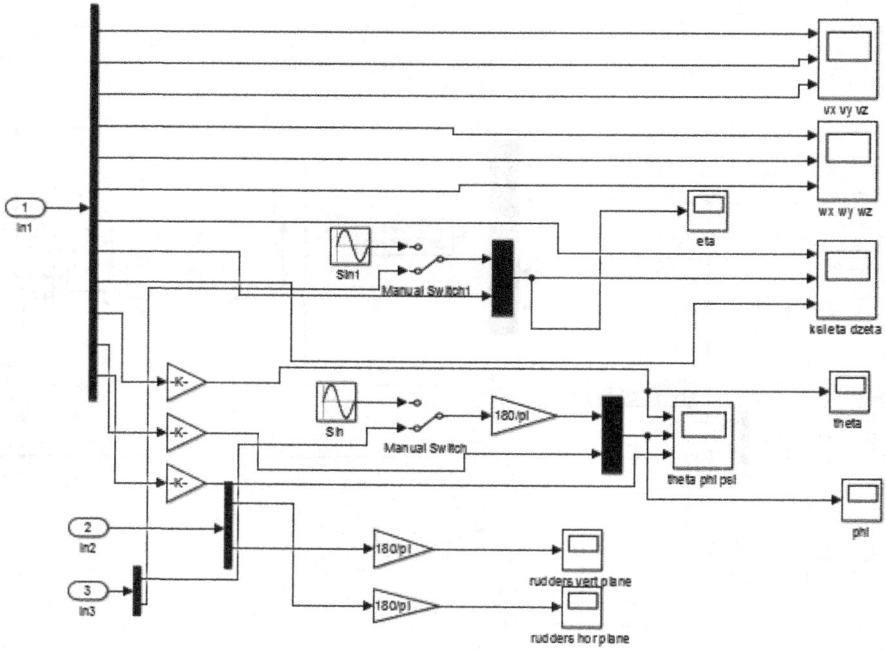

Fig. 7. The block **Visual Display**.

3 Example

Let's consider an example of the work of the developed software package using the motion control for the surface marine vessel motion. After linearization of nonlinear equations stated above we receive the following mathematical model for lateral motion

$$
\begin{aligned}
\dot{V}_z &= -a_{11}VV_z + a_{12}V\omega_y - b_1V^2\delta + d_1F_z,\\
\dot{\omega}_y &= a_{21}VV_z - a_{22}V\omega_y - b_2V^2\delta + d_2M_y,\\
\dot{\phi} &= \omega_d.
\end{aligned}
$$

The control takes place using two vertical rudders δ_l and δ_r which are deviated synchronously, $\delta = \delta_l + \delta_r$. Here ϕ is yaw angle, V is a fixed velocity, F_z, M_y are bounded external forces and moments.

If we will use the controller $u = k_1\beta + k_2\omega + k_3\phi + k_4\delta$, constructed for compensation of bounded disturbances with additional requirement to the degree of stability, for computer simulation in the developed software complex, we will obtain the graphs for main dynamic variables δ and ϕ shown in Figs. 8 and 9. The external disturbance is represented in Fig. 10. In the Fig. 8 blue line is the command signal $\phi_d = 15°$, yellow line is yaw changes of the vessel.

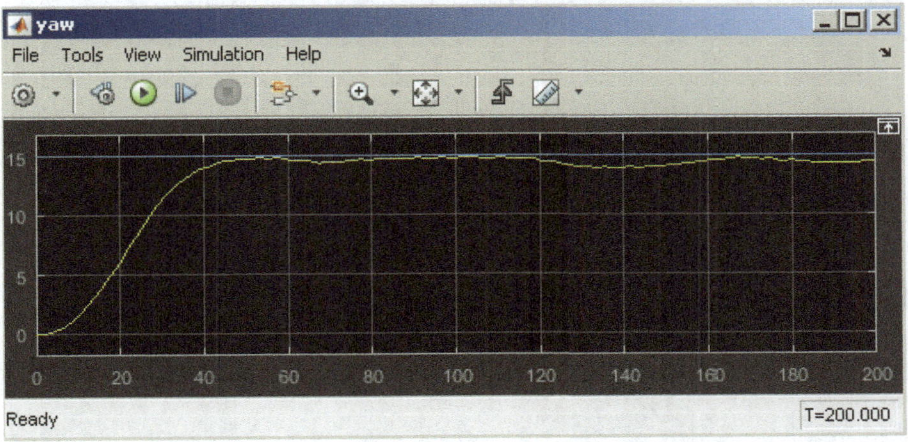

Fig. 8. Yaw graph.

In the Fig. 9 yellow line shows rudders angle change, blue and red line presents technical restrictions on rudder deviations. As we can see from the graph, the rudders do not reach the extreme positions. It is good result. Figure 10 illustrates the external disturbance used for simulation.

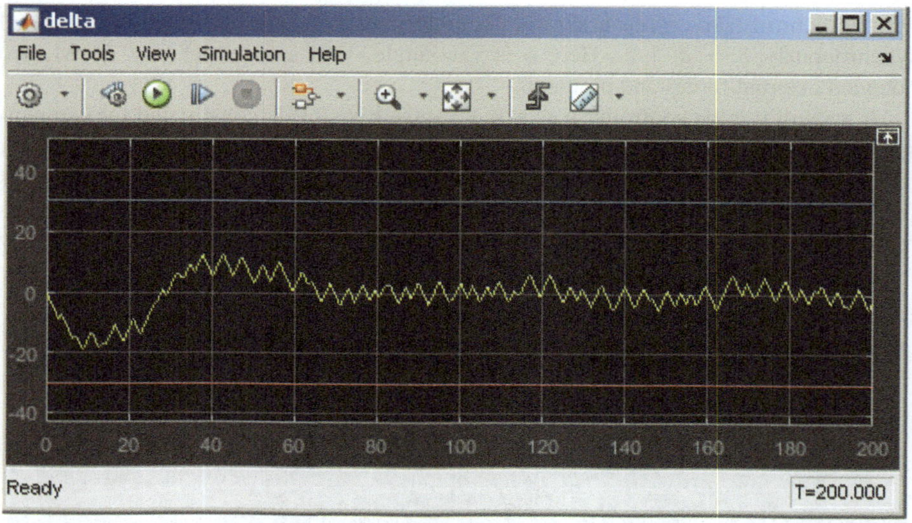

Fig. 9. Rudder deviation graph.

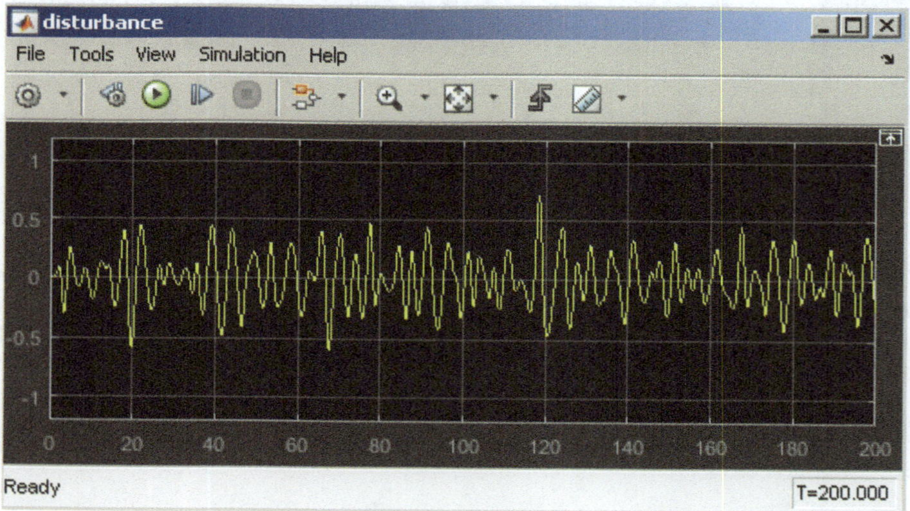

Fig. 10. Disturbance graph.

4 Conclusion

The paper demonstrates the software complex for analysis, simulation and modelling the control system for marine vessels. The complex is designed in MATLAB-Simulink package. It contains mathematical and computer model of the controlled object, mathematical and computer model of controller and tools for visualization. This software complex is universal for all objects with similar mathematical model and it can be easily modified for any marine vessel. In the paper the structure of software complex is given. Its work is illustrated by an example of real vessel.

References

1. Veremey, E.I., Korchanov, V.M, Korovkin, M.V., et al.: Computer Modeling of Control Systems for Marine Movable Objects. SPbSU, St. Petersburg (2002)
2. Lukomskiy, U.A., Korchanov, V.M: The Control of the Marine Movable Objects. Elmor, St. Petersburg (1996)
3. Veremey, E.I., Sotnikova, M.V.: Application of the H1-optimization method for synthesis of filters for marine disturbances. Gyroscopy Navig. **2**, 24–36 (2009)
4. Fossen, T.I.: Guidance and Control of Ocean Vehicles. John Wiley and Sons, New York (1999)
5. Veremey, E.I.: The synthesis of the laws of multipurpose control for marine movable objects. Gyroscopy Navig. **4**, 3–14 (2009)
6. Fossen, T.I., Strand, J.P.: Passive nonlinear observer design for ships using Lyapunov methods: experimental results with a supply vessel. Automatica **35**(1), 3–16 (1999)
7. Veremey, E.I.: Dynamical correction of positioning control laws. In: Proceedings of 9th IFAC Conference on Control Applications in Marine Systems, vol. 9, no. 1, pp.31–36 (2013)
8. Veremey, E.I., Smirnova, M.A., Smirnov, M.N.: Synthesis of stabilizing control laws with uncertain disturbances for marine vessels. In: Proceedings of the 2015 International Conference "Stability and Control Processes" in Memory of V.I. Zubov (SCP), pp. 1–3 (2015)
9. Smirnova, M.A., Smirnov, M.N.: Dynamic modeling and hybrid control design with image tracking for a quadrotor UAV. Int. J. Appl. Eng. Res. **12**(15), 5072–5077 (2017)
10. Smirnov, N.V., Smirnova, M.A., Smirnov, A.N., Smirnov, M.N.: Combined control synthesis algorithm. In: 2017 Constructive Nonsmooth Analysis and Related Topics (Dedicated to the Memory of V.F. Demyanov), CNSA 2017 - Proceedings, pp. 194–196 (2017)
11. Smirnova, M.A., Smirnov, M.N.: Questions of stabilization and control of unmanned aerial vehicles. Comptes Rendus de L'Academie Bulgare des Sciences **71**(1), 87–91 (2018)
12. Smirnova, M.A., Smirnov, M.N.: Control synthesis for marine vessels in case of limited disturbances. Telkomnika (Telecommun. Comput. Electron. Control) **16**(2), 648–653 (2018)

Evaluating the CO2 Storage Well Connectivity at the Kemper CO2 Storage Site Utilizing AI

Klemens Katterbauer(✉), Pramod Patil, Abdallah Al Shehri, and Ali Yousef

Saudi Aramco, Dhahran 31311, Saudi Arabia
Klemens.katterbauer@aramco.com

Abstract. Sustainability objectives have played an important role in order to minimize carbon emissions and create value from carbon dioxide (CO2). A crucial objective is to extract the emissions from industrial processes before being released into the atmosphere and then stored in subsurface geologic formations. In order to minimize value creation and reduce carbon dioxide emissions into the atmosphere, carbon collection, storage, and utilization is essential. Since carbon capture, utilization, and storage allows businesses to continue operating while emitting less greenhouse gases, it may represent an opportunity for enhancing sustainability. However, storage must be practical, affordable, and secure. Storage formations may be found in both onshore and offshore settings, and each kind of geologic formation has pros and cons of its own. CO2 is stored for longterm, which requires safe storage, and this process includes a number of parameters being considered to make sure the gas is adequately contained. To identify the appropriate formations, a variety of variables are often needed to be considered. These include saline formations, oil and natural gas reservoirs, impermeable coal seams, organic-rich shales, and basalt formations.

The Kemper CO2 Storage Project, which is being constructed in Kemper County, Mississippi, is essential to analyze and acquire experience in managing carbon storage projects as well as understanding how to sequester CO2 efficiently. It was first constructed by Mississippi Power, a branch of Southern Company, and is also known as the Kemper County Energy Plant or Plant Ratcliffe. The aim of the project was to reduce the amount of carbon emissions produced and released in the atmosphere by using CO2 sequestration. The location of the plant in Kemper County was selected to make use of the area's untapped brown coal potential and provide regional diversity, which would help to balance the state's energy demand and generate energy.

In order to establish well connections for carbon storage, the research offers a unique directed well connectivity approach that uses similarity learning. Similarity learning can assist with connection determination for CCUS when evaluating the connectivity between the injectors and producers by classifying and categorizing data as well as by adding several non-Euclidean metrics to evaluate similarity. Similarity learning may also be used to identify anomalies and take remedial measures to anticipate future CO2 sequestration problems.

The results demonstrated a strong connection between the injector and producer wells, supporting the relationship between increasing CO2 injection volumes and the subsequently detected CO2 levels in the characterization wells. The method might be a useful method for optimizing CCUS well placement and sequestration.

R. Silhavy and P. Silhavy (Eds.): CSOC 2024, LNNS 1120, pp. 412–427, 2024.
https://doi.org/10.1007/978-3-031-70518-2_37

Keywords: Artificial Intelligence · CO2 storage · LSTM · Optimization

1 Introduction

In order to promote sustainability goals, carbon capture and storage, or CCS, has garnered a lot of interest. It is a technique that involves extracting carbon dioxide (CO2) from emissions from industrial processes before it is released into the atmosphere and storing it underground in geologic formations. CCS is a powerful tool for reducing CO2 emissions from industry since it permits business as usual while reducing greenhouse gas (CHG) emissions. But storage needs to be secure, affordable, and efficient. Appropriate storage formations can exist in both onshore and offshore areas, and every kind of geologic formation has pros and cons of its own.

Geologic storage is the safe, long-term storage of CO2 underground; it necessitates the assessment of several variables to guarantee that the CO2 is adequately trapped. For the long-term storage of CO2, a variety of formations are often thought to be appealing, including saline formations, oil and gas reservoirs, unmineable coal seams, organic-rich shales, and basalt formations [1].

Usually, the subsurface CO2 is maintained in supercritical state for CO2 injection. The temperature at which CO2 displays favored features is described as 31.1 °C (88 °F) and 72.9 atm (approximately 1,057 psi), or its critical point. At such high temperatures and pressures, the CO2 exhibits both gaseous and liquid properties, such as having the density and viscosity of a liquid. The main advantage of supercritical CO2 storage is the dramatic reduction in storage area required compared to ordinary pressure settings (Fig. 1).

As one descends more below the Earth's crust, the pressure of the fluids—oil, gas, or brine—in the formations naturally rises. For most places on Earth, the natural temperature and fluid pressure are over the crucial CO2 threshold at depths below 800 m, or around 2,600 feet. This suggests that CO2 injected at this depth or below will remain in the supercritical form due to the existing pressures and temperatures.

Capturing carbon dioxide (CO2) is essential because it prevents CO2 from rising to the surface. The primary fundamental processes all have a major influence on keeping the subsurface environment's CO2 levels adequate. Structural trapping, or the physical trapping of CO2 in the rock, is the technique that captures the most CO2. The rock layers and faults above and inside the CO2 injection storage formation function as seals, preventing CO2 from escaping the formation.

Once injected, supercritical CO2 may float higher than other liquids in the surrounding pore area. As a result, the CO2 will move through the porous rocks until it comes into contact with an impermeable layer of seal rock, where it will get trapped [2].

The residual CO2 trapping is the CO2 that remains trapped in the pore space between the rock grains after the CO2 plume has gone through. Supercritical CO2 pushes through the porous rock in the formation and displaces the fluid that is already there. As it passes through the pore spaces, little quantities of CO2 could be left behind as disconnected, or residual, droplets. The various drops may not be moving through the porous rock space.

During the solubility trapping process, some of the injected CO2 will dissolve into the brine water that is already within the pore spaces of the rock. At the CO2/brine water

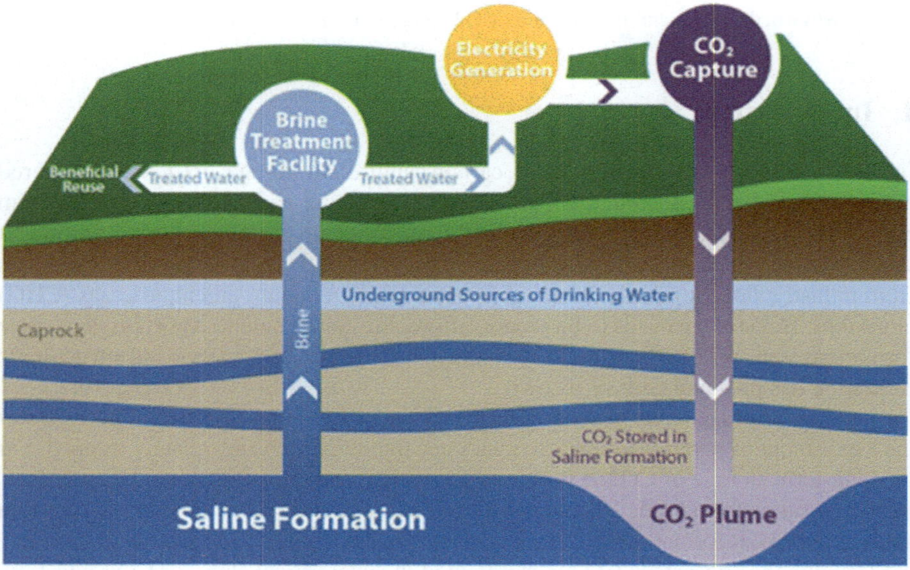

Fig. 1. Illustrative process overview of CO2 storage (Source: Department of Energy).

contact in the rock's pore cavity, some of the CO2 molecules dissolve into the brine water. A portion of that dissolved CO2 then combines with available hydrogen atoms to form HCO3-[3].

Mineral entrapment may result from the CO2 dissolved in the rock's brine water combining with the minerals in the rock. When CO2 dissolves in water, a weak carbonic acid (H2CO3) and ultimately bicarbonate (HCO3-) are generated. Some of the injected CO2 may be permanently trapped and stored if this weak acid reacts over time with the minerals in the surrounding rock to form solid carbonate minerals. When magnesium and CO3 in water combine, the mineral MgCO3 is formed on the surface of the rock grain. For a more extensive analysis, reservoir characteristics like as depth, injectivity, integrity, and storage resource are considered when choosing a storage site for long-term carbon dioxide storage.

The geologic storage process is considerably more complex. One or more caprocks, or seals—regionally continuous sealing formations—alongside a storage formation make up the full subsurface geologic storage complex. While the reactions that may occur in the formation's bigger or tinier pores are yet unknown, existing porosity-permeability equations may be able to predict reactions that are randomly distributed between tiny and huge pores or pore-through. The ability to store enormous amounts of compressed CO2 (millions of metric tons) at a storage facility is crucial. In particular, the storage resource for the majority of porous and permeable sedimentary formations is represented by a small amount of their accessible pore volume.

Another important consideration is the pace at which CO2 may be pumped underground. The injectivity of CO2 and the permeability of the formation are considerably coupled, as is the formation's integrity. One or more intact confining zones that seal above the injected formation and have no leakage channels are thus necessary in a storage complex in order to maintain the ability to securely confine the CO2 at a predetermined volume without leakage from the complex. To keep the CO2 in a supercritical condition, as was previously mentioned, the CO2 storage zone needs to be located at an appropriate depth and pressure.

Of special interest have been saline aquifers. These brine-filled, porous rocks known as saline formations have some intriguing properties that help them retain CO2. To optimize storage in these saline formations, brine deposits often have total dissolved solids (TDS) concentrations exceeding 10,000 ppm TDS and have demonstrated excellent storage capabilities.

Coal that is considered unmineable for geological, technical, or economic reasons might nonetheless contain CO2 (typically because it is too thin, too deep, or lacks internal continuity to be mined profitably). To be utilized for CO2 storage, the ideal coal seam has to be considered unmineable and have sufficient permeability. Coal seams may also contain methane (CH4), which can be produced in conjunction with CO2 injection through a process called enhanced coal bed methane (ECBM) recovery. In coal seams, CO2 may be injected and chemically held by connecting to the coal's surface and getting adsorbed, while CH4 is released and generated there. This trapping technique allows the CO2 to be stored over a long time [4, 5].

Another fascinating habitat is seen in basalt rocks, which are the result of large volcanic lava flows spreading, cooling, and eventually solidifying. Due to their unique physical and chemical properties, as well as those of the other formation types that may be found in between basalt layers, these basalts make good candidates for CO2 storage systems. Due to the chemistry of basalts, calcium and magnesium in the basalt may react with injected CO2 to generate the stable carbonate mineral forms of dolomite and calcite [6].

Shale formations are of interest because they are typically low porosity and low permeability rocks that function well as confining zones. Some shales share characteristics with coal, such as the ability to trap CO2 through adsorption (adherence to the surface), which releases methane and may make them attractive for storage [7].

At the moment, carbon dioxide (CO2) is being stored everywhere in the planet, including the US. For many years, large-scale, commercial projects such as the Sleipner CO2 Storage Site in Norway and the Weyburn-Midale CO2 Project in Canada have been injecting CO2. Every year, these programs store more than one million metric tons, or MMT, of CO2. Significant efforts are also being undertaken in Europe, Australia, and China. These massive commercial activities demonstrate that substantial volumes of CO2 may be permanently and securely stored.

The link between the reservoir compartments and wells is crucial to understanding the geologic CO2 storage analysis. The limited link between two zones that produce hydrocarbons and reservoir compartmentalization are important factors. Finding connections between various reservoir compartments is highly beneficial for plans including in-fill drilling, field rehabilitation, and well performance evaluation. Generally, two variables are used to quantify connectedness. The first are discontinuities in pressure gradients and variations in fluid similarity throughout the zones.

Connection can be ascertained using the results of reservoir fluid studies, pressure transient analysis, wireline formation testers such as RFT/MDT/XPT/RDT, and drill stem tests. Downhole fluid analysis (DFA) is another quick and effective method used for the purpose of vertical connection. When determining how several wells are connected to one another over time, these may be utilized as connection indicators [2].

However, judgments concerning the link between neighboring fields based solely on the evidence of fluid and pressure similarities are difficult to come by due to other geological characteristics. Firstly, it should be made abundantly evident that the mere fact that a number of zones exhibit initial pressure equilibrium does not imply that these zones would necessarily interact with one another during production. This is the result of the forces coming into balance throughout geologic time. It is possible for the various reservoir compartments to independently establish a pressure connection both during injection and over geological time. Secondly, it is possible that the fluid characteristics have altered throughout geological time in different ways, even if the reservoirs are linked.

The causes of the lateral and vertical fluid heterogeneity in the reservoirs include gravity segregation, temperature gradients, biodegradation, water washing, asphaltene precipitation during migration, charge from different source rocks, and continuous reservoir charging. Similarities in composition are often the basis for fluid comparisons. It is possible to compare fluid composition using a variety of methods. It is usual practice to use GC-MS to determine component distribution, biomarker distribution, and hydrocarbon fingerprint distribution. The two-dimensional gas chromatography technique is appropriate for this purpose. However, just comparing compositions is not an effective technique to establish correlations because of the inherent fluid heterogeneity [8].

Furthermore, if the hydrocarbons in the storage complex have accumulated in several zones inside a reservoir or many reservoirs within a single field, they may have been subjected to varying geological and chemical eras. A formation's stratigraphic shift and sealing faults may produce distinct filling histories or other secondary processes. Furthermore, vertical variations in the composition of the petroleum may arise via gradual intra-formation mixing. Therefore, discontinuities of the reservoir zones are not always ensured by fluid composition fluctuations, even when mixing of the whole reservoir occurs during fluid composition equilibration [9].

Even in the absence of compartmentalization, the fluid characteristics will vary across the reservoir because of things like ongoing charging. For assessing reservoir connectivity, sequence stratigraphy in conjunction with static modeling is the approach most commonly used. These systems do not, however, make use of the dynamic data from the reservoir and well. Static modeling and sequence stratigraphy results might be difficult to understand since subsurface data can have several meanings, particularly in the case of carbon dioxide storage. Although well data are restricted to the narrow region surrounding the shaft, they are more accurate and have higher resolution than seismic data. Furthermore, the accurate detection of reservoir dynamic changes during the hydrocarbon production process is made possible by 4D seismic monitoring [10].

However, the uncertainties in this data are higher than in well data due to issues with noise and repeating the seismic data acquisition. Non-uniqueness also occurs in the interpretation of seismic data sources. The most effective ways to get over these limitations are to develop varied models by combining many datasets, and to update or eliminate particular models as new data becomes available. Reservoir modeling is typically used to address difficulties that are more manageable, such as inter-well communication. To update the simulation model with the new data, however, requires time or might not be possible for carbon dioxide storage if the History Matching procedure is used. In order to incorporate the flow rate, BHP, or both data to evaluate inter-well connection, a direct link between the injector and the relief or observation wells is often constructed (Ezekiel, et al. 2020).

In order to correctly evaluate the connection between adjacent fields or the connectivity of various zones within a field, a number of geological and reservoir engineering criteria are studied simultaneously. Numerous new research projects have been launched that use machine learning and artificial intelligence to determine good connection estimates. Although they have been shown to work for traditional reservoir production, there are further difficulties when using them for carbon dioxide storage. In this article, we present a novel approach to well connectivity estimate for carbon dioxide storage, using simulated data to illustrate the framework on the CO2 Kemper storage site.

2 Methodology

The problem of finding similar items in a dataset is one that the similarity learning subfield of machine learning attempts to solve. It is often used in recommendation systems, where finding items similar to those a customer has already expressed interest in is the objective. The initial stages of similarity learning approaches frequently start with each item in the dataset being represented as a vector. The similarity between two items is then computed using the cosine of the angle produced by their vectors. There are several ways to represent items as vectors, and each has advantages and disadvantages of its own. For example, each item may be represented as a bag of words.

This method ignores the word order, which might have some negative effects even if it is simple and effective. Seeing everything as a collection of traits is an alternative approach. This way could be more descriptive, but it might also be more difficult to calculate how similar two items are. Similarity learning methods may be applied to a variety of tasks, including recommendation, grouping, and classification [11].

Applications of similarity learning for time series data, including pressure monitoring and carbon dioxide injection data, are numerous. The k-nearest neighbor approach, which compares a new data point to the k most comparable data points in the training set, is the most often used similarity learning technique. To determine similarity in such a case, the data points may take the form of subranges of the observation data [12, 13].

Another method for handling nonlinear similarity is support vector machines, which find a hyperplane that maximizes the distance between the closest data points of different classes. With the rise in computing power, neural networks—which can learn complex similarity rules—have attracted a lot of attention. A Siamese network has been widely used for similarity determination. It is made up of two identical neural networks trained to yield the same response for similar inputs. Autoencoders are a type of neural network that may be used to learn similarity and are a subclass of neural networks for similarity determination.

An autoencoder encodes a lower-dimensional version of an input before decoding it back to the original input. Because the autoencoder is trained to minimize the reconstruction error, it is compelled to acquire a condensed representation of the data [14].

To expedite similarity search, one approach that may be used is location-sensitive hashing. By avoiding comparing every data point, proximity-sensitive hashing may find related ones more quickly. It produces hash functions that are extremely likely to map related inputs to the same hash result. Among the benefits of similarity learning is that it can improve machine learning algorithms' efficiency and performance. Furthermore, similarity learning can aid in the comprehensibility of machine learning models. The curse of dimensionality, which occurs in high-dimensional environments when most points are far from one another and makes it difficult to learn effective similarity connections, is one of the main obstacles in similarity learning [15].

A similarity aware time-series classification framework was applied to analyze the CO2 similarity between wells. The framework operates as follows. Given the pair of wells, $\left(x_i^1, x_i^2\right)$, a function $g\left(x_i^1, x_i^2\right) = z_i$ is determined to estimate whether the wells are similar or not based on $z_i \in \{0,1\}$. Subsequently, with the similarity measure $y_i \in \mathbb{R}$, then the objective is to find a function, $f\left(x_i^1, x_i^2, z_i\right) \sim y_i$ that approximates the function for every quadruplet $\left(x_i^1, x_i^2, z_i, y_i\right)$ in order to determine their similarity.

Theoretically, this system addresses these problems by providing similarity information to deep learning models, such the LSTM. To correlate each time series with a node in a network and each pairwise similarity with a connection between nodes, the process is reformulated as a graph neural network (GNN). Next, a graph convolution layer is layered with a backbone (e.g., LSTM) to jointly model time-series similarities and extract features. The similarity is ascertained by employing both LSTM and dynamic time warping [11].

3 Results

In Kemper County, Mississippi, construction is already underway on the Kemper Project, a natural gas-fired power plant. Plant Ratcliffe, also known as the Kemper County Energy Plant, is another name for it. Mississippi Power, a Southern Company subsidiary, began construction on the project. The project's goal was to reduce its carbon emissions as much as possible by using CO2 sequestration. Kemper County is a county in eastern Mississippi that is situated around thirty miles north of Meridian. The decision to locate the plant in Kemper County was made to take use of the area's untapped potential for brown coal, or lignite, while also providing some geographical diversity to help balance the state's power demand and enhance energy reliability [9].

As part of the early environmental planning for the Kemper County energy complex, Mississippi Power Company (MPC) prepared an environmental impact statement (EIS) that focused on the quality of groundwater and adjacent aquifers. The five viable aquifers in Kemper County include the Lower Wilcox, which is Tertiary in age; the Upper Cretaceous Eutwa-McShan; and the Gordo, Coker, and Massive Sands, all of which are part of the Tuscaloosa Group (Fig. 2). The Black Warrior River aquifer is part of the Southeastern Coastal Plain aquifer system, which includes all of these divisions [16].

The Lower Wilcox, Gordo, and Eutaw-McShan are the three aquifers in Kemper County with the maximum transmissivity, ranging from 670 to 51,000 square feet per day (sfpd) for each. The Mississippi Department of Environmental Quality assessed the water quality in the region (MDEQ). Their investigation shows that the freshest water is often found in the updip (north to northeast) portions of all aquifers. The MDEQ Report states that the Lower Wilcox has a low concentration of total dissolved solids (TDS) (100–200 mg/L) and is mostly made up of calcium carbonate. The MDEQ data classifies the Eutaw-McShan, Gordo, and Coker aquifers as sodium chloride in type, with high TDS (> 1,000 mg/L). Iron concentrations in massive sands generally range from 0.5 to 20 mg/L, which is too much for intensive usage as potable water [17].

Based on data from the MDEQ from 2003, the Lower Wilcox aquifer is pumped at the highest rate in Kemper County, at 0.44 million gallons per day (Mgd). Nevertheless, in Mississippi, usage reporting is optional, and estimates place the comprehensiveness of the data at only 60–65%. Furthermore, data indicates that the combined Coker, Eutaw-McShan, and Massive Sands aquifers in Kemper County are pumped at a rate of 0.48 Mgd. Lauderdale County to the south shows that the Lower Wilcox is the main producing aquifer with 10 5 Mgd and just 0.21 Mgd coming from Middle and undifferentiated Wilcox aquifers, despite the fact that classification of Upper and Lower Wilcox is ambiguous and relies on the depth of wells.

It should be observed that the Lower Tuscaloosa Group 'Massive Sand' appears to have graded into a non-USDW in southwest Kemper County, close to the location of the energy complex. At the Kemper energy complex, a deep water well was constructed in 2008 to explore the idea of using the enormous sand as the facility's working fluid. Even though the unit had a high production capability (up to 500 gallons per minute per well), the total dissolved solids and chloride concentrations were above 20,000 mg/L and 10,000 mg/L, respectively.

Within ten miles of the intended injection site, there are sixteen deep well locations. Production casing may be found between 1,431 and 9,220 feet below the surface, whereas surface casing can be found between 41 and 4,832 feet below the surface. Three further deep characterization wells were drilled inside the study region in the summer of 2017 as part of the Project ECO2S characterization effort. The MPC 26-5, MPC 34-1, and MPC 10-4 are among them (Fig. 3).

Every casing in these three wells is cemented all the way down to the ground's surface [18]. There are 13 dry holes, seven that are still plugged in and abandoned, five that are dry and abandoned, one that seems to be plugged and abandoned but has no record, and one well location within a 10-mile radius that has expired, according to records from the Mississippi Oil and Gas Board (MSOGB). The wells that are closest to the planned injection location are the five-mile-distance three recently constructed ECO2S wells and the five older wells. Among these are the clogged and abandoned wells of Mrs. Etta H. Wilson, Mrs. Culpepper, Mrs. Ella M. Wilson, the MPC 26-5, the MPC 34-1, the MPC 10-4, and the W.T. Caldwell #1 expired location well.

A full total depth (TD) casing was present in the MPC 26-5, MPC 34-1, and MPC 10-4, and the TD to ground surface was continuously cemented. The MPC 265 well was drilled a total of 5,869 feet. It was installed with 9 5/8″ surface casing to a depth of 2,500 feet and 5 1/2″ long string casing from surface to depth. The depth of the MPC 34-1

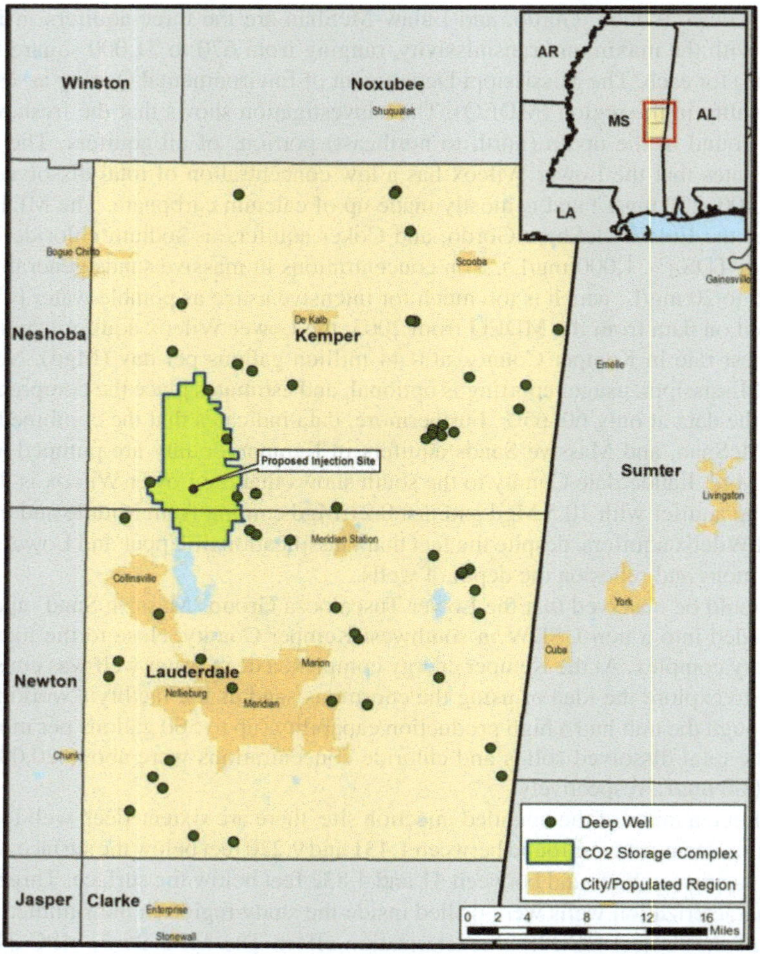

Fig. 2. Kemper Storage site overview (Source: Department of Energy).

well was 5,750 feet. It was installed with 9 5/8″ surface casing to a depth of 2,500 feet and 5 1/2″ long string casing from surface to depth. The proposed injection location is approximately 4 miles distant from the dry and abandoned W.T. Caldwell #1 well. In 1940, the well was excavated to a total depth of 3,309 feet. The 14-inch surface casing was completed at a depth of 40 feet, while the 7-inch production casing was positioned at 1,431 feet.

About five miles separate the Mrs. Etta H. Wilson 1 well—another dry, abandoned well—and the planned injection site. In 1956, the well was excavated to a total depth of 4,110 feet. 140 sacks of cement and 10 3/4-inch surface casing placed at a height of 309 feet completed the project. One deep water well is situated 10 miles from the proposed injection location. The well is 3,442 feet deep and located on the energy facility property, 4.5 miles from the site. Furthermore, the proposed injection location is located within a

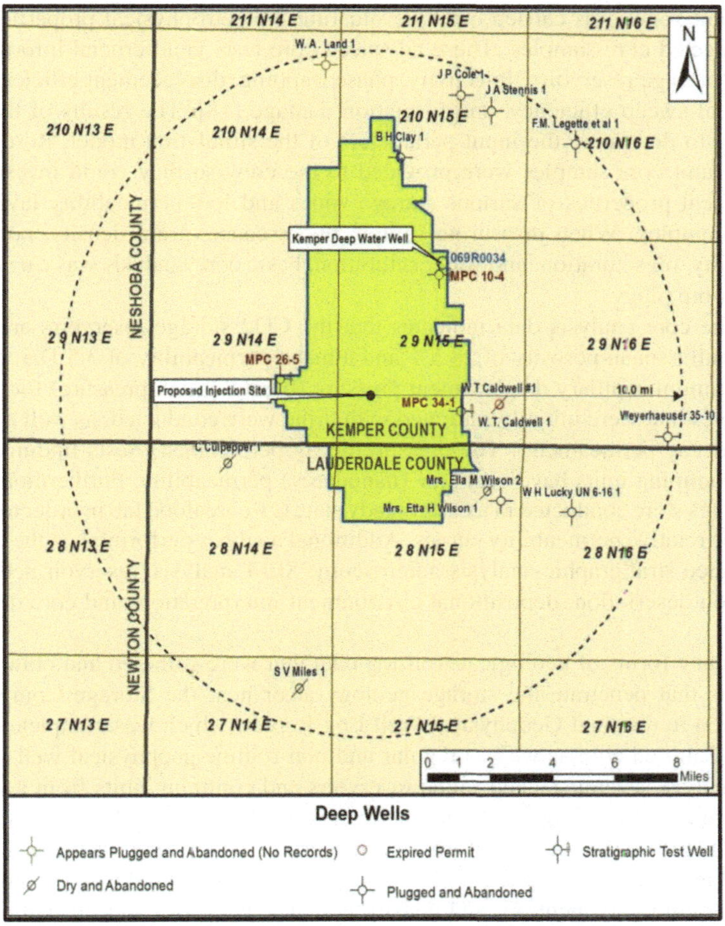

Fig. 3. Deep wells within 10 miles radius around the injection site (Source: Department of Energy).

10-mile radius of 42 shallow water wells. Currently, thirteen wells are used for various purposes, seven wells are not in use, three wells have been abandoned, five wells are used to monitor the quality of the groundwater, and twelve wells are used to provide water to the public. The wells are located between 1.5 and 10 miles from the proposed injection location, and they reach depths of 170 to 1,100 feet into the Lower and Middle Wilcox formations.

4 Geology

A geological model was developed for the Storage Complex based on data collected from the three Phase II characterization/monitoring wells. To get a better understanding of the target storage reservoirs, a comprehensive examination of the CO2 injection and confinement zones was carried out by evaluating the petrophysical properties present in the collected core samples. These advanced core tests yield crucial information on the target storage reservoirs' injectivity, phase trapping, displacement efficiency, aerial and vertical sweep efficiency, and formation damage [18]. The results of these trials were used to determine the input parameters of the simulation model. Reservoir and confining unit core samples were provided to the crew so they could investigate the petrophysical properties of various storage zones and low permeability layers at the Storage Complex. When measuring sample fluorescence, grain density, porosity, air permeability, oil saturation, and water saturation, basic core analysis was carried out on reservoir core plugs.

Routine core analysis data indicates that the CO2 storage reservoirs are of good quality, with a mean porosity of 28.5% and a mean permeability of 3.5 Darcy. Studies using minimum capillary displacement pressure (breakthrough pressure) for confining unit samples that were initially saturated with brine were conducted, as well as tests for pressure decay permeametry. The pressure decay permeametry tests findings suggest that the confining units have very low (nanodarcy) permeability. Furthermore, several core samples were conducted in a CO2-steady state CT core flood lab in order to generate CO2-brine relative permeability curves. Additional analysis performed on the core samples included stratigraphic analysis, microscopy, SEM analysis, reservoir petrography, thin section description, depositional environment interpretation, and core description [16].

The many forms of geological well log data that were gathered and obtained from deep wells that penetrate the storage geology at or near the Storage Complex were summarized in the final Geophysical Well Log Report, which was completed on June 29, 2020. Limited core as well as regular and non-routine geophysical well logs were used to analyze several stacked saline reservoirs and confining units from a variety of legacy oil/gas wells, a deep-water production test well, and three Project ECO2S wells finished in 2017. Among other forms of subsurface data, standard logging methods employed in regular geophysical well logs include the differential caliper, spontaneous potential, gamma ray, neutron, and density porosity. Well logs were usually available because Kemper County had a number of older gas and oil wells nearby. Furthermore, three further wells (MPC 26-5 #1, MPC 34-1 #1, and MPC 10-4 #1) were drilled in 2017 and yielded new Triple Combo logs and mud logs of the storage geology on the project site.

Well logs are based on well logging equipment measurements that determine petrophysical properties from the target formation. Non-standard geophysical logging including Formation Micro-Imaging (FMI), Dipole Sonic, and Combined Magnetic Resonance (CMR) were used in the new ECO2S wells. The saline storage reservoirs' porosity and permeability were determined by combining core, density/neutron porosity, and CMR. Significantly, the potential for CO2 storage at the Storage Complex was ascertained and reservoir performance during further CO2 injection was predicted using these evaluations of reservoir quality. Previous study has identified three porous, high-quality reservoirs: the Upper Cretaceous Lower Tuscaloosa group Massive Sand; the Lower Cretaceous Paluxy Formation; and the Washita-Fredericksburg interval. Each reservoir is layered with the massive regional seal known as the Marine Tuscaloosa Shale, which has lateral large-scale shale seals. Seals from the Tertiary era may also be discovered in the Porters Creek Clay and the Selma Group, which are located above the Marine Tuscaloosa. The project team received vital subsurface data for their updated stratigraphic studies, geologic modeling, and geologic mapping through the drilling of three characterization wells in 2017.

Depending on the mineralogy of the rock, CO2 undergoes a variety of additional geochemical reactions in the subsurface formation. Numerous simulations have been run by researchers to evaluate the processes of mineral dissolution and precipitation, as well as any potential impact on pore-throat widths. SEM and X-ray diffraction (XRD) methods are used in conjunction with these simulations, which are typically carried out at the pore and continuum scale, to examine the mineral reactions dependent on the rock [17]. Quartz is typically the predominant mineral phase in all samples, with smaller amounts of carbonates, feldspars, and clay minerals present. According to reach transport models, after CO2 injection, carbonates quickly dissolve in acidic environments whereas non-carbonate minerals maintain their stable mineral reactivity. As a result, the rock's overall porosity rises. It is evident that non-carbonate minerals dissolve over longer times (7,300 days). A rapid increase in rock porosity is prevented by amorphous silica precipitation. To create pore network models, 3D X-ray Computed Tomography (X-ray CT) scans of individual samples were used to collect data on the size distribution of pores and pore-throats as well as connection.

These models allowed us to simulate the CO2 volumes that should be detected at the three observation wells based on CO2 injection rates at the ECO2S well (Fig. 4). Throughout the course of the observation period, different CO2 injection rates were used to imitate the well injection rate, which was set to be between 5 and 28 MMSCF per day. The anticipated peak production availability of CO2 causes many spikes in CO2 injection. The preliminary simulation findings show that MPC 26-5 and ECO2S are correlated, since the measured CO2 levels at MPC 26-5 show a rise over the whole simulation period, which extends until 2030.

The directed well connectivity was then ascertained and shown in Fig. 5 using the simulated data. The coordinates x and y, which are in metric coordinates, must be seen in relation to one another. The normalization of the well connectivities results in a total of one. The findings show a significant well link between ECO2S and MPC 26-5, supporting the rise in CO2 levels reported at MPC 26-5 as a result of the CO2 injection in the ECO2S well. The results also show that ECO2S affects MPC34-1, and that ECO2S may have less of an effect on the well due to MPC10-4's close vicinity. An important part of the framework is the directed connection given the potential asymmetry in the impact of the between the various observation wells between each other.

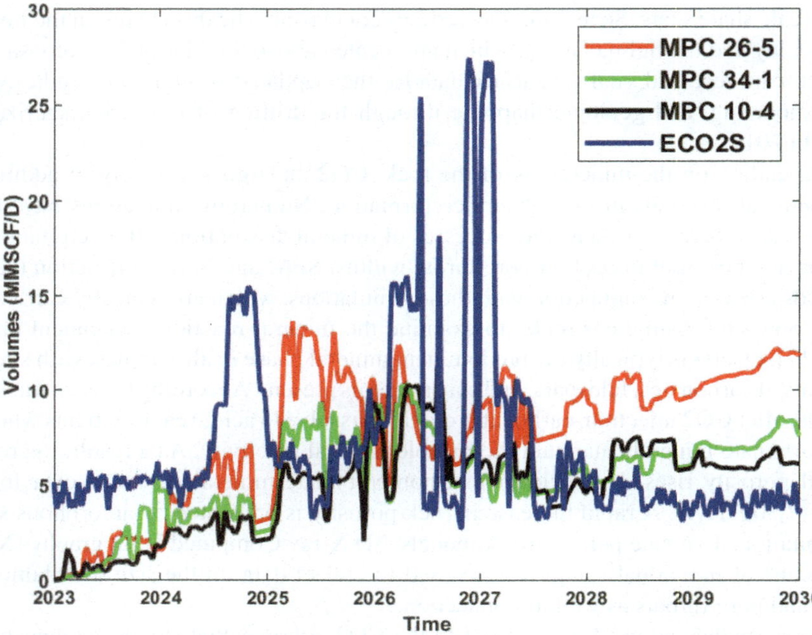

Fig. 4. Simulated CO2 injection and potential CO2 volume from the three observation wells.

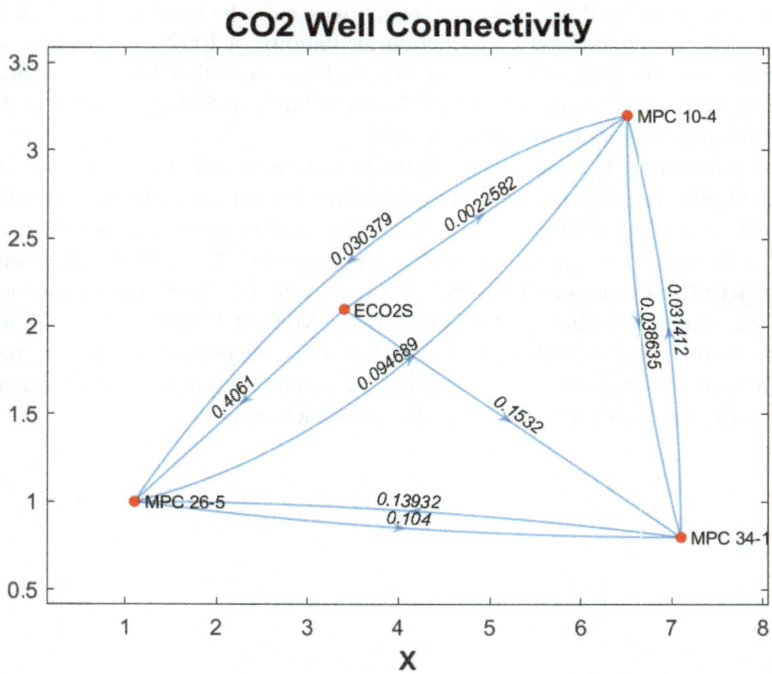

Fig. 5. Directed CO2 well connectivity between the injection and observation wells.

5 Conclusion

Carbon dioxide from emissions from industrial processes is extracted, stored in subterranean geologic formations in order to support sustainability goals. One important factor in lowering carbon emissions into the environment is CCS. Carbon Capture and Storage (CCS) is a useful instrument for reducing carbon dioxide emissions from businesses since it allows industry to operate while releasing less greenhouse gases into the atmosphere. Storage needs to be safe, inexpensive, and useful, though. Both onshore and offshore regions can include storage formations, and each kind of geologic formation has advantages and disadvantages of its own. Placing CO2 underground for safe, long-term storage is known as geologic storage, and it entails evaluating a variety of factors to ensure that the gas is sufficiently trapped.

For the long-term storage of CO2, a range of formations are generally considered suitable, including saline formations, oil and natural gas reservoirs, impassable coal seams, organic-rich shales, and basalt formations. The Kemper CO2 Storage Project, which is being built in Kemper County, Mississippi, is crucial to learning more about the effectiveness of CO2 sequestration and gaining experience managing carbon storage projects. Originally created by Mississippi Power, a division of Southern Company, it is sometimes referred to as Plant Ratcliffe or the Kemper County Energy Plant. The objective of the project was to lower its carbon emissions by using CO2 sequestration. In order to use the region's untapped brown coal potential and offer regional variance, Kemper County was chosen as the plant's location. This would assist in balancing the state's energy demand and ensure energy reliability.

The paper presents a novel directed well connectivity method that uses similarity learning to determine well connection for carbon storage. Comparing the relationships between the injectors and producers may be facilitated by using similarity learning, which has several benefits and can help with connection determination for CCUS. It can also be used to classify and organize data, as well as expand to non-Euclidean areas. Moreover, anomaly detection using similarity learning may be used to anticipate future CO2 sequestration problems and take corrective action.

The findings showed a robust association between the producer and injector wells, validating the hypothesis that higher CO2 injection volumes are correlated with later observations of CO2 quantities in the characterization wells. The methodology could serve as a valuable toolkit for optimizing the determination of CCUS wells and sequestration.

References

1. Bachu, S., Bonijoly, D., Bradshaw, J., Burruss, R., Holloway, S., Christensen, N.P., Mathiassen, O.M.: CO2 storage capacity estimation: methodology and gaps. Int. J. Greenh. Gas Control **1**, 430–443 (2007)
2. Ghiat, I., Al-Ansari, T.: A review of carbon capture and utilisation as a CO2 abatement opportunity within the EWF nexus. J. CO2 Util. **45**, 101432 (2021)
3. Hepburn, C., Adlen, E., Beddington, J., Carter, E.A., Fuss, S., Mac Dowell, N., Williams, C.K.: The technological and economic prospects for CO2 utilization and removal. Nature **575**, 87–97 (2019)

4. Luis, F., Al Hammadi, K., Tanakov, M.: Case study of CO2 injection to enhance oil recovery into the transition zone of a tight carbonate reservoir. In: Abu Dhabi International Petroleum Exhibition & Conference, 7–10 November, Abu Dhabi, UAE, Abu Dhabi (2016)
5. Soltani, R., Rosen, M.A., Dincer, I.: Assessment of CO2 capture options from various points in steam methane reforming for hydrogen production. Int. J. Hydrog. Energy **39**(35), 20266–20275 (2014)
6. Fleming, M.R., Adams, B.M., Kuehn, T.H., Bielicki, J.M., Saar, M.O.: Increased power generation due to exothermic water exsolution in CO2 plume geothermal (CPG) power plants. Geothermics **88**, 101865 (2020)
7. Rani, S., Padmanabhan, E., Prusty, B.K.: Review of gas adsorption in shales for enhanced methane recovery and CO2 storage. J. Pet. Sci. Eng. **175**, 634–643 (2019)
8. Adams, B.M., Kuehn, T.H., Bielicki, J.M., Randolph, J.B., Saar, M.O.: A comparison of electric power output of CO2 plume geothermal (CPG) and brine geothermal systems for varying reservoir conditions. Appl. Energy **140**, 365–377 (2015)
9. Ezekiel, J., Ebigbo, A., Adams, B.M., Saar, M.O.: Combining natural gas recovery and CO2-based geothermal energy extraction for electric power generation. Appl. Energy **269**, 115012 (2020)
10. Breen, S., Carrigan, C., La Brecque, D., Detwiler, R.: Bench-scale experiments to evaluate electrical resistivity tomography as a monitoring tool for geologic CO2 sequestration. Int. J. Greenh. Gas Control **9**, 484–494 (2012)
11. Katterbauer, K., Yousif, A.: Enhancing sustainability in reservoir management via a deep learning well connectivity determination framework. In: 83rd EAGE Annual Conference & Exhibition (2022)
12. Katterbauer, K., Arango, S., Sun, S., Hoteit, I.: Multi-data reservoir history matching for enhanced reservoir forecasting and uncertainty quantification. J. Pet. Sci. Eng. **128**, 160–176 (2015)
13. Katterbauer, K., Al Shehri, A., Marsala, A.: A novel deep reinforcement sensor placement method for waterfront tracking. In: SPE Middle East Oil & Gas Show and Conference, Manama (2021)
14. Romanenkova, E., Rogulina, A., Shakirov, A., Stulov, N., Zaytsev, A., Ismailova, L., Katterbauer, K.A.A.: Similarity learning for wells based on logging data. J. Pet. Sci. Eng. **215**, 110690 (2022)
15. Katterbauer, K., Qasim, A., Marsala, A., Yousef, A.: A data driven artificial intelligence framework for hydrogen production optimization in waterflooded hydrocarbon reservoir. In: Abu Dhabi International Petroleum Exhibition & Conference (2021)
16. Riestenberg, D.: CarbonSAFE: establishing an early CO2 storage complex in Kemper County, Mississippi: Project ECO2S. In: Southern States Energy Board, Peachtree Corners, GA (2018)
17. Bensinger, J., Beckingham, L.E.: CO2 storage in the Paluxy formation at the Kemper County CO2 storage complex: Pore network properties and simulated reactive permeability evolution. Int. J. Greenh. Gas Control **93**, 102887 (2020)
18. Wethington, C., Pashin, J., Wethington, J., Esposito, R., Riestenberg, D.: Mudstone baffles and barriers in lower cretaceous strata at a proposed CO2 storage hub in Kemper County, Mississippi, United States. Front. Energy Res. **10**, 904850 (2022)
19. Firozjaii, A.M., Saghafi, H.R.: Review on chemical enhanced oil recovery using polymer flooding: fundamentals, experimental and numerical simulation. Petroleum **6**(2), 115–122 (2020)
20. Azad, M.S., Trivedi, J.J.: Quantification of the viscoelastic effects during polymer flooding: a critical review. SPE J. **24**(6), 2731–2757 (2019)
21. Jouenne, S.: Polymer flooding in high temperature, high salinity conditions: Selection of polymer type and polymer chemistry, thermal stability. J. Petrol. Sci. Eng. **195**, 107545 (2020)

22. Zhao, Y., Yin, S., Seright, R.S., Ning, S., Zhang, Y., Bai, B.: Enhancing heavy-oil-recovery efficiency by combining low-salinity-water and polymer flooding. SPE J. **26**(3), 1535–1551 (2021)
23. Katterbauer, K., Marsala, A., Sofi, A., Yousif, A.: Carbon footprint forecasting of IOR activities via an intelligent NARX framework for promoting greener reservoir management. In: IOR 2021, Vienna (2021)
24. Katterbauer, K., Marsala, A.: A novel sparsity deploying reinforcement deep learning algorithm for saturation mapping of oil and gas reservoirs. Arab. J. Sci. Eng. **46**, 1–7 (2020)
25. Katterbauer, K., Al Sofi, A., Marsala, A., Yousif, A.: An innovative artificial intelligence framework for reducing carbon footprint in reservoir management. In: Annual Technical Conference and Exhibition, Dubai (2021)
26. Chen, S., Han, M., AlSofi, A.M.: Synergistic effects between different types of surfactants and an associating polymer on surfactant–polymer flooding under high-temperature and high-salinity conditions. Energy Fuels **35**, 14484–14498 (2021)
27. Han, M., Ayirala, S., Al-Yousef, A.: Review of offshore chemical flooding field applications and lessons learned. In: SPE Improved Oil Recovery Conference (2022)
28. Zhang, X., Han, M., Xu, L., AlSofi, A.M.: Long-term stability prediction of polyacrylamide-type polymers at harsh conditions via thermogravimetric analysis. Chem. Phys. Lett. **795**, 139538 (2022)
29. Hou, J., Han, M., Alsofi, A.M.: Optimization of surfactant-polymer formulations for improving oil production. Petrol. Sci. Technol. **42**, 1–17 (2022)
30. Zhang, X., Han, M., Fuseni, A., Alsofi, A.M.: An approach to evaluate polyacrylamide-type polymers' long-term stability under high temperature and high salinity environment. J. Petrol. Sci. Eng. **180**, 518–525 (2019)
31. Alkinani, H.H., Al-Hameedi, A.T.T., Dunn-Norman, S., Alsaba, M.T., Amer, A.S.: Review of the applications of decision tree analysis in petroleum engineering with a rigorous analysis. In: SPE Middle East Oil and Gas Show and Conference, Manama (2019)
32. Fuad, I.I.M., Rosli, L.H , Husni, H.: Application of hybrid ARIMA - decision tree algorthim in reservoir pressure surveillance. In: Abu Dhabi International Petroleum Exhibition & Conference, Abu Dhabi (2019)
33. Nikitin, N.O., Revin, I., Hvatov, A., Vychuzhanin, P., Kalyuzhnaya, A.V.: Hybrid and automated machine learning approaches for oil fields development: the case study of Volve field, North Sea. Comput. Geosci. **161**, 105061 (2022)
34. Saha, S., Vishal, V., Mahanta, B., Pradhan, S.P.: Geomechanical model construction to resolve field stress profile and reservoir rock properties of Jurassic Hugin Formation, Volve field, North Sea. Geomech. Geophys. Geo-Energy Geo-Resour. **8**, 68 (2022)

Movie Recommendation Using Collaborative Filtering Method and K-Nearest Neighbours: A Case Study on Netflix

Slamet Riyadi, Nia Audita(✉), and Febriyanti Azahra Abidin

Department of Information Technology, Muhammadiyah University of Yogyakarta, Kasihan, Indonesia
niaaudita01@gmail.com

Abstract. Streaming platforms like Netflix have become very popular in today's digital age, but users often need help finding movies or TV shows that suit their preferences. The Collaborative Filtering method has been used in recommendation systems to look for patterns of similarity among users with similar preferences. However, this method has limitations in situations with little user data or quick changes in preferences. To overcome this problem, this study proposes using the Collaborative Filtering method with the K-Nearest Neighbors (K-NN) algorithm. This research aims to develop a better recommendation system on Netflix by providing more personalized and accurate recommendations. The K-NN method uses user similarities based on genre preferences to provide relevant recommendations. Hopefully, this recommendation system can enhance the user experience by providing appropriate recommendations even in situations with little user data or quick changes in preferences. By implementing the Collaborative Filtering method with the K-NN algorithm, this research hopes to provide more relevant and interesting recommendations, helping users find content that matches their interests more efficiently.

Keywords: Netflix · Movies · Recommendation System · Collaborative Filtering · K-NN

1 Introduction

In the digital era, streaming platforms like Netflix have become global, offering a vast library of films and TV shows. Netflix, founded in 1999 as an online video store, has become the most widely used and still powerful, growing American online streaming provider specializing in video-on-demand distribution [1]. However, this abundance of content often overwhelms users, who struggle to find content that matches their interests. Despite its status as the most popular American online streaming provider, Netflix still faces the challenge of providing accurate and personalized recommendations to its users. This underscores the need for a solution to help users easily discover content that aligns with their preferences.

R. Silhavy and P. Silhavy (Eds.): CSOC 2024, LNNS 1120, pp. 428–436, 2024.
https://doi.org/10.1007/978-3-031-70518-2_38

Currently, the Collaborative Filtering method has been used to provide users with film and TV show recommendations. Collaborative Filtering is the process of filtering or evaluating items using opinions from other people [2]. This method focuses on the interconnections between user preferences and looks for similar patterns among users with similar interests. As one of the leading streaming platforms, Netflix has implemented the Collaborative Filtering method in its recommendation system. However. Although this method has provided good results in many cases, some limitations still must be overcome. One of the main limitations of Collaborative Filtering methods is the need for more accuracy when there is little user or item data. When there is more information about user preferences, recommendation systems may need help providing relevant recommendations. Additionally, when rapid changes in user preferences occur, these methods may need help to accurately keep up with those changes.

The Collaborative Filtering method using the K-Nearest Neighbors (K-NN) algorithm is proposed to overcome these limitations. The K-NN algorithm exploits user similarities based on genre preferences to provide more personalized and accurate recommendations. Using this approach, the recommendation system can provide recommendations that align with user interests, even when there is little user data or rapid changes in preferences. This research aims to develop a better recommendation system and improve the user experience on Netflix. By implementing the Collaborative Filtering method with the K-Nearest Neighbors (K-NN) algorithm, users will get more relevant and interesting recommendations, helping them find content that matches their interests more efficiently.

Research has developed a web-based film recommendation system using the prototype method, which allows users to test and evaluate the film recommendations provided. Implementing the KNN algorithm is used to predict film recommendations based on similar films the user has previously liked. This research implemented the content-based filtering method in a film recommendation system. This system provides recommendations for film titles based on similarities in genre and rating. However, because the system only uses these two variables, the evaluation results show that this recommendation system has a precision value of 45% and a recall value of 60% [3].

Other research designs film recommendation websites using the User-Base Collaborative Filtering method. This research's dataset contains information about film titles and user ratings collected. Then, a similarity calculation process is carried out between users based on the ratings given to certain films. This process uses distance calculation methods like Euclidean Distance or Cosine Similarity. The research results explain that the designed website can provide accurate film recommendations based on other users' rating patterns. This is achieved by applying the User-Based Collaborative Filtering method, where the rating patterns from one user are calculated to provide accurate recommendations based on available rating patterns from other users [4].

A. *Recommendation System*

A recommendation system is a mechanism used to provide suggestions or recommendations to users based on their preferences and profiles. The main goal of recommendation systems is to help users find movies and TV shows that match their interests and improve their viewing experience on platforms like Netflix. In developing recommendation systems, two approaches are commonly used, namely collaborative filtering and content-based filtering [5].

B. *Collaborative Filtering*
Collaborative filtering makes predictions by filtering item information from other people's opinions [6]. In this context, user rating data is used to discover similarities in preferences between similar users. If two users have similar preferences, they will likely have similarities in the movies and TV shows they like.

C. *User-Based Collaborative Filtering*
User-based collaborative filtering allows users to make recommendations based on similar preferences. This method looks for other users who have similar genre preferences to the user being given recommendations. User rating pattern information is used to identify neighboring users with similar preferences and suggest movies or TV shows that the user still needs to watch based on the preferences of those closest neighbors. Based on the items selected by the user's closest neighbors, this method makes predictions about items likely to be selected by the user in the future [7].

D. *Algoritma K-Nearest Neighbors (K-NN)*
The k-nearest neighbors' algorithm is an algorithm commonly used in collaborative filtering. The K-Nearest Neighbor (K-NN) model is a model that calculates the distance of the test data to each training data and then takes the k nearest neighbors of the initial data [8]. In recommending movies and TV shows on Netflix, the K-NN algorithm is used to find the nearest neighbors of users based on similar genre preferences. In this case, a rating matrix containing user rating information on movies and TV shows calculates the distance or similarity between users. Users closest to the user being recommended are considered the closest neighbors, and films or TV shows that these neighbors like will be recommended to the user.

E. *Genre sebagai Fitur Rekomendasi*
Genre is a key feature in movie and TV show recommendations on Netflix. A film recommendation system is built to make film selection easier for fans. This recommendation system aims to enable film fans to choose films through rating recommendations or choosing the genre and age limits the users choose [9]. Every film and TV show has a genre that describes its characteristics and themes. In the Collaborative Filtering method with K-NN, the user's genre preferences focus on finding similarities with other users. By leveraging genre as a feature, the recommendation system can identify neighboring users who have similar genre preferences and recommend a movie.

2 Methodology

A. *Data Collection*
The platform Kaggle, which is well-known for offering a variety of datasets to support research and data initiatives, provided the data for this study. The two CSV files, tmbd_5000_movies.csv and tmbd_5000_credits.csv, make up the "TMBD 5000

Movie Dataset," which was obtained from Kaggle with success. For analysis and algorithm implementation, each file includes vital data including movie scores, titles, release dates, genres, cast, crew, and other details.

B. *Data Pre-processing*

Data pre-processing is the next stage after data collection and is crucial to getting the data ready for analysis. Pre-processing deals with a number of data-related issues that could negatively impact system output, such as missing values, duplicate data, outliers, and inconsistent data formats [10]. Data cleaning is necessary to address problems like missing, erroneous, or inconsistent data, while data exploration is done to provide some early insights into the dataset.

Several essential Python modules are required in order to implement the Collaborative Filtering technique using K-Nearest Neighbors (K-NN) for genre-based Netflix movie and TV show recommendations. Specifically, the Pandas library makes data manipulation and analysis easier, making it possible to handle user preference historical data that is needed for Collaborative Filtering techniques in an efficient manner. The Scikit-learn toolkit then provides an abundance of tools and techniques for modeling and data analysis, such as the K-Nearest Neighbors algorithm that determines user similarity based on genre preferences. Additionally, NumPy is a foundational library for Python scientific computing, providing numerical operations necessary for executing Collaborative Filtering. Lastly, the Matplotlib package makes it easier to visualize the suggestions produced by the Collaborative Filtering technique.

To recommend movies and TV series on Netflix depending on genre, the Collaborative Filtering approach using the K-Nearest Neighbors algorithm is applied after the important steps of data exploration and cleaning. Understanding the dataset, spotting possible problems or inconsistencies, and cleaning the data to comply with analysis specifications are the goals of this iterative process.

C. *Algorithm Implementation*

The K-Nearest Neighbors (K-NN) method is used in Collaborative Filtering as part of the implemented algorithm. After the data pre-processing step, a ranking matrix is created based on user ratings for content. Then, based on similarity scores, nearest neighbors are determined by calculating user similarity using metrics like Pearson Correlation or Cosine Similarity. Movie and TV show recommendations are created by choosing content that is liked by neighbors who live close by but have not yet been watched by the user. Predicted ratings will determine the final recommendations. Users can get recommendations based on their preferences for genres through integration with the Netflix platform. Regular optimization and maintenance are essential for Netflix viewers to receive satisfactory recommendations and for algorithms to perform properly.

3 Result and Discussion

A. *Data Exploration and Cleaning*

Exploring and cleaning data in the DataFrame 'movies.head()' and 'credits.head()' is the first step in analyzing and cleaning data before applying the Collaborative Filtering method with the K-Nearest Neighbors algorithm to recommend films and

TV shows on Netflix. In the Data Exploration stage, we look at the first few rows of the two DataFrames to understand the structure and content of the data. Next, in the Data Cleaning stage, we look for missing values and address any problems, such as removing unnecessary columns, changing data formats, dealing with outliers, and correcting incorrect data formats. The goal is to ensure that the data used in the analysis is valid, accurate, and ready to be used to recommend movies and TV shows based on user preferences.

B. *Converting JSON to String*
The next process is converting JSON to string. The conversion from JSON to string is done because JSON stores data with complex structures such as objects and nested arrays. However, more advanced data analysis and processing is more practical and efficient if the data is in string format. The main reason why conversion from JSON to string is required in both DataFrames is to facilitate data exploration. By converting columns containing JSON data into strings, you can easily see and understand the contents of the data comprehensively. In addition, this conversion allows various functions and string-processing methods commonly used in programming languages. Thus, we can manipulate data such as searching, splitting, merging, and analyzing based on relevant patterns or attributes.

C. *Genre Column Analysis*

Table 1. Top 10 Genres

No	Genre	Number of Movies
1	Drama	2297
2	Comedy	1722
3	Thriller	1274
4	Action	1154
5	Romance	894
6	Adventure	790
7	Crime	696
8	Science Fiction	535
9	Horror	519
10	Family	513

Table 1 is the result of data visualization, which shows the 10 most popular film genres and their number of appearances in the Dataset.

D. *Cast Column Analysis*
The graph in Fig. 1 displays a list of actors with the most film appearances in the DataFrame. The number of appearances is calculated based on the 'cast' column data.

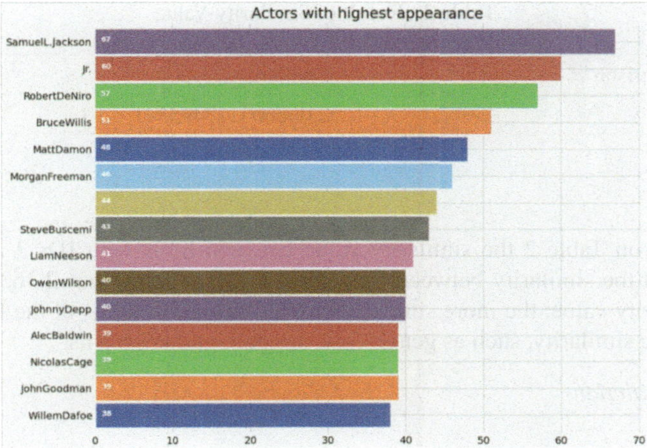

Fig. 1. Actors with Highest Appearance

E. *Director's Column Analysis*

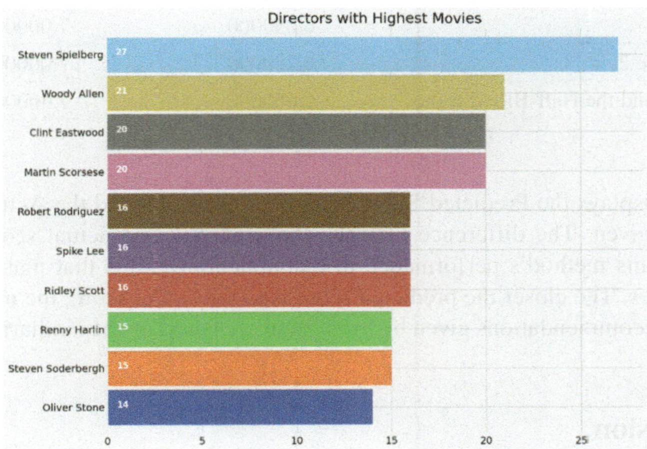

Fig. 2. Directors with Highest Movies

The graph in Fig. 2 displays a list of directors with many films in the DataFrame. The table shows directors with many films in the 'movies' DataFrame. The number of films is calculated based on the 'director' column data.

F. *Calculating of Similarities Between Films*
By measuring the similarities between the films in the dataset, you can identify films with similar or related genres. This makes it possible to recommend films or TV shows that have similar characteristics or are similar in terms of genre.

Table 2. Result of Similarity Value

Movie Comparison	Similarity Level
(3, 160)	2.068603127495341
(6, 100)	2.3636363636363638

Based on Table 2 the similarity value between films with IDs 3 and 160 was 2.068, and the similarity between films with IDs 6 and 100 was 2.363. The higher the similarity value, the more similar the two films are in terms of the features used to measure similarity, such as genre, cast, director, and keywords.

G. *Score Prediction*

Table 3. Score Predictor

Film	Score Prediction	Actual Score
The Godfather: Part III	6.950000	7.100000
Donnie Darko	6.010000	7.700000
Notting Hill	6.650000	7.000000
Despicable Me 2	6.280000	7.000000
Harry Potter and the Half-Blood Prince	6.650000	7.400000

Table 3 displays the Predicted Score results for each film and the Actual Score the user has given. The difference between the predicted and actual scores can help evaluate this method's performance in recommending films that match the user's preferences. The closer the predicted score is to the actual score, the more accurate the film recommendations given by the system are based on the similarity of the film genre.

4 Conclusion

Based on the results and discussion above, the Collaborative Filtering method with the K-Nearest Neighbors algorithm has been successfully used to recommend films and TV shows on Netflix. The data exploration stage ensures the validity and accuracy of the data used in this research, while the data cleaning stage helps address problems and change the data format to make it more ready for use. Calculating the similarities between films in a dataset helps identify similar films regarding genre and other elements. The score prediction results also show how accurate the system is in recommending films based on user preferences. The Collaborative Filtering method with K-NN is the first step to providing more accurate and personalized recommendations for Netflix users. By considering further developments, such as the application of model-based methods and optimal parameter selection, the recommendation system can continue to be improved. Developing these recommendation systems will help improve the user experience on streaming platforms and ensure they find films that match their interests and preferences.

Acknowledgements. The researcher would like to thank Muhammadiyah University of Yogyakarta for its extraordinary contribution and support while writing this article. Thank you for creating an inspiring academic environment, providing adequate facilities, and fully supporting the development of this research. Thank you also for your dedication in encouraging researchers to explore ideas and realize them through valuable knowledge.

References

1. Bagkar, P., Borude, A., Aga, Z.: Sentiment analysis on Netflix. IRE J. **4**, 121 (2021)
2. Dzumiroh, L., Saptono, R.: Application of the collaborative filtering method using implicit ratings in the film selection recommendation system in VCD rental. Jurnal ITSMART **1**(2) (2012). (in Bahasa)
3. Fitrianti, A.R., Rohmani, A., Widjanarto, W.: Website based film recommendation system using prototype method using K-nearest neighbors (KNN) algorithm. JOINS (J. Inf. Syst.) **5**(2), 278–287 (2020). https://doi.org/10.33633/joins.v5i2.4168. in Bahasa
4. Wahyudi, I.S.: Film Recommendation Engine Uses Genre Similarity Method Based on Collaborative Filtering (2017). (in Bahasa)
5. Mutiasari, H., Purboyo, T.W., Nugrahaeni, R.A.: Movie Recommendation System Using K-Means Clustering Method (n.d.). (in Bahasa)
6. Prasetyo, V.R., Hartanto, B., Mulyono, A.A.: Determining final assignment supervisors for students of the Informatics Engineering Department, University of Surabaya Using the Dice Coefficient Method. Teknika **8**(1), 44–51 (2019). https://doi.org/10.34148/teknika.v8i1.147. in Bahasa
7. Raghavendra, C.K., Srikantaiah, K.C.: Similarity based collaborative filtering model for movie recommendation systems. In: Proceedings - 5th International Conference on Intelligent Computing and Control Systems, ICICCS 2021, pp. 1143–1147 (2021). https://doi.org/10.1109/ICICCS51141.2021.9432354
8. Ricci, F., Shapira, B., Rokach, L.: Recommender systems: introduction and challenges. In: Ricci, F., Rokach, L., Shapira, B. (eds.) Recommender Systems Handbook, 2nd edn., pp. 1–34. Springer US, New York (2015). https://doi.org/10.1007/978-1-4899-7637-6_1
9. Bode, A. (2019). Feature Selection for Hollywood Movie Rating Prediction Using the K-Nearest Neighbor Model. (in Bahasa)
10. Ryana Agustian, E., Prasetyo Nugroho, E.: Film Recommendation System Using Collaborative Filtering Method and K-Nearest Neighbors, vol. 3, no. 1 (2020). https://ejournal.upi.edu/index.php/JATIKOM. (in Bahasa)
11. Salloum, S., Rajamanthri, D.: Implementation and evaluation of movie recommender systems using collaborative filtering. J. Adv. Inf. Technol. **12**(3), 189–196 (2021). https://doi.org/10.12720/jait.12.3.189-196
12. Sharma, P., Yadav, L.: Movie recommendation system using item based collaborative filtering. Int. J. Innov. Res. Comput. Sci. Technol. **8**(4), 266 (2020). https://doi.org/10.21276/ijircst.2020.8.4.2
13. Surendra, M., Babu, P., Raja, B., Kumar, S.: An Implementation of the User-based Collaborative Filtering Algorithm (n.d.)

14. Wardah, M.I., Putra, S.D.: Implementation of machine learning for film recommendations on IMDB using collaborative filtering based on IMDB sentiment analysis. Jurnal Manajamen Informatika Jayakarta **2**(3), 243 (2022). https://doi.org/10.52362/jmijayakarta.v2i3.868. in Bahasa
15. Wibowo, K., Studi Teknik Informatika STMIK TIME Medan, P.: Designing a film recommendation website using the user based collaborative filtering method. In: Jurnal Ilmiah Teknik Informatika, vol. 1, no. 2 (2021). http://ojs.fikom-methodist.net/index.php/. (in Bahasa)

Studying the Influence of Vietnamese Social Media on Vietnamese Stock Market to Forecast Market Trends

Cuong Bui Van, Trung Do Cao(✉), Tuyet Nguyen Thi, Ha Pham Dung, Hoang Do Minh, Le Nguyen Ha An, and Thai Pham Hong

DATX Vietnam, Hanoi, Vietnam
trung.do@datx.com.vn

Abstract. In the Vietnamese market, stock investors exchange information on popular social media platforms such as Zalo, Facebook, or forums like F247, F319. Here, market developments are reflected, analyzed, and evaluated from various perspectives and different viewpoints. This includes information about the overall market as well as various industries or specific stocks. Therefore, information on these platforms is clearly influenced by the stock market's movements. Conversely, does the information on these platforms have a reverse impact on the stock market's developments? Research will be conducted to identify this correlation.

Keywords: Vietnamese stock market · Stock prediction · Sentiment analysis · Machine learning · Decision tree

1 Introduction

To exchange information about the stock market, investors in the Vietnamese market have various channels available. One of the most popular platforms is Zalo, a Vietnamese messaging app [1]. On this platform, investors create chat groups to directly exchange information about stock market movements, with the primary exchange time typically occurring during trading hours from 9:00 AM to 2:45 PM daily from Monday to Friday. Additionally, Facebook is also widely used, with numerous stock groups regularly exchanging information about stock market trends during trading sessions as well as before and after trading hours.

In addition to these two platforms, in Vietnam, there are specialized stock forums where investors exchange market information, namely F247 [2] and F319 [3]. These forums allow investors to create accounts and post articles about their opinions on a stock or the general market trends. Information about individual stocks is particularly popular. Beneath each article, there are usually comments from interested investors expressing their personal opinions. Unlike Zalo, F247 and F319 are often used for information exchange even after the stock market trading hours have ended. The information exchanged on these platforms reflects the stock market trends in Vietnam quite clearly, promptly, and authentically.

R. Silhavy and P. Silhavy (Eds.): CSOC 2024, LNNS 1120, pp. 437–452, 2024.
https://doi.org/10.1007/978-3-031-70518-2_39

Almost all Vietnamese stock investors use at least one of these media platforms to research, exchange information, and guide their trading activities in the market. This means they are all to some extent influenced by the information shared on these forums. Simultaneously, the information on these forums is also clearly affected by the movements of the stock market. In [4], a group of authors analyzed the information status gathered from stock investor groups on the Zalo platform. This subsequent study builds upon and extends the research in [4], examining the reverse impact of the information status collected on the developments of the Vietnamese stock market, particularly on how the prices of impacted stocks are influenced by the status of the collected information. The information gathering platform is also expanded, including.

- Stock-related groups on social media platforms such as Facebook and Zalo in Vietnam.
- Stock forums F247 and F319.

Data will be collected from the beginning of 2021 to the present. Specifically, data from the year 2023 will be retained to validate observations, predictions, and conclusions drawn from the preceding data.

2 Related Work

The study [4] conducted data collection on stock market chat information from groups on the Zalo social media platform by using accounts to join these groups and developing software to crawl and process the data. Subsequently, they evaluated the information into 'positive,' 'neutral,' or 'negative' states corresponding to labels assigned as '1,' '0,' or '−1.' The results indicate that the information states accurately reflect the market trends as well as the movements of specific stock codes. This next study aims to expand data collection to other platforms such as Facebook, F247, F319, in addition to Zalo data, to assess the influence of these data states on the Vietnamese stock market trends.

The impact of information on the stock market has been extensively studied, with many results already published. In [5], the authors utilized information shared on the social media platform StockTwits to predict stock prices on the New York Stock Exchange using machine learning models, achieving an accuracy rate of approximately 65%. Research [6] leveraged data collected from social media platforms regarding the stock market to analyze positive or negative states, subsequently using these results to predict the prices of certain stocks. It concluded that "users' activity on StockTwits overnight positively correlates with stock trading on the next business day". Various studies [7–15] have also forecasted the states of stock markets and stock prices based on analyzing states on social media platforms. Meanwhile, [16] predicted stock market trends from the opening prices of two stocks, GOOGL and NKE, on the New York Stock Exchange, whereas [17] forecasted market trends based on analyzing 'hope' and 'fear' signals collected on Twitter (X).

3 Approach

3.1 Income and Social Media Data Processing

- ***Data Source:***

  - Posts, comments, and messages within approximately 200 groups related to the Vietnamese stock market on Facebook and Zalo.
  - Posts on the F247 and F319 forums.

- ***Text Data Processing:***

  - Extraction of stock codes (Named Entity Recognition - NER).
  - Assessment of the author's sentiment towards each mentioned stock code (Sentiment Analysis).

    - ***Aggregation of Statistical Results for Each Stock Code at Different Time Points:***

  - Popularity level assessment

$$\text{Popularity} = \sum_{i=1}^{n} w_i \tag{1}$$

- n represents the number of mentions within a week
- w is the time weight, where the closer to the current time being analyzed, the higher its value.

  - Sentiment

$$\text{Sentiment} = \frac{\sum_{i=1}^{n} w_i * s_i}{Popularity} \tag{2}$$

- is the processed sentiment score obtained in the previous step

After this process, we obtain a dataset containing the popularity and sentiment of each stock code at different time points.

3.2 Filtering and Preprocessing Data

- Currently, there are over 1600 stock codes in the Vietnamese stock market. Some of these codes are rarely mentioned on social media or have almost no trading activity. These stocks are excluded from the dataset to focus on the remaining ones.

  - Exclude 10% of the stocks with the lowest average popularity index or with the highest number of sessions with popularity = 0.
  - Exclude all stocks with more than 40% of trading sessions with unchanged prices.

- After filtering, the dataset contains 1403 stock codes.

- Calculate the change in sentiment and popularity indices for each time frame by measuring the difference between the scores at the current time and in the past. For example, the change in sentiment compared to the previous session:

$$\text{sentiment\_1d} = \text{current sentiment} - \text{sentiment of the immediate previous session} \tag{3}$$

$$\text{sentiment\_next\_1d} = \text{sentiment of the next immediate session} - \text{current sentiment} \tag{4}$$

- Similarly, calculate the percentage change in price and trading value concerning the Exponential Moving Average (EMA) for each stock code at each time point. For example, the price change of a stock compared to the price three sessions ago:

$$\text{price\_3d} = \text{current price}/\text{EMA}(\text{price of the immediate previous three sessions}) - 1 \tag{5}$$

- Calculate the percentage change in future price and the current price for comparison. For example, the price change of a stock in the next three sessions:

$$\text{price\_next\_3d} = \text{price of the next three sessions}/\text{current price} - 1 \tag{6}$$

3.3 Methodology for Evaluating Predictive Capability

- As the primary objective of the research is to assess the predictive ability of one or more social media indices on stock price changes, an appropriate evaluation method is necessary.
- Criteria for an Optimal Evaluation Method:

  - Minimize the influence of the distribution and magnitude of indices.
  - Ability to identify non-linear effects, as it is necessary to test non-linear assumptions, for example, such as 'extremely low or high sentiment causes a decrease in stock price'.
  - Ability to examine interactions between indices, for instance, when testing the hypothesis 'if sentiment and popularity are both high, the price increases, but if sentiment is low and popularity is high, the price decreases; however, if popularity is low, sentiment, whether high or low, has no effect'.
  - Limit the impact of data quantity to enable comparisons among different stock groups and the entire stock population.

- Applying Decision Tree Theory to Build the Method:

  - Treat each data point as a coordinate in a multidimensional space, with each dimension representing an index.
  - Iteratively select and split a dimension in the space to divide the data into different segments.

- Select dimensions and values for splitting to minimize the overall differentiation within each partition of divided data points.

$$m_i = \frac{\sum_{j=1}^{k_i} l_{ij}}{k_i} \tag{7}$$

$$\text{Loss} = \sum_{i=1}^{n} \sum_{j=1}^{k_i} l_{ij} - m_i \tag{8}$$

- n is the number of partitions created at the current step
- k_i is the number of data points in the i-th partition.
- L_{ij} is the label value of the j-th data point in the i-th data partition
- Limit the minimum number of data points (min_leaf_sample) for each data partition to 1% of the total data points to avoid overfitting.

  - When applied to subdatasets, the range of values for each index will be smaller than the range of values for the entire dataset. Therefore, to ensure that the method can effectively split these subdatasets into segments similar to when applied to the entire dataset, increase the min_sample_leaf value based on the average ratio of the range of values in the subset compared to the entire dataset

$$\text{min\_sample\_leaf} = 0.01 * \left(\frac{\sum_{i=1}^{n} \frac{\max(vi) - \min(vi)}{\max(Vi) - \min(Vi)}}{n} \right)^{-1} \left(\frac{\sum_{i=1}^{n} \frac{\max(vi) - \min(vi)}{\max(Vi) - \min(Vi)}}{n} \right)^{-1} \tag{9}$$

- n is the number of indices
- v_i is the i-th index of the subset.
- V_i is the i-th index of the entire dataset

  - Additionally, when reducing the number of indices, increase the min_sample_leaf value proportionally to the ratio of the number of indices
  - When further data splitting is no longer possible due to the constraints of min_sample_leaf, calculate the determination coefficient R_2 for the current partitioning.
  - The R_2 coefficient measures the total separation of each partitioned data. Each data partition with low separation indicates that the input indices can be relatively accurately used to predict the label.
  - The determination coefficient has a range from 0% to 100%. A value of 0% indicates that the input data has no influence on the output data. A value of 100% indicates that the output data is entirely dependent on and can be accurately predicted by the input data.

3.4 Individual Predictive Ability on the Entire Dataset

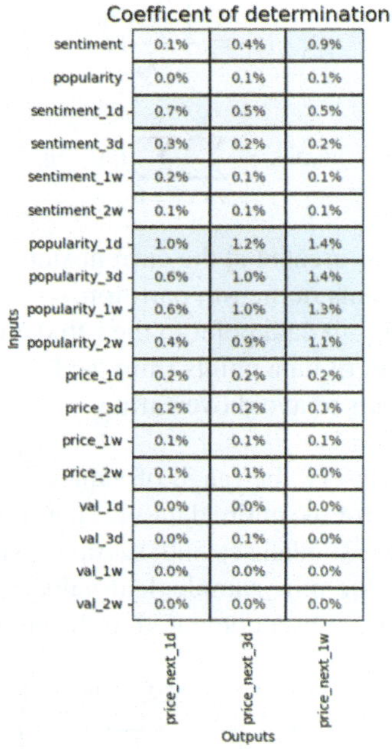

Fig. 1. The individual predictive ability of indicators for future prices

Data in Fig. 1 shows that individual indices almost have no predictive ability for future stock prices (Fig. 2).

- Individual social media metrics have a better capability to predict the future attitudes of social networks, however, their accuracy remains relatively low. This predictive ability stems from the fact that if popularity or sentiment is too high or too low, it can only change in the opposite direction. For example, when popularity has a value of 0, it can only increase in the future.
- Price and transaction value metrics do not reflect the influence on future social network attitudes. It is possible to discard the hypothesis that if a stock rises, the social network will have a positive attitude towards that stock.

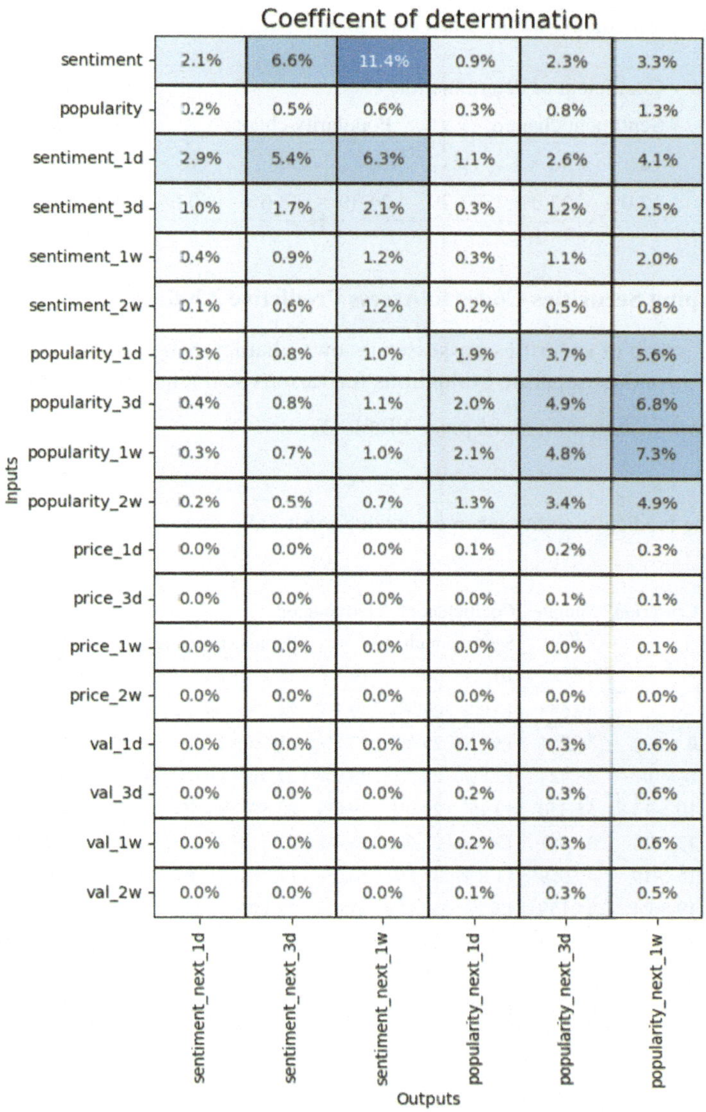

Fig. 2. The individual predictive ability of indicators for social media sentiment in the future

3.5 Predictive Ability When Using Multiple Input Variables Simultaneously

See Table 1.

- Combining input metrics results in significantly better segmentation outcomes. However, the ability to predict price changes remains very low.
- Overall, predicting the influence in the next week is easier than predicting the influence immediately after the trading session.

Table 1. The predictive ability when combining input indicators across the entire dataset

Sample size	Coefficient of determination								
	Sentiment change			Popularity change			Price change		
	1d	3d	1w	1d	3d	1w	1d	3d	1w
472,475	12.0%	27.0%	37.7%	24.3%	38.6%	40.6%	5.4%	5.6%	6.0%

3.6 Grouping Securities Codes to Assess Predictive Ability

Since each group of securities possesses its own characteristics and investor base, it is necessary to conduct separate evaluations for various segmentation methods.

a) *Segmentation based on stock price* (Table 2)

Table 2. The predictive ability when combining input indicators within groups categorized by stock prices

Cate-gory	Cri-terion	Sample size	Coefficient of determination								
			Sentiment change			Popularity change			Price change		
			1d	3d	1w	1d	3d	1w	1d	3d	1w
Price (thou-sand VND)	<6	49,658	13.0%	27.3%	39.1%	24.1%	39.9%	43.6%	11.1%	9.5%	9.2%
	6->8	36,251	13.9%	29.8%	37.3%	21.6%	36.8%	45.4%	6.7%	7.5%	9.1%
	8->10	38,424	12.8%	28.3%	38.9%	27.8%	40.6%	43.0%	6.5%	7.1%	8.0%
	10->13	55,350	13.1%	29.0%	39.0%	20.8%	41.4%	44.0%	6.1%	6.9%	7.9%
	13->15	31,751	12.7%	27.0%	36.4%	22.0%	41.2%	42.0%	6.9%	7.8%	9.3%
	15->19	49,059	11.8%	27.5%	36.0%	17.2%	29.9%	38.9%	6.6%	7.4%	8.4%
	19->24	44,159	12.2%	27.3%	37.0%	22.1%	35.8%	38.9%	6.4%	7.1%	8.5%
	24->31	41,148	11.9%	25.5%	35.3%	12.0%	31.7%	35.7%	7.7%	9.0%	9.2%
	31->45	43,129	11.6%	26.4%	36.2%	9.8%	30.9%	31.5%	7.4%	7.5%	8.4%
	>45	44,377	10.2%	21.9%	30.5%	6.5%	14.9%	29.5%	5.8%	6.2%	7.1%

b) *Segmentation based on market capitalization* (Table 3)

Table 3. The ability to predict when combining input indices from groups segmented by market capitalization

Category	Criterion	Sample size	Coefficient of determination								
			Sentiment change			Popularity change			Price change		
			1d	3d	1w	1d	3d	1w	1d	3d	1w
Value (billion VND)	<60	42,647	18.5%	38.3%	48.9%	46.4%	50.3%	51.2%	8.3%	7.2%	6.9%
	60->120	41,840	14.4%	33.1%	41.2%	33.9%	45.2%	43.3%	5.9%	6.2%	6.8%
	120->210	45,413	13.9%	31.9%	42.6%	28.9%	43.8%	45.1%	6.0%	6.8%	8.3%
	210->340	43,697	13.5%	30.0%	39.4%	27.2%	38.6%	37.6%	6.5%	7.2%	8.5%
	340->540	43,583	12.1%	28.0%	35.4%	22.5%	34.3%	35.2%	6.7%	7.5%	8.6%
	540->890	43,132	11.3%	25.2%	35.8%	14.3%	34.2%	34.6%	7.1%	8.9%	10.6%
	890->1600	44,184	10.9%	25.7%	34.4%	9.8%	17.4%	28.2%	7.9%	9.6%	11.0%
	1600->2900	42,093	10.3%	22.5%	30.8%	7.8%	15.9%	23.9%	9.1%	10.0%	11.8%
	2900->8100	43,202	8.5%	17.8%	25.5%	4.3%	9.4%	12.2%	11.2%	11.0%	13.1%
	>8100	43,515	7.7%	16.1%	22.3%	2.8%	7.4%	9.4%	13.1%	13.2%	14.9%

c) *Further subdividing at larger market capitalization ranges* (Table 4)

Table 4. The ability to predict when combining input indices from large-cap stock groups

Category	Criterion	Sample size	Coefficient of determination								
			Sentiment change			Popularity change			Price change		
			1d	3d	1w	1d	3d	1w	1d	3d	1w
Value (billion VND)	2900->3500	8,714	11.8%	23.9%	31.7%	8.9%	14.7%	17.3%	13.9%	17.0%	19.5%
	3500->4200	8,666	10.4%	20.5%	27.5%	5.3%	9.4%	12.1%	14.0%	15.1%	17.5%
	4200->5200	8,706	10.6%	18.2%	25.7%	4.9%	10.2%	12.6%	12.6%	13.4%	14.4%
	5200->6300	8,755	10.0%	19.0%	26.8%	6.1%	9.7%	11.8%	14.3%	14.4%	15.3%
	6300->8100	8,361	9.0%	19.9%	25.1%	5.3%	9.6%	13.3%	14.3%	14.8%	15.3%
	8100->11000	8,019	9.7%	20.4%	26.0%	4.1%	9.5%	12.8%	15.0%	14.2%	18.4%
	11000->16000	9,556	9.7%	19.5%	28.1%	5.2%	11.5%	13.2%	16.4%	18.1%	19.5%
	16000->30000	8,686	10.0%	19.8%	26.8%	6.7%	11.2%	13.7%	14.5%	15.9%	19.1%
	30000->67000	8,630	8.2%	15.4%	22.9%	3.6%	7.6%	11.1%	14.7%	14.0%	17.7%
	>67000	8,624	9.1%	14.0%	19.0%	10.0%	10.9%	13.7%	22.2%	17.8%	19.5%

- In general, the correlation coefficient when grouping stocks is higher than this coefficient across the entire dataset for price changes, but lower for sentiment changes on social media. The relationship between social media and the stock market varies among different stock groups.

- Immediate impact on both stock price changes and social media sentiment changes is more difficult to predict than the impact after a period.
- Social media sentiment is also more predictable for stocks with smaller market capitalization or value. Stock price changes tend to be more predictable for large-cap stocks, although the differences between groups are quite small.

d) *Segmenting based on DATX industry categories* (Table 5)

Table 5. The ability to predict when combining input indices for stocks within the DATX list

Category	Criterion	Sample size	Coefficient of determination								
			Sentiment change			Popularity change			Price change		
			1d	3d	1w	1d	3d	1w	1d	3d	1w
Sector	Materials	13,948	8.1%	19.7%	26.5%	5.8%	10.3%	13.7%	11.2%	11.8%	13.9%
	Financial	16,949	8.7%	19.2%	25.9%	5.0%	12.2%	15.2%	15.4%	14.9%	19.0%
	Consumer goods	13,361	9.5%	19.7%	27.7%	6.4%	10.6%	12.8%	120%	12.7%	15.2%
	Healthcare	4,217	12.8%	22.4%	34.3%	6.6%	14.0%	15.8%	19.5%	22.2%	22.6%
	Banking	9,780	8.1%	16.6%	22.8%	10.4%	12.7%	15.7%	16.0%	16.0%	19.0%
	Textile	5,661	12.1%	20.3%	28.4%	7.0%	14.9%	22.6%	16.0%	17.9%	20.0%
	Construction	20,604	9.1%	20.4%	29.0%	7.8%	16.5%	24.2%	12.4%	13.7%	14.6%
	Fisheries	4,716	9.6%	21.3%	27.7%	14.6%	21.3%	26.4%	18.6%	20.2%	22.2%
	Real estate	28,897	8.4%	17.0%	25.1%	5.1%	9.4%	12.5%	14.0%	16.0%	18.2%
	Energy	7,809	9.2%	18.3%	26.5%	10.3%	15.2%	19.0%	16.5%	16.2%	17.1%
	Agriculture	11,111	8.8%	19.6%	28.6%	4.9%	10.0%	12.9%	14.3%	14.6%	17.5%
	Industrial	6,558	12.8%	26.7%	35.0%	6.4%	14.5%	17.0%	16.3%	18.0%	21.8%
	Chemical	5,231	8.7%	17.9%	24.9%	5.8%	16.3%	23.7%	19.3%	20.7%	23.5%
	Insurance	2,598	14.4%	25.7%	33.0%	6.8%	11.6%	15.9%	24.1%	23.7%	22.2%
	State banking	1,572	16.9%	23.9%	28.4%	17.2%	18.8%	24.1%	35.8%	30.7%	35.7%
	Utilities	8,789	9.8%	22.0%	29.9%	6.2%	14.6%	16.0%	14.2%	15.1%	16.3%
	Public investment	10,955	7.6%	15.7%	23.9%	6.8%	9.2%	11.2%	14.6%	16.3%	19.3%
	Information technology	4,704	13.6%	23.3%	27.0%	7.9%	12.8%	14.9%	17.1%	17.9%	21.6%
	Industrial real estate	6,195	11.1%	18.9%	26.6%	11.7%	17.2%	18.2%	17.2%	18.7%	22.2%
	Shipping	4,679	12.4%	25.4%	38.8%	7.1%	13.9%	18.9%	18.2%	20.4%	22.7%
	Steel	7,588	10.7%	22.2%	29.0%	6.3%	11.3%	14.0%	14.3%	15.8%	19.8%

- Segmenting stocks by industry yields the best results. However, some industries have too little data, leading to statistical analyses on small datasets that may lack reliability.

3.7 Visualizing the Impact of Social Media on Stock Prices

- By representing several top layers of the decision tree built using the method in Sect. 3 on the chart, we can observe the segmentation approach as well as the influence of indicators on price changes statistically.
- Price changes in one week for the entire dataset (determination coefficient 6%) (Fig. 3).

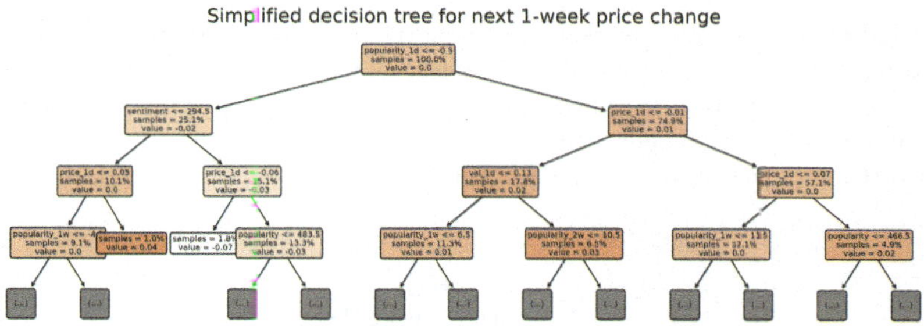

Fig. 3. A simple decision tree that statistically analyzes price changes over one week for the entire dataset

- Short-term popularity changes are the most crucial indicator for the entire dataset. Stocks with short-term popularity decrease incur an average loss of 2% after one week, while other stocks show an average gain of 1%.
- The portion of data with the largest average gain after one week at the top layers consists of stocks with moderate price increases compared to the previous session, lower sentiment, but a decrease in popularity.
- Price changes in one week for stocks with a market capitalization above 67,000 billion VND (determination coefficient 19.5%) (Fig. 4).

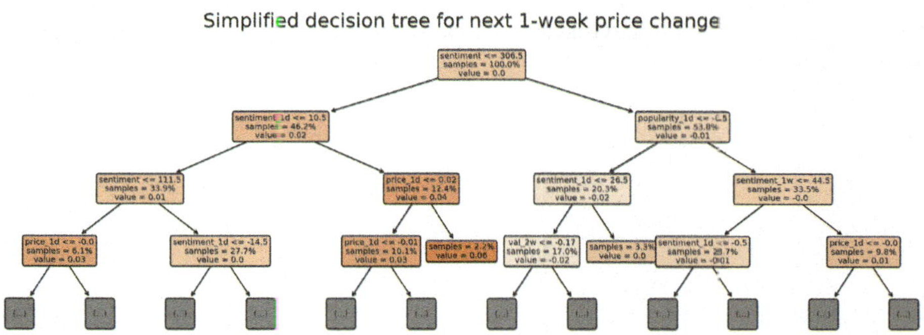

Fig. 4. A simple decision tree statistically analyzing price changes over one week for stocks with a market capitalization exceeding 67,000 billion VND

- Sentiment is the most crucial indicator for these stocks. Stocks with sentiment greater than 306.5 incur an average loss of 1% after one week, while other stocks show an average gain of 2%.
- The portion of data with the largest average gain after one week at the top layers of the decision tree consists of stocks with moderate sentiment and no decrease compared to the previous session, while the price increases by more than 2%.

3.8 Building Some Basic Models

- Utilized for stocks with a market capitalization exceeding 2,900 billion VND.
- Predicting outputs categorized into 20 levels, with higher levels being more positive for ease of comparison. Results will be presented using box plots to represent the distribution of price changes at each level.
- GAM (Generalized Additive Models) (Figs. 5 and 6).

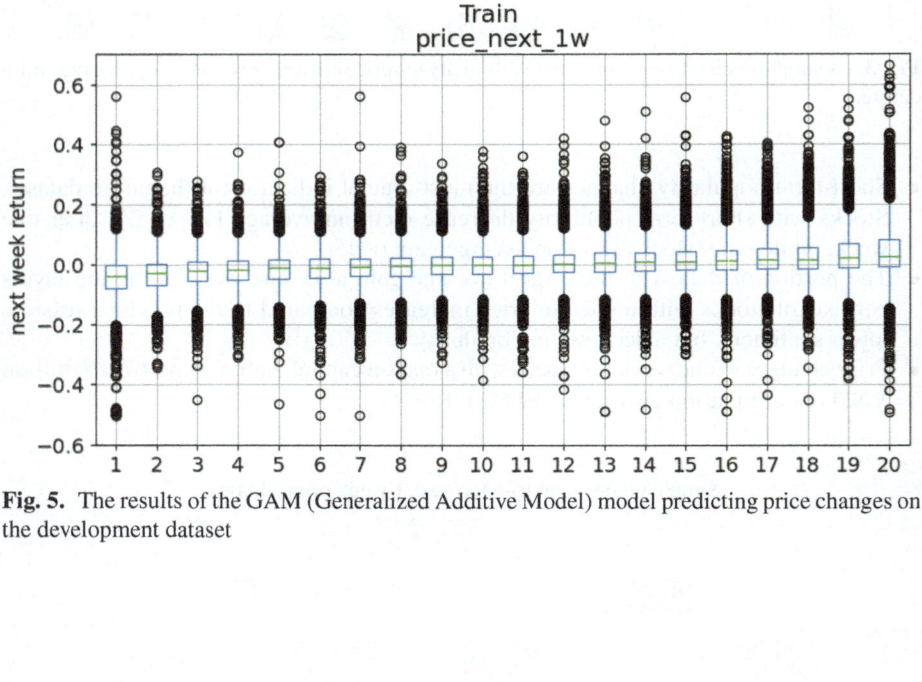

Fig. 5. The results of the GAM (Generalized Additive Model) model predicting price changes on the development dataset

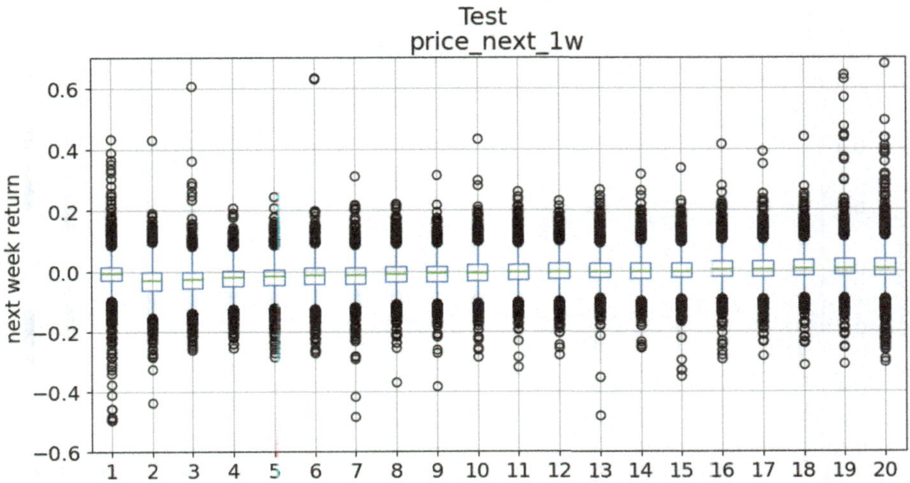

Fig. 6. The results of the GAM (Generalized Additive Model) model predicting price changes on the test dataset

- LGBM (Light Gradient Boost Machine) (Figs. 7 and 8)

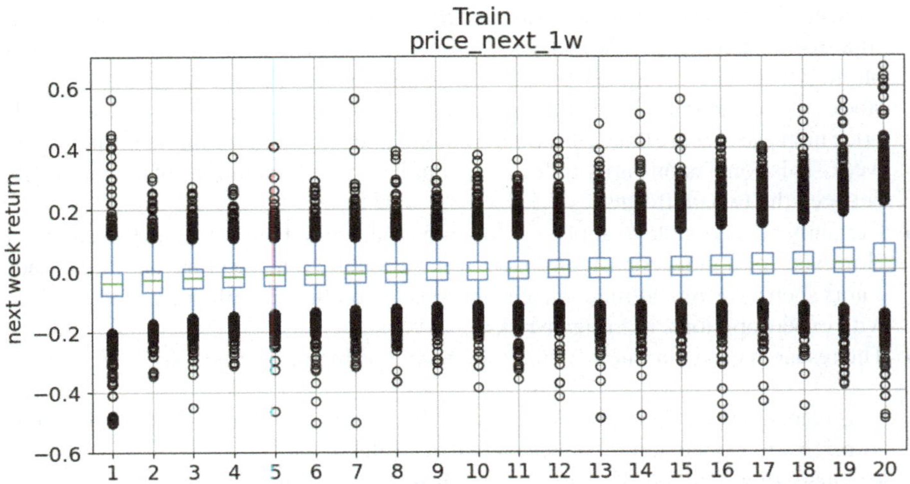

Fig. 7. The results of the LGBM (Light Gradient Boosting Machine) model predicting price changes on the development dataset

- The prediction results show clear separation at the highest/lowest levels. However, predictions at the mid-levels exhibit relatively similar price changes. The predicted levels also have a broad distribution, lacking concentration, which poses risks when applied to actual trading.

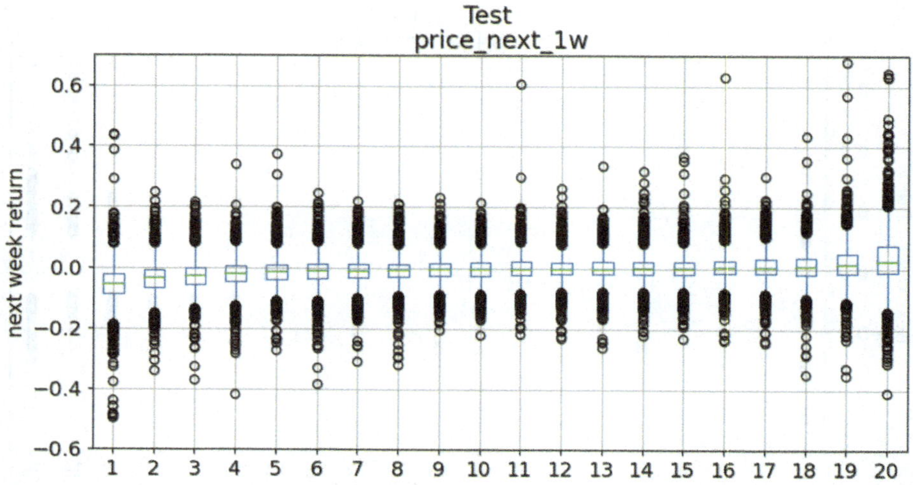

Fig. 8. The results of the LGBM (Light Gradient Boosting Machine) model predicting price changes on the test dataset

4 Conclusions and Developments

- Social media and the stock market influence each other; however, this interaction is quite complex and challenging to accurately determine. Absolute conclusions such as 'if a stock is popular, it will increase in price' cannot be unequivocally drawn. Based on the visualizations, a reasonable statistical-based conclusion could be 'large-cap stocks with moderate sentiment, no decrease compared to the previous session, and a price increase of more than 2% have the potential for significant gains in the upcoming week. This conclusion partly reflects the complexity and limitations of the interaction between the two platforms.
- Certainly, it is possible to apply a prediction model based on social media sentiment. However, for a more robust approach, it is advisable to incorporate additional indicators such as technical/fundamental analysis or combine it with other models, along with expert opinions, to mitigate risks.
- The research can continue with the following development directions:

  – Enhancing Data: Searching for and integrating additional data sources from various origins to expand the scope and accuracy of the model.
  – Integrating Multiple Factors: Research may extend to integrate various factors such as technical data, fundamentals, and expert opinions to enhance the predictive power of the model.
  – Advanced Model Development: Research may focus on developing advanced machine learning and artificial intelligence models to handle complex relationships between social networks and the stock market.
  – Risk Management: Adjust the model to minimize errors and clearly identify risk factors when applying it in practice.

– In-Depth Industry Research: Conduct thorough analysis on the impact of social networks within each specific industry to attain detailed insights and high accuracy.
– Integrating Investment Strategies: Develop investment strategies based on predictive outcomes, with support from financial experts and investors.
– Time Evaluation: Research can be expanded to assess the impact of social networks over time, aiming to better understand long-term and short-term trends.
– Testing and Real-World Application: Apply the model in real-world settings, test, and evaluate its performance before deploying it widely in the financial market.

References

1. Zalo – Apps on Google Play
2. https://f247.com
3. https://f319.com
4. Van, C.B., Cao, T.D., Manh, H.D., Dung, H.P., An, L.N.H., Trung, K.D.: Data mining natural language processing and sentiment analysis in Vietnamese stock market. In: Proceedings of 12th Computer Science On-line Conference 2023, Artificial Intelligence Application in Networks and Systems, vol. 3 (2023)
5. Coyne, S., Madiraju, P., Coelho, J.: Forecasting stock prices using social media analysis. In: IEEE 15th International Conference on Dependable, Autonomic and Secure Computing, 15th International Conference on Pervasive Intelligence and Computing, 3rd International Conference on Big Data Intelligence and Computing and Cyber Science and Technology (2017)
6. Xu, F., Keelj, V.: Collective sentiment mining of microblogs in 24-hour stock price movement prediction. In: IEEE 16th Conference on Business Informatics, Geneva, Switzerland, 2014, pp. 60–67 (2014)
7. Meesad, P., Li, J.: Stock trend prediction relying on text mining and sentiment analysis with tweets. In: 4th World Congress on Information and Communication Technologies (WICT 2014) (2014)
8. Gilbert, E., Karahalios, K.: Widespread worry and the stock market. Proc. Int. AAAI Conf. Web Soc. Med. **4**(1), 58–65 (2010)
9. Chen, S., Gao, T., He, Y., Jin, Y.: Predicting the stock price movement by social media analysis. J. Data Anal. Inf. Process. **7**, 295–305 (2019)
10. Schumaker, R.P., Chen, H.: Textual analysis of stock market prediction using breaking financial news: the AZFin text system. ACM Trans. Inf. Syst. (TOIS) **27**, 12 (2009)
11. Pagolu, V.S., Reddy, K.N., Panda, G., Majhi, B.: Sentiment analysis of Twitter data for predicting stock market movements. In: 2016 International Conference on Signal Processing, Communication, Power and Embedded System, Paralakhemundi, India, 3–5 October 2016, pp. 1345–1350 (2016)
12. Karabulut, Y.: Can Facebook predict stock market activity? In: Working Paper, University of Frankfurt, Germany (2011)
13. Mustansar, A.G., Madiha, A., Sin, W.L., Nadeem, Q., Amin, K., Jhanjhi, N.Z., Ali, J.: Stock market prediction based on machine learningand social sentiment analysis. Researchgate.net (2023)
14. Coelho, J., D'almeida, D., Coyne, S., Gilkerson, N., Mills, K., Madiraju, P.: Social media and forecasting stock price change. In: IEEE 43rd Annual Computer Software and Applications Conference (COMPSAC) (2019)

15. Shi, P., Song, S., Lang, Q., Liu, X.: Futures price forecasting based on LSTM model using public opinion sentiment analysis. In: 2023 42nd Chinese Control Conference (CCC), Tianjin, China, pp. 8683–8689 (2023)
16. Moghar, A., Hamiche, M.: Stock market prediction using LSTM recurrent neural network. Procedia Comput. Sci. **170**, 1168–1173 (2020)
17. Zhang, X., Fuehres, H., Peter, A.G.: Predicting stock market indicators through Twitter "I hope it is not as bad as I fear. Procedia Soc. Behav. Sci. **26**, 55–62 (2011)

Fiber Optic Temperature Sensor for Rigid Deformable Media

Tatiana Murashkina(✉), Elena Badeeva, Evgeniy Dudorov, and Alexsandr Badeev

Penza State University, Penza, Russia
timurashkina@mail.ru

Abstract. The safety of a nuclear power plant as a whole depends on the accuracy of measuring the temperature of the cement base of the dome of a nuclear power plant. The used measuring instruments should not create electromagnetic interference during measurements, therefore, fiber optic measuring instruments have been preferred recently. The subject of this study is scientific and technical solutions of fiber optic temperature sensors with compensation channel, which are part of the monitoring systems of nuclear power plant containment. The purpose of the work is to develop design and technological solutions of fiber optic temperature sensors, providing the possibility of its application in rigid deformable media with a minimum dynamic error due to inertial processes of transfer of thermal fields in the body of the sensor. The main approach to achieve the goal is adaptation of previously developed technical solutions of fiber optic temperature sensors to the conditions of measurement in rigid deformable media, which are the supporting parts of the dome of a nuclear power plant. As a result of research the structure and design of fiberoptic temperature sensors with a compensating channel of reflective type are determined. The ratios between the coefficients of thermal expansion of materials from which the structural elements of fiberoptic temperature sensors are made have been determined, providing the reduction of deformation of the protective housing and the sensitive element of fiberoptic temperature sensors at temperature change. The initial distance between the mirror surface and the ends of optical fibers opposite to the mirror surface, at which the sensor conversion function is linear, and the maximum sensitivity (up to 30%) and depth of modulation of optical signal conversion (up to 50%) are determined. Design-technological and structural solutions of fiberoptic temperature sensors with compensating channel of reflective type for systems of monitoring of stress-strain state of nuclear power plant containment placed in rigid deformable media have been developed. Application of the developed fiberoptic temperature sensor provides: reduction of internal stresses inside the deformable medium of the protective shell when its temperature changes in the range from minus 50 to +90 °C; reduction of dynamic error of temperature measurement caused by inertial processes in the body of the sensor due to the choice of materials and determination of relations between the coefficients of thermal expansion of structural elements of the fiberoptic temperature sensor; increasing the accuracy of temperature measurement of the protective shell due to the exclusion of deformation of the protective housing of the sensor and compensatory conversion of optical signals directly in the optical system of the sensing element #CSOC1120.

R. Silhavy and P. Silhavy (Eds.): CSOC 2024, LNNS 1120, pp. 453–463, 2024.
https://doi.org/10.1007/978-3-031-70518-2_40

Keywords: Fiber Optic Sensor · Temperature · Rigid Deformable Medium · NPP · Protective Shell · Micro-Optical System · Compensation Channel · Inertial Process · Coefficient of Thermal Expansion · Thermal Conductivity

1 Introduction

One of the main problems in the construction and operation of nuclear power plants (NPPs) is the accurate determination of the stress-strain state (SSS) of the NPP dome [1]. Therefore, the systems for monitoring the stress-strain state of the containment necessarily include sensors of strain, pressure, linear and angular microdisplacements, force, temperature, etc. One of the main requirements for the above sensors is the use of sensors in intrinsically safe design, the parameters of which do not depend on electromagnetic interference of any intensity [2].

The disadvantages inherent to electrical sensors and systems based on them are eliminated in fiberoptic systems combining fiberoptic sensors (FOS) of various physical quantities (Fig. 1) [3–7].

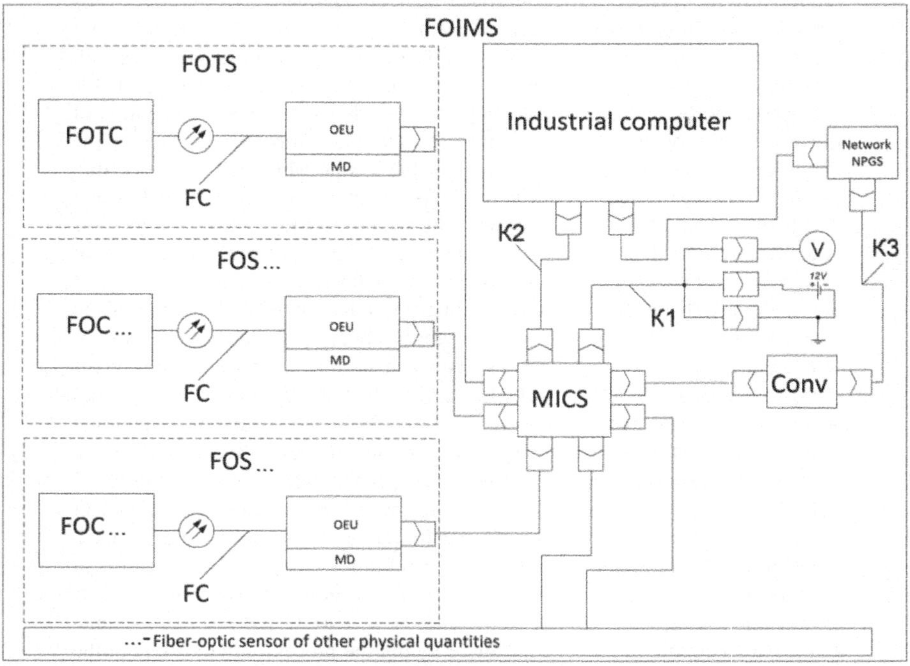

Fig. 1. Structural scheme of stress-strain state monitoring systems for nuclear power plant.

Fiberoptic sensors are designed to measure SSS of the containment of NPP power units, to measure the contact pressure of soil on the edge of concrete structures, temperature in the body of the containment of NPP power units and their environment (soil, concrete, air) during design periods of construction and operation [8, 9].

In order to measure various parameters of the NPP containment during its construction and operation as part of the systems for monitoring the stress-strain state, it is mandatory to use fiberoptic temperature sensors (FOTS). For example, in the process of NPP construction the temperature of cement in the base of the containment varies in the range of 0…+90°C. There are situations when the temperature changes in the range of minus 30…+90°C.

Nuclear power plant designers assume different operating modes: normal operation, mode of normal operation violation (modes of "small leakage", "big leakage", violation of heat removal) and emergency mode [10]. Under normal operation conditions, the temperature in the inter-shell space is from 10 to 60 °C with absorbed dose rate up to $1.19–10^{-6}$ Gy/h. When normal operation is disturbed, the concrete temperature rises to +85…90 °C with absorbed dose rate up to 1.0 Gy/h.

Measuring the temperature of large volume cement in the process of its hardening and then in the process of operation of such a large construction as NPP is a big problem connected with unreliability of known technical solutions of temperature sensors [11, 12]. The use of high-pressure protective sleeves for installation of known sensors in cement leads to a large (up to 5…10%) measurement error. At installation of such sensors in hardening cement their case and internal elements will experience large deformations, for example, the level of deformation of the case of FOTS, described in [12], is comparable to the measuring range of the sensor, which is determined by changing the distance between the reflector and the end of optical fibers, respectively, all metrological characteristics obtained in the process of grading or calibration of the sensor, will change significantly.

The design of the FOTS should provide its installation in the body of the shell (concreting blocks) or on the surface of concrete.

The developed FOTS generally consists of a fiberoptic temperature transducer (FOTT), a fiberoptic cable (FOC) and an optoelectronic unit (OUB) (see Fig. 1) [7]. The OUB includes a matching device (MD) docked with an electrical connector to the module of information conversion и storage (MICS). The MICS is connected to the power supply and measuring instrument (e.g., voltmeter V) by means of electric cable K1 and, if necessary, to an industrial computer for information visualization by means of cable K2, and to the NPP network by means of RS485 or RS232 CON interface convector by means of cable K3.

The use of a FOC up to 1 km long makes it possible to conduct measurements in the zone of elevated temperatures and radiation [2].

The paper [1] presents the results of temperature measurements inside and outside the containment of power unit No. 3 of Rostov NPP during the acceptance test period. These results were used in the clarifying calculations of the stress-strain state of the containment from the impact of loads acting during the acceptance test period. The temperature in concrete was determined by the average value of the temperature in the containment, as well as by the output signals of string temperature sensors installed on the body of the containment. The temperature during the tests varied from 29 to 39 °C.

In works [13, 14] the data obtained during concreting on samples of the base of the protective shell are given, according to which in the first day after concreting from the exotherm of concrete the temperature rises to 56 °C and then within 2.5 days it decreases to 25 °C. It is assumed that the data to be obtained on the in-situ structure will be different.

The results obtained cannot be considered accurate, as the measurements were not carried out in the body of cement, but on its samples or on the surface due to the lack of appropriate temperature sensors capable of withstanding the high pressure of a rigid deformable medium.

From the above it follows that for real estimation of the temperature inside the concrete structures of the NPP dome, appropriate means of measurements of high accuracy and reliability are required.

The aim of the work is to develop design and technological solutions of FOTS, providing the possibility of its application in rigid deformable media with minimal dynamic error due to inertial temperature processes in the sensor housing.

2 Methods and Approaches

The main approach to achieve the set goal is adaptation of previously developed technical solutions of FOTS to measurement conditions in rigid deformable media, which are the support parts of the NPP dome.

On the basis of the analysis of mechanical reliability of known FOS [15] it was concluded that it is necessary to develop FOTS in which optical fibers are not deformed [16]. On the basis of such a conclusion, the choice of the principle of operation of FOTS was made. We excluded technical solutions based on Bragg gratings [6, 17] and considered basic measuring transducers, in which the modulation of the optical signal can be realized with the help of various optical-modulating elements: mirror-reflecting surface, limiting attenuator, spherical or cylindrical lens [16].

A number of FOTS were considered, whose adaptation to temperature measurement in a rigid deformable medium is the most probable. In particular, a FOTS containing illuminating and receiving optical fibers (OF), the first ends of which are connected to the light source and receiver, and the second ends - to a capsule, in the cavity of which a mirror is installed, made on the end of the rod, the length of which varies under the influence of temperature change [11]. The fibers are attached to the capillary using glue or fusible glass. The main disadvantage of this FOTS is the lack of protection of its components from external mechanical effects, first of all, from the squeezing effect of cement.

The most developed for application in a rigid deformable environment can be considered as a FOTS, containing a capillary with an optical fiber and a light-reflecting element fixed in it, installed at a distance from the end of the optical fiber, a body with a through hole and two coaxial blind holes, one of which has a thread for a screw, the shank of which is installed in the through hole and has a light-reflecting end [12]. However, in this FOTS: first, the coefficients of thermal expansion of materials from which the sensor elements are made are not taken into account, which will lead to a significant error of measurements, as well as at large temperature changes (for example, in the range of minus 50…+90 °C) to large internal stresses and, accordingly, to a decrease in the mechanical reliability of the device. Secondly, when installing such a sensor in the hardening cement, the case will experience deformations, the levels of which are comparable in value to the measuring range of the sensor, which is determined by the change in the distance between the reflector and the end of optical fibers. Consequently, all metrological characteristics obtained in the process of grading or calibrating the sensor will change significantly. And, thirdly, the construction of the sensor is not hermetic, so in conditions of possible humidity on the reflecting surface and on the ends of optical fibers there may occur at high temperatures water droplets (acting as lenses), and at low temperatures - frost, which will introduce a significant additional error.

3 Results

The design of FOTS, which includes a protective case with a sensing element (SE) placed inside it (Fig. 2) has been developed [18]. The case 1 is made of a material with good thermal conductivity (for example, copper alloy) to reduce the inertia in temperature transfer from the environment to the sensing element 2 and, accordingly, to reduce the dynamic error. The thickness of the case is chosen from the consideration of ensuring the strength of the structure, if the sensor will be located in a rigid deformable medium, for example, in the concrete of the NPP containment base.

Sensing element 2 is a cylinder made of material with a high coefficient of thermal expansion. At both ends of the SE 2 a shank 3 and a sleeve 4 made of materials with small coefficients of thermal expansion are hermetically installed. On the narrow end of the shank 3 a reflective surface 5 is formed by polishing the metal from which the shank is made.

The sleeve 4 is necessary for fixing the working supplying 6 and working withdrawing 7 optical fibers, relative to the reflective surface 5 formed on the narrow end of the shank 3. The wide part of the sleeve 4 is rigidly fixed to the inner wall of the SE 2 by means of threads.

A cover 8 made of the same material as the case is fixed in the case 1 on the side of the wide part of the shank 3 with a gap.

The coefficients of linear expansion of the materials of the case α_1, shank α_3, sleeve α_4 and cylinder α_2 are determined by the equation:

$$\alpha_2 > \alpha_1 > (\alpha_3 \approx \alpha_4). \quad (1)$$

A through hole is made in the wall of the sensing element to fill the free space between the shank and the sleeve with inert gas, in which the sealing part is then hermetically installed. The free space between the inner surfaces of the case, the cover, the outer surfaces of the SE, the sleeve and the wide part of the shank is filled with a heat conducting composition 9.

To exclude deformation of the case at temperature change due to the difference of thermal expansion coefficients of the case and the SE, it is necessary that the change in length ΔL_1 of the case is equal to the change in length of the CE (cylinder) Δl_2, i.e. the condition is fulfilled:

$$\Delta l_2 = \Delta L_1. \quad (2)$$

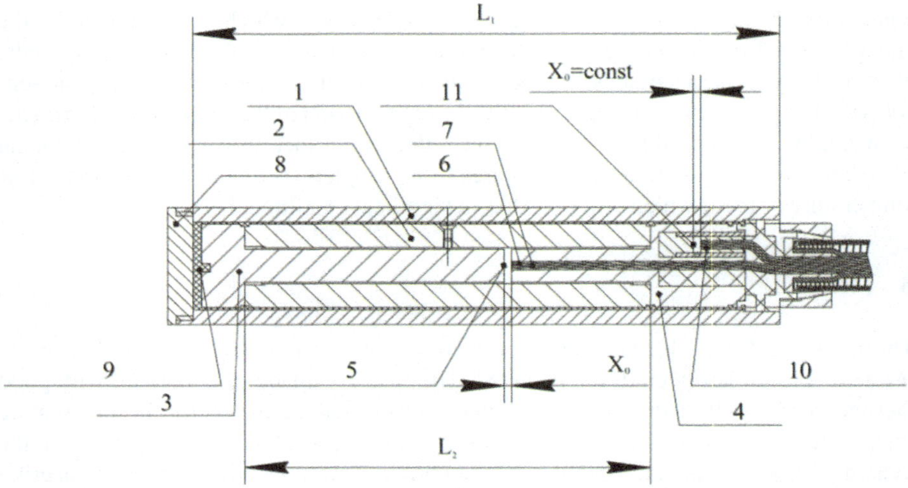

a)

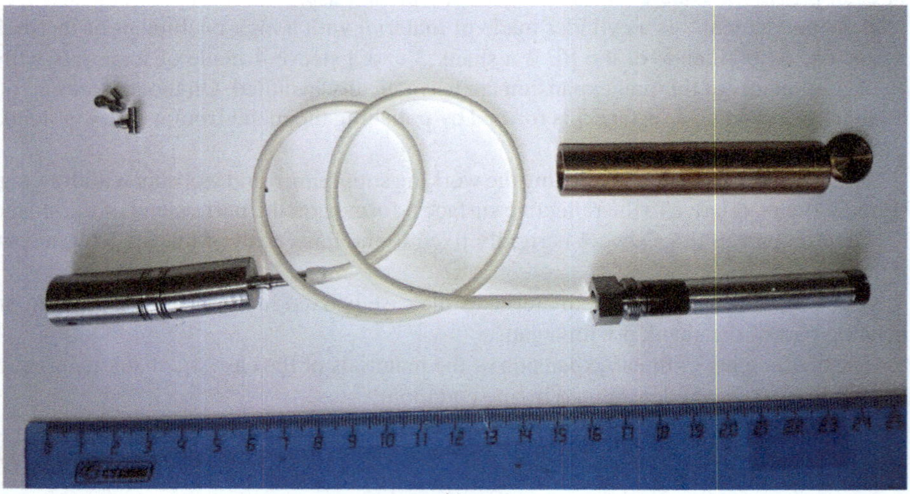

b)

Fig. 2. General view and photo of the fiber optic temperature sensor at the final stage of assembly.

If the ambient temperature changes, the length of the sensing element l_2 and the length L_1 of the case will change according to the expressions:

$$L_2 = l_{20}(1 + \alpha_2 \Delta T), \tag{3}$$

$$L_1 = L_{10}(1 + \alpha_1 \Delta T), \tag{4}$$

where l_{20}, L_{10} are the lengths of the cylinder and the body, respectively, under normal conditions (graduation temperature); ΔT is the range of temperature change. Substituting expressions (3) and (4) into (2) we obtain: $L_1 = l_2\alpha_2/\alpha_1$. But due to the fact that in the drawings linear dimensions should correspond to the standard values given in GOST 6639-69, a correction factor equal to (0.95…1) is introduced, i.e. $L_1 = (0,95\ldots1)l_2\alpha_2/\alpha_1$.

To reduce additional errors caused by bends of optical fibers, power fluctuations of the radiation source, an additional compensation channel 10 [19] is provided in the FOTS [19] (Fig. 3), the structural elements of which are located in the lateral blind hole in the sleeve 4, for which purpose a thickening is made in its end part. A fixed mirror 11 (FM) is fixed in the blind hole at a distance X0 relative to the end face of additional (compensating) supply 12 and withdrawal 13 optical fibers.

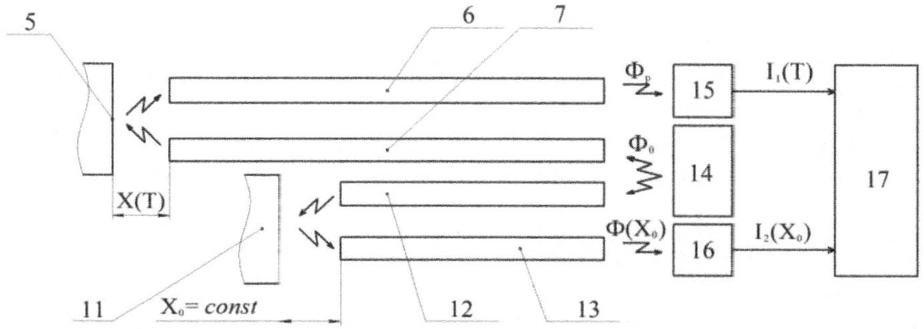

Fig. 3. Simplified structural diagram of fiberoptic temperature sensor with compensation channel.

The initial distance between the reflecting surface and the ends of optical fibers of both working and compensation channels at normal temperature is chosen from the ratio $X_0 = (0.3\ldots0.7)\ d_C$, where d_C is the diameter of the optical fiber core.

Optical fibers of the working and compensation channels are located in a single cable, for which purpose they are pulled through the shank 3. Such a constructive design of the fiberoptic cable allows to reduce additional errors from their bending during logometric or amplitude-phase conversion of the sensor signals [19].

The sensor works as follows.

The ambient temperature is perceived by the sensing element due to the process of heat transfer through the body made of material with good thermal conductivity, heat conductive composition to the cylinder. If the ambient temperature changes, the length l_2 of the cylinder will change according to expression (3). The initial length l_2 is chosen in such a way that its change in the measurement range provides a large modulation depth of the optical signal (up to 30%) and a linear function of light flux conversion from the change of distance X.

The fiberoptic temperature sensor is designed for measurement of either negative or positive temperature, so the initial distance X_0 can be shifted either towards smaller values of microdisplacements or towards larger values (Fig. 4).

The light flux ϕ_0 from the radiation source 14 through the working supply optical fiber 6 is fed into the measurement area, falls on the mirror surface of the shank, is reflected from it and arrives at the receiving end of the working diverting optical fiber 7 (see Fig. 2). On the working diverting optical fiber 7 the light flux is directed to the

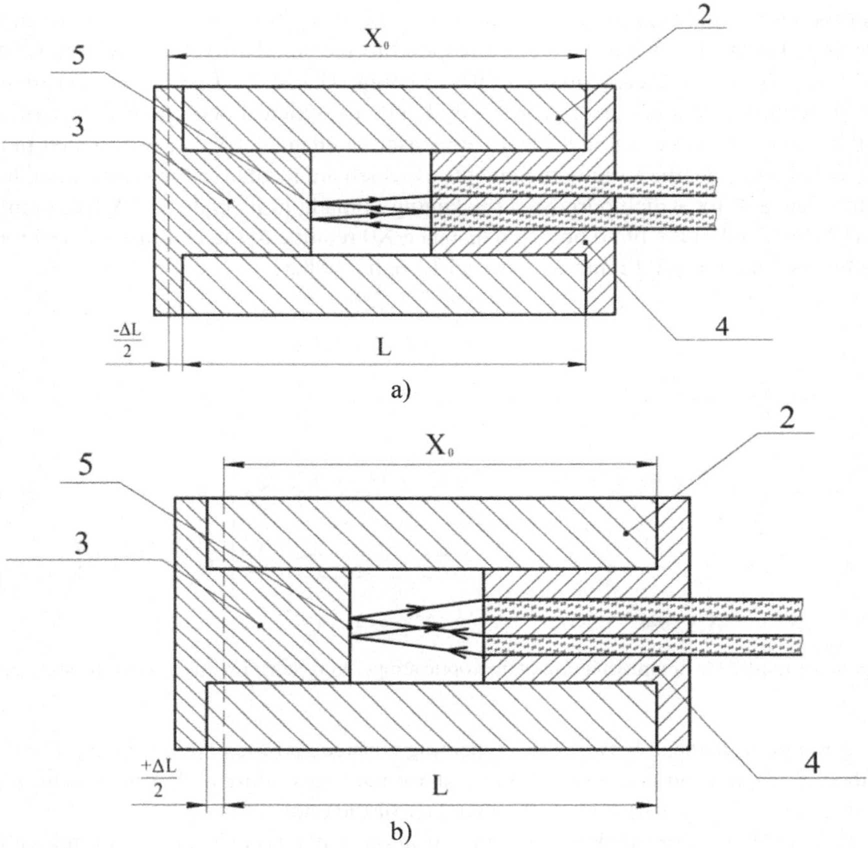

Fig. 4. To determination of the initial distance X_0: a) at temperature less than zero; b) at temperature less than zero.

working radiation receiver 15, where it is converted into an electrical signal $I_1(T)$, the value of which is proportional to the intensity of the reflected optical signal $K(X) = \phi(X)/\phi_0$, which in turn is proportional to the distance X between the ends of optical fibers and the mirror surface of the shank (Fig. 5).

Figure 5 shows as an example the results of calculating the dependence $K(X) = \phi(X)/\phi_0$ for different X_0 for an optical fiber with core diameter $d_C = 200$ μm, with aperture angle $\Theta_{NA} = 12°$.

Similar transformations occur in the compensation channel, but since the fixed mirror is mounted in a short part of the sleeve with a small coefficient of thermal expansion outside the sensing element, the distance X_0 remains unchanged during measurements, i.e., $X_0 = \text{const}$. Accordingly, $\phi_K(X_0 = \text{const})$ and the output signal from the compensating radiation receiver $I_k(T) \sim \phi_K\ (X_0 = \text{const})$.

In the working channel, a linear conversion function K(X) can be achieved if the WP is moved in the range of 0.25 d_C…0.75 d_C along the X axis relative to the ends of the

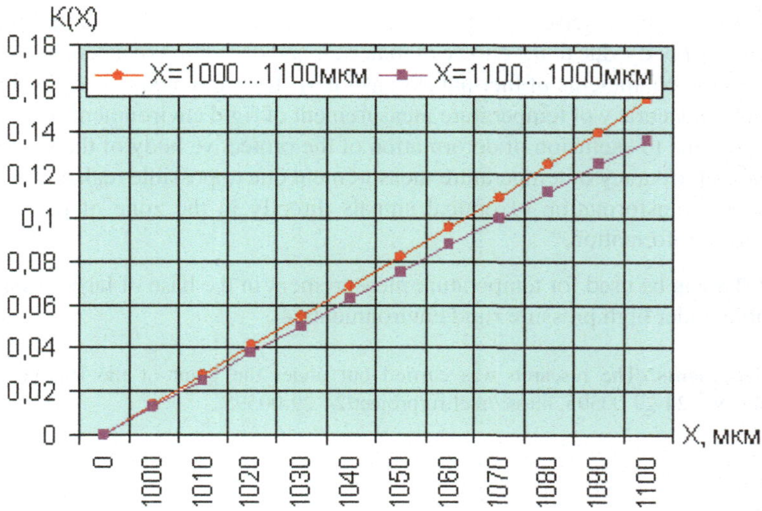

Fig. 5. Graphical dependencies $K(X) = \phi(X)/\phi_0$.

working optical fiber. For example, when $d_C = 200$ µm, the movement along the X axis will be 100 µm. Changing the initial distance X_0 leads to a change in the conversion sensitivity up to 30%, the dependence $K = f(X)$ is linear, the depth of modulation of the optical signal is more than 50%.

Since the temperature can both decrease and increase, the initial distance X_0 should be in the middle of the range 0.25 d_C...0.75 d_C (for example, for optical fiber $d_C =$ 200 µm the range of X distance variation will be 50...150 µm), i.e. 0.5 d_C (for example, for optical fiber with $d_C = 200$ µm - $X_0 = 100$ µm).

The introduction of the compensation channel makes it possible to implement, for example, the logometric transformation by the electronic module of information acquisition and conversion 16, described by the expression $I(T) = k\frac{I_1(T)-I_2(X_0)}{I_1(T)+I_2(X_0)}$, where k is the proportionality coefficient, allowing to reduce additional errors due to: bends of optical fibers during assembly and operation; the impact of external influencing factors; changes in the parameters of the power supply, sources and receivers of radiation, etc. [19].

4 Conclusions

The design and technological solutions of FOTS with compensating channel of reflective type for NPP containment placed in rigid deformable media have been developed.

Application of the developed FOTS provides:

- decrease of internal stresses and, accordingly, increase of mechanical reliability of the containment body in large temperature change ranges (for example, in the range of minus 50 OS...+90 °C);

- reduction of dynamic error of temperature measurement, caused by inertial processes in the sensor body, due to the choice of materials and derivation of calculated relations between the coefficients of thermal expansion of structural elements of FOTS;
- increase of accuracy of temperature measurement of rigid environment (for example, concrete) due to exclusion of deformation of the protective body of the sensor;
- increase of accuracy of temperature measurement due to possible realization of compensatory transformation of optical signals directly in the zone of perception of measuring information.

A FOTS can be used for temperature measurement in the base of large structures in deformable under high pressure rigid environments.

Acknowledgments. The research was carried out under the grant of the Russian Science Foundation No. 24-29-00595, https://rscf.ru/project/24-29-00595/.

References

1. Medvedev, V.N., Kiselev, A.S., Kiselov, A.S., Ulyanov, A.N., Strizhov, V.F., Salnikov, A.A.: Results of the containment temperature measurement during the acceptance test period. Glob. Nuclear Saf. **2**(15), 71–82 (2015)
2. Murashkina, T.I., Badeeva, E.A., Badeev, A.V., Serebryakov, D.I., Brostilova, TYu., Motin, A.V.: Radiation and resistant fiber and optical sensors with open mikro-optiko-mehanicheskim channel for information and measuring systems. In: Lukina, L.I., Bezhina, N.A., Lyamina, N.V. (eds.) Ecological, Industrial and Energy Security – 2018. SevSU, Sevastopol (2018)
3. Zelenskiy, V.A.: Theory Development and Development of Multiplexed FOIMS for Monitoring of the Complex Technical Objects. Moscow, Russia (2010)
4. Kolomiets, L.N.: Fiberoptic sensors in information-measuring systems. Sensors Syst. **1**, 8–14 (2006)
5. Egorov, F.A., Kolomiets L.N., Neugodnikov, A.P., Potapov, V.T., Pospelov, V.I., Rubtsov, I.V.: Theoretical and experimental aspects of construction monitoring on the basis of FOIMS. In: Proceedings of the International Conference "Sensors and Systems-2005", pp. 133–141. Penza, Russia (2005)
6. Buymistriuk, G.Y.: Monitoring the technical health of composite structural elements using the integrated intellectual optical fiber sensor system. Compos. Nanostruct. **12**(1), 21–24 (2020)
7. Murashkina, T.I., Badeeva, E.A., Kuznetsova, M.V., Kukushkin, A.N., Badeev, V.A.: Fiberoptic system for measuring large tilt angles of large-size test benches of rocket-space and aviation equipment. University Proceedings. Volga Region. Engineering Sciences **4**, 21–24 (2023)
8. Badeeva, E.A., Shcherbakova, A.A., Polyakova, E.A., Murashkina, T.I.: Evaluation of explosion and fire safety of information-measuring systems based on fiberoptic sensors with an open optical channel. In: Proceedings of the 11th International Scientific and Technical Conference "Ensuring the safety of NPPs with VVER", pp. 148–150. Hydropress, Podolsk, Russia (2019)
9. Murashkina, T.I., Badeeva, E.A.: New fiberoptic sensors for NPPs developed in STC "Nanotechnologies of fiberoptic systems" of Penza State University. In: Proceedings of the 11th International Scientific and Technical Conference "Ensuring the safety of NPPs with VVER", pp. 150–151. Hydropress, Podolsk, Russia (2019)

10. Komyagin, V.V., Makarov, V.V., Seleznev, A.V., Klimov, N.N., Bolvanchikov, S.N., Alekseev, Y.V., Safonova, E.A.: Vibration tests of block heat insulation of equipment and pipelines of VVER-1000 reactor unit. In: Proceedings of the 3rd Scientific and Technical Conference "Ensuring the safety of NPPs with VVER", pp. 114–123. Hydropress, Podolsk, Russia (2003)
11. Meshkovsky, I.K., Popkov, O.S., Voznesenskaya, A.O.: Fiberoptic temperature sensor. Patent RF 2256890, published 20.07.2005, Bulletin No. 20 (2005)
12. Veryaskin, M.E.: Fiberoptic temperature sensor (variants). Patent RF 2393431, published 27.06.2010, Bulletin No. 18 (2010)
13. Medvedev, V.N., et al.: To the question about the control of the NPP protective shells during the acceptance test. Glob. Nuclear Saf. **2**(35), 42–54 (2020)
14. Medvedev, V.N., Skorikova, M.I.: Influence of the concrete rheological characteristics on the SSS of the NPP containment. At. Energ. **216**(6), 317–320 (2019)
15. Badeeva, E.A., Gorish, A.V.: Analysis of mechanical reliability of the fiberoptic cable for sensors. Inf. Meas. Technol. Ecol. Monit. **6**, 243–254 (2003)
16. Murashkina, T.I., Badeeva, E.A.: Fiberoptic Devices And Systems: Scientific developments of STC "Nanotechnology of Fiberoptic Systems". Penza State University, Politechnika, Saint-Petersburg, Russia (2018). https://doi.org/10.25960/7325-1132-1
17. Butov, O.I., Tomyshev, K.A.: Fiberoptic sensors based on Bragg gratings with tilted strokes. Special Issue Photon Exp. Sci. **6**, 22 (2019). https://doi.org/10.24411/2308-6920-2019-16006
18. Murashkina, T.I., Badeeva, E.A., Serebryakov, D.I., Dudorov, E.A., Khasanshina, N.A., Badeev, V.A.: Fiberoptic temperature sensor. Patent RF 2795841, Published 12.05.2023, bul. no. 14 (2023)
19. Badeeva, E.A., Murashkina, T.I., Polyakova, E.A., Slavkin, I.E., Kukushkin, A.N.: Implementation of the principle of dual-channel in the fiberoptic information-measuring systems. University Proceedings. Volga Region. Engineering Sciences **2**(58), 87–98 (2021)

Proposal of a Smart Control System Using Edge Computing and Deep Learning Technologies

Lenka Halenarova(✉), Igor Halenar, and Pavol Tanuska

Faculty of Materials Science and Technology in Trnava, Slovak University of Technology in Bratislava, Bratislava, Slovakia

{lenka.halenarova,igor.halenar,pavol.tanuska}@stuba.sk

Abstract. The core of this paper is a proposal of a control and monitoring system of the production process using smart sensors and Edge Computing (EC) technology. A solution proposed in this work provides the possibility of obtaining a wider spectrum of information than a normal control. Using the Multilayer perceptron neural network (MPL) and EC a faster data collection can be achieved without limiting the computing capacity and transfer services of the server. Additional benefits include availability of indirect data in process control, higher level of security by data collection and transfer, and, on top of that, low cost of components used. The proposed system reflects the current state of technology in the field, reflects on actual forms of the collection and usability of data, and offers opportunities to improve the quality of process management in general. This proposal shows that through EC, we are able to process a large amount of data (Big Data) from a number of devices/sensors (within proximity to the data), which we can analyze and process in a relatively short period of time. Thanks to analyzes and the use of AI (Deep Learning), we are able to react to errors or bad equipment conditions "Just in Time".

Keywords: Data collection · RevPi · Edge Computing · Industry 4.0 · Innovations

1 Introduction

Today's technological progress moves us to use key elements of information and technical development in all areas of life. The rise of technological development brings solutions that can be used not only in everyday routine activities, and the development affects the research environment. The implementation of modern approaches is needed mainly in the industrial area and companies. It is necessary to find new solutions for increasing the efficiency of production, optimizing processes, and automatization of operations. By deploying digitalization, data collection and monitoring the status of individual devices a better overview of the overall progress of the production process can be gained. The data are a key article to decision processes for support, plan, and operations in real time. New approaches in the collection and evaluation framework bring to companies not only the possibility of prediction, setting activities based on the analysis of their previous activities, but also the overall optimization of the company's functioning.

R. Silhavy and P. Silhavy (Eds.): CSOC 2024, LNNS 1120, pp. 464–475, 2024.
https://doi.org/10.1007/978-3-031-70518-2_41

Data collection and processing has become an important part of our lives and it is necessary to reflect on developing trends in every area and pay our attention to this direction.

The primary reason for collecting data in production processes is to obtain enough information about ongoing processes, production quality, equipment efficiency, and an adequately fulfilled production plan. Data can be obtained directly from devices, sensors, or indirectly, which must pass through another processing process, and only then can they provide us with adequate and usable information.

2 Related Work

The industrial sector and the companies working in it focus their activities on the successful achieving of the desired goal. This must be done in the shortest possible time and with the least possible effort. The deployment of modern technologies in production results in adequate streamlining of processes, cost reduction, and productivity increase. Economic, technological, and time indicators are used to examine, compare, and analyze key attributes of the company and to establish and engage in further development and trends. The Industry 4.0 area according to [1–3] provides a considerable number of methods to bring activities within the company to the level of advanced digitalization and simplify the access to data and their analysis.

The Internet of Things (IoT) and Industrial Internet of Things (IIoT) technologies within these principles [4–7] enable the expansion of the production system infrastructure by providing access to additional intelligent devices. However, it is not possible to obtain data completely directly, from the device, from the sensors or additionally from the environment. The increase in data demand has led to the development and search for other solutions. And that is why the possibilities of communication with devices that are remote, but not directly connected, are coming to the fore, through which we get higher computing possibilities for further data processing.

In the past and still today, a commonly used way to overcome the difficulties of data collection, storing, and moving data is to redirect tasks to remote computing centers, cloud servers, and centralized data centers. Developments [8–10] within this area indicate that if we move tasks to remote computing systems that are connected via a network, we get options that are helpful for more demanding operations and task distributions. There are even problems that arise as a result of the pressure on the computing capacity and possible delays in the transfer of data between the cloud and the field.

One of the technologies currently used is Edge Computing [11–13]. The concept is not completely new and has been discovered, and its principle follows from activities that have been repeated and researched for decades. EC [7, 12, 13] brings the possibility to process data at the edge of the network, as close as possible to the data source. After several experiences [7, 14–17], it was found to be effective not to move the data to the data center, but to process where it is collected. As a result, central storage is not overloaded, and data processing is more reliable and efficient. This principle provides significant benefits such as autonomy, data sovereignty, and edge security. The importance is increasing in cases where the connection is not working well, bandwidth is limited, or the network is overloaded. The principle is suitable as a powerful tool for data that we cannot obtain directly, and the collected amount of data causes impracticality, price inaccessibility or the violation of fundamental rules (poor data quality, data security, scalability). Data collection and transfer of data are ensured by the MQTT protocol (Message Queuing Telemetry Transport), which enables communication between edge devices and the cloud and directs the operation, storage, and forwarding of data.

A part of the collection, completion of data, and their immediate use for further analyses, it is possible to use Edge Computing together with microcontrollers. In the past and today too, kinds of microcontrollers have been used – Arduino, Raspberry Pi, Revolution Pi. They are versatile, powerful, space-saving, and simple to operate.

The current trend now, together with more affordable means for communication, acquisition, and collection of data from remote sources and for the management control of production system operation, is the use of group RevPi (Revolution Pi) type devices [18]. This group of microcontrollers is included in the industrial field as the modules within this system enable communication with sensors, actuators, or PLC devices. The basic module can be expanded with I/O modules, PROFIBUS, PROFINET, EtherCat industrial bus gateways, or radio modules for M-Bus. The IIoT gateway can be extended to a small industrial controller as part of the available software.

The merging of EC with AI [19, 20] into AIEC models allows the possibility of processing and analyzing data at the local level, and in addition faster feedback can be provided. This gives us the opportunity to increase efficiency, save costs, and strengthen security during the data transporting and processing. The authors [21–24] point to the uniqueness of using AIEC models in data collection and transmission due to faster decision-making in real time. AIEC offers opportunities for further development in this area – 5G networks, Edge-to-Cloud, Edge Intelligence, data protection and security.

3 The Design of a Control System Using Edge Computing Methods

The existing equipment in the laboratory is used as an object of control. The automatic production line (Fig. 1) - Festo AFB Factory [25] is located in the Laboratory of control systems, which is under the ownership of The Institute of Applied Informatics, Automation and Mechatronics of the Faculty of Materials Science and Technology in Trnava. The modular hybrid automatic production system consists of the basic manufacturing systems stated:

- production,
- manipulation with seals,
- packing line,
- filling line,
- system of conveyors,
- automatic shelf stacker / storage,
- inputs / outputs.

The system also contains procedural and factory automatization. The whole device is able to work like a complex entity but can also be separated into individual systems. As a testing device for the implementation of EC has been chosen, the vibrating bulk material conveyor (MPS PA Mixing Station). In its present state, the control system

Fig. 1. Automatic production line – Festo AFB Factory [26]

of separate modules contains industrial PLC Siemens S7 mounted on a bar connected to regulated technology via IEEE 488/24 pin connectors and to a visualisation through an external interface. Operator panels (HMI) are mounted on a plate with outsourced connectors for communication interfaces.

The entire design of the solution can be divided into individual steps, which can be summarized as follows:

1. replacement of the present control system represented by the PLC Simatic S7–300 with the RevPi device, i.e., connecting all of the sensors and active elements;
2. creation of a control program of production process using RevPi to ensure the production process;
3. propose and install additive sensors for the sensorics layer and EC implementation;
4. proposes the way of additional data processing and evaluation for a higher level of control of the production process using new and modern technologies.

The first and second points are clear. The process control must be available and must be compatible with the rest of the production line. The advantage is that the process itself is quite simple. The station is a simple vibration conveyor with few sensors – level of material storage tank, sensor at the conveyor output, and two active elements – air control valves. One for Festo Fluidic Muscle and one for material transfer to the next station.

The third point can be more complicated due to the number of options for measurement - physical quantities, i.e., pressure, power, temperatures, voltage, current, … Serious research can be carried out in this area. According to the controlled process and the simplicity of the testing purposes, only one additive sensor was used. As an example to gain additive data from the production process the vibration sensor was used in this proposal. This information may provide additional feedback to the process control unit.

The fourth point is benefit of EC implementation and is the main quest and represents key difference against clasic process control using PLC. The device (RevPi) is able to run other application processes, in addition to managing the production process. In this proposal an intelligent monitoring application for dynamic Fluiding Muscle (FM) movement tracking is used to run.

3.1 Device Installation

The RevPi device is mounted on the device panel on the production line, as shown in Fig. 2.

Fig. 2. Connected Revolution Pi device

RevPi (Revolution Pi) is an open, modular industrial PLC based on Raspberry Pi. The base unit can be extended with other I/O modules (analog or digital) and many communication interfaces, as Profinet, Profibus, RS-485. Thanks to platform (Linux) it is possible to use technologies such a Node-RED, Phyton, or C language and run applications with support of MQTT and OPC UA communication.

The idea is that the RevPi controller fully replaces the older system S7–300 and the new control system is fully compatible with the rest of the line. Due to educational purposes, the old PLC will remain mounted on panel, but the process control can be switched between these two devices. The configuration (in the testing phase) is shown in Figs. 3 and 4.

Fig. 3. System control using PLC

Fig. 4. System control using RevPi

The sensor used for movement tracking is a compact high-performance 3D digital linear acceleration sensor based on the ADXL345 chip. The accelerometer is firmly mounted to the conveyor. The entire situation is shown in Fig. 5.

3.2 Data Collection and Processing

The sensor connection is done using the I2C communication interface. The sensor is able to obtain 3-axis of acceleration. ADXL 345 is a small, thin, ultra-low power consumption 3-axis accelerometer with high resolution (13-bit) measurement at up to ± 16 g. The digital output data are formatted as a 16-bit two-s complement and are accessible through I2C(addr:0x53). The application for communication with sensor running in RevPi has been written using Python language. The collected data (separate acceleration values for the X/Y/Z axes) are stored in memory for further processing (Fig. 6). Due to computational limitations, data collection is carried out at regular intervals.

The data stream recording is continuous, and the data is stored in memory. The plan is to use a neural network to identify that process is continuously running and values are in the given tolerance. Obviously, continuous data are not suitable for such use. The solution is to cut continuous data stream to patterns of the same length and then use a neural network to evaluate with output TRUE/FALSE [0/1]. In application is the output from the neural network processed as icon with GREEN/RED color. This gives the information to the operator, that the process is continuously working.

The examples of patterns submitted as input to neural network are shown in Fig. 7.

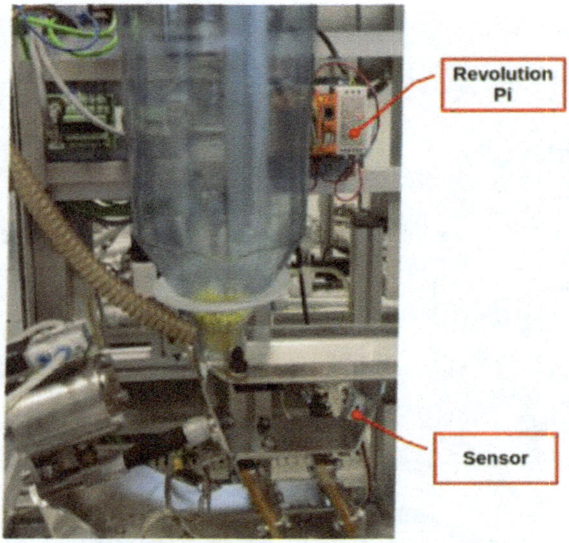

Fig. 5. Acceleration sensor installation

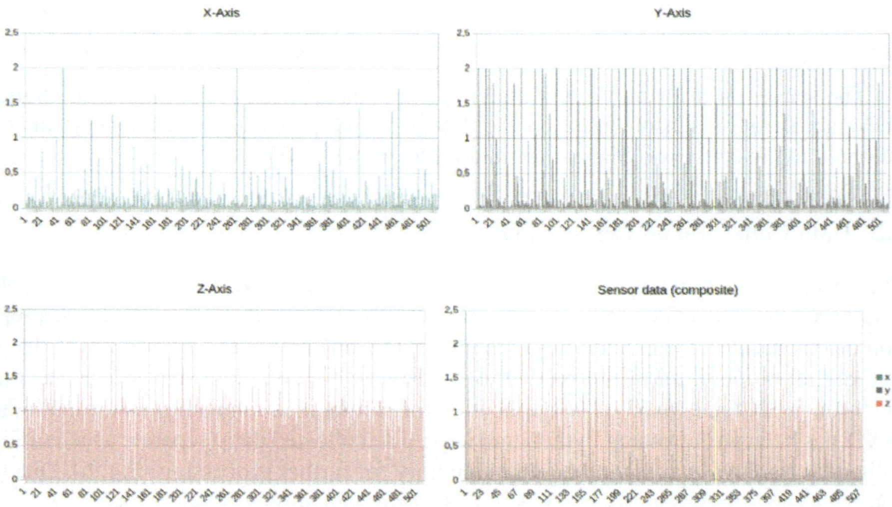

Fig. 6. Recorded data example – separate X/Y/Z axis and overall

The data stream divided into patterns according to the given example can be easily evaluated using a suitable neural network. Data shown in Fig. 7 are from the Y-axis. Due to the way the sensor is mounted, the Y-axis data stream provides sufficiently appropriate data on belt movement. Of course, it is also possible to process separate data from all axes and process them independently.

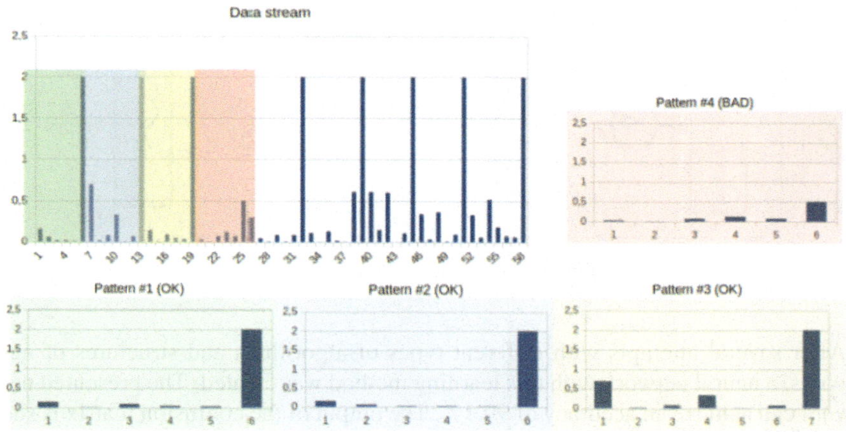

Fig. 7. Example of data patterns (input to MLP)

3.3 Design of the Neural Network

Neural networks are, for their characteristics, widely used as classifiers for image recognition. There are many types of neural networks. In this case, is used Multilayer perceptron neural network. This type of neural network can be defined as a feedforward artificial neural network that generates a set of outputs from a set of inputs. An MLP is characterized by several layers of input nodes connected as a directed graph between the input and output layers. MLP uses backpropagation to train the network. MLP is a deep learning method. MLP is widely used to solve problems that require supervised learning, as well as research into computational neuroscience and parallel distributed processing. Applications include speech recognition, image recognition, and machine translation.

For simplicity in implementation in laboratory conditions, Matlab software was used, using built-in modules for MLP. The advantage is that the trained neural network (program) can be deployed in the used computer architecture (Raspberry Pi). The control device used in this proposal (Revolution Pi) is an open, modular, and inexpensive industrial PC based on the well-known Raspberry Pi. Thus, the programs created and exported from Matlab for Raspberry Pi are fully compatible with Revolution Pi, and directly runnable.

Because the device has relatively small computing power, the neural network must be simple and effective. The idea is to make the entire neural network as simple as possible so that the 3-layer neural network is used. The input layer is made up of 7 neurons. These correspond to patterns prepared from the recorded data stream. The network contains one hidden layer of 10 neurons. The network basically works as a simple two-class classifier and as output is just one neuron/bit. The identification of a successful FM stroke in the data pattern is represented by the output neurons: true, false [0,1]. After many tests we considered that the best algorithm (Fig. 8) seems the scaled conjugate gradient algorithm (trainscg).

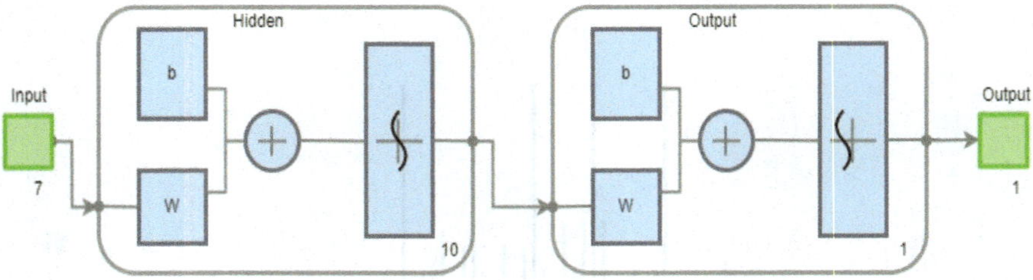

Fig. 8. Neural network architecture

After several attempts with different types of algorithms and structures of neural networks, a neural network with this learning method was created. The presented neural network can achieve an accuracy of 99.1%. The output of the confusion matrix is shown in image (Fig. 9).

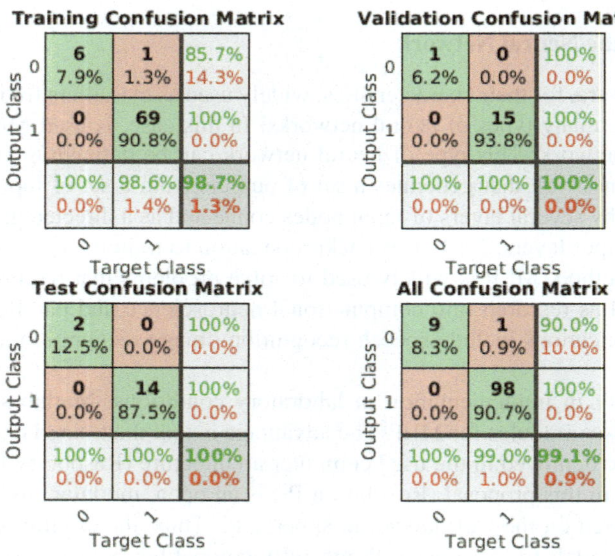

Fig. 9. Confusion matrix of the trained neural network

4 Conclusion

The core of this paper is the improvement of production line in laboratory environment where a classic process control system is replaced by a new system with edge computing possibilities. All this while maintaining compatibility and functionality. The process state monitoring is enhanced using the application of well-known technologies of neural networks. This configuration can give a higher level of control. This work describes the initial state of the research, and additional information from the production process is represented by a single accelerometer. In further work, the system can be extended with additional sensors, and so provide comprehensive information about the process.

A significant benefit of the proposed solution is that the system is fully functional using a Revolution Pi device, such as a classic PLC, but with the possibility to implement new approaches to process enhanced data from the production system. The source of additional information from production process is the sensorics layer and data are processed directly at the control level using Edge Computing. The system can be implemented for other stations/devices of the used production line, where it is not limited by the computing capacity, it offers potential with guarantees of latency and reliability during data collection. We can apply the proposed solution for process monitoring of equipment status, detection of anomalies, predictive maintenance through modern data processing techniques such as machine learning, artificial intelligence, genetic algorithms, and others.

In conclusion, the whole idea offers a solution that is open, efficient in application, and easy to implement in practice.

Acknowledgment. This publication is the result of the implementation of the VEGA project 1/0193/22: "Proposal of identification and monitoring of production equipment parameters for the needs of predictive maintenance in accordance with the concept of Industry 4.0 using Industrial IoT technologies." supported by the VEGA.

References

1. Tupa, J., Simota, J., Steiner, F.: Aspects of risk management implementation for industry 4.0. In: Procedia Manufacturing, vol. 11, pp. 1223–1230 (2017), ISSN 2351–9789, https://doi.org/10.1016/j.promfg.2017.07.248
2. Hassan Reza, M.N., Agamudai Nambi Malarvizhi, C., Jayashree, S., Mohiuddin, M.: Industry 4.0–technological revolution and sustainable firm performance. In: 2021 Emerging Trends in Industry 4.0 (ETI 4.0), Raigarh, India, pp. 1-6 (2021). https://doi.org/10.1109/ETI4.051663.2021.9619363
3. Joshi, S.J., Mamaniya, S., Shah, R.: Integration of intelligent manufacturing in smart factories as part of industry 4.0-a review. International Conference on Industry 4.0 – Nascent Technologies and Sustainability for "Make in India" Initiative, in 2022 , Mumbai, India, pp. 1–5 (2022). https://doi.org/10.1109/SPICON56577.2022.10180471
4. Talwana, J.C., Hua, H.J.: Smart world of internet of things (IoT) and its security concerns. In: 2016 IEEE International Conference on Internet of Things (iThings) and IEEE Green Computing and Communications (GreenCom) and IEEE Cyber, Physical and Social Computing (CPSCom) and IEEE Smart Data (SmartData), Chengdu, China, pp. 240–245 (2016). https://doi.org/10.1109/iThings-GreenCom-CPSCom-SmartData.2016.64
5. Narayanan, R.G., Gunasekera, J.S., (2023). Sustainable manufacturing processes - references. Elsevier. Retrieved from https://app.knovel.com/hotlink/pdf/id:kt013CP6V1/sustainable-manufacturing/waste-ener-references
6. Khan, F., Pasman, H.J., Yang, M.: Methods in chemical process safety - methods to assess and manage process safety in digitalized process system - 2.3.1.1 industrial internet of things (IIoT). Elsevier (2022). Retrieved from https://app.knovel.com/hotlink/pdf/id:kt0131QID3/methods-in-chemical-process/industrial-internet-things

7. Lin, Z., Liu, J., Xiao, J., Zi, S.: A survey: resource allocation technology based on edge computing in IIoT. In: 2020 International Conference on Communications, Computing, Cybersecurity, and Informatics (CCCI), Sharjah, United Arab Emirates, pp. 1–5 (2020). https://doi.org/10.1109/CCCI49893.2020.9256663
8. Konishi, T., Nakamichi, T., Kamikawa, R., Yamasaki, Y.: Spectroscopic inspection optimization for edge computing in industry 4.0. In: 2020 22nd International Conference on Transparent Optical Networks (ICTON), Bari, Italy, pp. 1–4 (2020). https://doi.org/10.1109/ICTON51198.2020.9203553
9. Satyam, Geetha, P., Shashikala, K.S., Kumar, N.A.: AI-enabled edge computing models: trends, developments, and future implications. In: 2023 2nd International Conference on Edge Computing and Applications (ICECAA). Namakkal, India, pp. 63–67 (2023). https://doi.org/10.1109/ICECAA58104.2023.10212294
10. Awaysheh, F.M., Aladwan, M.N., Alazab, M., Alawadi, S., Cabaleiro, J.C., Pena, T.F.: Security by design for big data frameworks over cloud computing. IEEE Trans. Eng. Manage. **69**(6), 3676–3693 (2022). https://doi.org/10.1109/TEM.2020.3045661
11. Taheri, J., Shuiguang, D., (2020). Edge computing - models, technologies and applications. In: Institution of Engineering and Technology (The IET). Retrieved from https://app.knovel.com/hotlink/toc/id:kpECMTA007/edge-computing-models/edge-computing-models
12. Cao, J., et al.: Edge computing: a primer, springer international publishing. 90s. Available online (2018). ISBN 978-3-030-02082-8
13. Papcun, P., Kajáti, E., Cupková, D., Zolotová, I.: Smart/Intelligent edge – cloud a edge platform provides. In ATP J. **07**, 50–51 (2019)
14. Cao, K., Liu, Y., Meng, G., Sun, Q.: An overview on edge computing research. IEEE Access, year. **8**, 85714–85728 (2020)
15. Zhang, X., Cao, Z., Dong, W.: Overview of edge computing in the agricultural internet of things: key technologies, applications, challenges. IEEE Access **8**, 141748–141761 (2020). https://doi.org/10.1109/ACCESS.2020.3013005
16. Bourechak, A., Zedadra, O., Kouahla, M.N., Guerrieri, A., Seridi, H., Fortino, G.: Na sútoku umelej inteligencie a okrajových výpočtov v aplikáciách založených na IoT: Prehľad a nové perspektívy. (At the Confluence of Artificial Intelligence and Edge Computing in IoT-Based Applications: An Overview and New Perspectives). Sensors **23**, 1639 (2023)
17. Kong, X., Wu, Y., Wang, H., Xia, F., Edge computing for internet of everything: a survey. In: IEEE Internet of Things Journal, vol. 9, num. 23, pp. 23472–23485, (2022)
18. The Revolution Pi module documentation. Available online: https://revolutionpi.com/revolution-pi-series
19. Huh, J.H., Seo, Y.S.: Understanding edge computing: engineering evolution with artificial intelligence. IEEE Access **7**, 164229–164245 (2019)
20. Hua, H., Li, Y., Wang, T., Dong, N., Li, W., Cao, J.: Edge computing with artificial intelligence: a machine learning perspective. In: ACM Compput. Surv , year 55, num. 9, pp. 35 (2023)
21. Satyam, Geetha, P., Shashikala, K.S., Kumar, N.A.: AI-enabled edge computing models: trends, developments, and future implications. In: 2023 2nd International Conference on Edge Computing and Applications (ICECAA), Namakkal, India, pp. 63–67 (2023). https://doi.org/10.1109/ICECAA58104.2023.10212294
22. Al-Doghman, F., Moustafa, F., Khalil, I., Sohrabi, N., Tari, Z., Zomaya, A.Y.: AI-enabled secure microservices in edge computing: opportunities and challenges. In: IEEE Transactions on Services Computing, vol. 16, no. 2, pp. 1485–1504 (2023). https://doi.org/10.1109/TSC.2022.3155447
23. Ramachandran, P., Ranganath, S., Bhandaru, M., Tibrewala, S.: A survey of ai enabled edge computing for future networks. In: 2021 IEEE 4th 5G World Forum (5GWF), Montreal, QC, Canada, pp. 459–463 (2021). https://doi.org/10.1109/5GWF52925.2021.00087

24. Fragkos, G., Tsiropoulou, E.E., Papavassiliou, S.: Artificial intelligence enabled distributed edge computing for internet of things applications. In: 2020 16th International Conference on Distributed Computing in Sensor Systems (DCOSS), Marina del Rey, CA, USA, pp. 450–457 (2020). https://doi.org/10.1109/DCOSS49796.2020.00077
25. The Festo documentation. Available online: https://www.festo-didactic.com/ov3/media/customers/1100/afb_en_monitor_056961.pdf
26. The Festo Stack magazine module documentation. Available online: https://www.festo-didactic.com/int-en/learning-systems/mps-the-modular-production-system/stack-magazine-module.htm?fbid=aW50LmVuLjU1Ny4xNy4xOC41ODUuNDA5Ng
27. The Revolution Pi module documentation, Available online: https://revolutionpi.com/wp-content/uploads/manuell/revolution-pi-booklet.pdf

Analysis of Early Cancer Diagnosis Using Machine Learning

Marija Gjosheva, Zlate Bogoevski, Bojana Velichkovska, Danijela Efnusheva(✉), Goran Jakimovski, and Sanja Atanasova

Dedicated Computer Systems Department, Faculty of Electrical Engineering and Information Technologies, Ss. Cyril and Methodius University, Skopje, North Macedonia
{bojanav,danijela,goranj,ksanja}@feit.ukim.edu.mk

Abstract. Cancer is a group of diseases with similar symptoms, all involving uncontrolled growth and reproduction of cells. With around 8 million deaths each year, it is the second leading cause of death worldwide in developing countries and the first in the developed world. In contemporary medicine, early cancer diagnosis for every known type is essential. Machine learning has the potential to completely transform the process and increase the number of lives saved. In order to make predictions, computers develop complex data models and search for patterns. Early cancer diagnosis could undergo a revolution because of machine learning.

This research's goal is to outline the issue surrounding cancer diagnoses in patients and all the difficulties they experience. A suitable strategy will be to model the risk of cancer and patient outcomes given the growing trend of employing machine learning technics in cancer research. A specific model has been developed that, if applied appropriately, can reduce the number of lost lives and, at the same time, increase the number of individuals capable of coping with this disease. The results indicate that the created model can be used by professionals to identify lung cancer with efficiency. If the prediction is accurate, the doctor may be able to develop a better treatment plan and provide the patient with an early diagnosis. The study's findings show that the number of patients has been rising recently, yet early detection is crucial because it can help avert serious complications.

Keywords: machine learning · diagnosis · analysis · cancer · transformation · lungs

1 Introduction

Research on cancer has undergone constant change throughout the last few decades. Researchers used a variety of techniques, such as early screening, to identify cancer types before they manifest symptoms. Additionally, they have created fresh approaches to early cancer treatment outcome prediction. The medical research community has access to a vast amount of cancer data that has been gathered since the introduction of new medical technologies [1].

On the other hand, one of the most fascinating and difficult jobs for doctors is making accurate predictions about the course of a disease. Consequently, ML techniques

R. Silhavy and P. Silhavy (Eds.): CSOC 2024, LNNS 1120, pp. 476–485, 2024.
https://doi.org/10.1007/978-3-031-70518-2_42

have gained popularity as a tool among medical researchers. These methods are good at predicting the future course of a particular cancer type and can find patterns and correlations among them in large datasets.

One of the key priorities of the World Health Organization is to identify the cause of the increasing percentage of cancer patients [2]. The diagnosis of many tumor subtypes has been improved through remote screening programs, but the two challenges of the process - patient selection and risk categorization - remain.

A subfield of artificial intelligence called machine learning (ML) allows computers to "self-learn" from training data and get better over time without needing to be explicitly programmed. In the field of cancer diagnostics, machine learning has become a revolutionary force, providing previously unheard-of improvements in early detection.

Machine learning models can sort through enormous volumes of medical data by utilizing complex algorithms and data analysis. This allows them to spot minute details and trends that conventional diagnostic techniques can miss [3]. These models are quite good at identifying possible cancer indicators, which helps medical practitioners catch the disease early on, when treatment can be most successful. The application of machine learning to cancer diagnosis is a promising development in the ongoing search for more dependable and effective medical treatments.

This paper is organized as follows: Sect. 2 gives an overview of the benefits and techniques for machine learning in general. In the next section lung cancer is discussed with emphasis on diagnostics. In Sect. 4 the proposed machine learning system is described. The focus is on the database, its parameters, and their visualization. Section 5 presents the implemented system design and discusses the results. Section 6 concludes the paper.

2 Benefits and Techniques for Machine Learning

Machine learning algorithms are able to detect patterns in data and learn from them, in order to make their own predictions. In short, machine learning algorithms and models learn through experience [4].

Each learning process includes two stages: (i) using a given dataset to estimate unknown dependencies in a system, and (ii) using the estimated dependencies to forecast new system outputs. In biomedical research, machine learning (ML) has also shown to be an intriguing field with a wide range of applications. Using various approaches and algorithms, an appropriate generalization is found by searching an n-dimensional space for a given set of biological samples [5]. ML techniques can be broadly classified into two categories: supervised learning and unsupervised learning. A labeled collection of training data is used in supervised learning to estimate or map the input data to the intended output. On the other hand, no labeled examples are given when using unsupervised learning techniques, and there is no concept of the output when learning [6, 7].

Consequently, the learning scheme/model is responsible for identifying the groups or patterns in the incoming data. This process can be conceptualized as a classification problem in supervised learning. The process of learning to classify the data into a limited number of classes is referred to as the classification task. The two other popular ML tasks are grouping and regression [8]. A learning function converts the data into a real-value variable in regression situations. Based on this procedure, the value of a predictive variable can then be estimated for every new sample. Finding categories or clusters to characterize the data items is a typical unsupervised task called clustering.

Every fresh sample can be categorized into one of the identified clusters based on this approach, taking into account the shared traits.

Artificial intelligence and machine learning have significantly transformed business processes and everyday life. Recent developments in this field have demonstrated extraordinary progress and possibilities for easing the workload of healthcare professionals, while enhancing precision, predictability, and the quality of treatment. Learning strategies developed for and within these industries have immense potential to improve clinical care and medical research, especially as more healthcare practitioners utilize electronic health records.

Diagnostics and outcome prediction are two areas where the use of machine learning approaches in the medical industry can be advantageous. As can be observed, machine learning technologies have a broad spectrum of potential applications in healthcare, ranging from improving patient data, medical research, diagnosis, and treatment to reducing costs and enhancing patient safety.

3 Lung Cancer: Predictions and Diagnosis

A type of cancer that originates in the lungs is lung cancer. When cells in the body start to multiply uncontrollably, cancer develops. Figure 1 illustrates the anatomy of the lungs. The lungs contain two organs resembling sponges called the lungs. Because the heart occupies more space on the left side of the body, the left lung is smaller [9].

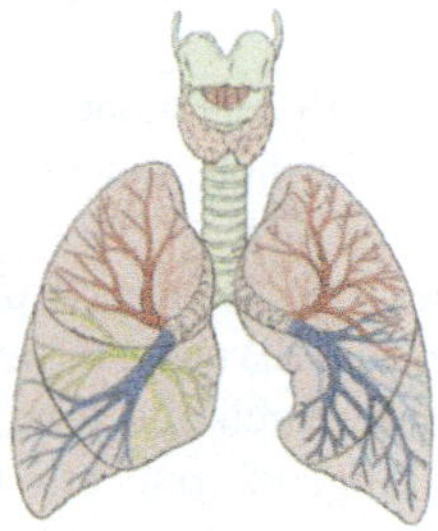

Fig. 1. Anatomy of the Human Lungs

The term "early detection of lung cancer" refers to methods that can identify the disease when it can still be treated with surgery or stereotactic body radiation therapy (SBRT).

The type of lung cancer, the overall health of the patient, and the lung cancer's response to treatment are some of the factors that influence the prognosis for any patient with lung cancer, regardless of when lung cancer is detected. On the other hand, lung cancer is easiest and most successfully treated when detected early [10]. According to statistics, those with lung cancer who receive a diagnosis in the early stages have a significantly greater chance of surviving for at least five years afterward.

Using cutting-edge algorithms and computing capacity, machine learning is a key component in changing the lung cancer diagnosis environment. In order to find possible signs of lung cancer, these systems are particularly good at identifying intricate patterns within medical data, such as radiological images and patient records. Machine learning models' capacity to identify minute details and anomalies helps with early detection, which is essential for enhancing treatment results. These algorithms improve over time by continuously learning and adapting, which increases their diagnostic accuracy. This technology helps healthcare practitioners make better decisions about treatment strategies in addition to helping with the timely identification of possible instances. Machine learning promises to improve lung cancer detection efficiency and accuracy, which could lead to better patient outcomes and more individualized and efficient healthcare interventions.

4 Discussion

In this type of research, where the outcome could potentially save human lives, it is crucial to select a dataset that meets the requirements of the analysis. This would facilitate the creation of an acceptable analysis and a model that would be useful [11].

4.1 Data Set Selection

The analysis was conducted based on an already structured dataset, publicly available for research, which consolidates the following parameters: gender, age, smoking, yellow fingers, anxiety, peer pressure, chronic disease, fatigue, allergy, wheezing, alcohol consuming, coughing, shortness of breath, swallowing difficulties and chest pain. Through the use of machine learning, the presence of a disease in the patient is clearly determined. Based on the acquired parameters considered to most effectively and efficiently provide an indication of the presence of the disease, we have two types of output results. The ultimate goal of the system is to obtain an output parameter that indicates whether the patient has lung cancer or not according to the contained results. Specifically, if the value is "YES," it means that the person has lung cancer, and if it is "NO," it means they do not. If the value of the gender parameter is "F," the subject is female, and if "M," it refers to a male subject. Regarding other factors, a value of "2" signifies a positive value, indicating that the subject is a smoker, has yellow fingers, is anxious, perceives environmental pressure, etc. The value would be "1" if not. Age is the only parameter with a numerical value, while all other values are nominal.

For a better understanding of the presented facts, it is first necessary to assess the already available data. If a detailed analysis of the available data is conducted (Fig. 2), the following conclusions can be drawn:

The database primarily consists of older individuals, with an average and median age of 62 years. However, despite this, there are also lower age groups present, with a minimum age of 21 years. Another parameter of significant importance is the "smoking" parameter, from which it can be noted that the database is predominantly composed of smokers. Considering that the values can only be 1 or 2, the mean value would be 1.5. Regarding the parameter "chest pain" and the conclusion drawn from the mean value, it is observed that the majority of participants in the database experience pain.

```
In [7]: df.describe().loc[['mean','std','min','max']] # Анализа на нумеричките податоци
Out[7]:
```

	AGE	SMOKING	YELLOW_FINGERS	ANXIETY	PEER_PRESSURE	CHRONIC DISEASE	FATIGUE	ALLERGY	WHEEZING	ALCOHOL CONSUMING	COUGHING	SHORTNE OF BRE/
mean	62.673139	1.563107	1.569579	1.498382	1.501618	1.504854	1.673139	1.556634	1.556634	1.556634	1.579288	1.640
std	8.210301	0.496806	0.495938	0.500808	0.500808	0.500787	0.469827	0.497588	0.497588	0.497588	0.494474	0.480
min	21.000000	1.000000	1.000000	1.000000	1.000000	1.000000	1.000000	1.000000	1.000000	1.000000	1.000000	1.000
max	87.000000	2.000000	2.000000	2.000000	2.000000	2.000000	2.000000	2.000000	2.000000	2.000000	2.000000	2.000

Fig. 2. Detailed analysis of the available data

Within the analysis of characteristics, it is important to calculate the importance score of each feature relevant to the target class. The Random Forest method is used as a ranking method [12]. Random Forest calculates the Gini impurity to measure a feature's ability to optimally discriminate examples in both classes.

Table 1 shows the ranking estimates of the parameters in descending order based on their importance. Values close to 0 and/or negative indicate features of little or no importance. Therefore, from the table, it can be concluded that the parameter with the highest significance is age. Regarding gender, men and women are equally likely to be diagnosed with lung cancer. Therefore, this parameter is not of great significance. Furthermore, with a detailed analysis of the distribution of participants concerning feature values and class labels in balanced data, it can be inferred that each of the explored features appears in 26% to 35% of lung cancer patients.

A significant amount of patients reported these symptoms even before receiving a lung cancer diagnosis. Monitoring risk factors and symptoms, followed by clinical assessment, can prevent or mitigate the negative effects of the disease. However, models will be trained and validated considering all attributes since, although not of the same rank of importance, they are significant indicators for the development and management of lung cancer by healthcare professionals.

4.2 Visualization of Data

The easiest way to analyze research data is to visualize it [13]. Greater understanding of the relevance of the data, interdependence, and whether all factors in this form of research are equally important can be achieved through data visualization.

This type of analysis begins with categorizing the categorical (nominal) and numerical data in the database. Considering that initially only age is numerical data, with its visualization, as shown in Fig. 3, it can be concluded that the most common age in the database ranges from 50 to 70 years.

4.3 The Architecture of the Machine Learning Model

In Fig. 4, the proposed architecture illustrating the workflow of the planned work is presented.

Table 1. Ranking of the parameters in the data set

Parameter	Rang
Age	0.3462
Allergy	0.2809
Alcohol consuming	0.2665
Wheezing	0.2567
Coughing	0.2442
Swallowing difficulties	0.2327
Peer pressure	0.2245
Chronic disease	0.1662
Chest pain	0.0758
Anxiety	0.0774
Smoking	0.0753
Yellow fingers	0.0725
Shortness of breath	0.0431
Gender	−0.0053
Fatigue	−0.0334

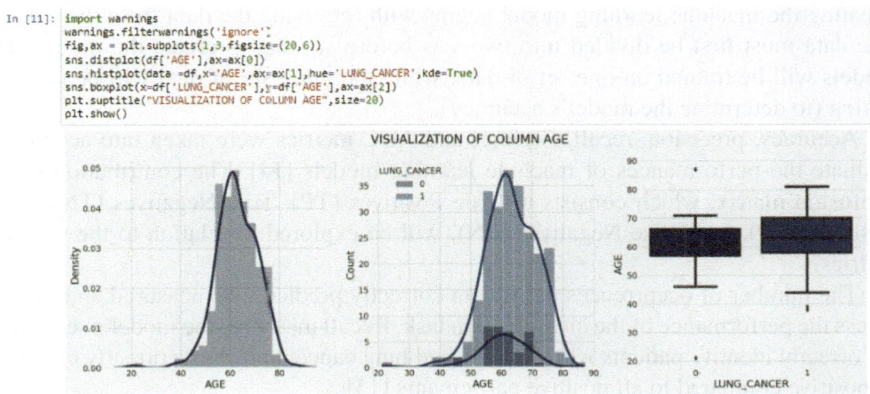

Fig. 3 Visualization of column AGE

Firstly, the data from the database undergoes the necessary processing. Once processed, the data is divided into training and testing sets. 25% of the data will be used for testing, while the majority of the data, approximately 75%, will be used for training.

After the data division, a model for data classification needs to be selected. In this case, a variety of models will be proposed, and based on their performances, the final chosen model will be determined.

With the help of the models, the data will be classified. The target class will indicate whether the individual being tested has lung cancer or not.

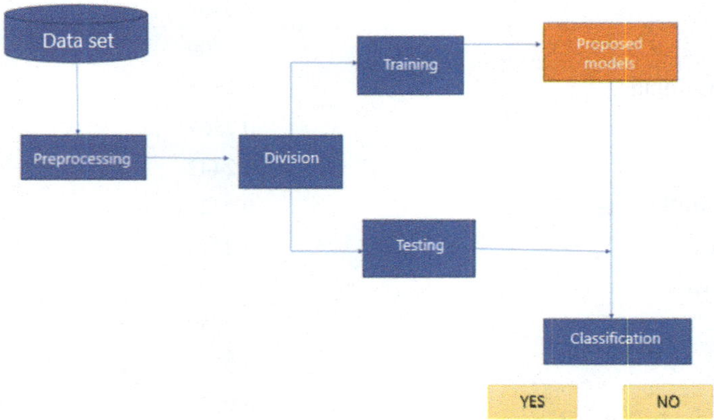

Fig. 4 Detailed architecture of the machine learning model

5 Results

Creating the machine learning model begins with retrieving the data from the dataset. The data must first be divided into two sets before starting to create the model. The models will be trained on one set of data, while the other set of data will be used for testing (to determine the model's accuracy).

Accuracy, precision, recall, F-score, and AUC metrics were taken into account to evaluate the performances of machine learning models [14]. The contribution of the confusion matrix, which consists of True Positives (TP), True Negatives (TN), False Positives (FP), and False Negatives (FN), will be explored in relation to the required metrics.

The number of occurrences of all data correctly predicted is measured and used to assess the performance of the classification task. Recall measures the model's sensitivity to correctly identify patients who actually had lung cancer and were correctly classified as positive compared to all positive participants [15].

$$Recall = \frac{Truepositives}{(TruePositives + FalseNegatives)} \quad (1)$$

After analyzing and training several models (KNN classifier,Support Vector Machine, Logistic regression, Random Forest classifier, Gradient boosting classifier, LGBM classifier), Support Vector Machine (SVM) classifier was selected as the best one. Table 2 describes the evaluation for each of the 6 models. They are assessed based on their performance, depending on each of the metrics used.

Table 2. Evaluation of the performance of machine learning algorithms.

	Accuracy	Recall	Precision	F-score
KNN	0.89	0.78	0.82	0.88
SVM	0.99	0.98	0.98	0.99
LR	0.94	0.95	0.90	0.90
RFC	0.94	0.88	0.90	0.94
GBC	0.94	0.88	0.90	0.94
LGBM	0.95	0.90	0.91	0.95

In this research, the Recall metric is of significant importance because it is necessary to reduce the number of false negatives. It is necessary to minimize the number of patients who have cancer but are classified as not having it.

SVMs are particularly effective in dealing with binary classification problems, which involve dividing data points into two groups, as is the case in the specific database [16].

The goal of the SVM algorithm is to identify the optimal decision boundary or line, which separates the data points into multiple data classes.

SVMs are helpful in deriving inferences from intricate datasets that exhibit deviations from a simple linear model. They accomplish this by applying a mathematical technique called non-linear Support Vector Machines (SVM), which converts the data into a higher-dimensional space in which border location becomes easier.

The Receiver Operating Characteristic (ROC) curve, often referred to as the ROC curve, is a representation similar to a graph indicating how well the classification model performs [17]. It enables understanding of how the model makes decisions at different levels of confidence. The curve consists of two lines, one representing the portion where the model correctly identifies positive cases (true positives), and the other representing the portion of the time when it incorrectly classifies positive cases as negative cases (false positives). By examining this graph, the model's performance can be assessed, and a threshold that achieves the ideal balance between true and false predictions can be chosen.

A tool for measuring binary classification problems is the ROC curve. Essentially, it separates the "signal" from the "noise" by plotting True Positive Rate (TPR) against False Positive Rate (FPR) with different threshold values. In other words, it shows how well the categorization model performs. The Area Under the Curve (AUC), which serves as a summary of the ROC curve, is a measure of the binary classifier's ability to distinguish between classes.

The classifier can successfully distinguish between all positive and negative class values when AUC = 1. However, if AUC were 0, the classifier would predict that all positives are negatives and all negatives are positives. There is a high probability that the classifier will be able to distinguish positive class values from negative class values when the AUC value is between 0.5 and 1. This is the case when most true positives and true negatives can be detected by the classifier compared to false positives and false negatives. The classifier cannot distinguish between positive and negative class values when AUC = 0.5, meaning the classifier predicts either a random class or a fixed class for each data point.

Therefore, the classifier's ability to discriminate between positive and negative classes improves with a higher AUC value.

In the case of the SVM model we selected for diagnosing lung cancer, the AUC value is 0.9915. This value is very close to the AUC = 1 value. This means that the diagnostic model successfully distinguishes between positive and negative values in 99% of cases. The AUC ROC curve for the selected model is shown in Fig. 5.

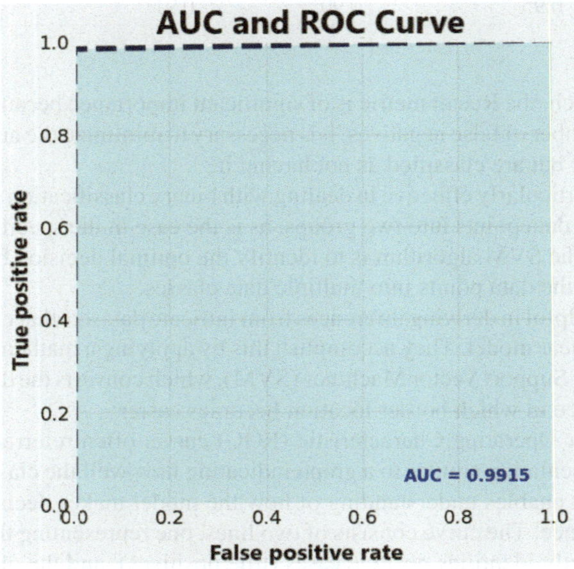

Fig. 5. AUC & ROC curve

The positive results from the performance assessment of the proposed approach indicate that oncologists can use the SVM model to identify lung cancer with efficiency. If the prediction is accurate, the doctor can develop a better prescription and provide the patient with an early diagnosis. As the quality of studies improves, the use of machine learning classifiers will increase in many clinical and hospital settings.

6 Conclusion

Predicting lung cancer is one of the most difficult medical challenges to address because cancer cells are so complicated. In the past, determining whether a patient has lung cancer or not required the doctor to do multiple tests on them, making the diagnosis extremely difficult and time-consuming. Delaying treatment for lung cancer significantly increases the likelihood of dying from the disease. Cancer can be cured if it is found and treated early enough.

For the analysis and visualization of data, machine learning is a crucial tool. It enables the extraction of conclusions and patterns from significant data sets that can be used to understand complex systems and make informed decisions. The field of machine learning is rapidly expanding, with many interesting new developments and prospects for exploration.

According to the accuracy matrices for classification and confusion, the SVM model outperforms other machine learning algorithms with an accuracy of 99 percent. SVM is included in this study to predict the onset of lung cancer. The main goal of this system is to provide patients with early warnings so they can save time and money.

Acknowledgements. This paper was supported by the project "Нид-ca-02-1902/8-Interdisciplinary application of data processing" financed by Faculty of Electrical Engineering and Information Technologies, Skopje, Ss. Cyril and Methodius University, Republic of N. Macedonia.

References

1. Green, D.R.: Cancer and apoptosis: who is built to last? Cancer Cell **31**(1), 2–4 (2017)
2. Ruger, J.P., Yach, D.: The global role of the world health organization. 2–5 (2009)
3. Toh, C., Brody, J.P.: Applications of machine learning in healthcare (2021)
4. Ayodele, T.: Machine learning overview, 10–11 (2010)
5. Raschka, S.: Model evaluation, model selection, and algorithm selection in machine learning, pp. 4–8 (2018)
6. Lauretto, M.S.: Introduction to Supervised Learning, pages 2–7
7. Dridi, S.: Unsupervised learning – A systematic literature (2021)
8. Saleh, H.: Machine Learning -Regression (2022)
9. Kim, J., Lee, H.: Lung cancer: diagnosis, treatment principles, and screening (2022)
10. Aggarwal, A., Lewison, G., Idir, S.: The state of lung cancer research: a global analysis (2016)
11. Sidey-Gibbons, J.A.M., Sidey-Gibbons, C.J.: Machine learning in medicine: a practical introduction (2019)
12. Ali1, J., Khan, R., Ahmad, N., Maqsood, I.: Random forests and decision trees (2012)
13. Leban, G.: Information visualization using machine learning, pp 109–110 (2013)
14. Zaman, M., Lung, C.H.: Evaluation of machine learning techniques for network intrusion detection, pp. 1–5. Taiwan (2018)
15. Goutte, C., Gaussier, E.: A probabilistic interpretation of precision, Recall and F -score, with Implication for Evaluation (2005)
16. Awad, M., Khanna, R.: Support vector machines for classification, pp 39–54 (2015)
17. Kumar, R., Indrayan, A.: Receiver operating characteristic (ROC) curve for medical researchers (2011)

Automated Quiz Generation Tool Using Artificial Intelligence Techniques

Richard Fox(✉) and Peyton Hayes

School of Computing and Analytics, Northern Kentucky University, Highland Heights, KY 41099, USA
foxr@nku.edu

Abstract. Artificial intelligence has been used to construct numerous intelligent tutorial systems over the years with varying degrees of complexity and capability. They generally are constructed on top of a domain model of knowledge, some form of model of the student's learning, a pedagogical strategy that presents material to the student, and an interface. Automated question generation is often considered a subproblem within the larger tutorial domain (although not all automated question generation systems use artificial intelligence). Perhaps simpler to construct, the ability to generate questions to test a student is a very valuable tool. As with tutorial systems, question generation approaches vary widely. In recent years, approaches have concentrated on the use of large language models. Between tutorial system and question generation attempts, research indicates a significant investment in time and effort to construct such systems. This paper presents an AI-based approach to automated question generation that foregoes the language model/language interface and uses a much shallower domain model to simplify construction, yet provides beneficial results to the student. This paper introduces the QuizGenie system to demonstrate the approach taken. The work is still in progress but shows promise as a study resource for students.

Keywords: Artificial intelligence · tutorial system · question generation · tool

1 Introduction

Artificial Intelligence (AI) applied to education dates back to such expert systems as LISP Tutor [1], Sophie [2], GUIDON [3] and SPICE [4]. These intelligent tutorial systems (ITS) were built on top of existing knowledge-based systems whereby the knowledge embedded in the problem-solving system was used to create lesson plans, to pose questions and respond to answers. Later development concentrated on building systems without first having a fully functioning expert system which led to a field called computer-aided instruction (CAI).

CAI refocused the approach by building a domain model (the material to be taught), a pedagogical model (encoding one or more ways to teach the material) and a model of the student's current state of understanding of the material [5]. As CAI research progressed into the 1990s and 2000s, CAI systems relied more heavily on a natural language interface

R. Silhavy and P. Silhavy (Eds.): CSOC 2024, LNNS 1120, pp. 486–496, 2024.
https://doi.org/10.1007/978-3-031-70518-2_43

to accept both questions and answers from the student. Today, CAI/ITS approaches are commonly built on top of large language model-based generative AI systems (e.g., ChatGPT). The obvious drawback of these approaches is the time needed to construct the underlying knowledge-base or domain and pedagogical models along with a natural language interface, and/or a large language model.

While these efforts are producing very useful results, a simplified approach may still be valuable. Concentrating just on the subproblem of question generation, when question types are limited to those expecting specific answers (e.g., multiple choice, true/false, fill in the blank), the need for a language interface is eliminated and construction of the question generating system can concentrate on the domain model with a modest pedagogical component. This paper presents QuizGenie, an AI-based quiz generation tool. The system uses a simple set of rules to create questions from a database of concepts that represent the domain model. The system operates in two modes, domain model input and quiz generation. AI techniques are applied in the latter mode to generate questions, explain answers and diagnose student performance.

In this paper, Sect. 2 examines some notable ITS/CAI approaches. Section 3 introduces the QuizGenie system and how it operates, followed by some examples demonstrating its use in Sect. 4. As QuizGenie is in its early stages of construction, the final section offers some brief conclusions and discusses planned future work.

2 Related Work

Two of the earliest known AI tutorial systems are LISP Tutor and GUIDON. LISP Tutor was based on the ACT (advanced computer tutoring) Theory [6], whereby the system had the appropriate computational model to solve problems that students would be asked to solve, and would solve them using strategies that the students should be applying. ACT theory differentiates between process and domain knowledge, applying production rules (if-then statements) both to solve problems and to model the student's current understanding within the domain. As the name of the system implies, the theory was implemented on a system to teach students how to write LISP code.

GUIDON [4] was built on top of the already existing MYCIN [7] expert system. As with LISP Tutor, the underlying knowledge was in the form of production rules. And like LISP Tutor, GUIDON had both declarative knowledge (how to perform diagnosis) and domain knowledge (the infectious disease domain). MYCIN itself comprised some 650 rules encoding infectious disease diagnosis and treatment knowledge but GUIDON included a model of the student's understand as well as some modest natural language facilities to respond to questions from the user (student).

Another early system was HERON [8], developed by a cognitive psychologist to solve mathematical word problems using the simplex method [9]. HERON was designed to improve a child's ability to work through word problems. It operated using a graphics-based interface that let the student add to a growing model of the problem. HERON would generate subgoals for the student to pursue in breaking the larger problem into smaller problems. The result was a tree structure of goal/subgoal nodes.

CAI/ITS systems progressed from those built on top of already existing knowledge bases to those that implemented one or more pedagogical approaches using one or more underlying models of the domain and one or more problem solving techniques applied to the domain. As noted in [5], most of these systems comprised an explicit domain model, a student model, and one or more pedagogical models. In addition, some form of interface was applied so that the system could receive natural language input from the student.

The domain model might apply any number of AI approaches such as semantic networks, production rules, scripts or frames. One form of student model was a differential approach, comparing student answers to expected result and using this to critique the student's understanding. Another approach was to use an overlay model in which the portions of the domain model that the student understood were noted so that the focus could shift to portions of the domain that the student had yet to master. With a disturbance model, the CAI attempted to diagnose student misconceptions. With a simulation model, a computational model was run to test if the student's responses matched the results of the simulation. Two other approaches were to generate a belief model, which explicitly stated what the system believed the student knew and did not know, and a constraint-based model in which numeric constraints were proposed to suggest where the student was in his or her understanding. Any number of approaches might be employed to implement these models including computational modeling, fuzzy logic reasoning, production rules and Bayesian networks.

A subset of ITS concentrates solely on question generation, although not all automated question generation (AQG) systems use AI approaches and not all AQG are used for tutorial purposes (for instance, AQG is commonly used in chatbots). AQG requires a means of generating meaningful, correct and syntactically and semantically valid questions from a base of knowledge. As a tutorial tool, AQG can be used as a study resource for students to test their current knowledge and perhaps pinpoint areas where they need to study further.

AQG dates back to the 1970s [10] but gained significant attention with the popularity of MOOCs in the 2010s [11]. AQG systems are as diverse as CAI/ITS systems. They generally divide into domain-specific and generic systems with the primary goal of assessing a student's understanding. AQG might generate questions using template-based, rule-based, semantic network-based or neural network-based approaches. Aside from neural networks, these approaches use an explicit model and techniques which select a piece of knowledge from the domain and generate a question from the selected knowledge. Questions are commonly in the form of multiple choice, true/false and fill-in-the-blank. With a significant language model and natural language processing techniques, questions demanding longer answers can be created. Within the mathematics domain, one approach tackled generating word problems [12]. Today, with the popularity of generative AI, most new CAI/ITS and AQG systems are being developed to use large language models implemented using deep neural networks [13–15].

Most of the significant AQG systems require a significant investment in time and effort. Whether built on top of existing knowledge bases, existing large language models, or constructed entirely from scratch, these systems comprise several components that collectively must work well together for success. The research presented in this paper instead focuses on a greatly simplified approach that shows potential as a study tool for students.

3 The QuizGenie Tool

QuizGenie is a GUI-based app written in Python (using PyQt5) which operates in one of two modes. In the first mode, the user enters (or loads) text that represents the domain of study. From this text, the user then creates a database of concepts. In this mode, the user defines categories, terms, term definitions and grammatical categories of terms. The user is able to link terms together to indicated related ideas from which QuizGenie can form questions.

The second mode is the quiz generation mode. QuizGenie generates a series of questions for the user to respond to. The student attempts to answer the questions and QuizGenie responds with the student's results, including an analysis of what the student missed. This analysis is a form of diagnosis that might help the student learn from a mistaken understanding of concepts. This mode also provides a facility for the user to request explanations for answers of individual questions. The remainder of this section explores the two modes in more detail.

3.1 QuizGenie Operating in Database Mode

Mode 1, domain model manipulation, does not employ AI. Instead, it presents a GUI editor, as shown on the left-side of Fig. 1, from which the user is able to create categories of concepts and define terms. The editor may be preloaded with text that contains all the domain concepts or the user can enter or edit the text. The user then defines categories of concepts.

Focusing on the left-side of Fig. 1, categories are listed in the left pane. Here, there are four categories: devices, networks, protocols and technology. Under each category, defined terms of that category are listed (notice that a term may appear in multiple categories, such as Ethernet which is both a device and a technology in the figure). Selecting a term in the left pane highlights all instances of the term in the central pane.

With a term selected in the text of the central pane, the user may select the Extract Term button. This opens a second window, as shown on the right-side of Fig. 1, from which the user is able to indicate the category that the term belongs in, define keywords that are related to, or link to, this term, and select the term's grammatical category. Additionally, from the main pane, the user can connect related terms together. Term entry consists of 6 specific parts as described below.

First, the term itself is selected (or entered). This term will be a word or phrase used in questions like fill-in-the-blank or multiple choice. It is also the identifying phrase used by the user to select the entry for modification when needed using the term definition window (as shown in the right-side of Fig. 1).

Second, the term is assigned a grammatical role. The default is as a noun, but other categories are verb and "other" (e.g., verb phrase or clause). Terms of the same grammatical type can be used as alternatives in questions like multiple choice and true/false.

Third, the user selects the definition of the term by highlighting the definition in the text. Definitions may be as long as needed but are typically one sentence so that they can be used as the basis of a question. For instance, an entire sentence can be used for a true/false question or used as a fill-in-the-blank question with a keyword removed. There is no restriction on having longer definitions if desired.

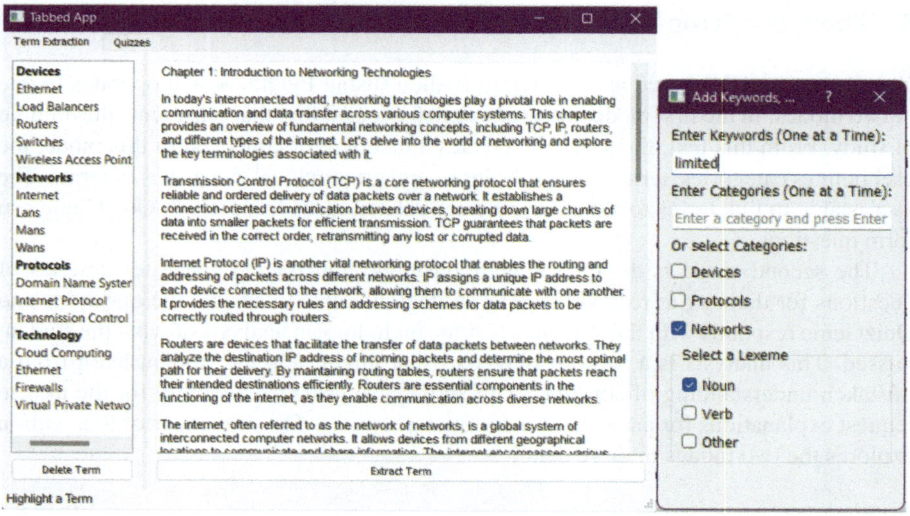

Fig. 1. QuizGenie operating in input mode.

Fourth, the user can highlight key phrases within the definition that are related to the term. These linked phrases can then be drawn upon to form a question. For instance, a true/false question might exchange the term for one of the highlighted phrases, or linked phrases can be reordered within a definition for a term. A highlighted term can also be removed to form a fill-in-the-blank question.

The fifth and sixth forms of input provide the user with the chance to link this new term to terms already in the database and to place a term within a defined category. Linking terms together gives QuizGenie the ability to find related terms in the formation of some question types. For instance, a true/false question may have the term exchanged with a linked term. Categorizing terms allows QuizGenie to identify other terms that are related to be used as potential answers, for instance in multiple questions.

The domain model used by QuizGenie is the collection of the original text, the terms with their definitions, grammatical categories and linked phrases, and the categories. The entire domain model is stored in an SQL database.

Notice in this section the word "user" has been used instead of "student". The reason for this is that someone other than the student may set up the domain model. The tool is envisioned as being used by both a teacher and the teacher's students. The teacher can set up the database and distribute it to the students. A student then loads the database to use in mode 2, as described in the next subsection. Alternatively, a student could define the domain model and/or modify the domain model. When the student sets up the domain model, it helps reinforce the material as the student hopefully learns while creating and modifying the model.

A portion of one of the implemented databases is shown in Fig. 2 to illustrate how these concepts, terms and definitions are stored. The database stores the term, its definition, keywords, the category and the grammatical role (listed as lexeme here).

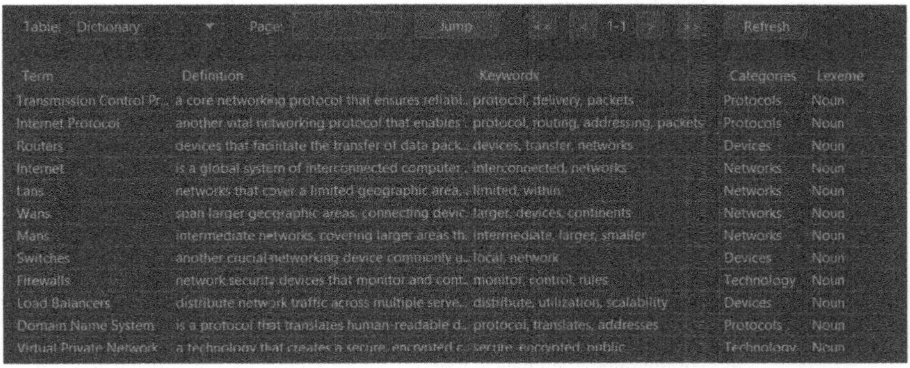

Table: Dictionary Page: Jump 1-1 Refresh

Term	Definition	Keywords	Categories	Lexeme
Transmission Control Pr...	a core networking protocol that ensures reliabl...	protocol, delivery, packets	Protocols	Noun
Internet Protocol	another vital networking protocol that enables ...	protocol, routing, addressing, packets	Protocols	Noun
Routers	devices that facilitate the transfer of data pack...	devices, transfer, networks	Devices	Noun
Internet	is a global system of interconnected computer...	interconnected, networks	Networks	Noun
Lans	networks that cover a limited geographic area...	limited, within	Networks	Noun
Wans	span larger geographic areas, connecting devic...	larger, devices, continents	Networks	Noun
Mans	intermediate networks, covering larger areas th...	intermediate, larger, smaller	Networks	Noun
Switches	another crucial networking device commonly u...	local, network	Devices	Noun
Firewalls	network security devices that monitor and cont...	monitor, control, rules	Technology	Noun
Load Balancers	distribute network traffic across multiple serve...	distribute, utilization, scalability	Devices	Noun
Domain Name System	is a protocol that translates human-readable d...	protocol, translates, addresses	Protocols	Noun
Virtual Private Network	a technology that creates a secure, encrypted c...	secure, encrypted, public	Technology	Noun

Fig. 2. Sample (partial) database for the networking domain model.

3.2 QuizGenie Operating in Quiz Generation Mode

In Mode 2, QuizGenie creates a quiz. The student may select a specific category to be quizzed over, or all categories of the domain. The number of quiz questions generated is based on the number of terms stored within the selected category(ies). For a small domain model, the number of questions defaults to three per category.

A question is generated randomly by selecting a term from the database. Simple AI-based rules are applied to generate a quiz question based on the stored information of the selected term. At this point, QuizGenie can generate one of five types of questions. These are described in table 1, along with the rules used to determine if this question type is available for this particular term.

Although the rules are simplistic, they are easy to implement and provide a range of question types that can challenge the student. Along with the question generation, the answer is stored. QuizGenie opens a new window displaying the question and the question type along with possible answers for multiple choice questions or drop-down menus for the order type.

Figure 3 shows how the quiz questions are stored. Each question is stored by its term, the type of quiz question, and the question itself. Notice how the question is not just text but also references to words/phrases and other terms. Not shown in the figure is the answer to the question but this is part of the storage.

Upon answering a question, the student is given a response of correct or incorrect. In either case, the student may ask for an explanation of the answer. The explanation is drawn from the domain model by displaying the text from which the term was defined.

QuizGenie waits until the entire quiz is completed before providing the user a report of their performance. The student may ask for an analysis of their performance which provides the student feedback in three forms. First, if the student consistently missed one type of question (e.g., true/false), this is reported. The rationale is that a student may struggle with one type of question rather than the content of the question. Students often find true/false questions to be tricky, for example.

Table 1. Quiz Generation Rules

Type	Rule for Generation	Generating the Question
Fill in the blank	Does the definition contain at least one highlighted word/phrase?	Randomly select either the term or one of the highlighted words/phrases to remove
Matching	Are there at least four terms linked together that are of the same grammatical type and does each have a short definition?	Select the terms and their definitions and randomly arrange the definitions to create a matching question; limited to no more than seven terms
Multiple choice	Is there at least one other term in the database linked to this term and is that term of the same grammatical role?	Randomly select four other terms in the database with the same grammatical role, including at least one that is linked to this term, and display the definition along with this term's choice and others in random order; randomly the actual term can be removed in favor of "none of the above"
Order	Does the definition contain at least two linked terms/phrases?	Remove at least two of the linked words/phrases and display the term along with its definition replacing the selected words/phrases with blank spaces, then list the removed words/phrases in a random order for the user to decide which word/phrase fits in which blank
True/false	Does the definition contain a word like "is", "is not", "does", "does not"?	Randomly select true or false and if false, replace the selected word with its opposite. For instance, change "is" to "is not"

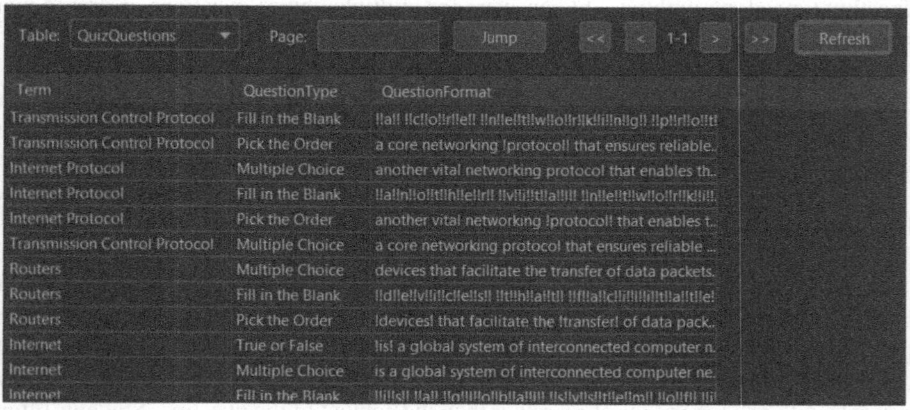

Fig. 3. Sample questions as stored in QuizGenie

Second, QuizGenie examines the terms used to generate each question. If the student missed a question because they confused terms, QuizGenie will report this. By illustrating to the student that he or she confused one term with another, it can help the student better remember the appropriate term for a given definition.

Third, if the student consistently missed questions within a specific category, QuizGenie reports this. The student can then concentrate on studying material corresponding to that (or those) category(ies). This is only available when the student requests a quiz across multiple categories.

In quiz mode, QuizGenie serves as a tutorial system that can diagnose issues that the student has either with the domain or the type of question. QuizGenie can pinpoint a specific topic or linked topics that should be revisited. At the time of this writing, only some of the reporting facilities have been implemented. Work continues to improve this mode.

One additional facility of quiz mode is that QuizGenie stores the results of the quiz by the domain (that is, the database loaded), score, number of questions asked, and date. The student is then able to track progress over hours or days of test results. Although a simple accounting process, the student can use this to track progress over time.

4 QuizGenie in Operation

QuizGenie's usefulness is limited to domain areas that comprise concepts and terms. To illustrate this, two databases have been created to test QuizGenie: categories of species within biology (specifically cell study) and TCP/IP (computer networks). QuizGenie is unlikely to be of use when applied to an area of study that revolves around applied problem solving, as with algebra or computer programming. In such domains, concepts may be important but problem solving is essential. QuizGenie's methods for generating questions is limited to definitions of terms. A different approach would be required to extend QuizGenie to other types of domains.

This section demonstrates QuizGenie's second mode in action with examples from the computer network domain. Figures 4 and 5 show QuizGenie quizzing a student. As can be seen in Fig. 4, the quiz is 3 questions and the student has asked for questions in all categories. The student is given a multiple-choice question and has answered Transmission Control Protocol. Now, the student selects Check Answer and is told the answer is wrong. Whether or correct or wrong, the pop-up window that contains the answer also has an explanation button.

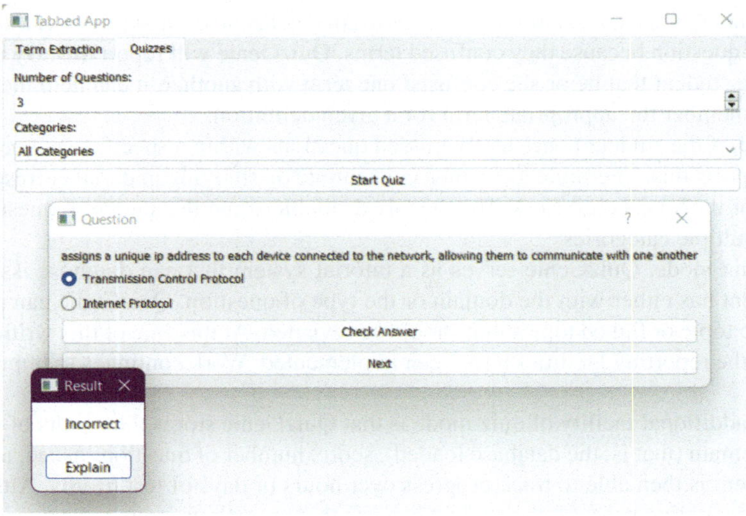

Fig. 4. QuizGenie in Quiz Mode

Selecting the Explain button brings up an explanation window, as shown in Fig. 5. The explanation is drawn directly from the database that defined the term, in this case, Internet Protocol. To move on to the next quiz question, the user selects the Next button on the question's pop-up window.

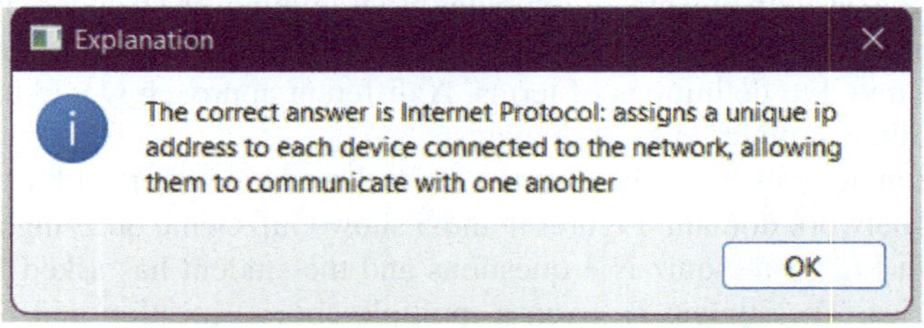

Fig. 5. QuizGenie Providing an Explanation

Upon completing the quiz, QuizGenie displays the results of the student's performance. This is shown in Fig. 6. A diagnostic report can be generated although that is not shown here in this figure. As noted earlier, work continues to expand the capabilities of this report.

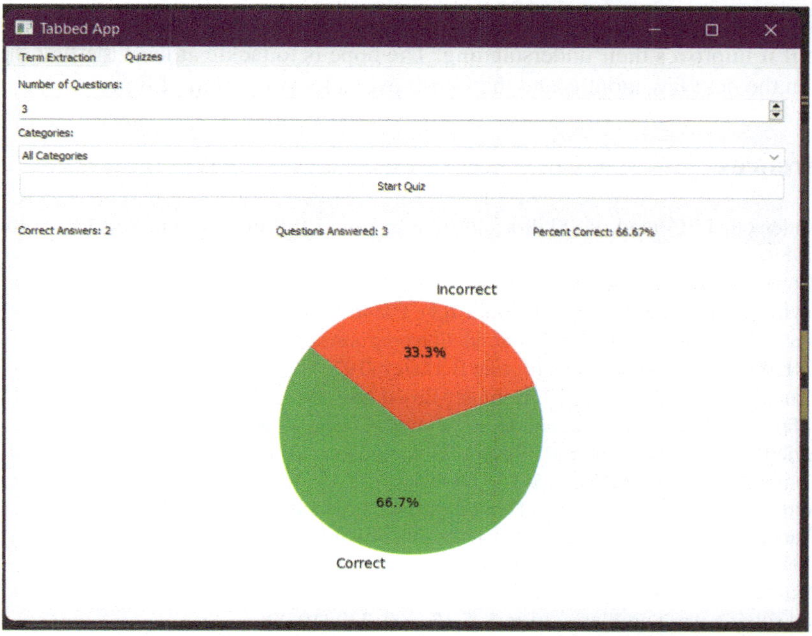

Fig. 6. QuizGenie Quiz Results

5 Conclusions and Future Work

While generative AI systems have captured the imagination of the public and shown great potential, and while AI tutorial systems have been around for decades, the amount of work needed to create one of these systems is immense. The payoff, in terms of the capability of these systems, is equally immense. However, within the limited context of question generation, the approach can be dramatically simplified.

QuizGenie takes a far more tractable approach, eliminating the need of a language model/natural language interface. The pedagogical model is simplified by applying AI rules to generate questions. Only the domain model remains as a component that requires building and here, the editing capabilities makes it an easy model to develop. With additional AI rules, the tool can provide useful feedback to the student learning the material of the selected domain.

This paper introduced the mechanisms behind QuizGenie. As noted, this work has only just begun. QuizGenie is still under construction although early returns are encouraging that this can be a very useful tool. Future work centers around four endeavors. First, to date only two domains have been partially implemented (biology and computer networks). Fleshing out both databases and creating additional databases would further demonstrate the tool's utility. Second, the ability to diagnose possible student learning issues is only partially implemented. The more significant diagnosis (determining errors because of a misunderstanding within the domain) is the more important part of the diagnosis, and at the time of this writing, it has not yet been implemented. Third, the interface continues to be improved. Fourth, the tool has not yet been used by students to see if it improves their understanding. The hope is to tackle at least the first three of these in the next few months and the fourth over a longer period of time.

References

1. Anderson, J., Conrad, F.: Skill acquisition and the LISP tutor. Cogn. Sci. **13**(4), 467–505 (1989)
2. Brown, J., Burton, R., Bell, A.: SOPHIE: a step toward creating a reactive learning environment. Int. J. Man Mach. Stud. **7**(5), 675–696 (1975)
3. Nagel, L., Pederson, D.: SPICE (simulation program with integrated circuit emphasis), Memo ERL-M382, University of California Berkeley (1973)
4. Clancey, W.: GUIDON. Int. J. Comput.-Based Inst. **10**(1), 8–15 (1983)
5. Alfaro, L., Rivera, C., Castaneda, E., Zuniga-Cueva, J., Rivera-Chavez, M., Fialho, F.: A review of intelligent tutorial systems in computer and web based education. Int. J. Adv. Comput. Sci. Appl. **11**(2), 775–763 (2020)
6. Anderson, J.: The architecture of cognition. Harvard University Press, Cambridge, MA (1983)
7. Clancey, W.: Use of MYCIN's knowledge-base for tutoring. Int. J. Man Mach. Stud. **11**, 25–49 (1979)
8. Reusser, K.: Problem-solving beyond the logic of things: Contextual effects on understanding and solving word problems. Instr. Sci. **17**, 309–339 (1988)
9. Dantzig, G: Origins of the simplex method, a history of scientific computing, pp. 141–151. Association for Computing Machinery, New York, NY (1990)
10. Alsubait, T., Parsia, B., Sattler, U.: Next generation of e-assessment: automatic generation of questions. Int. J. Technol. Enhanc. Learn. **4**(3–4), 156–171 (2012)
11. Qayyum, A., Zawacki-Richter, O.: Distance Education in Australia, Europe and the Americas. Springer, Berlin (2018)
12. Kurdi, G., Leo, J., Parsia, B., Sattler, U., Al-Emari, S.: A systematic review of automatic question generation for educational purposes. Int. J. Artif. Intell. Educ. **30**, 121–204 (2020)
13. Bhat, S., Nguyen, H., Moore, S., Stamper, J., Sakr, M., Nyberg, E.: Towards automated generation and evaluation of questions in educational domains, Proceedings of the 15th International Conference on Educational Data Mining, vol 701 (2022)
14. Wang, Z., Valdez, J., Basu Mallick, D., Baranuick, R.: Towards human-like educational question generation with large language models, International Conference on Artificial Intelligence in Education, pp. 153–166. Cham: Springer International Publishing (2022)
15. Tsai, D., Huang, A., Lu, O., Yang, S.: Automated question generation for repeated testing to improve student learning outcome. In: 2021 International Conference on Advanced Learning Technologies, pp. 339–341 (2021)

Proposal of a Computational Vision Model for the Pre-diagnosis of Anemia Based on the Image of the Ocular Conjunctiva

Alejandro Ildefonso[1] and Wilfredo Ticona[1,2](✉)

[1] Universidad ESAN, Lima, Perú
13200125@ue.edu.pe, wmamani@esan.edu.pe
[2] Universidad Tecnológica del Perú, Lima, Perú

Abstract. Anemia is a persistent public health problem in Peru, with significant repercussions on individual quality of life and socio-economic progress at the national level. Although the hemogram is considered the reference method for diagnosing anemia, its need for time and a laboratory setting to be performed can represent a considerable limitation, especially in remote or less developed areas. The aim of the present research is to implement a computer vision model for the pre-diagnosis of anemia from the image of the ocular conjunctiva. The applied methodology was based on 5 phases: obtaining dataset, preprocessing, modeling and classification, feature extraction and implementation of CNN architectures. Several models were run with the classifiers: SVM, RF, MLP, and RNN and feature extraction techniques: SIFT, SURF, ORB and HOG. The model with RF and HOG extractor obtained the highest accuracy with 79%. Finally, deep learning models were explored, adjusting parameters such as the number of neurons, epochs, and samples in each model. Although initially the custom model obtained the highest accuracy of 93.18%, the Inception-ResNet-v2 model, supported by existing studies, was finally chosen and demonstrated a robust accuracy of 90.69% and loss of 0.2788, which is better than the loss of 0.3563 obtained with the comparison model.

Keywords: Computer Vision · Anemia · Machine Learning · Predictive Accuracy · Convolutional Neural Network Architectures · Ocular Physiology · Pre-diagnosis

1 Introduction

Health and happiness are essential in any family, so it is important that all members are healthy and able to grow and develop to their full potential, especially children. During childhood, care is critical to a child's future and quality of life, so it is important to protect their health and nutrition. Without proper care at home, various disorders and diseases can develop during childhood [1]. Iron deficiency anemia is characterized by a low level of hemoglobin in the blood due to a lack of iron in the body [2]. The disease affects all stages of life, but women and pre-school children are most susceptible [3]. It should be

R. Silhavy and P. Silhavy (Eds.): CSOC 2024, LNNS 1120, pp. 497–518, 2024.
https://doi.org/10.1007/978-3-031-70518-2_44

noted that if a child is anemic, he or she is likely to have developmental delays, which can negatively affect the child's cognitive, psychomotor and language development. In addition, anemia can have long-term effects on a child's health, possibly affecting their physical and mental abilities as an adult. Compared to people with adequate nutrition, people with anemia perform poorly, as a deficiency reduces the amount of oxygen in the blood and reduces the energy available to increase productivity [4] and [5].

Anemia remains a serious problem in Peru today. According to recent studies, the prevalence of anemia in miners is 40%, which indicates a reduction of only 0.1 percentage point from the values recorded in 2019 [6]. Data show that the anemia rate in rural areas (48.4 per cent) is higher than in urban areas (36.7 per cent). Although anemia has decreased by 3.5 per cent year-on-year, many people are still affected [7].

In Peru, provinces located far from the capital lack the human and material resources necessary to accurately identify patients suffering from anemia. According to [8], a clear decentralization policy is needed; some regions do not have the human resources they really need, and there is a great disparity in the distribution of health human resources in the country. However, there are different ways to have indicators that a person may have anemia, paleness in various regions of the body is a crucial indicator for a detection of this disease, this color change can occur in the nail bed, conjunctiva, and palmar creases [9].

2 Related Work

In the work of Bauskar *et al.* [10], the authors emphasize the global importance of anemia, a blood disorder caused by a lack of healthy red blood cells or hemoglobin. The aim of their research is to present a non-invasive computerized technique that uses images of the ocular conjunctiva to de- detect anemia. This would allow early diagnosis and timely intervention for the benefit of patients. In this study, the process begins with the collection of images of the ocular conjunctiva of the patients, which are then analyzed. The acquisition of the database was performed manually, following a specific procedure. To capture the images, the lower eyelid of the eye is gently pulled using the index finger and the picture is taken. It is important to note that the images were taken without the use of flash to avoid excessive light. The dataset used in this research is composed of a total of 99 images, of which 51 images correspond to healthy patients and 48 images correspond to patients with anemia. The study aims to build a classifier to detect anemia based on a non-invasive method that can provide a more accurate estimate of red blood cell concentration by analyzing the ocular conjunctiva, the results of this experiment had 93% accuracy using the SVM technique, it is still recommended that an invasive technique be performed in order to confirm this pre-diagnosis, it recommends and stresses that imaging is a crucial point for the study to be successful, as in this case the level of brightness has caused some images to be in error.

Similarly, Dhanasekaran and Shanker [11], in the introduction to the study, detail what anemia is, the different types of anemia and what characteristics are shown when a patient suffers from this condition. On the one hand, it talks about what is the purpose of evaluating the cells in the blood, what is sought within the analysis and what are the factors that determine the health of the patient. On the other hand, mention is made of how technology has been perfecting the accuracy of this diagnosis. The study highlights that there are hardware's such as the hematological counter, which has the function of analyzing the main parameters: Leukocytes, Erythrocytes/Platelets, and hemoglobin; thus, this counter allows to provide an accurate diagnosis. The method divides the original image into different classes and regions where pixels in the same region share some common characteristics. It is not convenient to work with the whole image, so it is segmented considering the color, all pixels are represented in the image under the color red, green and blue. The idea of this segmentation method allows the analysis work to be simpler and the computational load to be reduced. Image segmentation for this case also introduces techniques such as histogram-based methods, optimization techniques, thresholding, clustering, data augmentation, etc. Finally, PSO algorithms are recommended for this type of process, however, the author proclaims that these are not effective, so we will use Darwinian Particle Swarm Optimization (DPSO), this technique is superior to the recommended one, for this case each swarm works separately as a PSO algorithm. The DPSO algorithm makes these swarms traverse the image space, the particle goes through different positions until it finds the best location or until the velocity is close to 0.

In the work developed by Heredia and Ticona [12], they indicate that anemia is present in more than 40% of Peru's population, is common in both children and adolescents, and predominates in pregnant women and children under two years of age, which seriously compromises their development. To diagnose anemia, it is necessary to carry out laboratory tests by analyzing the blood, where hemoglobin levels are determined. However, most hospitals do not have adequate equipment to perform the tests, resulting in delays in the delivery of diagnoses. The aim of this research is to compare techniques based on computer vision and machine learning for the early detection of anemia from fingernail analysis, so that doctors can use it to support the detection of anemia for early ruling out. With early diagnosis, patients will be spared from suffering from the different stages of this disease and especially those in the chronic stage when the consequences are severe, because anemia can indicate the presence of another underlying disease. The proposed methodology contains 5 phases: dataset construction; preprocessing; feature vector extraction (SIFT, SURF and HOG), machine learning (SVM and ANN) and CNN architectures (AlexNet, VGG16, VGG19, ResNet50 and proprietary model). Several experiments were performed, and the best results were Accuracy 0.989, precision 0.98, recall 0.98 and F1-score of 0.98, with the VGG19 architecture as feature extractor in combination with the support vector machines (SVM) classifier. The research demonstrated that it is possible to detect anemia, without the need for a blood test, more quickly and with reliable results.

On the other hand, in the work developed by [13], diagnosis and Identification of cells and disease infected cells are considered as very important in medical science all over the world. There are health implications which can often be identified by observing the morphological changes of cells. The traditional methods of identifying such issues can be expensive and time consuming to the extent that only certain medical centers can perform the task at hand or take days to receive a report of. However, deep Learning with convolutional neural networks (CNNs) can take over most of this tedious process. This paper aims towards creating a custom CNN model that can quickly classify different kinds of peripheral blood cells such as different white blood cell times and platelets.

Such a model can be used in blood cell counts which can be used to identify cases like leukemia. Moreover, such a method can be extended into other fields such as red blood cell detection or even infected cell detection, which includes identifying diseases from Sickle Cell Anemia to cells affected by COVID19. For this research, a total of 32768 images were used. Eventually, the custom CNN model has performed exceptionally well, achieving accuracies as high as 99.1% and 98.9% in training and validation respectively, which is significantly higher than using pretrained models such as DenseNet or NasNet. The result also proves to be higher compared to the previously completed similar tasks to the best of our knowledge.

In the work of Bauskar *et al.* [14], this paper proposes the development of a mobile application for anemia detection using YOLO v5. The app aims to bring an easy diagnosis as support to health-care professionals to know the presence of anemia in children in places where mobile signals are missing (isolated). Here, we present the implementation and inclusion of the YOLO v5 neural network into the app. We used an image dataset obtained from Universidad Peruana Cayetano Heredia, which contains several pictures of children under five years old and their respective prognosis with the blood test. Although YOLO v5 gets good results using a computer, its implementation in a mobile app reduces its performance. Despite this, the app shows a sensitivity of 0,71 and a specificity of 0,89 to predict anemia.

Finally, in the work of Sujatha *et al.* [15], the proposed online-based malnutrition-induced anemia detection smart phone app is built, to remotely measure and monitor the anemia and malnutrition in humans by using a non-invasive method. This painless method enables user-friendly measurements of human blood stream parameters like hemoglobin (Hb), iron, folic acid, and vitamin B12 by embedding intelligent image processing algorithms which will process the photos of the fingernails captured by the camera in the smart phone. This smart phone app extracts the color and shape of the fingernails, will classify the anemic and vitamin B12 deficiencies as onset, medieval, and chronic stage with specific and accurate measurements instantly. On the other dimension, this novel technology will place an end to the challenge involved in the disposal of biomedical waste, thereby offering a contactless measurement system during this pandemic Covid-19 situation.

3 Methodology

The methodology of this study, based on frameworks referenced in [10, 12], and [13], unfolds across five pivotal phases: dataset acquisition, preprocessing, feature extraction, modeling and classification, and CNN architecture implementation. Each phase comprises critical activities fundamental to the research's development, with a graphical representation provided in Fig. 1.

Initially, the study sourced a comprehensive dataset from IEEE DataPort, pivotal for anemia detection. Subsequent preprocessing included the delineation of the region of interest and data augmentation to reinforce the training process. Feature extraction capitalized on established techniques such as SIFT, HOG, and ORB, including a variant of SIFT with padding to address image heterogeneity. The study evaluated various machine learning algorithms—SVM, KNN, Random Forest, and MLP—to ascertain their predictive efficacy. Concluding with the implementation of sophisticated CNN architectures like EfficientNet, ResNet, and Inception-ResNet, the study sought to harness deep learning for detailed feature identification and classification.

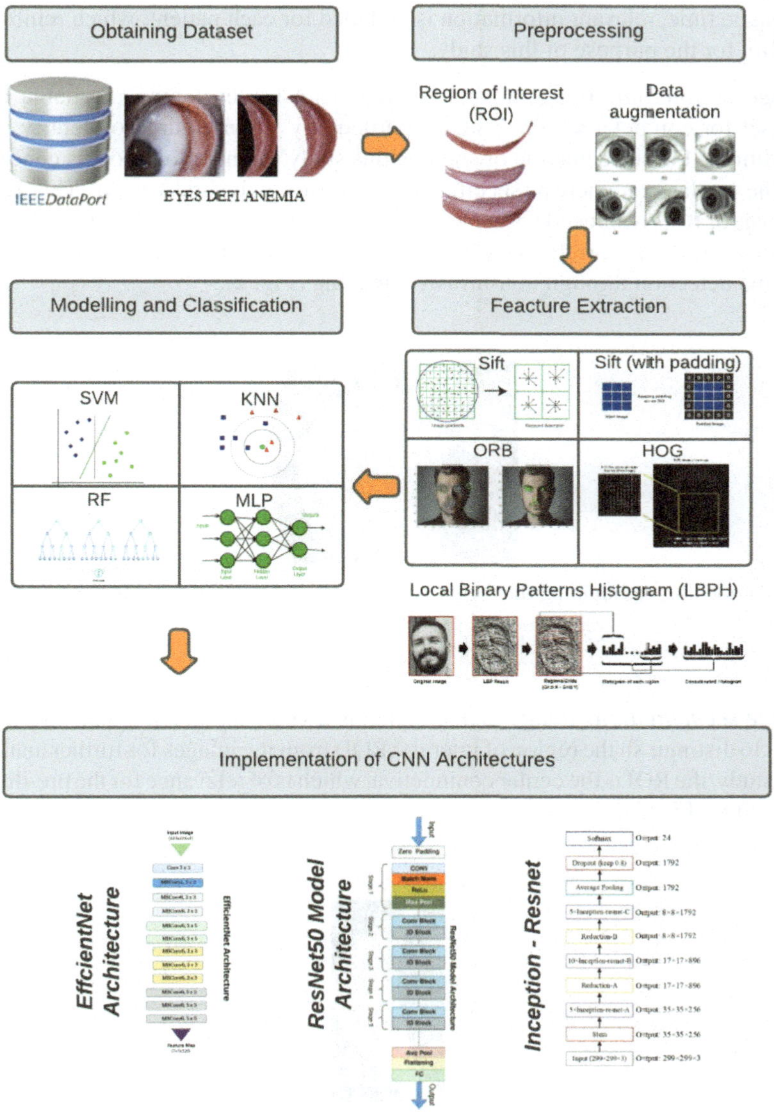

Fig. 1. Proposed methodology.

3.1 Obtaining Dataset

The dataset obtained from EYES-DEFI-ANEMIA available in IEEE-DATAPORT (https://org/documents/eyes-defy-anemia) was used for this research. The dataset contains specific features of the ocular conjunctiva images of the patients and consists of 217 images, all obtained under similar circumstances and focusing on the ocular conjunctiva. At the same time, relevant information is included for each patient, which reinforces its suitability for the purpose of this study.

- ***Image annotation:*** Image annotation is crucial in developing a reliable training dataset for computer vision systems, particularly when aiming to identify medical conditions. For the detection of anemia, this study's annotation process concentrates on the ocular conjunctiva, specifically the fornix palpebral region. This region is known for its pronounced vascular patterns and tendency to exhibit distinct color and texture variations correlating with hemoglobin levels, making it a critical area for anemia detection through non-invasive imaging (Fig. 2).

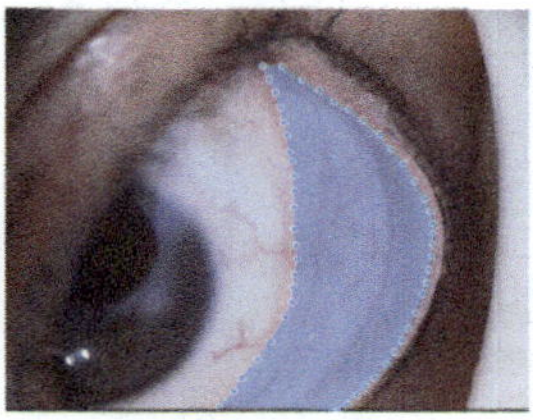

Fig. 2. Annotation highlighting the fornix palpebral of the ocular conjunctiva, a key region for anemia detection.

- ***Image segmentation:*** Image segmentation is a vital step in computer vision, which aims to distinguish the region of interest (ROI) from the images for further analysis. In this study, the ROI is the ocular conjunctiva, which is of relevance for the pre-diagnosis of anemia (Fig. 3).

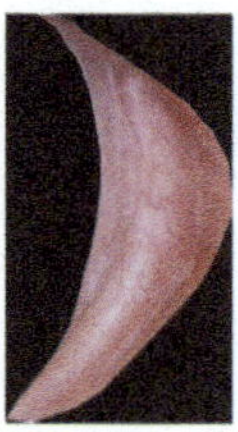

Fig. 3. Segmented ocular conjunctiva, displayed against a black background to emphasize the ROI, with extraneous elements removed to facilitate precise analysis.

3.2 Preprocessing

Table 1 shows the activities of the preprocessing phase.

Table 1. Pre-processing activities.

Activities	Description	Tasks
Evaluation of image classification	The images acquired from the database should be assigned to the relevant categories depending on the patient's condition, namely whether the patient is anemic or not	Classifying patients with anemia
		Determination of the group to which each image belongs based on additional patient information
Image labelling	The images are classified in the groups "Positive Result" and "Negative Result" according to the results of the initial evaluation	Assignment of the labels "Positive Result" and "Negative Result" to the images
		Data cleaning to correct typographical errors and manage missing data
		Substitution of these textual labels by numerical representations, i.e., 0 and 1
Evaluation of predictor variables	The additional data are used to confirm the classification of the images and to ensure their correct cataloguing	Correlation analysis between the predictor variables and the variable to be predicted
		Evaluation and interpretation of the results of the correlation analysis

Table 1 delineates the structured pre-processing activities essential for preparing the dataset for subsequent analysis. The initial stage involves the evaluation of image classification, ensuring images are categorized based on patient anemia status. Followed by image labeling, where images are sorted into 'Positive Result' or 'Negative Result' categories. The final activity includes the evaluation of predictor variables to validate the classification of images and affirm their accurate cataloging. This comprehensive pre-processing ensures the integrity of the dataset for reliable machine learning application.

- ***Evaluation of image classification***: Image classification assessment is a critical process in the pre-processing of medical image data for machine learning-based diagnosis. This activity ensures accuracy in associating each image with its respective diagnostic outcome "Positive Result" or "Negative Result". This process comprises several essential tasks to ensure the integrity of the data and ultimately the effectiveness of the machine learning model (Table 2).

Table 2. Original and Segmented Conjunctiva Images for Anemia Analysis, showing both the unprocessed and processed states of the ocular region.

Image original	Palpebral	Fornix palpebral	Fornix
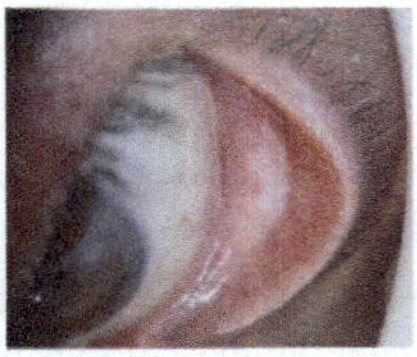		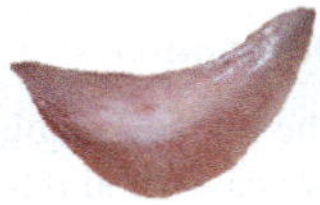	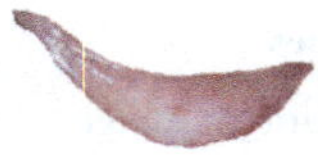
No anemia	Gender: Female	Hgb: 11.7	Age: 73

The table illustrates the segmentation process, which refines the original image to highlight regions of interest — the palpebral and fornix sections of the conjunctiva. These segments are critical for the machine learning algorithm to accurately classify the presence of anemia, as demonstrated in the non-anemic 73-year-old female subject with a hemoglobin level of 11.7.

- ***Image labeling:*** In this section we proceeded to label the images according to their condition, the corresponding labels were assigned: "1" for patients diagnosed with anemia and "0" for those without anemia.
- ***Evaluation of predictor variables:*** Correlation analysis is a statistical technique that examines and quantifies the association between two or more variables [16]. Applied to the diagnosis of anemia, a correlation is made between predictor variables (e.g., age, gender, medical history) and the target variable (presence or absence of anemia). In this way, patient characteristics that may be correlated with an increased risk of anemia can be identified, potentially increasing the accuracy and reliability of image classification.

3.3 Feature Extraction

In this phase, five feature extraction techniques were applied: LBPH, HOG, SIFT, SIFT Pading and ORB. Table 1 shows the activities of the feature extraction phase.

Table 3 presents a systematic outline of the feature extraction methodologies utilized within the machine learning classification model. It enumerates five distinct activities: feature extraction employing Local Binary Patterns Histograms (LBPH), Histogram of Oriented Gradients (HOG), Scale-Invariant Feature Transform (SIFT), an augmented SIFT approach with padding, and the Oriented FAST and Rotated BRIEF (ORB) method. Each technique is delineated in terms of image preprocessing, specific method application, and subsequent result validation, thus providing a comparative framework for the various feature extraction strategies implemented in the context of anemia detection imaging.

Table 3. Modeling and classification.

Activities	Description	Task
Feature extraction with LBPH	Feature extraction for each record using the Local Binary Patterns Histograms (LBPH) method	Preprocess the image Apply LBPH Store and validate results
Feature extraction with HOG	Feature extraction for each record using the Histogram of Oriented Gradients (HOG) method	Preprocess the image Apply HOG Store and validate results
Feature extraction with SIFT	Feature extraction for each record using the Scale-Invariant Feature Transform (SIFT) method	Preprocess the image Apply SIFT Store and validate results
Feature extraction with SIFT Padding	As a variation to the model, a Padding process is added to the SIFT feature extractor	Apply Sift padding Store and validate results
Feature extraction with ORB	Feature extraction of each record using the Oriented FAST and Rotated BRIEF (ORB) method	Preprocess the image Apply ORB Store and validate results

- ***Feature extraction with LBPH:*** The Local Binary Patterns Histograms (LBPH) method was used for feature extraction from the images. This method is effective in describing the local textures of an image, which is particularly useful in image classification tasks. Initially, a conversion of the original image to its greyscale equivalent is performed, this step simplifies the complexity of the image and provides a suitable representation for texture analysis. After this conversion, the LBPH algorithm is employed, which analyses each pixel of the image in relation to its surrounding neighbors. This evaluation process generates a local binary pattern for each central pixel, based on whether the intensity of each neighboring pixel is greater or less than that of the central pixel. This procedure results in a matrix of local binary patterns for the entire image. Finally, to compress the dimensionality of the feature space and eliminate non-uniform variability, a uniform local binary pattern strategy is employed, which counts only patterns that contain a maximum number of transitions from 0 to 1 or vice versa [17].
- ***Feature extraction with HOG:*** The process of feature extraction with the Histogram of Oriented Gradients (HOG) method is broken down into a series of steps that effectively capture the shape and appearance of objects in an image. The procedure begins with the transformation of the original image to a greyscale version. This conversion leads to a simplified representation of the image, which is suitable for the detailed gradient analysis that HOG performs. The HOG algorithm, at its core, evaluates the direction and magnitude of the gradient for each pixel in the image.

This gradient calculation identifies rapid changes in pixel intensity and provides critical information about the boundaries and textures of objects contained in the image. Once the gradients are determined, histograms of the oriented gradients are constructed. This is done within specific regions of the image called cells. These histograms are used to count the frequency of gradients at different orientations within each cell. The next step of the HOG procedure is blocking normalization. This step improves the robustness of the HOG descriptor by adjusting the histograms of each cell with information from adjacent cells, thus increasing the invariance to changes in illumination and shadows [18].

- ***Feature extraction with SIFT:*** The procedure for feature extraction using the Scale-Invariant Feature Transform (SIFT) method is broken down into a series of steps that ensure the efficient identification and description of key points in an image. Initially, the image is transformed to a greyscale version, thus reducing its complexity, and preparing it for the texture and structure analysis performed by SIFT. Subsequently, the construction of the Gaussian scale pyramid is carried out. This process generates different levels of "blurring" of the original image, each of which captures details of the image structure at a specific scale. The Gaussian difference scale space is obtained by differencing adjacent Gaussian images, thus capturing key details at a wide range of scales. From this Gaussian difference scale space, key points are detected by identifying local maxima and minima in scale space and orientation [19].
- ***Feature extraction with SIFT Padding:*** In feature extraction using the Scale-Invariant Feature Transform with Padding (SIFT Padding) method, an additional step is introduced to the original SIFT procedure which is used to ensure that key points located near the edge of the image are properly handled. The first step is to convert the image to grayscale, then the standard SIFT procedure is applied, which includes the construction of the Gaussian scale pyramid, the detection of key points and the assignment of orientations and descriptors to these key points. The main difference in SIFT Padding is the incorporation of a padding step before the construction of the Gaussian scale pyramid. Padding is applied to extend the original image, adding extra pixels at the edges. These extra pixels are commonly generated by duplication of edge pixels or by mirroring, although other more sophisticated methods exist. The aim of this step is to ensure that key points located near the edges of the image can have sufficiently large neighborhoods to compute their descriptors [19].
- ***Feature extraction with ORB:*** Feature extraction using the Oriented FAST and Rotated BRIEF (ORB) method begins with the conversion of the original image to greyscale. This initial step simplifies the image analysis by focusing only on the light intensity of each pixel, resulting in a more effective study of textural and structural patterns. Keypoint detection is the next component of the ORB process. This procedure uses the Features from Accelerated Segment Test (FAST) algorithm. As a corner detector, FAST examines the circular patterns of pixels around the target pixel, identifying key points through a process of comparing pixel intensities [20]. For the calculation of the orientation of the detected key points, ORB applies a centered intensity measure, providing them with rotation invariance. This aspect is critical to allow the same features to be recognized in the image even when it has been rotated. In the final phase of ORB feature extraction, descriptors are generated for the detected key points using the Binary Robust Independent Elementary Features (BRIEF) method

with the assigned orientation. BRIEF creates a highly efficient binary feature vector, which enables fast and memory-efficient matching [21]. For this purpose, BRIEF performs comparative pixel intensity tests and compiles the results into a feature vector.

3.4 Modelling and Classification

In this phase, four techniques were used for classification: SVM, Random Forest, KNN and MLP. Table 4 shows the activities of the modelling and classification phase.

Table 4. Modelling and classification activities.

Activities	Description	Task
Model building with SVM	Implementation of a classification model using the Support Vector Machine (SVM) algorithm, which constructs an optimal hyperplane for class separation	Define the hyperparameters of the SVM model
		Train the SVM model using the training dataset
		Evaluate the accuracy of the SVM model using the test dataset
Model building with Random Forest	Development of a classification model using the Random Forest algorithm, which operates by generating a set of decision trees during training	Configure the hyperparameters of the Random Forest model
		Train the Random Forest model using the training dataset
		Evaluate the accuracy of the Random Forest model using the test data set
Model building with KNN	Implementation of a classification model using the K-Nearest Neighbors (KNN) algorithm, which classifies an observation based on the labels of its 'k' nearest neighbors	Select the optimal value of 'k' for the KNN model
		Train the KNN model using the training data set
		Evaluate the accuracy of the KNN model using the test data set
Model building with MLP	Implementation of multilayer perceptron (MLP) network structures, with different number of hidden layers and different neurons	Define the MLP structures

(*continued*)

Table 4. (*continued*)

Activities	Description	Task
		Train the MLP using the training dataset
		Evaluate the accuracy of the MLP using the test data set

Table 4 delineates the core classification models employed in this study, along with their corresponding implementation tasks. Four distinct model-building activities are elaborated: SVM (Support Vector Machine), which involves constructing an optimal hyperplane for class distinction; Random Forest, which entails creating a forest of decision trees; KNN (K-Nearest Neighbors), based on proximity to 'k' nearest data points; and MLP (Multilayer Perceptron), which is a neural network characterized by multiple layers and neurons. Each model undergoes a process of hyperparameter definition, training with respective datasets, and evaluation of accuracy against a test dataset. This structured approach ensures the rigor and reproducibility of the classification outcomes pertinent to the machine learning framework of anemia detection.

- ***Model building with SVM:*** The creation of a Support Vector Machine (SVM) model in this specific implementation follows several essential steps. Initially, a grid of parameters to be explored is defined. In the case of SVMs, two critical parameters are 'C', which controls the trade-off between achieving a wide margin and minimizing margin violations (i.e., correctly classifying all training points); and 'gamma', which defines the extent of influence of a single training example, with low values meaning 'far' and high values meaning 'near' [22].
- ***Model building with Random Forest:*** The construction of a model with the Random Forest method follows a particular set of steps. This method, which is a type of machine learning ensemble algorithm, is based on the creation of numerous decision trees during the training phase and the production of a classification prediction as the mode of the individual tree predictions [23]. Initially, the parameters for the creation of the forest are defined. These parameters include the number of trees to create (n_estimators), the maximum depth of the trees (max_depth), the minimum number of samples required to split an internal node (min_samples_split), among others.
- ***Model building with KNN:*** The creation of a classification model using the k-Nearest Neighbors (KNN) method is carried out following a defined process. This method, which is a type of supervised learning algorithm, bases its predictions on a proximity parameter, i.e., an unlabeled example is classified according to the labels of the 'k' closest examples in the training set [24]. The first step is to determine the optimal value for 'k', the number of nearest neighbors to be considered. This can be done through a series of experiments or by cross-validation.
- ***Model building with MLP:*** The number of hidden layers, number of neurons in the hidden layers, the activation function, and the optimization algorithm to be used are

defined. The hidden layer sizes tested are (50,), (100) and (50,100,), with activation functions 'identity', 'logistic', 'tanh' and 'relu', and solvers 'lbfgs', 'sgd' and 'adam'.

3.5 Implementation of CNN Architectures

Five deep learning algorithms were applied in this phase: Customized CNN Model, CNN with EfficientNet, CNN with ResNet and CNN with Inception-ResNet. Table 5 shows the activities of deep learning modelling.

Table 5 presents the development and customization of deep learning models used in the study, including a custom CNN and modifications to pre-trained networks like EfficientNetB0, ResNet50, and InceptionResNetV2. The table summarizes the configuration, layer enhancement, and regularization steps applied to each model, culminating in their training over several epochs for optimized performance.

- ***Creation of the Customized CNN Model:*** The creation of a customized CNN (Convolutional Neural Networks) Model is based on the implementation of a series of specific layers that correspond to the fundamental structure of a CNN. The first step is the addition of convolutional layers to the model using the Conv2D class. These layers have the function of applying a series of filters to the input to extract features from it. In this case, three convolutional layers with different numbers of filters and kernel sizes are implemented, each followed by a batch normalization layer (BatchNormalization) that helps speed up the training and brings some regularization to the model [25]. The MaxPooling layer (MaxPooling2D) is then applied to reduce the dimensionality of the resulting matrices without losing the most important features. The MaxPooling layers take the maximum value of each sliding window, thus preserving the most salient features and reducing the risk of overfitting [26].
- ***CNN with EfficientNet:*** In this case, EfficientNet, a pre-trained network architecture, is used. EfficientNet is a family of CNN models that have been shown to be more efficient in terms of performance and resource consumption compared to traditional architectures, thanks to its composite scaling strategy that adjusts the depth, resolution, and width of the network in a coordinated manner [27].
- ***CNN with ResNet:*** For this implementation, the Convolutional Neural Network (CNN) architecture known as ResNet50 is used, a prominent model in the field of computer vision for its ability to train very deep networks without falling into the problem of gradient fading. The procedure starts by loading the pre-trained ResNet50 model using the TensorFlow library, excluding its last classification layer. This technique, known as transfer learning, allows us to benefit from the knowledge ResNet50 has already acquired from huge datasets and adapt it to our specific task.
- ***CNN con Inception-ResNet:*** The implementation of a Convolutional Neural Network (CNN) using the Inception-ResNetV2 architecture is an advanced and highly efficient approach to image processing and analysis. Inception-ResNetV2 is a robust architecture that combines the efficient features of the Inception architecture with the residual connections of ResNet, designed to alleviate the problem of gradient fading in deep networks.

Table 5. Deep learning modelling activities.

Activities	Description	Task
Creation of the Customized CNN Model	Design of a customized CNN architecture with convolutional, flattening, and dense layers	Initialization of the CNN architecture
		Incorporation of convolutional layers
		Application of regularization techniques
		Flattening of the convolutional output
		Addition of dense layers
CNN with EfficientNet	Use of the pre-trained EfficientNetB0 model, customizing it with additional layers	Loading the EfficientNetB0 base model
		Customization of the base model
CNN with ResNet	Application of the pre-trained ResNet50 model, adapted with additional layers	Loading the ResNet50 base model
		Customization of the base model
CNN con Inception-ResNet	Implementation of the pre-trained InceptionResNetV2 model, customized with additional layers	Loading the InceptionResNetV2 base model
		Customization of the base model
Training of CNN Models	Training of each model using the input data and labels	Training of the model in several epochs
		Saving of model weights after each epoch
Evaluation of the Trained Model	Evaluation of the model's performance on a test set	Loading of saved model weights
		Performing predictions on the test set

(*continued*)

Table 5. (*continued*)

Activities	Description	Task
		Calculating performance metrics such as accuracy, recall and F1 score
		Documentation of performance metrics for each epoch

- ***Training of CNN Models:*** Once the Convolutional Neural Network (CNN) architectures are configured and customized for the problem at hand, the next step is to train these models. This process consists of feeding the networks with training data so that they can learn the underlying patterns in the data. During training, use is made of a set of training data (X_train, Y_train), which is presented to the model in batches. These data consist of labelled images, where X_train contains the images and Y_train the corresponding labels. The training strategy employed is based on running multiple epochs (iterations through the entire training dataset), specified by the variable epochs. During each epoch, the model is adjusted to minimize a loss function, which is a measure of how well the model is performing the classification task. This adjustment is achieved by updating the model weights as a function of the gradient of the loss function with respect to those weights.
- ***Evaluation of the Trained Model:*** Table 6 shows the activities of the Evaluation of the trained model.

Table 6. Evaluation of the Trained Model.

Activities	Description	Task
Comparison of model results based on metrics	Examine each model based on accuracy, precision, recall and f1-score metrics	Determine the best performing model for each metric
		Develop visual representations to contrast the metrics of the different models
		Save the trained model and results

Table 6 summarizes the evaluation phase of the trained models. It details the assessment of model performance through key metrics such as accuracy, precision, recall, and F1-score. Tasks include identifying the top-performing model for each metric, creating visual representations for comparative analysis, and documenting the final model configurations and their results.

- ***Comparison of model results based on metrics***: In this activity, all models are compared in terms of accuracy, precision, recall and f1-score. While it is possible that some models may excel in certain individual metrics, our choice of the optimal model will be based primarily on the accuracy and f1-score results. The f1-score is particularly significant in this evaluation as it provides a balance between accuracy and recall, providing a more holistic measure of model performance. In cases where several models show comparable results in terms of accuracy and f1-score, additional metrics will be considered to break the tie and determine which model is superior.

4 Results

4.1 Analysis of Optimal Parameters

For each of the models evaluated, a set of parameters was determined that maximized their performance. These parameters, selected after a rigorous trial and error process, represent the optimal conditions under which each model provided the most accurate results. Table 7 shows the optimal parameters identified for each model and the corresponding training and test results.

The structure presented in the table reflects an array of machine learning models applied to the classification of anemia from conjunctival images. The feature extraction methods employed—SIFT, HOG, ORB, and others—process and transform the input images to highlight the essential characteristics pertinent to the classification tasks. Each model's hyperparameters, detailed in the 'Parameters' column, such as the regularization parameter 'C' in SVMs and the number of estimators in Random Forests, are finely tuned to enhance the model's ability to discern patterns within the data. The 'Train' and 'Test' columns quantify the models' performance metrics, with the accuracy on the test set being indicative of each model's generalization potential. This tabulated data not only aids in model comparison but also serves as a foundation for selecting a model that balances precision and robustness in real-world applications.

4.2 Model Results

The effectiveness of the models, assessed in terms of accuracy, is presented in Tables 8 and 9. These results reflect the ability of each model to make accurate predictions when trained with the optimal parameters identified in the previous section.

Table 9 provides a clear comparison of how each model (SVM, KNN, Random Forest and MLP) performs depending on the feature extractor used (SIFT, SIFT(Padding), LBPH, HOG, ORB). It is worth noting that the best result was obtained when using the Random Forest model with the HOG feature extractor, reaching an accuracy of 79%.

Table 9 shows a summary of the results of various deep learning models with tuned parameters. Although the custom model achieved the highest accuracy (93.18%) using 1024 neurons, 50 epochs and 1000 samples. But not only accuracy is crucial, the loss and robustness of the architecture used must also be considered. Despite its high accuracy, the custom model lacks the empirical validation provided by recognized architectures, such as In-ception-ResNet-v2, which showed the lowest loss, indicating possible better adaptability to new data.

Table 7. Results with optimal parameters for SVM, KNN, Random Forest and MLP

	Feature Extractors	Parameters	Train	Test
SVM	SIFT	{'C': 0.5, 'gamma': 0.1, 'kernel': 'linear'}	1	0.6463
	SIFT (Padding)	{'C': 0.5, 'gamma': 0.1, 'kernel': 'poly'}	1	0.6506
	LBPH	{'C': 0.5, 'gamma': 100, 'kernel': 'poly'}	0.7568	0.6875
	HOG	{'C': 0.5, 'gamma': 0.1, 'kernel': 'poly'}	1	0.7656
	ORB	{'C': 0.5, 'gamma': 0.1, 'kernel': 'sigmoid'}	0.6149	0.6250
KNN	SIFT	{'algorithm': 'auto', 'n_neighbors': 21}	0.6411	0.6675
	SIFT (Padding)	{'algorithm': 'auto', 'n_neighbors': 21}	0.6340	0.6807
	LBPH	{'algorithm': 'auto', 'n_neighbors': 61}	0.6149	0.6950
	HOG	{'algorithm': 'auto', 'n_neighbors': 21}	0.6214	0.7507
	ORB	{'algorithm': 'auto', 'n_neighbors': 21}	0.6352	0.6394
Random Forest	SIFT	{'n_estimators': 10}	0.9798	0.6531
	SIFT (Padding)	{'n_estimators': 30}	1	0.7063
	LBPH	{'n_estimators': 30}	1	0.6719
	HOG	{'n_estimators': 30}	0.9932	0.7969
	ORB	{'n_estimators': 1200}	1	0.6425
MLP	SIFT	{'hidden_layer_sizes': (100,), 'activation': 'identity', 'solver': 'adam'}	0.7018	0.6925
	SIFT (Padding)	{'activation': 'logistic', 'hidden_layer_sizes': 100, 'solver': 'lbfgs'}	0.6901	0.6263
	LBPH	{'activation': 'identity', 'hidden_layer_sizes': (50, 100), 'solver': 'lbfgs'}	0.6685	0.6625

(*continued*)

Table 7. (*continued*)

	Feature Extractors	Parameters	Train	Test
	HOG	{'activation': 'relu', 'hidden_layer_sizes': 100, 'solver': 'lbfgs'}	0.7097	0.7141
	ORB	{'activation': 'tanh', 'hidden_layer_sizes': 100, 'solver': 'sgd'}	0.7097	0.6425

Table 8. Accuracy of models using various feature extractors.

Model	Feature extractor				
	SIFT	**SIFT(Padding)**	**LBPH**	**HOG**	**ORB**
SVM	64%	65%	68%	76%	62%
KNN	67%	68%	65%	75%	63%
Random Forest	65%	70%	67%	**79%**	64%
MLP	69%	62%	66%	71%	64%

Table 9. Accuracy of deep learning models using different parameters.

Model	Neurons	Epochs	Samples	Accuracy	Loss
Inception-ResNet-v2	2048	50	1500	**90.69%**	**0.2788**
Resnet	512	50	500	84.09%	0.6131
EfficientNet	2048	50	1500	86.04%	0.4077
Personalized	1024	50	1000	**93.18%**	0.3563

The Inception-ResNet-v2 model, with 2048 neurons, 50 epochs and 1500 samples, showed the lowest loss (0.2788) despite a slightly lower accuracy (90.69%). This low loss suggests a possible better generalization to new data, a crucial factor for the practical application of the models. The choice of the best model will therefore require a trade-off between accuracy and loss.

Figure 4 presents an analysis of the Inception-ResNet v2 model's accuracy over a series of epochs. The graph plots the trajectory of accuracy rates during the model's training phase, illustrating the learning consistency and volatility. Peaks in accuracy suggest epochs where the model predictions were highly aligned with the validation data, whereas troughs indicate periods of less predictive performance. Such fluctuations are crucial for understanding model behavior and potential overfitting or underfitting at various stages of the learning process.

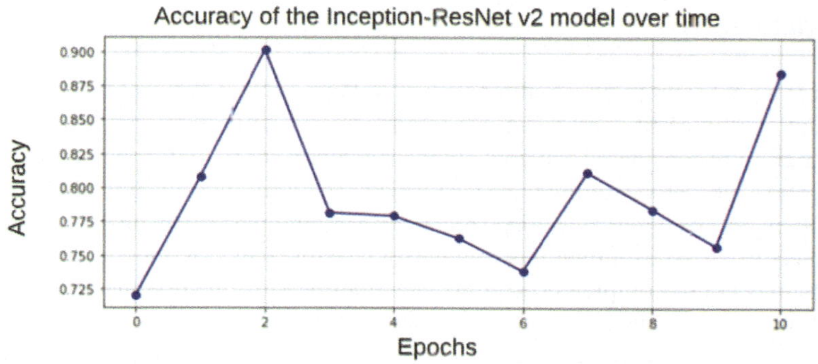

Fig. 4. Accuracy trend of Inception-ResNet v2 across epochs.

Figure 5 portrays the loss experienced by the Inception-ResNet v2 model across successive training epochs. Loss metrics serve as an indicator of the model's optimization, with lower values representing better performance. The depicted oscillation in loss values across epochs provides insight into the model's adjustment to the data over time. Sharp increases in loss may reflect the model's struggle with particular data complexities or represent an epoch where the model's generalization capabilities were challenged.

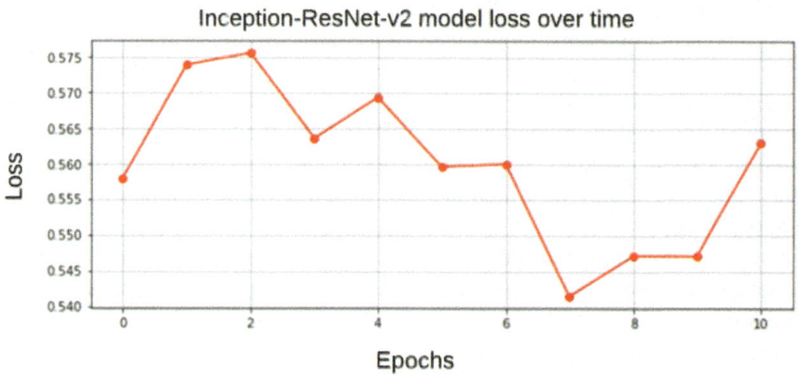

Fig. 5. Loss progression of the Inception-ResNet v2 model during training epochs.

The implementation of the Inception-ResNet v2 model is predicated upon its proven ability to generalize effectively across the dataset in question, as evidenced by its consistent performance throughout the testing phases. This performance aligns with previous research endorsing the efficacy of Inception-ResNet v2 in medical image classification tasks [28]. The balance between high accuracy, evidenced by a rate of 90.69%, and low loss, recorded at 0.2788, underscores the model's capacity for efficient learning without overfitting, indicative of its robustness and potential for practical application in the pre-diagnosis of anemia.

5 Conclusions

Anemia is a persistent public health problem in Peru, with significant repercussions on individual quality of life and socio-economic progress at the national level. Although the hemogram is considered the gold standard method for diagnosing anemia, its need for time and a laboratory setting to be performed can represent a considerable limitation, especially in remote or less developed areas. As a common practice, health professionals often resort to visual examinations to identify possible signs of anemia. However, this approach may not be completely accurate and relies heavily on the subjective experience of the clinician. This study aims to overcome such limitations by offering an innovative, accessible, and reliable pre-diagnostic solution designed to be used by anyone, thereby increasing the early detection and timely treatment of anemia.

The methodology of this work began with the collection of images of the ocular conjunctiva of patients with and without anemia, thus fulfilling the first objective. Subsequently, these images were preprocessed to optimize their analysis. Preprocessing included image segmentation into four different regions of interest (palpebral conjunctiva, fornix conjunctiva, the sum of palpebral conjunctiva and fornix, and the original image), greyscale transformation and resizing. In addition, data augmentation techniques were implemented to compensate for the limitation of a small initial database.

As for the data analysis, several feature extractors such as SIFT, SURF, ORB and HOG were used to obtain the characteristic vectors of each image. With the application of machine learning techniques such as SVM, RF, MLP, and RNN, these features were classified. The Random Forest model with HOG extractor obtained the highest accuracy with 79%. In addition, deep learning models were explored, adjusting parameters such as the number of neurons, epochs, and samples in each model. Although initially the custom model obtained the highest accuracy of 93.18%, the Inception-ResNet-v2 model, supported by existing studies, was finally chosen and demonstrated a robust accuracy of 90.69% and loss of 0.2788, which is better than the loss of 0.3563 obtained with the comparison model.

In summary, this study has demonstrated that machine learning and deep learning techniques can provide a reliable prediction of anemia from images of the ocular conjunctiva. The implementation of the Inception-ResNet-v2 model, with its 90.69% accuracy, offers a promising tool for the preliminary diagnosis of anemia. It is hoped that these findings will effectively contribute to the fight against anemia in Peru, and in doing so, improve public health and the overall well-being of the population.

References

1. Wallander, J.L., Koot, H.M.: Quality of life in children: a critical examination of concepts, approaches, issues, and future directions. Clin. Psychol. Rev. **45**, 131–143 (2016). https://doi.org/10.1016/J.CPR.2015.11.007
2. Gordon, N.: Iron deficiency and the intellect. Brain Develop. **25**(1), 3–8 (2003). https://doi.org/10.1016/S0387-7604(02)00148-1
3. Organización Mundial de la Salud. Carencia de micronutrientes [Internet]. Ginebra: OMS; 2020 [citado el 12 de agosto de 2020]. Disponible en: https://www.who.int/nutrition/topics/ida/es/

4. Falkingham, M., Abdelhamid, A., Curtis, P., Fairweather-Tait, S., Dye, L., Hooper, L.: The effects of oral iron supplementation on cognition in older children and adults: a systematic review and meta-analysis. Nutr. J. **9**(1), 4 (2010)
5. Thompson, J., Biggs, B.-A., Pasricha, S.-R.: Effects of Daily Iron Supplementation in 2- to 5-Year-Old Children: Systematic Review and Meta-analysis. Pediatrics. **131** (4):739–753 (2013). https://doi.org/10.1542/peds.2012-2256
6. Nacional y Regional (ENDES 2018). Lima: INEI; 2019 [acceso 10/12/2019] Disponible en: https://www.inei.gob.pe/media/MenuRecursivo/publicaciones_digitales/Est/Lib1656/index1.html
7. Instituto Nacional de Estadística e Informática (2020). La prevalencia de anemia en la población peruana. Lima: INEI. https://www.inei.gob.pe/media/MenuRecursivo/investigaciones/prevalencia_anemia.pdf
8. Ministro de Salud destaca necesidad de una mejor distribución del recurso humano. (2021, 18 diciembre). Andina. https://andina.pe/agencia/noticia-ministro-salud-destaca-necesidad-una-mejor-distribucion-del-recurso-humano-873910.aspx
9. Thaver, I. H., Baig, L.: Anemia in Children: ¿Part I. Can Simple Observations by Primary Care Provider Help in Diagnosis?. J. Pak. Med. Assoc. **44**, 282–284 (1994)
10. Bauskar, S., Jain, P., Gyanchandani, M.: A Noninvasive Computerized Technique to Detect Anemia Using Images of Eye Conjunctiva. Pattern Recognit Image Anal. **29**(3), 438–446 (2019). https://doi.org/10.1134/S1054661819030027
11. Dhanasekaran, S., Shanker, N.R.: Detection of anemia disease using DPSO algorithm. ARPN Journal of Engineering and Applied Sciences, **14**(7), 1362–1369 (2019). https://www.scopus.com/inward/record.uri?eid=2-s2.0-85064879188&partnerID=40&md5=c0910c4db8aa7281fe5fcfc8b988acf5
12. Heredia-Menor, K.A., Mamani-Ticona, W.: Comparison of computer vision and machine learning based techniques for early detection of anemia from fingernail analysis. En Universidad de Lima (Ed.), Actas del III Congreso Internacional de Ingeniería de Sistemas (pp. 151–164), Lima, 17 y 20 de noviembre del 2020. Universidad de Lima, Fondo Editorial (2021)
13. Aadi, O.A., Akash, M.M.H., Hossain, A., Ishraq, F., Fahim, A.A., Karim, D.Z.: (2023). CellIdentifier: Classification of Peripheral Blood Cell Images using Deep Learning. International Conference on Electrical, Computer, Communications and Mechatronics Engineering, ICECCME 2023, https://doi.org/10.1109/ICECCME57830.2023.10253397
14. Rivero-Palacio, M., Alfonso-Morales, W., Caicedo-Bravo, E.: Mobile application for anemia detection through ocular conjunctiva images. In: 2021 IEEE Colombian Conference on Applications of Computational Intelligence, ColCACI 2021 - Proceedings, art. no. 9469593 (2021). https://doi.org/10.1109/ColCACI52978.2021.9469593
15. Sujatha, K., et al.: Indigenous non-invasive detection of malnutrition-induced anemia from patient-sourced photos using smart phone app. Lect. Notes Netw. Syst. **400**, 431–440 (2023). https://doi.org/10.1007/978-981-19-0095-2_41
16. Dormann, C.F., et al.: Collinearity: a review of methods to deal with it and a simulation study evaluating their performance. Ecography **36**(1), 27–46 (2013). https://doi.org/10.1111/j.1600-0587.2012.07348.x
17. Pietikäinen, M., Hadid, A., Zhao, G., Ahonen, T.: Computer Vision Using Local Binary Patterns. Springer, London (2011)
18. Dalal, N., Triggs, B.: Histograms of oriented gradients for human detection. In: Pro-ceedings of the IEEE Computer Society Conference on Computer Vision and Pattern Recogni-tion, vol. 1, pp. 886–893. IEEE (2005).
19. Lowe, D.G.: Distinctive image features from scale-invariant keypoints. Int. J. Comput. Vision **60**(2), 91–110 (2004)

20. Rosten, E., Drummond, T.: Machine learning for high-speed corner detection. In: Euro-pean Conference on Computer Vision, pp. 430–443. Springer, Berlin, Heidelberg (2006).
21. Calonder, M., Lepetit, V., Strecha, C., Fua, P.: BRIEF: binary robust independent elementary features. In: European Conference on Computer Vision, pp. 778–792. Springer, Ber-lin, Heidelberg (2010).
22. Cortes, C., Vapnik, V.: Support-vector networks. Mach. Learn. **20**(3), 273–297 (1995)
23. Breiman, L.: Random Forests. Mach. Learn. **45**(1), 5–32 (2001)
24. Cover, T., Hart, P.: Nearest neighbor pattern classification. IEEE Trans. Inf. Theory **13**(1), 21–27 (1967)
25. Ioffe, S., Szegedy, C.: Batch normalization: accelerating deep network training by reducing internal covariate shift. In: Proceedings of the 32nd International Conference on Ma-chine Learning. (2015). https://arxiv.org/abs/1502.03167
26. Scherer, D., Müller, A., Behnke, S.: Evaluation of pooling operations in convolutional architectures for object recognition. In: Artificial neural networks–ICANN 2010, pp. 92–101. Springer, Berlin, Heidelberg (2010)
27. Tan, M., Le, Q.V.: EfficientNet: rethinking model scaling for convolutional neural networks. In: Proceedings of the 36th International Conference on Machine Learning (2019). http://proceedings.mlr.press/v97/tan19a.html
28. Litjens, G., et al.: A survey on deep learning in medical image analysis. Med. Image Anal. **42**, 60–88 (2017)

Linear Regression with PM2.5 and PM10 Concentration for Air Quality in East Lima, Peru

Juan J. Soria, Angelo Ordoñez Cardenas(✉), and Lidia Segura Peña

Universidad Tecnológica del Perú, Lima, Perú
c20723@utp.edu.pe

Abstract. This research paper provides an in-depth analysis of the impact of meteorological factors on the regression of PM2.5 and PM10 concentrations and their correlation with SARS-COV-2 transmission in Peru, employing heteroscedasticity considerations. The findings of the study reveal significant OLS estimators and intercepts for both PM2.5 and PM10, with remarkable comparisons with previous research underlining the robustness of the developed models. In particular, regression models designed for different settings (rural, urban and industrial) highlight how PM10 concentrations are predictive of PM2.5 levels, with coefficients of determination exceeding those found in previous studies. In addition, the article uses the Kruskal-Wallis test and the Holm-Bonferroni method to demonstrate significant variations in particulate matter in different study areas, lending further credence to the comprehensive nature of the analysis performed. Given the rigorous quantitative analysis, the clear significance of the findings in multiple settings, and the contribution of the study to the understanding of the complex relationship between air pollution and virus transmission, this article is a valuable addition to the field. The detailed comparison with existing models, along with the innovative use of statistical methods to validate the results, positions this research as a significant advance in environmental science and public health. A regression equation between PM2.5 and PM10 was determined; in rural area with linear coefficient of 0.66105 $\mu g/m^3$ and intercept of 5.57229 $\mu g/m^3$, for urban it has a coefficient of 0. 672768 $\mu g/m^3$ and intercept of 1.149877 $\mu g/m^3$, and industrial a coefficient of 0.567541 $\mu g/m^3$ and an intercept of 4.485924.

Keywords: PM10 · PM2.5 · Temtop Sensor · Linear Regression · Concentration · Air Pollutants

1 Introduction

The Air pollution by particulate matter (PM10 y PM 2.5), It's one of the biggest problems that exist today [1]. Because they affect human health and the environment in different cities around the world, since the deposition of this pollutant in the upper respiratory tract depends on the inhalation flow, causing cardiovascular diseases and lung cancer mortality. [2]. One of the main pollutants that affect the atmosphere is the PM10, PM 2.5 y CO2.

R. Silhavy and P. Silhavy (Eds.): CSOC 2024, LNNS 1120, pp. 519–533, 2024.
https://doi.org/10.1007/978-3-031-70518-2_45

According to [3] show that almost the entire world population breathes air that exceeds the limits recommended by the World Health Organization, containing large amounts of pollutants, and according to 2021 estimates, air pollution in cities and rural areas around the world causes 2 million premature deaths per year.

Urban air pollution of the last decades and its impact on human health is just one example of the ongoing ecological degradation and its links to a certain type of society [4]. At the urban level, the effects of air pollution especially fossil fuel products both in energy production and in the burning of automobiles and other organic matter demonstrate how the gases sulfur oxides (SOx) and nitrogen oxides (NOx) are directly related to respiratory tract, illness and hospitalization; something similar happens with sulfur dioxide (SO2) [5].

Hazardous air pollutants are substances known to cause serious health effects, and the effects of these pollutants have also been studied in conjunction with climate change. There are many sources of these pollutants, but industrial emissions attract interest because they are considered important sources of PM10 and PM2.5 particles, gases, and heavy metals. Residents living around industrial complexes are exposed to higher levels of pollution than those living far from these areas [6].

In addition[7], collected data for the year 2022, where measurements were taken to determine the presence of particulate matter (PM10) and (PM2.5). Therefore, these results have led the World Health Organization to stress the importance of curbing the use of fossil fuels and taking other measures to mitigate air pollution levels worldwide.

In global ecological protection and urban development, air pollution has become a crucial problem affecting human health and the environment, along with climate change, has attracted increasing attention from academics and government administrators [8]. In Latin America, at least 100 million people are exposed to air pollution levels above those recommended by the World Health Organization [9]. The groups most vulnerable to the harmful effects of poor air quality include children, the elderly, people with health problems, and people of low socioeconomic status.

According to [10] a recent study noted that Lima Metropolitana (LIM) is prone to high concentrations of PM10 and PM2.5 due to accelerated industrialization and economic growth and it is 29% of the 34 105 243 of the total Peruvian population.

The PM10 and PM2.5 parameters exceed the Peruvian National Ambient Air Quality Standard (ENCA) by more than 333% in 25 zones located in the north and east of the metropolitan area of Lima and Callao, with the highest values of PM10 and PM2.5 in the north, south and east of Lima, respectively [11].

As shown by Kazi et al. [12] the prediction of PM2.5 and PM10 air pollutants using a linear regression model $y = 3.413 + 0.698x$ finds a significant correlation of $r = 0.7895$, a Std error de 0.0405 with a p-value less than 2×10^{-6} which makes the predictive model relevant.

Gulati et al. [13] shows a predictive model of PM2.5 and PM10 particulate air pollutants in support of air quality, finding a linear regression model $y = 11.721 + 0.7201x$, with a moderate correlation $r = 0.7201$ between particulate pollutants and RMSE of 11.7585.

Li et al. [14] found a multivariate model to predict the spatial distribution of air pollutants in North America and China with respect to PM2.5, where $y = 105.1 -$

$1.4948e-06(G5-400)-6.894e-04(RE-5000)+1.964(PE-5000)$, where G5–400 indicates green spaces, RE-5000 indicates the restaurant area and PE-5000 are the polluting companies, with, $R^2 = 0.718$, un $RMSE = 8.355\frac{\mu g}{m^3}$ and a p-value less than 2×10^{-6} making the predictive model significant.

In the study by Maher et al. [15] twelve schools in Palestine with natural ventilation were analyzed and correlations were found between air pollutants and relative humidity inside and outside the schools with a multivariate regression model $PM10(in) = 0.09 - 0.49PM2.5(out) + 0.03RH(in) + 1.35PM10(out)$ with a coefficient of determination $R^2 = 0.72$ and a p-value of less than 0.01 which makes the model significant.

As clarified by [16], particulate matter is an atmospheric pollutant that affects human health conditions and air quality; which analyzed PM10 concentrations in the Tababela airport sector of Quito using the geostatistical analysis method found the relationship $y = 0.072 + 12000x$ with a coefficient of determination of $R^2 = 0.9733$ and a p-value of less than 0.01 that makes the model significant.

In the research of Alvarez et al. [4], a quantitative contribution of PM2.5 and PM10 particulate matter was determined in the Mantaro valley in the urban area of Jauja, Concepción and Huancayo, finding a PM2.5 of 21.8 (Jauja, s = 13.8), 15.8 (Concepción, s = 7.8) and 34.5 (Huancayo, s = 14.7) and for PM10 particulate matter, 31.7 (Jauja, s = 11.6), 28.2 (Concepción, s = 13.1) y 64.5 (Huancayo, s = 30.7) were found.

In Maryam's research [17] mentions that the increase of air and dust pollution has caused human health problems and environmental hazards in which PM2. 5 and PM10 are the indicators to evaluate the air concentration, finding a linear regression model with aerosol optical depth (AOD) as dependent variable and PM2.5 as independent variable $PM10 = \alpha_1 + \alpha_{1AOD}(AOD) = 133.3 + 218.4(AOD)$ with a coefficient of determination $R^2 = 0.62$ and p-value less to 0.01 which makes the model significant.

2 Materials and Methods

2.1 Linear Regression Model

The generalization of the k-variable population regression model (FRP) with dependent variable Y and k-1 explanatory variables $X_2, X_3, X_4, X_5, \cdots, X_k$ is written by [18]:

$$Y_i = \beta_1 + \beta_2 X_{2i} + \beta_3 X_{3i} + \beta_4 X_{4i} + \cdots + \beta_k X_{ki} + u_i;\ i = 1, \cdots, n \tag{1}$$

where β_1 is the intercept, β_2 a β_k are the partial slope coefficients, u is the stochastic disturbance term and i is the ith observation, with n as the population size.

Equation (1) is a shorthand expression for the set of n si-multaneous equations shown in Eq. (2) [19]

$$\begin{aligned}
Y_1 &= \beta_1 + \beta_2 X_{21} + \beta_3 X_{31} + \beta_4 X_{41} + \cdots + \beta_k X_{k1} + u_1 \\
Y_2 &= \beta_1 + \beta_2 X_{22} + \beta_3 X_{32} + \beta_4 X_{42} + \cdots + \beta_k X_{k2} + u_2 \\
Y_3 &= \beta_1 + \beta_2 X_{23} + \beta_3 X_{33} + \beta_4 X_{43} + \cdots + \beta_k X_{k3} + u_3 \\
&\cdots\cdots\cdots\cdots\cdots\cdots\cdots\cdots\cdots\cdots\cdots\cdots \\
Y_n &= \beta_1 + \beta_2 X_{2n} + \beta_3 X_{3n} + \beta_4 X_{4n} + \cdots + \beta_k X_{kn} + u_n
\end{aligned}$$

The system of equations is written in matrix form of the form:

$$\begin{bmatrix} y_1 \\ y_2 \\ y_3 \\ y_4 \\ \vdots \\ y_n \end{bmatrix} = \begin{bmatrix} 1 & x_{21} & x_{31} & x_{41} & \cdots & x_{k1} \\ 1 & x_{22} & x_{32} & x_{42} & \cdots & x_{k2} \\ 1 & x_{23} & x_{33} & x_{43} & \cdots & x_{k3} \\ \vdots & \vdots & \vdots & \vdots & \vdots & \vdots \\ 1 & x_{2n} & x_{3n} & x_{4n} & \cdots & x_{kn} \end{bmatrix} \begin{bmatrix} \beta_1 \\ \beta_2 \\ \beta_3 \\ \beta_4 \\ \vdots \\ \beta_k \end{bmatrix} + \begin{bmatrix} u_1 \\ u_2 \\ u_3 \\ \vdots \\ u_n \end{bmatrix}$$

$$Y_{nx1} = X_{nxk} . \beta_{kx1} + u_{nx1} \tag{2}$$

where Y is the nx1 column vector of observations on the dependent variable Y, the variable X is the matrix nxk, with n observations on k-1 variables X2 to Xk and the first column of numbers 1 represents the intercept term, the value of β is the kx1 column vector of the unknown parameters $\beta_1, \beta_2, \cdots, \beta_k$ and the value of u is the nx1 column vector of n perturbations ui.

2.2 Data Collection

Currently, Lima and Callao form the largest metropolitan area in Peru. Of which about 31.1 million inhabitants, about 9.8 million live in the capital.[11].

The study area for monitoring was established in the rural zone (Carapongo), urban zone (Alameda) and Ñaña (industrial zone). Monitoring was planned for 19 days from February 11 to February 29, 2023 with 5 monitoring points. Each day was monitored in the three locations shown in Table 1.

Table 1. Geographical location of air quality monitoring zones

Monitoring zone	Monitoring hours	Coordinates
Carapongo (Zona rural)	7:30 a.m	12°00′10" S y 76°52′ 55" W
Alameda (Zona Urbana)	12:00 m	11°59′ 17" S y 76°50′ 06" W
Ñaña (Zona Industrial)	18:30 p.m	11°59′ 33" S y 76°49′ 39" W

Source: Own elaboration

Within each monitoring zone, 5 monitoring points were determined as shown in Fig. 1, where the parameters that contribute to the concentration of air pollution (PM 10 y PM 2.5).

For data collection was based on the measurement of air quality through the TEMTOP M2000 [20] air quality monitor PM2.5 PM10 formaldehyde particles air quality detector with audio alarm, temperature, and humidity screen recording curve of easy calibration, as shown in Fig. 2. This type of automatic equipment allowed us to obtain information automatically and allowed us to evaluate the behavior of air pollutants (PM 10 y PM 2.5), These measurements were taken minute by minute to obtain the concentration level of

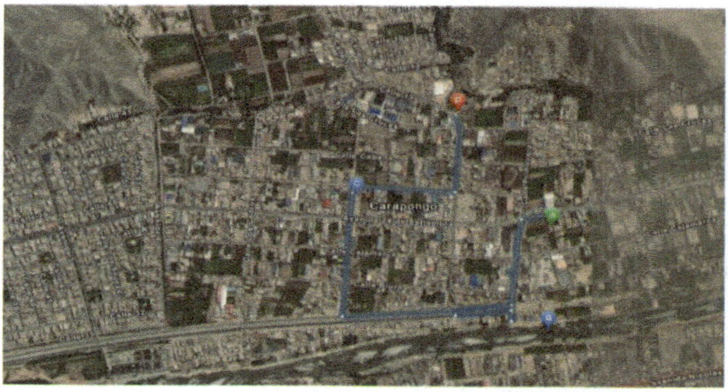

Fig. 1. Location of air quality measurement points

Fig. 2. TEMTCP M2000 sensor for air quality measurement monitor

PM 10 and PM 2.5, temperature, relative humidity and CO2. Between 7:00 a.m, 12:30 a.m. and 18:30 p.m.

A total of 285 data were collected at different times (day, afternoon, evening, night), between February 11, 2023 and February 29, 2023. The data were analyzed using the ANOVA statistical method with repetitions [21] with r Studio [22], as well as linear regression analysis [21].

3 Results

3.1 Descriptive Statistics

The study analyzed air pollutants in East Lima, finding an average PM2.5 de 50.8 $\mu g/m^3$ (S = 2.08) with a minimum of 10.1 $\mu g/m^3$ and a maximum de 238 $\mu g/m^3$, un PM10 an average 73.3 $\mu g/m^3$ (S = 3.08) with a minimum of 12.8 $\mu g/m^3$ and a maximum of 318 $\mu g/m^3$, a CO2 an average 635 $\mu g/m^3$ (S = 4.04) with a minimum of 501 $\mu g/m^3$ and a maximum of 832 $\mu g/m^3$, an average temperature of 26.3 °C (S = 0.162) with a minimum of 20 °C and a maximum of 33 °C and an average relative humidity of 51.5% (S = 0.492) with a minimum 27% and a maximum of 73%.

Table 2, shows the analysis of air pollutants where the rural area (Carapongo), an average PM2.5 de 83.0 $\mu g/m^3$ (S = 3.65) with a minimum of 43.9 $\mu g/m^3$ and a maximum of 238 $\mu g/m^3$, an average PM10 of 117.0117.0 $\mu g/m^3$ (S = 5.34) with a minimum de 61.9 $\mu g/m^3$ and a maximum of 318 $\mu g/m^3$, an average CO2 on average 668 $\mu g/m^3$ (S = 6.75) with a minimum of 501 $\mu g/m^3$ and a maximum of 832 $\mu g/m^3$, an average temperature of 24.3 °C (S = 0.17) with a minimum of 20 °C and a maximum of 27 °C and an average relative humidity of 59.6% (S = 0.445) with a minimum of 52% and a maximum of 76%. The analysis of atmospheric pollutants in the urban area (Alameda), found an average of PM2.5 de 37.3 $\mu g/m^3$ (S = 1.42) with a minimum of 14.0 $\mu g/m^3$ and a maximum of 76.8 $\mu g/m^3$, PM10 an average 53.8 $\mu g/m^3$ (S = 2.1) with a minimum of 21.0 $\mu g/m^3$ and a maximum of 113 $\mu g/m^3$, a CO2 on average 609 $\mu g/m^3$ (S = 6.36) with a minimum of 503 $\mu g/m^3$ and a maximum de 809 $\mu g/m^3$, an average temperature of 29.4 °C (S = 0.164) with a minimum of de 26 °C and a maximum of 33 °C and an average relative humidity of 43.3% (S = 0.631) with a minimum of 27% with a minimum of 55%. Analysis of atmospheric pollutants in the industrial zone (Ñaña), an average of PM2.5 de 32.3 $\mu g/m^3$ (S = 2.69) with a minimum of 10.1 $\mu g/m^3$ and a minimum of 169.0 $\mu g/m^3$, a PM10 on average 49.0 $\mu g/m^3$ (S = 4.73) with a minimum of 12.8 $\mu g/m^3$ and a minimum of 296 $\mu g/m^3$, a CO2 on average 627 $\mu g/m^3$ (S = 6.55) with a minimum of 521 $\mu g/m^3$) and a minimum of 825 $\mu g/m^3$, an average temperature of 25.3 °C (S = 0.143) with a minimum of de 23 °C and a maximum of 31 °C and an average relative humidity of 51.7% (S = 0.423) with a minimum of 45% and a maximum of 64%.

Table 2. .

Zonas	PM2.5	PM10	CO2	Temperatura
Rural	$\bar{x}$ = 83.0 S = 3.65	$\bar{x}$ = 117.0 S = 5.34	$\bar{x}$ = 668.0 S = 6.75	$\bar{x}$ = 24.3 S = 0.17
Urban	$\bar{x}$ = 37.3 S = 1.42	$\bar{x}$ = 53.8 S = 2.01	$\bar{x}$ = 609.0 S = 6.36	$\bar{x}$ = 29.4 S = 0.164
Industrial	$\bar{x}$ = 32.3 S = 2.69	$\bar{x}$ = 49.0 S = 4.73	$\bar{x}$ = 627.0 S = 6.55	$\bar{x}$ = 25.3 S = 0.143

Source: Own elaboration

Figure 3a shows the measurement of PM2.5 in the three planned zones, with the rural zone having the highest concentration of PM2.5. Figure 3b shows the measurement of PM10 in the three measurement zones, with the rural zone having the highest concentration of PM10.

Table 3 shows that the highest concentration of PM2.5 particulate matter is in rural areas with a median of 72.2 $\mu g/m^3$, Likewise, for the particulate matter PM10 is mostly concentrated in the rural area with a median of 100 $\mu g/m^3$.

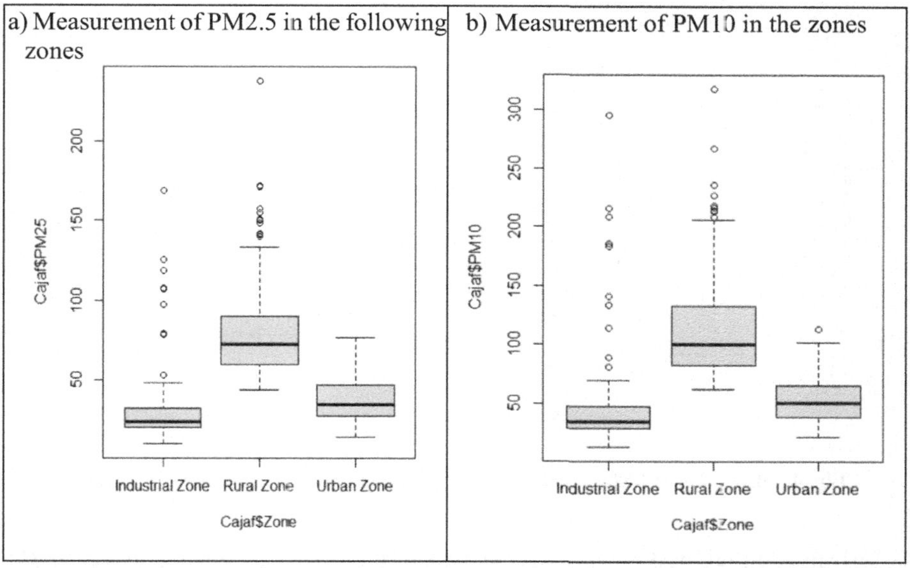

Fig. 3. Measurement of particulate matter in rural, urban and industrial areas of eastern Lima

Table 3. Concentration of particulate matter in the measurement areas.

Zone	PM2.5	PM10
Rural	*Md* = 72. 2 Min = 43.9 Max = 238	*Md* = 100 Min = 61.9 Max = 318.0
Urban	*Md* = 34. 9 Min = 14.0 Max = 76.6	*Md* = 50. 3 Min = 21.00 Max = 113.00
Industrial	*Md* = 23.8 Min = 10.1 Max = 169.0	*Md* = 34.2 Min = 12.8 Max = 296

Source: Own elaboration

In Fig. 4 we observe that the median measurement of particulate matter in the PM2.5 in the rural area, the 50% of the measurements are less than 72.2 $\mu g/m^3$ and the other 50% is higher than 72.2 $\mu g/m^3$, In the urban area as well, the 50% of measurements in less than 34.9 $\mu g/m^3$ and the other 50% is greater than 34.9 $\mu g/m^3$, in addition, in the industrial zone, the 50% of the measurements is less than 23.8 $\mu g/m^3$ and the other 50% is greater than 23.8 $\mu g/m^3$. In addition, we observed that the median measurement of PM10 particulate matter in the rural area, the 50% of the measurement is less than 100 $\mu g/m^3$ and the other 50% is greater than 100 $\mu g/m^3$, likewise, in the urban area, the 50% of measurements in less than 50.3 $\mu g/m^3$ and the other 50% is greater to 50.3

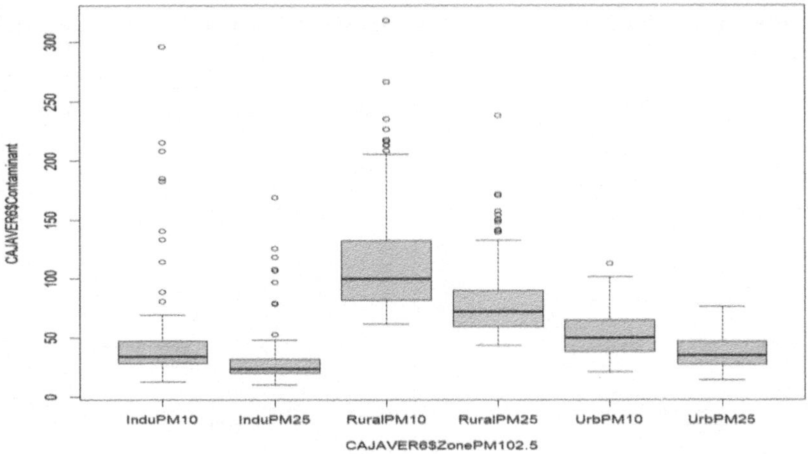

Fig. 4. Box plots of PM2.5 and PM10 particulate matter measurements by zones.

$\mu g/m^3$, In addition, in the industrial the 50% of the measurements are less than 34.2 $\mu g/m^3$ and the other 50% are greater than 34.2 $\mu g/m^3$.

3.2 Inferential Statisticians

Analysis of Variance for PM2.5

The study applied the Kruskal-Wallis test for PM2.5 particulate matter and found a chi-square of 156.34 with a p-value of less than 2.2e-16 in which significance was found in the difference of at least two study zones for particulate matter. Likewise, a post-hoc comparison was made, applying the Holm-Bonferroni method, finding a significant difference in the measurements of PM2.5 particulate matter between the rural and industrial zones with a p-holm value less than 2e-16; likewise the urban zone and the industrial zone with a p-holm less than 6.9e-07 and the urban zone with the rural zone with a p-holm less than 2e-16.

In Fig. 5a) we observe that the sample data for PM2.5 is not normally distributed, which is a limitation in using the ANOVA test, so the study performed the Kruskal-Wallis test to verify if there is a difference in medians between the PM2.5 particulate matter in the rural, urban and industrial areas of East Lima. En Fig. 5b) we observed that 50% of the PM2.5 particulate matter measurements in the rural area is less than 72.2 $\mu g/m^3$ and the other 50% has a measurement greater than 72.2$\mu g/m^3$. Also, in the urban zone 50% of the PM2.5 particulate matter measurements are less than 34.9 $\mu g/m^3$ and the other 50% has a measurement greater than 34.9 $\mu g/m^3$.

Likewise, in the industrial zone, 50% of the PM2.5 particulate matter measurements are less than 23.8 $\mu g/m^3$ and the other 50% have a measurement greater than 23.8 $\mu g/m^3$.

Analysis of Variance for PM10

The study applied the Kruskal-Wallis test for PM10 particulate matter and found a

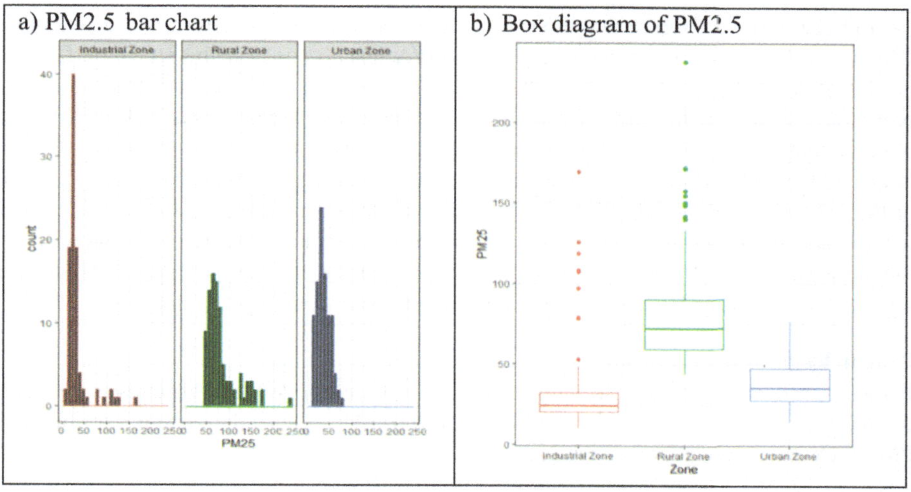

Fig. 5. Histogram and box plot of PM2.5 measurements by zones.

chi-square of 146.80 with a p-value of less than 2.2e-16 in which significance was found in the difference between at least two study zones for particulate matter. A post-hoc comparison was also performed, applying the Holm-Bonferroni method, finding a significant difference in the measurements of PM10 particulate matter between the rural and industrial zones with a p-holm value less than 2e-16; likewise the urban zone and the industrial zone with a p-holm less than 2e-06 and the urban zone with the rural zone with a p-holm less than 2e-16 (Fig. 6).

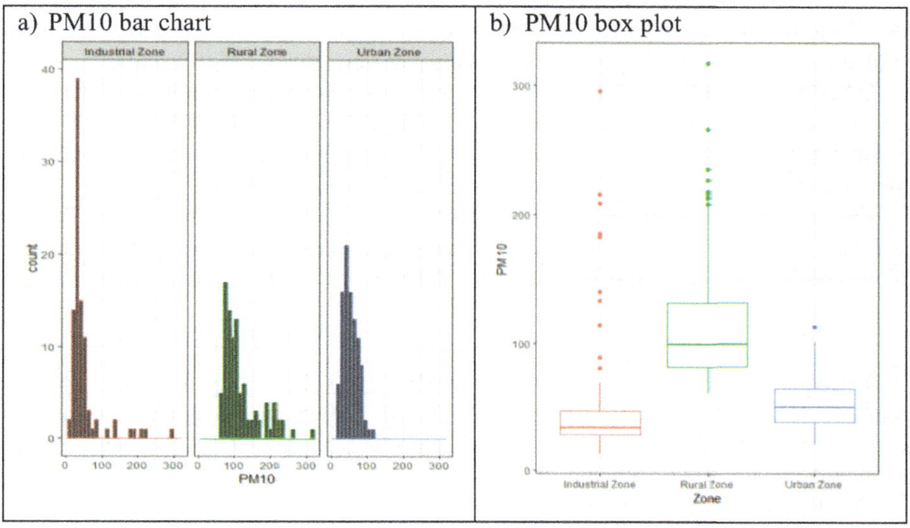

Fig. 6. Histogram and box plot of PM10 measurements by zones.

In Figure a) we observe that the sample data for PM10 are not normally distributed, which is a limitation in employing the ANOVA test, for which the study performed the Kruskal-Wallis test to verify if there is a difference in medians between the PM10 particulate matter in the rural, urban and industrial areas of East Lima. In Figure b) we observe that 50% of the PM10 particulate matter measurements in the rural area is less than 100 $\mu g/m^3$ and the other 50% has a measurement greater than 100 $\mu g/m^3$. Also. in the urban zone the of the PM10 particulate matter measurements are less than 50.3 $\mu g/m^3$ and the other 50% have a measurement greater than 50.3$\mu g/m^3$. Likewise, in the industrial zone, 50% of the PM10 particulate matter measurements are less than 34.2 $\mu g/m^3$ and the other 50% have a measurement greater than 34.2 $\mu g/m^3$

Linear Regression Equation

In the scientific research article, a linear regression equation by study zones was obtained for the measurements of particulate matter PM10 (X) as an independent variable and PM2.5 (Y) as a dependent variable, represented in Eq. (3)

$$Y = \beta_0 + \beta_1 X + \varepsilon \tag{3}$$

For the measurements made in the rural area of East Lima, a high Pearson correlation of 0.9669877 was found, with a very significant coefficient of determination R^2 = 0.9351 in which for the data collected, the equation shown in Eq. (4) was obtained.

$$Y_{PM2.5} = 5.57229 + 0.66105 X_{PM10} \tag{4}$$

The equation shows that the coefficient for PM10 es de 0.66105 $\mu g/m^3$. The coefficient indicates that for each additional $\mu g/m^3$ in PM10 can be expected to increase by an average of PM2.5 increases by an average of 0.66105 $\mu g/m^3$. The intercept value 5.57229 of the line, indicates the average level of PM2.5 when PM10 is zero.

The linear regression equation has a p-value of less than 2.2e-16 which makes the model significant, so we affirm that PM10 and PM2.5 are significantly correlated in the rural area and allow predicting PM2.5 values over time, given PM10 values.

Likewise, an analysis of variance was applied to validate the reliability of the linear regression model, obtaining a calculated F of 1339.2 with a Df = 1 and that at 5% significance is in the rejection region with a pvalue less than 2.2e-16, which confirms a good accuracy of the linear regression model, as shown in Fig. 7.

For the measurements carried out in the urban area of East Lima, a high Pearson correlation of 0.9838981 was found, with a very significant coefficient of determination $R^2 = 0.9878$ in which for the data collected, the equation shown in Eq. (5) was obtained.

$$Y_{PM2.5} = 1.149877 + 0.672768 X_{PM10} \tag{5}$$

The equation shows that the coefficient for PM10 is 0.672768 $\mu g/m^3$. The coefficient indicates that for each additional $\mu g/m^3$ in PM10, PM2.5 can be expected to increase by an average of 0.672768 $\mu g/m^3$. The intercept value 1.149877 of the line indicates the average level of PM2.5 when PM10 is zero.

The linear regression equation has a p value less than 2.2e-16 which makes the model significant, by which we affirm that PM10 and PM2.5 in this area are significantly correlated, allowing us to predict PM2.5 values over time, given PM10 values.

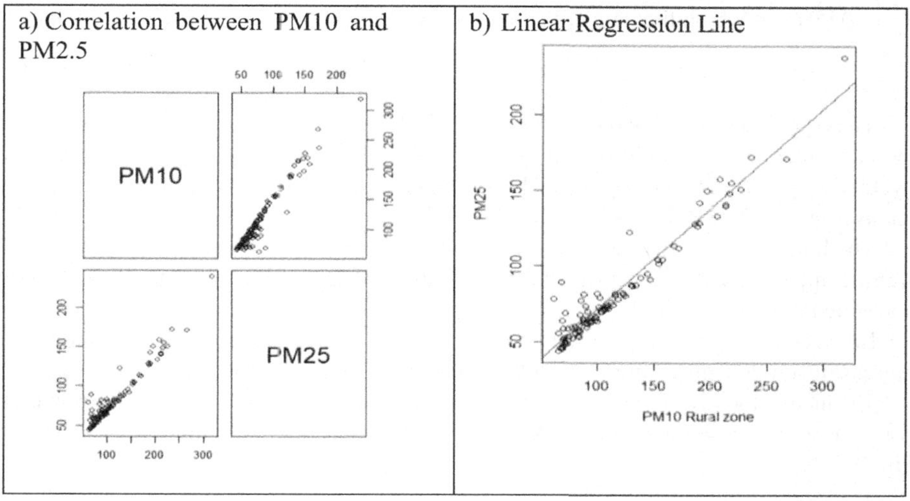

Fig. 7. Correlation diagram and regression line between PM2.5 and PM10 in rural áreas.

Likewise, an analysis of variance was applied to validate the reliability of the linear regression model, obtaining a calculated F of 7550.9 with a Df = 1 and 5% significance is located in the rejection region with a p value less than 2.2e-16, which confirms a good accuracy of the linear regression model as shown in Fig. 8.

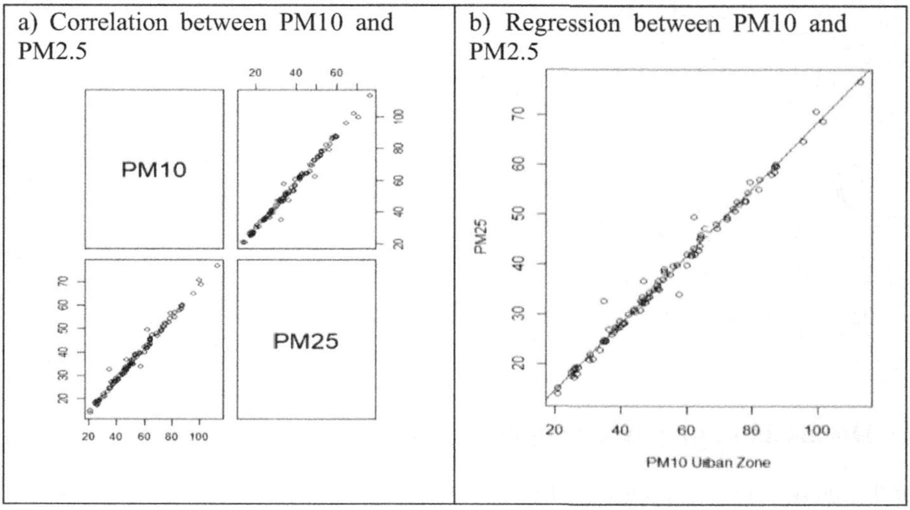

Fig. 8. Correlation diagram and regression line between PM2.5 and PM10 in urban áreas.

For the measurements carried out in the industrial zone of East Lima, a high Pearson correlation of 0.9959157, was found, with a very significant coefficient of determination

$R^2 = 0.9918$ in which the data collected, the equation shown in Eq. (6) was obtained.

$$Y_{PM2.5} = 4.485924 + 0.567541 X_{PM10} \tag{6}$$

The equation shows that the coefficient for PM10 is 0.567541 $\mu g/m^3$. The coefficient indicates that for each additional $\mu g/m^3$ in PM10, PM2.5 can be expected to increase by an average of 0.567541 $\mu g/m^3$. The intercept value 4.485924 4 of the line, indicates the average level of PM2.5 when PM10 is zero.

The linear regression equation has a p-value less than 2.2e-16 hich makes the model significant, by which we affirm that PM10 and PM2.5 in this zone are significantly correlated allowing to predict in time PM2.5 values, given PM10 values.

Likewise, an analysis of variance was applied to validate the reliability of the linear regression model, obtaining a calculated F of 11315 with a Df = 1 and that at 5% significance is located in the rejection region with a p value of less than 2.2e-16, which confirms a good accuracy and performance of the linear regression model, as shown in Fig. 9.

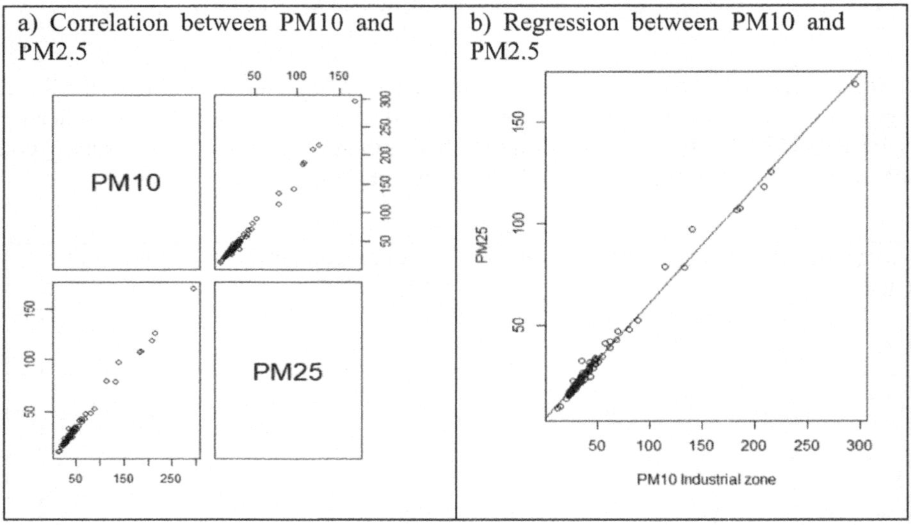

Fig. 9. Correlation plot and regression line between PM2.5 and PM10 in Industrial zone.

4 Discussion and Conclusion

In the analysis of meteorological factors of the regression taking into account the heteroscedasticity in Chile on PM2.5 and PM10 in the SARS-COV-2 transmission, they found an OLS estimator of 0.00191 and an intercept of 4.673 for PM2.5 and an OLS estimator of −0.00176 and an intercept of 1.700 for PM10 were highly significant [22]; however, the investigation found an OLS estimator of0.66105 and an intercept of5.57229 for PM2.5 and an OLS estimator of -0.00176 and an intercept of 1.700 for PM10.

A regression model of PM2.5 concentration for single-family houses was found with slope 0.301682 and an independent term of -1.942038 where the analysis of variance allowed estimating the directional coefficients of the regression model [5], in the article the model was found with slope 0.66105 and independent term of 5.57229.

A linear regression model was found with data from 2005 to 2009 in industrial and urban areas in Taiwan relating PM10 and PM2.5 with a coefficient of 0.90, which makes the model reliable [23], compared to the scientific paper which had a value of 0.93.

A linear regression model $y = 60.219 - 1.7493x$ with a coefficient of determination $R^2 = 0.2836$ was found on the concentration of PM2.5 and PM10 in Lima and Callao in which they used SENAMHI data [11], compared to the one found in the research of 0.93 which makes it more significant.

The quantitative analysis conducted in the study, found air pollutants for PM2.5 in the rural area of Carapongo an average of 83.0 $\mu g/m^3$, in the urban area of Alameda an average of 37.3 $\mu g/m^3$ and in the industrial area an average of 32.3 $\mu g/m^3$. Air pollutants for PM10 were found in the rural zone with an average of 83.0 $\mu g/m^3$, in the urban zone with an average of 37.3 $\mu g/m^3$ and in the industrial zone with an average of 32.3$\mu g/m^3$.

In addition, the Kruskal-Wallis test found a chi-square of 156.34 with a p-value of less than 2.2e-16 finding significance in at least two particulate matter study zones; the Holm-Bonferroni method found a significant difference in the measurements of PM2.5 between the rural and industrial zones with a p-holm value less than 2e-16; likewise the urban zone and the industrial zone with a p-holm less than 6.9e-07 nd the urban zone with the rural zone with a p-holm less than 2e-16.

In the rural area, a regression equation with independent variable PM10 and dependent variable PM2.5 was found, in which the OLS estimator for PM10 is 0.66105 $\mu g/m^3$, indicating that for each additional $\mu g/m^3$ additional in PM10, PM2.5 can be expected to increase by an average of 0.66105 $\mu g/m^3$ and with an intercept 5.57229 from the line indicating the average level of PM2.5 when PM10 is zero. For the urban area the OLS estimator for PM10 is 0.672768 $\mu g/m^3$, indicating that for each additional $\mu g/m^3$ additional in PM10, PM2.5 is expected to increase by an average of 0.672768 $\mu g/m^3$ and with an intercept 1.149877 from the line indicating the average level of PM2.5 hen PM10 is zero. Likewise in the industrial zone the OLS estimator for PM10 is 0.567541$\mu g/m^3$, indicating that for each additional $\mu g/m^3$ in PM10 PM2. 5 increase by an average of 0.567541 $\mu g/m^3$ and with an intercept 4.485924 from the line indicating the average level of PM2.5 when PM10 is zero.

References

1. Santillán, P., Rodríguez, M., Orozco, J., Ríos, I., Bayas, K.: Evaluación de la concentración y distribución espacial de material particulado en los campus de la UNACH - Riobamba. REVISTA DIGITAL DE CIENCIA, INGENIERÍA Y TECNOLOGÍA **4**(2), 111–126 (2021). https://doi.org/10.37135/ns.01.08.07
2. Sturm, R.: Modeling the deposition of bioaerosols with variable size and shape in the human respiratory tract - A review. J. Adv. Res. **3**(4). Elsevier B.V., 295–304 (2012). https://doi.org/10.1016/j.jare.2011.08.003

3. OMS, "Calidad del Aire Ambiente." Accessed: Feb. 09, 2023. [Online]. Available: https://www.paho.org/es/temas/calidad-aire-salud/calidad-aire-ambiente
4. Álvarez-Tolentino, D., Suárez-Salas, L.: Apportionment of emission sources of PM10 and PM2.5 at urban sites of Mantaro Valley, Peru. Revista Internacional de Contaminacion Ambiental **36**(4), 875–892 (2020). https://doi.org/10.20937/RICA.53473
5. Załuska, M., Gładyszewska-Fiedoruk, K.: Regression model of PM2.5 Concentration in a single-family house. Sustainability (Switzerland), **12**(15) (2020). https://doi.org/10.3390/su12155952
6. Rojano, R., Mendoza, Y., Arregoces, H., Restrepo, G.: Dispersión de Contaminantes del Aire (PM10, NO2, CO, COV y HAP) emitidos desde una estación modular de compresión, tratamiento y medición de gas natural. In: Informacion Tecnologica, Centro de Informacion Tecnologica, pp. 99–110 (2016). https://doi.org/10.4067/S0718-07642016000500012
7. OMS: Miles de millones de personas siguen respirando aire insalubre: nuevos datos de la OMS. Accessed: 09 Feb 2023. [Online]. Available: https://www.paho.org/es/noticias/4-4-2022-miles-millones-personas-siguen-respirando-aire-insalubre-nuevos-datos-oms
8. Tao, T., Shi, Y., Gilbert, K., Liu, X.: Spatiotemporal variations of air pollutants based on ground observation and emission sources over 19 Chinese urban agglomerations during 2015–2019. Sci. Rep. **12**(1) (2022). https://doi.org/10.1038/s41598-022-08377-9
9. Green, J., Sánchez, S.: La calidad del aire en América Latina: Una visión panorámica. (2012)
10. Cordova, C., et al.: Air quality assessment and pollution forecasting using artificial neural networks in Metropolitan Lima-Peru. Sci. Rep. **11**(1) (2021). https://doi.org/10.1038/s41598-021-03650-9
11. Pacsi Valdivia, S.A.: Analisis temporal y espacial de la calidad del aire determinado por material particulado pm10 y pm2,5 en lima metropolitan. Anales Científicos **77**(2), 273 (2016). https://doi.org/10.21704/ac.v77i2.699
12. Kazi, Z., Filip, S., Kazi, L.: Predicting PM2.5, PM10, SO2, NO2, NO and CO Air Pollutant Values with Linear Regression in R Language. Applied Sciences (Switzerland) **13** (6) (2023). https://doi.org/10.3390/app13063617
13. Gulati, S., Bansal, A., Pal, A., Mittal, N., Sharma, A., Gared, F.: Estimating PM2.5 utilizing multiple linear regression and ANN techniques. Sci. Rep. **13**(1) (2023) https://doi.org/10.1038/s41598-023-49717-7
14. Han, L., Zhao, J., Gao, Y., Gu, Z., Xin, K., Zhang, J.: Spatial distribution characteristics of PM2.5 and PM10 in Xi'an City predicted by land use regression models. Sustain Cities Soc. **61** (2020). https://doi.org/10.1016/j.scs.2020.102329
15. Elbayoumi, M., Ramli, N.A., Yusof, N.F.F.M.: Development and comparison of regression models and feedforward backpropagation neural network models to predict seasonal indoor PM2.5–10 and PM2.5 concentrations in naturally ventilated schools. Atmos. Pollut. Res. **6**(6), 1013–1023 (2015). https://doi.org/10.1016/j.apr.2015.09.001
16. Ramírez-Cando, L.J., Armijos, M., Crespo, M., Pino-Casignia, S.P., Álvarez-Mendoza, C.I.: Modelamiento geoestadístico de mediciones de concentración de material particulado (PM10) para la validación de un método simplificado. Anales Científicos **79**(1), 81 (2018). https://doi.org/10.21704/ac.v79i1.1143
17. Gharibzadeh, M., Saadat Abadi, A.R.: Estimation of surface particulate matter (PM2.5 and PM10) mass concentration by multivariable linear and nonlinear models using remote sensing data and meteorological variables over Ahvaz, Iran. Atmos Environ X **14** (2022). https://doi.org/10.1016/j.aeaoa.2022.100167
18. Álvarez, A.P.E.: Temas de Álgebra Lineal para Administración y Dirección de Empresas (2016)
19. G.D.: Econometría, Quinta. **13**(1) McGRAW-HILL, (1959)
20. Temtop M2000. Manual de Temtop M2000. (2000)

21. Forsyth, C.H., West, C.J.: Introduction to Mathematical Statistics., **25**(9) (1918). https://doi.org/10.2307/2974342
22. Scapini, V., Torres, S., Rubilar-Torrealba, R.: Meteorological, PM2.5 and PM10 factors on SARS-COV-2 transmission: The case of Southern Regions in Chile. Environ. Pollut. **322**, 120961 (2023). https://doi.org/10.1016/j.envpol.2022.120961
23. Chu, H.J., Huang, B., Lin, C.Y.: Modeling the spatio-temporal heterogeneity in the PM10-PM2.5 relationship. Atmos. Environ. **102**(1), 176–182 (2015). https://doi.org/10.1016/j.atmosenv.2014.11.062

Machine Learning Models for Salary Prediction in Peruvian Teachers of Regular Basic Education

Tinoco Ramos José(✉), Yupanqui Arellano Jhoset, Juan J. Soria, and Nemias Saboya

Escuela Profesional de Ingeniería de Sistemas, Universidad Peruana Unión, Lima, Perú
joselo18@upeu.edu.pe

Abstract. This paper presents an analysis of machine learning (ML) models to predict the salaries of 108,317 appointed teachers in Ventanilla, Lima, Peru, using recent data. The focal point of the study is appointed teachers, deliberately excluding salaries of hired teachers' from the analysis. A significant result of this research is the identification of a new ML model capable of predicting teacher salaries with considerable accuracy, based on regressor variables closely related to salary. This finding is noteworthy because it fills a gap in existing ML applications for salary prediction, indicating a promising direction for future research in this area. The methodology used to analyze the wage data, while comprehensive, does not account for gender differences, which may affect wage variation over the five-year period considered. This oversight suggests that future research should include a wider range of variables, including gender, to improve the accuracy and applicability of salary predictions for both appointed and contract faculty. Such an approach could provide more nuanced information on the factors influencing teacher salaries and help develop more equitable and effective salary models. One of the key contributions of the article is the detailed examination of the factors influencing salaries of appointed teachers, including age, educational level, length of service, teaching scale, and hours worked. The use of linear regression, Ridge, Lasso, and Elastic Net models yielded accurate metrics for choosing the best model for salary prediction. This research not only advances our understanding of the determinants of teacher salaries in Peru, but also provides a valuable framework for similar studies in other contexts. Comparison with other research highlights the robustness of the chosen ML models, underscoring the potential of ML in educational administration and policymaking.

Keywords: Regularization · Machine learning · Teaching Salary · Lasso · Ridge · Elastic net

1 Introduction

The salary of a professional plays a fundamental role in the existence of the employees of any organization, since it is the only compensation for their work, covering all aspects that represent an economic benefit. This income is the exclusive source that makes it possible to meet with dignity the essential needs of various kinds, both their own

R. Silhavy and P. Silhavy (Eds.): CSOC 2024, LNNS 1120, pp. 534–552, 2024.
https://doi.org/10.1007/978-3-031-70518-2_46

and those of their families [1]. In addition, income disparities have become a major problem throughout the world [2]. In that sense, there are several reasons why nowadays employees change jobs, one of the main factors is the salary, even more in a competitive world where employees have high aspirations and defined objectives. which causes losses in companies or organizations [3].

In Peru, the salary of employees is no stranger to economic constraints. According to a survey conducted in 2019 by the National Education Council (CNE) [4], between 65% and 68% of teachers are dissatisfied with their salary; public teachers are between 78% and 80%. The salaries of teachers in the Peruvian state are governed by eight pay scales [5], the status of the teacher, the type of service provided by the state, the years of service as a teacher, training, and research, which allow for increased salary incentives for teachers in the Peruvian teaching profession. Currently there are predictive models of machine learning [2, 6], in which the linear regression model, lasso, ridge and elastic net, allow to organize large amounts of data with lambda(λ) penalties, generating predictions with a very high degree of accuracy, they can also adapt quickly to changes, especially when it is required to predict the salary of Peruvian teachers. In addition, multilevel hierarchical mixed models allow associating teacher salary with its regressor variables to find predictions. There is little research that uses machine learning techniques and models to pre-predict the salary level of a teacher knowing the factors that include the salary. [6].

Numerous research use machine learning models to predict salaries in organizations using predictive models, as shown in the research [7], which analyzed data from 2003 to 2006 on teacher salaries from ''Education at a Glance'' and national data on mathematics performance in 30 countries and found a multiple regression model relating the change in salaries of new teachers and teachers with 15 years of experience to national mathematics performance. The model found was: Salario = 2.86(New teacher salary change) + 1.28(Educational expenditure) + 0.21(GDP per capita), where the coefficient of determination R^2 is 0.29, which means that 29% of the regressor variables explain teacher salary. In the same context McKinley L. Blackburn [8], analyzed data from the years 2012 to 2016 from an American Community Survey (ACS) to determine salary differences between teachers and non-teachers, for which a logit regression model was used.

Emma Garcia and Eunice S. Han [9], used nationally representative data between the years 2009 to 2015 to examine the relationship in teachers' salaries and students' academic performance, for which they used linear regression models and multilevel mixed-effects models, obtaining as a result that the model shows a positive significant association between teachers' base salary and students' mathematics performance.

Yasser T. Matbouli [10], conducted a study using machine learning models to predict wages based on occupational and organizational characteristics of the entire economy of Saudi Arabia, using data from 2010 to 2017 and more than two thousand economic activities, obtaining as a result that the Bayesian Gaussian regression showed an improvement over the multiple linear regression (from 0.50 to 0.98). Likewise, low levels of mean squared error were shown, where it decreased by 80% compared to the multiple linear regression, which was reduced by 90%. Similarly Rupashri Barik [11], conducted a study to predict the salary of a person in a given period of time, and then develop a system that allows to show all the data of the daily work and salary growth of a person, the data to be used are the history of salary records of an employee of an organization, for which linear regression and polynomial regression models were used. Also At present industries are using machine algorithms to solve various problems, Sarala, V. [3], addresses the salaries of employees of recent graduates and aims to predict the salary, according to a particular field and their qualifications, for the present research used prediction algorithms such as linear regression, also used algo-rhythms of decision tree and random forest regressor, the linear regression algorithm gives better accuracy, the model predicted an accuracy of 95.68% in training, and an accuracy of 95.33% in the test ones.

The literature covers few wage predictions using Linear regression, Ridge regression and Elastic Net regression algorithms, but they use these models to solve various concrete situations, as mentioned by Xiangning Dong [12], where he presents a new technique to predict the incidence of ground delay schedules due to bad weather or airport capacity problems, using machine learning techniques. These methods could improve passenger safety and the economic benefits of flights, for which flight operations data were taken from January 1, 2021, to June 30, 2021, from Nanjing Lukou inter-national airport, the second largest airport in China. Linear regression models such as Ridge and LASSO were used, resulting in regression models showing better performance in quantifying and generating a score.

Devesh Singh [13] to attract public investment regional factors play an important role, this research uses 18 years data at the county level of Hungary, finding a Ridge, LASSO and Elastic Net regression model, the results show that the Elastic Net model is the best method to determine the prediction of public investment at the regional scale.

On the other hand, Fadhil M. Basysyar [14], conducted a research, with a range of housing prices, seeks to forecast the value of housing accurately, the models used for this research are linear regression, Ridge regression, Lasso Regression, and Elastic Net Regression, 1460 records and 81 housing characteristics were used, obtaining as a result showed that the Lasso regression model had a better performance when performing the analysis compared to the other models.

Frank Emmert [15], in his research studies and compares the regression models Lasso, Elastic Net, Ridge, the data used for the analysis consists of 93 variables of 156 samples of inflation indexes of the economy in the country of Brazil, regularization terms were discussed to decrease coefficients that lead to better performance metrics, it was found that the re-gression models are very efficient in analyzing high dimensional data, each of them faces particular problems, and that currently these models are very popular for data analysis.

This paper aims to predict the salary of Peruvian regular basic education teachers using regression models with machine learning, using teacher salary as predictor variables and age, educational level, length of service, teaching scale and hours worked as regressor variables.

The study will be a precedent and theoretical reference for future research in predictive models with machine learning because it helps to analyze the relationships, concepts and scope of the regressor variables of the study with the predictor variable that is the teacher's salary under linear regression models, Lasso, Ridge and Elastic Net, which are very accurate in their prediction. The results will help us to determine the future salary of teachers who meet their scale standards, since the larger the scale, the higher the salary.

2 Methodology

2.1 Collection of Information

To have the information of the data of the remunerations of the teachers of the Ugel Ventanilla - Callao, a letter was sent to the legal representative of the institution requesting to use the data of the payroll of the regular basic education teachers appointed, from 2018 to 2023 to be studied and analyzed. Likewise, the research method was carried out as shown in Figure 1.

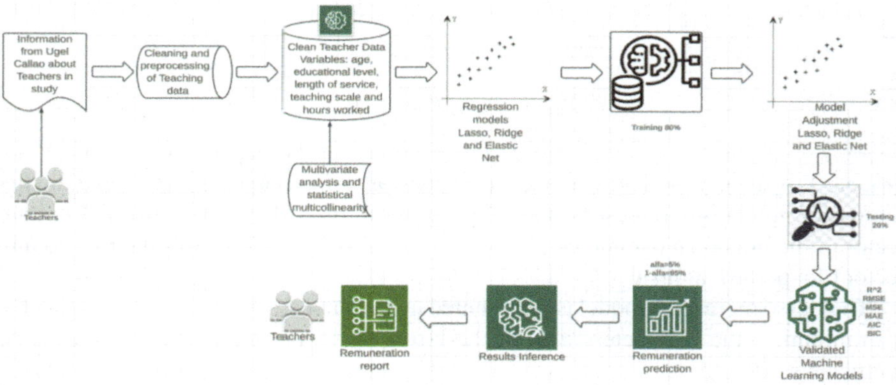

Fig. 1. Research Method.

2.2 Machine Learning Models

The machine learning predictive models used in the research, linear regression, Lasso regression, Ridge regression and Elastic Net regression, allowed for a very effective analysis in the prediction of teachers' salaries at Ugel Ventanilla, Peru, showing efficiency and excellent performance through the metrics.

Linear regression is a generalization of the k-variable population regression model (PRF) with dependent variable Y and k-1 explanatory variables $X_2, X_3, X_4, X_5, \cdots, X_k$ [16] and is written as shown in Eq. 1.

$$Y_i = \beta_1 + \beta_2 X_{2i} + \beta_3 X_{3i} + \beta_4 X_{4i} + \cdots + \beta_k X_{ki} + u_i; i = 1, \cdots, n \quad (1)$$

where β_1 is the intercept, β_2 to β_k are the partial slope coefficients, u is the stochastic disturbance term and i is the ith observation, with n as the population size.

Equation (1) is a shorthand expression for the set of n simultaneous equations[17] shown in Eq. 2 and 3.

$$\left.\begin{array}{l} Y_1 = \beta_1 + \beta_2 X_{21} + \beta_3 X_{31} + \beta_4 X_{41} + \cdots + \beta_k X_{k1} + u_1 \\ Y_2 = \beta_1 + \beta_2 X_{22} + \beta_3 X_{32} + \beta_4 X_{42} + \cdots + \beta_k X_{k2} + u_2 \\ Y_3 = \beta_1 + \beta_2 X_{23} + \beta_3 X_{33} + \beta_4 X_{43} + \cdots + \beta_k X_{k3} + u_3 \\ \qquad \cdots \\ Y_n = \beta_1 + \beta_2 X_{2n} + \beta_3 X_{3n} + \beta_4 X_{4n} + \cdots + \beta_k X_{k3} + u_n \end{array}\right\} \tag{2}$$

The system of Eqs. 2 is written in matrix form as shown in Eq. 3 and 4.

$$\begin{bmatrix} y_1 \\ y_2 \\ y_3 \\ y_4 \\ \vdots \\ y_n \end{bmatrix} = \begin{bmatrix} 1 & x_{21} & x_{31} & x_{41} & \cdots & x_{k1} \\ 1 & x_{22} & x_{32} & x_{42} & \cdots & x_{k2} \\ 1 & x_{23} & x_{33} & x_{43} & \cdots & x_{k3} \\ \vdots & \vdots & \vdots & \vdots & & \vdots \\ 1 & x_{2n} & x_{3n} & x_{4n} & \cdots & x_{kn} \end{bmatrix} \begin{bmatrix} \beta_1 \\ \beta_2 \\ \beta_3 \\ \beta_4 \\ \vdots \\ \beta_k \end{bmatrix} + \begin{bmatrix} u_1 \\ u_2 \\ u_3 \\ \vdots \\ u_n \end{bmatrix} \tag{3}$$

$$Y_{nx1} = X_{nxk} \cdot \beta_{kx1} + u_{nx1} \tag{4}$$

where Y is the nx1 column vector of observations on the dependent variable Y, the variable X is the matrix nxk, with n observations on k-1 variables X_2 to X_k and the first column of numbers 1 represents the intercept term, the value of β is the kx1 column vector of the unknown parameters $\beta_1, \beta_2, \cdots, \beta_k$ and the value of u is the nx1 column vector of n perturbations u_i.

The Lasso regression model is represented mathematically by Eq. 5, whose objective is to minimize the parameters α; β in the errors by taking the penalty of the model coefficients [18].

$$\underset{\alpha,\beta}{Minimize}\left\{\frac{1}{2n}\sum_{i=1}^{N}\left(y_i - \alpha - \sum_{j=1}^{p} x_{ij}\beta_j\right)^2\right\} \tag{5}$$

Therefore, the Lasso Model considering the variables mentioned above, is represented by Eqs. 6 and 7.

$$\begin{bmatrix} y_1 \\ y_2 \\ y_3 \\ y_4 \\ y_5 \end{bmatrix} = \begin{bmatrix} 1 & x_{11} & x_{12} & x_{13} & x_{14} & x_{15} \\ 1 & x_{21} & x_{22} & x_{23} & x_{24} & x_{25} \\ 1 & x_{31} & x_{32} & x_{33} & x_{34} & x_{35} \\ 1 & x_{41} & x_{42} & x_{43} & x_{44} & x_{45} \\ 1 & x_{51} & x_{52} & x_{53} & x_{54} & x_{55} \end{bmatrix} \begin{bmatrix} \alpha \\ \beta_1 \\ \beta_2 \\ \beta_3 \\ \beta_4 \\ \beta_5 \end{bmatrix} + \begin{bmatrix} u_1 \\ u_2 \\ u_3 \\ u_4 \\ u_5 \end{bmatrix} \tag{6}$$

$$Y_i = X_{kx5} \cdot \beta_{5x1} + u_i; \quad i = \overline{1,5} \tag{7}$$

The Lasso regression model is shown in Eqs. 8 and 9.

$$Y_i = \alpha + \beta_1 X_{i1} + \beta_2 X_{i2} + \beta_3 X_{i3} + \beta_4 X_{i4} + \beta_5 X_{i5} + u_i \tag{8}$$

$$u_i = Y_i - \alpha - \beta_1 X_{i1} - \beta_2 X_{i2} - \beta_3 X_{i3} - \beta_4 X_{i4} + \beta_5 X_{i5} \tag{9}$$

where u_i is the error to be minimized

The least squares technique performs the minimization of the errors and its representation is as follows:

$$D = \sum_{i=1}^{n} u_i^2$$

In performing the Lasso regression, we add a penalty factor to the least squares, which reduces the loss function S to a minimum value, represented by Eq. 10.

$$S = \underset{\alpha,\beta_1,\beta_2,\beta_3,\beta_4,\beta_5}{Min}\left[u_i^2 + \lambda(|\beta_1| + |\beta_2| + |\beta_3| + |\beta_4| + |\beta_5|)\right] \tag{10}$$

$$S = \underset{\alpha,\beta_1,\beta_2,\beta_3,\beta_4,\beta_5}{Min}\left[(Y_i - \alpha - \beta_1 X_{i1} - \beta_2 X_{i2} - \beta_3 X_{i3} - \beta_4 X_{i4} - \beta_5 X_{i5})^2 + \lambda(|\beta_1| + |\beta_2| + |\beta_3| + |\beta_4| + |\beta_5|)\right]$$

The normal equations by least squares theory for the Lasso multiple regression are shown in Eq. 11.

$$\begin{cases} \alpha n + \beta_1 \sum_{i=1}^{n} X_{i1} + \beta_2 \sum_{i=1}^{n} X_{i2} + \beta_3 \sum_{i=1}^{n} X_{i3} + \beta_4 \sum_{i=1}^{n} X_{i4} + \beta_5 \sum_{i=1}^{n} X_{i5} = \sum_{i=1}^{n} y_i \\ \alpha \sum_{i=1}^{n} X_{i1} + \beta_1 \sum_{i=1}^{n} X_{i1}^2 + \beta_2 \sum_{i=1}^{n} X_{i1}X_{i2} + \beta_3 \sum_{i=1}^{n} X_{i1}X_{i3} + \beta_4 \sum_{i=1}^{n} X_{i1}X_{i4} + \beta_5 \sum_{i=1}^{n} X_{i1}X_{i5} = \sum_{i=1}^{n} X_{i1}y_i - \lambda n \\ \alpha \sum_{i=1}^{n} X_{i2} + \beta_1 \sum_{i=1}^{n} X_{i2}X_{i1} + \beta_2 \sum_{i=1}^{n} X_{i2}^2 + \beta_3 \sum_{i=1}^{n} X_{i2}X_{i3} + \beta_4 \sum_{i=1}^{n} X_{i2}X_{i4} + \beta_5 \sum_{i=1}^{n} X_{i2}X_{i5} = \sum_{i=1}^{n} X_{i2}y_i - \lambda n \\ \alpha \sum_{i=1}^{n} X_{i3} + \beta_1 \sum_{i=1}^{n} X_{i3}X_{i1} + \beta_2 \sum_{i=1}^{n} X_{i3}X_{i2} + \beta_3 \sum_{i=1}^{n} X_{i3}^2 + \beta_4 \sum_{i=1}^{n} X_{i3}X_{i4} + \beta_5 \sum_{i=1}^{n} X_{i3}X_{i5} = \sum_{i=1}^{n} X_{i3}y_i - \lambda n \\ \alpha \sum_{i=1}^{n} X_{i4} + \beta_1 \sum_{i=1}^{n} X_{i4}X_{i1} + \beta_2 \sum_{i=1}^{n} X_{i4}X_{i2} + \beta_3 \sum_{i=1}^{n} X_{i4}X_{i3} + \beta_4 \sum_{i=1}^{n} X_{i4}^2 + \beta_5 \sum_{i=1}^{n} X_{i4}X_{i5} = \sum_{i=1}^{n} X_{i4}y_i - \lambda n \\ \alpha \sum_{i=1}^{n} X_{i5} + \beta_1 \sum_{i=1}^{n} X_{i5}X_{i1} + \beta_2 \sum_{i=1}^{n} X_{i5}X_{i2} + \beta_3 \sum_{i=1}^{n} X_{i5}X_{i3} + \beta_4 \sum_{i=1}^{n} X_{i5}X_{i4} + \beta_5 \sum_{i=1}^{n} X_{i5}^2 = \sum_{i=1}^{n} X_{i5}y_i - \lambda n \end{cases} \tag{11}$$

The Ridge regression model, unlike the Lasso regression, reduces the coefficients of the correlated predictors, which allows them to borrow the strength of each other. Also, from a Bayesian perspective, the penalty of the Ridge model is appropriate in the case where there are several predictors and all have non-zero coefficients, i.e., they are drawn from a Gaussian distribution [15]. In addition, it is of priority to consider in the analysis the properties of the mean square error of the Ridge regression such as the variance and bias of the estimator, the theorem on the mean square function and the comments made on the mean square error function [19]. The mathematical representation of the Ridge regression mode is shown in Eq. 12.

$$\beta^{ridge} = \underset{\beta\epsilon\mathbb{R}}{argmin}\|Y - X\beta\|_2^2 + \lambda\|\beta\|_2^2 \tag{12}$$

In performing the Ridge regression, we add a penalty factor to the least squares, which reduces the loss function S to a minimum value, represented in Eq. 13.

$$S = \underset{\alpha,\beta_1,\beta_2,\beta_3,\beta_4,\beta_5}{Min}\left[\frac{1}{2n}\sum_{i=1}^{N} u_i^2 + \lambda\left(\beta_1^2 + \beta_2^2 + \beta_3^2 + \beta_4^2 + \beta_5^2\right)\right] \tag{13}$$

$$S = \underset{\alpha,\beta_1,\beta_2,\beta_3,\beta_4,\beta_5}{Min}\left[\frac{1}{2n}(Y_i - \alpha - \beta_1 X_{i1} - \beta_2 X_{i2} - \beta_3 X_{i3} - \beta_4 X_{i4} + \beta_5 X_{i5})^2 + \lambda\left(\beta_1^2 + \beta_2^2 + \beta_3^2 + \beta_4^2 + \beta_5^2\right)\right]$$

The normal equations of the least-squares theory for the Ridge multiple regression are represented in 14.

$$\begin{cases} \alpha n + \beta_1 \sum_{i=1}^{n} X_{i1} + \beta_2 \sum_{i=1}^{n} X_{i2} + \beta_3 \sum_{i=1}^{n} X_{i3} + \beta_4 \sum_{i=1}^{n} X_{i4} + \beta_5 \sum_{i=1}^{n} X_{i5} = \sum_{i=1}^{n} y_i \\ \alpha \sum_{i=1}^{n} X_{i1} + \beta_1 \sum_{i=1}^{n} X_{i1}^2 + \beta_2 \sum_{i=1}^{n} X_{i1}X_{i2} + \beta_3 \sum_{i=1}^{n} X_{i1}X_{i3} + \beta_4 \sum_{i=1}^{n} X_{i1}X_{i4} + \beta_5 \sum_{i=1}^{n} X_{i1}X_{i5} = \sum_{i=1}^{n} X_{i1}y_i - 2\lambda n\beta_1 \\ \alpha \sum_{i=1}^{n} X_{i2} + \beta_1 \sum_{i=1}^{n} X_{i2}X_{i1} + \beta_2 \sum_{i=1}^{n} X_{i2}^2 + \beta_3 \sum_{i=1}^{n} X_{i2}X_{i3} + \beta_4 \sum_{i=1}^{n} X_{i2}X_{i4} + \beta_5 \sum_{i=1}^{n} X_{i2}X_{i5} = \sum_{i=1}^{n} X_{i2}y_i - 2\lambda n\beta_2 \\ \alpha \sum_{i=1}^{n} X_{i3} + \beta_1 \sum_{i=1}^{n} X_{i3}X_{i1} + \beta_2 \sum_{i=1}^{n} X_{i3}X_{i2} + \beta_3 \sum_{i=1}^{n} X_{i3}^2 + \beta_4 \sum_{i=1}^{n} X_{i3}X_{i4} + \beta_5 \sum_{i=1}^{n} X_{i3}X_{i5} = \sum_{i=1}^{n} X_{i3}y_i - 2\lambda n\beta_3 \\ \alpha \sum_{i=1}^{n} X_{i4} + \beta_1 \sum_{i=1}^{n} X_{i4}X_{i1} + \beta_2 \sum_{i=1}^{n} X_{i4}X_{i2} + \beta_3 \sum_{i=1}^{n} X_{i4}X_{i3} + \beta_4 \sum_{i=1}^{n} X_{i4}^2 + \beta_5 \sum_{i=1}^{n} X_{i4}X_{i5} = \sum_{i=1}^{n} X_{i4}y_i - 2\lambda n\beta_4 \\ \alpha \sum_{i=1}^{n} X_{i5} + \beta_1 \sum_{i=1}^{n} X_{i5}X_{i1} + \beta_2 \sum_{i=1}^{n} X_{i5}X_{i2} + \beta_3 \sum_{i=1}^{n} X_{i5}X_{i3} + \beta_4 \sum_{i=1}^{n} X_{i5}X_{i4} + \beta_5 \sum_{i=1}^{n} X_{i5}^2 = \sum_{i=1}^{n} X_{i5}y_i - 2\lambda n\beta_5 \end{cases} \tag{14}$$

The Elactic Net regression penalty is an adaptation of least squares and allows addressing the estimation problem by producing a biased β-estimator but with small variances [20]. The mathematical representation of the Elactic Net regression model is represented in Eq. 15 and is defined as follows:

$$\beta^{EN} = \underset{\beta \in \mathbb{R}}{argmin} \frac{1}{2n} \|Y - X\beta\|_2^2 + \lambda\left[(1-\alpha)\|\beta\|_2^2 + \alpha\|\beta\|_1\right] \tag{15}$$

When performing the Elastic Net regression, we add a penalty factor to the least squares, which reduces the loss function S to a minimum value, represented in Eq. 16.

$$S = \underset{\alpha,\beta_1,\beta_2,\beta_3,\beta_4,\beta_5}{Min} \left[\frac{1}{2n} \sum_{i=1}^{N} u_i^2 + \lambda\left[(1-\alpha)\left(\beta_1^2 + \beta_2^2 + \beta_3^2 + \beta_4^2 + \beta_5^2\right)\right] + \alpha\left[|\beta_1| + |\beta_2| + |\beta_3| + |\beta_4| + |\beta_5|\right] \right] \tag{16}$$

$$S = \underset{\alpha,\beta_1,\beta_2,\beta_3,\beta_4,\beta_5}{Min} \left\{ \tfrac{1}{2n}\left(Y_i - \alpha - \beta_1 X_{i1} - \beta_2 X_{i2} - \beta_3 X_{i3} - \beta_4 X_{i4} + \beta_5 X_{i5}\right)^2 + \lambda\left[(1-\alpha)\left(\beta_1^2 + \beta_2^2 + \beta_3^2 + \beta_4^2 + \beta_5^2\right)\right] + \alpha\left[|\beta_1| + |\beta_2| + |\beta_3| + |\beta_4| + |\beta_5|\right] \right\}$$

Then the normal equations of the least squares theory for the Elastic net regression is represented in Eq. 17.

$$\begin{cases} \alpha n + \beta_1 \sum_{i=1}^{n} X_{i1} + \beta_2 \sum_{i=1}^{n} X_{i2} + \beta_3 \sum_{i=1}^{n} X_{i3} + \beta_4 \sum_{i=1}^{n} X_{i4} + \beta_5 \sum_{i=1}^{n} X_{i5} = n\lambda \\ \quad [(\beta_1^2 + \beta_2^2 + \beta_3^2 + \beta_4^2 + \beta_5^2)] - [|\beta_1| + |\beta_2| + |\beta_3| + |\beta_4| + |\beta_5|] + \sum_{i=1}^{n} y_i \\ \alpha \sum_{i=1}^{n} X_{i1} + \beta_1 \sum_{i=1}^{n} X_{i1}^2 + \beta_2 \sum_{i=1}^{n} X_{i1}X_{i2} + \beta_3 \sum_{i=1}^{n} X_{i1}X_{i3} + \beta_4 \sum_{i=1}^{n} X_{i1}X_{i4} + \beta_5 \sum_{i=1}^{n} X_{i1}X_{i5} = \sum_{i=1}^{n} X_{i1}y_i + \lambda n[2\beta_1(\alpha - 1) - \alpha] \\ \alpha \sum_{i=1}^{n} X_{i2} + \beta_1 \sum_{i=1}^{n} X_{i2}X_{i1} + \beta_2 \sum_{i=1}^{n} X_{i2}^2 + \beta_3 \sum_{i=1}^{n} X_{i2}X_{i3} + \beta_4 \sum_{i=1}^{n} X_{i2}X_{i4} + \beta_5 \sum_{i=1}^{n} X_{i2}X_{i5} = \sum_{i=1}^{n} X_{i2}y_i + \lambda n[2\beta_2(\alpha - 1) - \alpha] \\ \alpha \sum_{i=1}^{n} X_{i3} + \beta_1 \sum_{i=1}^{n} X_{i3}X_{i1} + \beta_2 \sum_{i=1}^{n} X_{i3}X_{i2} + \beta_3 \sum_{i=1}^{n} X_{i3}^2 + \beta_4 \sum_{i=1}^{n} X_{i3}X_{i4} + \beta_5 \sum_{i=1}^{n} X_{i3}X_{i5} = \sum_{i=1}^{n} X_{i3}y_i + \lambda n[2\beta_3(\alpha - 1) - \alpha] \\ \alpha \sum_{i=1}^{n} X_{i4} + \beta_1 \sum_{i=1}^{n} X_{i4}X_{i1} + \beta_2 \sum_{i=1}^{n} X_{i4}X_{i2} + \beta_3 \sum_{i=1}^{n} X_{i4}X_{i3} + \beta_4 \sum_{i=1}^{n} X_{i4}^2 + \beta_5 \sum_{i=1}^{n} X_{i4}X_{i5} = \sum_{i=1}^{n} X_{i4}y_i + \lambda n[2\beta_4(\alpha - 1) - \alpha] \\ \alpha \sum_{i=1}^{n} X_{i5} + \beta_1 \sum_{i=1}^{n} X_{i5}X_{i1} + \beta_2 \sum_{i=1}^{n} X_{i5}X_{i2} + \beta_3 \sum_{i=1}^{n} X_{i5}X_{i3} + \beta_4 \sum_{i=1}^{n} X_{i5}X_{i4} + \beta_5 \sum_{i=1}^{n} X_{i5}^2 = \sum_{i=1}^{n} X_{i5}y_i + \lambda n[2\beta_5(\alpha - 1) - \alpha] \end{cases} \tag{17}$$

2.3 Model Validation Metrics Under Study

The metrics used in the research give the performance of the pre-dictive models [21] which were the MAE (Mean Absolute Error), which measured the average magnitude of the errors in the predictions, without considering their direction. The MAE is the sample average of the absolute differences between the prediction and the actual observation, where all individual differences have the same weight and is shown in Eq. 18.

$$Mean\ Absolute\ Error : MAE = \frac{1}{n}\sum_{i=1}^{n} \left|y_i - \widehat{y_i}\right| \tag{18}$$

Also, the RMSE (root mean squared error) is the square root of the average of the squared differences between the prediction and the actual observation (SSE/n) and is defined by Eq. 19.

$$Root\ Mean\ Square\ Error : RMSE = \sqrt{\frac{1}{n}\sum\nolimits_{i=1}^{n} (y_i - \widehat{y_i})^2} \quad (19)$$

The coefficient of determination R^2 is another metric used in predictive models, which measures the variance in the data explained by the model and is determined as one minus the ratio of the sum of squares of the errors (SSE) to the total (SST) as shown in Eq. 20.

$$\text{Coefficient of determination} : R^2 = 1 - \frac{\sum_{i=1}^{n} (y_i - \widehat{y_i})^2}{\sum_{i=1}^{n} (y_i - \overline{y_i})^2} \quad (20)$$

2.4 Data Normalization

The research used 108317 teaching records with single variable data, where Shanker et al. [22] suggest that it is necessary to normalize the characteristics of the variables, since they produce better results in general. In addition, the requirements of the algorithms used require normalization of the data, where Atlas et al. [23] mentions that it should be used in the cases of the return and predictor variables, so we proceeded to normalize the data using the min-max normalization technique to guarantee the homogeneity of the variables concentrated in a continuous interval of [0, 1] mentioned by M-Dawam [24] as shown in Eq. 21.

$$\widehat{X}[:, i] = \frac{X[:, i] - min(X[:, i])}{max(X[:, i]) - min(X[:, i])} \quad (21)$$

3 Results

3.1 Descriptive Statistics

The research conducted an exploratory descriptive analysis, finding an average teacher salary of 2721.80 soles, with a minimum of S/. 302.2 and a maximum of S/. 10,760.10, as well as teachers with an average age of 53.24 years, two months and 26 days with a minimum of 23 years and a maximum of 71 years, the educational level had teachers of level 1, a maximum of 8 with an average of 2. 372 Also the time of teaching service with an average of 14 years. also an average teaching scale of 2 and the hours worked with an average of 31H+8M+24S with a minimum of 26 hours and a maximum of 40 hours, as shown in Table 1.

An exploratory analysis of the salaries per year of the appointed teachers was also carried out and it was found that the salary was increased from 2018 to 2023 as shown in Fig. 2.

Table 1. Descriptive statistics of the variables under study.

Study variables	Min	Median	Means	Max
Teacher Salary	302.2	2620.1	2771.8	10760.1
Age	23	53	53.24	71
Educational Level	1	2	2.372	8
Service time	1	14	13.88	44
Teacher Scale	1	2	1.919	7
Hours Worked	26	30	31.35	40

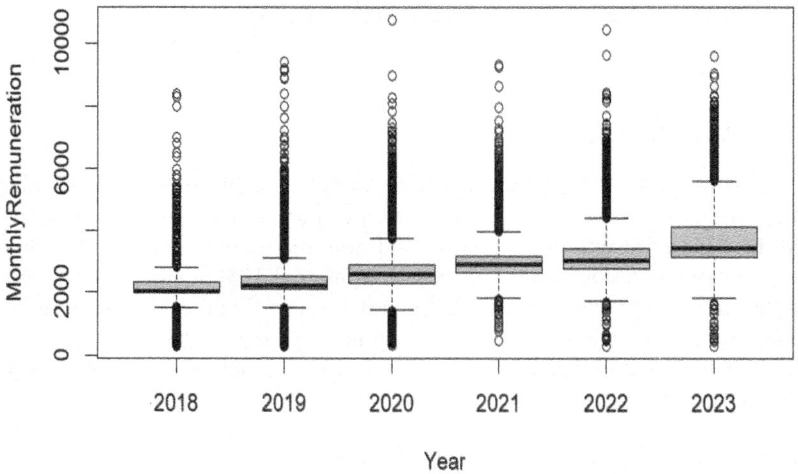

Fig. 2. Five-year cash flow chart of teachers' salaries.

In the research we found the remunerations by teaching scale, with Scale 4 in first place, followed by Scale 5, then Scale 6, followed by Scale 2, then Scale 1, then Scale 3 and finally Scale 7, which over time was increasing the salaries of teachers as shown in Fig. 3.

3.2 Inferential Statistics

In the investigation of the 108317 observations, 80% (86654) of the observations were taken for training and 20% (21663) for testing, finding in the training a resampling cross validation with 5 repetitions a RMSE=895.3793, a Rsquared=0.2872618 and a MAE=619.7701.

The multivariate linear regression analysis with 5 regressor variables analyzed, which are age (Age), educational level (NEduc), length of service (Tserv), teaching scale (EsDoc) and hours worked (HoLab) by teachers, as well as the predictor

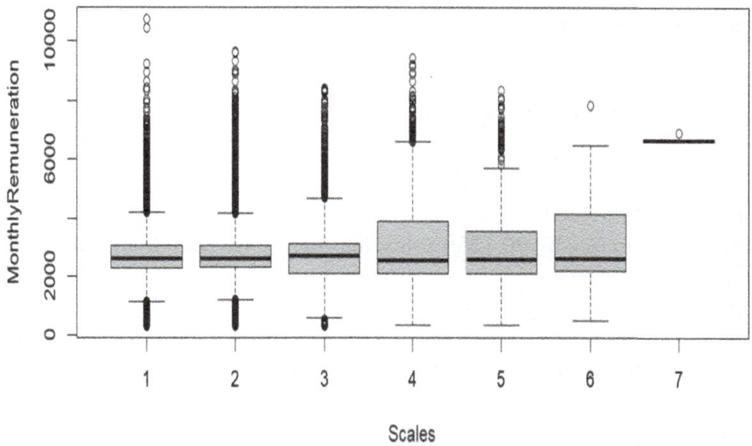

Fig. 3. Box diagram of the eight scales of teachers' salaries.

variable teacher salary, fulfilling the statistical prerequisites, yielded the results in the R Studio software shown in Fig. 4.

```
> summary(lm)

Call:
lm(formula = .outcome ~ ., data = dat)

Residuals:
    Min      1Q  Median      3Q     Max 
-4283.0  -344.6    31.1   428.6  8114.1 

Coefficients:
              Estimate Std. Error t value Pr(>|t|)    
(Intercept) -3109.7302    38.2512 -81.298   <2e-16 ***
Edad           -5.2337     0.5378  -9.731   <2e-16 ***
NEduc          24.1794     3.2491   7.442    1e-13 ***
TServ          18.0115     0.4056  44.410   <2e-16 ***
EsDoc         -90.1259     3.0967 -29.104   <2e-16 ***
HoLab         192.1820     1.1054 173.859   <2e-16 ***
---
Signif. codes:  0 '***' 0.001 '**' 0.01 '*' 0.05 '.' 0.1 ' ' 1

Residual standard error: 895.4 on 86690 degrees of freedom
Multiple R-squared:  0.2873,	Adjusted R-squared:  0.2873 
F-statistic:  6989 on 5 and 86690 DF,  p-value: < 2.2e-16
```

Fig. 4. Results of R Studio codes for multiple linear regression. *** Indicates the contribution of significance to the predictor variable

These results shown in Fig. 1, where the significance for the regressor variables were highly significant with p-value less than 2×10^{-16} allowed us to propose the multiple regression model of teacher salaries shown in Eq. 22.

$$Y = -3109.7302 - 5.2337(Edad) + 24.1794(NEduc) + 18.0115(TServ) \\ -90.1259(EsDoc) + 192.1820(HoLab) \quad (22)$$

This means that for each year of a teacher, the teacher's salary decreases by 5.2337 soles, likewise for each educational level the salary increases by 24.1794 soles, and for each year of service the salary increases by 18.0115 soles, however for each teaching scale the salary decreases by 90.1259 soles, and finally for each hour worked the salary increases by 192.1820 soles.

The graphs shown in Fig. 5 show the normalized residuals and fitted values, with residual quantiles and the root mean square error, for which the linear regression model meets all the specifications and prerequisites for prediction.

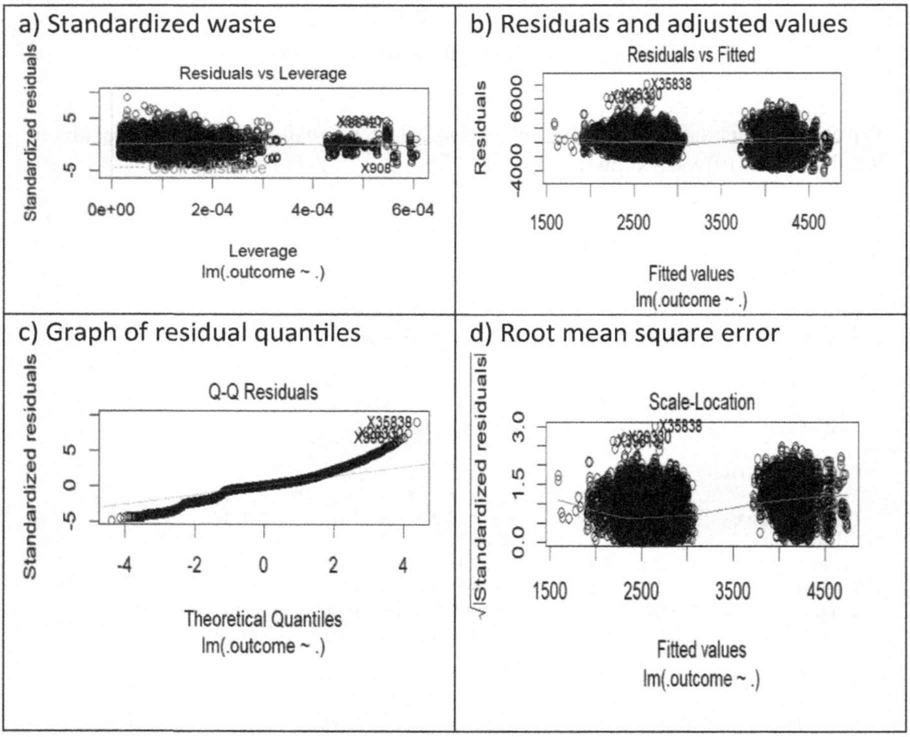

Fig. 5. Multiple linear regression model performance graphs

The results for the Ridge regression model, with a training of 86654 observations and 21663 observations for testing, found that the metrics of the resampling across fit parameters are as shown in Table 2.

Figure 6 shows the graphs for the Ridge regression model with cross-validation and the logarithms of the Lambdas penalty values, as well as the values of the model coefficients and the significance of the regressor variables.

Table 2. Ridge regression model performance metrics.

Lambda	RMSE	Rsquared	MAE
0.000100	896.5545	0.2866115	622.6167
0.250075	896.5545	0.2866115	622.6167
0.500050	896.5545	0.2866115	622.6167
0.750025	896.5545	0.2866115	622.6167
1.000000	896.5545	0.2866115	622.6167

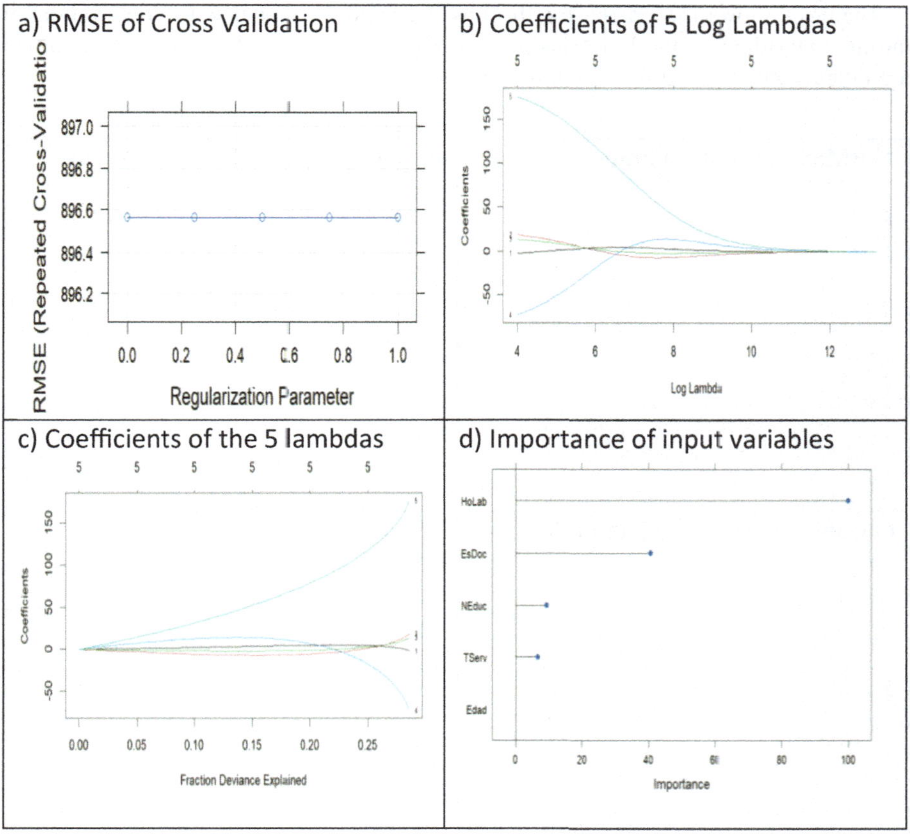

Fig. 6. Ridge regularization model plots.

The results for the Lasso regression model, with a training of 86654 observations and 21663 observations for testing, found that the metrics of the resampling through the fit parameters are as shown in Table 3.

Table 3. Lasso regression model performance metrics.

Lambda	RMSE	Rsquared	MAE
0.000100	895.3873	0.2872513	619.851
0.050075	895.3873	0.2872513	619.851
0.100050	895.3873	0.2872513	619.851
0.150025	895.3873	0.2872513	619.851
0.200000	895.3873	0.2872513	619.851

Figure 7 shows the graphs for the Lasso regression model with cross-validation and the logarithms of the Lambdas penalty values, as well as the values of the model coefficients and the significance of the regressor variables.

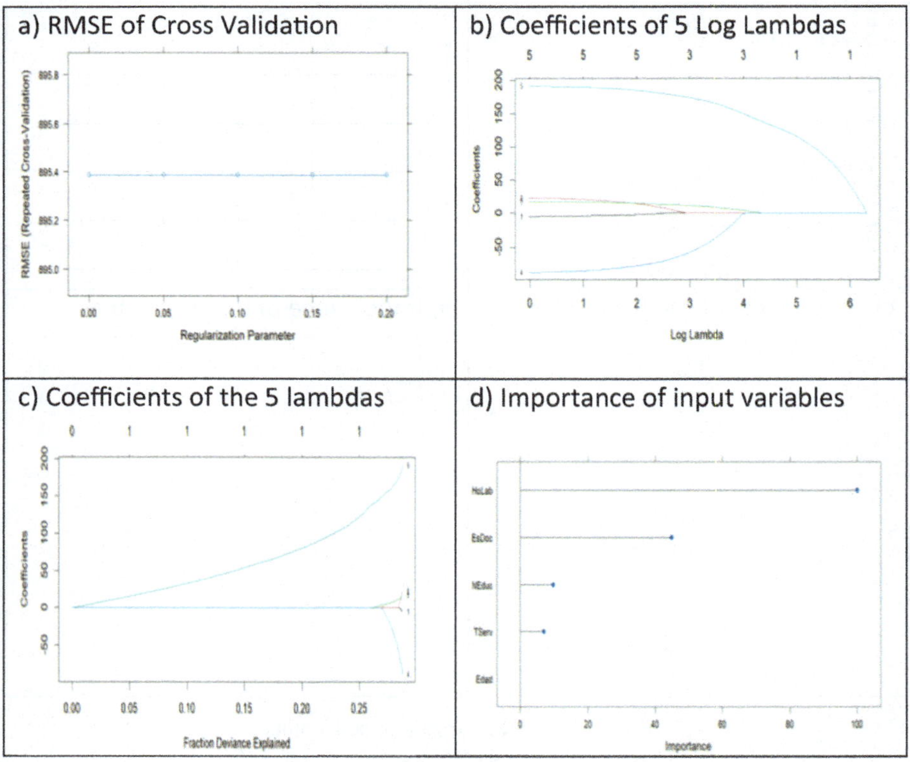

Fig. 7. Lasso regulatory model graphs.

The results for the Elastic Net regression model, with a training of 86654 observations and 21663 observations for testing, found that the resampling metrics MAE, RMSE and Rsquared through fit parameters are as shown in Table 4, 5 and 6 respectively.

Table 4. MAE metric for model resampling.

MAE	Min	Mean	Max
LinearModel	607.2400	619.7701	632.1245
Ridge	610.6784	622.6167	633.5756
Lasso	607.3899	619.8510	632.1050
ElacticNet	607.4004	619.8605	632.1108

Table 5. RMSE metric for model resampling.

RMSE	Min	Mean	Max
LinearModel	877.5731	895.3793	912.9543
Ridge	879.1204	896.5645	912.8855
Lasso	877.6042	895.3873	912.8773
ElacticNet	877.6047	895.3870	912.8733

Table 6. Rsquared metric for model resampling.

Rsquared	Min	Mean	Max
LinearModel	0.2558271	0.2872618	0.3196392
Ridge	0.2560430	0.2866115	0.3197616
Lasso	0.2559246	0.2872513	0.3196918
ElacticNet	0.2559196	0.2872524	0.3196939

Figure 8 shows the graphs for the Elastic net regression model with the cross-validation and the logarithms of the Lambdas penalty values, as well as the values of the model coefficients and the significance of the regressor variables.

The best model that allowed predicting the teacher salary is the thirteenth one with $\alpha = 0.5555556$ and a Lambda $\lambda = 0.2$ whose modeling is shown in Eq. 23.

$$Y = -3092.582975 - 4.824496(Edad) + 22.972778(NEduc) + 17.623234(TServ) - 88.511756(EsDoc) + 191.104877(HoLab) \quad (23)$$

This means that for each year of a teacher, the teacher's salary decreases by 5.2337 soles, likewise for each educational level the salary increases by 24.1794 soles, in addition for each year of service the salary increases by 18.0115 soles, however for each teaching scale the salary decreases by 90.1259 soles, finally for each hour worked the salary increases by 192.1820 soles.

Finally, the prediction with the best model shown in Eq. (2) with the training data obtained an RMSE of 895.3383 and with the test data an RMSE of 906.8914, being a

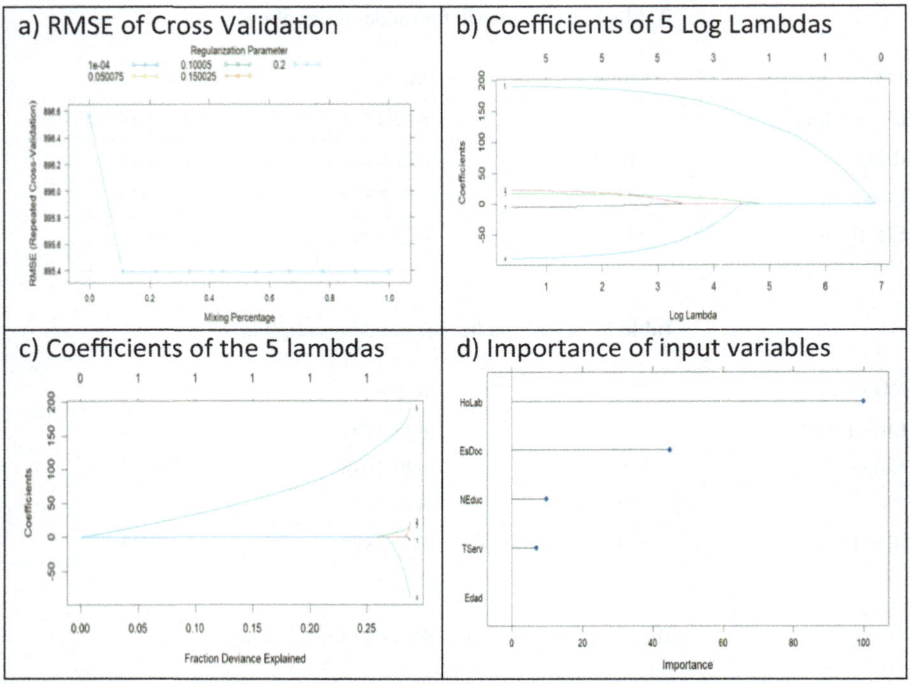

Fig. 8. Elastic Net regularization model graphs.

difference of 11.5531 that represents 1.2739% of error with a 98.7261% of efficiency in the prediction of the model.

4 Discussion and Conclusion

The analysis of Machine Learning models in the salary prediction of 108317 teachers with more recent data provided by the Ugel de Ventanilla, Lima Peru produced new results on the accuracy of the salary of appointed teachers. Before discussing the findings, it is important to point out that the study only analyzed only the salary of appointed teachers, but not the salary of hired teachers. Currently, there is no Machine Learning predictive model with high accuracy that allows the prediction of teacher salary, which is why in future studies the predictive model found helps to predict the salary based on regressor variables that are related to its predictor variable.

Secondly, the data analyzed with average teacher salary does not take into account the gender invariance of the teacher, so the amount of salary per month differs in the five years analyzed where the average variation is very marked for a more accurate comparison of the prediction. Future studies should examine the variation in salary for both appointees and contra-appointees to fully understand how the accuracy works in terms of the predictive machine learning models analyzed.

One of the key findings of the study is that a regular basic scale teacher has an average salary of S/. 2771.8 and is framed as a function of age, educational level, length of service, teaching scale and hours worked. The machine learning models linear regression, Ridge, Lasso and Elastic Net were accurate in finding the penalty factor of the model taking its best metrics such as the root mean square error (RMSE), the mean absolute error (MAE) and the coefficient of determination (R2) that allowed taking the best model that predicted the salary of the teachers appointed by the Ugel de Ventanilla.

The research [25] found a machine learning model for graduate salary as a function of field performance using a linear regression model with an accuracy of 95.68% for the training data and 95.33% for the test data, obtaining a very small difference affirming a good performance of the model. The research determines the teaching salary with a linear regression model with 5 regressor variables with a precision of 98.7261% and a determination coefficient $R^2 = 0.2072618$ which allows a deep analysis for other future research.

Research [26] predicted employee salary classes using machine learning, finding accuracy and correlation of feature extraction using a PCA System and a DNN deep neural network model with accuracy of 94.9%, in addition Decision Tree with 89.6% and Random Forest with 76.4% compared to the research linear regression accuracy of 98.7261% there is a difference of 3.8226% which is minimal.

The article [9] found regression models with association between teachers' base salary and mathematics performance in different districts, where the correlation coefficient between the logarithm of base salary and mathematics test re-scores was approximately 10.5 indicating that a 10% increase in teachers' base salary is associated with 1.05 higher average mathematics test scores. In contrast to the research model, an association was found between teacher salary and time of service, finding that for each year of service provided by the teacher, the salary increases by 18.0115 soles.

Dreher G. et al. [27] found a regression model of salary satisfaction as a function of company perception and maintenance variables such as years of service (β = 0.01), training (β = -0. 07), company valuation (β = 0.17), potential (β = 0.00), gender (β = 0.23), monthly salary (β = 0.29), and percentage of salary increase (β = 0.01), with a coefficient of determination R^2 of 12.00%. The research found a linear regression model of teacher salary as a function of the variables of age (β = -5.2337), Educational level (β = 24.1794), length of service (β = 18.0115), Teaching scale (β = -90.1259), and hours worked (β = 192.1820) with an R^2 coefficient of determination of 28.72618%.

Mohamed A. et al. [28] found the factors of remuneration of fresh engineering graduates in the job markets of Bharat, India with machine learning algorithms Naïve Bayes (RMSE = 0.308), Random Forest (RMSE = 0.306) and Support Vector Machine (RMSE = 0.305) identifying the factors like demographics, academic success, personality traits and score obtained in exams which were significant in starting salary. In the research, machine learning models were found lasso regression, Ridge and Elastic net with a lambda penalty λ = 0.5555556 and a value of α = 0.20 taking as regressors the teacher's age (β = -4. 824496), educational level (β = 22.972778),, length of service (β = 17.623234), teaching scale (β = 88.511756), and hours worked (β = 191.104877) with an RMSE for the best model of 895.3870.

Dutta et al. [29] found a prediction engine for the appropriate salary of a job using machine learning models such as decision tree (MSE = 389.64) and joint model (MSE = 329.12) with an accuracy of 0.844 and 0.873 respectively. The research found for the

Lasso (MAE = 619.8510), Ridge (MAE = 622.6167) and Elastic net (MAE = 619.8605) models under study.

Geraldo-Campos et al. [30] found Lasso and Ridge regularization models for the prediction of credit risk in reactiva peru with 501,298 companies analyzed, with economic sector, granting entity, amount covered and department as regressors and risk level as predictor, determining a Lasso regression model with $\lambda_{60} = 0.00038$ and an RMSE = 0.3573685; as well as a Ridge regression with $\lambda_{100} = 0.00910$ and an RMSE = 0.3573812, represented in Eq. 24 and 25 respectively.

$$\begin{aligned} Y_{Lasso} &= 0.51487 + 0.05878(Economicsector) \\ &-0.19292(Creditgrantingentity) \\ &+1.29671(Amountcovered) + 0.03115(Departament) \end{aligned} \tag{24}$$

$$\begin{aligned} Y_{Ridge} &= 0.51408 + 0.05849(Economicsector) \\ &-0.19071(Creditgrantingentity) + 1.27321(Amountcovered) \\ &+0.03198(Departament) \end{aligned} \tag{25}$$

In the research, the lasso model (RMSE = 895.3873), Ridge (RMSE = 896.5645) and Elastic net (RMSE = 895.3870) were found, where the best model found with a $\alpha = 0.5555556$ and a Lambda $\lambda_{13} = 0.2$ is represented in Eq. 26.

$$\begin{aligned} Y &= -3092.582975 - 4.824496(Edad) + 22.972778(NEduc) \\ &+17.623234(TServ) - 88.511756(EsDoc) + 191.104877(HoLab) \end{aligned} \tag{26}$$

In conclusion, the research obtained a new regularization model that allows predicting teacher salary based on regressive and predictor variables, which obtained efficiency in its metrics as shown in Table 7.

Table 7. Metrics of Machine Learning models.

	Linear Model	Ridge	Lasso	Elastic Net
MAE	619.7701	622.6167	619.8510	619.8605
RMSE	895.3793	896.5645	895.3673	895.3870
Rsquared	28.72618%	28.66115%	28.72513%	28.72524%

Likewise, the models obtained with their respective coefficients are shown in Table 8.

This means that for each year of a teacher, the teaching salary decreases by 4.824496 soles, likewise for each level of teacher education the salary increases by 22.972778 soles, in addition for each year of service provided the salary increases by 17.623234 soles, however For each teaching scale the salary decreases by 88.511756 soles, finally for each hour worked the salary increases by 191.18104877 soles, which, represented mathematically, are shown in Eqs. 27 and 28 respectively.

Multiple linear regression model

$$\begin{aligned} Y = -3109.7302 - 5.2337(Edad) + 24.1794(NEduc) + 18.0115(TServ) \\ -90.1259(EsDoc) + 192.1820(HoLab) \end{aligned} \tag{27}$$

Table 8. Coefficients of the linear regression model and Elastic Net.

Model coefficients	Lineal Model	Elastic Net
Intercept	-3109.7302	-3092.582975
Age	-5.2337	-4.824496
Educational level	24.1794	22.972778
Service time	18.0115	17.623234
Teaching Scale	-90.1259	-88.511756
Hours labored	192.1820	191.104877

Elastic net regression model

$$Y = -3092.582975 - 4.824496(Edad) + 22.972778(NEduc) + 17.623234(TServ) \\ -88.511756(EsDoc) + 191.104877(HoLab) \quad (28)$$

References

1. Vieira, E.C., Debenedetti, B., Reyna, L., Santisteban, S., Lorena, R., Aguirre, S.: Políticas remunerativas Grupo de Investigación ius et veritas
2. Zaid, M., Rajendran, T.: Higher classification accuracy of income class using decision tree algorithm over naive bayes algorithm. In: Advances in Parallel Computing, IOS Press BV, pp. 555–561 (2022). https://doi.org/10.3233/APC220079
3. Sarala, V., Ganesh, K.: Employee salary prediction system using machine learning
4. "6 DE CADA 10 DOCENTES DE ESCUELAS ESTATALES REPORTAN QUE ELLOS MISMOS SE PROVEEN DE MATERIALES PARA EL AULA 8 DE CADA 10 DOCENTES DEL ÁREA RURAL INDICAN QUE NO CUENTAN CON SALA DE PROFESORES."
5. "Ley de Reforma Magisterial LEY N° 29944 (*) De conformidad con la Única Disposición Complementaria Final del Decreto Supremo N° 227–2013-EF."
6. Siswanto, J.V., Castilani, L.A., Winata, N.H., Nugraha, N.C., Sagala, N.T.M.: Salary classification & prediction based on job field and location using ensemble methods. In: ICCoSITE 2023 - International Conference on Computer Science, Information Technology and Engineering: Digital Transformation Strategy in Facing the VUCA and TUNA Era, Institute of Electrical and Electronics Engineers Inc. pp. 325–330 (2023). https://doi.org/10.1109/ICCoSITE57641.2023.10127828
7. Akiba, M., Chiu, Y.L., Shimizu, K., Liang, G.: Teacher salary and national achievement: a cross-national analysis of 30 countries. Int. J. Educ. Res. **53**, 171–181 (2012). https://doi.org/10.1016/j.ijer.2012.03.007
8. Blackburn, M.L.: Are U.S. teacher salaries competitive? Accounting for geography and the retransformation bias in logarithmic regressions. Econ. Educ. Rev. **84** (2021) https://doi.org/10.1016/j.econedurev.2021.102169
9. García, E., Han, E.S.: Teachers' base salary and districts' academic performance: evidence from national data. Sage Open. **12**(1) (2022). https://doi.org/10.1177/21582440221082138

10. Matbouli, Y.T., Alghamdi, S.M.: Statistical machine learning regression models for salary prediction featuring economy wide activities and occupations. Information (Switzerland). **13**(10) (2022). https://doi.org/10.3390/info13100495
11. Sayan, D., Rupashri, B., Ayush, M.: Salary prediction using regression techniques. pp. 1–5 (2015)
12. Dong, X., Zhu, X., Hu, M., Bao, J.: A methodology for predicting ground delay program incidence through machine learning. Sustainability (Switzerland), **15**(8) (2023). https://doi.org/10.3390/su15086883
13. "10.2478_danb-2022-0017" (1)
14. Basysyar, F.M., Dwilestari, G.: House price prediction using exploratory data analysis and machine learning with feature selection. Acad. Trans. AI Mach. Learn. **1**(1), 11–21 (2022). https://doi.org/10.56578/ataiml010103
15. Emmert-Streib, F., Dehmer, M.: High-dimensional LASSO-based computational regression models: regularization, shrinkage, and selection. Mach. Learn. Knowl. Extr. **1** (1), 359–383, MDPI (2019). https://doi.org/10.3390/make1010021
16. Yan, X., Gang Su, X.: Linear regression analysis theory and computing
17. Gujarati, D.: Econometría. Quinta. **5** (1) McGRAW-HILL (2010)
18. James, G., Witten, D., Hastie, T., Tibshirani, R.: An Introduction to Statistical Learning. (2014). [Online]. Available: http://www.springer.com/series/417
19. Das, K., Das Chatterjee, N., Jana, D., Bhattacharya, R.K.: Application of land-use regression model with regularization algorithm to assess PM2.5 and PM10 concentration and health risk in Kolkata Metropolitan. Urban Clim. **49** (2023). https://doi.org/10.1016/j.uclim.2023.101473
20. Härdle, W.K., Hlávka, Z.: Multivariate Statistics (2015)
21. Montgomery douglas, peck elizabeth, and vining Geoffrey. Introduction to Linear Regression Analysis. Arizona State University (2012)
22. Shanker, M.: Effect of data standardization on neural network training. Int. J. Mgmt Sci **24**(4), 385–397 (1996)
23. Atlas, L., et al.: Performance comparisons between backpropagation networks and classification trees on three real-world applications
24. Dawam, S.R.M., Ku-Mahamud, K.R.: Reservoir water level forecasting using normalization and multiple regression. Indonesian Journal of Electrical Engineering and Computer Science. **14**(1), 443–449 (2019). https://doi.org/10.11591/ijeecs.v14.i1.pp443-449
25. Sarala, V., Ganesh, K.: Employee salary prediction system using machine learning. pp. 1–6 (2023)
26. Aminu, H., Imam Yau, B., Umar Zambuk, F., Ramsom Nanin, E., Abdullahi, A., Zahraddeen Yakubu, I.: Salary prediction model using principal component analysis and deep neural network algorithm. (2023). [Online]. Available: www.ijisrt.com
27. Dreher, G.F.: Predicting the salary satisfaction of exempt employees. Pers. Psychol. pp. 34 (1981)
28. Mohamed Saeed, A.K., Abdullah, P.Y., Tahir, A.T.: Salary prediction for computer engineering positions in india. J. Appl. Sci. Technol. Trends. **4**(01), pp. 13–18 (2023) https://doi.org/10.38094/jastt401140
29. Samanda, D., Airiddha, H., Kousik, D.: Design of a novel prediction engine for predicting suitable salary for a job. pp. 1–5, (2018)
30. Geraldo-Campos, L.A., Soria, J.J., Pando-Ezcurra, T.: Machine learning for credit risk in the reactive peru program: a comparison of the lasso and ridge regression models. Economies. **10**(8) (2022). https://doi.org/10.3390/economies10080188

Model Proposal for the Detection of Infected Potato Leaves Using Deep Learning

Hernán Marecos[1], Joaquin Delgado[1], Sebastian León[1], and Wilfredo Ticona[1,2](✉)

[1] Universidad ESAN, Lima, Perú
{21100534,21100165,21100310}@ue.edu.pe, wmamani@esan.edu.pe
[2] Universidad Tecnológica del Perú, Lima, Perú

Abstract. In Peru, the cultivation of potatoes holds significant economic importance, serving as a cornerstone of the agricultural sector. However, the potato crop faces persistent threats from various diseases and pests, impacting both yield and quality. Addressing this issue is crucial for both the economy and food security in agricultural communities. Therefore, the study aimed to find the best classifier of infected potato leaves in the Peruvian context. For this purpose, the dataset titled'Data for: Identification of Plant Leaf Diseases Using a 9-layer Deep Convolutional Neural Network' was chosen due to its relevance in the agricultural domain. This dataset comprises 39 different classes of plant leaf and background images, collected from various locations worldwide, totaling 61,486 images. Naturally, we applied preprocessing, which consisted of three phases: Preparation of Processing Tools, Pixel Values Normalization, Organization of Dataset. These phases were necessary to prepare the data and tools required for subsequent analysis, ensuring that the data are in an appropriate format and that the algorithms can work more efficiently and accurately. Finally, different Deep Learning models were implemented: InceptionV3, Resnet, and Vision Transformer. The results were evaluated according to the Accuracy, Precision, Recall and F1-score metrics. The best model resulted in Vision Transformer, whose metrics were superior to the others with 96.00% Accuracy, 96.18% Sensitivity, 96.00%, Precision and 95.94% F1-score.

Keywords: Deep Learning · Convolutional Neural Networks · Detection · Infected Potato Leaves · Vision Transformer · InceptioV3

1 Introduction

The potato, scientifically known as Solanum tuberosum, stands as one of the most crucial and widely consumed crops globally. Its nutritional value and culinary versatility make it an essential food for millions of people. Peru, known for its rich culinary heritage and geographical diversity, is a country that plays a prominent role in global potato production. It stands as one of the world's leading potato producers, cultivating approximately 4.5 million tons of potatoes per year [1], which represents a significant pillar in its agriculture and economy. With a diverse range of varieties cultivated in different regions and climates, the potato has become an essential part of the daily diet of Peruvians.

R. Silhavy and P. Silhavy (Eds.): CSOC 2024, LNNS 1120, pp. 553–562, 2024.
https://doi.org/10.1007/978-3-031-70518-2_47

The versatility of this tuber, alongside its cultural importance, makes it a fundamental resource for the country's food security and economy.

However, potato production in Peru faces a constant challenge: foliar diseases that affect the plant's leaves. These diseases have a significant impact on yields and the quality of harvested tubers. It is estimated that in Peru, up to 80% of economic losses are caused by these diseases [2], this not only poses a threat to the economy of agricultural communities but also to food security.

The issue lies in the delayed detection of infected leaves, leading to significant production losses. Currently, the detection of infected potato leaves is done manually, entailing high costs in terms of time, resources, and labor. Moreover, relying on farmers and agricultural experts for visual inspections of leaves to identify signs of disease results in delayed detection and diagnostic errors. This manual methodology is also limited in terms of scalability, as it's not feasible for monitoring large crop areas.

This study focuses on implementing deep learning models, such as Convolutional Neural Networks (CNNs), to develop a highly accurate detection system for infected potato leaves. The model's performance will be thoroughly evaluated using precision and sensitivity metrics to ensure its effectiveness in detecting this pest, which is crucial in safeguarding potato crops.

2 Related Work

In the work of Barón S., Portilla L., y Fernández A. [3], the authors presented a study on the application of Convolutional Neural Networks (CNNs) for late blight detection. The study utilized an augmented dataset from PlantVillage, comprising images of healthy and late blight-affected potato leaves. Several CNN models, including AlexNet, ZFNet, VGG16, and VGG19, were trained to classify images and detect the disease. The models were evaluated using performance metrics like accuracy, sensitivity, F1 score, and precision. Results showed that the VGG16 model achieved the highest accuracy at 96%, followed by AlexNet at 94%. VGG16 also demonstrated a sensitivity of 92%, indicating its efficiency in detecting samples with the disease.

The study, conducted by Edmundo J. [4], focuses on researching the use of unmanned aerial vehicles for pest detection. The initial phase involves adapting the drone, comprising a flight structure and a camera system, allowing for the capture of aerial images. Subsequently, flight route planning was conducted to capture images limited to the olive cultivation area under evaluation. Next, image processing was undertaken using various processing algorithms for the detection and description of key points, including SIFT, SURF, BRISK, and FREAK. Finally, the detection of the Orthezia Olivícola pest in the processed images was executed. Additionally, a comparison was made regarding the total time in milliseconds taken by each algorithm: SIFT (152.45), SURF (432.14), BRISK (15.67), and FREAK (57.49). The results indicate that the SIFT algorithm showed the best relationship considering the total time and the detection of points of interest, facilitating the detection of the olive pest.

The study, carried out by Ishengoma F. [5], it focuses on the importance of precision agriculture technologies to ensure a stable food supply. The automatic detection of corn leaves infected by corn armyworms is achieved using convolutional neural network-based recognition algorithms such as VGG16, VGG19, InceptionV3, and MobileNetV2.

These algorithms were applied to images captured by unmanned aerial vehicles (UAVs) and further enhanced through Shi-Tomasi corner detection. The results demonstrate that these CNN models significantly outperform previous proposals in terms of accuracy. Particularly, the performance improved significantly when using modified images, achieving precision rates close to 100%. This approach holds promise for increasing farmers' yield and profits while reducing input costs and time.

The article "Multi-Level Deep Learning Model for Potato Leaf Disease Recognition" [6]. It's a study on potato leaf disease detection based on Convolutional Neural Networks (CNN). The performance of a model called PDDCNN was evaluated in two configurations: one with data augmentation and one without. Without data augmentation, the PDDCNN model achieved an average accuracy of 93.90% in detecting three classes of potato leaves: early blight, healthy, and late blight. The model also showed accuracies of 87%, 94%, and 90% in detecting these classes, respectively. The performance was evaluated on a dataset cross-referenced between PlantVillage and PLD, demonstrating that the environment and regional varieties can affect the model's accuracy. In a PLD dataset, the PDDCNN model achieved an accuracy of 86.38Compared to other deep learning models and previous studies, the PDDCNN outperformed several of them, achieving an accuracy of 99.75% with fewer trainable parameters and higher computational efficiency. In summary, the PDDCNN model excelled in detecting diseases in potato leaves, proving to be effective and efficient, especially when data augmentation was used during training.

In the following research conducted by Ishengoma F., Idris A., and Rutabayiro S. [7], various detection techniques are proposed to improve the speed and accuracy of models for pest detection in corn. A hybrid model of Convolutional Neural Networks (HCNN) is suggested for the swift detection of fall armyworm in corn crops. This system involves the use of UAVs for capturing photos of corn leaves and a hybrid CNN model based on a parallel structure that benefits from the advantages of VGG16 and InceptionV3 models. Additionally, the certainty performances of other CNN models: VGG16 (91.0%), InceptionV3 (92.13%), XceptionNet (90.26%), and Resnet50 (89.8%) are compared. The results demonstrate that the hybrid VGG16- InceptionV3 model achieves the highest certainty (96.98%).

The latest research conducted by Ishengoma F., Idris A., and Gatare I. [8], aimed to implement an autonomous system to locate the fall armyworm pest in corn crops. This system utilizes images captured by unmanned aerial vehicles (UAVs) and machine learning techniques to accurately detect infected corn plants and the affected area. This enables farmers to make informed decisions based on the location of the infected plants on their farms. The system performs three tasks: firstly, it crops the UAV photos into smaller images while saving the coordinates; secondly, it extracts the coordinates of infested corn plants and counts the number of infected images per UAV photo, indicating the percentage of infection and size; finally, it sends a report to the farmer specifying the area covered by the infection in each photo. This report enables farmers to take prompt action by spraying pesticides in the affected areas, thereby saving them money and time.

3 Methodology

The proposed methodology was based on [7, 8, 14, 16]. It consists of four phases: Dataset Selection, Dataset Augmentation, Model Implementation, and Evaluation. Each stage involves general activities that are crucial for the research development. The methodology used is depicted graphically below (Fig. 1):

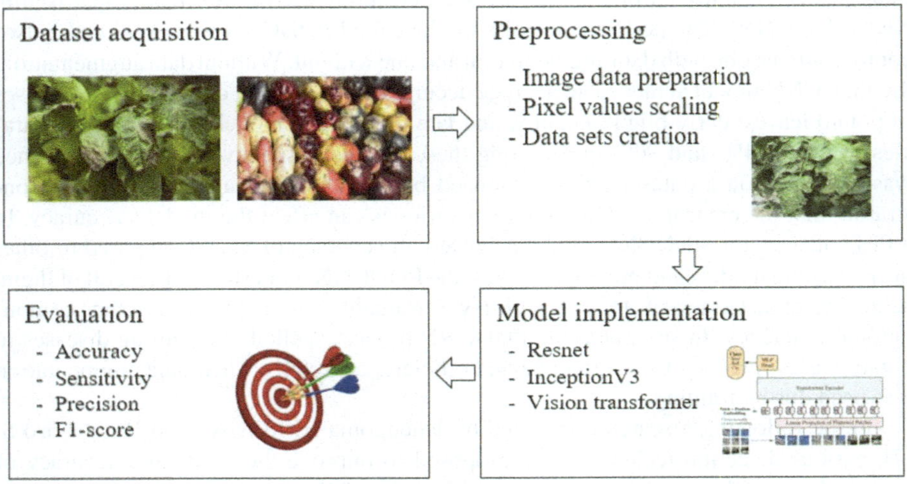

Fig. 1. Proposed Methodology

3.1 Dataset Acquisition

In the present investigation, the dataset entitled "Data for: Identification of Plant Leaf Diseases Using a 9-layer Deep Convolutional Neural Network" [9] was selected for its pertinence within the agricultural domain. Considering agriculture's pivotal role in ensuring global food security, directing attention to a prominent crop such as the potato (Solanum tuberosum) yields valuable perspectives on the adversities encountered by farmers. This dataset comprises 39 different classes of plant leaf and background images, totaling 61,486 images. Six data augmentation techniques were applied, including image flipping, gamma correction, noise injection, color augmentation via PCA, rotation, and scaling. For this research, three specific classes have been chosen:

- Potato early blight
- Potato healthy
- Potato late blight

This choice is based on agricultural relevance, considering the crucial role agriculture plays in global food security. Focusing on such a significant crop as the potato (Solanum tuberosum) provides valuable insights into the challenges faced by farmers.

3.2 Preprocessing

In this phase, three fundamental steps are followed. Table 1 displays the activities of the preprocessing phase.

Table 1. Preprocessing activities

Activities	Description	Task
Preparation of specific tools	Setup and arrangement of various Software or tools specifically designed for handling and manipulating images	Tool Setup
		ImageDataGenerator configuration
		Functionality customization
Pixel value normalization	This process adjusts the pixel values to a suitable range	Scaling pixel values
		Parameter configuration
Organization of datasets	Structures and prepares datasets specifically tailored for the training and validation	Dataset preparation
		Data flow generator configuration

- ***Preparation of Processing Tools:*** Specific tools are prepared to handle the images. This involves setting up data generators using ImageDataGenerator to ease image manipulation and processing.
- ***Pixel Values Normalization:*** An essential step involves scaling the pixel values of the images. This process adjusts the pixel values to be within a suitable range, typically between 0 and 1. This is achieved by setting the parameter rescale = 1/255 in the data generators. This normalization aids the model in learning more efficiently during training.
- ***Organization of Dataset:*** Specific datasets are prepared for training and validation. These datasets are created using data flow generators, responsible for loading and organizing images from specific directories. For each set (training and validation), the desired image size (target size), batch size (batch size), and class classification type (class mode = "categorical") are set.

3.3 Model Implementation

The three selected models will be trained with the chosen dataset. The chosen models are as follows:

- ***Resnet:*** A pivotal model in the realm of convolutional neural networks (CNNs), ResNet introduces the concept of residual learning via skip connections. These connections allow the network to skip certain layers by adding the original input to the output of these layers, easing the training of extremely deep networks. ResNet's architecture comes in various depths, such as ResNet-18, ResNet-50, and even deeper variants, facilitating the capture of intricate patterns in images. ResNet's innovative design tackles the vanishing gradient problem, boosting its efficiency in image recognition, object detection, and classification tasks while enabling the training of significantly deeper neural networks compared to previous architectures. Its success has influenced the development of subsequent CNN designs by emphasizing the importance of skip connections in enhancing network training and performance [10] (Fig. 2).

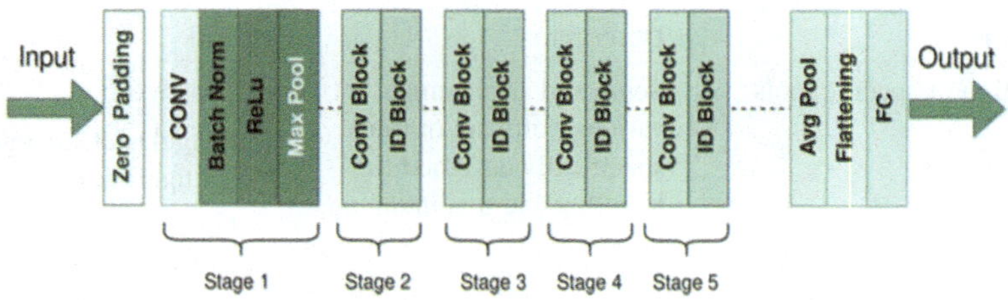

Fig. 2. ResNet5o model architecture

- ***InceptionV3:*** This model is part of the'Inception' series developed by Google. InceptionV3 utilizes'Inception' modules that combine different sizes of convolutional filters in parallel to capture features at various spatial scales. It has a deeper and more sophisticated structure than its predecessor, InceptionV1. This network has shown great effectiveness in image classification tasks and object detection, balancing accuracy with computational load [11] (Fig. 3).

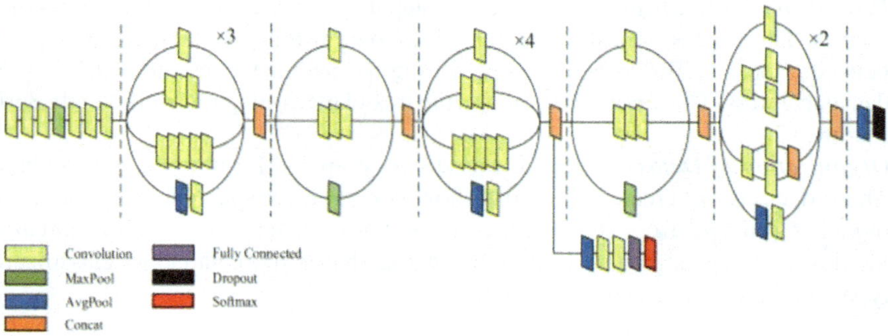

Fig. 3. The architecture of Inception-V3 model.

- ***Vision Transformer (ViT):*** This model represents a significant breakthrough in the field of computer vision by combining elements of convolutional neural networks (CNNs) with the Transformer architecture, originally developed for natural language processing. Unlike traditional CNNs that rely on convolutions to extract local features, ViT revolutionizes this approach by splitting the image into patches and processing them through attention operations in Transformer blocks [12] (Fig. 4).

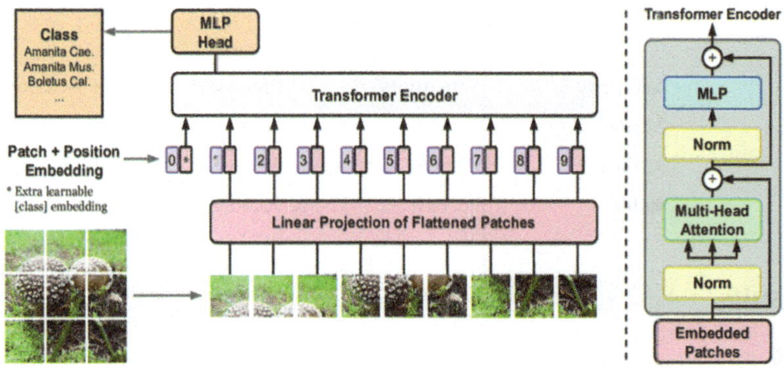

Fig. 4. Vision Transformer architecture.

3.4 Evaluation

The classification models will be evaluated using the parameters of Accuracy, Sensitivity, Specificity, Precision, and F1-score:

- ***Accuracy:*** Refers to how close a sample statistic is to a population parameter. It's the sum of correct predictions divided by all predictions [13].

$$Accuracy = \frac{TP + TN}{TP + FP + FN + TN} \quad (1)$$

- ***Sensitivity:*** It's the ratio of true positives divided by the sum of true positives plus the number of false negatives. It expresses the ability to identify all relevant instances in a dataset [7].

$$Sensitivity = \frac{TN}{TP + FN} \quad (2)$$

- ***Precision:*** It's defined as the number of true positives divided by the sum of true positives plus the number of false positives [13].

$$Precision = \frac{TP}{TP + FP} \quad (3)$$

- ***F1-Score:*** Balance between Precision and Sensitivity. It provides a better measure of incorrectly classified cases than the Precision metric [15].

$$F1 = 2 \times \frac{Precision \times Recall}{Precision + Recall} \quad (4)$$

where:

- TP: TruePositives
- FP: FalsePositives
- TN: TrueNegatives
- FN: FalseNegatives

Table 2. Model results.

Model	Accuracy	Sensitivity	Precision	F1-Score
Inception	74.66%	74.66%	74.67%	74.67%
Resnet	92.66%	92.66%	93.39%	92.61%
ViT	**96.00%**	96.18%	96.00%	95.94%

4 Results

Below is a summary table of the models with their parameters and the results of the metrics (Table 2):

Accuracy, a metric indicating the proportion of correct predictions made by a model relative to the total predictions, serves as a gauge of overall precision. The results showcase ViT achieving the highest accuracy at an impressive 96.00%, followed by ResNet at 92.66%, and finally Inception at 74.66%.

Sensitivity, also known as recall, focuses on a model's ability to correctly identify real positive cases. It measures the proportion of true positive cases correctly identified by the model. In this study, all models exhibit quite solid sensitivity, with ViT leading again at 96.18%, followed by ResNet at 92.66%, and Inception at 74.66%.

Precision, on the other hand, zeroes in on the proportion of cases identified as positive that are positive. This metric is particularly important in scenarios where correctly identifying positive cases is crucial. The findings demonstrate ResNet demonstrating the highest precision at 93.39%, followed by ViT at 96.00%, and Inception at 74.67%.

Lastly, the F1-score combines precision and sensitivity into a single metric, providing a more balanced measure of model performance. This metric is especially useful when there's an imbalance between prediction classes. In this study, the ViT model garnered the highest F1-score at 95.94%, closely followed by ResNet at 92.61%, and Inception at 74.67%.

5 Conclusions

This current study integrates various methodologies and techniques from prior research to address a persistent issue in Peru: foliar diseases impacting plant leaves. Over the years, a substantial portion of the Peruvian population has grappled with this problem. A comprehensive preprocessing of data was undertaken, focusing on preparing data for training Deep Learning models, and this meticulous approach is evident in the model results.

Different Deep Learning models were implemented: InceptionV3, Resnet, and Vision Transformer. The results were evaluated according to the Accuracy, Precision, Recall and F1-score metrics. The best model resulted in Vision Transformer, whose metrics were superior to the others with 96.00% Accuracy, 96.18% Sensitivity, 96.00%, Precision and 95.94% F1-score. In conclusion, it is suggested that an increased quantity and diversity of images could enhance results. Although the models can already distinguish infected potato leaves, their robustness could be further improved with additional training data. Further-more, this research provides an initial insight for future endeavors that may use it as a reference and adopt a similar methodology for the detection of infected potato leaves.

References

1. José Carlos León Carrasco: Más de 700 mil familias cultivan papa en Perú. Agencia Agraria de Noticias (2017)
2. Ministerio de Agricultura y Riego: Informando sobre el desarrollo agrario y la seguridad alimentaria. SENASA (2016)
3. Suarez-Barón, M.J., Lozada-Portilla, W.A., Avendaño-Fernández, E.: Aplicación de redes neuronales convolucionales para la detección del tizón tardío phytophthora infestans en papa solanum tuberosum. Revista U.D.C.A Actualidad Divulgación Científica **24**(2) (2021)
4. Torres, J.E.G., Condori, G.E.A.: Identificación de plagas en el cultivo de olivo utilizando percepción remota con un vehículo aéreo no tripulado en el valle la yarada los palos. Ingeniería Investiga **1**, 81 (2019)
5. Ishengoma, F.S., Rai, I.A., Said, R.N.: Identification of maize leaves infected by fall armyworms using uav-based imagery and convolutional neural networks. Comput. Electron. Agric. **184**, 106124 (2021). https //doi.org/10.1016/j.compag.2021.106124
6. Rashid, J., Khan, I., Ali, G., Almotiri, S.H., Alghamdi, M.A., Masood, K.: Multi-level deep learning model for potato leaf disease recognition. Electronics (Switzerland) **10**(17), 2064 (2021). https://doi.org/10.3390/electronics10172064
7. Ishengoma, F.S., Rai, I.A., Ngoga, S.R.: Hybrid convolution neural network model for a quicker detection of infested maize plants with fall army- worms using UAV-based images. Ecol. Inform. **67**, 101502 (2022). https://doi.org/10.1016/j.ecoinf.2021.101502
8. Ishengoma, F.S., Rai, I.A., Gatare, I.: Autonomous system for locating the maize plant infected by fall armyworm. In: Lecture Notes in Networks and Systems, 724 LNNS, pp. 106–113 (2023). https://doi.org/10.1007/978-3-031-35314-7_10
9. Geetharamani, G., Arun Pandian, J.: Identification of plant leaf diseases using a nine-layer deep convolutional neural network. Comput. Electr. Eng. **76**, 323–338 (2019). https://doi.org/10.1016/j.compeleceng.2019.04.011
10. Jha, P., Dembla, D., Dubey, W.: Implementation of transfer learning based ensemble model using image processing for detection of potato and bell pepper leaf diseases. Int. J. Intell. Syst. Appl. in Eng. **12**(8s), 69–80 (2024)
11. Ahad, M.T., Li, Y., Song, B., Bhuiyan, T.: Comparison of CNN-based deep learning architectures for rice diseases classification. Artif. Intell. Agric. **9**, 22–35 (2023). https://doi.org/10.1016/j.aiia.2023.07.001
12. Gole P., Bedi P., Marwaha S.: Automatic diagnosis of plant diseases via triple attention embedded vision transformer model. In: Lecture Notes in Networks and Systems, 731 LNNS, pp. 879–889 (2024). https://doi.org/10.1007/978-981-99-4071-4_67
13. Menze, B.H., Kelm, B.M., Masuch, R., Himmelreich, U., Bachert, P.: Petrich W (2009) Comparison of random forest and its Gini importance with standard chemometric methods for the feature selection and classification of spectral data. BMC Bioinform. **10**, 213 (2009)
14. Khalifa, N.E.M., Taha, M.H.N., Abou El-Maged, L.M., Hassanien, A.E.: Artificial intelligence in potato leaf disease classification: a deep learning approach. In: Hassanien, A.E., Darwish, A. (eds.) Machine Learning and Big Data Analytics Paradigms: Analysis, Applications and Challenges. Studies in Big Data, vol. 77. Springer, Cham (2021)

15. Tiwari, D., Ashish, M., Gangwar, N., Sharma, A., Patel, S., Bhardwaj, S.: Potato leaf diseases detection using deep learning. In: 2020 4th International Conference on Intelligent Computing and Control Systems (ICICCS), pp. 461–466 (2020)
16. Wang, H., Li, G., Ma, Z., Li, X.: Application of neural networks to image recognition of plant diseases. In: 2012 International conference on systems and informatics (ICSAI2012), pp. 2159–2164 (2012)

Software Patterns for Creating Computer Simulations

Yuriy Sobolev, Roman Kreinin, Tatiana Kudinova(✉), and Farida Nanay

MIREA—Russian Technological University, Moscow 119454, Russia
Kudinova@mirea.ru

Abstract. This article provides an overview of Object-oriented, Entity Component System and Functional Reactive Programming software patterns for implementing real-time simulations with numerous interacting components. Three former patterns are used to implement a simple scene in Unity Engine, a comparison of flexibility and applicability is made #CSOC1120.

Keywords: Real-Time Simulations · Entity Component System · Functional Reactive Programming · Monadic Stream Functions

1 Introduction

Real-time simulation is an application of a computer for running a virtual physical system at the same rate as in the real world. Real-time simulations are widely used for computer gaming, engineering, construction, statistical analysis. This problem typically involves interaction of numerous stateful agents processed in discrete steps, which drastically differs from a common set of problems solved using a computer that typically reduce to synchronous computations and a few agents.

Object-oriented approach, in which a scene is modelled as a set of objects, where each object contains proper state and update method, and interaction between objects is performed via direct mutable update of others' objects data. It's a straightforward approach that naturally fits into most imperative, object-oriented languages. It utilizes the principle of hierarchical organization, so agents are able to extend the functionality of other simpler agents. Object-oriented model is being adopted by numerous mainstream simulation platforms such as Godot, Unity Engine, Unreal Engine, Flax Engine. But this approach has disadvantages and might not be suitable for all use cases.

Entity Component System is a software pattern in which data and update method are strictly separated. In ECS, each entity is a composition of components. Components are mere data container that possesses no logic. On each simulation step, a set of methods called systems are being run. Each system traverses a subset of entities, loads data from their components, performs necessary computations and saves the state back to the components. ECS is not as widely adopted as object-oriented approach and is more complicated, but is more efficient in terms of parallelizability and utilization of processor cache. ECS also localizes interactions between agents of the system, easing debugging and reasoning about the model.

R. Silhavy and P. Silhavy (Eds.): CSOC 2024, LNNS 1120, pp. 563–573, 2024.
https://doi.org/10.1007/978-3-031-70518-2_48

Functional Reactive Programming is an implementation of reactive programming by using primitives from functional programming. FRP uses functions as building blocks and composes simple functions to create complex behaviour. In FRP, each function may possess internal state, preserved over discrete steps. In this representation, the entirety of real-time simulation with all behaviours and state is stored as a part of a modular functional reactive computation. The state of agents is encapsulated by the functions that operate on them, and interaction between agents is performed by directly passing data between functions (hence reactive programming is considered a form of dataflow programming). FRP is not a specific pattern, rather a collection of patterns with these common principles. In this paper, Monadic Stream Functions implementation of FRP implementation is being used, which introduces monad-aware MSF data structure as the base building block of the system and defines rules for composing these blocks to create complicated models. The major disadvantage of FRP is that it typically requires some level of support from the used programming language in order to be practical.

2 Methods

2.1 Theoretical Analysis

Object-Oriented Pattern. In object-oriented patterns, the simulation is represented as a set of object, each object attached to multiple components. Components are typically represented as classes in Object-Oriented Programming and contain both behaviour of the agent and data related to it. On each step of the simulation, all components of all agents in the scene are traversed and updated by calling the method associated with them (Fig. 1).

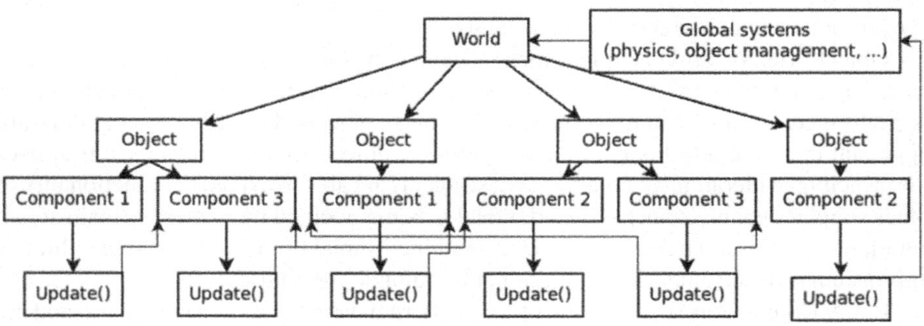

Fig. 1. Simplified example of simulation structure in object-oriented pattern.

Object-oriented pattern encourages the programmer to describe common logic in self-contained components and then attach these components to any number of objects.

This approach permits incremental development of new simulations by adding more agents and attaching existing components alongside with new custom components to them, with minimum maintenance work required. Therefore, object-oriented approach is applicable for quick prototyping, but large systems with numerous components are difficult to manage due to distributed simulation step logic and implicit relations between objects.

Although all components of all agents could be processed in parallel, parallelizability of object-oriented systems is limited in practice. This happens due to implicit relations between objects and shared mutable locations, which potentially leads to race conditions if synchronization is not managed correctly.

Interaction between different agents is usually implemented as direct communication between components, in which a component of some object A interacts with a component of some object B through an interface provided by the component of object B.

Entity Component System Pattern. Entity Component System is a pattern that consists of three main concepts.

Entities are abstract objects that hold no real data or behaviour. Each entity represents a single agent existing in the simulation and is associated with multiple components.

Components are data containers for entities. They are used to store the state of the simulation. As an example, components could be used to store position, rotation, velocity, mass of a physical object. Each component type typically represents a single trait of the associated entity, and each entity is associated only with applicable components.

Systems are processes run on each discrete step of the simulation. Each system requests entities with some set of components, performs required calculations for the step, and stores the result of the steps back in the components. Systems are not restricted in the amount of types of components they request at each step, but it's recommended to keep it at minimum for greater performance and organization (Fig. 2).

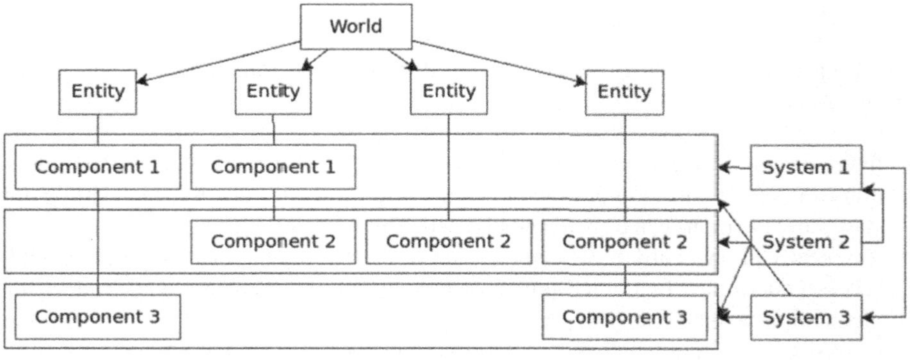

Fig. 2. Simplified example of simulation structure in entity component system pattern.

Extending an existing simulation likely requires the programmer to create both new components and systems, as well as some maintenance work on a global level (e.g. registering new systems and components, creating new global event queues, instantiation of new entities). Therefore, ECS is usually less useful for quick prototyping than object-oriented design.

While the object-oriented approach distributes step logic across all agents, ECS centralizes the logic in a few systems. This approach is practical at handling large amounts of entities with the same behaviour, but is not as elegant when numerous different behaviours are required from different entities, forcing the programmer into applying the strategy pattern (by selecting how the entity should be handled based on the data its components hold) or splitting the processing into many independent systems.

A major advantage of systems is their parallelizability. Each system is able to process all entities in parallel without threat of colliding with systems that operate over other components. It is also notable that components of the same type in ECS are located in a single contiguous memory location, so accessing component data by the system likely results in a CPU cache hit. Systems are also able to make use of SIMD operations to perform some operation over multiple components at a time.

In ECS, each system is only allowed to write the data to components it requested or global variables. Interaction between different systems can be performed via global event queues or shared components.

Functional Reactive Programming Pattern. Exist numerous approaches to functional reactive programming that differ in approach to external event processing and state preservation.

Approaches to external event processing:

- Push-based. Push-based systems eagerly push all external inputs to the network.
- Pull-based. Pull-based systems do not immediately react to external inputs. Instead, systems of these type wait for the result to be requested and only then perform the computations required. This type is applicable in models with a numerous observable outputs that are mostly independent.
- Hybrid push-pull approach. Systems of this type combine features of both push-based and pull-based systems.

In our research, we also found useful to distinguish two main approaches to state preservation:

- Internal state management. In this approach, each stateful part of the FRP computation is assigned with proper mutable location to preserve state. On each simulation step, these locations can be read and updated. This method is easier to implement, and it typically allows burdenless composition of FRP computations, but such mutable access undermines functional purity of the system and therefore may lead to further complications. For example, aborting such computation possibly leaves it in an inconsistent state, in which only a part of the network's state was updated.
- External state management. In this approach, the preservation of the state is performed by an external system, while the parts of the FRP network only have a restricted interface to this system. Some ways of implementing such interface include using State algebraic effect or mere modelling of stateful parts of the computation as functions that accept old state as part of the input and return new state as part of the output. External state management enables sophisticated manipulations with simulation state presented later, but requires the programmer to maintain the interface.

This article will focus on the Monadic Stream Functions implementation of FRP. This implementation is push-based and employs external state management.

The core concept of Monadic Stream Functions is MSF data structure. In Haskell programming language, it is defined as:

$$newtype\ MSF\ m\ a\ b = MSF\ \{unMSF :: a- >m(b, MSF\ a\ b)\}$$

In simple words, this definition defines "MSF m a b" computation that accepts value of type "a" as input. It operates in context of monad m and returns value of type "b" as well as a new "MSF m a b" that should be used for subsequent simulation steps. The state of the FRP network is entirely stored within the returned MSF object.

This definition allows MSF object to implement Arrow type class that defines how MSF object over the same monad can be composed.

">>>" (inherited from Category type class) allows composing "MSF m a b" with "MSF m b c", yielding "MSF m a c". This is analogous to sequential computation.

"***" allows composing "MSF m a b" and "MSF m c d", yielding "MSF m (a, c) (b, d)". This is analogous to parallel computation. It is important to note that the term "parallel" in this case only refers to the fact that two inputs "a" and "b" are processed by independent computations, whether both computations are performed in parallel on multi-core system depends on the implementation and the monad used (some monads such as State may inhibit parallelism, because performing the second computation requires access to the state produced by the first one).

This allows to express the simulation as a sequence of steps, such as:

$$getUserInput\ >>> (updateObjectPositions * * * updateGUI) >>> render$$

Visualisation of such composition is presented in Fig. 3.

Each step can also be a composition of other steps.

Monadic Stream Functions are monad-aware. This enables use of monads to simplify implementation of complicated patterns. Some examples of usable monads include:

- Reader monad. Allows the computation to access read-only environmental variable and override it locally. This monad could be used to store immutable data about the simulation step, such as current simulation time.
- Writer monad. Allows the computation to output messages that can be captured by the outer computation. This monad could be used to store debug log or listen to the events that happened in the inner computation and react to them.
- State monad. Allows to emulate mutable global state.
- List monad. Allows to express numerous independent computations, where each computation is allowed to halt or fork. This monad could be used to process independent agents of the simulation or maintain multiple simulation scenes, each generated by some choice in the past.
- Maybe monad. Allows to express a computation that potentially stops yielding results at some point. This monad could be used to express temporary environmental factors. Once the Maybe computation has finished, it is not possible to continue it, but it is possible to switch it to a new computation.
- Either/Except monad. Allows to express parts of the network that can fail irrecoverably at some point or switch itself to a new computation.

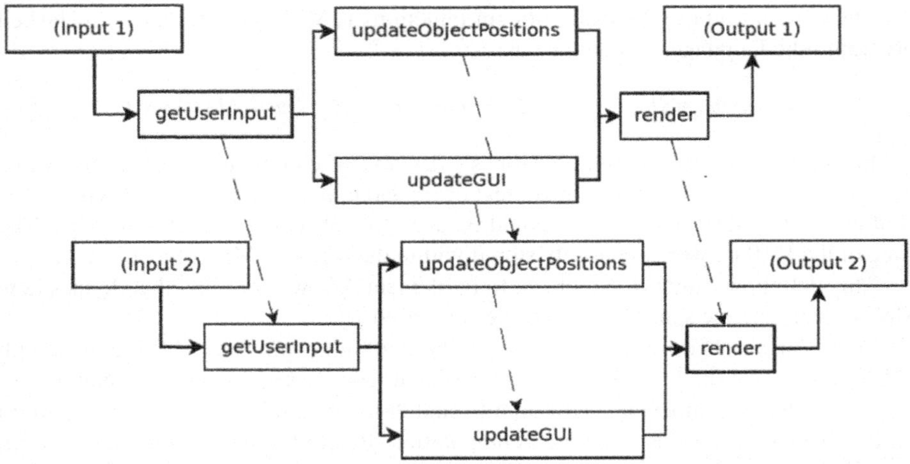

Fig. 3. Visualisation of a simple MSF computation. On each step, the computation accepts some input and returns output alongside with a new MSF computation to execute on the next step.

All side effects of MSF computations are typically represented with monads.

While object-oriented and ECS-based simulations require a global system that manages agents and their components, in MSF-based simulations agents and components are encoded as part of the MSF computation. New state and behaviour can be introduced by the programmer at any place of the computation, making MSF highly-extensible.

MSF are based on lambda calculus and therefore have a potential for massive parallelism in theory. Parallelizing MSF-based simulation reduces to parallelizing an underlying monadic computation, which can be proven challenging in practice, especially if complex monadic stacks are present: some monads (such as State) could inhibit parallelism, while some (such as List monad) could be beneficial for it.

In MSF, interaction between agents is mainly implemented by passing outputs of previous computation steps as inputs to subsequent computation steps via ">>>" and related operators. Inter-agent communication could also be carried by the monadic context. Manually composing MSF computations in systems with large amount of interactions can quickly become a burden and can be simplified by using Arrow notation [1].

2.2 Implementation of Sample Simulation

For our comparison, we will implement a simple simulation in these three models in Unity Engine. The object of our comparison is ease of use, maturity, and flexibility (or extensibility), parallelism of the models. In this paper, we do not focus on making three implementations of the simulation strictly equivalent and the result of the simulations may differ.

We use a physical simulation involving multiple colliding spheres on a horizontal plane. On each new collision, each sphere may change behaviour.

Sphere 1: Not affected by the collision.
Sphere 2: Creates a new sphere.
Sphere 3: Changes visible colour.
Sphere 4: Changes mass.
Sphere 5: Changes world's gravity.
Sphere 6: Increases restitution on first hit, ignores subsequent collisions.

Expected results:

- *Object-oriented pattern* is expected to be the simplest to prototype and implement. Due to the fact that Unity Engine mostly executes components in single thread (called main thread) [2], this solution cannot benefit from multi-threading environment.
- *Entity Component System pattern* is expected to be similar to object-oriented pattern in terms of ease of use and some parts of the implementation, but be more suitable for multi-threading environment.
- *Functional Reactive Programming pattern* is not native to Unity Engine and is expected to require more work to implement from the ground up. The result simulation is expected to be parallelizable, but this could require extra work and would not arise naturally. We expect this implementation to be the most flexible and have the ability to be easily extended by new functionality on the level of concrete agents or entire simulation world.

Implementation of Sample Simulation can be accessed and downloaded by everyone on Github page [3].

Object-Oriented Pattern. For this pattern, we implement OnHitBehaviourChange base abstract class, which extends MonoBehaviour (base class for all components in Unity Engine).

OnHitBehaviourChange class exposes OnCollisionEnter method that is executed by Unity Engine each time collision occurs.

To describe actual spheres, we introduce several more components that extend OnHitBehaviourChange, overriding CollisionAction method to specify behaviour on collision. Each such component either modifies other components of the same object, or accesses Unity Engine's global API to mutate the world.

While quick to prototype, this implementation heavily relies on the internal structure of the engine and implicit connections between components. For example, Rigidbody component, which is responsible for mass and impulse of bodies in Unity Engine, is tightly connected to Collider component [4]. While concepts of mass and impulse are not applicable for all simulations and, therefore, Ridigbody may not be sensible in such simulations, lack of it disrupts built-in collision detection. Such design problems are not direct consequence of object-oriented design, but implicit relations between objects, endorsed by object-oriented pattern, may be one of the contributing factors to such limitations.

Entity Component System Pattern. Detection of collisions through separate methods, employed in Object-oriented implementation of the simulation, is not native to Entity Component System design. Instead, a single system is used to handle all collisions occurring in the world. It is also required to manually detect ongoing collisions and only fire changes of behaviour if a new collision occurs.

For this implementation, we create CollisionsComponent that is attached to all spheres and stores a list of ongoing and new collisions, and CollisionsSystem that is executed for all collisions in the world and maintains correct values of CollisionsComponent. All different types of sphere are handled by different systems, that are able to detect collisions through accessing CollisionsComponent. This way, inter-system communication in this paper is implemented by using a shared component.

Alternatives to this include using event queues to notify each system of new collisions or directly calling on collision logic from CollisionsSystem.

Note that the code, besides System and Component, features Authoring and Baker. Authoring and Baker are concepts specific to Unity Engine and are needed to attach specific components to respective entities.

Functional Reactive Programming Pattern. This pattern is not native to Unity Engine, implementing it requires more work. Implementation is done in F# functional programming language that targets Mono (and therefore is compatible with Unity). F# code loads initial variables from Unity Engine scene, and, on each simulation step, executes the MSF-based simulation and updates Unity Engine scene from its outputs, storing new MSF to execute on the next step.

The immutable nature of MSF is not completely compatible with mutable behaviour of physics of Unity Engine. To preserve correctness, simplified physics are implemented from the ground up.

F# language, used to implement this simulation, does not have built-in support for higher-kinded polymorphism [5], while definition of MSF data type utilizes this concept to abstracts over monad. This limitation was circumvented by introducing multiple definitions for MSF over different monads, but this solution increases the complexity of the code exponentially as more monadic stacks are utilized.

Also, F# does not have built-in support for arrow notation, making managing inputs and outputs of computations more challenging.

We were able to use Reader and List monads to achieve greater modularity and simplify composition of steps, but use of List monad introduced a problem later in the development. The collision system needs to access the position of all objects to calculate ongoing collisions, but the position of each object in our simulation is encapsulated in the respective MSF computation. List monad, which is used to handle multiple objects, does not support any mechanism for MSF computations to communicate positions of respective objects, perform global collisions computation and then to continue independent execution with collision data—such inter-process synchronization is not part of List monad design and therefore some approach other than desirable (see Fig. 4) needs to be implemented. Our implementation (see Fig. 5) exposes position of objects as outputs of respective computations and perform collisions detections at the end of the step, which is not a natural solution.

3 Results

As expected, native support of object-oriented design by Unity Engine makes it simple to use for models of small scale, such as presented in this paper. Behaviour of spheres is expressed in components attached to them, and these components are able to receive

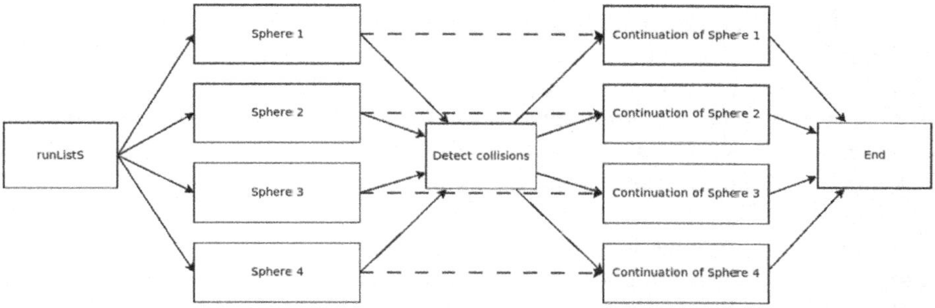

Fig. 4. Desirable implementation of collision system.

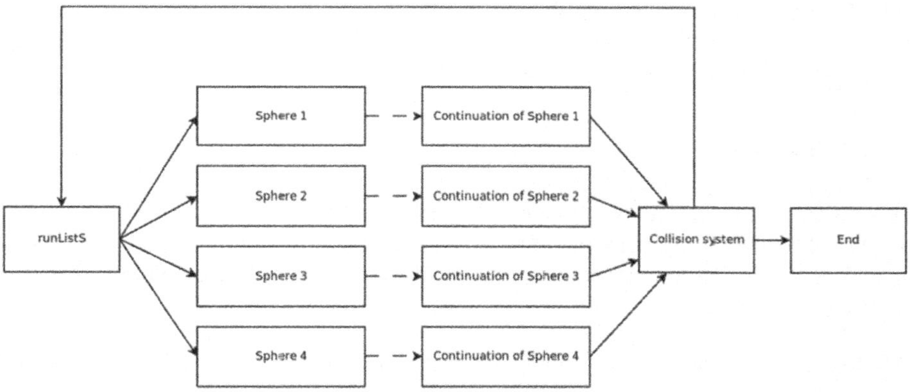

Fig. 5. Final implementation of collision system.

collision events by providing respective method, influence other components, edit global variables and access global API to mutate the world.

ECS implementation of the simulation was more complicated than initially expected. New collisions weren't reported directly, and instead a new system that handles all the collisions needed to be introduced. Most of the created systems could be scheduled for parallel execution, meaning it was possible to process multiple spheres of the same type in parallel, but the Collisions System itself could not be easily parallelized due to mutable access to components of other entities. Such random access could possibly result in race conditions and is not thread-safe.

MSF implementation of the simulation could be considered more flexible than other attempts due to lack of implicit mutable relations, ability to easily introduce new state and extend existing objects, control over both global and local processes. In our case, it was easy to introduce a stateful stepIntegral MSF later in the development to achieve better approximation of position calculation, which would require additional work in other patterns.

But MSF in its current state is not perfect by any means, and more research is required to identify and resolve its limitations. In our attempt at implementing the simulation in MSF, we encountered numerous issues, related to both F# and the pattern itself.

One such problem was the limitation imposed by List monad covered in the previous section. This problem demonstrates that the ecosystem around MSF might be underdeveloped and new concepts need to be introduced in order to make this approach practical.

Limitations of tooling arose as well. As expected, although the result consists of numerous independent computations that could be executed simultaneously in theory, multi-threading did not arise naturally, and additional work is required for the code to make use of multiple threads. Lack of support for higher-kinded polymorphism and arrow notation made the model less easy to create and maintain. It is important to note that those problems are not specific to F#, because the level of support for higher-kinded polymorphism, arrow notation and parallel reduction remains low for mainstream languages at the time of writing of this article.

4 Conclusion

Theoretical and practical comparison of object-oriented, entity component system and functional reactive programming software patterns guides us to the conclusion that all three models have proper advantages and disadvantages and are applicable in different contexts. Advantages and disadvantages of these approaches are informally summarized in Table 1, clarifications follow.

Object-oriented pattern is highly suitable for quick prototyping of new simulations, but often has limitations due to relying on implicit mutable relations between object and is not as flexible as other approaches. It also requires additional work to be parallelizable. This approach is adopted by numerous real-time simulation platforms (such as Godot, Unity Engine, Unreal Engine, Flax Engine) and has a vast ecosystem.

Entity Component System pattern is not as suitable for quick prototyping due to separation of data and logic. Implicit relations are specifically limited in this software pattern, making it easier to reason about the simulation and extend it. It also allows for a massive degree of parallelism and efficient optimizations by processing many entities simultaneously.

Functional Reactive Programming is not as suitable for prototyping as object-oriented design due to immutable nature and requires advanced tooling in order to be a viable approach, but showed great flexibility in the context of our research. More work is required to eliminate existing limitations of this approach and explore its full potential.

Table 1. Informal comparison of object-oriented, ECS and FRP software patterns for implementing real-time simulations.

	Object-oriented pattern	Entity Component System pattern	Functional Reactive Programming pattern (Monadic Stream Functions)
Suitability for prototyping	High	Medium	Medium
Flexibility	Low	Medium	High
Parallelizability	Low	High	Potentially high, but low in practice
Maturity of the ecosystem	High	High	Low

References

1. Hughes, J.: Programming with Arrows. https://www.cse.chalmers.se/~rjmh/afp-arrows.pdf
2. Unity User Manual 2022.3 (LTS). Job System Overview (2022). https://docs.unity3d.com/Manual/JobSystemOverview.html
3. Project Repository. https://github.com/gardspirito/Entity-Component-System-vs-Functional-Reactive-Programming-vs-Object-Oriented
4. Unity Documentation: Rigidbody. https://docs.unity3d.com/ScriptReference/Rigidbody.html
5. Yallop, J., White, L.: Lightweight Higher-Kinded Polymorphism. https://www.cl.cam.ac.uk/~jdy22/papers/lightweight-higher-kinded-polymorphism.pdf
6. Kockelke, F.: Overview of Functional Reactive Programming (2021)
7. Wan, Z., Hudak, P.: Functional Reactive Programming from First Principles (2001)
8. Barenz, M., Perez, I., Nilsson, H.: On the Mathematical Properties of Monadic Stream Functions (2016)
9. Dunai: Generalised reactive framework supporting classic, arrowized and monadic FRP, https://hackage.haskell.org/package/dunai. Accessed 03 Jan 2024
10. Unity-Technologies Entity Component System Samples. https://github.com/Unity-Technologies/EntityComponentSystemSamples. Accessed 03 Jan 2024

Application of Modern Programming Languages in Solving the Problem of Emulator Development for Embedded Systems

Andrey Vlasov(✉), Aleksey Gladkikh, and Kirill Kutaev

Bauman Moscow State Technical University, 2-nd Baumanskaya 5, Moscow 105005, Russia
vlasov@iu4.ru

Abstract. The work is devoted to the use of modern programming languages in solving the problem of developing a hardware platform emulator used for debugging the developed software of embedded systems. The authors perform a research of the performance characteristics of modern programming languages in performing certain operations and algorithms most often used in the process of emulating embedded systems using the software developed as part of the research. The paper discusses the characteristics of existing emulators of built systems and offers recommendations for developing an emulator that has higher performance characteristics and more efficient support potential for it as a software product. The results obtained can be useful for engineers and developers working in the field of embedded systems design and software development for them #CSOC1120.

Keywords: Embedded System · Emulator · Programming Language · Processor · Microcontroller

1 Introduction

In the modern world, embedded systems are an integral part of most branches of human activity [1]. An embedded computing system is any computing hardware and software system that is not a general-purpose system such as a personal computer or server. Their applications cover a wide variety of fields including automotive, medical, energy, security systems, industrial manufacturing, home appliances, etc. Internet of Things (IoT) technology also relies heavily on embedded systems to create a network of devices that can be remotely monitored and controlled [2].

In the process of developing software for embedded systems, there is often a need to debug and test it. It is not always possible to perform these operations on the embedded system under development due to lack of physical access to it or lack of necessary hardware parts on it that are still in the process of design or production. Even with a fully designed hardware part of the embedded system in some cases there is no possibility to get detailed information about all processes occurring in it, which must be debugged and tested. This is due to the fact that modern embedded systems use many hardware components, between which there is various communication, and it is difficult to monitor

R. Silhavy and P. Silhavy (Eds.): CSOC 2024, LNNS 1120, pp. 574–598, 2024.
https://doi.org/10.1007/978-3-031-70518-2_49

software states in all components at the same time [3]. These communication processes may also depend on external events, and then the complexity of debugging increases significantly [4].

Running the software of embedded systems on the developer's computing system (personal computer) can also be difficult. One of the reasons is the difference between processor architectures of the developer's computing system and the target system for which the development is carried out. Embedded systems often use specialized processors or microcontrollers that are oriented to perform highly specialized tasks with low power consumption [1]. One way to solve this problem is to use an emulator - a software tool that simulates (emulates) the operation of a computing system and its environment in another computing system, so that the simulated behavior corresponds as closely as possible to the behavior of the real computing system.

The work is dedicated to the application of modern programming languages to solve the problem of developing a hardware platform emulator used for debugging the developed software of embedded systems in conditions when debugging the system physically is difficult or impossible. Embedded computing systems have become widespread in recent decades as more and more processes in both everyday life and industry require a high degree of automation and integration [2]. Thus, the efficiency of software development for embedded systems becomes a critical aspect in their production.

The object of the research is modern tools and methods of software development, in particular imperative programming languages that were popular at the time of writing this article. Imperative programming language is a formal language used for writing computer programs that define the sequence of commands executed by the processor. The basic idea of imperative programming is to tell the computer system what operations to perform and in what order.

The subject of the research is the fast performance of modern imperative programming languages in the context of the emulation task of embedded systems software, as well as their capabilities for efficient support and development of new emulator components for embedded systems.

The purpose of the work is to investigate the characteristics of the speed performance of modern programming languages in the execution of certain operations and algorithms, most often used in the process of emulation of embedded systems. The result is a software tool to conduct a research of programming languages to determine the speed of execution of specified algorithms, as well as proposals for the use of software development tools to implement a modern emulator of embedded systems that meets the criteria of speed and efficiency of supporting it as a software product.

The relevance of the research is due to the development of information technology and the increasing integration of embedded systems in many areas of industry and everyday life. Embedded software becomes more and more complex and gets more and more extensive functionality. Because of this, the number of processes occurring in embedded systems software is also increasing. This leads to an exponential increase in the number of possible paths of software execution [5]. Testing, debugging, and analyzing the execution of such software becomes more difficult. With the increasing integration of embedded systems into a wide range of automation processes, efficient software development and testing become critical components of developing reliable, fault-tolerant, and secure embedded systems [6].

It is also important to mention that in industry, where embedded systems are widely used, small deviations in the expected behavior of devices can lead to serious consequences, including damage to critical infrastructure and harm to humans.

One of the possible ways to increase the efficiency of testing and debugging is to run software in an emulator. Emulators have a wide range of applications [7]. They are used, in particular, for dynamic analysis of software execution within the framework of certification tests [8], for performance testing [9], for confirmation of correct behavior of devices or programs [10], and for solving reverse engineering tasks [11].

The use of emulation as a method of software testing and debugging has existed for quite a long time, since the first computing systems [12]. It has a number of advantages over the use of physical hardware. When emulation is started, the software to be tested is loaded by writing it directly into the emulated memory. Thus, there is no need to load the software over a serial line or network and start monitoring the state of the target embedded system. The emulated embedded system is controlled over the same command transfer interface as the real physical system. Consequently, there is no need for physical access to the target embedded system.

Due to the fact that the emulator runs on a developer workstation, it is possible to create test scenarios and automate their launch. Since testing is performed by running a normal process (emulator) on the operating system of a workstation, test suites (multiple individual test cases) can be run simultaneously on multiple workstations. Emulation determinism simplifies regression testing, which is a check of a previously tested part of the software to make sure that the changes made did not cause defects in the part of the software that was not changed. Any change in performance results compared to the results of previous versions of the same software can only be caused by differences in the software under test and not by hardware failures or other random changes [13].

In general, the use of emulator of embedded systems with the characteristics of speed and efficiency of its support as a software product represents an important step in improving the quality of developed systems for both industrial and everyday applications.

2 Background and Related Works

Development of emulator of embedded systems is a non-standard task due to its low prevalence and the small number of existing solutions. When designing such software, the choice of programming language is an important task, because the characteristics of the developed software product and the effectiveness of its support will depend on this choice. Many programming languages are used for modern software development. According to the rating on the TIOBE index, about 50 programming languages currently occupy 70% of the market share [14]. Each of them has its own application area, philosophy, features, different dialects, frameworks and many other characteristics by which they can be compared.

Studies show that the right choice of tools and instruments in software development has a significant impact on the development process, the quality of the result and the ability to further support the product, especially for various critical systems [15].

Although most of the programming languages are universal in terms of producing the required result, some of them are more suitable for specific purposes than others [16]. In order to reduce the number of choices, when selecting a programming language, it is necessary to start from the goals that are achieved by implementing the given software. For this purpose, it is necessary to evaluate a set of criteria that the developed product should possess. Criteria such as speed of operation, memory efficiency, dependence or independence from the platform, availability of user interface, execution environment, etc. can be used.

Since a programming language is an abstract description of a computational process, only the process itself and its results can be objectively measured. The essence of such methods of comparing programming languages is that the code written in different languages and executing the same process to get the same result is evaluated. It is necessary to take into account the difference between the programming language and the program when using this method and interpreting the results [16]. When comparing programming languages, one should use only basic syntactic constructions of the language to achieve the most objective performance evaluation. This is due to the presence of unique syntactic abstractions, the principle of operation of which in the internal structure of the language is not obvious or unknown and thus cannot be accurately reproduced in other programming languages.

Thus, the choice of a programming language for writing modern software is a complicated task that requires a complex approach and the use of various methods of analysis. However, the development of methods and software tools for comparing the performance of programming languages is an important direction for the design of modern software that meets the criteria of speed and efficiency of its support as a software product.

When designing a modern emulator of embedded systems, it is necessary to identify the criteria it should meet. Below is a review and analysis of existing solutions and analogs of emulator of embedded systems.

Qemu (Quick Emulator) is an open-source tool that is used to emulate various architectures [17]. Qemu has a large developer community and as a consequence, good support. It supports two emulation modes: user mode, in which a complete emulation of the system, including processor and peripherals, takes place, and system mode, in which only translation of instructions and system calls of the emulated system takes place. Figure 1 shows an example of Qemu in system mode, where the operating system ReactOS is emulated.

Qemu has many positive features like RSP GDB (Remote Serial Protocol GNU Debugger) interface, modeling based on logical bus coupling and high performance, since it is written in C language. On the other hand, it creates a big disadvantage in the form of high complexity of support and extension, in particular writing new modules (architectures, peripherals, etc.).

Kopycat is a Russian emulator of hardware platforms that allows low-level program emulation of arbitrary hardware systems and their debugging through the GDB interface supported by most development tools [18]. It is implemented in the Kotlin (Java) programming language, the main advantage of which is the efficiency of developing new solutions. Figure 2 shows the emulator architecture for emulating the STM32 series microcontroller.

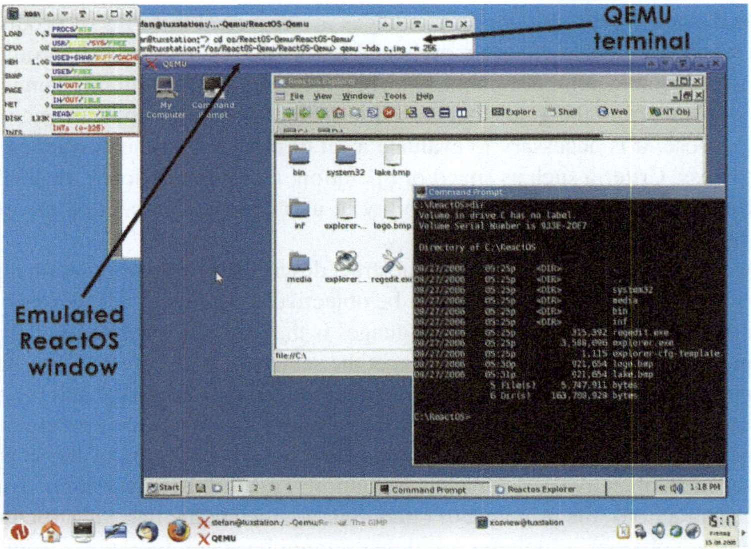

Fig. 1. Emulation of the ReactOS operating system in the Qemu emulator.

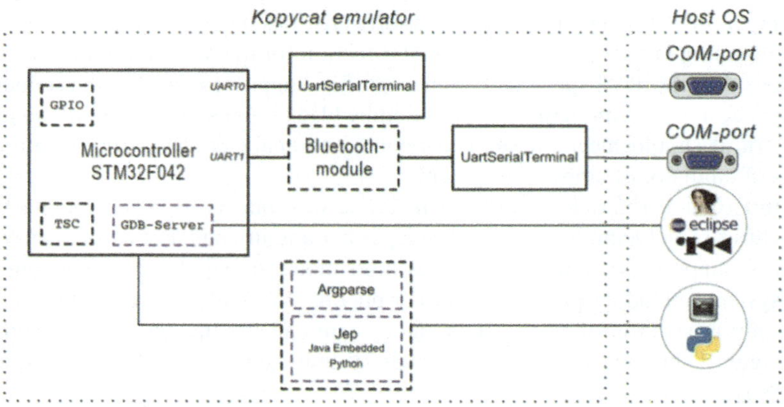

Fig. 2. Architecture of Kopycat emulator for emulation of STM32 series microcontroller.

One of the competitive advantages of Kopycat is the availability of a convenient software development kit and documentation for creating new computing cores and peripherals [19]. There are two possible approaches for creating new modules: the first is a structural approach, the second is a behavioral description. During structural description, the installed modules and connections between them are specified in a file in JSON (JavaScript Object Notation) format. Structural description can also be performed using the Kotlin/Java programming language. In behavioral description, handlers for writing or reading from certain addresses can be added to the module. Inside the handler it is specified how the module should handle this event. Behavioral description can be specified only in the Kotlin/Java programming language. Among the disadvantages, we should note the average performance when emulating embedded systems that do not have full-fledged general-purpose operating systems (for example, distributions based on the Linux kernel). In case of emulation of such systems, performance is significantly degraded.

Unicorn is a lightweight, multi-platform and multi-architectural processor emulator. This emulator differs in its architecture from the others in the list by having only basic functionality [20]. It does not emulate the operation of all software or an entire operating system.

It does not support system commands (such as opening a file, outputting a character to the console, etc.). The developer has to partition the memory of the emulated software and load data into it himself. Figure 3 shows the graphical interface of BIOS (BasicInput/OutputSystem) emulation in Unicorn.

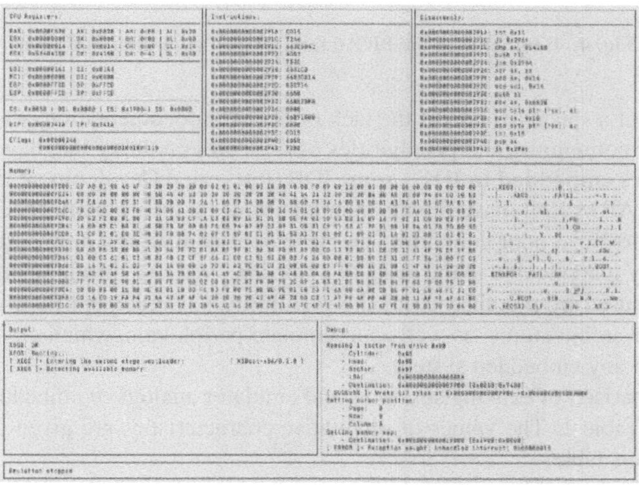

Fig. 3. BIOS emulation in Unicorn.

The disadvantages of this emulator include the lack of built-in abstractions and tools for software operation, configuration and use can be difficult for inexperienced users, the lack of system commands, the complexity of implementing new modules of the emulator. Thus, working with Unicorn requires high qualification of a specialist.

Proteus is a software package for computer-aided design of electronic circuits, the distinctive feature of which is the ability to simulate the operation of programmable devices [21]. Emulation of microcontroller of PIC16 family is presented in Fig. 4. Emulation in Proteus, unlike Qemu, takes place at the level of electrical signals, which significantly slows down its speed.

Proteus is proprietary software, unlike the emulators described above, but there is a possibility to write your own modules in C++. Their implementation requires high

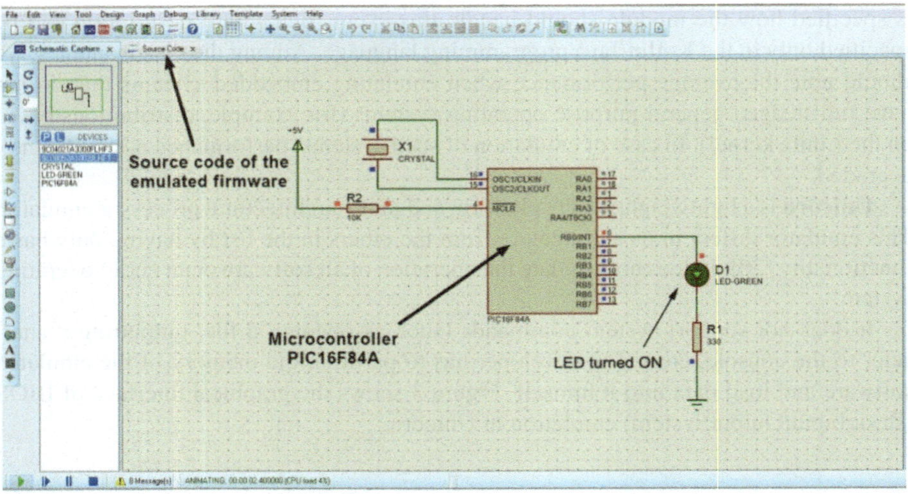

Fig. 4. Emulation of the PIC16 family microcontroller in Proteus.

qualification of a specialist due to the lack of detailed documentation, weak support in the developer community and peculiarities of C++ programming language.

Emulators embedded in IDE - some IDE (Integrated Development Environment), for example, ARM Keil, have a built-in emulator of hardware platforms (Fig. 5) [22]. Such emulators have poor support of the developer community, low performance and are practically inapplicable in the tasks of embedded systems development.

This is caused by the absence of the possibility to implement their own modules and processor architectures, as well as to emulate peripherals, which are an obligatory component of any embedded system.

The comparison of existing solutions and emulator analogs of embedded systems is presented in Table 1. The values of qualitative characteristics are given relative to the elements of this table.

As a result of consideration of this issue, we can conclude that existing solutions for emulation of embedded systems do not have the characteristic of speed and the ability to easily implement new modules at the same time. At high speed of the C programming language, they have a small number of built-in tools and abstractions and are poorly developed.

On the other hand, C++ programming language has a large number of built-in facilities and abstractions, but the complexity of development in it is extremely high and the resulting outputs often have non-intuitive behavior. Thus, there is a need to develop an emulator that will have high performance characteristics and ease of implementation of new modules.

3 Materials and Methods

The key method of the research is to test the speed of execution of given algorithms by different programming languages and then compare the results of the experiments.

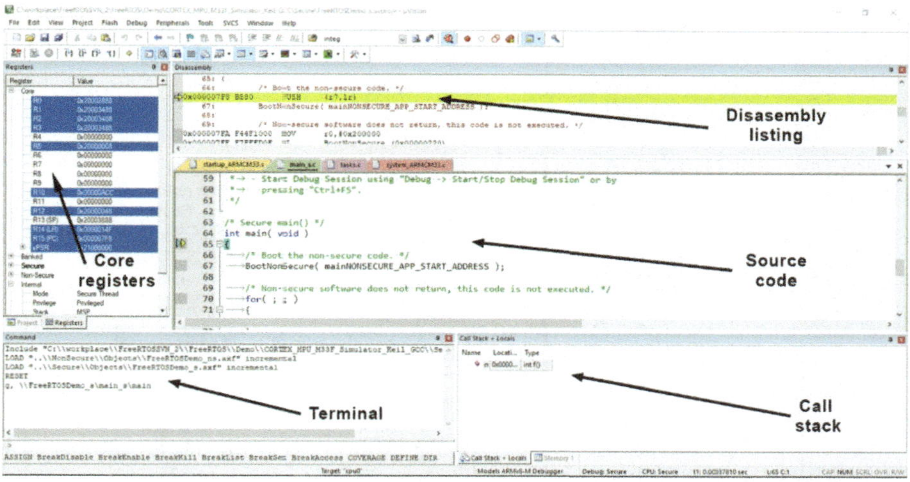

Fig. 5. Emulation of ARM Cortex-M33 core operation in KeiluVision IDE.

Table 1. Comparison of embedded system emulation and circuit simulation tools.

Software	Performance	Difficulty in implementing new modules	GDB interface	Modeling of electrical signals
Qemu	High	High	+	–
Kopycat	Average	Low	+	–
Unicorn	High	High	+	–
Proteus	Low	High	–	+
Keil IDE	Low	None	+	–

In order to test the performance of programming languages, we developed "FPLB" (Flexible Programming Language Benchmarking) [23] with the necessary environment for testing, in the design of which we used the research presented in the work "Cross-Language Compiler Benchmarking" [15]. The flowchart of the algorithm of the "FPLB" software is presented in Fig. 6.

The "FPLB" software allows to run various pre-prepared algorithms for specified programming languages in an automated mode. It also allows to log test results, draw a bar graph based on test results, generate the necessary input data for testing, as well as run a sequence of identical tests and calculate the average value of the test results. The output of the reference data on the use of the "FPLB" software is shown in Fig. 7.

The Python programming language was chosen for the development of the FPLB software because it is easy to develop, debug, and maintain software in. It has a large and active community of developers, which provides many resources, documentation, libraries and development tools. The Docker container platform was chosen as the test

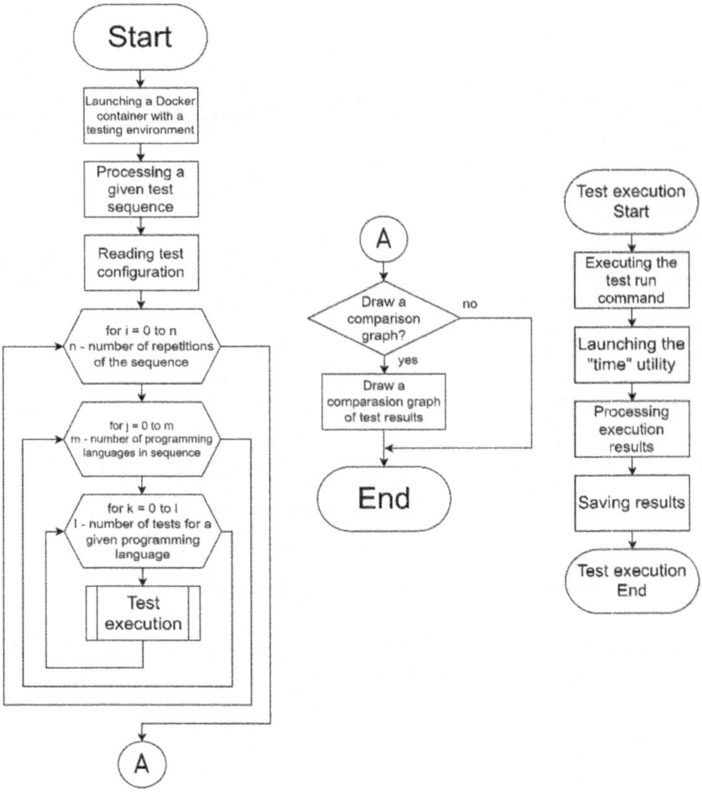

Fig. 6. The flowchart of the algorithm of the "FPLB" software operation.

execution environment. It is a common lightweight and self-contained runtime container that includes all the necessary environment to run the software, including libraries, system tools, and runtime environment. It keeps the environment of the algorithms under test in the same stable state, which reduces the impact of different processes running on the computer system under test.

```
> python . --help
usage: . [-h] -p {full,hello-world,local,matmul,prime-number,qsort,versions} [-o OUTPUT_FILE] [-i] [-g]
         [-c COUNT] [-d]

FPLB - Flexible Programming Language Benchmarking. Tool for comparing the speed of programming languages

options:
  -h, --help            show this help message and exit
  -p {full,hello-world,local,matmul,prime-number,qsort,versions}, --pipeline {full,hello-world,local,matmul,pr
ime-number,qsort,versions}
                        Pipeline for testing
  -o OUTPUT_FILE, --output-file OUTPUT_FILE
                        Path to file with test result
  -i, --image           Creates an image graph.jpeg with a graph comparing execution speeds
  -g, --generate        Generates a new data for tests
  -c COUNT, --count COUNT
                        Number of time to repeat tests
  -d, --debug           Enables debug mode
```

Fig. 7. Output of reference data on the use of the "FPLB" software.

To estimate the execution speed of the tested software, we used the method proposed in the research "Comparative Testing of Programming Languages" [16]. The value of the execution time of the tested software in a certain programming language is measured using the "time" software. This is a standard UNIX utility that outputs a statistical message about the time spent to execute the software passed to the utility as input.

The choice of algorithms, the speed of execution of which will be tested, was based on the work "Research of Speeds of Performance of Basic Mathematical Problems of Popular Programming Languages" [24]. Since the purpose of the work is to research the characteristic of programming languages' performance in the execution of algorithms most often used in the emulation of embedded systems, it is necessary to choose those algorithms that contain the operations most often used in the emulator's work. These operations are reading and writing data to memory, as well as basic arithmetic operation sand comparison operations. This is due to the specificity of modern processor architectures [25]. Thus, algorithms containing a set of such operations are required.

Visualization of the results was performed using the method used in the research "Cross-Language Compiler Benchmarking" [15]. For this purpose, the library for the Python language "MatPlotLib" was used. It allows visualization of data by two-dimensional and three-dimensional graphs. Thus, it is possible to visually evaluate the ratio of algorithm execution speeds of different programming languages.

3.1 Analysis of Software Execution

From the processor's point of view, software is a sequence of instructions describing the actions to be performed on a set of data. These processor instructions can be written as mnemonic codes in Assembly language. All Assembly languages are machine-specific, that is, they are designed for a particular type of processor architecture. Examples of mnemonics in Assembly language for the ARM (Advanced RISC Machine) architecture are shown in Fig. 8.

```
.global _start
_start:
    MOV R0, #0xFFFFFFFF
    MOV R1, #5
    MOV R2, #4
    MOV R3, #3

    ADDS R4,R0,R1
    ADC R5,R2,R3
```

Fig. 8. Instruction mnemonics in ARM assembly language.

From a software developer's point of view, it is a syntactic unit that follows the rules of a particular programming language, consisting of definitions, operators or instructions necessary for a particular function, task or problem solving. It is represented as a text file with a set of source code. An example of C source code is shown in Fig. 9.

In order for the processor to execute the actions specified in the software source code, they must be converted into processor instructions. There are many ways of conversion. Let us consider the most basic and common ones.

Compilation is the process of converting software source code into machine code (a set of processor instructions) that can be directly executed by the processor. During compilation, the compiler analyzes all the source code and creates an executable file that contains the machine code of the software. The executable file is machine-specific, thus a separate compilation process must be performed for each processor architecture. Programming languages that use compilation as a way to convert source code into processor instructions are called compiler-based languages.

One of the main advantages of compiler-based languages is their ability to generate time- and memory-optimized machine code. Most compiler-based programming languages require explicit declaration of variable data sizes and types. This allows to use computing system resources more efficiently and to detect typing errors at early stages of development, which helps to improve software reliability and security.

Software development in compiler-based languages can be more time-consuming because such languages usually require more rigorous syntax and language constructs. An error in the source code can lead to hard-to-detect compilation errors, which requires additional time to debug, find and fix the cause of the error. Compiler-basedprogramming languages require the program code to be recompiled each time a change is made, which can be a time-consuming and costly process when dealing with large software. This results in reduced flexibility and speed of software development. Common compiler-based programming languages include C, C++, Rust, Go.

Interpretation is the process of executing source code by analyzing it and interpreting it sequentially by an interpreter during software execution. Unlike compiler-based programming languages, where the source code is pre-transformed into processor instructions and stored as an executable file, interpreted programming languages analyze and execute each syntactic unit of the source code directly at runtime. Programming languages that use interpretation as a way of translating source code into machine code are called interpreted languages.

One of the main advantages of program interpretation is its platform independence. Interpreted programming languages in most cases can be executed on different platforms without the need to modify the source code. When interpreting software, it is easier to track and correct errors because the interpreter executes the code line by line. Interpreted programming languages usually support dynamic typing, which means that variable types can be defined at runtime.

However, software interpretation also has its disadvantages. One of the main disadvantages is low performance compared to compiler-based programming languages. Since the interpreter analyzes and executes the source code, the execution time can be longer than that of software compiled into machine code. This is especially noticeable when processing large amounts of data or performing highly loaded calculations. Common interpreted programming languages include Python, JavaScript, PHP, Ruby, SQL and Bash.

```c
int main(void){
    srand ( time(NULL) );
    initWeights();
    initData();
    for(int j = 0;j <= numEpochs;j++){
        for(int i = 0;i < numPatterns;i++){
            patNum = rand() % numPatterns;
            calcNet();
            WeightChangesHO();
            WeightChangesIH();
        }
        calcOverallError();
    }
    displayResults();
    return 0;
}
```

Fig. 9. Source code in the C programming language

There are also **bytecode programming languages** that use both compilation and interpretation to convert source code into machine code. Bytecode programming languages represent the source code of a program in an intermediate form. The compiler converts the source code into a sequence of instructions that are low-level and intermediate for execution on a virtual machine or interpreter. This intermediate code is called bytecode because it is represented as a sequence of bytes.

One of the main advantages of using bytecode is platform independence. Building a virtual machine or interpreter that executes bytecode allows software to run on different platforms without changing the source code. Bytecode is also often used to improve software performance. It can be compiled in the most efficient way for a particular virtual machine or interpreter, resulting in more efficient software execution. Common bytecode programming languages include Java, C#, Kotlin and Scala.

Using the results of the analysis, a mind map of programming languages classification by the criterion of source code conversion into processor instructions was made, presented in Fig. 10.

According to the results of the analysis we can conclude that each of the presented types of programming languages has its own advantages and disadvantages. As a consequence, each of them is used in different areas of software development. The choice programming language type is determined by the task requirements.

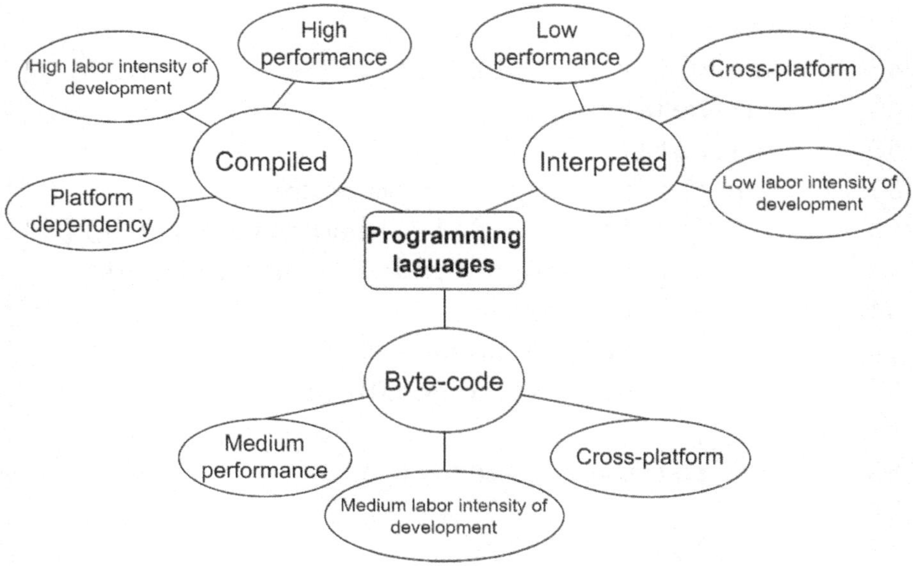

Fig. 10. Mind map of programming languages classification.

3.2 Analysis of the Programming Languages Under Research

Based on the conducted analysis of programming languages it is necessary to select those that potentially have the characteristics of speed and efficiency of development in their application.To solve the programming language selection problem, the evaluation criteria were compiled:

- performance - how fast the software developed with the use of programming language is executed on the computer system;
- relevance - how up-to-date the programming language is and what potential it has in the future for further prospective development of the software product;
- prevalence - the popularity of the language among the developer community and, as a consequence, the amount of available information and documentation on the Internet;
- number of integrated abstractions and syntax - a set of constructs, abstractions, paradigms and special tools that increase the efficiency of development using a programming language;
- ecosystem - availability of IDEs with automation of typical repetitive processes, availability of a compiler or interpreter without program errors, availability of an effective debugging system for the developed software product.

According to the compiled criteria the selection of modern and relevant programming languages was made. Below is an overview and analysis of the selected programming languages to evaluate their performance.

Python is an interpreted script programming language developed in the late 1980s by Guido van Rossum. It has a simple syntax and is the most popular programming language at the moment according to the TIOBE index [14]. One of the key advantages of Python is its ease of use. Python offers extensive built-in abstractions, which makes development in it highly efficient. Python is an interpreted language and as a consequence it is much slower compared to compiler-based languages, but there are various ways to optimize for speed. Python is known for its many libraries and frameworks that make it a very effective tool for a wide range of tasks. It has a well-developed ecosystem with a large number of development tools.

Cython is a programming language that is an add-on to Python with the ability to explicitly define data types. It allows developers to write high-performance code close to C that can be easily integrated with existing Python code. One of the main advantages of Cython is its performance. By explicitly defining data types and using static typing, Cython can synthesize CPU instructions that are optimized for performance. Cython is used extensively in projects that require a tradeoff between ease of development and performance. Cython is not as widespread as Python, but still has an audience and is growing in popularity. Cython provides the same integration and syntax as Python, adding the ability to use static typing.

JavaScript is a programming language developed by Netscape Communications Corporation in 1995. It is widely used to create interactive web pages and web applications and, along with HTML and CSS, is one of the primary languages for web development. JavaScript can also be used to develop server-side web applications using the Node.js platform. This allows to develop full-fledged software that can run on both client-side and server-side.

JavaScript is an interpreted language, but many built-in optimizations improve its performance. JavaScript is one of the most widely used programming languages and has a huge audience of developers. JavaScript provides many integrated abstractions and has an easy-to-understand syntax.

C# was developed by Microsoft in 2000 and is a key language for the ".NET" platform. C# combines the strengths of C++ with the convenience of Java, making it very popular among developers. One of the main advantages of C# is its object-oriented nature. It provides inheritance, encapsulation and polymorphism mechanisms to build modular and extensible programs. In addition, C# has built-in tools for automatic memory management, which makes development easier and eliminates many errors associated with memory misuse.

C# is slower relative to compiler-based languages, but modern solutions to optimize for performance make it quite efficient. C# is used to develop applications under Windows operating system and computer games on Unity platform. C# offers a wide set of integrated abstractions and has a syntax that is a combination of C++ and Java. C# is accompanied by an extensive ecosystem, including the integrated Visual Studio development environment and many libraries for various purposes.

Kotlin is a programming language developed by JetBrains in 2011 specifically for developing applications on the Java platform. Kotlin is a statically typed language that has a simple and clear syntax and is an extension of Java. One of the main advantages of Kotlin is its compatibility with Java. This allows developers to easily integrate Kotlin into existing Java projects and reuse existing source code. It has many improvements over Java. In addition, Kotlin provides extensive syntax features, making it easy to write safe and efficient code.

It is a bytecode language, compiles to Java bytecode and exhibits good performance but still lacks to the level of compiler-based languages. Kotlin is not as widely used as Java, but is getting popular rapidly.

Rust is a relatively new system programming language developed by Mozilla Research in 2010. First of all, Rust is known for its ability to improve security and avoid many common programming mistakes. One of the key advantages of Rust is its memory access control system, which prevents errors related to inconsistent memory accesses. This makes Rust especially safe for system software development, where low-level errors can lead to serious problems.

Rust also has an efficient type system and static analysis, which allows errors to be detected at compile time. This greatly simplifies debugging and improves the quality of the code being developed. In addition, Rust provides a mechanism for automatic memory management. Rust is a compiler-based programming language that shows excellent performance. Rust is still evolving and is not as widely used as other languages, but has potential applications in specialized areas. Has an evolving ecosystem with libraries and development tools.

C is a programming language developed in the early 1970s and is one of the most widely used languages in the software development industry. It is known for its capabilities and efficiency, as well as its broad support for various architectures and operating systems. One of the main advantages of C is its low level of abstraction from hardware, which allows developers to have full control over memory and the processor. This makes C particularly suitable for developing system software, embedded software, operating systems, and device drivers.

In this research, the C programming language is used as a benchmark programming language with the fastest performance and the widest distribution. Thus, it is not considered for the development of emulator of embedded systems due to the lack of support for object-oriented programming and a small number of integrated abstractions.

As a result of the analysis, seven programming languages were chosen, which will be used to implement the algorithms under research and comparative analysis of the results of their execution. It should be noted that some of the selected programming languages are used to implement emulators of the embedded systems discussed above. Thus, in further studies it will be possible to carry out a comparative analysis by the criterion of performance of the developed emulator of embedded systems on the programming language selected by the results of this research and existing solutions of emulators of embedded systems.

3.3 Analysis of the Researched Algorithms

Three algorithms were selected for the research of programming language performance. These operations are the most common operations performed by the processor. Hence, they will be performed most often in the emulator of embedded systems. It is necessary that the software developed in the programming language, in which the emulator will be implemented, performs these operations as fast as possible. The given algorithms do not consider auxiliary operations such as memory allocation, data reading, and output of results. Let us consider the selected algorithms in details.

Matrix multiplication algorithm is a process that consists of multiplying two matrices to obtain a third matrix, called the product matrix. The algorithm is implemented using loops, where each element of the product matrix is matched by the sum of the products of the elements of the corresponding rows and columns. The difference in the implementation is that the second matrix is transposed for more efficient memory usage. The flowchart of the matrix multiplication algorithm is shown in Fig. 11.

The matrix multiplication algorithm has a complexity of $O(n^3)$, where n is the matrix side size, because it contains three nested loops. The time complexity of this algorithm depends on the size of the input matrices and can be significant for large matrices.

Quicksort is one of the most efficient methods of sorting data arrays. The quicksort algorithm is implemented by selecting a reference element from the original array and then dividing the array into two subgroups: elements smaller than the reference and elements larger than the reference. These subgroups are recursively subjected to a sorting process using quicksort until a final ordering is achieved. This process is called "partitioning".

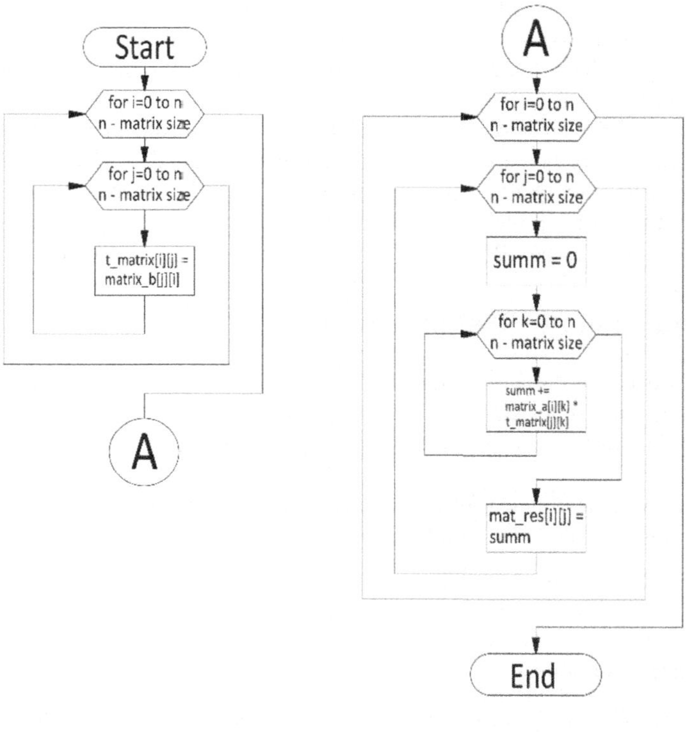

Fig. 11. Flowchart of the matrix multiplication algorithm.

The quicksort algorithm is highly efficient, having an average time complexity of $O(n * log(n))$, where n is the number of elements in the array. However, in the worst case, its time complexity can reach $O(n^2)$, which happens when the reference element is unsuccessfully chosen or when the array is already ordered in ascending or descending order.The flowchart of the quicksort algorithm is shown in Fig. 12.

The algorithm for finding prime numbers solves the problem of finding all prime numbers up to a given integer N. Prime numbers are natural numbers that are divisible by themselves and by 1. The rest of the natural numbers are called composite numbers. The only exception, 1, is not referred to either prime or composite. The solution of this problem is widely used in cryptography, data analysis and other applied problems.

There are many algorithms optimized by the criterion of speed to find prime numbers: Sieve of Eratosthenes, Sieve of Sundaram, Sieve of Atkin, Wheel factorization, Base 2 Segmented factorization. For this study, the simplest algorithm for finding prime numbers was chosen - searching all numbers up to a given N and checking them for primality. This algorithm lacks obvious optimizations in terms of speed, such as discarding all even numbers after the number "2". The algorithm has a time complexity of $O(n * sqrt(n))$, where n is the number of prime numbers sought.

The justification of this choice is to test the performance of software developed in a certain programming language. It is necessary to perform the task of loading the computer system to perform certain mathematical operations, but not the task of searching for prime numbers. For this reason, the developed algorithm does not save the found numbers except the last one. The flowchart of the developed prime number search algorithm is presented in Fig. 13.

As a result of the analysis, three algorithms were selected that contain the operations most frequently performed in the emulator of embedded systems. For each of the algorithms, the flowchart of the algorithm and its time complexity are presented. These algorithms were implemented in the programming languages discussed above and integrated into the "FPLB" software to test the speed of their execution.

4 Description of the Research Method

The method for the research was designed to create a stable environment in which the effect of the programming language on the speed of execution of the algorithms could be evaluated. Standard scripts for testing programming languages for performance were used as a basis.

The software for testing was installed on a virtual machine with the distribution of the operating system of the Linux family Ubuntu 22.04. Parameters of the virtual machine:

- Processor AMD Ryzen 5 5600U with Radeon Graphics 2 cores;
- RAM 4 GB;
- ROM 60 GB;
- Video adapter is not used.

The testing is performed in a Docker container based on the Ubuntu Linux 22.04 family operating system. It allows the environment to be kept in the same stable state, which reduces the impact of various processes occurring on the computing system on which the testing takes place.

The "FPLB" testing software provides for the configuration of a sequence of tests in certain programming languages and certain algorithms. Thus, three configurations

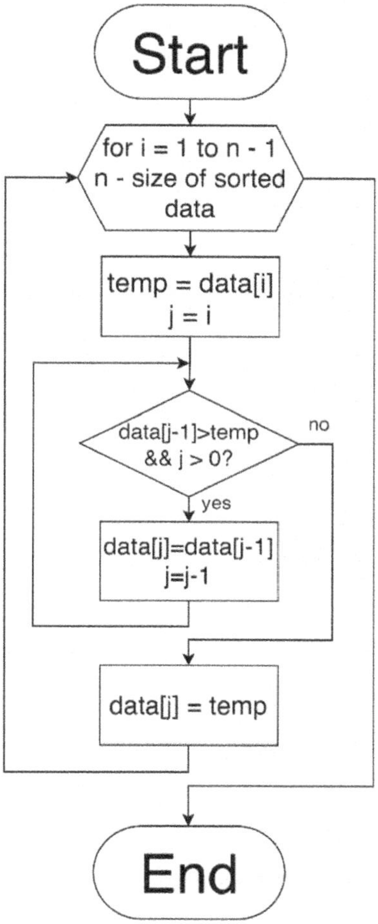

Fig. 12. Flowchart of the algorithm of quicksort.

are created for each of the researched algorithms. The "FPLB" software also has the functionality of running a sequence of a specified number of the same tests. This allows to run many identical tests and calculate the arithmetic mean of the obtained results. The result is rounded to 2 decimal places. For each test a new set of input data of the same specified dimensionality is generated. Tests were carried out at 20 cycles of test repetitions. Based on statistical methods, this method of testing will give results close to real values.

For the brute force prime number search algorithm, "FPLB" software generates the maximum amount of prime numbers to search in the range from 10000 to 15000. For the matrix multiplication algorithm, matrices of size 2000 by 2000 elements with values in the range from 0 to 65536 are generated. For the quicksort algorithm, data arrays of 1000000 elements with numbers in the range of 0 to 4294967294 are generated.

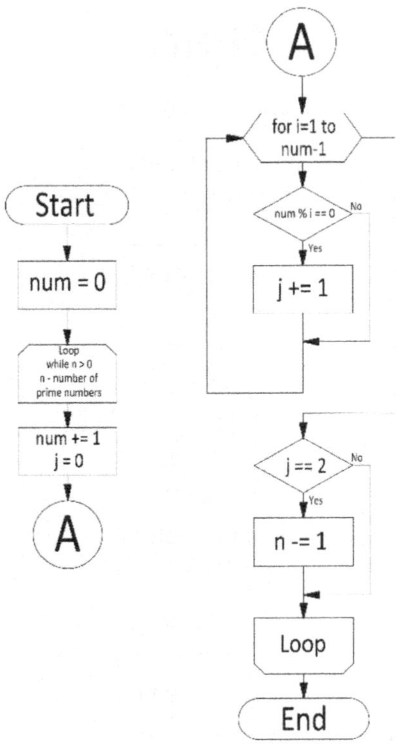

Fig. 13. Flowchart of the brute force prime number search algorithm.

5 Analysis of the Research Results

The result of this research is a set of bar graphs, with the vertical axis representing time in seconds and the horizontal axis representing programming languages. The resulting bar charts from the "FPLB" software for each of the studied algorithms are presented in Figs. 14, 15 and 16.

The results of the research correlate with the results of the analysis of programming languages and their classification by the criterion of source code conversion into processor instructions. Compiler-based Rust and C languages show the best performance in the conducted research. Their results are approximately equal taking random error into account. The Cython programming language also shows high performance as it has a compilation mechanism similar to that used in the C programming language.

Bytecode programming languages Kotlin and C# show average speedup values. In the research of the algorithm for finding prime numbers, the Kotlin programming language managed to get a result comparable in speed to compiler-based programming languages. On the other hand, in the research of the algorithm of quicksort the C# programming language yielded to the interpreted programming language JavaScript

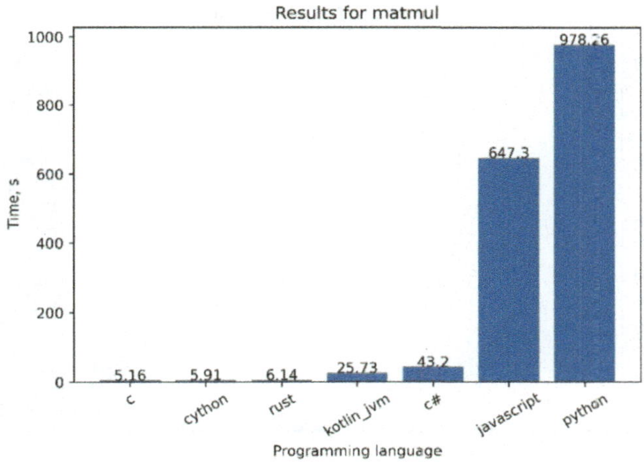

Fig. 14. Matrix multiplication algorithm evaluation results.

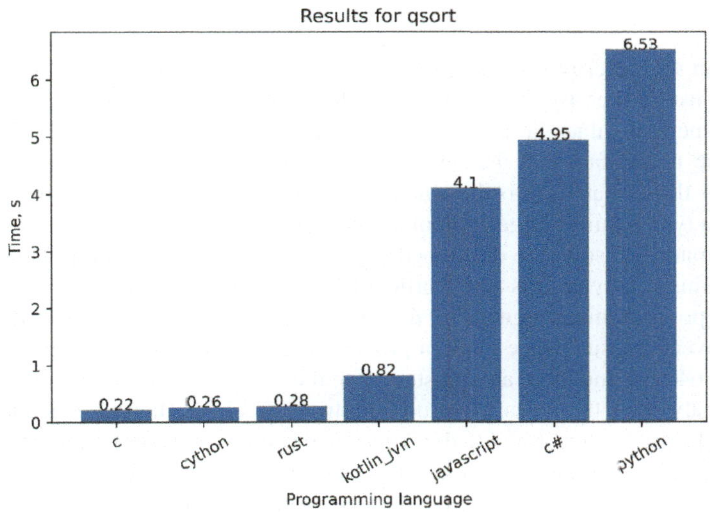

Fig. 15. Quicksort algorithm evaluation results.

with a small lag. Based on this, we can conclude that bytecode programming languages can show performance results in a wide range depending on the task at hand.

The interpreted programming languages Python and JavaScript showed the worst performance results in this research. The results of other programming languages are an order of magnitude higher than those of interpreted ones. It should be noted that in the research of the algorithm of searching for prime numbers by brute force Javascript showed the result almost 3 times slower than Python. At the same time, in other studies

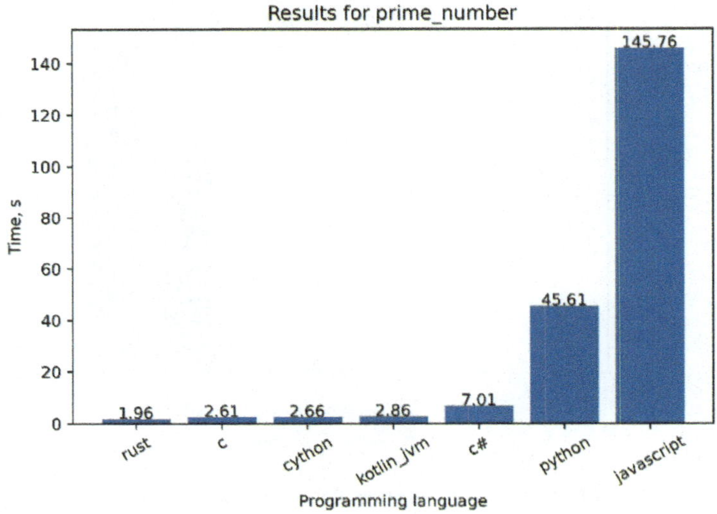

Fig. 16. Enumerative prime numbers search algorithm evaluation results.

of algorithms JavaScript showed better results than Python. This result is presumably due to the use of the "BigInt" data type in the implemented algorithms, which allows storing numerical values larger than $2^{53}-1$, as opposed to the standard Int data type. This is due to the fact that the investigated algorithms can obtain numerical values larger than those that fit into the standard Int data type. Presumably, the use of the BigInt data type requires large computational resources and, as a consequence, reduces the performance of software developed in the JavaScript programming language with the use of this data type. It is also worth noting that in this case the applicability of the JavaScript programming language for the development of emulator of embedded systems is a controversial issue, since modern processor architectures use machine words with the size of 64 bits, and there are registers with the size of 128 bits and more.

The analysis of the research results allowed to confirm the classification of programming languages according to the criterion of source code conversion into processor instructions. The peculiarities of the behavior of software developed in some programming languages were also revealed. It follows from the results of the research that the speed of programming languages can vary significantly depending on their peculiarities and optimizations. However, it should be noted that the speed of operation is not the only criterion when choosing a programming language, usability, availability of developers and other factors should also be taken into account.

It was revealed that the choice of programming language for software development should be carried out taking into account the specific requirements and tasks that the developed software product solves. However, it should be noted that the results of this research are relative and may depend on many factors, including the specific implementation of the programming language, code optimization, hardware and other conditions.

6 Discussion

The results of the research have contributed to the understanding of the impact of programming language choice on the execution speed of the software being developed for the emulator embedded systems development task. This section analyzes the results obtained, places them in the context of existing knowledge and current technological trends, and discusses possible practical applications of the study and future research directions.

An important aspect of the obtained results is the confirmation of the fact that software developed with the use of compiler-based programming languages has a performance characteristic that is orders of magnitude higher than that of software developed with the use of interpreted languages. Thus, interpreted programming languages cannot be applied to implement a modern emulator of embedded systems that meets the criteria of speed and efficiency of its support as a software product.

It is worth noting that the Cython programming language showed impressive results. Despite the fact that in the process of execution of some parts of the software there are calls to the Python virtual machine, which reduces its performance, the results at the level of compiler-based C programming language Rust were demonstrated. Thus, the Cython programming language is a promising language for the development of emulator of embedded systems due to its high performance, which is a consequence of the compilation of the source code, simplicity, prevalence, the number of integrated abstractions and syntax, as well as the presence of an extensive ecosystem that is present around the Python language.

It is also worth noting that JIT compilation was not considered or utilized in the conducted research. It is an effective tool to entrain the performance of interpreted and bytecode programming languages. This is due to the complex behavioral algorithm of JIT-compilation, which requires in-depth research and analysis. In this way, it will be possible to explicitly make an evaluation of the execution speed of software in different programming languages considering JIT-compilation. Also, object-oriented programming techniques have not been addressed in the research. This is due to the different implementation of object-oriented programming in the internal architecture of different programming languages. The impact of applying object-oriented programming techniques on the performance of software in different programming languages requires a separate research.

The results of this research also correlate with the results obtained in other works [15, 16, 24]. Software execution performance is crucial to the success of computer systems and applications in the modern world. With the increasing number of workloads on modern computing systems, it is important to execute tasks successfully and timely. Thus, the importance of the speed of programming languages in the modern world can hardly be overemphasized, as the further development of information technology directly depends on it.

It is noteworthy that some programming languages showed slightly different results executing different algorithms. This opens prospects for new studies using more algorithms, which will reveal more accurate patterns in understanding performance of software developed in the studied programming languages. In general, the conducted research opens new prospects for the development of modern emulators of embedded systems and the development of new methods to improve their performance.

7 Conclusion

In this paper a research aimed at analyzing the performance of programming languages in the context of the task of emulation of embedded systems software has been performed. The problem of selecting a programming language for software product development has been revealed. Thus, the choice of programming language is a complicated and multifaceted task that requires a complex approach and the use of various methods of analysis.

The comparative analysis and research of existing solutions of emulators of embedded systems was carried out. The need to develop an emulator that will have high performance characteristics and ease of implementation of new modules has been revealed. The need for comparative testing of the performance of different programming languages has been revealed. To test the performance of programming languages, a software "FPLB" in the Python programming language has been developed. The algorithm of the developed software is described in the paper in details.

The analysis of the software execution on the computer system has been carried out. The classification of programming languages by the criterion of conversion of the source code into processor instructions has been made. To choose programming languages, for which the research was carried out, the criteria for evaluation of programming languages have been made. As a result of the analysis, seven programming languages have been chosen for the research, for which the implementation of the algorithms under research was carried out and the comparative analysis of the results of their execution has been made.

To conduct the research, we selected and analyzed the algorithms that were implemented using the selected programming languages. As a result of this analysis, three algorithms were selected that contain operations that are most frequently performed in the emulator of embedded systems. For each of the algorithms, a block diagram of the algorithm and its time complexity is presented. The method of conducting the research using the developed software "FPLB" is developed and described.

The results of the research were analyzed. The analysis of the research results allowed to confirm the classification of programming languages according to the criterion of conversion of source code into processor instructions. Based on the results of this research it was revealed that the choice of programming language for software development should be carried out taking into account specific requirements and tasks that the software product being developed solves.

The results of this research can be useful for engineers and developers working in the field of embedded systems design and software development for them. The obtained recommendations will help to increase the efficiency of software development and testing for embedded systems, as well as to reduce possible risks and problems associated with incorrect emulator behavior.

In the future, based on the results of this work, it is possible to conduct additional research in the field of emulator optimization for embedded systems by the criterion of performance. In general, the presented work is an important contribution to the field of emulator development for embedded systems and can become a useful reference material for specialists engaged in this problem.

Acknowledgements. The results were partially obtained within the project «PRIOR/SN/NU/22/SP2/5» under the Development Program of Bauman Moscow State Technical University as part of the Priority-2030 federal academic leadership program.

References

1. Artemiev, B.V., Vlasov, A.I., Juravleva, L.V.: Quality estimation method in advanced software systems. AIP Conf. Proc. **2467**, 050011 (2022)
2. Zobov, O.V., Vlasov, A.I., Juravleva, L.V.: The building data collection devices problems analysis in the industrial IoT paradigm. AIP Conf. Proc. **2402**, 050041 (2021). https://doi.org/10.1063/5.0072134
3. Efimov, S.N., Terskov, V.A., Galushin, P.V., Yarkov, K.V.: Reliability model of multiprocessor hardware-software complexes of real-time control systems with multi-version software. Herald Bauman Moscow State Techn. Univ. Ser. Instrum. Eng. **4**(137), 41 (2021). https://doi.org/10.18698/0236-3933-2021-4-41-58
4. Zaborovskiy, N.V., Tormasov, A.G.: Modeling of multithreaded program execution and the method of static code analysis for race conditions. J. Appl. Inform. **4**(34), 105–110 (2011)
5. Boonstoppel, P., Cadar, C., Engler, D.: RWset: attacking path explosion in constraint-based test generation. In: Lecture Notes in Computer Science Book Series, vol. 4963, pp. 351–366 (2008)
6. Vlasov, A.I., Selivanov, K.V., Kovalev, I.O.: Investigation of quality and identity of similar electronic components from different manufacturers. In: International Conference on Industrial Engineering, Applications and Manufacturing, ICIEAM 2023, pp. 333–338. IEEE, Sochi, Russian Federation (2023). https://doi.org/10.1109/ICIEAM57311.2023.10139051
7. Sepulveda Rodriguez, L.E., Chavarro-Porras, J.C., Sanabria-Ordonez, J.A., et al.: A survey of virtualization technologies: towards a new taxonomic proposal. Ingenieria E Investigacion **42**(3), e97363 (2022)
8. Selvesyuk, N.I., Ostrovskiy, A.S., Rusanov, P.V., Komakhin, M.O.: Object-oriented design of the system of emulation of program interface of the application in the tasks of ensuring information security. Inf. Sci. Control Syst. **2**(56), 3 (2018). https://doi.org/10.22250/isu.2018.56.3-13
9. Zhang, B., et al.: A survey on I/O virtualization and optimization. In: Fifth Annual China Grid Conference 2010, pp. 117–123. IEEE, Guangzhou, China (2010). https://doi.org/10.1109/ChinaGrid.2010.54
10. Dileep, K.P., Devesh, G., Raghavendra, R., Suman,A., Srikanth, S.: Verification of Linux device drivers using device virtualization. In: 2nd International Conference on Computing for Sustainable Global Development, INDIACom2015, pp. 694–698. IEEE, New Delhi, India (2015)
11. Gladkikh, A.A., Vlasov, A.I., Uzenkov, D.A., Fatkhutdinov, T.M.: Dynamic modification of embedded devices internal firmware for solve reverse engineering problems. In: All-Russian Scientific and Technical Conference "Problems of Development of Advanced Micro- and Nanoelectronic Systems", vol. 2, pp. 63–69 (2021). https://doi.org/10.31114/2078-7707-2021-2-63-69

12. Gill, S.: The diagnosis of mistakes in programmes on the EDSAC. Proc. R. Soc. Lond. Ser. A Math. Phys. Sci. **206**(1087), 538–554 (1951)
13. Engblom, J., Girard, G., Werner, B.: Testing embedded software using simulated hardware. In: Conference ERTS'06, Jan 2006, Toulouse, France. ffhal-02270472f (2006)
14. Index TIOBE. https://www.tiobe.com/tiobe-index/. Accessed 03 Mar 2024
15. Marr, S., Daloze, B., Mössenböck. H.: Cross-language compiler benchmarking: are we fast yet? In: Proceedings of the 12th Symposium on Dynamic Languages, DLS 2016, pp. 120–131. Association for Computing Machinery, New York (2016)
16. Ivanov, S., Ilin, D., Bolshakov, I.: Benchmark of programming languages. Bull. Chuvash Univ. **3**, 222–227 (2017)
17. Qemu Emulator. https://www.qemu.org/. Accessed 03 Mar 2024
18. Kopycat Emulator. https://kopycat.ru/. Accessed 03 Mar 2024
19. Aristov, R.S., Gladkikh, A.A., Davydov, V.N., Komakhin, M.O.: Developing software Kopycat platform for emulation of complex computing systems. Nanoindustry **S**(89), 350–352 (2019). https://doi.org/10.22184/NanoRus.2019.12.89.350.352
20. Unicornemulator. https://www.unicorn-engine.org/. Accessed 03 Mar 2024
21. CAD Proteus. https://www.labcenter.com/. Accessed 03 Mar 2024
22. Integrated development environment Keil. https://www.keil.com/. Accessed 03 Mar 2024
23. Software FPLB. https://github.com/eleven5150/mips-emulation/tree/main/benchmarking. Accessed 03 Mar 2024
24. Korovin, I.V., Pulkin, I.A., Veranyan, A.S.: Research of speeds of performance of basic mathematical problems of popular programming languages. Econ. Qual. Commun. Syst. **3**(13), 63–68 (2019)
25. Frolov, V.A., Galaktionov, V.A., Sanzharov, V.V.: Investigation of the RISC-V. Trudy ISP RAN/Proc. ISP RAS **32**(2), 81–98 (2020). https://doi.org/10.15514/ISPRAS-2020-32(2)-7

Model Proposal for Phishing Detection in Text Messages Using Machine Learning

Jeferson Sandoval[1], Victor Ormeño[1], and Wilfredo Ticona[1,2](✉)

[1] Universidad ESAN, Lima, Perú
{20100714,20100835}@ue.edu.pe, wmamani@esan.edu.pe
[2] Universidad Tecnológica del Perú, Lima, Perú

Abstract. In the dynamic landscape of digital communication, text messages are seamlessly integrated into everyday interactions, but they have become a prime target for cybercriminals using phishing attacks. With around 55% of users falling victim to such fraudulent tactics, addressing the growing threat of SMS phishing is imperative. This article presents an in-depth analysis of the central role of machine learning in cyber security, focusing on the detection and prevention of SMS phishing. Based on several works such as Moncada's methodology, Dueñas' study of AI applications, ALMahadin's optimization based SVM model, Sheik's email classification and Pandiyan's algorithms to distinguish between legitimate and malicious websites, the study presents a robust methodology. This approach includes data collection, preprocessing, feature extraction using CountVectorizer and Tf-Idf, and data classification using various machine learning algorithms. The results show that the Random Forest Classifier is the optimal model, achieving an excellent accuracy of 97.94% in distinguishing between legitimate and potentially malicious text messages. The study concludes by highlighting the invaluable contribution of machine learning techniques in strengthening cyber security measures against the threat of SMS phishing.

Keywords: Phishing · Machine Learning · Natural Language Processing · Accuracy · Phishing Message

1 Introduction

The works of Pandiyan [1] argue that in the digital age, the ease of sending text messages has become an integral part of everyday life. However, cybercriminals see SMS as a new playground for phishing attacks, exploiting people's trust in this form of communication to defraud and steal sensitive information. Recent studies show that around 55% of users have fallen victim to phishing attacks, exploiting the trust inherent in text messages. That's why it's important to address the growing threat landscape. In addition, Dueñas [2] emphasizes that artificial intelligence and machine learning are important allies in the fight against phishing, allowing organizations to identify and neutralize threats more effectively. This article discusses the critical role of machine learning in cybersecurity, particularly in the detection and prevention of text message phishing. It explores strategies, challenges, and technological advances to protect users from this growing cyber

R. Silhavy and P. Silhavy (Eds.): CSOC 2024, LNNS 1120, pp. 599–610, 2024.
https://doi.org/10.1007/978-3-031-70518-2_50

threat. Recent cases highlighted by Pandiyan [1] show a 20% increase in SMS phishing in the last year alone. This alarming growth underlines the urgent need for improved technical solutions. The importance of machine learning is that it can significantly help detect and prevent SMS phishing. The integration of intelligent algorithms facilitates the detection of malicious patterns with high accuracy, which strengthens the defense against cyber threats in the field of text messages.

2 Related Work

Moncada et al. [3] used a comprehensive 5-step methodology that included data collection, pre-processing, extraction, classification, and results. The study focused on different machine learning techniques to identify phishing websites. Tree classifiers such as Decide Tree and Random Forest have achieved significant accuracy and efficiency of 97% to 99%.

Similarly, Dueñas et al. [2] conducted a state-of-the-art study on cyber security applications of machine learning. Their work involved implementing a supervised learning-based classification model for decision trees using Python and Scikit-learn. Analyzing the results of the CSIC-20102 dataset, the proposed model showed exceptional accuracy, reaching up to 100% in the classification of HTTP requests.

This study is based on ALMahadin et al. [6] who focused on detecting phishing attacks using machine learning. Their research included innovative machine learning approaches and integrated optimization strategies into the Gray Wolf model and Artificial Bee Colony Optimization-Based Support Vector Machine (GWABC SVM). This approach significantly improved detection and detection of Botnet DDoS attacks, achieving an impressive detection rate of 99.62% compared to traditional methods.

The study by Sheik and Neware [7] conducted a study of phishing attacks designed to steal sensitive data. Information A model was developed to classify emails into legitimate, phishing and spam categories. Regression logistic, random forest and support vector machines show high accuracy. The focus is on continuous improvement to provide better protection against phishing attacks.

The paper by Pandiyan et al. [5] discusses the impact of phishing attacks and highlights the use of machine learning algorithms such as decision trees, random forest, and SVM. Detect and block phishing attempts. The study emphasizes the importance of these techniques to distinguish between legitimate and malicious websites, thus improving overall security measures.

Finally, Vennam et al. [8] aimed to improve the detection of phishing sites using machine learning techniques. Using the Random Forest method, they achieved a detection rate of 97.14% with few false positives. The study also highlighted that the performance of the classifier improved when there were more training examples. Future plans include using a hybrid technique that combines the blacklist method and Random Forest to more effectively detect phishing sites.

3 Methodology

My method is inspired by [1, 2, 4], integrating the main elements of each source into a structural framework consisting of five essential steps: data collection, data processing, feature extraction, data classification and results, and Analysis. Each phase contains specific activities that are critical to the successful progress of the study.

This graphical representation serves as a comprehensive guide that visually illustrates the complex interplay of activities in each phase. It aims to increase clarity and facilitate a deeper understanding of the methodological approach used in this study by providing readers and researchers with a visual road map [12] (Fig. 1).

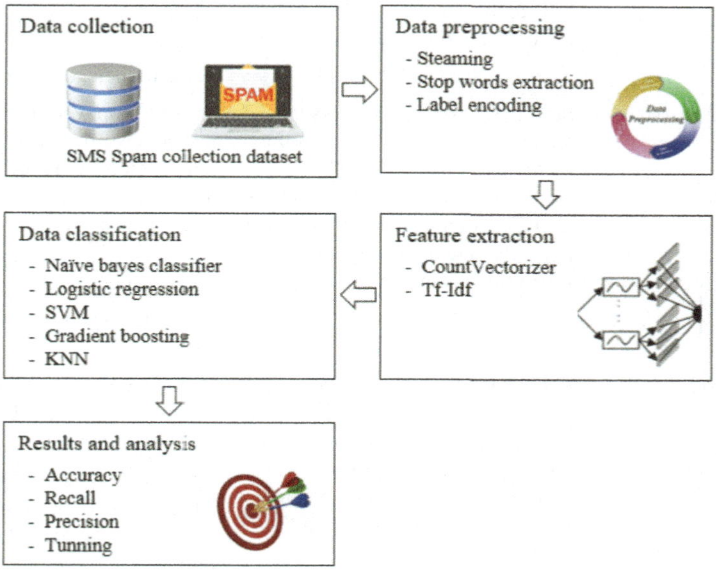

Fig. 1. Proposed methodology

3.1 Data Collection

First, data sources were carefully selected to start the investigation and focused on text messages due to their prevalence in phishing incidents. The Kaggle dataset contains 5572 rows and contains two variables: "type" (normal or phishing) and "message" (content).

3.2 Data Preprocessing

In this phase, moving to the data preprocessing step, a series of steps were applied to refine the textual data. This involved removing mentions and emoticons using regular expressions, and then removing Spanish characters such as the umlaut and tilde using the

Unicode library. A later normalization converted all content to lowercase. Tokenization then removed non-alphabetic characters, including punctuation, numbers, and special characters, leaving only words. To increase the meaning of the content, common ending words were systematically removed [13].

Table 1. Data preprocessing activities

Activities	Description	Tasks
Stemming	Stemming is a process of reducing words to their base or root form by removing suffixes and prefixes. This helps simplify words for more effective analysis and comparison	Tokenization: Split the text into words
		Apply a stemming algorithm: Use algorithms like the Porter or Snowball algorithm to reduce words to their root form
Stop Word Extraction	Stop word extraction involves identifying and removing common words that usually do not contribute valuable meaning to text analysis as they are highly frequent in a language but have little semantic content	Create a stop word list: Identify common words that will be considered stop words in the context of the analysis, in this case, Spanish stop words
		Removestop words: Apply the stop word list to filter and eliminate these words from the text
Label Encoding	Label encoding is a technique for transforming categorical data into integer values. It as- signs a unique number to each category, facilitating processing in machine learning algorithms that require numerical data	Map categories to numbers: Assign a unique number to each category in the dataset
		Apply the encoding: Replace the categories with the assigned numbers in the dataset, creating a numerical representation of categorical variables

Table 1 briefly describes the steps taken in the preprocessing stage and provides a clear overview of the careful steps taken. The resulting corpus, shown in Table 2, reflects the transformation of the original content into a refined dataset. The study used the CountVectorizer and TfidfVectorizer techniques to represent the text. These tools convert text documents into numbers by counting the frequency of words in each document. Parameters such as 500, 750 and 1000 were chosen for each method.

3.3 Feature Extraction

Seeing the results, applying the methods CountVectorizer and TfidfVectorizer produced a data frame that represented the values based on the functionality of each algorithm. This integrated approach to data processing and presentation creates a strong foundation

Table 2. Original content and the resulting corpus

Content
Entrada gratuita en 2 competiciones semanales para ganar los ensayos de la final de la Copa FA el 21 de mayo de 2005. Envíe un mensaje de texto con FA al 87121 para recibir la pregunta de inscripción (tasa de texto estándar). Se aplican los términos y condiciones 08452810075 para mayores de 18 años
Content Preprocessed
entrad gratuit competicion semanal gan ensay final cop fa may envi mensaj text fa recib pregunt inscripcion tas text estand aplic termin condicion mayor año

for subsequent machine learning analysis, contributing to the overall effectiveness of SMS phishing detection [14] (Table 3).

Table 3. Feature selection activities

Activities	Description	Tasks
Perform feature extraction with CountVectorizer	Extraction of record characteristics using CountVectorizer, a technique that converts text data into a bag of words representation by counting the occurrences of each word	Identify relevant parameters for CountVectorizer, such as tokenization options and ngram range
		Apply CountVectorizer to the corpus to create feature vectors representing the word frequencies
Perform feature extraction with Tf-Idf	Extraction of record characteristics using Tf-Idf (Term FrequencyInverse Document Frequency), a technique that measures the importance of each word in a document relative to a collection of documents	Identify relevant parameters for Tf-Idf, including adjustments for document frequency and term weighting
		Apply Tf-Idf to the corpus to create feature vectors capturing the importance of words across documents

When you use the CountVectorizer and TfidfVectorizer methods, a data frame becomes available and displays the values assigned to the operations of each algorithm. Parameters, specifically 500, 750 and 1000 features, were chosen for both methods to improve the representation of textual data.

3.4 Data Classification

3.4.1 Model Selection

During model selection, different machine learning algorithms were evaluated, each of which was evaluated. Clear advantages:

- ***Naive Bayes Classifier:*** According to Pandiyan [1], it is one of the most widely used algorithms in machine learning tasks. Among its advantages, it stands out for its fast-training speed.
  Parameters: 'alpha': 0.5
- ***Logistic regression:*** Sheik [5] says, is a data analysis technique that uses mathematics to find the relationships between two data factors. It then uses this relationship to predict the value of one of those factors based on the other. The prediction usually has a finite number of outcomes, like yes or no.
  Parameters: 'C': 10.0, 'penalty': 'l2'
- ***SVM (Support Vector Machine):*** Chen [8] defines it as a type of supervised learning algorithm used in machine learning to solve classification and regression tasks; SVMs are particularly good at solving binary classification problems, which require classifying the elements of a data set into two groups.
  Parameters: 'C': 10.0, 'kernel': 'rbf'
- ***Random Forest Classifier:*** According to Dueñas [2], decision trees have the ability to visually represent a set of rules for the purpose of making decisions related to the assignment of an output value.
  Parameters: 'max-depth': None, 'n-estimators': 100
- ***Gradient Boosting:*** Akanbi [7] defines it as a method standing out for its prediction speed and accuracy, particularly with large and complex datasets.
  Parameters: 'learning-rate': 0.2, 'n-estimators': 200
- ***KNN (K-Nearest Neighbors):*** ALMahadin [4] says, this algorithm, known as k-NN (k-Nearest Neighbors), assigns a label to each new example by measuring the distance between that example and all the examples in the training set.
  Parameters: 'n-neighbors': 3, 'p': 2

The most appropriate algorithms were selected according to the type of problem and the available data.

3.4.2 Training and Testing

In agreement with ALMahadin [4], we decided that the data were divided into training and test sets (90% training and 10% test) to train and evaluate the models (Table 4).

3.5 Results and Analysis

- **Accuracy:** This metric represents the total percentage of values correctly classified, both positive and negative [9].

$$Accuracy = \frac{TP + TN}{TP + FP + FN + TN} \tag{1}$$

Table 4. Result of models

Model	Description	Accuracy	Precision	Recall
Naive Bayes	CV 1000	0.964158	0.938772	0.931183
	CV 750	0.962366	0.937179	0.933333
	CV 500	0.964158	0.932386	0.939785
	tfidfV 1000	0.967742	0.965758	0.916129
	tfidfV 750	0.964158	0.963267	0.905376
	tfidfV 500	0.967742	0.965758	0.916129
Logistic Rgression	CV 1000	0.973118	0.974149	0.927957
	CV 750	0.974910	0.975320	0.933333
	CV 500	0.974910	0.975320	0.933333
	tfidfV 1000	0.978495	0.977664	0.944086
	tfidfV 750	0.971326	0.963786	0.931183
	tfidfV 500	0.976703	0.967698	0.947312
SVM	CV 1000	0.974910	0.980200	0.929032
	CV 750	0.973118	0.979106	0.923656
	CV 500	0.967742	0.975835	0.907527
	tfidfV 1000	0.974910	0.975320	0.933333
	tfidfV 750	0.976703	0.976492	0.938710
	tfidfV 500	0.982079	0.980013	0.954839
Gradient Boosting	CV 1000	0.978495	0.982397	0.939785
	CV 750	0.976703	0.986402	0.930108
	CV 500	0.980287	0.983499	0.945161
	tfidfV 1000	0.956989	0.969344	0.875269
	tfidfV 750	0.967742	0.970640	0.911828
	tfidfV 500	0.971326	0.963786	0.931183
KNN	CV 1000	0.931900	0.962227	0.795699
	CV 750	0.937276	0.965000	0.811828
	CV 500	0.939068	0.965932	0.817204
	tfidfV 1000	0.919355	0.938279	0.766667
	tfidfV 750	0.928315	0.960396	0.784946
	tfidfV 500	0.946237	0.969697	0.838710
Random Forest	CV 1000	0.978495	0.982397	0.939785
	CV 750	0.974910	0.970725	0.937634

(*continued*)

Table 4. *(continued)*

Model	Description	Accuracy	Precision	Recall
	CV 500	0.982079	0.980013	0.954839
	tfidfV 1000	0.982079	0.989474	0.946237
	tfidfV 750	0.978495	0.982397	0.939785
	tfidfV 500	0.985663	0.991563	0.956989

- **Precision:** This metric represents the count of accurate positive out- comes divided by the count of positive outcomes that the classifier anticipated, situations where False Positives are considered more harmful than False Negative [10].

$$Precision = \frac{TP}{TP + FP} \quad (2)$$

- **F1-Score:** Enable the equilibrium between precision and recall, optimizing their in-terplay for improved performance [10].

$$F1 = 2 \times \frac{Precision \times Recall}{Precision + Recall} \quad (3)$$

where:

- TP: True Positives
- TN: True Negatives
- FP: False Positives
- FN: False Negatives

A summary of the results of the algorithms with their respective metrics is presented in Table 5.

Table 5. Algorithm results.

Algorithms	Accuracy	Precision	Recall
Naive Bayes (tfidfV 500)	0.967742	0.965758	0.916129
Logistic.R (tfidfV 500)	0.976703	0.967698	0.947312
SVM (tfidfV 500)	0.982079	0.980013	0.954839
Random Forest (tfidfV 500)	0.985663	0.991563	0.956989
Gradient Boosting (CV 500)	0.980287	0.983499	0.945161
KNN (tfidfV 500)	0.946237	0.969697	0.838710

The best parameters of the algorithms obtained from the experiments are presented below (Table 5).

- Naive Bayes Classifier
  Parametes: 'alpha': 0.1
- Logistic Regression
  Parametes: 'C': 10.0, 'penalty': 'l2'
- SVM (Support Vector Machine)
  Parametes: 'C': 10.0, 'kernel': 'rbf'
- Random Forest Classifier
  Parametes: 'max-depth': None,'n-estimators':100
- Gradient Boosting
  Parametes: 'learning-rate': 0.2, 'n-estimators': 200
- KNN (K-Nearest Neighbors)
  Parametes: 'n-neighbors': 3, 'p': 2

3.5.1 ROC Curve

The receiver operating characteristic (ROC) curve is a valuable tool for evaluating the ability of a model to discriminate between different classes, especially in the context of binary classes. This graphical representation effectively represents the trade-off between true positive value (sensitivity) and false positive percentage between different thresholds. Basically, the ROC curve provides a visual representation of the model's performance at different decision thresholds, which helps. Understand its sensitivity to true positives and false positives. The larger the area under the ROC curve (AUC-ROC), the more effectively the model separates the classes it intends to classify [11]. A higher AUC-ROC means a better ability to accurately distinguish between positive and negative cases, which reflects the overall performance of the model in binary classification scenarios. For the ROC curve, the Random Forest model with a TfidfVectorizer of 500 features was used. AUC value is 0.96 (Fig. 2).

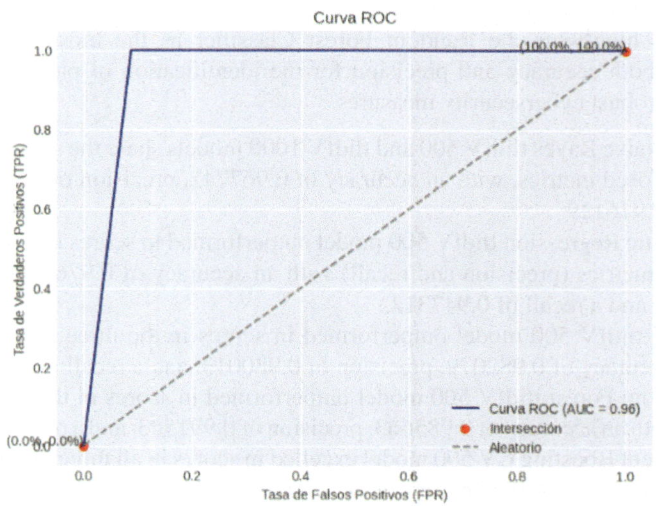

Fig. 2. Roc curve

3.5.2 Confusion Matrix

The confusion matrix of the text message classification model, where class 0 represents phishing messages and class 1 normal messages, shows a nuanced performance [15]. The model effectively detected 85 phishing messages (true positives), demonstrating its ability to detect real phishing cases. However, it showed 8 false negatives that misclassified phishing messages as normal, indicating that phishing events were detected to some extent. In contrast, the model correctly classified 465 normal messages (true negatives), demonstrating its ability to detect non-phishing. Events content However, the 0 false positives represented cases where normal messages were incorrectly flagged as phishing. This finding suggests that the model needs to be improved, particularly to reduce the number of false negatives and positives, thereby improving its overall accuracy and reliability in detecting phishing messages. These reviews highlight the iterative nature of model development and encourage modifications of the model achieves an optimal balance between sensitivity and specificity, which is crucial for robust and reliable classification in real-world scenarios (Table 6).

Table 6. Confusion matrix

	Predicted Class 0	Predicted Class 1
Class 0	85	8
Class 1	0	465

3.5.3 Analysis

The analysis highlights the Random Forest Classifier as the most reliable model, excelling in both accuracy and precision for the identification of phishing messages, essential for robust cybersecurity measures.

- Both the Naive Bayes tfidfV 500 and tfidfV 1000 models share the same scores in the three proposed metrics, with an accuracy of 0.967742, precision of 0.965758, and a recall of 0.916129.
- The Logistic Regression tfidfV 500 model outperformed in scores in two of the three proposed metrics (precision and recall) with an accuracy of 0.976703, precision of 0.967698, and a recall of 0.947312.
- The SVM tfidfV 500 model outperformed in scores in the three proposed metrics with an accuracy of 0.982079, precision of 0.980013, and a recall of 0.954839.
- The Random Forest tfidfV 500 model outperformed in scores in the three proposed metrics with an accuracy of 0.985663, precision of 0.991563, and a recall of 0.956989.
- The Gradient Boosting CV 500 model excelled in scores in all three proposed metrics with an accuracy of 0.980287, precision of 0.983499, and a recall of 0.945161.
- The KNN tfidfV 500 model excelled in scores in all three proposed metrics with an accuracy of 0.946237, precision of 0.969697, and a recall of 0.838710.

3.5.4 Model Performance Metrics

To evaluate the performance of the models, metrics such as accuracy, recall, F1-score, confusion matrix and ROC curves were used. These metrics provided a detailed understanding of how the models performed in detecting threats and suspicious activities.

3.5.5 Key Finding

Significant findings related to cybersecurity were found, including effective identification of cyber threats, reduction of false positives, and improved early detection of malicious activity.

3.5.6 Discussion

The results were discussed in the context of cybersecurity, the implications of the findings were explored, and possible improvements in cybersecurity strategies were suggested.

4 Conclusions

This study addresses the common problem of text message phishing and proposes a solution combining different methods and techniques, with a special focus on using machine learning to detect text message phishing. The spread of misinformation, especially in the context of phishing, has contributed to many people falling prey to misleading messages.

Building the database presented challenges that included both online data collection and manual tagging and validation. Some records marked with sarcastic undertones or unrelated tendencies were considered inappropriate for the classification model and were carefully cleaned. Effective preprocessing, as described in the cutting-edge technology, played a key role in achieving successful model results. By adding Peruvian slang at the end words aimed at the informal language usually used in the country, while the built-in models showed promising vectorizations that hinted at their possible scale with larger corpora.

Machine learning experiments with different algorithms culminated in a robust result. Classification model with an impressive 98.56% accuracy. A key factor in its success was the Random Forest technology. However, hyperparameter tuning required careful consideration to avoid overfitting or underfitting, which required cross-validation, which lengthened training runs.

Although the limited size of the dataset limited the duration of the exercise, increasing the size of the dataset may challenge the execution time. Although metadata such as URL, authors and date were not included in the modeling in the current study, it is recommended, that these factors should be considered in future work. Adding metadata can improve the reliability of the classifier by providing insight into the credibility of the source. The structure of URLs and authors of reputable sources can help distinguish them from less reliable sources that use attractive links. Combining metadata with content analysis offers the potential for more powerful classifiers to identify phishing data.

Finally, increasing the size and diversity of the dataset can push the model's performance beyond the 2023 records. Although the model discriminated phishing data effectively, more data would improve its reliability. This study lays the initial basis for future research to reference false information and use similar methods.

In conclusion, the evaluation of machine learning models for phishing message prediction highlights the Random Forest tfidfV 500 model as the best performer. With an accuracy of 0.985663, a precision of 0.991563, and a recall of 0. 956989, this model is a strong candidate for cybersecurity applications. Its solid performance across all metrics suggests its potential reliability in practical applications in cybersecurity environments. It prioritizes minimizing both false positives and false negatives to accurately and comprehensively identify potential phishing threats.

References

1. Selvaraj, P., Burugari, V.K., Benadit, J., Kanmani, P.: Phishing attack detection using machine learning. Measurement **24**, 100476 (2022)
2. Dueñas Quesada, J.M.: Aprendizaje supervisado para la detección de amenazas web mediante clasificación basada en árboles de decisión. Tesis para optar el Título Profesional de Ingeniero de Sistemas, Universidad Oberta de Catalunya (2020)
3. Moncada Vargas, A.E.: Comparación de técnicas de machine learning para detección de sitios web de phishing. Tesis para optar el Título Profesional de Ingeniero de Sistemas, Universidad de Lima (2021)
4. ALMahadin, G., Hiari, M.O., Hussein, A.H., Turab, N.M.M., Alkhresheh, A., Al-Tarawneh, M.A.B.: Performance evaluation of an intelligent and optimized machine learning framework for attack detection. Int. J. Commun. Netw. Inf. Secur. **14**(3), 358–371 (2022)
5. Sheik, A., Neware, S.: Phishing detection in email using machine learning. J. Algebr. Stat. **13**(3), 2268–2273 (2022)
6. Vennam, V., Hafeez, R.I.A., Khan, P.S., Fareeduddin, F.M., Naveed, S.: Using machine learning to find phishing websites. J. Algebr. Stat. **13**(3), 2373–2378 (2022)
7. Akanbi, O.A., Amiri, I.S., Fazeldehkordi, E.: A machine-learning approach to phishing detection and defense (2014)
8. Chen, Y., Chen, J.: Machine learning mechanisms for cyber phishing attack. IEICE Trans. Inf. Syst. **E102D**(5), 878–887 (2019)
9. Buriro, A., Buriro, A.B., Ahmad, T., Buriro, S., Ullah, S.: MalwD&C: a quick and accurate machine learning-based approach for malware detection and categorization. Appl. Sci. (Switzerland) **13**(4), 2508 (2023). https://doi.org/10.3390/app13042508
10. Menze, B.H., Kelm, B.M., Masuch, R., Himmelreich, U., Bachert, P., Petrich, W.: Comparison of random forest and its Gini importance with standard chemometric methods for the feature selection and classification of spectral data. BMC Bioinform. **10**, 213 (2009)
11. Chatterjee, S., Das, S., Banerjee, S., Biswas, U.: An approach towards development of a predictive model for female kidnapping in India using R programming. Int. Ethical Hack. Conf. (2018). https://doi.org/10.1007/978-981-13-1544-2_40
12. Akinyelu, A.A.: Advances in spam detection for email spam, web spam, social network spam, and review spam: ML-based and nature-inspired-based techniques. J. Comput. Secur. **29**(5), 1–57 (2021)
13. Anonymous: Machine learning for email spam filtering: review, approaches and open research problems. Heliyon (2019)
14. Mohammad, R.M.A.: A lifelong spam emails classification model. Appl. Comput. Inform. **20**(1/2), 35–54 (2024)
15. AbdulNabi, I., Yaseen, Q.: Spam email detection using deep learning techniques. Procedia Comput. Sci. **184**, 853–858 (2021)

Analysis and Forecasting of Russell 2000 Stock Price Dynamics

Roman Dzerjinsky(✉), Timur Anosov, and Sofia Daeva

Applied Mathematics Department, MIREA – Russian Technological University, 78, Vernadsky pros., Moscow 119454, Russia
9015111295@mail.ru

Abstract. The paper presents the results of the study of hidden periodicities in the dynamics of the Russell 2000 index stock using anamorphosis for Gompertz models. Boundaries of ranges of automodel growth are interconnected with local minima of functions and together with them determine the spectral composition of time series of world prices on the index fund "Russell 2000". It is shown that at the moment the price is between critical levels, when the next critical level is reached, the price of the index fund will go down to break through the value of the critical level in the future, continuing growth #CSOC1120.

Keywords: Time Series · Trend Exclusion · Anamorphosis · Near-Period

1 Introduction

The stock exchange rate is a perfect example of specific time series. Their peculiarity lies in the possibility of analyzing the interaction of processes using a variety of time scales. One way to analyze the stock price is to divide the original time series into components characterized as "fast" and "slow" elements of the process. Studying fast movements helps to identify a hierarchy of time intervals that are close to near-periods. Knowledge of near-periods allows to use them as smoothing intervals of the original series to determine the hierarchy of trends corresponding to the obtained values of near-periods.

Slow motions in this case are considered as trends of the main characteristics of the process and are analyzed together with fast motions using the apparatus of anamorphosis construction, displaying the original data in the form of piecewise linear dependencies.

In addition, there is a problem of near-proportions [1–3], which can be detected both by means of algorithms for finding near-proportions and can be reproduced through the ratios of near-periods.

This paper examines the stock price performance of the Russell 2000 index fund. Russell 2000 is a stock index of the lower 2000 funds from the Russell 3000 index. The index is backed by FTSE Russell, a branch of the London Stock Exchange Group.

The Russell 2000 is currently the best-known starting point for small-cap mutual funds. The Russell 2000 is the most widely used measure for determining the overall stock performance of small and mid-capitalization companies. The index covers approximately 8% of the total market capitalization of the Russell 3000 index [4–11].

R. Silhavy and P. Silhavy (Eds.): CSOC 2024, LNNS 1120, pp. 611–628, 2024.
https://doi.org/10.1007/978-3-031-70518-2_51

The purpose of the study is to determine the main characteristics of the process, as well as to forecast the stock price of the index fund and determine the current phase of its development.

2 Analysis of the Stock Price of the Russell 2000 Fund

To analyze the Russell 2000 fund's share price the initial data will be needed - the MICEX Index throughout the entire interval under study. Let us analyze the chart of dynamics of the share rate during the period under consideration. On the axis of abscissas trading days starting from the first one are plotted.

To make it convenient to work with the graphs, the days starting from the first one will be put on the abscissa axis (Fig. 1).

Fig. 1. Shares of Russell 2000 during the interval under consideration.

Looking at the raw data, one can see a general upward trend. Let us take the natural logarithm of the stock price and plot its dependence on the number of trading days in order to better visualize the available data. Let us proceed to the allocation of critical levels (Fig. 2).

We will select critical levels of development by changing the step in order to identify the period of formation of the critical level T_k.

Fig. 2. Allocation of critical levels with a step of 1/e.

Analytically determine the development cells on the basis of critical levels. Let us draw graphs of dependence of the natural logarithm of the natural logarithm of the fund's share price on the trading day with allocation of steps $T_k = 3000$ and $T_k = 1500$ respectively.

A change in the value of the price logarithm by one corresponds to a change in the initial data by e times. The lines parallel to the x-axis corresponding to the values 1, 2, 3 and 4 of the natural logarithm of the stock price are support/resistance lines for the corresponding areas of growth/decline of the stock price, respectively (Figs. 3, 4 and 5).

3 Intensive Growth Model

Let us proceed to analyze the intensive growth model. For this purpose, let us define the general form of the structure of this model.

$$\frac{dy}{dt} = k(y, t) * y \tag{1}$$

In order to obtain an anamorphosis that straightens the graph where there is intensive growth, we need to plot $1/y$ ~ t and mark a trend line on it. Let us denote the intervals of intensive growth in the coordinates ($1/y \sim t$) and ($\ln(y) \sim t$).

It is worth noting that any intensive growth is bounded. On a graph with the original data, this would look like a quadratic function tending to a vertical asymptote. In order to find this asymptote, the end of the period of intensive growth must be known. To do

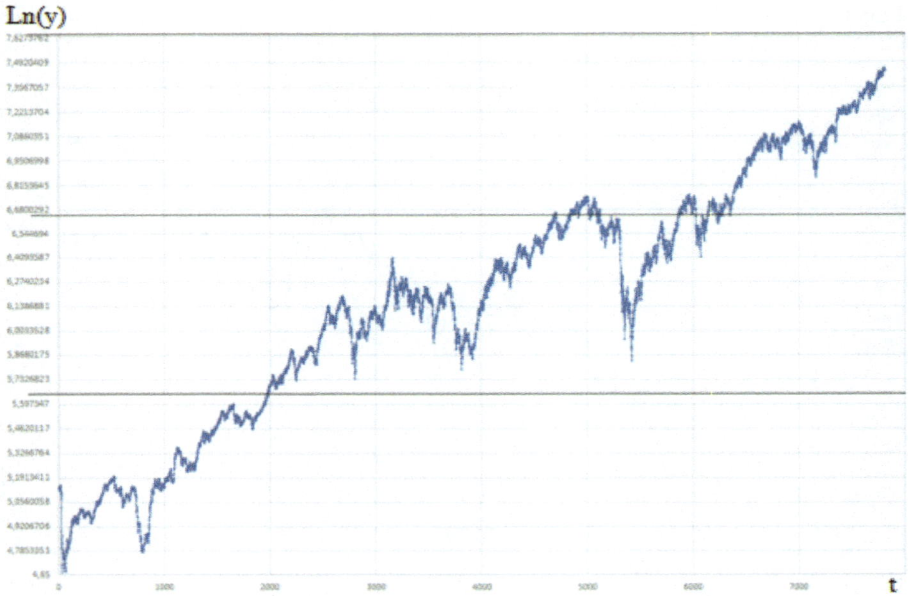

Fig. 3. Allocation of critical levels with step 1.

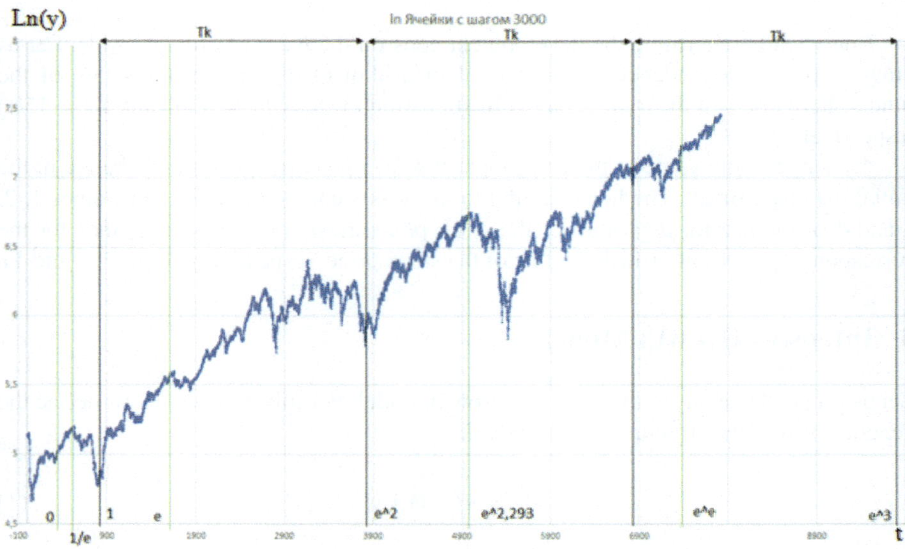

Fig. 4. Allocation of critical levels at $T_k = 3000$.

this, we need to draw lines at all plots where we can see a clear straightening of the graph.

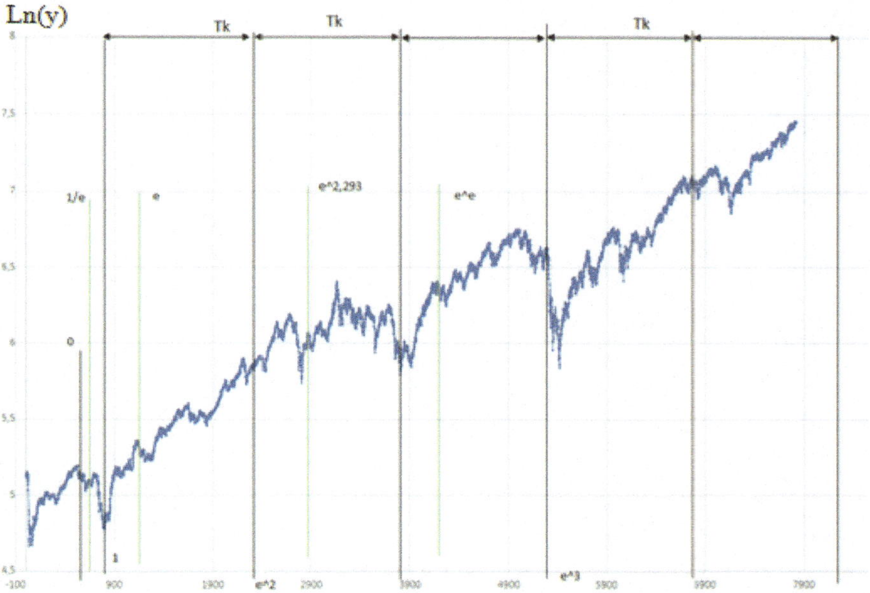

Fig. 5. Allocation of critical levels at $T_k = 1500$.

The values of 1/y in the graph of Fig. 6 decrease rather sharply, so it will be expedient to consider the dynamics of intensive growth in some parts of the graph where the straightening occurs. Let us consider the selected intervals 2845–3235 and 4180–5070. Let us find the growth limit in these intervals and denote it.

Let us construct an anamorphosis to intense growth for the trading days interval from 2845 to 3285 and analyze the data (Fig. 7).

From the chart we can see that the downward trend is slowing down - in order to analyze the dynamics of the stock price, let us build a corresponding graph and analyze it (Fig. 8).

Let us plot a similar graph as a function of t^2 (Fig. 9).

The required coefficients of the intensive growth model are found for this plot:

$$k = 9.72 \times 10^{-8}$$
$$a = 5.199694$$
$$y^* = 219.8878$$
$$t^* = 4255.426491$$

Let us do the same operations for the interval 4180–5070 (Figs. 10, 11 and 12).

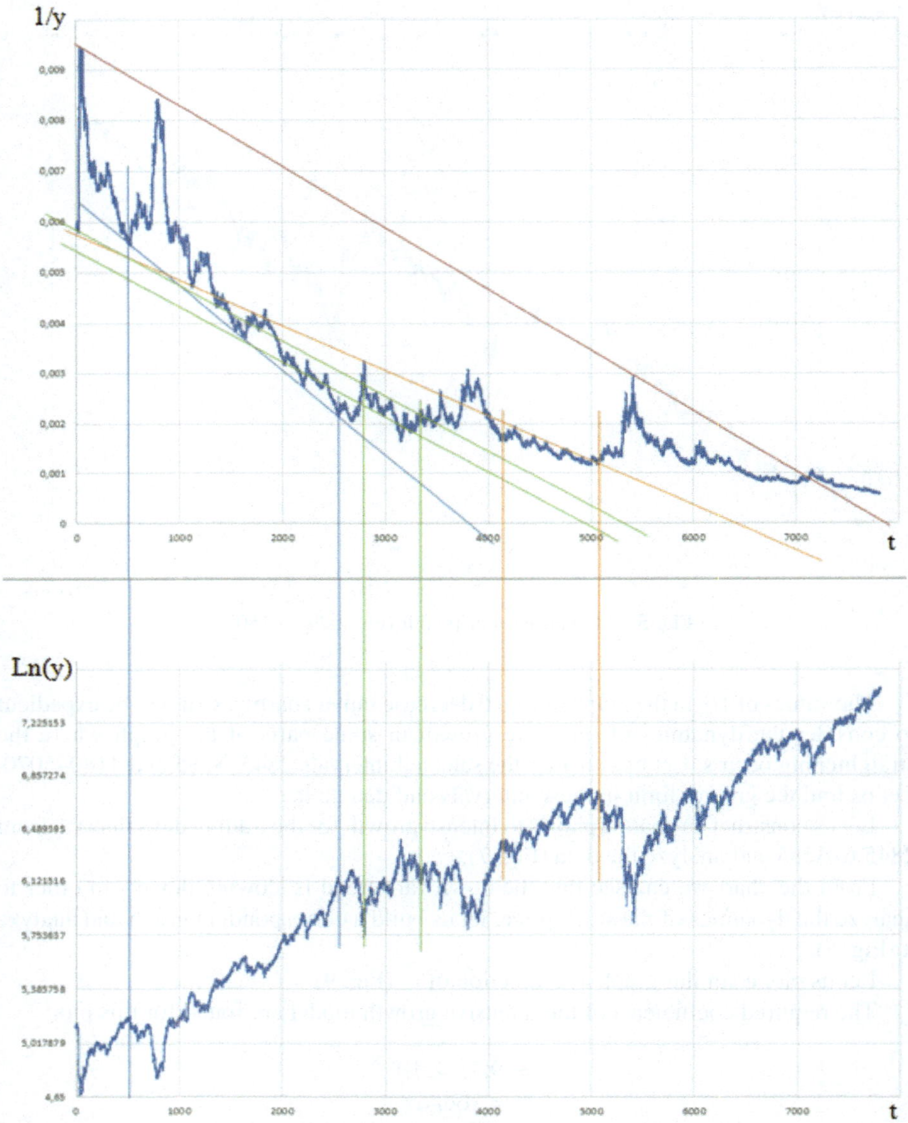

Fig. 6. Anamorphosis to intensive growth in coordinates 1/y ~ t and dynamics of share price in logarithmic scale corresponding to this period.

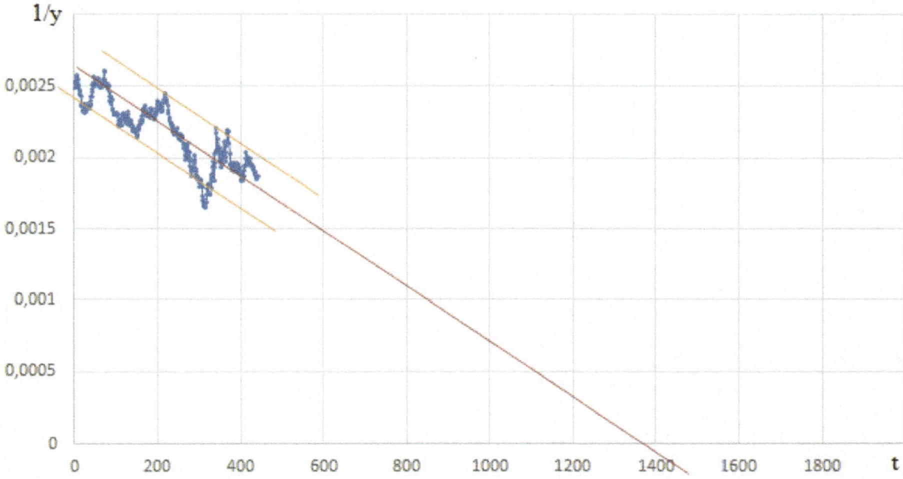

Fig. 7. Anamorphosis to the intensive growth on the interval 2845–3285.

Fig. 8. Dynamics of the share price on the interval 2845–3285 in logarithmic scale.

The required coefficients of the intensive growth model are found for this plot:

$$k = 4.85 \times 10^{-8}$$
$$a = 5.484182$$
$$y^* = 334.2535$$
$$t^* = 6778.350958$$

Let us construct a table containing the results obtained for intensive growth (Table 1).

Fig. 9. Dynamics of the share price on the interval 2845–3285 in logarithmic scale from t^2.

Fig. 10. Anamorphosis to intense growth on the interval 4180–5070.

4 Constrained Growth Model

Let us proceed to analyze limited growth. For a model to be a limited growth model, k must have the form of a decreasing function. For example, it can be a Gompertz model that decreases according to an exponential law or a logistic model that decreases from some limit. In order to see whether a plot belongs to one of these models, we need to plot the graphs in coordinates corresponding to the anamorphosis (Figs. 13 and 14).

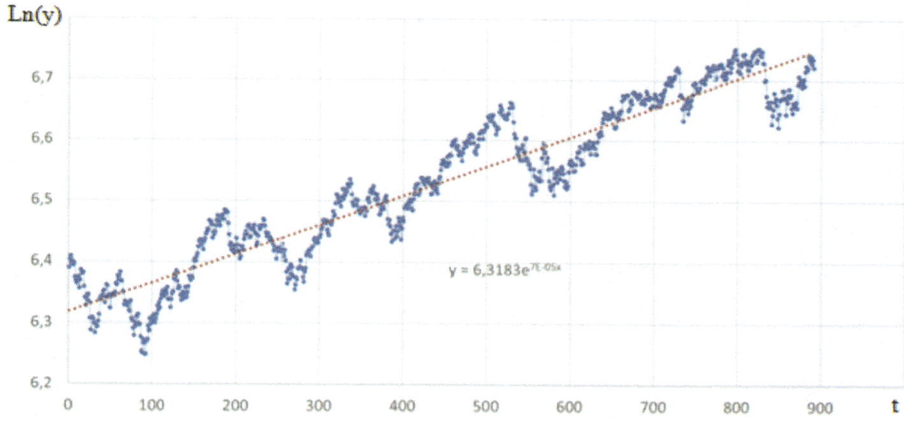

Fig. 11. Dynamics of the stock price on the interval 4180–5070 in logarithmic scale.

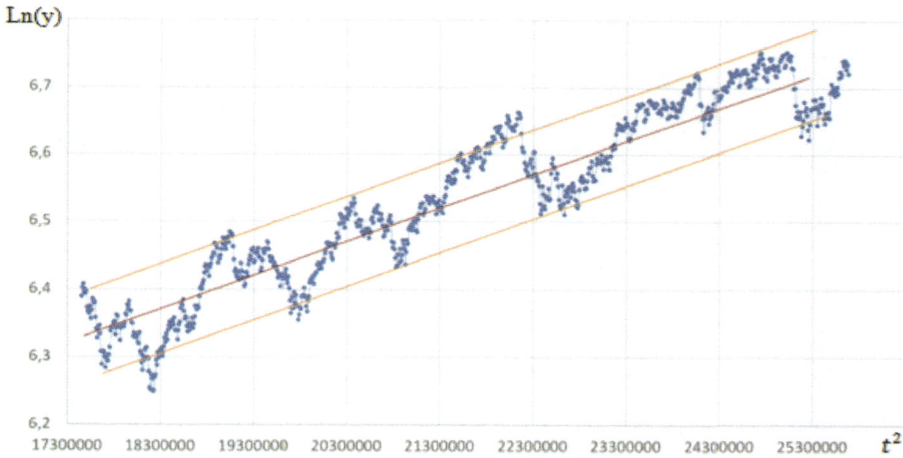

Fig. 12. Dynamics of share price on the interval 4180–5070 in logarithmic scale from t^2.

Table 1. Results for intensive growth.

Interval	a	k	y^*	t^*
2845–3285	5.199694	9.72E−08	219.8878	4255.426491
4180–5070	5.484182	4.85E−08	334.2535	6778.350958

Let us build the Gompertz model. The equation of this model has the following form:

$$\frac{dy}{dt} = A * e^{-k*t} * y$$

Fig. 13. Dependence of the ratio of current quotes to accumulated quotes from time t.

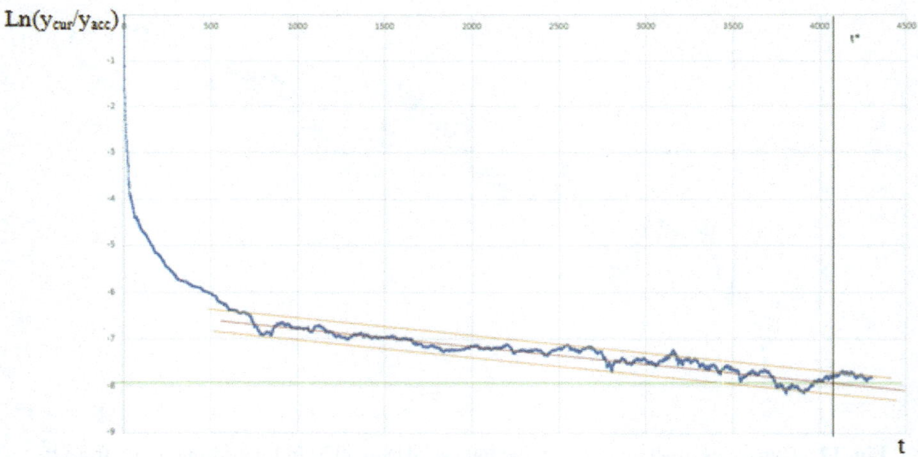

Fig. 14. Dependence of the ratio of current quotes from accumulated quotes from time t on a logarithmic scale.

Thus the anamorphosis is of the form:

$$\ln\left(\frac{\dot{y}}{y}\right) = \ln(A) - k * t$$

By analyzing the Gompertz model, an inflection point can be identified. This is done by finding the coefficient k through the equation of the straight line:

$$y = kx + b.$$

We find k = −0.000363346191307722; the inflection point is $t^* = 4085$.

To analyze the horizontal limit to which the system tends, it is necessary to build a logistic model. The anamorphosis to the logistic model is of the form:
$\frac{\dot{y}}{y} = k * \ln(y\infty) - k * \ln(y)$.

Let us find the coefficients k and b through the straight line equation $y = kx + b$ (Fig. 15).

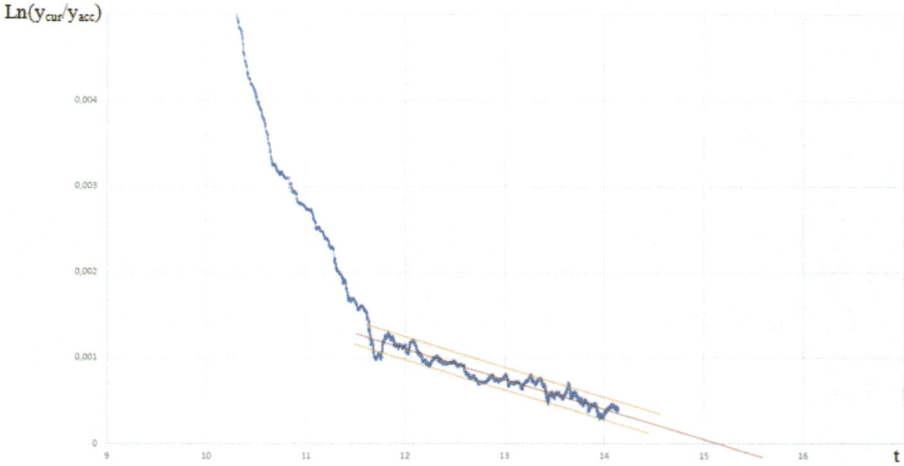

Fig. 15. Dependence of the ratio of current quotes to accumulated quotes from time t on a logarithmic scale.

To further analyze the bounded growth, it is necessary to denote the growth boundary. To do this, let us find the coefficients k and b:

$$k = -0.00034517$$
$$b = 0.00525061$$

From the straight line equation, we find x, raise the exponent to the degree obtained, and obtain the corresponding growth boundary.

$$x = 15.2118456$$
$$e^x = 4040365.12$$

Let us denote this growth boundary in the graph (Fig. 16).

This bounded growth is associated with the period $T_k = 3000$ discussed earlier. Consider the bounded growth with period $T_k = 1500$ (Fig. 17).

We find $k = -0.00027$; the inflection point $t^* = 6017.202$ (Fig. 18).

To further analyze the limited growth, it is necessary to denote the growth boundary. To do this, let us find the coefficients k and b:

$$k = -0.00065$$
$$b = 0.008915$$

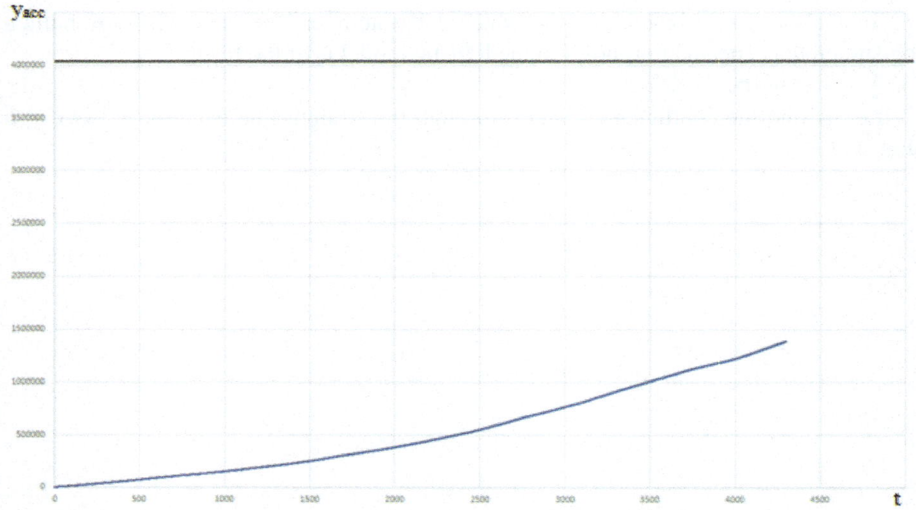

Fig. 16. Constrained growth with corresponding limit $y = 4040365.12$.

Fig. 17. Dependence of the ratio of current quotes to accumulated quotes from time t ($T_k = 1500$).

From the straight line equation we find x, raise the exponent to the degree obtained and obtain the corresponding growth boundary.

$$x = 13.74948$$
$$e^x = 936105.9$$

Let us denote this growth boundary on the graph (Fig. 19).

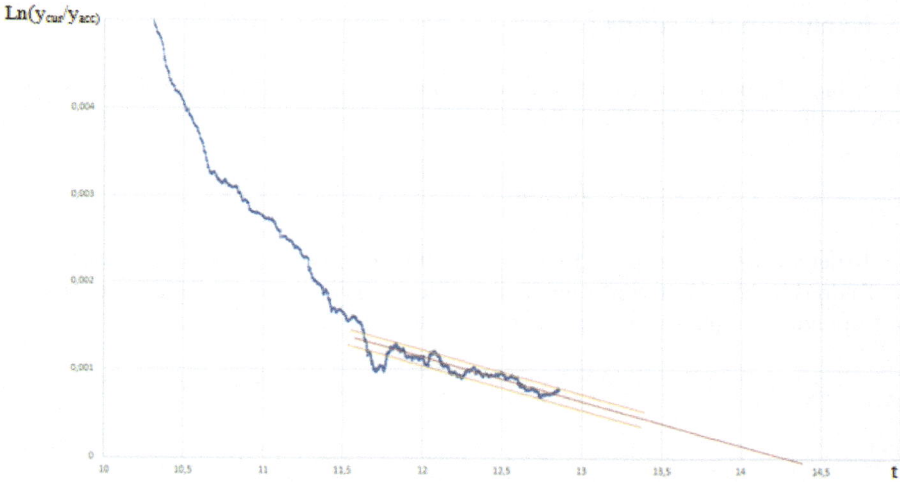

Fig. 18. Dependence of the ratio of current quotes to accumulated quotes from time t on a logarithmic scale ($T_k = 1500$).

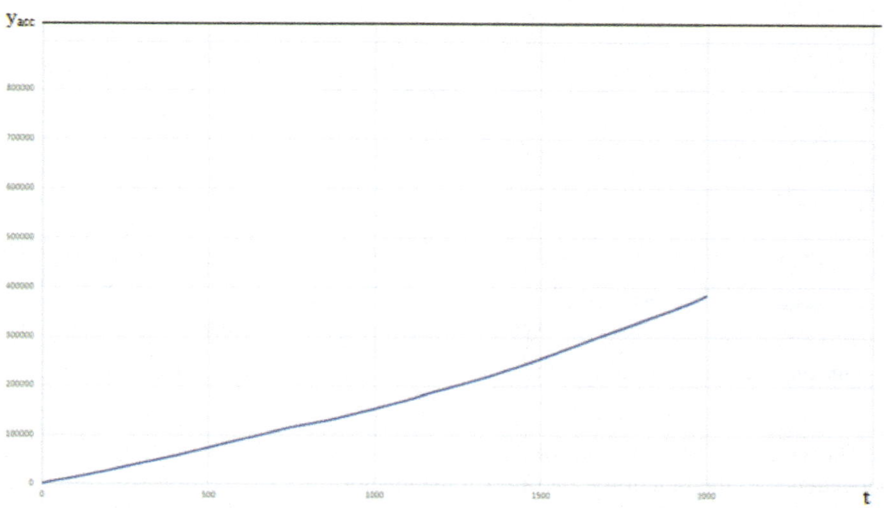

Fig. 19. Restricted growth with corresponding limit $y = 936105.9$ ($T_k = 1500$).

This bounded growth is associated with the period $T_k = 1500$ discussed earlier.

5 Exclusion of Trends

To exclude the trend, we will use the geometric proportion, which we will prologarithmize to plot the corresponding dependencies.

$$R = \ln\left(\frac{y_{t-\Delta t} * y_{t+\Delta t}}{y_t^2}\right).$$

To determine the near-period and construct the shift function, we must first determine *dt*. To do this, let us determine the mathematical expectation of the function - plot the corresponding graph and find the values closest to 0 (Fig. 20).

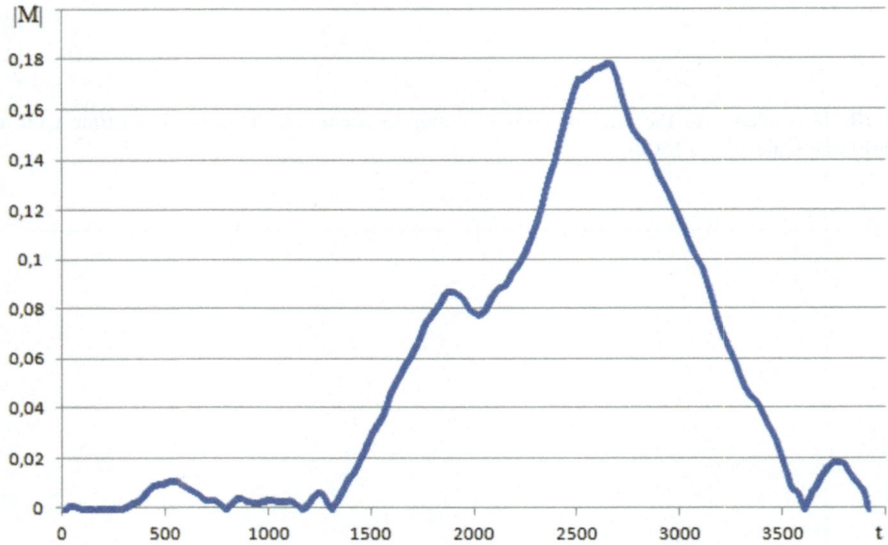

Fig. 20. Graph of dependence of the mathematical expectation of the function on time *t*.

The values closest to 0 are observed at $t = 100$. Let us plot the trend exception (dependence of the function R on time) at the given t.

After determining the near period, we proceed to the construction of the shift function. It can be determined by the formula (Figs. 21 and 22):

$$a(\tau) = \frac{1}{n - \tau} * \sum_{i=1}^{n-\tau} |f(t + \tau) - f(t)|.$$

It can be seen that in the graph of the shift function, the values $t = 1500$ and $t = 3000$ correspond to its local minima.

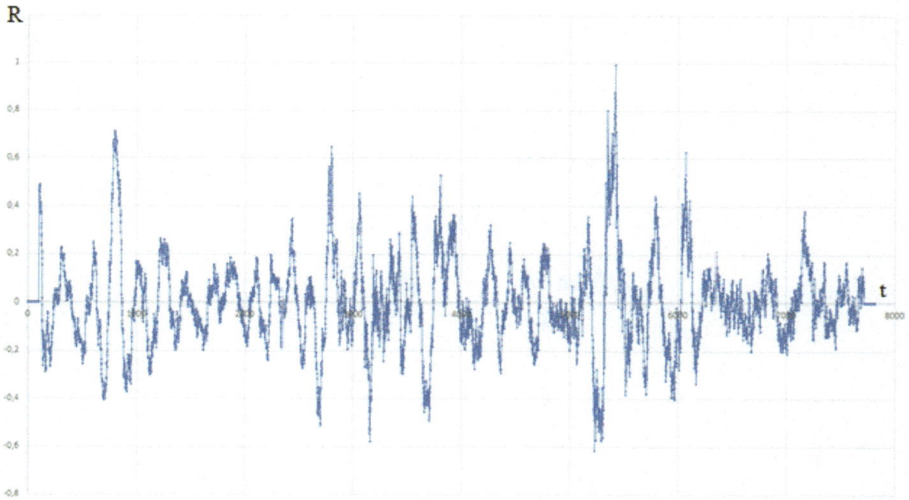

Fig. 21. Trend exclusion (dependence of function R on time).

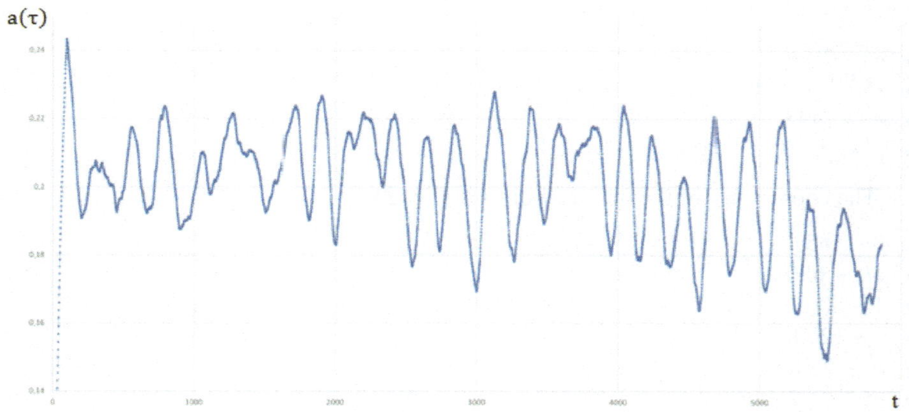

Fig. 22. Plotting the shear function.

6 Development Cells

Let us determine the optimal values of the shear function and find other points through the exponent. Let us do this operation for periods $T_k = 3000$ and $T_k = 1500$ respectively (see Figs. 23 and 24).

The coordinates of the optimal points in Fig. 23:

- 0 ~ 330;
- 1/e ~ 500;
- 1 ~ 800;
- e ~ 1600;

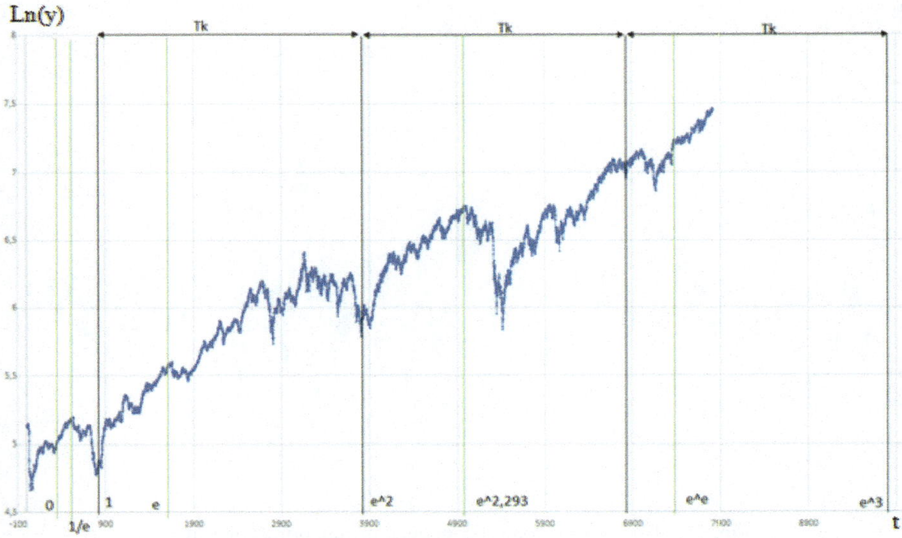

Fig. 23. Construction of the shift function ($T_k = 3000$).

- e^2 ~ 3800;
- $e^{2.293}$ ~ 4980;
- e^e ~ 7445;
- e^3 ~ 9760.

The coordinates of the optimal points in Fig. 24:

- 0 ~ 565;
- 1/e ~ 650;
- 1 ~ 800;
- e ~ 1200;
- e^2 ~ 2300;
- $e^{2.293}$ ~ 2890;
- e^e ~ 4125;
- e^3 ~ 5280.

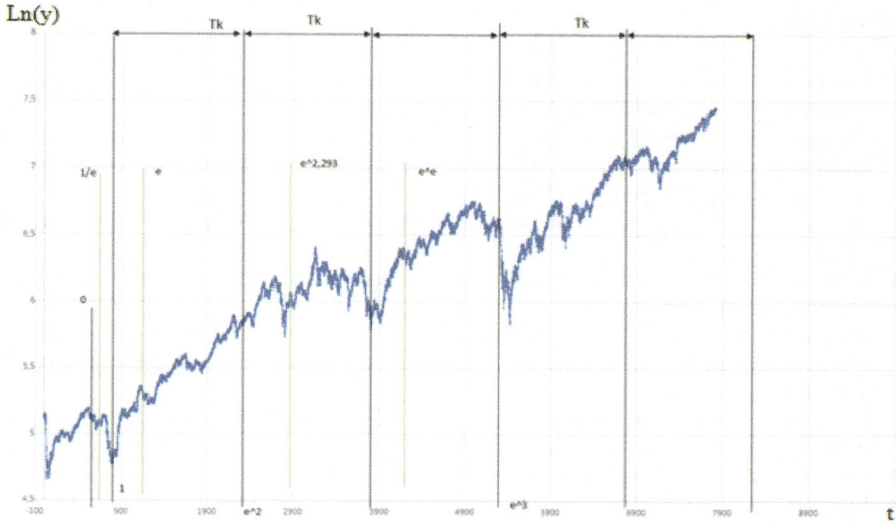

Fig. 24. Construction of the shear function ($T_k = 1500$).

7 Conclusion

Based on the results obtained from the description of the model of intensive and limited growth, it can be said that the stock quotes of the fund have a significant potential for growth. Based on the results obtained in the construction of development cells, it is possible to say that at the moment the price for this asset is between the critical levels. This indicates that when the next critical level is reached, the price may correct to further overcome this critical level and subsequent growth.

At the moment Russell 2000 shares are in correction after reaching the local maximum in the middle of 2021. Based on the analysis, it can be predicted that the next local maximum will be reached in 2025–2027 in case of absence of global geopolitical changes.

It is worth noting that the Russell 2000 includes companies with capitalization from $200 million to $10 billion. Investments in Russell 2000 are characterized by:

- high volatility - the volatility of asset prices of small businesses is higher than that of large businesses;
- wider diversification compared to other indices, which include only a couple of dozens or hundreds of companies;
- higher growth potential - it is easier for small companies to double or even triple their capitalization than large companies.

Thus, due to its broad diversification and high growth potential, the Russell 2000 is an attractive fund both in the medium and long term investment perspective.

References

1. Dzerjinsky, R.I., Sidorov, S.V., Anosov, T.E.: The problem of large local fluctuations appearance. In: Lecture Notes in Networks and Systems, vol. 723, pp. 573–580 (2023)
2. Dzerjinsky, R.I., Krynetsky, B.A., Chernorizova, N.V.: The changes dynamics analysis in the Japanese stock exchange Nikkei^225 index in the latest time. In: Lecture Notes in Networks and Systems, vol. 231, pp. 102–126 (2021)
3. Kuzmin, V.I., Gadzaov, A.F.: Methods of Model Building on Empirical Data. MIREA, Moscow, Russia (2012)
4. Serebrennikov, M.G., Pervozvansky, A.A.: Detection of Hidden Periodicities. Nauka, Moscow, USSR (1965)
5. Kuzmin, V.I., Gadzaov, A.F.: Mathematical methods of periodic components of nonlinear processes and predict the dynamics of the limited growth based on them. Russ. Technol. J. **4–2**(9), 94–104 (2015)
6. Investing.com, service for obtaining data on stock quotes of the Russell 2000 fund. https://ru.investing.com/. Accessed 03 Mar 2024
7. Levitan, B.M.: Near-Periodic Functions. Gostekhizdat, Moscow, USSR (1953)
8. Evgrafov, M.A.: Analytical Functions. Nauka, Moscow, USSR (1968)
9. Kendel, M.: Time Series. Finance and Statistics, Moscow, Russia (1981)
10. Varga, R.: Functional Analysis and Approximation Theory in Numerical Analysis. Mir, Moscow, USSR (1974)
11. Gubanov, V.A.: Selection of non-stationary cyclic component from time series. Econ. Math. Methods **39**(1), 25–39 (2003)

Evaluation of Highly Compressed Semantic Features for Efficient Image Representation

Stanisław Łażewski(✉) and Bogusław Cyganek

Faculty of Computer Science, Electronics and Telecommunications,
AGH University of Krakow, Al. Mickiewicza 30, 30-059 Krakow, Poland
{slazewsk,cyganek}@agh.edu.pl
http://agh.edu.pl

Abstract. Visual information means large amounts of data. Therefore, methods of representing images using sparse features are constantly being researched, which are effective in at least two aspects: they enable accurate classification and are lightweight. In the paper, we present a number of this type of sparse features, which arise from various combinations of features originating from the semantic layers of the ResNet-50 network and compressed with one of two dominant component analysis methods. The first is principal component analysis and the second is neighborhood component analysis. The obtained results confirm our initial assumptions that various combinations of features and component analysis enable both classification and content-based image retrieval at the level of the best performing methods.

Keywords: Sparse features · neighborhood component analysis · ResNet-50 · image classification · content-based image retrieval

1 Introduction

Visual information contained in static images or video is characterized by enormous data redundancy. Hence, research has been ongoing for a long time aimed at extracting substitute features from the image, which would be much smaller than the image itself, and at the same time would enable high-quality image classification or the search for visual content. Many methods of this type have been created [2,3,11,15,17–19] and perhaps the most famous among the older methods is SIFT [13] and modifications, such as [1,10] to name a few. Nevertheless, the emergence of deep neural networks has opened up much greater possibilities. In the context of extracting short and highly representative features, the deeper semantic layers of these networks are very suitable. There are many solutions here. An interesting idea is, for example, combining several semantic layers of the ResNet-50 network to obtain features called ResFeats [14]. We recently proposed a modification of this method, which includes extracting only dominant

R. Silhavy and P. Silhavy (Eds.): CSOC 2024, LNNS 1120, pp. 629–640, 2024.
https://doi.org/10.1007/978-3-031-70518-2_52

components using PCA, as well as additional compression of floating-point representations [12]. Thanks to this, it is possible to represent images as small as several dozen bytes, which, however, allows classification at a level not worse than 5pp compared to features without compression. In this paper, we extend this idea by presenting further possibilities of both, combining features from semantic layers, as well as their subsequent compression not only using PCA, but also by Neighbourhood Components Analysis (NCA) proposed by Goldberger et al. [6]. The new improved versions of our method presented in this article can find numerous applications in the classification of images or videos with various contents, also due to their universality [5,9].

2 Method Description

2.1 Composition of Deep Semantic Features

The scheme of semantic feature extraction in the deep neural network ResNet-50 is shown in Fig. 1.

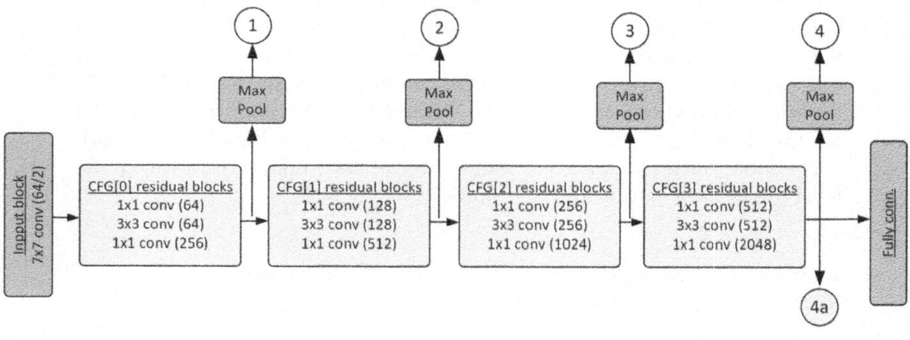

Fig. 1. Semantic feature extraction and concatenation in ResNet-50.

The scheme of feature combination for a given method is explained in Table 1.

Table 1. Description of the feature combination symbols used in this article.

Feature combination	Description
L4a	Output of the 4-th ResNet-50 residual unit after average-pooling i.e. ResNet-50 output
LX_LY_LZ	Concatenation of outputs of the X-th, Y-th and Z-th ResNet-50 residual units after max-pooling

2.2 Component Analysis for Feature Post-processing

Both of the methods we are examining assume a special transformation of data to extract latent information. The first one, PCA, has been known for years [16]. It is an unsupervised method and its main idea is to decorrelate data to find such new directions of the coordinates in which the data variance is maximized [4].

The second method we use, i.e. NCA, is slightly less known. It was proposed 20 years ago by Goldberg et al. as a supervised nearest-neighbor like classification method, but with appropriately transformed data [6]. Namely the goal of NCA is to learn a distance metric from the data by computing a linear transformation such that the average accuracy of the stochastic variant of the leave-one-out (LOO) classification is maximized in such transformed space. To explain NCA let's assume $\{\mathbf{x}_i\}_{1..N} \in \Re^L$ is a set of N L-dimensional labeled features. Now the goal of NCA is to compute a distance metric that maximizes the accuracy of the nearest neighbour classification. As an approximating setup, we concentrate on optimizing performance of the leave-one-out classification on the training data. LOO means that all data except one exemplar is taken for model building, and then accuracy is checked with that left exemplar data. Let $\mathbf{A}$ denote the data transformation matrix. Then, the quadratic (Mahalanobis) distance can be expressed as follows:

$$d(\mathbf{x}, \mathbf{y}) = (\mathbf{x} - \mathbf{y})^T(\mathbf{A}^T\mathbf{A})(\mathbf{x} - \mathbf{y}) \tag{1}$$

To compute $\mathbf{A}$ Goldberger et al. propose a differentiable cost function built on stochastic neighbour assignment in the transformed space [6]. For this purpose, the probability p_{ij} is introduced which describes a probability of a point i selecting a point j as its neighbour. This also entails assigning the class label from j to i. These probabilities are defined as follows:

$$p_{ij} = \frac{\exp(-\|\mathbf{A}\mathbf{x}_i - \mathbf{A}\mathbf{x}_j\|^2)}{\sum_{k \neq i} \exp(-\|\mathbf{A}\mathbf{x}_i - \mathbf{A}\mathbf{x}_k\|^2)}, \quad p_{ii} = 0 \tag{2}$$

Now it is possible to compute the probability p_i that point i is correctly classified,

$$p_i = \sum_{j \in C_i} p_{ij} \tag{3}$$

where $C_i = \{j | c_i = c_j\}$ denotes a set of points in the same class as that of the point i. With the above definitions, the objective function θ is formed to maximize the expected number of points that are correctly classified, that is

$$\theta(\mathbf{A}) = \sum_i \sum_{j \in C_i} p_{ij} = \sum_i p_i \tag{4}$$

Now the goal is to find out such $\mathbf{A}$ that θ in Eq. (4) is maximized. Since θ is differentiable this can be done using for example the conjugate gradient algorithm.

Note that it is also possible to restrict **A** to be a nonsquare matrix of dimensions $q \times L$, where $q < L$. Thanks to this, the low rank metric can be learned and the transformed data points will lie in lower dimensional space $\Re^q$.

We also tightened the accuracy loss condition at the expense of reducing feature length. We consider acceptable variants for which the accuracy is not worse than 1pp than the variant for which we received the best accuracy (1pp condition).

3 Results

3.1 Image Classification

To evaluate our proposed methods we selected five benchmark datasets: Caltech-256, Caltech-101, EuroSAT, Oxford Flowers, and Dogs vs. Cats. They vary in terms of both difficulty and number of samples. A detailed description of the parameters of these datasets has been given in [12]. They were divided into training and test sets using 5-fold cross-validation. The tables below show accuracy (in percents) averaged over all folds. Each feature combination was also reduced using one of two post-processing methods: PCA and NCA with a different number of components (not larger than the size of the features in a given combination). The values of the tested number of components are: [30, 60, 100, 200, 400, 600, 800, 1000, 1200]. Additionally, for each feature variant, SVM classification with a linear or radial basis kernel function was performed, and the better result was selected. Tables 2, 3, 4, 5 and 6 present the results obtained using feature connected as in the reference methods developed in [12] and [14] for the tested datasets. L2_L3_L4 connection of features classified using the SVM with the linear kernel is called ResFeats in [14]. On the other hand, L2_L3_L4 connection of features, post-processed with PCA to 60 components, and classified using the SVM with the linear kernel is called PCA-ResFeats in [12]. Finally, L4a denotes only the final layer of ResNet-50.

Table 2. Caltech-256 - Comparison of results for specific variants of reference features described in previous studies. L2_L3_L4 features classified with the linear SVM is called ResFeats [14]. L2_L3_L4 PCA compressed to 60 components, and classified with the linear SVM is called PCA-ResFeats [12]. L4a denotes only the final layer of ResNet-50.

Feature combination	Post-processing method	Classifier type	Feature length	Accuracy
L2_L3_L4	–	SVM_lin	3584	**86.77**
L4a	–	SVM_rbf	2048	85.92
L2_L3_L4	PCA	SVM_rbf	**60**	81.61

Table 3. Caltech-101 - Comparison of results for specific variants of reference features described in previous studies.

Feature combination	Post-processing method	Classifier type	Feature length	Accuracy
L2_L3_L4	–	SVM_lin	3584	**94.48**
L4a	–	SVM_lin	2048	94.01
L2_L3_L4	PCA	SVM_rbf	**60**	92.00

Table 4. EuroSAT - Comparison of results for specific variants of reference features described in previous studies.

Feature combination	Post-processing method	Classifier type	Feature length	Accuracy
L2_L3_L4	–	SVM_lin	3584	**96.23**
L4a	–	SVM_rbf	2048	95.47
L2_L3_L4	PCA	SVM_rbf	**60**	95.04

Table 5. Oxford Flowers - Comparison of results for specific variants of reference features described in previous studies.

Feature combination	Post-processing method	Classifier type	Feature length	Accuracy
L2_L3_L4	–	SVM_lin	3584	**96.76**
L4a	–	SVM_lin	2048	94.31
L2_L3_L4	PCA	SVM_rbf	**60**	93.44

Table 6. Dogs vs. Cats; Comparison of results for specific variants of reference features described in previous studies.

Feature combination	Post-processing method	Classifier type	Feature length	Accuracy
L2_L3_L4	–	SVM_rbf	3584	**99.25**
L4a	–	SVM_rbf	2048	99.24
L2_L3_L4	PCA	SVM_rbf	**60**	99.23

Tables 7, 8, 9, 10 and 11 show the best result achieved for all tested methods (depending on the dataset) in the first row. The following rows contain the ten shortest variants of features for which the accuracy achieved is not worse than 1pp from the best result in the dataset (first row of the tables). The most effective methods are in bold. We achieved at least a 9-fold reduction in feature length for all five tested datasets without reducing the classification accuracy by more than 1pp. Four variants with the length of 400 were found for the most difficult Caltech-256. The Caltech-101, which has 101 classes, can be represented with high success (93.60% accuracy) by features of length 100 based on components from the PCA transformation applied to the concatenated outputs of the third and fourth residual unit. For this dataset, we obtained the largest, over 30-fold

Table 7. Caltech-256 - Comparison of results for the ten shortest feature variants, for which the accuracy was not worse than 1pp from the best result, which was placed in the first row.

Feature combination	Post-processing method	Classifier type	Feature length	Accuracy
L2_L3_L4	–	SVM_lin	3584	**86.77**
L3_L4	PCA	SVM_rbf	**400**	85.93
L2_L3_L4	NCA	SVM_rbf	**400**	85.87
L4a	PCA	SVM_rbf	**400**	85.86
L3_L4	NCA	SVM_rbf	**400**	85.85
L3_L4	PCA	SVM_lin	600	86.07
L2_L3_L4	PCA	SVM_lin	600	86.02
L3_L4	NCA	SVM_lin	600	85.96
L4a	PCA	SVM_rbf	600	85.90
L1_L3_L4	NCA	SVM_lin	600	85.90
L2_L3_L4	NCA	SVM_lin	600	85.83

Table 8. Caltech-101 - Comparison of results for the ten shortest feature variants, for which the accuracy was not worse than 1pp from the best result, which was placed in the first row.

Feature combination	Post-processing method	Classifier type	Feature length	Accuracy
L3_L4	–	SVM_lin	3072	**94.58**
L3_L4	PCA	SVM_rbf	**100**	93.60
L3_L4	PCA	SVM_rbf	200	94.30
L2_L3_L4	NCA	SVM_rbf	200	94.23
L2_L3_L4	PCA	SVM_rbf	200	94.21
L1_L3_L4	PCA	SVM_rbf	200	94.00
L3_L4	NCA	SVM_rbf	200	93.88
L1_L2_L3_L4	PCA	SVM_rbf	200	93.77
L1_L3_L4	NCA	SVM_rbf	200	93.72
L4a	PCA	SVM_rbf	200	93.72
L2_L4	PCA	SVM_rbf	200	93.67

reduction in the length of the features (comparing the combination giving the best accuracy to the shortest variant of the features for which the result achieved is not worse than 1pp from the best accuracy). On the EuroSAT dataset, the most optimal variant of features resulting from the application of NCA on a combination of outputs from the first, second, and third residual units of the ResNet-50 network (i.e., a combination of two novelties that we introduce in this article) turned out to be the most optimal. The shortest features that meet the accuracy condition for the Oxford Flowers are 200 long. The simplest

Table 9. EuroSAT - Comparison of results for the ten shortest feature variants, for which the accuracy was not worse than 1pp from the best result, which was placed in the first row.

Feature combination	Post-processing method	Classifier type	Feature length	Accuracy
L2_L3	PCA	SVM_rbf	400	**96.77**
L1_L2_L3	NCA	SVM_rbf	**30**	96.00
L2_L3	NCA	SVM_rbf	**30**	95.92
L1_L2_L3	NCA	SVM_rbf	60	96.41
L2_L3	NCA	SVM_rbf	60	96.27
L2_L3	PCA	SVM_rbf	60	96.13
L1_L2_L3	PCA	SVM_rbf	60	96.06
L1_L3	NCA	SVM_rbf	60	96.04
L1_L3	PCA	SVM_rbf	60	95.93
L1_L2_L3	NCA	SVM_rbf	100	96.47
L2_L3	NCA	SVM_rbf	100	96.39

Table 10. Oxford Flowers - Comparison of results for the ten shortest feature variants, for which the accuracy was not worse than 1pp from the best result, which was placed in the first row.

Feature combination	Post-processing method	Classifier type	Feature length	Accuracy
L2_L3_L4	–	SVM_lin	3584	**96.76**
L1_L2_L3_L4	PCA	SVM_lin	**200**	95.93
L2_L3_L4	PCA	SVM_lin	**200**	95.92
L1_L2_L3_L4	NCA	SVM_lin	**200**	95.88
L2_L3_L4	NCA	SVM_lin	**200**	95.80
L2_L3	PCA	SVM_lin	**200**	95.77
L2_L3_L4	PCA	SVM_lin	400	96.51
L1_L2_L3_L4	PCA	SVM_lin	400	96.41
L1_L2_L3_L4	NCA	SVM_lin	400	96.36
L2_L3_L4	NCA	SVM_lin	400	96.32
L1_L3_L4	PCA	SVM_lin	400	96.26

Dogs vs. Cats dataset can be represented very effectively (99.31% accuracy) using just features of length 30, which, with the *numpy.float16* type used, gives only 60 bytes for the image representation.

Due to diversity of the datasets, we also checked which shortest feature variants are the most stable (they achieve results not worse than 1pp from the best on all five benchmark datasets), regardless of the classifier kernel. The ten shortest variants that meet this condition are presented in Table 12. Two variants

Table 11. Dogs vs. Cats - Comparison of results for the ten shortest feature variants, for which the accuracy was not worse than 1pp from the best result, which was placed in the first row.

Feature combination	Post-processing method	Classifier type	Feature length	Accuracy
L1_L2_L3_L4	NCA	SVM_rbf	400	**99.31**
L3_L4	NCA	SVM_rbf	**30**	99.18
L2_L3_L4	NCA	SVM_rbf	**30**	99.18
L2_L3_L4	PCA	SVM_rbf	**30**	99.17
L1_L2_L4	NCA	SVM_rbf	**30**	99.17
L1_L3_L4	NCA	SVM_rbf	**30**	99.17
L2_L4	PCA	SVM_rbf	**30**	99.16
L1_L3_L4	PCA	SVM_rbf	**30**	99.16
L3_L4	PCA	SVM_rbf	**30**	99.15
L1_L4	PCA	SVM_rbf	**30**	99.15
L1_L4	NCA	SVM_rbf	**30**	99.15

Table 12. The shortest universal variants whose accuracy was not worse than 1pp compared to the best results achieved on all benchmark datasets.

Feature combination	Post-processing method	Feature length	Accuracy				
			Caltech-256	Caltech-101	EuroSAT	Oxford Flowers	Dogs vs. Cats
L3_L4	PCA	**400**	85.93	94.42	95.77	96.10	99.24
L2_L3_L4	NCA	**400**	85.87	94.30	95.98	96.32	99.22
L3_L4	PCA	600	86.07	**94.53**	95.82	96.35	99.26
L3_L4	NCA	600	85.96	94.40	95.79	96.23	99.25
L2_L3_L4	PCA	600	86.02	94.48	**96.32**	96.51	**99.27**
L2_L3_L4	NCA	600	85.83	94.32	96.15	96.46	**99.27**
L1_L3_L4	PCA	600	85.82	94.26	96.14	**96.52**	**99.27**
L1_L3_L4	NCA	600	85.90	94.33	95.95	96.36	99.24
L3_L4	PCA	800	**86.25**	94.48	95.81	96.45	99.25
L2_L3_L4	PCA	800	86.16	94.50	96.27	**96.52**	99.26

with a length of 400 turned out to be the most optimal. Due to the large diversity of datasets, we would start testing on other datasets by checking these two variants.

3.2 Content-Based Image Retrieval

We checked the ten shortest variants, achieving the accuracy not worse than 1pp than those giving the best accuracy (shown in Table 12) in the content-based image retrieval (CBIR) task. We performed experiments for the most complicated Caltech-256 dataset. As before, the dataset was divided into training (examples) and test (query) sets using 5-fold cross-validation. The $MAP_{overall}@20$ and $MAP_{over\_class}@20$ metrics were averaged over the test sets [12]. No query image was used to determine the compact representation of the set. The distances of examples from queries were determined based on the Euclidean distance between their representations. The results are shown in Fig. 2. Additionally, the first variant ($L2\_L3\_L4; PCA; 60$) was checked in our previous work [12]. The last variant ($L2\_L3\_L4; -; 3584$) was proposed in [14]. The best results were achieved by the variant ($L4a; PCA; 400$) with $MAP_{overall}@20 = 66.26\%$ and $MAP_{over\_class}@20 = 63.07\%$. This result is better than the previously tested variant by about 9pp (however, on features over six times longer). Figure 3 shows one of the queries and examples whose representations were closest to that query representation. As many as 17 of the 20 most related representations were images from the same class as query (snake).

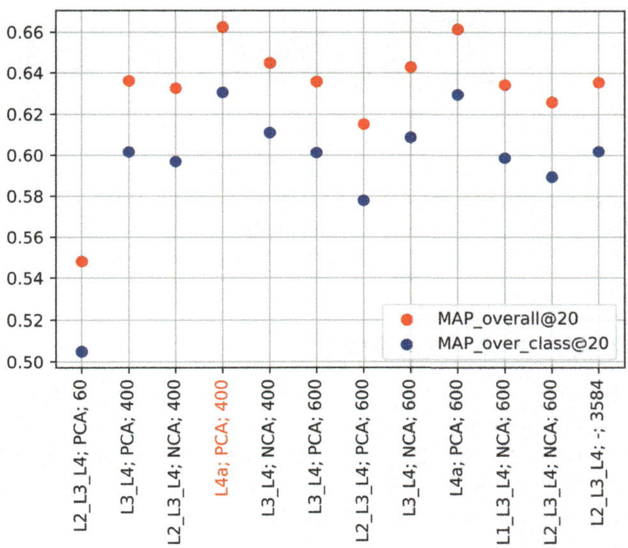

Fig. 2. The $MAP_{overall}@20$ and $MAP_{over\_class}@20$ metrics in the CBIR task on Caltech-256. The variants are L2_L3_L4+PCA with 60 components (first variant), the ten shortest variants meeting the 1pp condition (see Table 7), and L2_L3_L4 (last variant). The variant with the highest metrics is marked in red.

Fig. 3. The query image (190_0077.jpg) from the "190.snake" class in Caltech-256 is on the left. On the right are 20 examples whose representations determined by the $(L4a; PCA; 400)$ method are closest to the query image representation. Above the examples is the distance to the query and the class of the example.

3.3 Data and Code Availability

All results containing averaged accuracy values, as well as further classification reports and CBIR results are available at: https://aghedupl-my.sharepoint.com/:f:/g/personal/slazewsk_agh_edu_pl/ErljGLuQHjZGjIdSCccAD5QBOVDB8EOhF4FswU7yinK-XQ?e=SSov8. We also provide a code repository: https://github.com/Stanislaw-Lazewski/PCA-ResFeats/tree/nca.

4 Conclusions and Future Work

This article presented a significant extension of the methods we studied in [12]. First, we used NCA as a post-processing method, which turned out to be a good alternative to the previously used PCA. In the feature combinations, we also included the output from the first residual unit of the ResNet-50 network. Combining these two new features allowed us to obtain even better results for the EuroSAT dataset (see Table 9). Since in the case of real datasets we rarely have to deal with linear data separability, we classified proposed features using SVM endowed with linear or radial kernel basis functions. We evaluated the features in two ways: first, we checked for which feature variants we achieved the highest accuracy. Secondly, we reduced the number of components during feature post-processing to make them as short as possible. Of course, as the size of the features decreases, the classification accuracy also decreases. Therefore, as a trade-off, we assumed the acceptable accuracy should be no more than 1pp worse than the highest accuracy we obtained for any method for a given dataset. For most datasets, we obtain better results for the shortest variants meeting the 1pp condition for the newly tested SVM radial basis function kernel. We used 5-fold cross-validation to confirm the reliability of the results.

We also evaluated the feature variants that performed best on the five benchmark datasets in the CBIR task. For this purpose, we chose the most difficult dataset - Caltech-256. The best method achieved $MAP_{overall}@20 = 66.26\%$ and $MAP_{over\_class}@20 = 63.07\%$ which is a significant improvement (by about 9pp).

In the future, we will explore the application of here proposed features for medical image classification. We also want to test neural network architectures other than ResNet-50 for feature extraction (such as DenseNets [8] or MoblieNets [7]).

Acknowledgments. This work has been supported by the AGH University of Krakow under subvention funds no. 16.16.230.434.

References

1. Arandjelović, R., Zisserman,A.: Three things everyone should know to improve object retrieval. In: 2012 IEEE Conference on Computer Vision and Pattern Recognition, pp. 2911–2918. IEEE (2012)
2. Bay, H., Tuytelaars, T., Van Gool, L., Surf: speeded up robust features. In: European Conference on Computer Vision, pp. 404–417. Springer (2006)
3. Calonder, M., et al.: Brief: binary robust independent elementary features. In: European Conference on Computer Vision, pp. 778–792. Springer (2010)
4. Cyganek, B.: Object Detection and Recognition in Digital Images: Theory and Practice. John Wiley & Sons (2013)
5. Cyganek, B., Wozniak, M.: Tensor-based shot boundary detection in video streams. New Gener. Comput. **35**(4), 311–340 (2017)
6. Goldberger, J., et al.: In: Distinctive Image Features from Scale-Invariant Keypoints, NIPS'04, pp. 513–520. MIT Press, Cambridge (2004)
7. Howard, A.G., et al.: Mobilenets: Efficient Convolutional Neural Networks for Mobile Vision Applications (2017). arXiv:1704.04861
8. Huang, G., et al.: Densely connected convolutional networks. In: Proceedings of the IEEE Conference on Computer Vision and Pattern Recognition, pp. 4700–4708 (2017)
9. Jankowski, D., Jackowski, K., Cyganek, B.: Learning decision trees from data streams with concept drift. In: Altintas, I., Norman, M., Dongarra, J., Krzhizhanovskaya, V.V., Lees, M., Sloot, P.M.A. (eds.) International Conference on Computational Science 2016 (ICCS 2016) of Procedia Computer Science, vol. 80 , pp. 1682–1691 (2016)
10. Ke, Y., Sukthankar, R.: PCA-SIFT: a more distinctive representation for local image descriptors. In: Proceedings of the 2004 IEEE Computer Society Conference on Computer Vision and Pattern Recognition, 2004. CVPR 2004, vol. 2, pp. II–II. IEEE (2004)

11. Knapik, M., Cyganek, B.: Comparison of sparse image descriptors for eyes detection in thermal images. In: Proceedings of the 14th International Joint Conference on Computer Vision, Imaging and Computer Graphics Theory and Applications - Volume 5: VISAPP, pp. 638–644. INSTICC, SciTePress (2019)
12. Łażewski, S., Cyganek, B.: Highly compressed image representation for classification and content retrieval. In: Integrated Computer-Aided Engineering, pp. 1–18 (2023). Preprint
13. Lowe, D.G.: Distinctive image features from scale-invariant keypoints. Int. J. Comput. Vis. **60**(2), 91–110 (2004)
14. Mahmood, A., et al.: Resfeats: residual network based features for underwater image classification. Image Vis. Comput. **93**, 103811 (2020)
15. Mikolajczyk, K., Schmid, C.: Scale & affine invariant interest point detectors. Int. J. Comput. Vis. **60**(1), 63–86 (2004)
16. Pearson, K.: LIII. On lines and planes of closest fit to systems of points in space. Lond. Edinb. Dublin Philos. Mag. J. Sci. **2**(11), 559–572 (1901)
17. Rosten, E., Drummond, T.: Machine learning for high-speed corner detection. In: European Conference on Computer Vision, pp. 430–443. Springer (2006)
18. Rublee, E., et al.: ORB: an efficient alternative to sift or surf. In: 2011 International Conference on Computer Vision, pp. 2564–2571. IEEE (2011)
19. Tola, E., Lepetit, V., Fua, P.: DAISY: an efficient dense descriptor applied to wide-baseline stereo. IEEE Trans. Pattern Anal. Mach. Intell. **32**(5), 815–830 (2010)

Statistical Filtering of Images in the Spectral Domain

Valery V. Vasilevskiy(✉)

Moscow Aviation Institute (National Research University), 4, Volokolamskoe Highway, Moscow 125993, Russia
echinops777@rambler.ru

Abstract. The problem of increasing the efficiency of search and identification of anomalies in multispectral images obtained in observation systems is investigated. The approach based on the use of the spectral form of the mathematical description of the observation system model and statistical filtering of video information performed in the spatial-frequency domain is considered. The method is based on finding the spectral characteristic of the image and filtering operators, transition to solving the system of linear algebraic equations of generalized characteristic functions and estimation of the posterior distribution density of the image state vector. To perform optimal statistical filtering and image anomaly detection, an algorithm of recurrent processing of video information by a multichannel frequency filter is implemented, providing maximum suppression of noise additively distorting the useful image signal #CSOC1120.

Keywords: Observation System · Image Model · Statistical Filtering · Distribution Density Function · Spectral Response · Filter Transition Function · Computational Algorithm

1 Introduction

Observation system technologies have recently been widely used in natural resource management, cartography and many other sectors of the economy. Ensuring the automation of obtaining results taking into account the significant volume and speed of information update is an urgent and important scientific problem.

In this paper we investigate the possibilities of using an adaptive algorithm for statistical filtering of low-contrast images of the observation system in the spectral domain taking into account the uncertainty of the initial a priori information.

The class of low-contrast images here will include images whose state vector (brightness, color purity, color tone) is a random function.

The observation system realizing the processes of image acquisition, measurement and evaluation, search and recognition of individual objects (anomalies) with random energy characteristics of radiations is represented by mathematical models of stochastic control systems [1, 4].

The electromagnetic radiation field of image anomalies, i.e. the field of the observed space, can be represented by a function of the form $U_t = U(x, \lambda, t)$, describing the

R. Silhavy and P. Silhavy (Eds.): CSOC 2024, LNNS 1120, pp. 641–647, 2024.
https://doi.org/10.1007/978-3-031-70518-2_53

distribution of light energy by spatial coordinates $x = (X, Y, Z)$, wavelengths λ and the current time of image formation t.

Since the results of measurements of the distribution of intensities (amplitudes) of the image field radiations are random, let us call the algorithm of processing and estimation of their state a *statistical filter* - a dynamic (inertial) system that transforms the input signal $U(t)$ of measurements into the output signal $\hat{Z}(t)$ of the image state vector estimation.

The approach used is based on the use of the spectral method of filtering and estimation of the image state in the spatial-frequency domain [1, 3].

A random process of anomaly registration on the current image of the observation system can be characterized by an extended state vector $Z(t) = z(t, u)$ as a function of the time argument t and spatial coordinates (variables) $u = u(x, y)$, defined on the set $L([t_0, T] \times \Omega)$, $\Omega = \left\{u(x, y) : x = \overline{1, m},\ y_j = \overline{1, n}\,; m = n = a\right\}$ [1, 4].

It is required at discrete time instants $t = t_0 + T$ of all measurements $U_{t_0}^T = \{u(x, y),\ t_0 \le t \le T\}$ to obtain the image state estimation vector $\hat{Z}(t)$ in the form of a functional [1, 4]:

$$\hat{Z}(t) = \psi\left(t, U_{t_0}^T\right), \tag{1}$$

where $\psi\left(t, U_{t_0}^T\right)$ is the estimating functional of the spectrozonal information of the image.

This expression defines the input-output relation of the synthesizable *statistical non-linear filter* of the observation system, which performs the evaluation of the extended image state vector $z(t, u)$ and decision-making about the presence of an anomaly by the spectral method [2, 5].

2 Image Model

The basis for solving the problem by the spectral method is to find the spectral characteristic of the image and filtering operators, and to pass to the system of linear algebraic equations of *generalized characteristic functions* to determine the posterior distribution density $p(t, z)$ of the current image state vector $z(t, u)$ [1, 4].

We will call the spectral characteristic of the image state vector function $z(t, u)$ a $(n + 1)$-dimensional matrix $Z(n + 1, 0)$, whose elements represent the coefficients of the expansion of this function in a series of orthonormalized functions of the basis system $\{e_{i0\ldots in}(t, u)\}_{i_0,\ldots,i_n=0}^{\infty}$:

$$z_{i0\ldots in} = \int_{t0}^{T} \int_{\Omega} e_{i_0,\ldots i_n}(t, u)\, z(t, u)\, du dt, \quad i_0, \ldots, i_n = 0, 1, 2, \ldots \tag{2}$$

As a basis system of state vector functions $e(t, u)$ orthonormalized on the set Ω, we use trigonometric functions [1].

To represent the image state vector function $z(t, u) \in \Omega$, as a series over the selected basis system of functions, we use the two-dimensional discrete Fourier transform [3]:

$$Z\left(v_x, v_y\right) = \frac{1}{MN} \sum_{x=0}^{M-1} \sum_{y=0}^{N-1} u(x, y) \cdot e^{-j2\pi\left(v_x x/M + v_y y/N\right)}, \tag{3}$$

$$u(x, y) = \frac{1}{MN} \sum_{u=0}^{M-1} \sum_{v=0}^{N-1} Z(v_x, v_y) \cdot e^{j2\pi (v_x x/M + v_y y/N)},$$

where $Z(v_x, v_y)$ is the spectral characteristic of the image state vector $z(t, u)$ on the variables (v_x, v_y), which, in general, is a complex function;

$u(x, y)$ are the coefficients of the image brightness function decomposition by the basis system;

(x, y) are the spatial image variables;

(v_x, v_y) are the spectral image variables;

$\Delta v_x = 1/M\,\Delta x$, $\Delta v_y = 1/M\,\Delta y$ is the discretization step on spectral variables;

Δx, Δy is the discretization step on spatial variables, respectively;

$W(v_x, v_y) = |Z(v_x, v_y)|^2$ is the energy spectrum of the image;

$\phi(v_x, v_x)$ is the phase spectrum of the image.

The functions of the amplitude and phase spectrum of the image are bilateral and periodic, which allows to use the procedure of centering the spectral characteristic of the image:

$$\Psi\left[u(x, y)(-1)^{x+y}\right] = Z(v_x - M/2, v_y = N/2). \tag{4}$$

The spectral characteristic $Z(v_x, v_y)$ of the image state vector function in the presence of an anomaly is characterized by the presence and parameters of the corresponding components - constant frequency component ($v_x = v_y = 0$), low and high values of harmonic components.

To perform optimal image filtering we use an adaptive algorithm of recurrent multichannel video information processing aimed at maximum, in the sense of the selected criterion, suppression of noise distorting the useful image signal [5].

As filters performing multichannel processing and estimation of the image state in the spectral domain, we use the following types.

1. Butterworth low-pass filter (LPF), the transition function of which has the form (Fig. 1):

$$F(v_x, v_y) = \frac{1}{1 + \left[G(v_x, v_y)/G_0\right]^{2n}}, \tag{5}$$

where $G(v_x, v_y) = \left[(v_x - M/2)^2 + \left(v_y - N/2\right)^2\right]^{\frac{1}{2}}$.

2. A Gaussian low-pass filter whose transition function is of the form (Fig. 2):

$$F(v_x, v_y) = \exp\left(-G^2(v_x, v_y)/2G_0^2\right), \tag{6}$$

where G_0 is the cutoff frequency of the filter.

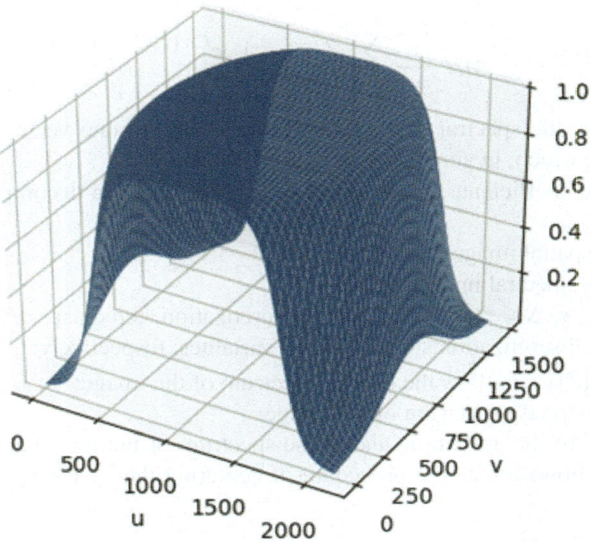

Fig. 1. Transient function of the Butterworth low-pass filter.

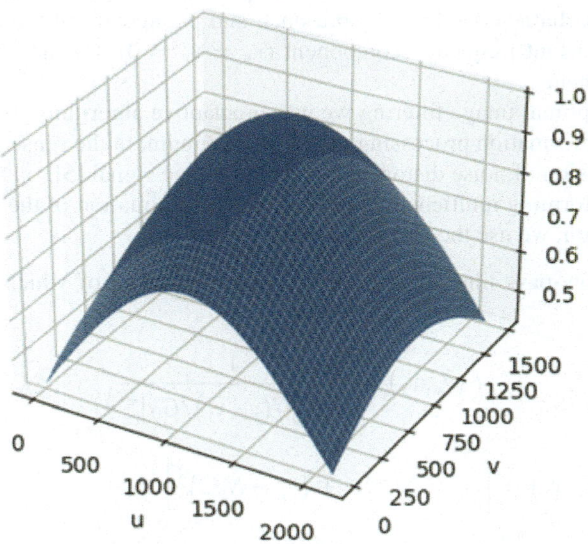

Fig. 2. Transient function of a Gaussian low-pass filter.

3. The Butterworth high-pass filter (HPF), the transition function of which has the form (Fig. 3):

$$F(v_x, v_y) = \frac{1}{1 + \left[G_0/G(v_x, v_y)\right]^{2r}}, \tag{7}$$

where $G(v_x, v_y) = \left[(v_x - M/2)^2 + \left(v_y - N/2\right)^2\right]^{\frac{1}{2}}$; G_0 is the filter cutoff frequency; r is the filter order.

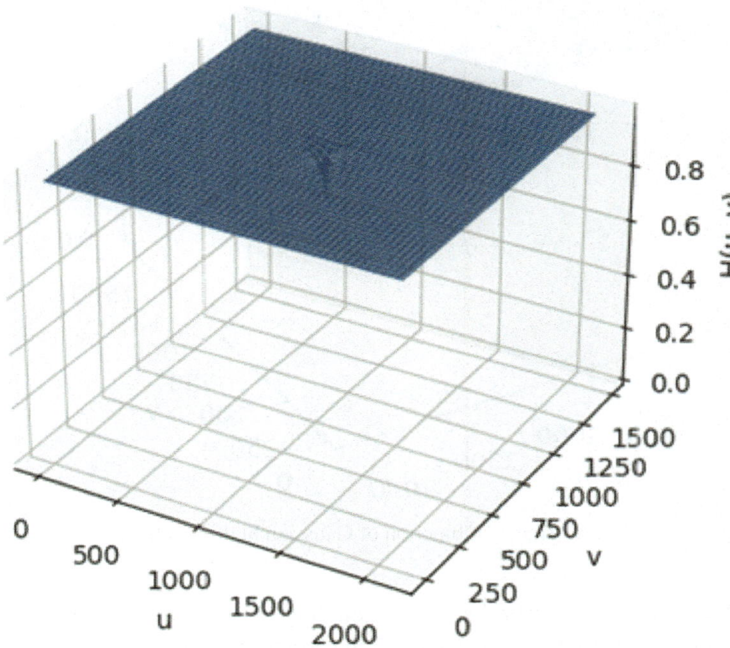

Fig. 3. Transient function of Butterworth high-pass filter.

4. Gaussian high-pass filter whose transition function is of the form (Fig. 4):

$$F(v_x, v_y) = 1 - \exp\left(-G^2(v_x, v_y)/2G_0^2\right), \tag{8}$$

where $G(v_x, v_y) = \left[(v_x - M/2)^2 + \left(v_y - N/2\right)^2\right]^{\frac{1}{2}}$; G_0 is the cutoff frequency of the filter.

The multichannel filtering operation is realized by the multiplication operator of the spectral characteristic function $Z(v_x, v_y)$ of the original image and the transition function of the corresponding filter $F(v_x, v_y)$:

$$\overline{Z}(v_x, v_y) = Z(v_x, v_y) \cdot F(v_x, v_y), \tag{9}$$

where $\overline{Z}(v_x, v_y)$ is the spectral response of the enhanced image, for which an image can be found using the inverse two-dimensional discrete Fourier transform:

$$\overline{u}(x, y) = \frac{1}{MN} \sum_{u=0}^{M-1} \sum_{v=0}^{N-1} \overline{Z}(v_x, v_y) \cdot e^{j2\pi (v_x x/M + v_y y/N)}. \tag{10}$$

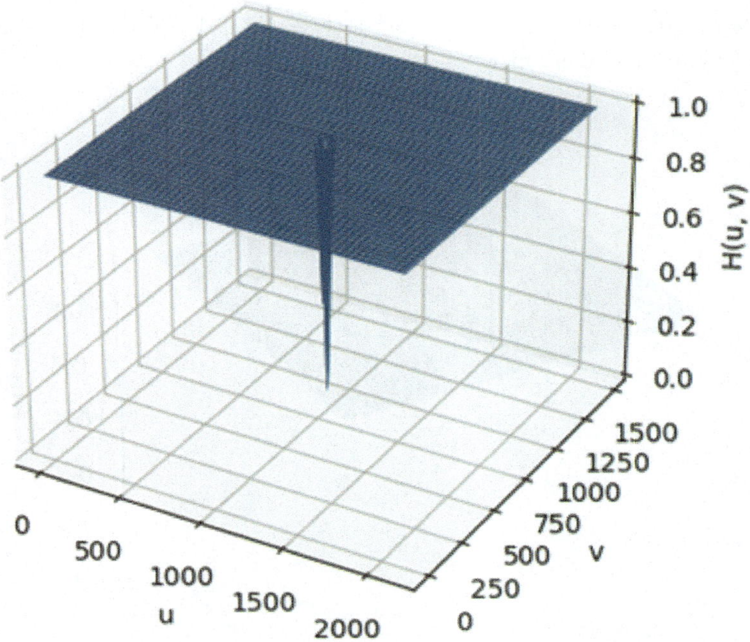

Fig. 4. Transient function of Gaussian high-pass filter.

Then the solution to the problem of estimating the image state vector $\hat{Z}(t)$ when searching for an anomaly in the form of the functional (1) can be reduced to the deterministic problem of finding the partial minimum of the *loss function*:

$$\psi\left(t, \overline{U}_0^T\right) = \underset{\psi \in R^n}{\arg\min}\, J\left(t, \overline{U}_0^T, \psi\right), \tag{11}$$

where $J\left(t, U_0^T, \psi\right) = \int\limits_{R^n} (z-\psi)^T C(t)(z-\psi) p\left(t, z\middle|\overline{U}_0^T\right) dx$ is the conditional risk function expressed through the posterior density $p\left(t, x|U_0^T\right)$; $C(t) > 0$ is the symmetric positively defined matrix of weight coefficients; $p\left(t, x|U_0^T\right)$ is the posterior probability density function of the image state vector $z(t, u)$ obtained by spectral processing method.

3 Computational Filtering Algorithm

Taking into account the adopted model of the observation system, the control law of spectral channels of video information and the random nature of background textures and anomalies, a computational algorithm for adaptive multichannel filtering of images in the frequency domain is constructed [4, 5].

1. Perform the convolution operation of the function $u(x, y)$ with the function $(-1)^{x+y}$.
2. Compute the spectral response of the image state vector $Z(v_x, v_y)$ using relation (3).

3. Set parameters and calculate transfer functions $F(\nu_x, \nu_y)$ of filters according to expressions (5)–(8).
4. Apply the multiplication operator of the spectral characteristic of the image state vector $Z(\nu_x, \nu_y)$ by the transition function $F(\nu_x, \nu_y)$ according to expression (9).
5. Obtain and extract the real part of the brightness function $\overline{u}(x, y)$ of the enhanced image according to expression (10).
6. Find the estimate of the state vector $\hat{Z}(t)$ of the enhanced image according to criterion (11).
7. Perform convolution of the function $\overline{u}(x, y)$ with the function $(-1)^{x+y}$.

From the analysis of the results of testing the algorithm of adaptive multichannel filtering on model problems of test image processing we can conclude that the estimation of the state vector $\hat{z}(t, u)$ of the enhanced image with an anomaly can be provided with higher reliability compared to the original image due to the suppression of harmonic components of the background texture, on the one hand, and the amplification of spectral components of the anomaly field.

4 Conclusion

The problem of improving the efficiency of low-contrast image processing algorithms for anomaly search is considered on the basis of investigating the possibility of applying statistical filtering in the spatial-frequency domain.

The solution of the problem of image state vector estimation is reduced to the transition to the spatial-frequency domain, performing multichannel adaptive spectral filtering and finding the state vector estimation of the improved image in accordance with the quality criterion.

The efficiency of the statistical filtering algorithm in the spectral domain is determined by the errors in the initial data on the properties of images and identifiable anomalies, parameters of transformation and filtering operators.

References

1. Vasilevskiy, V.V.: Spectral method of analysis and complex processing of video information from multi-zone images. In: Lecture Notes in Networks and Systems, vol. 910, pp. 330–335 (2024). https://doi.org/10.1007/978-3-031-53552-9_28
2. Fukunaga, K.: Introduction to the Statistical Theory of Pattern Recognition. Nauka, Moscow, USSR (1979)
3. Ditkin, V.A., Prudnikov, N.P.: Integral Transformations and Operational Calculus. Fizmatgiz, Moscow, USSR (1974)
4. Pugachev, V.S., Sinitsyn, I.N.: Theory of Stochastic Systems. Logos, Moscow, Russia (2000)
5. Fomin, V.N.: Recurrence Estimation and Adaptive Filtering. Nauka, Moscow, Russia (1984)

Machine Learning Approaches for Predicting Individual's Financial Inclusion Status with Imbalanced Dataset

Elliot Mbunge[1(✉)], Stephen G. Fashoto[1], Boluwaji A. Akinnuwesi[1(✉)], Andile S. Metfula[1], James Sicelo Manyatsi[2], Shamsudeen A. Sanni[1], John Mahlalela[1], Mzabalazo Lupupa[1], Jeremiah Olamijuwon[3], Prudence Mirriam Mnisi[2], Ntando Nkambule[2], Faith-Michael E. Uzoka[4], Dianabasi Nkantah[5], and Madoda A. Nxumalo[1]

[1] University of Eswatini, Kwaluseni, Eswatini
mbungeelliot@gmail.com, moboluwaji@gmail.com
[2] Center for Financial Inclusion, Mbabane, Eswatini
[3] Etihuku Pty Limited, Johannesburg, South Africa
[4] Mount Royal University, Calgary, Canada
[5] Coventry University, Coventry, UK

Abstract. Financial inclusion plays an important role in designing financial products and promoting sustainable development and economic growth aimed at reducing poverty. Designing financial products and services that meet individual's financial needs requires the integration and development of advanced data-driven machine learning (ML) models. However, there is a dearth of literature on the application of intelligent ML models in predicting individual's financial inclusion status. Therefore, this study applied random forest and decision trees to predict individual's financial inclusion (FI) status using data collected from Hhohho, Lubombo, Manzini and Shiselweni regions in Eswatini. Among the four regions, Hhohho had the highest number of participants (27.71%), followed closely by Lubombo (25.16%), Manzini (24.20%) and 22.93% from Shiselweni region. The findings revealed that 61.15% of the participants are partially included, 33.75% are financially included, and 5.10% are financially excluded. Moreover, the synthetic minority oversampling technique (SMOTE) was used to address the class imbalance problem and further applied random forest and decision trees to predict FI status. The study results revealed that random forest performed better with a high accuracy of 98.73%, precision of 98.76%, recall of 98.73% and F1-score of 98.67%. Decision trees also performed better with an accuracy of 97.47%, precision of 98.19%, recall of 97.47% and F1-score of 97.65%. Random outperformed the decision trees across all metrics in predicting individual's FI status. Implementing such models can aid policymakers and financial institutions in developing strategies to improve financial access and support financial inclusion initiatives.

Keywords: Financial Inclusion · Machine Learning · Eswatini

R. Silhavy and P. Silhavy (Eds.): CSOC 2024, LNNS 1120, pp. 648–658, 2024.
https://doi.org/10.1007/978-3-031-70518-2_54

1 Introduction

Financial inclusion (FI) plays an important role in poverty eradication while improving social inclusion and contributing to inclusive economic growth. FI refers to the delivery of financial services and products [1] and ensuring that they are available, accessible and affordable to all segments of society in the formal financial system [2]. Policymakers around the world have embraced financial inclusion as the key driver to economic empowerment [3], reducing poverty levels and bridging the existing financial exclusion, especially in rural and marginalized communities. This can be achieved by improving financial literacy [1], providing equal access to financial services [4], products and resources, and most importantly opening avenues for secure and safe saving practices guided by the policies that support sustainable inclusive financial systems [2]. Furthermore, reducing the growth of exploitative informal financial credit sources can bring socio-economic benefits [3]. Thus, improving access to financial services such as credit, savings, insurance, payments and remittances, provided by diverse financial service providers aligns with 6 out of the 17 goals of the sustainable development goals (SDGs) and main objectives of the International Labour Organization (ILO) which are instruments for promoting economic growth and sustainable development [5]. Despite financial inclusion being regarded as a key indicator of inclusive growth, financial products and schemes which are means of facilitating financial facilities in societies are not effectively captured in many developing countries. This consequently leads to the dearth of financial inclusion literature and the development of data-driven models for assessing and predicting an individual's financial inclusion status using intelligent predictive models such as machine learning and deep learning algorithms.

Globally, an estimated 1.7 billion adults are deprived of formal financial services [6] but Eswatini has made significant progress in improving access to financial products and services to build a sustainable society and inclusive economy. Financial inclusion in Eswatini is mainly characterized by banking services (deposits, withdrawals, remittances of funds and loan facilities) [7], mobile money [8], savings and credit and Insurance [9]. However, other emerging important variables such as mobile money services and digital banking influence financial inclusion through digital financing. This resulted in a further decline in the financially excluded population from 27% in 2014 to 13% in 2018 [9], bringing the financially included population to more than 75% of the bankable population of the nation [8]. The increased financial inclusion is associated with mobile money services, yet the use of digital financial services is low [10].

There are still barriers that cause certain individuals to be structurally excluded from participation in the formal financial system. These impediments include demand-side barriers, supply-side barriers, regulatory barriers, and access barriers as indicated in the National Financial Inclusion Strategy (NFIS) for Eswatini [7]. A study by Anthanasius Fomum & Opperman [11] also indicated that Micro, Small & Medium Enterprises (MSMEs) lack access to finance, and there is a dearth of literature on the enabling role of FI on MSMEs' performance. In addition, Nkambule [12] also highlighted that some studies overlooked certain key factors for effectively investigating financial inclusion, especially in Eswatini rural communities where mobile money services are extensively used for transactions [13]. In addition, the lack of data-driven intelligent applications for re-engineering financial inclusion and determining an individual's FI status in real-time is also a barrier. To harness the power of machine learning predictive models to build a sustainable society and inclusive economy in the Kingdom of Eswatini, it is imperative to develop a reliable and dependable financial inclusion (FI) scheme for citizens and businesses to improve access to financial services and contribute to economic growth. Therefore, this study sought to address this gap by predicting an individual's degree of financial inclusion using machine learning algorithms.

2 Related Work

Financial inclusion is a key driver for sustainable development goals. Increasing access to finance services and products for individuals is paramount to facilitate secure payments [11], including for basic services and trade; smooth cash flows [4]; offer financial protection; and improve capital allocation while enabling investments. Several techniques including the generalized ordered logit model [11], and propensity score matching (PSM) methods [13] have been utilized to assess individual's financial inclusion status. However, these techniques lack predictive capacity, hence the need for applying machine learning techniques.

Machine learning techniques have made remarkable progress in the financial sector to solve different tasks, including fraud detection, anomaly detection, pattern recognition, credit score and risk assessment [14]. Several studies applied machine learning algorithms to analyse financial transaction data and subsequently transform customers' financial services [15] and their interaction with financial ecosystems. For instance, machine learning models such as logistic regression (LR), support vector machines (SVM), Naïve Bayes, decision trees (DT), AdaBoosting, random forest, Bagging, XGBoosting, voting Ensemble, K-Nearest Neighbour (KNN), Stack, and Gradient Boosting Classifier were applied by Ismail et al. [16] to predict the state of FI in Africa using data from Zindi. Their study revealed that the XGBoost model achieved a high accuracy of 89.23%. Akinnuwesi et.al [17] analysed the financial inclusion dataset in Eswatini to determine the extent to which MSMEs have access to financial services using support vector machines (SVM) and logistic regression (LR). Their study revealed that FI differs with regions and the performance of SVM and LR was relatively affected by the large number of missing data. To validate the performance of SVM and LR, [5] applied the same ML models to the recent FI scheme from 2018 and the FI dataset extracted from Finscope. The performance was relatively improved, and SVM achieved optimal recall of 73% and 77.8% for LR. Despite the limited literature on the application of ML in financial inclusion, this section indicates that there is significant progress. However, [18] also highlighted that there is a lack of data-driven FI predictive tools that can determine an individual's FI status which exacerbates inequalities and financial exclusion, especially in a society that survives through an informal economy. Furthermore, most of the studies applied machine learning algorithms to determine FI status as included or excluded, none of the researchers predicted the degree of financial inclusion using multiclass classification models.

3 Materials and Methods

A. Study Area and Data Source. Eswatini is a landlocked country with a population of 1.2 million and an area of only 17,373 km^2 [11]. The country is divided into four administrative regions: Hhohho, Lubombo, Manzini and Shiselweni. Eswatini's economy is diverse, with mining, forestry, and agriculture contributing more than 13%, manufacturing contributing 37% and the services subsector making up more than 50% of the gross domestic product (GDP) [10]. Researchers collected data under the guidance of the Centre for Financial Inclusion (CFI), a policy-advocacy unit within the Ministry of Finance that coordinates and monitors the implementation of the National Financial Inclusion Strategy (NFIS) implementation [10]. In each region, constituencies were carefully selected, and participants were randomly selected to participate in the study. A cross-tabulation of Eswatini Access to Finance Strand by region was used to purposively select a constituency per region (1*4 Constituencies) that had the highest percentage of excluded population. Data were collected through survey questionnaires in the following constituencies Timphisini (Hhohho), Hlane (Lubombo), Mangcongco (Manzini) and Shiselweni (Shiselweni). The survey questionnaire collected data on access to financial services, usage of financial services and quality of the financial products and service delivery as well as sociodemographic data. To calculate the sample size, the study used an online Check Market Calculator, at a 5% margin error and 95% confidence interval. Therefore, a total of 362 respondents participated in this study. The participants were purposively selected at each constituency after liaising with the leadership of each constituency.

B. Data Description- Financial Inclusion Indicators. Data collection instruments were designed and developed guided by the following key indicators: employment status, source of income, ownership of insurance policy, ownership of bank account, types of insurance owned, ownership of mobile money, ownership of mobile money account, types of mobile money account, mode of sending and receiving money. The study also included other demographic characteristics to assess FI. A study by Ajetomob et al. [19] also highlighted that there is a link between gender and financial inclusion and the importance of women in sustainable economic growth and encouraged financial institutions to capacitate and mentor female-owned firms, adjust tax and collateral requirements for access to credit in Eswatini. Improving access to financial services increases opportunities for communities to unlock economic potential and financial inclusion. Also, a study by Anthanasius Fomum & Opperman [11] indicated that developing inclusive financial systems that provide access to financial services such as formal accounts, savings, credit, insurance, and payment options to all eligible populations may foster inclusive growth.

C. Measuring Financial Inclusion. Generally, access to financial services, usage of financial services and quality of the financial products and service delivery are the three important dimensions for measuring financial inclusion [20]. Access to financial services represents the possibility for individuals to use them. Usage indicator was measured by holding at least one financial product, keeping savings, and having a loan in a formal financial institution. Quality is the ability of the financial service or product to meet the needs of the consumer. It describes whether financial products and services match clients' needs, or the range of options available to customers, and clients' awareness and understanding of financial products [21]. These dimensions are, at the same time, determined by several demand-side individual-level indicators for the cases of usage and quality, and supply-side country-level indicators for access [20]. Therefore, to determine the financial inclusion status, a two-stage Principal Component Analysis (PCA) was used as an indexing strategy. Access is key since it represents a necessary but not sufficient condition for using formal financial services. Therefore, we determined the degrees of FI and classified them into four classes: banked, financially excluded, non-banked and informally served. Banked refers to consumers with bank accounts or digital wallets in deposit money (or microfinance) banks and mobile money operators. Unbanked (non-banked) are individuals, without a bank account or digital wallet, who conduct all their financial transactions independently. Furthermore, in our study, we combined unbanked and informally served together because of the following reasons:

1. Since there were only 3 samples for the informally served. There are three ways to handle this; one is to ignore it, secondly, leave it as a class and thirdly, add it to another class based on a justifiable reason.
2. When the informally served is added to the non-banked. The authors refer to those who are non-banked as informally served because we consider the banking system as a formal form of financial service. However, to guide against mere assumptions, the authors decided to go with partial inclusion as either way, those who are non-banked or informally served have some form of financial service hence they are classified as partially included.

The ones who are banked are deemed financially included. The ones who do not interact with the financial system are considered financially excluded. The ones that are non-banked or informally served are considered partially included.

D. Data Preprocessing. The collected data had some missing values. To address this, the authors deleted samples with missing values and remained with 314 samples. Furthermore, the study applied label encoder to change the categorical data into numerical values. Figure 1 shows the class distribution of the target class. The samples in the target class are highly imbalanced, therefore, we applied SMOTE to address the class imbalance problem.

The data was first divided into a ratio of 80:20. The train data was 80% and the test data was 20%. The train data was then divided into x_train (X) and y_train (Y) while test data was divided into x_test and y_test. The train data was cleaned and encoded to a categorical variable after which it was split into X and Y. X being the independent variable(s) with Y being the dependent variable, in this case, the label column with either Financially Included, Partially Included, or Financially Excluded item. To balance the data, the SMOTE method was applied to both X and Y to make sure the values were evenly distributed across the data. The data was then scaled using the standard scaler library in Python. A new set of data (x_test) without the labeled column is encoded and transformed using the same scaler used for X in the training part. Then the trained model was used to predict the outcome of this new data stating either Financially Included, Partially Included, or Financially Excluded. It has been done in such a way that when a set of dependent variables (X) is passed in the ML model, it returns an outcome (Y).

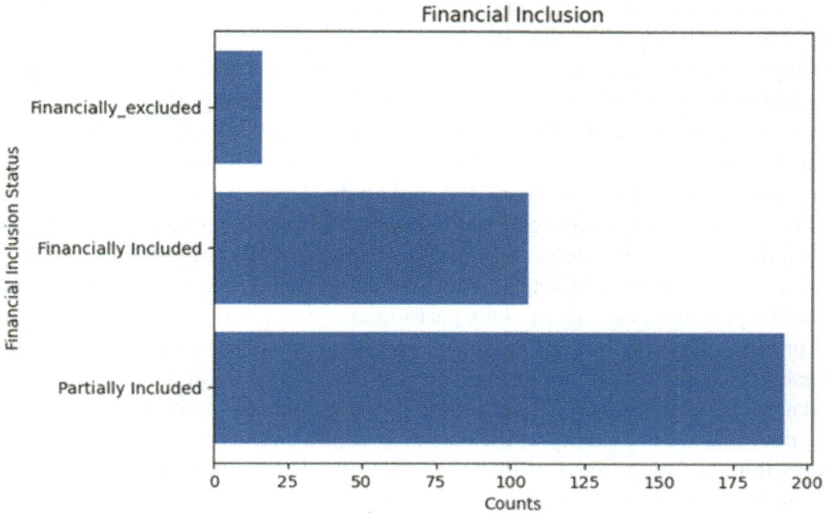

Fig. 1. Class Imbalance

E. FI Status Predictive Models and Evaluation Metrics. After applying the feature engineering techniques, two machine learning classifiers, namely, decision trees (DT) and random forests (RF) were used to predict the financial inclusion status of individuals. These models are explained in the subsequent section.

A Decision Tree is a predictive model that maps features to outcomes by making a series of decisions based on the values of those features [22]. DT is a versatile and interpretable algorithm that can be used to predict the financial inclusion status of individuals. DT was applied by Ismail et al. [16] to predict the state of financial inclusion in Africa. Similarly, DT was used to predict stock price movements through the Spark big data framework by Hasan et al. [22]. The DT model inherently performs feature selection by choosing the most informative features to split on at each node. During model training, the algorithm recursively splits the data based on the selected features until the stopping criterion is met. Hyperparameter tuning helped to prevent overfitting. The model performance was evaluated based on accuracy, precision, recall, and F1 score. The results are presented in Table 1. Decision Tree is advantageous for its simplicity, interpretability, and ability to handle both numerical and categorical data. However, it is prone to overfitting, and the interpretability might decrease as the tree becomes more complex. Therefore, ensemble methods like Random Forests can be considered to mitigate overfitting while maintaining interpretability.

Random Forest (RF) is an ensemble learning technique that can be employed to predict the financial inclusion status of individuals. It can measure multiple trees with the same datasets and calculate the value of the forecast of every single tree [22]. Random Forest combines the predictions of multiple decision trees to create a robust and accurate model. The features included in the model were: demographic information, income, education, employment status, and other features that might influence financial inclusion status. Feature importance was assessed using the feature importance scores provided by the RF algorithm and this helps to identify the most influential features for predicting financial inclusion status. While training the RF model on the training dataset, the algorithm built multiple decision trees by randomly selecting subsets of features and data points. Each tree in the forest "votes" on the predicted outcome. With the view to optimize the model performance, we fine-tune the hyperparameters of the Random Forest, such as the number of trees, the maximum depth of each tree, and the minimum number of samples required to split a node. We implemented the StratifiedKFold cross-validation to ensure the generalization performance and robustness of the model. The cross-validation helps to detect overfitting and provides a more reliable estimate of the model's performance. The model's performance was evaluated using the testing set based on the following metrics: accuracy, precision, recall, and F1-score. The results are presented in Table 1. Random Forest is known for its ease of use, ability to handle large datasets, and resistance to overfitting. However, like any machine learning model, its success relies on the quality of the data and feature engineering.

After implementing financial inclusion status predictive models, their performance was evaluated by using a confusion matrix, as shown in Figs. 2 and 3. Therefore, accuracy, precision, recall and F1-score were used to measure the ability of the model to predict the financial inclusion status of individuals, as shown in Table 1.

Table 1. FI status predictive models

FI status predictive models	Accuracy	Precision	Recall	F1-score
Random Forest	0.9873	0.9876	0.9873	0.9867
Decision Trees	0.9747	0.9819	0.9747	0.9765

The interpretation of the confusion matrices results. For the confusion matrices, the three outcomes are coded as follows: Financially excluded, financially included, and partially included. They are denoted by 0, 1 and 2 respectively, as shown in Figs. 2 and 3.

4 Results Analysis

After data preprocessing, a total of 314 samples were considered for further analysis. Among the four regions, Hhohho has the highest number of participants, with 87 (27.71%), followed closely by Lubombo with 79 participants (25.16%). Manzini has 76 participants (24.20%), and 72 participants (22.93%) were from Shiselweni region. However, the gender distribution shows a slight imbalance, with a slightly higher number of females than males. Females account for 171 (54.46%), while males account for 143 (45.54%) of the total number of participants. The study also reveals that most participants are married 143 (45.54%), with 119 singles (37.90%) being the next largest group, followed by 44 individuals (14.01%) who are widowed, 4 individuals (1.27%) are divorced, and another last 4 individuals (1.27%) who are separated. The study further

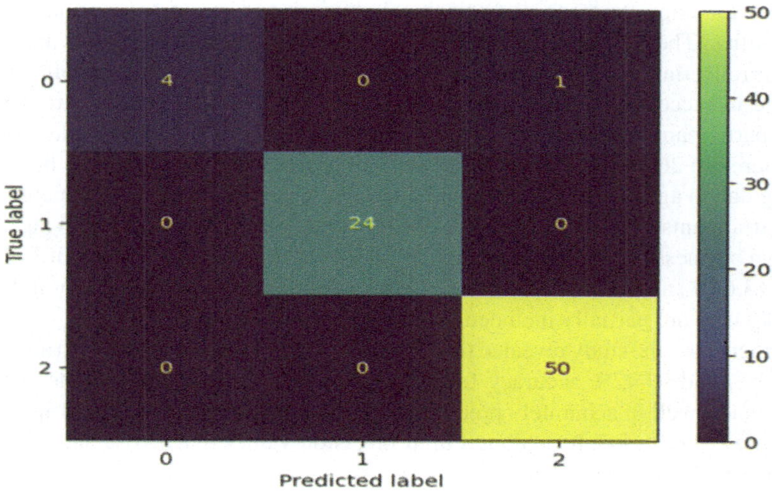

Fig. 2. Random Forest confusion matrix

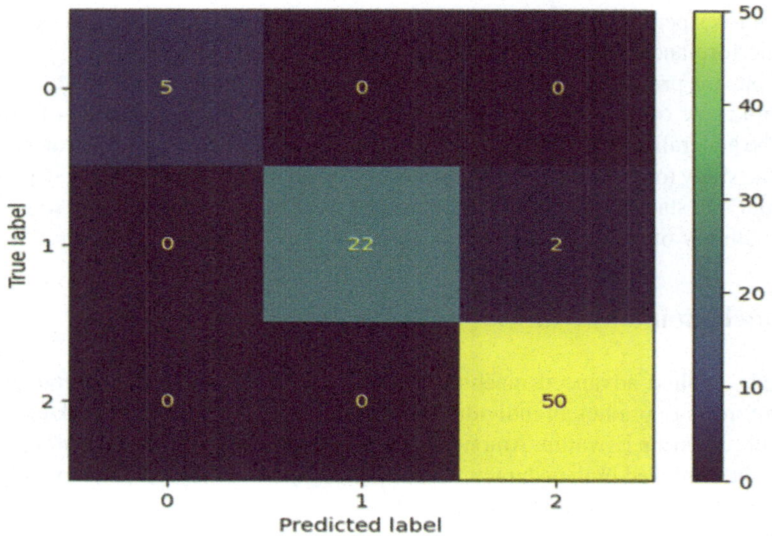

Fig. 3. Decision Trees' confusion matrix

revealed that 192 individuals (61.15%) are partially included, 106 individuals (33.75%) are financially included, and 16 individuals (5.10%) are financially excluded. This shows that a significant number of the participants are not fully financially included. However, there is still a subset of individuals who are financially excluded (5 10%) and may not have access to adequate financial services and opportunities. There is a need to develop strategies for improving access to financial services and products, especially in rural communities. The findings reveal that a significant number of participants are partially included reflecting the need for creating awareness and sensitizing communities about available and accessible financial products and services. Furthermore, the study revealed that 24 participants (7.64%) are formally employed or receive a regular salary or wage. Also, there are 26 individuals (8.28%) who are retired, implying that they have ceased working due to age or other reasons and no longer receive regular employment income. Most participants, 127 (40.45%) are self-employed, probably making a living through their own business or professional practice. However, a significant number of 137 participants (43.63%) are unemployed. This probably contributes to the highest of individuals (61.15%) who are partially included in financial inclusion status.

Furthermore, the study revealed that random forest performed better with an accuracy of 98.73%, and 97.47% accuracy for decision trees. Thus, random forests performed exceptionally well in accurately predicting the financial inclusion status of individuals. Furthermore, both models exhibited high precision values of 98.76% and 98.19% for random forest and decision trees, respectively. In terms of recall, random forest and decision trees also achieved recall scores of 98.73% and 97.47%, respectively. In terms of F1-score, random forest recorded an F1-score of 98.67% and decision trees achieved an F1-score of 97.65%. Overall, random forest performed better than decision trees in terms of accuracy, precision, recall, and F1-score, achieving perfect scores in all metrics. Decision trees perform slightly lower compared to random forests but still demonstrate strong performance overall.

Despite the promising results of these models, the study has some limitations. Firstly, the sample size of the study was relatively small. Secondly, the findings of this study cannot be generalized to represent the entire population in the country but can be used as a baseline study to understand the importance of applying machine learning in predicting individual's FI status. Therefore, there is a need to increase the sample size to ensure the generalizability of the findings.

5 Conclusion

The study applied advanced machine learning classifiers such as decision trees and random forests to predict an individual's financial inclusion status using data collected from four regions in Eswatini. Among the classifiers, random forests outperformed other models and achieved high accuracy, recall, precision and F1-score. Implementing such models in financial inclusion can potentially assist Eswatini in providing and promoting financial products and services in building a sustainable society for inclusive economic growth, poverty reduction and employment creation. Our findings reveal that machine learning can be used to develop data-driven applications that can analyze financial data and predict an individual's FI status. This can tremendously assist policymakers, individuals and financial service providers with effective ways of providing and managing financial services across different population groups to ensure inclusive participation, economic growth and financial inclusion. This can help regulators make informed decisions on financial inclusion and monitor the financial inclusiveness of various households in the four regions of Eswatini.

Acknowledgements. Authors would like to thank Mr Abel Motsomi for helping determine the FI status. We also, extend our appreciation to the Royal Academy of Engineering (RAE), London, UK under the Distinguished International Associates – Round 2 programme. RAE is incorporated by Royal Charter of 3 Carlton House Terrace, London SW1Y 5DG, Registered Charity 293074. RAE provided financial support for this project through a research grant with Grant Reference Number: DIA-2022-178.

References

1. Khan, F., Siddiqui, M.A., Imtiaz, S.: Role of financial literacy in achieving financial inclusion: a review, synthesis, and research agenda. Cogent. Bus. Manag. **9**, 2034236 (2022). https://doi.org/10.1080/23311975.2022.2034236
2. Ozili, P.K.: Financial inclusion research around the world: a review. Forum Soc. Econ. **50**, 457–479 (2021). https://doi.org/10.1080/07360932.2020.1715238
3. Sarma, M., Pais, J.: Financial inclusion and development. J. Int. Dev. **23**, 613–628 (2011). https://doi.org/10.1002/JID.1698
4. Fakudze, S., Tsegaye, A., Sibanda, K.: The relationship between financial development and economic growth in Eswatini (formerly Swaziland). Afr. J. Econ. Manag. Stud. **13**, 15–28 (2022). https://doi.org/10.1108/AJEMS-06-2021-0291/FULL/XML
5. Fashoto, S.G., Akinnuwesi, B.A., Mbunge, E., Metfula, A.S.: Financial inclusion dataset classification in Eswatini using support vector machine and logistic regression. Int. J. Bus. Inf. Syst. **43**, 507–527 (2023). https://doi.org/10.1504/IJBIS.2023.132811
6. Yasir, A., Ahmad, A., Abbas, S., Inairat, M., Al-Kassem, A.H., Rasool, A.: How artificial intelligence is promoting financial inclusion? A study on barriers of financial inclusion. In: 2022 International Conference on Business Analytics for Technology and Security ICBATS (2022). https://doi.org/10.1109/ICBATS54253.2022.9759038
7. Eswatini Ministry of Finance: National Financial Inclusion Strategy for Swaziland 2017–2022 (2017)
8. Eswatini State of Financial Inclusion Report. Eswatini State of Financial Inclusion Report. Eswatini State Financ Incl Rep (2022). https://finmark.org.za/system/documents/files/000/000/391/original/Eswatini_Financial_Inclusion_Refresh.pdf?1614849476 (accessed November 28, 2023)
9. Mabuza, Z.: The Impact of Financial Inclusion on Monetary Policy Effectiveness in Eswatini. Cent Bank Eswatini (2023)
10. Central Bank of Eswatini: Eswatini FinTech Working Group, Olayinka David-West Consulting. Eswatini-Fintech-Landscape-Report (2023)
11. Anthanasius Fomum, T., Opperman, P.: Financial inclusion and performance of MSMEs in Eswatini. Int. J. Soc. Econ. **50**, 1551–1567 (2023). https://doi.org/10.1108/IJSE-10-2020-0689/FULL/PDF
12. Nkambule, M.B.: The Financial Inclusion Status of Rural Households in Eswatini. University Pretoria, pp. 2003–2005 (2022)
13. Myeni, S., Makate, M., Mahonye, N.: Does mobile money promote financial inclusion in Eswatini? Int. J. Soc. Econ. **47**, 693–709 (2020). https://doi.org/10.1108/IJSE-05-2019-0310/FULL/PDF
14. Mhlanga, D., Boubaker, S.: Financial inclusion in emerging economies: the application of machine learning and artificial intelligence in credit risk assessment. Int. J. Financ. Stud. **9**, 39 (2021). https://doi.org/10.3390/IJFS9030039

15. Kshetri, N.: The role of artificial intelligence in promoting financial inclusion in developing countries. J. Glob. Inf. Technol. Manag. **24**, 1–6 (2021). https://doi.org/10.1080/1097198X.2021.1871273
16. Ismail, Q.F., Al-Sobh, E.S., Al-Omari, S.S., Bani Yaseen, T.M., Abdullah, M.A.: Using machine learning algorithms to predict the state of financial inclusion in Africa. In: 2021 12th International Conference Information Communication System ICICS 2021, pp. 317–323 (2021). https://doi.org/10.1109/ICICS52457.2021.9464590
17. Akinnuwesi, B.A., Fashoto, S.G., Metfula, A.S., Akinnuwesi, A.N.: Experimental application of machine learning on financial inclusion data for governance in Eswatini. In: Lectutre Notes Computer Science (Including Subser Lect Notes Artif Intell Lect Notes Bioinformatics) LNCS, vol. 12067, pp. 414–425 (2020). https://doi.org/10.1007/978-3-030-45002-1_36/TABLES/6
18. Duvendack, M., Mader, P.: Impact of financial inclusion in low- and middle-income countries: a systematic review of reviews. J. Econ. Surv. **34**, 594–629 (2020). https://doi.org/10.1111/JOES.12367
19. Ajetomob, J.O., Dlamini, S.G., Dlamini, D.V.: Sustainable Financial Inclusion for Women Owned Firms in Eswatini (2022)
20. Cáamara, N., Tuesta, D.: Measuring Financial Inclusion: a Multidimensional Index Measuring Financial Inclusion, vol. 2 (2014)
21. Pearce, D., Ortega, C.R.: Financial Inclusion Strategies Reference Framework, vol. 60. World Bank (2012)
22. Hasan, M.M., Popp, J., Oláh, J.: Current landscape, and influence of big data on finance. J. Big Data **7**, 21 (2020). https://doi.org/10.1186/S40537-020-00291-Z

Proposal of a Machine Learning Model for the Early Detection of Depression in University Students

Samir Aguilar, Antony Huaman, and Wilfredo Ticona(✉)

Universidad Tecnológica del Perú, Lima, Perú
{U17305076,U18203356}@utp.edu.pe, wmamani@esan.edu.pe

Abstract. Mental health problems such as depression in university students has increased with the passing of time, it is a topic little investigated by governmental institutions, but of great importance for the development of a country. The importance of early detection of mental illness would help to optimize the treatment process. Therefore, the overall objective of this research work is to detect depression early in college students by using a machine learning model, applying newer technologies and methodologies to develop a more robust and consistent model. The methodology used in this research consists of 5 phases: Dataset Extraction, Preprocessing, Feature Extraction, Model Implementation and Evaluation. Each phase was divided into activities and tasks to understand the problem-solving processes. In the results, the logistic regression (LR) technique obtained a superior performance with the metrics: precision of 0.79, sensitivity of 0.85, accuracy of 0.81 and f1 score of 0.81. In conclusion, machine learning techniques can detect depression in college students.

Keywords: Machine learning · Depression · University students · Mental health · Methodology · SVM

1 Introduction

The mental health of university students is a concern for society in general, as they represent the 'future' of the country. Therefore, it is important to detect if the student is suffering from any psychological disorder that may affect their personal growth. According to data from the World Health Organization (WHO), 3.8% of the world's population experiences depression, with approximately 280 million people worldwide suffering from depression. Each year, more than 700,000 people commit suicide, making suicide the fourth leading cause of death in the 15–29 age group [1]. In addition, a WHO study revealed that more than 22% of Latin Americans suffer from mental disorders such as severe depression and anxiety [2]. On the other hand, multiple studies related to depression in university students report that between 25% and 50% of the university population suffers from depression, partly determined by the different instruments used in each research for measurement [3]. In addition, a National University Mental Health

R. Silhavy and P. Silhavy (Eds.): CSOC 2024, LNNS 1120, pp. 659–672, 2024.
https://doi.org/10.1007/978-3-031-70518-2_55

Survey conducted in Chile found that 46% of university students show depressive symptoms and signs of anxiety, 54% suffer from stress and 30% present all three problems simultaneously [4].

In Peru, the mental health of high school students has been an issue that has received little attention from the country's mental health institutions, as they only collect data on those with psychological disorders instead of implementing solutions to address this problem. According to data from the Ministry of Health (MINSA), more than 300,000 cases of people with depression were treated in 2021, with a total of approximately 313,455 cases of people with depression receiving treatment nationwide in various mental health services [5]. According to a study conducted by a consortium of universities in Peru, at least 39% of university students show symptoms of severe depression, 32% experience an unmanageable level of stress, and 30.8% have contemplated suicide [6]. According to MINSA, 30% of the university population in Peru suffers some type of mental disorder due to academic pressure [7]. According to the National Institute of Health (INS), in the context of the pandemic, 30% to 40% of young people aged 19 to 26 years in Peru showed symptoms of anxiety and depression. The causes of these symptoms included family dysfunction, loss of family members, and economic adversity [8]. Depression in university students is increasingly prevalent, affecting their mental health, academic performance, and even their personal lives.

2 Related Work

Early detection of depression has become essential in the treatment of this condition, as it prevents more serious long-term problems for those suffering from this disease. In one study, an attempt is made to detect depression using a machine learning (ML) model in the context of COVID-19. First, data collection was conducted through an online survey composed of 21 questions, resulting in 1694 records. In evaluating the survey results, the Hamilton rating scale, which is used to detect symptoms of depression, was used. Then, decision tree (DT), Naive Bayes (NB) and k-nearest neighbor (KNN) algorithms were used to train the model. As a result, it was found that 56% of women experience depression compared to men, and KNN had the best performance among all machine learning models with an accuracy of 92.32% [9].

Similarly, according to this study aiming to detect maternal depression and anxiety during COVID-19 pandemic using ML techniques, data collection was initiated involving 3569 pregnant or postpartum women. Various machine learning techniques such as Gradient Boosting (GB), Random Forest (RF), Extreme Random Trees (XRT), NB, Support Vector Machines (SVM), Multilayer Neural Networks (MNN) and DT were used to train the model. The results indicated that 20.3% of the women had moderate to severe maternal depression, while 30.1% had moderate to severe anxiety. In addition, the GB model achieved the highest accuracy rate with 82.9% [10].

In addition, another study focused on detecting depression using ML algorithms with data collected from social networks through data mining algorithms. Data collection involved the use of data mining techniques to gather a total of 10,000 depression and non-depression data points. ML algorithms such as RF, Logistic Regression (LR), DT, SVM, KNN and Naive Bayes Multinomial were employed to classify the collected data. Notably, the SVM model achieved the highest accuracy rate with 75.15% [11].

Additionally, another study focused on diagnosing depression through ML and determining whether it is possible to identify groups at high risk of suicide based on the words uttered by participants in a semi-structured interview. Data collection involved a total of 166 participants with their respective analysis types, and the Naive Bayes algorithm was used to classify the data. The final model achieved an accuracy rate of 83.1% [12].

On the other hand, this study aims to detect childhood depression using ML methods. Data collection involved a survey with a total of 6310 participants, and machine learning methods such as RF, Boruta, TPOTclassifier, Extreme Gradient Boosting (XGB), DT and Gaussian Naive Bayes (NB) were employed. Although RF, XGB, DT and Gaussian NB were used to train the model, XGB performed the best in terms of accuracy, although the other models were not far behind, with an accuracy rate of 95.35% [13].

3 Methodology

The proposed methodology was based on [9, 11, 13]. It consists of five phases: Dataset Extraction, Preprocessing, Feature Extraction, Model Implementation y Evaluation. The methodology used is shown graphically below (Fig. 1):

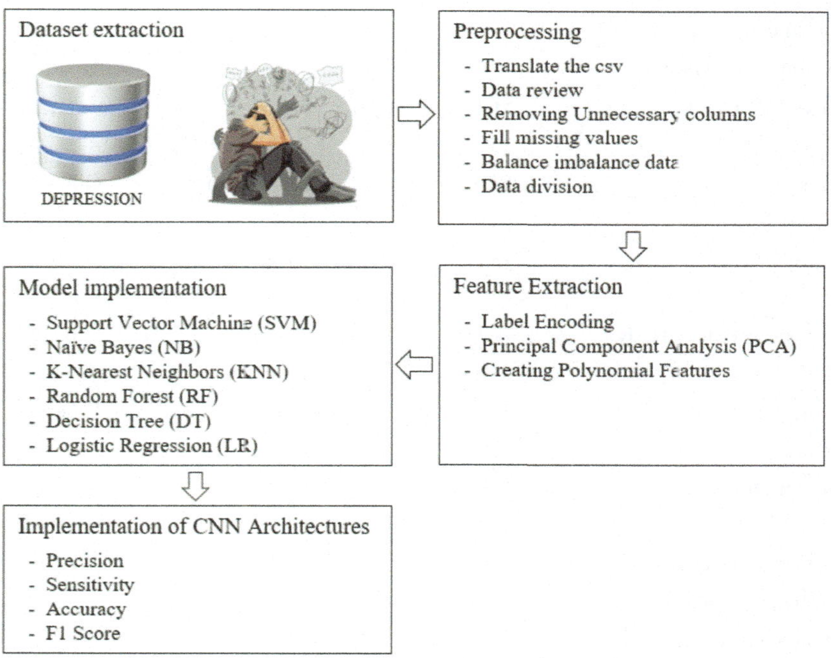

Fig. 1. Proposed Methodology

In Fig. 1, the process of developing the Machine Learning model is depicted, highlighting both the specific topic and the techniques that will be employed at each stage of the proposed methodology. From the beginning to the conclusion of the process, the established order to be followed for the development of the Machine Learning model is clearly shown.

3.1 Dataset Extraction

The dataset was obtained through the Kaggle platform, and the obtained dataset is called "Student Mental health.csv". This dataset was compiled through a survey conducted with university students via Google Forms to examine their current academic status and mental health. A total of approximately 101 students participated, responding to various questions in a questionnaire to obtain data on the student's mental health. An example of the survey results is shown in Table 1. The use of publicly available data from reliable sources ensures the credibility of the developed machine learning model. Coming from reliable sources, public data provides greater assurance of accuracy and consistency, thereby strengthening confidence in the results and predictions generated by the machine learning model [14].

Table 1. Example of questionnaire results

Total students surveyed (101)			
Question Description			
Choose your Gender	Male	Female	–
	26%	74%	–
What is your course?	BCS	Engineering	Others
	18%	17%	65%
Age	18-year-old	24-year-old	Others
	32%	23%	45%
Your current year of study	1ª year	3ª year	Others
	41%	19%	41%
What is your CGPA?	3,00–4,00	3,00–3,50	Others
	47%	43%	11%
Your marital status	Yes	No	–
	16%	84%	–
Do you have Depression?	35%	65%	–
Do you have Anxiety?	34%	66%	–
Do you have Panic Attack?	33%	67%	–

Table 1 provides a detailed analysis of the raw data extracted from the dataset, along with a classification and percentage-based division of the different types of data present. This report is of vital importance as it will serve as a guide to select the main and secondary variables in the Machine Learning model. Its prior analysis will allow for a precise definition of these variables before model development.

To understand the nature of the data, a small sample is presented in the following figure (Fig. 2):

Content
Timestamp 08/07/2020 12:02:00, ...
Choose your gender: Female, Male, Male, ...
Age: 18, 21, 19, ...
What is your course?: Engineering, Islamic education, BIT, ...
Do you have Depression?: Yes, No, ...
Your current year of Study: year 1, year 2, ...
......

Fig. 2. Sample of the data set obtained.

In Fig. 2, the raw content of the extracted dataset is displayed, providing an initial view of the data before any modifications are made. This initial visualization is crucial for understanding the original nature of the data and for facilitating comparison with the transformations made later in the preparation process for the Machine Learning model. Therefore, examining this data in its initial state is crucial for better understanding and appreciating the changes that will be implemented.

The content of Table 1 is an example of the questionnaire and in some questions not all the alternatives are shown, since only the alternates with the highest number of answers are shown.

Finally, before using this dataset another evaluation must be performed again to verify if it meets all the conditions mentioned above, it is important to clarify any doubts about the dataset before running the experiments with the Machine Learning models.

3.2 Preprocessing

At this stage, a comprehensive evaluation of the obtained dataset will be carried out to determine which preprocessing techniques will be employed to enhance data quality and optimize the performance of the ML models. Since these decisions will have a direct impact on the results of the ML models, this phase is of vital importance. Next, the proposal of the methods to be implemented is presented in Table 2.

In Table 2, the planned activities for the Machine Learning model are detailed, presenting both the tasks to be performed and their respective descriptions. This organized presentation is crucial as it provides a structured overview of the changes that will be implemented in the Machine Learning model at this phase of the methodology.

To carry out the preprocessing of the raw dataset, the Google Colab platform will be used, as it allows us to execute and write arbitrary Python code directly from a browser, which is essential when developing Machine Learning models. First, the necessary libraries will be imported to make the proposed modifications, and the dataset will

Table 2. Activities of the preprocessing methods.

I	Description	Tasks
Translate the content of the csv file	A complete translation of all the cells of the data set obtained from English to Spanish will be performed	Install the required libraries
		Load the raw csv file
		Translate the cells of the dataset
		Translate the header of the dataset
Review of the data	A review of the translated dataset will be performed to propose actions to improve the ML models	Display the first few rows of the csv file
		Review the information in the csv file
Elimination of unnecessary columns	Several columns of the revised dataset that do not contribute significantly to the performance of the ML models will be eliminated	Identify unnecessary columns
		Eliminate unnecessary columns
Fill in missing values	Missing values from the dataset will be filled in	Identify missing values
		Fill in the missing values with the mean of each column
Balancing unbalanced data	The dataset will be balanced to improve the performance of the ML models	Separate the features and the target column from the data set
		Apply one-hot encoding to all data except the target column
		Apply SMOTE to the features and the target column to balance the data
Data splitting	Separate the dataset into a training and test set	Separate dependent and independent variables
		Divide into training and test sets

be loaded from a CSV file. Both the data and the header of the dataset will be translated from English to Spanish using the "translate" library.

Subsequently, a data review will be conducted to identify and eliminate those that are not relevant for training the Machine Learning model, specifically the "Timestamp" and "What is your CGPA?" columns. Once the less relevant data has been removed, we will address the presence of missing values, filling them in with the mean of each column from the CSV file.

The next stage involves addressing data imbalance by applying one-hot encoding only to the features, limited to the target column "Do you have depression?" in the dataset. Subsequently, SMOTE will be implemented to balance the data, a technique especially useful in binary classification problems.

Finally, the data will be divided into training and testing sets to prepare them for training the Machine Learning models. Figure 3 shows a visual representation of some of the changes made during the preprocessing of the dataset.

Raw content
Timestamp: 08/07/2020 12:02:00, …. Choose your gender: Female, Male, Male, …. Age: 18, 21, 19, … What is your course?: Engineering, Islamic education, BIT, .. Do you have Depression?: Yes, No, … Your current year of Study: year 1, year 2, … ……
Pre-processed content
Elige tu género_Femenino: 1.00000, 0.00000, 0.00000, … Elige tu género_Masculino: 0.00000, 1.00000, 0.0000, … Edad_18.0: 1.00000, 0.00000, 0.00000 …. ¿Buscaste algún especialista para un tratamiento?_Sí: 0.00000, 0.00000, … ¿Tienes depresión?: Si, No

Fig. 3. Sample of the preprocessed data set.

Figure 3 illustrates both the raw content and the preprocessed content of the dataset, highlighting the changes made after implementing the activities in the Machine Learning model. This visual representation provides a clear comparison between the original data and the data after preprocessing, allowing for a visual understanding of the adjustments made to prepare the data for the Machine Learning model.

3.3 Feature Extraction

In this phase, feature extraction techniques will be applied to the dataset with the aim of improving the performance of the ML models. Next, the proposal of the techniques to be implemented will be presented in Table 3.

Table 3 details the extraction techniques to be applied to the Machine Learning model, providing a description of the tasks associated with each technique. This organized layout is crucial as it offers a clear and structured overview of the changes that will be implemented in the Machine Learning model at this specific stage of the methodology.

To begin with feature extraction, we will first apply label encoding to normalize the binary output of the dataset, then we will employ PCA to reduce the dimensionality of the ML models, and finally, we will proceed with the creation of polynomial features to enhance the performance of the ML model, using 'PolynomialFeatures' on all modified

Table 3. Activities of feature extraction techniques

Activities	Description	Tasks
Apply Label Encoding	Perform Label Encoding to scale the csv data to improve ML model performance	Insert the necessary libraries
		Load the preprocessed csv file
		Start the label encoder
		Apply Label Encoding to the target column
Principal Component Analysis (PCA)	Perform PCA to reduce the dimensionality of the csv data	Separate numerical and categorical csv characteristics
		Encode the categorical variables using one-hot encoding
		Concatenate the encoded numeric and categorical features
		Apply PCA to the concatenated data
Creation of polynomial features	Perform polynomial feature creation of csv files to improve the performance of ML models	Insert the necessary libraries
		Load the updated csv file with the PCA
		Identify the categorical and numerical features in the csv
		Apply PolynomialFeatures to all features

features of the dataset. These implemented changes have the potential to significantly improve the performance of the machine learning model. Although some changes are not directly visible in the dataset, their effects will be reflected in Fig. 4.

Figure 4 displays both the preprocessed content and the content with the implemented extraction techniques, highlighting some of the changes made to the dataset after applying these techniques to the Machine Learning model. This visual representation allows for a clear and concise observation of how the data has been transformed as a result of the implementation of the extraction techniques in the Machine Learning model.

3.4 Model Implementation

The utilization of algorithms such as KNN, LR, RF, among others, enhances the predictive capacity of Machine Learning models in binary classification. Additionally, the availability of additional methods to optimize existing Machine Learning models offers flexibility and potential to further improve their performance [15, 16]. In this section, the Machine Learning algorithms SVM, NB, KNN, RF, DT, and LR will be applied.

Pre-processed content
Elige tu género_Femenino: 1.00000, 0.00000, 0.00000, …   Elige tu género_Masculino: 0.00000, 1.00000, 0.0000, …   Edad_18.0: 1.00000, 0.00000, 0.00000   ….   ¿Buscaste algún especialista para un tratamiento?_Sí: 0.00000, 0.00000, …   ¿Tienes depresión?: Si, No
Content with feature extraction implemented
Elige tu género_Femenino: 1.00000, 0.00000, 0.00000, …   Elige tu género_Masculino: 0.00000, 1.00000, 0.0000, …   Edad_18.0: 1.00000, 0.00000, 0.00000   ….   ¿Buscaste algún especialista para un tratamiento?_Sí: 0.00000, 0.00000, …   **¿Tienes depresión?: 1, 0**

Fig. 4. Sample dataset with feature extraction used.

Subsequently, we will proceed to train these models using the previously modified dataset.

- ***Support Vector Machine (SVM):*** SVMs are excellent for classification and regression on complex or high-dimensional data, finding an optimal hyperplane that maximizes the margin between classes, even in nonlinear situations by using kernels [11]. For training the ML model the parameters used were 'C': 1, 'degree': 2, 'kernel': 'poly'.

$$f(x) = w^t x + b \tag{1}$$

where:

- f(x): Decision function or the score assigned to sample x.
- w: Vector of weights.
- t: Transposition of the vector or matrix.
- x: Vector of sample characteristics.
- b: Bias.

- ***Naïve Bayes (NB):*** NB is an efficient classification algorithm based on Bayes' theorem, excelling in applications such as text classification and sentiment analysis [11].

$$P(y|X) = \frac{P(X|y) * P(y)}{P(X)} \tag{2}$$

where:

- P(y|X): Conditional probability of class y given feature vector X.
- P(X|y): Conditional probability of feature vector X given class y.
- P(y): A priori probability of the class y.
- P(X): Marginal probability of X.

- ***K-Nearest Neighbors (KNN):*** KNN classifies or predicts by assigning labels or values based on the labels of nearby points in the feature space. Used in classification and regression, it considers that nearby points share similar labels or values [11]. For training ML models, the parameters used were 'n_neighbors': 3.

$$y = mode(y1, y2, \ldots, yk) \tag{3}$$

where:

- y: Class label we are trying to predict for the new sample.
- y1, y2,..., yk: Class labels of the k nearest neighbors of the Xtest. Sample.
- ***Random Forest (RF):*** RF uses multiple decision trees to improve accuracy and robustness in data prediction in both classification and regression problems using an ensemble technique [11]. For the training of the ML model the parameters used were 'max_depth': None, 'n_estimators': 100.

$$y^\wedge = mode(y1, y2, \ldots, yn) \tag{4}$$

where:

- y^: y1, y2, ..., yn: Predicted class label for the new sample.
- y1, y2, ..., yn: Predictions of the n decision trees in the forest.

- ***Decision Tree (DT):*** A DT is a hierarchical model that makes decisions by partitioning the dataset based on features, achieving accurate classifications or predictions [11]. For the training of the ML model the parameters used were 'max_depth': None.

$$y^\wedge = \arg\ max_y(P(y|x)) \tag{5}$$

where:

- y^: Predicted class label for sample x.
- P(y|x): Conditional probability of class label y given feature set x.

- ***Logistic Regression (LR):*** The LR uses the sigmoid function to model probabilities and perform binary or multiclass classifications from a linear combination of weights associated with the features [11]. For training the ML model the parameters used were 'C': 1.

$$P(y = 1|X) = \frac{1}{1 + e^{-(w^T X + b)}} \tag{6}$$

where:

- P(y = 1|X): Conditional probability of positive class.
- X: Vector of characteristics.
- W: Weights associated with each feature.
- B: Bias.
- wT: Transposition of the vector of weights.
- e: Base of the natural logarithm.

3.5 Evaluation

The utilization of evaluation metrics such as precision, sensitivity, accuracy, F1, among others, significantly enhances the ability to determine the performance of a Machine Learning model in binary classification [17]. In this phase of the methodology, to evaluate the performance of the ML models, precision, sensitivity, accuracy, and F1 score metrics will be employed.

- **Precision:** The precision metric in neural networks refers to the measure of fractions of positive predictions. In which, it is the number of correct positive results di-vided by the number of positive results predicted by the classifier [13].

$$Precision = \frac{TP}{TP + FP} \tag{7}$$

- **Sensitivity:** The sensitivity metric in neural networks measures the number of correct positive results divided by the number of all relevant samples [13].

$$Sensitivity = \frac{TP}{TP + FN} \tag{8}$$

- **Accuracy:** The accuracy metric in neural networks refers to the measure of how accurate a model is. It is measured by the number of correct predictions divided by the total number of predictions [13].

$$Accuracy = \frac{TP + TN}{TP + FP + FN + TN} \tag{9}$$

where:

- TP: True Positives.
- FP: False Positives.
- TN: True Negatives.
- FN: False Negatives.

- **F1 Score:** Allows a balance between precision and recovery [13].

$$F1 = 2 \times \frac{Precision \times Recall}{Precision + Recall} \tag{10}$$

4 Results

Table 4 shows the best results of the Machine Learning models with their respective parameters.

The results demonstrate the various levels of performance obtained through the techniques implemented in the Machine Learning model. Additionally, they graphically illustrate the wide differences that these techniques can present [18]. It is worth noting that logistic regression (LR) exhibited the best overall performance, with both precision and F1 score of 0.81 each. In contrast, the Naive Bayes (NB) model showed the worst

Table 4. Results of Machine Learning techniques

Techniques	Parameters	Precision	Sensitivity	Accuracy	F1 Score
SVM	'C': 1, 'degree': 2, 'kernel': 'poly'	0.83	0.77	0.81	0.80
NB	No parameters were adjusted	0.69	0.85	0.74	0.76
KNN	'n_neighbors': 3	0.62	**1.00**	0.70	0.76
RF	'max_depth': None, 'n_estimators': 100	**0.90**	0.69	0.81	0.78
DT	'max_depth': None	0.82	0.69	0.78	0.75
LR	'C': 1	0.79	0.85	**0.81**	**0.81**

performance, with a precision of 0.69 and an accuracy of 0.74. Although the K Nearest Neighbors (KNN) algorithm demonstrated outstanding sensitivity of 1.00, it also presented the lowest precision, with a value of 0.62. There is a considerable variability in the results among the different ML techniques, evidenced by the significant disparities in precision, sensitivity, accuracy, and F1 score. This highlights the diversity in the performance of the evaluated models.

5 Conclusions

The mental health of university students is a concern for society in general, since they represent the "future of the country", and it is important to detect if the student is suffering from any psychological disorder that may affect their personal growth. Therefore, the present work combines several methodologies and techniques from previous works for the early detection of depression in university students.

The research used a dataset called "Student Mental health", which was obtained from the Kaggle Platform. This dataset was obtained by surveying university students to examine their academic situation and their mental health. A total of 101 students participated in the survey, in which they answered several questions related to mental health.

Likewise, good preprocessing was carried out, which was reflected in the results of the model. This preprocessing was carried out based on the information presented in the state of the art and activities were developed, such as: elimination of unnecessary columns, completing missing data and balancing the data.

The diversity of existing techniques in the field of Natural Language Processing is a great advantage when it comes to vectorizing textual records since they facilitate the conversion of textual data to numerical data. Three techniques were chosen for the research: label coding, principal component analysis, and polynomial feature creation.

Likewise, different Machine Learning algorithms were used, such as: Support Vector Machines, Naive Bayes, K-Nearest Neighbors, Random Forest, Decision Tree, and Logistic Regression. Different parameters were used to carry out the experiments. The results were obtained with the logistic regression (LR) technique with the metrics: precision of 0.79, sensitivity of 0.85, accuracy of 0.81 and f1 score of 0.81. In conclusion, machine learning techniques can detect depression in college students.

References

1. Organización mundial de la salud (OMS): Depresión (2023). https://www.who.int/es/news-room/fact-sheets/detail/adolescent-mental-health
2. Banco Mundial: Bajo su superficie alegre, Latinoamérica también sufre de depresión (2017). https://www.bancomundial.org/es/news/feature/2017/04/07/latinoamerica-tambien-sufre-depresion-dia-mundial-salud-2017
3. Javier, A., et al., 2010, Depresión en estudiantes universitarios y su asociación con el estrés académico, Scielo, http://www.scielo.org.co/scielo.php?script=sci_arttext&pid=S0120-87052010000100002#:~:text=La%20prevalencia%20de%20depresi%C3%B3n%20reportada,medici%C3%B3n%20(13%2D16)
4. CNN Chile: Salud mental: El 44% de los estudiantes universitarios ha estado con tratamiento psicológico (2019). https://www.cnnchile.com/pais/salud-mental-estudio-estudiantes-universitarios-tratamiento-psicologico_20190429/
5. Ministerio de Salud (MINSA): Minsa: Más de 300 mil casos de depresión fueron atendidos durante el 2021 (2022). https://www.gob.pe/institucion/minsa/noticias/575899-minsa-mas-de-300-mil-casos-de-depresion-fueron-atendidos-durante-el-2021
6. Carolina, D.: El peso que cargan los universitarios, Saludconlupa (2022). https://saludconlupa.com/salud-mental/el-impacto-psicologico-de-la-pandemia-en-los-universitarios/
7. RPP: Hasta 30% de la población universitaria del Perú sufre de impactos en su salud mental por presión académica (2018). https://rpp.pe/vital/vivir-bien/hasta-30-de-la-poblacion-universitaria-del-peru-sufre-de-impactos-en-su-salud-mental-por-presion-academica-noticia-1151266?ref=rpp
8. Instituto Nacional de Salud (INS): Entre 40 y 30 % de jóvenes de 19 y 26 años en el Perú presentaron síntomas de ansiedad y depresión tras la llegada de la pandemia de la COVID-19 (2022). https://web.ins.gob.pe/es/prensa/noticia/entre-40-y-30-de-jovenes-de-19-y-26-anos-en-el-peru-presentaron-sintomas-de-ansiedad
9. Sofia, Arun, M., Mohammad, S., Evans, A.: Machine learning based model for detecting depression during Covid-19 crisis. Sci. Afr. **20**, e01716 (2023). https://doi.org/10.1016/j.sciaf.2023.e01716
10. Radwan, Q., et al.: Machine learning techniques for predicting depression and anxiety in pregnant and postpartum women during the COVID-19 pandemic: a cross-sectional regional study. F1000Research **11**, 390 (2022). https://doi.org/10.1268/f1000research.110090.1
11. Zannatun, V., Bidyut, S., Israt, E., Jabir, A., Johora, P.: Depression detection in social media comments data using machine learning algorithms. Bull. Electr. Eng. Inform. **12**, 987–996 (2023). https://doi.org/10.11591/eei.v12i2.4182
12. Daun, S., et al.: Detection of depression and suicide risk based on text from clinical interviews using machine learning: possibility of a new objective diagnostic marker. Front. Psychiatry **13**, 801301 (2022). https://doi.org/10.3389/fpsyt.2022.801301
13. Umme, H., Enamul, K., Rasheda, K.: Detection of child depression using machine learning methods. PLoS ONE **16**, 12 (2021). https://doi.org/10.1371/diario.pone.0261131
14. Leme, D., Cesar, O.: Machine learning models to predict future frailty in community-dwelling middle-aged and older adults: the ELSA cohort study. J. Gerontol. Ser. A Biol. Sci. Med. Sci. **78**, 2176–2184 (2023). https://doi.org/10.1093/gerona/glad127

15. Jun-Bo, T., et al.: Using machine learning techniques to predict the risk of osteoporosis based on nationwide chronic disease data. Sci. Rep. **14**, 5245 (2024). https://doi.org/10.1038/s41598-024-56114-1
16. Yuting, L., et al.: Machine learning based biomarker discovery for chronic kidney disease–mineral and bone disorder (CKD-MBD). BMC Med. Inform. Decision Making **24**, 36 (2024). https://doi.org/10.1186/s12911-024-02421-6
17. Amir Reza, N., Hamed, Z., Ahmad, S.: Machine learning approaches for early detection of non-alcoholic steatohepatitis based on clinical and blood parameters. Sci. Rep. **14**, 2442 (2024). https://doi.org/10.1038/s41598-024-51741-0
18. Muntequa Imtiaz, S., et al.: Impact of mobile connectivity on students' wellbeing: detecting learners' depression using machine learning algorithms. PLoS ONE **18**, e0294803 (2023). https://doi.org/10.1371/journal.pone.0294803

A Focused Systematic Review of the Regulation for the Crypto Asset in Peru and Its Limits

Renato Arias[1(✉)], Ricardo Arias[2], and Kelly Ochoa[3]

[1] Universidad Autónoma del Perú, Lima, Peru
renato.arias@me.com
[2] Universidad Tecnológica del Perú, Lima, Peru
[3] Universidad Pompeu Fabra, Barcelona, Spain
https://www.autonoma.pe/

Abstract. This research article evaluates the current legislative proposal for a crypto-assets law in Peru. It is noteworthy that various regulatory entities within the country's financial system have explicitly expressed their opposition to the legislative proposal. Despite this, the Peruvian Congressional Commission responsible for evaluating the approval has overlooked the pronouncements of regulatory entities, basing its decision on similar legislative initiatives in various South American countries. Therefore, the purpose of this research is to highlight the deficiencies of the legislative proposal and the potential consequences of its regulation under these terms, as well as to make recommendations and gather international experience in the regulation of crypto assets. In this regard, a systematic analysis of official documents from Peru and the content of 169 research articles retrieved from the ScienceDirect, Scopus and IEEE database has been conducted. It is worth noting that for this analysis, a systematic review of primary sources and the Preferred Reporting Items for Systematic Reviews and Meta-Analyses for Abstract (PRISMA-A) and Population, Intervention, Comparison and Outcome (PICO) guideline were used, enabling a rigorous and transparent investigation to showcase influential and important studies addressed from 2020 to the present. In summary, this research presents the deficiencies in the legislative proposal, along with the most recent trends in the regulation of crypto assets. Furthermore, we suggest that legislators be more critical when proposing such legislative initiatives, and we believe that this article could serve as a basis for future study on the regulation of crypto assets.

Keywords: Crypto assets · Cryptocurrencies · Regulation · Legal framework · Regulatory authorities

1 Introduction

Currently, several South American countries are undertaking legislative initiatives related to the potential regulation of crypto assets. For instance, the Peru-

R. Silhavy and P. Silhavy (Eds.): CSOC 2024, LNNS 1120, pp. 673–693, 2024.
https://doi.org/10.1007/978-3-031-70518-2_56

vian Congress of the Republic is currently debating the approval of Bill 01042-2021-CR, here the main objective of this bill is to establish a framework law that enables the purchase of cryptocurrencies with Peru's national currency, facilitating participation in various global transactions of goods and services. In this context, the bill has been crafted with reference to legislative approaches in Chile, Colombia, Argentina, and El Salvador; however, it relies heavily on regulatory proposals or intentions, neglecting optimal standards in terms of regulation and supervision, as well as recommendations provided by the "Financial Action Task Force" and the "International Organization of Securities Commissions" in this field. It is also relevant to evaluate about the terminology 'cryptocurrencies' and 'crypto-assets' have been used interchangeably in the Bill. While both share Distributed Ledger Technology (DLT), the term 'crypto-assets' has a broader scope than the commonly understood 'cryptocurrencies'.

Therefore, the legislator must be highly rigorous when proposing regulation for crypto assets, as regulatory and supervisory parameters will differ on the detail category of crypto asset intended for regulation. Also the fundamental idea behind blockchain technology and its many uses is that people submit to the authority of an immutable technical system [1], instead of the authority of centralized organizations that are seen to be unreliable [2].

It is worth noting, due to portfolio risks and decreasing diversification prospects brought about by the growing integration of traditional financial systems, investors are turning to alternative assets like cryptocurrency and green finance, and commodities [3]. In 2024, researchers demonstrate the software and hardware industries have lower exposure correlations of 0.53 and 0.42 for the long-only portfolio. It also published the effects of the equity prices of corporations with disclosure to "crypto-assets" through several industries. As a result, the asset management industry has a substantial portion of the variation in stocks for the fluctuations in this crypto-asset price with a variance by 9.4%, and it has lower variance compared with the Bitcoin price volatility [4]. Therefore, during periods of significant capital inflows or outflows from asset-backed stablecoins, there may be a rise in the instability, or beyond broadly of these reserve assets' tail risk. International organizations have expressed concerns about the financial stability of asset-backed stable coins due to the instability spillover from cryptocurrencies to old-style financial assets caused by reserve adjustments; extreme stable coin or cryptocurrency asset failures might lead to massive asset-backed stable coin redemptions and a fire sale of their reserve assets, which could have a significant effect on established financial markets like the money market this study examined [5,6].

The field of financial regulation is not the first to see the rise of illicit activities due to decentralized technology. In 2023, some researchers suggested shifting towards more flexible regulatory measures as a solution to the drawbacks of crypto-asset wallets. The question at hand is software suppliers' accountability for illegal activity committed by their users. Inspired by this effort, we look into alternative legal frameworks: under secondary copyright responsibility, countries that deal with decentralized software and its ideas might give the law the necessary flexibility [7].

About the cyber-attacks, Australia reported 3900 reports in 2015 regarding fraud and scams about identity problems with credit card, information, or complicated data breaches, even the hiding the hack produces more damage than its notification [8]. In that way, the advantage of "blockchain technology" [9] and expand the security of the transactions improves the defense with proactive and timely notification of a data breach.

This paper has the sections as follows: Sect. 2, introduces the methodology for the systematic review by combination of the methodology PRISMA-A and PICO, regarding the search strategy. Besides, in the Sect. 3, it develops the results of the quality evaluation criteria and selection process with the inclusion and exclusion aspects, the data strategy and finally, the evolution and trends in the enhancing of the crypto asset and its limits. In the Sect. 4, the discussion for the implementation of the Peruvian markets with the Blockchain challenges and the opportunity for the integration of digital information and management. Finally, in the Sect. 5, the conclusions and future works regarding the implementation of the crypto assets in the Peruvian economy.

2 Methodology

This research article was conducted using a systematic review by combining two methodologies: Population, Intervention, Comparison, Outcome (PICO) methodology for the identification of the research questions and quality, it was formulated to be not only relevant, however it is also capable of identifying and delimiting the scope of future investigations. In this regard, the research questions have been structured following the PICO method, allowing us to establish the focus of the research and thereby identify primary studies. The research questions are as follows:

- (RQ1): What would be the impact in Peru if the crypto-asset regulation Bill were approved, according to the current text?
- (RQ2): What modifications to the mentioned Bill should legislators consider?
- (RQ3): How effective would it be to consider the non-regulation of crypto-assets, as mentioned by regulatory entities in the financial system in Peru?
- (RQ4): What results have been generated by the regulation of crypto-assets in other countries?

2.1 Search Strategy

In the evaluation of the search string, it proposed to use the PICO methodology, the strategy used to obtain search terms is developed in the Table 1.

Table 1. PICO model for research questions.

Pico	Research Questions
P	The central issue is that the approval of the crypto-asset regulation bill, according to the current text, could impact society and the financial system as a whole
I	The intervention would eliminate ambiguity regarding the definition of the term 'crypto-asset,' and it could also help protect the financial market and prevent conflicts with regulatory authorities' competencies
C	The comparison is based on the pronouncements of regulatory entities in Peru and in accordance with the investigations of reviewed primary sources
O	The expected outcome is to verify whether, as of the present date, the regulation of crypto-assets by the Government is timely or, conversely, if it could adversely affect investors and/or the overall financial market

Determine the main search terms from the research questions by identifying the following terms in the Table 2.

Table 2. Main search terms.

Item	Terms
Population	Crypto-assets, Cryptocurrencies, Regulation, Policy
Intervention	Crypto-assets, Regulation, Financial market, legal framework, legislative framework
Comparison	Crypto-assets, Regulatory authorities, Policy, Legal framework, Legislative framework, Recommendations
Outcome	Crypto-assets, Regulation, Non-Regulation, Government, Policy, Recommendations, Impact

2.2 Resource Search

The systematic literature review entails a meticulous process of literature exploration, aiming to pinpoint pertinent studies aligned with the research topic. This method involves two pivotal stages: primary search and secondary search. Within the systematic review framework at hand, our exclusive emphasis has been on the primary search phase.

The initial stage of primary search in this systematic literature review has defined the relevance criterion, favoring research articles categorized in quartile 1 from reputable academic sources. Furthermore, the selection was constrained to scientific studies released within the preceding 4 years, guaranteeing the incorporation of the latest and pertinent research in the systematic review.

Similarly, the integrity of the selected research articles in this systematic review relies heavily on a meticulous selection of search engines. In this context, it was deemed relevant to employ ScienceDirect, IEEE and Scopus as the chosen search engine.

Based on the Sect. 2.1, the search strings are constructed from the terms identified in Table 3.

Table 3. Search strings for the systematic review.

Item	Search Strings
Population	((Crypto-assets) OR (Cryptocurrencies)) AND (Regulation) AND (Policy) NOT (Economics) NOT (Trading) NOT (Medicine) NOT (Psychology)
Intervention	(Crypto-assets) AND (Regulation) AND (Financial market) AND (legal framework OR legislative framework) NOT (Economics) NOT (Trading) NOT (medicine) NOT (psychology)
Comparison	(Crypto-assets) AND (Regulatory authorities) AND (Crypto-assets) AND (Regulatory authorities) AND (Policy) AND ((legal) OR (framework) OR (legislative) OR (framework)) AND (Recommendations) NOT (Trading)
Outcome	(Crypto-assets) AND (Regulation OR Non-Regulation) AND (Government) AND (Policy) AND (Recommendations) AND (Impact) NOT (economics) NOT (trading)

3 Results

The evaluation of the results for the quality, inclusion and exclusion criteria were undertaken with the aim of identifying primary studies that offer direct evidence pertaining to the research questions. Quality criteria were employed to ascertain the reliability and solidity of evidence in each study.

3.1 Quality Evaluation Criteria and Selection Process

The assessment of study quality criteria was undertaken with the aim of identifying primary studies that offer direct evidence pertaining to the research questions. Quality criteria were employed to ascertain the reliability and solidity of evidence in each study. In instances where the quality is deemed low, the study may be excluded due to insufficient evidence.

3.2 Inclusion and Exclusion Criteria (I&EC)

The establishment of I&EC, as outlined in Table 4, is aimed at gathering articles from library sources (search managers). Inclusion criteria pertain to those articles deemed suitable for the subsequent refinement phase. On the other hand, exclusion criteria are applicable to articles that do not align with the research objectives and will consequently be excluded from the case study.

Table 4. Inclusion and exclusion criteria in articles.

PICO	Search Strings	Inclusion criteria	Exclusion criteria
Population	(Crypto-assets) AND (Cryptocurrencies) AND (Regulation) AND (Policy) NOT (Economics) NOT (Trading) NOT (Medicine) NOT (Psychology)	Q1 Articles	Not economics, Not trading, Not medicine, Not psychology
Intervention	(Crypto-assets) AND (Regulation) AND (Financial market) AND (legal framework OR legislative framework) NOT (Economics) NOT (Trading) NOT (medicine) NOT (psychology)	Q1 Articles	Not economics, Not trading, Not medicine, Not psychology
Comparison	(Crypto-assets) AND (Regulatory authorities) AND (Policy) AND (legal framework OR legislative framework) AND (Recommendations) NOT (Trading) NOT (medicine) NOT (psychology)	Q1 Articles	Not economics, Not trading, Not medicine, Not psychology
Outcome	(Crypto-assets) AND (Regulation OR Non-Regulation) AND (Government) AND (Policy) AND (Recommendations) AND (Impact) NOT (economics) NOT (trading)	Q1 Articles	Not economics, Not trading, Not medicine, Not psychology

3.3 Eligibility Criteria

In this preliminary selection phase, specific inclusion and exclusion criteria were employed to choose articles for the primary studies. Research articles meeting the stipulated minimum requirements were incorporated into the reference; therefore, in the eligibility criteria the results for the identification are 169 research articles, in the screened section are 67 papers and included in the systematic review is 59 papers, regarding to the Table 5 and Fig. 1.

Table 5. Search strings for the systematic review.

Item	Identification	Screened	Included
Population	Science Direct: 2, IEEE: 64, Scopus: 103	Science Direct: 1, IEEE: 5, Scopus: 62	Science Direct: 1, IEEE: 3, Scopus: 56
Intervention	Science Direct: 2, IEEE: 13, Scopus: 48	Science Direct: 1, IEEE: 3, Scopus: 25	Science Direct: 1, IEEE: 3, Scopus: 16
Comparison	Science Direct: 1, IEEE: 0, Scopus: 3	Science Direct: 1, IEEE: 0, Scopus: 1	Science Direct: 1, IEEE: 0, Scopus: 1
Outcome	Science Direct: 3, IEEE: 0, Scopus: 3	Science Direct: 1, IEEE: 0, Scopus: 1	Science Direct: 1, IEEE: 0, Scopus: 1
Total	169	67	59

For the evaluation of the Table 5, the eligibility criteria is the following:

- Relevance to the research topic: This systematic review aims to provide a comprehensive analysis of regulating crypto-assets in the Peruvian market. By synthesizing existing literature, it addresses key questions surrounding the adoption, regulation, and impact of crypto-assets, with a specific focus on cryptocurrencies in the Peruvian context. The relevance of this topic lies in its potential to inform policymakers, investors, and researchers about the role and implications of crypto-assets in Peru's evolving financial landscape. It considered from 2011 to 2024, based on the analysis of the literature of the 13 last years.
- Methodological rigor: The systematic review follows a rigorous methodology, including comprehensive search strategies across multiple databases, predefined inclusion and exclusion criteria, and systematic screening and data extraction processes. The review adheres to established guidelines, such as

Population, Intervention, Comparison and Outcome (PICO) methodology for the research string identification and the PRISMA-A, guaranteeing transparency and replicability. Additionally, the evaluation may employ meta-analysis techniques to quantitatively synthesize findings from included studies, further enhancing methodological rigor. Selected papers should explain the research methodology, such as empirical studies, experimental designs, or comprehensive data analysis techniques.
- Data quality: Data quality is paramount in this systematic review, with a focus on selecting studies that meet predefined criteria for relevance, validity, and reliability. Quality assessment tools, such as Science Direct, IEEE and SCOPUS. Moreover, sensitivity analyses and subgroup analyses may be conducted to assess the robustness of findings across different study designs and quality levels. Papers should evaluate the evaluation with statistic with metrics and to provide reliable and well-documented data.
- Inclusion of relevant outcomes: The systematic review aims to include a comprehensive range of outcomes relevant to the study topic, such as the adoption rates of crypto-assets in Peru, the regulatory framework governing cryptocurrencies, the influence of virtual currencies on financial stability and security, and the socio-economic implications of crypto-assets. By synthesizing evidence on these outcomes, the review provides insights into the opportunities and challenges associated with cryptocurrencies as crypto-assets in the Peruvian market, thus informing future research directions and policy decisions. Selected papers should report outcomes relevant to the objectives of the study, with database and double blinded evaluation.
- Publication status: Papers should be peer-reviewed and published in Q1 international journal.

3.4 Source Database

The information sources are: ScienceDirect, IEEE and Scopus.

3.5 Bias Analysis

One potential risk of bias in the crypto-asset evaluation for the Peruvian market is the database and gray literature, for this problem, the bias analysis is composed according with the following steps:

- Selection bias mitigation: Implementing comprehensive search strategies across multiple databases and gray literature sources to identify all relevant studies. Clearly defining inclusion and exclusion criteria to minimize the risk of cherry-picking studies.
- Performance bias mitigation: Ensuring consistent and objective data extraction processes among reviewers. Utilizing standardized data extraction forms and training reviewers to enhance consistency.

- Detection bias mitigation: Conducting thorough quality assessments of included studies using established tools like the IEEE, Scopus and Science Direct database. Sensitivity analyses can be performed to assess the impact of including studies with varying levels of bias.
- Attrition bias mitigation: Tracking the reasons for exclusion of studies during the screening process to ensure transparency. Assessing the prospective effect of excluded papers on the robustness of the review's findings.
- Reporting bias mitigation: Addressing publication bias by searching for published studies and gray literature. Considering the impact of selective reporting within included studies on the overall findings of the review.

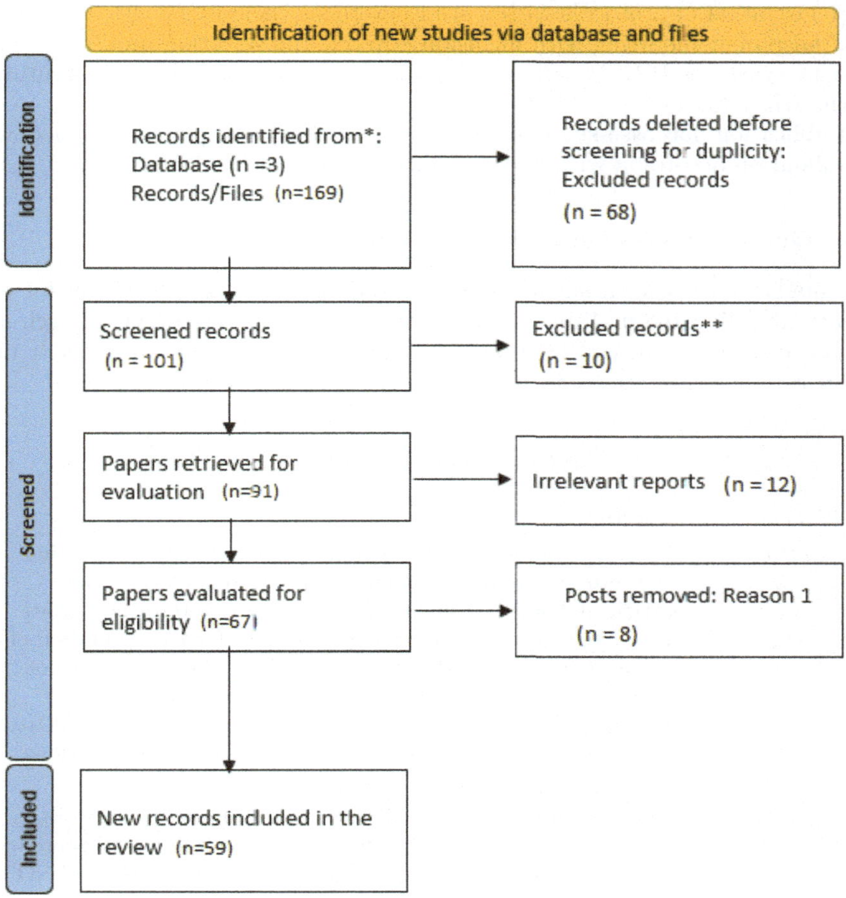

Fig. 1. PRISMA-A diagram.

To minimize the risk of publication bias in the crypto-asset evaluation for the systematic review, this paper has employed the following strategies:

- Selection of the search string with the evaluation with PICO methodology for the identification of population, intervention, comparison, and outcome.
- Comprehensive literature search with PRISMA-A methodology, with multiple databases as IEEE, ScienceDirect and Scopus with international journal Q1, to identify all relevant studies regardless of their publication status or outcomes.
- Exclusion of unpublished data, this paper didn't consider unpublished studies, conference abstracts, and reports to capture a broader range of research findings and prevent the exclusion of studies with negative or non-significant results.
- Registration of protocols with a study protocol, including eligibility criteria.
- Transparent reporting with inclusion criteria, and selection process in accordance with the PRISMA-A guidelines to facilitate replication and scrutiny of the study findings.
- Publication bias assessment of 169 papers, with and output of 59 papers with validation according with two systematic review methodologies.

3.6 Quality Assessment Checklist and Procedures

The quality checklist entails a series of questions, answers, and scores for each article, as delineated in Table 6. This structure enables the sorting of each article based on its score and facilitates filtering based on what we consider most relevant to the research topic of the systematic literature review.

Table 6. Evaluation verification.

Element	Description
Questions	Does the article establish definitions or differences regarding the term crypto-assets and/or cryptocurrency? Does the article elaborate on any distinct class of crypto-assets? Does this article incorporate recommendations for the regulation of crypto-assets? Does the article emphasize the importance of regulating crypto-assets? Does this article mention the regulatory processes in other countries for crypto-assets? Does the article detail any benefits of regulation or considerations for the non-regulation of crypto-assets?"
Answers	Assignment of weight to all questions evaluated. Therefore, the weight is one "IF" the answer is yes, it will have a score of 0.5 if the answer is "Partially" and a score of zero if it is "NO"
Cut-off record	The upper limit score is five, and low limit is 3.5, those investigations that have a score less than 3.5 will be discarded

3.7 Data Extraction Strategy

In the data extraction strategy, a detailed analysis was carried out and selected articles were grouped into their respective thematic categories, Table 7. This in-depth review involved an exploration of each study, regarding its objectives, methodologies, findings and conclusions.

Table 7. Data extraction from research articles.

Systematic review	Process output	Information cluster
Population	16 research articles	Q1 (16)
Intervention	56 research articles	Q1 (56)
Comparison	1 research articles	Q1 (1)
Outcome	1 research articles	Q1 (0)

In 2021, researchers have identified maturity level for CCBIT 2019 for the industry 4.0, especially for the crypto-asset related to Bitcoins, the challenges for the cyber security centered around to take advantage of the coin's irreversibility and pseudo-anonymity, which make it harder for government enforcement to find, by using the "ISO/IEC 15504" typical package for "engineering institute capability model integration" [10]. Therefore, the evaluation with fuzzy logic with the methodology hesitant z-cognitive mapping introduced by Torra [11], it allows to integrate different expert opinions from the evaluation in the ISO/IEC standards, with the membership evaluation with a triangle classification regarding the rules of the diffuse evaluation [12,13].

In trends, about the American's policy about the renewable energy, it has been increased and the invertors for the blockchain is growing in the last years from a case study of 80 from 83 power plants based on solar and wind technology with a profit of 7.68 million dollars per power plant; based on the Sustainable Development Goals 7, 9 and 13, by integrating the bitcoin mining as an incentive for future infrastructure [14], in this line, we can point out that blockchain-based innovations have grown tremendously since the creation of bitcoin in 2008, offering a valuable framework for comprehending scale-up [15].

On the other hand, in Kenya, the entrepreneurs are confident with the public-private business model with strong emphasis on community or transparency in the government structure [16], and blockchain with tokens, therefore, it requires an agile retailored and re-clarified with the latest technologies Calzada (n 6). , the two main problems are the following: Emerging business models and products offered by the industry are reflected in the technological advancements and affordances that are thought to satiate customer demand; and the core idea of the entire policy agenda, which is that greater competition leads to the supply of more and better services [1,17].

In the computer architecture, the trends are network configuration becomes simpler when network environments are dynamically programmed and reorganized from the Software-Defined Internet of Thing controller. Any data transaction may be tracked by trusted nodes inside a network using blockchain technology. It is not an option to rely on any third-party application or service provider to make decisions regarding information collecting, storage, and safety because IoT devices have limited resource capacity for carrying out intelligent tasks [18], through dockless bike schemes as the capital's avatar [19].

About the law enforcement, the financial intelligence unit has two potential improvements is the potential improvement, it allows to give them the necessary law-enforcement authority without having to create a new organization and legal administrative structure from scratch, and it is the shield among private sector and law enforcement [20].

4 Discussions

The regulatory intention regarding crypto assets in Peru has sparked opinions and stances in the financial market. It began with the issuance of several requests by the Commission of the Congress of the Republic to the Ministry of Economy of Peru and regulatory bodies in the financial market. The aim was to gather different opinions on the proposal for the regulation of crypto assets; in this case, the Peruvian government doesn't have explicit encryption laws or an emerging legal framework of encryption, as a common factor in the Latin America jurisdiction [30] (Tables 8 and 9).

Firstly, to explain the state structure of Peru in financial matters, the country has a Ministry of Economy and Finance, responsible for planning and executing the economic policy of the Peruvian state. It includes a technical body in charge of regulating and supervising the financial system, another technical body responsible for regulating and supervising the securities market, and finally, a Central Bank responsible for preserving the monetary stability of the country.

It is worth noting that all public bodies mentioned in the previous paragraph explicitly expressed their opposition to the regulation of crypto assets. Among their main arguments, they pointed out that: (i) There is no pressing need to develop that market at the moment; (ii) Recent experience is contradictory in favor of the market and its collateral benefits for the entire economy; (iii) There would not be a clear focus on managing the inherent risks of financial activity, which could lead to financial instability, market manipulation, financial fraud, and that this problem has not been resolved even in developed markets; and (iv) Cryptocurrencies do not constitute financial assets through which monetary policy is transmitted Comisión de Economía, Banca, Finanzas e Inteligencia Financiera, 'Dictamen Recaído en el Proyecto de Ley 1042/2021-CR, Que propone la "Ley Marco de Comercialización de Criptoactivos' (2023) Congreso de la República del Peru.

In the last year, the cybersecurity initiative was proposed for the Peruvian law No 27444 Peruvian Government 'Cybersecurity Policy' (2024) [33]; it considered the regulation from the European union from 2016, as national strategy [34].

Besides, the stance of the specialized bodies in the financial sector indicates one of the key impacts about the regulation of crypto assets: the impact on risk

Table 8. Evolution and trends in the crypto-assets.

Year	Methodology	Contribution	Limitation	Metrics
2020	Internet of Multimedia Things orchestration [21]	Varied collection of "domains as smart authentication"	Partial privacy, it means to complete partial encryption of zones	Theta platform with 1000+ TPS. Waltonchain with 100 TPS
2021	Ethereum to improve security with broken authentication [22]	From the DApp to Ethereum or Distributed Ledge [23]	In order to test the architecture's performance on a permissioned blockchain network, It also anticipates adding the step for rescinding credentials	20 transactions/second with 15 s with mined block
2022	The article does not focus on a specific methodology [24]	Several startup blockchain projects have the potential to grow quickly and transform supply networks, property rights frameworks, and capital markets	The underdeveloped world is probably where this scaling will happen most quickly	The article does not present any specific metrics
2022	Mixed method research design for infraestructure [24]	It presents the side effects of implementing data centers in Nordic countries	No specific limitations are mentioned in the article	600% of energy waste caused by the architecture and hydro power source in76%
2020	Qualitative evaluation for benefits and limits [25]	Explores the benefits of "blockchain" and "distributed ledger technology" A systematic review was not conducted		

management issues, which could lead to various economic crimes such as market manipulation and financial fraud. However, in order to curb these inherent risks in the financial sector, the Peruvian legislator proposed in this bill the creation of a Multisectoral Commission chaired by all the aforementioned public bodies. The

Table 9. Last trends in the crypto-assets.

Year	Methodology	Contribution	Limitation	Metrics
2022	Digital currency controlled by central bank	Central Banks' Interest in Digital Currencies [26]		
2023	Garch-Midas model	The frequent fluctuations in political uncertainty regarding [27] financial regulation contribute to a more stable cryptocurrency market	Tacit information for uncertainty evaluation	Econometric model known as GARCH-MIDAS
2023	TVP-VAR model	To stabilize the traditional payment markets, regulators should constantly track the movements of cryptocurrencies [28]	Adjust policies and regulations	
2023	The article does not focus on a specific methodology	Cyberattacks are possible because of the blockchain economy's financial accessibility and privacy aspects [29]	Game theory model where all players are risk-averse	The article does not present any specific metrics

purpose of this commission is to: (i) Develop policies related to the regulation, protection, surveillance, inspection, and control of cryptocurrency operations; (ii) Establish and maintain a national registry of cryptocurrency operations; (iii) Order the conduct of audits, inspections, and expert opinions related to the activity; (iv) Request reports and opinions from public or private entities specialized in cryptocurrencies; and (v) Support policies that encourage and promote the use of cryptocurrencies according to the "Congreso de la República del Perú" Congreso de la República de Perú, 'Proyecto de Ley 1042/2021-CR, Proyecto Ley Marco de Comercialización de Criptoactivos' (2021), Congreso de la Republica de Peru.

In this regard, the Peruvian legislator, motivated by legislative initiatives from neighboring countries, would be 'regulating' crypto assets, but focusing solely on 'cryptocurrencies,' without making a clear distinction between the two terminologies. As mentioned earlier, while they share Distributed Ledger Technology (DLT), also known as Blockchain, which allows currencies to function without a central authority Raza and others (n 26)., provides the prospect to incorporate "digital information, management, and contracts", enables peer-to-peer transactions by providing an immutable and transparent record of transactions Hunhevicz and Hall (n 24)., and even suggests a new way of transmitting financial information [23]. The term 'crypto assets' includes, in addition to cryptocurrencies, other types of distinct crypto assets such as Stablecoins, Non-Fungible Tokens, functionality tokens, safety, equity, and debt tokens, among others.

In the same vein, mention should be made of security tokens, which, having an underlying asset and granting rights similar to securities, raise the debate that such crypto assets may be considered under securities market regulation rather than capital market regulation. It is important to highlight the position taken by the Central Bank of Peru, which expressed disagreement with assigning regulatory functions regarding cryptocurrencies through a law. The central bank is also not in favor of presiding over a commission for the drafting of regulations in this matter. This is because it believes that cryptocurrencies do not constitute financial assets through which monetary policy is transmitted, and fulfilling the functions of a central bank through them is not feasible Comisión de Economía, Banca, Finanzas e Inteligencia Financiera (n 29).

On the other hand, the legislative proposal also establishes the creation of a Single Registry of Cryptocurrency Exchange Platforms, aimed at registering providers of cryptocurrency exchange services, also known as Exchanges. However, despite the pronouncements of the aforementioned public bodies, it should be noted that, according to the research conducted, it is likely that some of the many emerging blockchain companies will grow rapidly and revolutionize financial markets, as well as property law systems and supply chains Hans Rawhouser and others, "Scaling, Blockchain Technology, and Entrepreneurial Opportunities in Developing Countries" (2022) 18 Journal of Business Venturing Insights.

To conclude, at this point, the main impact of crypto-asset regulation under Peru's current legislative proposal could affect risk management due to its limited analysis of anti-money laundering and financial crime management. Additionally, it could potentially clash with the jurisdiction of different regulators, creating legal uncertainty in the country that approves a similar proposal. Regarding Blockchain technology, it should be emphasized that, both at the level of Peru and internationally, This technology is currently in its early stages of development and has limited real-world uses. For instance, 488 distributed ledger and blockchain initiatives were introduced globally in 2019, a 56% increase over the previous year, creating the value of 1045 projects in the last four years. Latin America has been extending gradually however progressively in this regard, with Argentina and Mexico making the most progress, while, for example, Colombia's use is still incipient. However, in all cases, regulatory barriers are the main obstacle, as there are legal difficulties or gaps hindering the implementation of blockchain technology [36].

Regarding the disadvantages of non-regulation, analyzed in this research, it should be noted that the growing incorporation of conventional financial sector has reduced branching out opportunities and grown group threats. Due to this, investors are now looking for alternative assets, such as cryptocurrency and green financing [37], the latter being highly volatile compared to traditional investments. It is important to mention the Metaverse, which integrates numerous technologies, potentially leading to security issues and privacy breaches. There may also be targeted attacks on blockchain users, including phishing, social engineering, malware, Distributed Denial of Service (DDoS), and NFT-related attacks [38], it should be noted, that criminals understand the opportunities offered by an online society, and methods for indetity theft, financial crime, and other criminal activities have adapted to the internet [8].

Similarly, it is critical to evaluate about the digital platforms could lead to harmful consequences, not only financially but also environmentally and in the pursuit of various crimes. For example, in Nordic countries, the implementation of data centers has resulted in a series of consequences, including equipment manufacturing, data cable installation, building construction, and issues related to the dark web, cryptocurrency mining, hacking, espionage, waste Sovacool, Upham and Monyei (n 22). As a complementary, the distributed ledger technology improved this hole for financial reporting standards or accounting information system [39].

Now, it has been determined that non-regulation is not entirely detrimental to the investor. For example, in the specific case of cryptocurrencies, contrary to expectations with regulation, it has been observed that greater uncertainty in financial regulation policy leads to a decrease in cryptocurrency volatility Syed Ali Raza and others, 'Uncertainty in the Financial Regulation Policy and the Boom of Cryptocurrencies' (2023) 52 Finance Research Letters. Therefore, in situations of uncertainty in cryptocurrency regulation, investment in them becomes more attractive.

Finally, this study has reviewed what would be the most suitable way to regulate crypto assets. For example, in the case of Colombia, according to the review conducted, there needs to be a focus on overcoming legal barriers that hinder the implementation of blockchain technology [36], considering that, From a privacy perspective, it is problematic as every transaction from every blockchain address is part of a publicly available dataset. Anytime an item is moved to a different address or a smart contract is engaged, the transaction is recorded on the blockchain indefinitely. As an alternative, the government and the private sector would be more confident to develop projects utilizing this technology if the state regulated three legal aspects: data protection, intellectual property, and anti-money laundering and counter-terrorism financing Marlys Herazo Narvaez and others, 'Legal Challenges for Blockchain Implementation in Colombia'.

In this sense, it is evident that, regarding crypto asset regulation, efforts need to be focused on improving consumer rights to alert them to the consequences and risks of dealing with unregulated and highly volatile assets. There is also a need to enhance systems for preventing money laundering and terrorism financing, as well as data protection, to prevent cybercrimes. Anti-money laundering laws, in particular, discourage cyber attackers from committing cyberattacks in a shared blockchain-based economy that complies with AMLKirimhan (n 27). , considering that cybercrimes are a key challenge for digital economies and society[40]also Cybercriminals see opportunities in the uncertainty that comes with altering daily routines, and the rise in virtual life is translated into more attack vectors [41]. The trends are the quantum-safe cryptography and transition phase for technical standardization with the risk of the Catch-22 loop with delays for the synchronization and governance between institutions [2, 42].

It should be noted that, according to the systematic review conducted, cryptocurrency regulation is not imperative, as there is a novel and digital means of currency exchange that is of global concern: Central Bank Digital Currencies. Many central banks have expressed interest in CBDCs, and several projects have emerged in this direction, as a financial technology enterprise, it found the latent to improve financial inclusion, growth the speed and competence of financial transactions, and decrease the cost regarding with old currency payment [43]. For example, the case of the Central Bank of Sweden, conceiving the payment system as a public good, stating that it is the government's role to ensure that the system is safe, inclusive, efficient, and ultimately, that trust in the monetary system is maintained. It should be noted, understanding the link between financial technology and financial stability is important for financial sector regulators[44].

Therefore, in 2017 the Central Bank of Sweden (CBDC) has been working on the e-Krona project, in this context, the issuance of a CBDC by the Swedish Central Bank aims to safeguard the Swedish crown (the currency of Sweden) as a stock of value and unit of account against the competition of "private digital" currenciesDe Portu (n 24)., it is an evidence of the methodology for acceptance of the population for the self-sovereign identity with technical domain of the crypto-asset [9], with social domain in the evaluation, of good governance [2] with institution domain [45].

5 Conclusion

Specific research questions (RQ) were formulated to emphasis on analyzing the influence of crypto asset regulation, proposed modifications to the bill, and the possible non-regulation of crypto assets in safeguarding consumer rights, seeking results generated in other economies.

For this purpose, the PRISMA-A guideline was employed to select and evaluate research articles that met the quality criteria for drafting this article. Initially, 169 research articles were obtained from the ScienceDirect, Scopus and IEEE database. In a second stage, a series of procedures (duplication, unrecovered, excluded, and those that did not meet the relevance criterion) were completed, allowing us to obtain a total of 59 Q1 quartile research articles, as well as official documents and specific proceedings related to this research. The PICO and

PRISMA-A research methods were useful for the current investigation because they allowed us to formulate the four research questions, and the latter helped find research articles that answered the already formulated questions.

The main contribution in the research is to conclude that strict regulation is not required for Cryptocurrencies; on the contrary, aspects such as the defense of consumer rights, money laundering prevention systems, and data protection need improvement to avoid cyber-attacks, for example, in the EU, there are recent amendments to the "General Data Protection Regulation" (GDPR), which aim to protect the rights and freedoms of natural persons. Every organization that processes personal data must comply with the GDPR to safeguard these rights while enhancing business models [39], considering that last cyber-attacks in the online platforms as LinkedIn between others, give an emphasize the irrefutable reality about the weakness of the digital security is concerning every person who participate in the chain and network at some point of time [44], with this, an ever-increasing dependence on protected data distributing has made significant information infrastructures crucial for authorities and industries [42].

Therefore, if the legislator focuses on the aspects mentioned in the paragraphs above, it can lay the groundwork for future legislative projects regarding different crypto assets, which, due to their characteristics, cannot be regulated through a general regulatory framework. This could contradict the functions of various regulatory bodies and generate legal uncertainty, contrary to the intended purpose of regulation. However, it is important to highlight the limitation regarding the obtained information, which has shown that, by applying the PICO and PRISMA-A methods, there are very few research articles on crypto assets other than cryptocurrencies. Researchers have mostly focused on analyzing DLT and CBDC, but not on other types of crypto assets, as mentioned in the previous points.

In summary, this research aimed to address a problem in the proposed regulation of crypto assets in Peru, demonstrating the importance of correctly delimiting the term 'crypto asset.' On the other hand, it was determined that regulation is not required, neither from an investor's perspective nor otherwise. Instead, legislators are recommended to focus their proposals on laying the groundwork (through regulation in AML, data protection, and consumer rights) for other types of crypto assets, such as virtual currencies like CBDC. In this regard, it is expected that the results will contribute to future legislative proposals and research on crypto assets, aiming to make better decisions. Additionally, we believe that this research can contribute to further investigations on the law of various crypto-assets and its impact, therefore it could have on a country's economy.

Finally, the limit of the uncertainty depends on the global economic policy uncertainty, as the pandemic of the COVID 19, a significant predictor of mining industry and its variation of the crypto-asset.

Acknowledgments. Thanks to the Universidad Autónoma del Perú, Universidad Tecnológica del Perú and Universidad Pompeu Fabra.

References

1. Calzada, I.: Disruptive technologies for e-diasporas: blockchain, DAOs, data cooperatives, metaverse, and ChatGPT. Futures **154**, 103258 (2023). https://doi.org/10.1016/j.futures.2023.103258
2. De Filippi, P., et al.: Blockchain as a confidence machine: the problem of trust & challenges of governance. Technol. Soc. **62**, 101284 (2020). https://doi.org/10.1016/j.techsoc.2020.101284
3. Abdullah, M., Chowdhury, M., Sulong, Z.: Asymmetric efficiency and connectedness among green stocks, halal tourism stocks, cryptocurrencies, and commodities: portfolio hedging implications. Resour. Policy **81**, 103419 (2023). https://doi.org/10.1016/j.resourpol.2023.103419
4. Mercik, A., et al.: Understanding crypto-asset exposure: an investigation of its impact on performance and stock sensitivity among listed companies. Int. Rev. Financ. Anal. **92**, 103070 (2024). https://doi.org/10.1016/j.irfa.2024.103070
5. Wu, G., Leung, P.: Do asset-backed stablecoins spread crypto volatility to traditional financial assets? Evidence from Tether. Econ. Lett. **229**, 111213 (2023). https://doi.org/10.1016/j.econlet.2023.111213
6. Birindelli, G., et al.: SFDR, investor attention, and European financial markets. Financ. Res. Lett. **56**, 104135 (2023). https://doi.org/10.1016/j.frl.2023.104135
7. Barbereau, T., Bodó, B.: Beyond financial regulation of crypto-asset wallet software: in search of secondary liability. Comput. Law Secur. Rev. **49**, 105829 (2023). https://doi.org/10.1016/j.clsr.2023.105829
8. Alazab, M., Hong, S.H., Ng, J.: Louder bark with no bite: privacy protection through the regulation of mandatory data breach notification in Australia. Future Gener. Comput. Syst. **116**, 22–29 (2021). https://doi.org/10.1016/j.future.2020.10.017
9. Buterin, V., Illum, J., Nadler, M., Schär, F., Soleimani, A.: Blockchain privacy and regulatory compliance: towards a practical equilibrium. In: Blockchain: Research and Applications, pp. 100176 (2023). https://doi.org/10.1016/j.bcra.2023.100176
10. Ariffin, K.A.Z., Ahmad, F.H.: Indicators for maturity and readiness for digital forensic investigation in era of industrial revolution 4.0. Comput. Secur. **105**, 102237 (2021). https://doi.org/10.1016/j.cose.2021.102237
11. Karaşan, A., et al.: A multicriteria decision making methodology based on two-dimensional uncertainty by hesitant Z-fuzzy linguistic terms with an application for blockchain risk evaluation. Appl. Soft Comput. **113**, 108014 (2021). https://doi.org/10.1016/j.asoc.2021.108014
12. Nguyen, P.H., et al.: Breaking ground in ESG assessment: integrated DEA and MCDM framework with spherical fuzzy sets for Vietnam's wire and cable sector. J. Open Innov. Technol. Mark. Complex. **9**(3), 100136 (2023). https://doi.org/10.1016/j.joitmc.2023.100136
13. Velásquez, R.M.A., Lara, J.V.M.: Principal components analysis and adaptive decision system based on fuzzy logic for power transformer. Fuzzy Inf. Eng. **9**(4), 493–514 (2017). https://doi.org/10.1016/j.fiae.2017.12.005
14. Lal, A., Niaz, H., Liu, J.J., You, F.: Can bitcoin mining empower energy transition and fuel sustainable development goals in the US? J. Clean. Prod. **439**, 140799 (2024). https://doi.org/10.1016/j.jclepro.2024.140799
15. Rawhouser, H., et al.: Scaling, blockchain technology, and entrepreneurial opportunities in developing countries. J. Bus. Ventur. Insights **18**, e00325 (2022). https://doi.org/10.1016/j.jbvi.2022.e00325

16. Sydow, A., Sunny, S.A., Coffman, C.D.: Leveraging blockchain's potential - the paradox of centrally legitimate, decentralized solutions to institutional challenges in Kenya. J. Bus. Ventur. Insights **14**, e00170 (2020). https://doi.org/10.1016/j.jbvi.2020.e00170
17. Ferrari, M.V.: The platformisation of digital payments: the fabrication of consumer interest in the EU FinTech agenda. Comput. Law Secur. Rev. **45**, 105687 (2022). https://doi.org/10.1016/j.clsr.2022.105687
18. Khalid, M., Hameed, S., Qadir, A., Shah, S.A., Draheim, D.: Towards SDN-based smart contract solution for IoT access control. Comput. Commun. **198**, 1–31 (2023). https://doi.org/10.1016/j.comcom.2022.11.007
19. Chambers, P.: O Bike in Melbourne: a plea for more scepticism about disruption and capital, based on what we can know about one Dockless bike scheme. Transp. Res. Part A Policy Pract. **140**, 72–80 (2020). https://doi.org/10.1016/j.tra.2020.07.016
20. Brewczyńska, M.: Financial intelligence units: reflections on the applicable data protection legal framework. Comput. Law Secur. Rev. **43**, 105612 (2021). https://doi.org/10.1016/j.clsr.2021.105612
21. Jan, M., et al.: Security and blockchain convergence with internet of multimedia things: current trends, research challenges and future directions. J. Netw. Comput. Appl. **175**, 102918 (2021). https://doi.org/10.1016/j.jnca.2020.102918
22. Mishra, R., et al.: Privacy protected blockchain based architecture and implementation for sharing of students' credentials. Inf. Process. Manag. **58**(3), 102512 (2021). https://doi.org/10.1016/j.ipm.2021.102512
23. Weigand, H., et al.: Shared ledger accounting - implementing the economic exchange pattern. Inf. Syst. **90**, 101437 (2020). https://doi.org/10.1016/j.is.2019.101437
24. Sovacool, B.K., et al.: The "whole systems" energy sustainability of digitalization: Humanizing the community risks and benefits of Nordic datacenter development. Energy Res. Soc. Sci. **88**, 102493 (2022). https://doi.org/10.1016/j.erss.2022.102493
25. Hunhevicz, J.J., Hall, D.M.: Do you need a blockchain in construction? Use case categories and decision framework for DLT design options. Adv. Eng. Inform. **45**, 101094 (2020). https://doi.org/10.1016/j.aei.2020.101094
26. De Portu, T.: New trends in retail payments: How technological changes are reshaping the payments system. Introducing a proposal for a new pan-European instant payment system. Lat. Am. J. Cent. Bank. **3**(4), 100075 (2022). https://doi.org/10.1016/j.latcb.2022.100075
27. Raza, S.A., et al.: Uncertainty in the financial regulation policy and the boom of cryptocurrencies. Financ. Res. Lett. **52**, 103515 (2023). https://doi.org/10.1016/j.frl.2022.103515
28. Yousaf, I., Goodell, J.W.: Linkages between CBDC and cryptocurrency uncertainties, and digital payment stocks. Financ. Res. Lett. **54**, 103765 (2023). https://doi.org/10.1016/j.frl.2023.103765
29. Kirimhan, D.: Importance of anti-money laundering regulations among prosumers for a cybersecure decentralized finance. J. Bus. Res. **157**, 113558 (2023). https://doi.org/10.1016/j.jbusres.2022.113558
30. Dizon, M., Upson, P.J.: Laws of encryption: an emerging legal framework. Comput. Law Secur. Rev. **43**, 105635 (2021). https://doi.org/10.1016/j.clsr.2021.105635
31. Arias Velasquez, R., et al.: Enhancing medium voltage underground circuit design: assessing limitations, thermal influence, and accurate modelling. Res. Eng. **20**, 101552 (2023). https://doi.org/10.1016/j.rineng.2023.101552

32. Velásquez, R.M.A.: Bearings faults and limits in wind turbine generators. Res. Eng. **21**, 101891 (2024). https://doi.org/10.1016/j.rineng.2024.101891
33. Maesschalck, S., et al.: Don't get stung, cover your ICS in honey: How do honeypots fit within industrial control system security. Comput. Secur. **114**, 102598 (2022). https://doi.org/10.1016/j.cose.2021.102598
34. Vandezande, N.: Cybersecurity in the EU: How the NIS2-directive stacks up against its predecessor. Comput. Law Secur. Rev. **52**, 105890 (2024). https://doi.org/10.1016/j.clsr.2023.105890
35. Velásquez, R.M.A., Lara, J.V.M.: Forecast and evaluation of COVID-19 spreading in USA with reduced-space Gaussian process regression. Chaos Solitons Fract. **136**, 109924 (2020). https://doi.org/10.1016/j.chaos.2020.109924
36. Narvaez, M.H., et al.: Legal challenges for Blockchain implementation in Colombia. Procedia Comput. Sci. **210**, 323–327 (2022). https://doi.org/10.1016/j.procs.2022.10.158
37. Raza, S.A., Masood, A., Benkraiem, R., Urom, C.: Forecasting the volatility of precious metals prices with global economic policy uncertainty in pre and during the COVID-19 period: novel evidence from the GARCH-MIDAS approach. Energy Econ. **120**, 106591 (2023). https://doi.org/10.1016/j.eneco.2023.106591
38. Yu, B., et al.: METAseen: analyzing network traffic and privacy policies in Web 3.0 based metaverse. Digit. Commun. Netw. (2023). https://doi.org/10.1016/j.dcan.2023.11.006
39. Rhahla, M., Allegue, S., Abdellatif, T.: Guidelines for GDPR compliance in big data systems. J. Inf. Secur. Appl. **61**, 102896 (2021). https://doi.org/10.1016/j.jisa.2021.102896
40. Ho, H., Ko, R., Mazerolle, L.: Situational crime prevention (SCP) techniques to prevent and control cybercrimes: a focused systematic review. Comput. Secur. **115**, 102611 (2022). https://doi.org/10.1016/j.cose.2022.102611
41. Saleous, H., et al.: COVID-19 pandemic and the cyberthreat landscape: research challenges and opportunities. Digit. Commun. Netw. **9**(1), 211–222 (2023). https://doi.org/10.1016/j.dcan.2022.06.005
42. Kong, I., Janssen, M., Bharosa, N.: Realizing quantum-safe information sharing: implementation and adoption challenges and policy recommendations for quantum-safe transitions. Gov. Inf. Q. **41**(1), 101884 (2024). https://doi.org/10.1016/j.giq.2023.101884
43. Gupta, S., Pandey, D.K., El Ammari, A., Sahu, G.P.: Do perceived risks and benefits impact trust and willingness to adopt CBDCs? Res. Int. Bus. Finance **66**, 101993 (2023). https://doi.org/10.1016/j.ribaf.2023.101993
44. Yudaruddin, R., et al. Financial technology and bank stability in an emerging market economy. Heliyon **9**(5), e16183 (2023). https://doi.org/10.1016/j.heliyon.2023.e16183
45. Weigl, L., Barbereau, T., Fridgen, G.: The construction of self-sovereign identity: extending the interpretive flexibility of technology towards institutions. Gov. Inf. Q. **40**(4), 101873 (2023). https://doi.org/10.1016/j.giq.2023.101873

Insights of Evolving Methods Towards Screening of AI-Enhanced Malware in IoT Environment

Ramesh Shahabadkar[1](✉), S. B. Nandeeswar[2], Rajashree Soman[3], and G. Sangeetha[4]

[1] Department of CSE, AMC Engineering College, Bangaluru, Karnataka 560083, India
ramesh.shahabadkar@amceducation.in

[2] Department of CSE (AIML), AMC Engineering College, Bengaluru, Karnataka, India
nandeeswar.basavaraju@amceducation.in

[3] Department of CSE, BNM Institute of Technology, Banashankari Stage II, Bangaluru, Karnataka 560070, India

[4] Department of Computer Science, Christ University, Bengaluru, Karnataka, India
sangeetha.g@christuniversity.in

Abstract. Internet-of-Things (IoT) has been encountering a series of potential form of threats since past half decades. Artificial Intelligence (AI), which is frequently seen to be adopted to solve various challenges in IoT operation, has now been adopted even by attackers for their malicious purposes. Of all forms of threats, AI-enhanced malwares are one of the most potential forms of threats which has its extensive effectiveness towards the complete operation of the entire IoT environment. Hence, this manuscript discusses existing detection and prevention approaches evolved in current literatures to understand various taxonomies of solution-based methodologies for circumventing such threats. The paper also contributes towards highlighting the potential open-ended issues that are yet to be addressed.

Keywords: Internet of Things · Artificial Intelligence · Malware · Security · detection · prevention · Threats

1 Introduction

IoT (Internet of Things) environments face a variety of threats due to their interconnected nature and the proliferation of connected devices. Hackers may attempt to gain unauthorized access to IoT devices or networks to steal sensitive data, manipulate device functionality, or launch further attacks within the network [1]. IoT devices often collect and transmit sensitive data, such as personal information or proprietary business data. Breaches of this data can lead to privacy violations, identity theft, financial losses, or reputational damage [2]. Malicious software can infect IoT devices, disrupting their operation, stealing data, or holding devices hostage for ransom. Ransomware attacks on IoT devices can result in financial losses or operational disruptions for individuals and organizations [3]. Compromised IoT devices can be enlisted into botnets, large networks of devices controlled by attackers. These botnets can be used to launch Distributed Denial of Service (DDoS) attacks, overwhelming targeted servers or networks with traffic and

R. Silhavy and P. Silhavy (Eds.): CSOC 2024, LNNS 1120, pp. 694–704, 2024.
https://doi.org/10.1007/978-3-031-70518-2_57

causing service disruptions [4]. In some cases, IoT devices control physical systems or machinery. Unauthorized access or manipulation of these devices can result in physical damage, injury, or even loss of life [5]. Attackers may compromise the supply chain of IoT devices, inserting malicious components or tampering with devices during manufacturing, shipping, or installation. This can lead to widespread security vulnerabilities in IoT ecosystems [6]. Weak authentication mechanisms or lack of encryption can make IoT devices vulnerable to interception or tampering of data during transmission and unauthorized access by malicious actors [7]. Misconfiguration of IoT devices or networks, such as using default passwords or leaving unnecessary ports open, can create opportunities for attackers to exploit vulnerabilities and gain access to sensitive data or device control [8]. There are various approaches to offering IoT security; however, less potential inference towards significance of this security concern has been discussed [9–12].

Hence, this paper presents a compact discussion associated with the AI-based security threats, their implications and suggest some possible mechanism to stop it thereby protecting IoT environment. The organization of the paper is as follow: Sect. 2 presents discussion of AI-based threats in IoT followed by discussion of varied essential operation being performed by AI-based threats in Sect. 3. Section 4 presents discussion of the characteristic of AI-enhanced malware followed by briefing of identification methods of AI-enhanced malware in Sect. 5. Further, Sect. 6 presents discussion on prevention techniques of AI-enhanced malware. The highlights of essential findings of study are given in Sect. 7, while conclusion in given in Sect. 8.

2 AI-Based Threats in IoT

The most dangerous attack in AI-based IoT (Internet of Things) would likely be a combination of AI-driven malware or AI-powered botnets targeting IoT devices. However, there are various forms of AI-based threats that are exhibited in Fig. 1

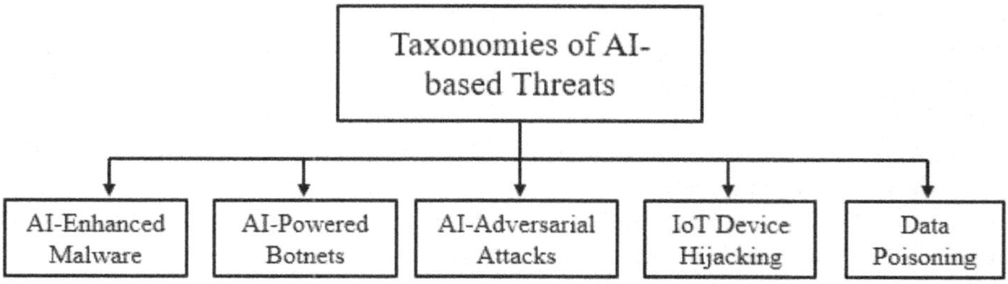

Fig. 1. Taxonomies of AI-based Threats

Figure 1 highlights various forms of attacks that can exploit vulnerabilities in IoT systems to cause significant damage. The first form of threat is known as *AI-Enhanced Malware.* Malicious software equipped with AI capabilities can adapt and evolve, making it harder to detect and mitigate by traditional cybersecurity measures. It could infiltrate IoT devices, steal data, or even take control of critical infrastructure [13]. The second form of threat is known as *AI-Powered Botnets.* Botnets are networks of compromised devices controlled by a single entity. With AI, these botnets can become more sophisticated, coordinating attacks, and dynamically adjusting their strategies to evade detection and maximize damage [14]. The third form of threat is known as *AI-Adversarial Attacks.* IoT systems often rely on AI algorithms for tasks like anomaly detection or decision-making. Adversaries can manipulate these AI models through adversarial attacks, leading to incorrect or malicious actions, potentially causing physical harm or disrupting essential services [15]. *IoT Device Hijacking* is another potential form of AI-based threat. Attackers could hijack IoT devices, such as connected vehicles or industrial control systems, using AI techniques to bypass authentication mechanisms, manipulate sensor data, or issue unauthorized commands, leading to safety hazards or operational disruptions [16]. *Data Poisoning* is another form of AI-based threat that has potentially negative effective on IoT networks. AI algorithms rely heavily on data for training and decision-making. Attackers could poison IoT data streams with malicious inputs, causing AI systems to make incorrect decisions or predictions, with potentially catastrophic consequences in critical applications like healthcare or autonomous vehicles [17].

Therefore, securing AI-based IoT systems requires a multi-layered approach, including robust authentication mechanisms, encryption, anomaly detection, regular software updates, and ongoing monitoring for unusual behavior. Collaboration between manufacturers, developers, and cybersecurity experts is crucial to stay ahead of emerging threats in this rapidly evolving landscape.

3 Operation of AI-Based Threats

AI-enhanced malware leverages artificial intelligence techniques to improve its capabilities and evade detection, making it more sophisticated and effective at compromising systems. Here's how it typically works: The primary operation of AI-based threat is to undertake *Adaptive Behavior*. AI-powered malware can adapt its behavior in real-time based on the environment it encounters. It can analyze the target system, identify vulnerabilities, and choose the most effective attack vectors. This adaptability allows it to bypass traditional security measures that rely on fixed patterns or signatures [18]. *Dynamic Evasion* is the next operation performed by AI-based threats in IoT. AI-driven malware can dynamically change its code or behavior to evade detection by antivirus software and intrusion detection systems. It can generate polymorphic code, which alters its structure with each infection, making it challenging for security tools to recognize and block [19]. The next potential form of threat is known as *Zero-Day Exploits.* AI algorithms can analyze software and systems to discover previously unknown vulnerabilities, known as zero-day exploits. AI-enhanced malware can exploit these vulnerabilities before security patches are available, giving attackers a significant advantage in launching targeted attacks [20]. Interestingly, *social engineering* is also represented as one of the most intrusive threats in IoT. AI can analyze massive amounts of data to create highly personalized phishing emails or social engineering attacks, increasing the likelihood of successful infiltration. By mimicking the writing style or behavior of legitimate users, AI-enhanced malware can trick individuals into revealing sensitive information or installing malicious software [21]. *Stealthy Operations* has also proven to be critical issue when it comes to IoT security related to AI-based threats: AI-powered malware can operate stealthily within a compromised system, minimizing its footprint and avoiding detection for extended periods. It can monitor user behavior, network traffic, and system processes to avoid triggering alarms or suspicion [22]. Finally, *automated propagation* is also reported to have an adversarial impact on communication in different scales of applications in IoT. AI can automate the propagation of malware across networks by identifying vulnerable devices and exploiting them to spread infection rapidly. This autonomous behavior enables AI-enhanced malware to establish botnets or create distributed denial-of-service (DDoS) attacks with minimal human intervention [23].

Overall, AI-enhanced malware represents a significant threat due to its ability to continuously learn and evolve, making it increasingly challenging for cybersecurity defenses to keep pace. Combatting these threats requires the development of AI-driven security solutions capable of detecting and mitigating advanced attacks in real-time.

4 Characteristic of AI-Enhanced Malware

AI-enhanced malware possesses several distinctive characteristics that differentiate it from traditional malware: The first characteristic of AI-enhanced Malware is *Adaptability*. AI-enhanced malware can adapt its behavior based on the environment it encounters, allowing it to evade detection and exploit vulnerabilities more effectively. *Self-Learning* is the next essential feature of such form of malware. AI-powered malware can learn from its interactions with target systems and security defenses, continuously improving its evasion techniques and attack strategies. This form of malware is also known for its adoption of *Dynamic Evasion Techniques***.** It employs dynamic evasion techniques such as polymorphic code generation, obfuscation, and encryption to evade traditional signature-based detection methods. Another important property of this attacker is *Zero-Day Exploits***.** AI-enhanced malware can discover and exploit previously unknown vulnerabilities (zero-day exploits) before patches or security updates are available, enabling it to launch targeted attacks with minimal risk of detection. *Stealthy Operations* is another essential characteristic of AI-enhanced malware. It operates stealthily within compromised systems, minimizing its footprint and avoiding detection by security tools and administrators. *Automated Propagation* follows as subsequent feature of this attacker. AI-enhanced malware can autonomously propagate across networks by identifying and exploiting vulnerable devices, enabling it to establish botnets or execute large-scale attacks without human intervention. It employs *sophisticated social engineering* techniques, leveraging AI to analyze and generate convincing phishing emails, messages, or websites tailored to specific targets, increasing the likelihood of successful infiltration. AI-enhanced malware utilizes *advanced persistence mechanisms* to maintain access to compromised systems, such as rootkit installation, privilege escalation, and backdoor creation. It leverages AI algorithms to make *data-driven decisions*, such as identifying high-value targets, selecting optimal attack vectors, and adapting its behavior in real-time based on feedback from the target environment. Due to its adaptability and dynamic nature, AI-enhanced malware exhibits *resilience to traditional security measures*, requiring advanced AI-driven security solutions capable of detecting and mitigating emerging threats in real-time.

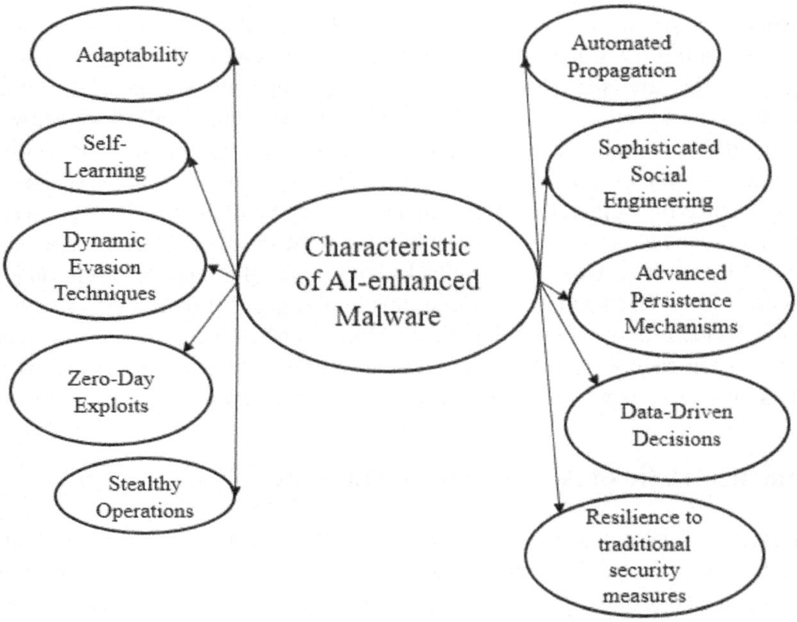

Fig. 2. Characteristic of AI-enhanced Malware

The characteristics shown in Fig. 2 make AI-enhanced malware highly sophisticated, challenging to detect, and potentially devastating in their impact on targeted systems and networks. As a result, defending against these threats requires a comprehensive cybersecurity strategy that integrates advanced AI-driven threat detection, proactive threat hunting, and incident response capabilities.

5 Identification Methods for AI-Enhanced Malware

This section discusses about the existing approaches towards identification of AI-based methods. Identifying AI-enhanced malware in IoT (Internet of Things) environments requires a combination of proactive measures, continuous monitoring, and the use of advanced cybersecurity solutions. Here are some strategies to help identify AI-enhanced malware in IoT (Fig. 3):

Existing system is witnessed to *implement behavioral anomaly detection* mechanisms that monitor the normal behavior of IoT devices and networks [24]. AI-enhanced malware may exhibit abnormal behavior patterns, such as unusual data traffic, unexpected system resource utilization, or irregular communication patterns. By analyzing deviations from normal behavior, anomaly detection systems can flag potential malware infections.

Existing system is also witnessed to perform deep packet inspection and *analysis of network traffic* to identify suspicious communication patterns indicative of malware activity [25]. AI-enhanced malware may utilize encrypted channels, command-and-control (C2) communication, or peer-to-peer networking to evade detection. Advanced network traffic analysis tools equipped with AI capabilities can help identify these malicious patterns and anomalies. Existing literature is also reported to deploy *endpoint monitoring* solutions on IoT devices to detect signs of compromise, such as unauthorized system modifications, unusual process execution, or suspicious network connections [26]. AI-powered endpoint detection and response (EDR) solutions can analyze device telemetry data in real-time to identify indicators of AI-enhanced malware activity. The sensor data and device telemetry are analyzed for *anomalous patterns* that may indicate AI-enhanced malware manipulation [27]. For example, sudden changes in environmental sensor readings, abnormal device behavior, or unexpected data fluctuations could be indicative of malware interference. Machine learning algorithms can analyze historical data patterns to identify anomalies and detect potential malware infections.

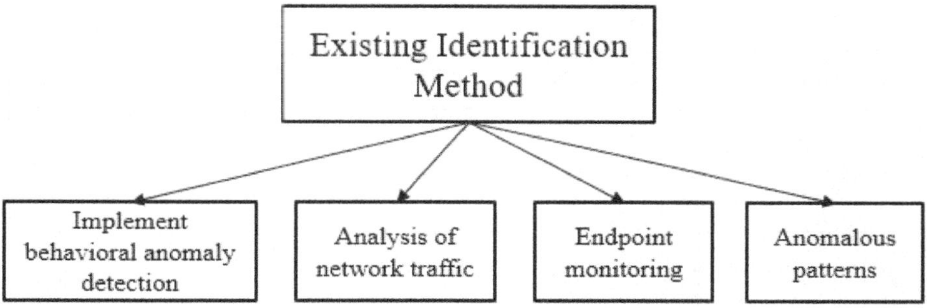

Fig. 3. Existing Identification Method

Existing system also conduct *code analysis* and *reverse engineering* of IoT firmware and software to identify signs of AI-enhanced malware [28]. The obfuscated code, polymorphic malware variants, or evidence of AI algorithms embedded are looked for within the malware code. Automated static and dynamic analysis tools equipped with AI capabilities can assist in identifying and analyzing suspicious code patterns. There are certain research works where system participate in *threat intelligence* for sharing communities and collaborate with industry peers to stay informed about emerging AI-enhanced malware threats targeting IoT devices [29]. Sharing threat intelligence data on malware signatures, behavior patterns, and indicators of compromise (IOCs) can help organizations proactively identify and mitigate potential threats. Finally, *continuous monitoring* and *incident response* procedures to detect and respond to AI-enhanced malware infections in real-time [30]. An incident response plans that outline the steps for isolating infected devices, containing the spread of malware, and restoring affected systems to a known-good state.

By leveraging a combination of these strategies and deploying advanced AI-driven cybersecurity solutions, organizations can enhance their ability to identify and mitigate AI-enhanced malware threats targeting IoT environments. Additionally, maintaining a proactive approach to cybersecurity hygiene, including regular software updates, patch management, and employee training, can help minimize the risk of malware infections and strengthen overall security posture.

6 Prevention of AI-Based Malware

This section presents discussion of existing literatures associated with prevention of AI-based malwares. Stopping and preventing AI-enhanced malware intrusion in a large-scale, heterogeneous IoT network requires a comprehensive and multi-layered approach to cybersecurity. Here's a strategy that combines various techniques and measures to mitigate the risk as shown in Fig. 4:

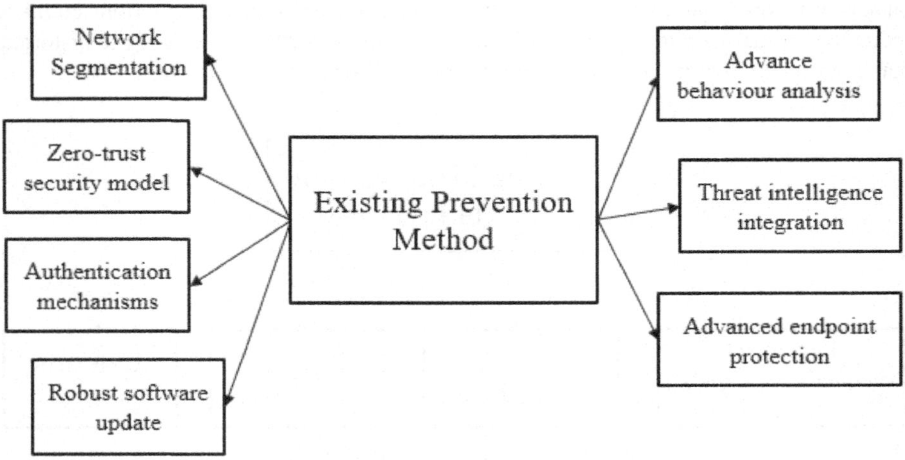

Fig. 4. Existing Prevention Method

The IoT network is divided into segments based on device type, function, or security requirements. *Network segmentation* is implemented to isolate critical assets and restrict lateral movement of malware within the network [31]. This helps contain potential infections and prevents malware from spreading across the entire network. A *zero-trust security model* is implemented in existing literatures that assumes no device or user can be trusted by default, regardless of their location within the network [32]. Implement strict access controls, authentication mechanisms, and least privilege principles to enforce granular access policies and prevent unauthorized access to IoT devices and resources. Existing study models are also found to implement strong *authentication mechanisms*, such as *Multi-Factor Authentication* (MFA) and *biometric authentication*, to verify the identity of users and devices accessing the IoT network [33]. Enforce access control policies based on the principle of least privilege to limit the privileges granted to users and devices, reducing the attack surface for potential malware intrusion.

Existing literatures also deployed *advanced behavioral analysis* and *anomaly detection* solutions to monitor the behavior of IoT devices and detect unusual or suspicious activities indicative of AI-enhanced malware intrusion [34]. Leverage machine learning algorithms to analyze telemetry data, network traffic, and device behavior patterns in real-time, enabling early detection and response to potential threats. The existing study has also presented technique where threat intelligence feeds and security information and event management (SIEM) systems to enhance threat detection capabilities and stay informed about emerging AI-enhanced malware threats targeting IoT environments. Leverage *threat intelligence integration* data on malware signatures, IOCs, and attack techniques to proactively identify and block malicious activity in the network [35]. Existing study also deployed *advanced endpoint protection* solutions on IoT devices to detect and prevent AI-enhanced malware infections at the device level [26]. The AI-driven endpoint detection and response (EDR) capabilities are utilized to analyze device telemetry data, detect malicious behavior, and respond to threats in real-time, even in heterogeneous IoT environments. A *robust software update* and *patch management process* is also used in existing system to ensure that IoT devices are running the latest firmware and security patches. Regularly update device firmware and software to address known vulnerabilities and mitigate the risk of exploitation by AI-enhanced malware [36].

By implementing these preventive measures and adopting a proactive approach to cybersecurity, organizations can strengthen their defenses against AI-enhanced malware intrusion in large-scale, heterogeneous IoT networks. It's essential to continuously monitor the evolving threat landscape, adapt security controls accordingly, and collaborate with industry peers to stay ahead of emerging threats.

7 Essential Findings of Study

IoT security faces a plethora of challenges, particularly when it comes to defending against AI-based attacks. Here are some of the key challenges:

- **Complexity of IoT Ecosystem**: IoT environments are diverse, comprising a wide range of devices with varying capabilities and security features. Coordinating security measures across this complexity is challenging.
- **Resource Constraints**: Many IoT devices have limited computational resources, making it difficult to implement robust security measures such as encryption and complex authentication protocols.
- **Legacy Systems and Compatibility**: Numerous IoT devices are built on outdated platforms with limited support for security updates. Compatibility issues make it challenging to implement standardized security protocols across the IoT ecosystem.
- **Data Privacy Concerns**: IoT devices collect vast amounts of sensitive data, raising concerns about privacy breaches if these data fall into the wrong hands. AI-based attacks can exploit vulnerabilities to access and misuse this data.
- **Lack of Standardization**: The absence of uniform security standards across IoT devices complicates efforts to develop comprehensive security solutions. Varying standards lead to inconsistencies and vulnerabilities in IoT ecosystems.
- **Interconnectedness**: IoT devices are often interconnected, creating a ripple effect where a security breach in one device can compromise the entire network. AI-based attacks can exploit these interconnections to propagate across the IoT ecosystem.
- **Insufficient Authentication Mechanisms**: Many IoT devices rely on weak or default authentication mechanisms, making them susceptible to brute-force attacks and unauthorized access by AI algorithms.
- **Cyber-Physical Risks**: IoT devices control physical systems and infrastructures, posing significant risks if compromised. AI-based attacks targeting IoT devices can disrupt critical operations and cause physical harm.

- **Adversarial AI Techniques**: Attackers can leverage AI algorithms to develop sophisticated attack strategies, such as adversarial examples to bypass security measures, or using AI to automate and scale attacks on IoT systems.
- **Inadequate Update Mechanisms**: Updating IoT devices with security patches and fixes is often challenging due to factors like limited connectivity, vendor negligence, or device obsolescence. This leaves devices vulnerable to known exploits that AI-based attacks can exploit.

Addressing these challenges requires a multi-faceted approach involving collaboration between stakeholders, including device manufacturers, cybersecurity experts, policymakers, and end-users, to develop and implement robust security measures tailored to the unique characteristics of IoT environments.

8 Conclusion

This chapter has presented a snapshot of research work effectiveness towards measuring its capabilities for resisting AI-enhanced malware which is considered as one of the most lethal threats in IoT. The contribution of this paper are as follows: i) this paper presents a compact discussion of detection and prevention approaches towards resisting AI-enhanced malwares explicitly, ii) the paper also highlights the open-end challenges associated with both the approaches to showcase that this specific security problem has yet not been solved. The outcome of the paper offers a compact and clear information for readers planning to address the AI-enhanced malware problems to undertake more fruitful solution. The future direction of current work will emphasize towards addressing the identified challenges in evolving a robust countermeasure that can optimize the resources to make better resiliency against such threats.

References

1. Aljabri, M., et al.: Machine learning-based detection for unauthorized access to IoT devices. J. Sens. Actuator Netw.Netw. **12**(2), 27 (2023)
2. Vojković, G., Milenković, M., Katulić, T.: IoT and smart home data breach risks from the perspective of data protection and information security law. Bus. Syst. Res. Int. J. Soc. Adv. Innov. Res. Econ. **11**(3), 167–185 (2020)
3. Sharma, P., Zawar, S., Patil, S.B.: Ransomware analysis: Internet of Things (Iot) security issues challenges and open problems inthe context of worldwide scenario of security of systems and malware attacks. In: International Conference on Recent Innovation in Engineering and Management, vol. 2, no. 3, pp. 177–184 (2016)
4. McDermott, C.D., Majdani, F., Petrovski, A.V.: Botnet detection in the internet of things using deep learning approaches. In: 2018 International Joint Conference on Neural Networks (IJCNN), pp. 1–8. IEEE (2018)
5. Yang, X., Shu, L., Liu, Y., Hancke, G.P., Ferrag, M.A., Huang, K.: Physical security and safety of IoT equipment: a survey of recent advances and opportunities. IEEE Trans. Ind. Inf. **18**(7), 4319–4330 (2022)
6. Hiromoto, R.E., Haney, M., Vakanski, A.: A secure architecture for IoT with supply chain risk management. In: 2017 9th IEEE International Conference on Intelligent Data Acquisition and Advanced Computing Systems: Technology and Applications (IDAACS), vol. 1, pp. 431–435. IEEE (2017)

7. Diro, A., Reda, H., Chilamkurti, N., Mahmood, A., Zaman, N., Nam, Y.: Lightweight authenticated-encryption scheme for Internet of Things based on publish-subscribe communication. IEEE Access **8**, 60539–60551 (2020)
8. Hussain, S.R., Nirjon, S., Bertino, E.: Securing the insecure link of Internet-of-Things using next-generation smart gateways. In: 2019 15th International Conference on Distributed Computing in Sensor Systems (DCOSS), pp. 66–73. IEEE (2019)
9. Khan, M.A., Salah, K.: IoT security: review, blockchain solutions, and open challenges. Futur. Gener. Comput. Syst.. Gener. Comput. Syst. **82**, 395–411 (2018)
10. Hassija, V., Chamola, V., Saxena, V., Jain, D., Goyal, P., Sikdar, B.: A survey on IoT security: application areas, security threats, and solution architectures. IEEE Access **7**, 82721–82743 (2019)
11. Al-Garadi, M.A., Mohamed, A., Al-Ali, A.K., Du, X., Ali, I., Guizani, M.: A survey of machine and deep learning methods for internet of things (IoT) security. IEEE Commun. Surv. Tutor. **22**(3), 1646–1685 (2020)
12. Alaba, F.A., Othman, M., Hashem, I.A.T., Alotaibi, F.: Internet of Things security: a survey. J. Netw. Comput. Appl.Netw. Comput. Appl. **88**, 10–28 (2017)
13. Farooq, U.: Cyber-Physical Security: AI Methods for Malware/Cyber-Attacks Detection on Embedded/IoT Applications. Doctoral Dissertation, Politecnico di Torino (2023)
14. Memos, V.A., Psannis, K.E.: AI-powered honeypots for enhanced IoT botnet detection. In: 2020 3rd World Symposium on Communication Engineering (WSCE), pp. 64–68. IEEE (2020)
15. Ibitoye, O., Shafiq, O., Matrawy, A.: Analyzing adversarial attacks against deep learning for intrusion detection in IoT networks. In: 2019 IEEE Global Communications Conference (GLOBECOM), pp. 1–6. IEEE (2019)
16. Shobana, M., Rathi, S.: IoT malware: an analysis of IoT device hijacking. Int. J. Sci. Res. Comput. Sci. Comput. Eng. Inf. Technol. **3**(5), 2456–3307 (2018)
17. Zhu, Y., Wen, H., Wu, J., Zhao, R.: Online data poisoning attack against edge AI paradigm for IoT-enabled smart city. Math. Biosci. Eng.Biosci. Eng. **20**(10), 17726–17746 (2023)
18. Fuller, T.R., Deane, G.E. IoT applications in an adaptive intelligent system with responsive anomaly detection. In: 2016 Future Technologies Conference (FTC), pp. 754–762. IEEE (2016)
19. Jeon, J., Park, J.H., Jeong, Y.S.: Dynamic analysis for IoT malware detection with convolution neural network model. IEEE Access **8**, 96899–96911 (2020)
20. Popoola, S.I., Ande, R., Adebisi, B., Gui, G., Hammoudeh, M., Jogunola, O.: Federated deep learning for zero-day botnet attack detection in IoT-edge devices. IEEE Internet Things J. **9**(5), 3930–3944 (2021)
21. Hijji, M., Alam, G.: A multivocal literature review on growing social engineering based cyber-attacks/threats during the COVID-19 pandemic: challenges and prospective solutions. IEEE Access **9**, 7152–7169 (2021)
22. Layton, P.: Fighting artificial intelligence battles: operational concepts for future AI- enabled wars. Network **4**(20), 1–100 (2021)
23. Kollu, P.K., Kumar, K., Kshirsagar, P.R., Islam, S., Naveed, Q.N., Hussain, M.R., Sundramurthy, V.P.: Development of advanced artificial intelligence and IoT automation in the crisis of COVID-19 Detection. J. Healthc. Eng. **2022**, 1 (2022)
24. Arrington, B., Barnett, L., Rufus, R., Esterline, A.: Behavioral modeling intrusion detection system (BMIDS) using Internet of Things (IoT) behavior-based anomaly detection via immunity-inspired algorithms. In: 2016 25th International Conference on Computer Communication and Networks (ICCCN), pp. 1–6. IEEE (2016)
25. Azath, H., Devi Mani, M., Prasanna Venkatesan, G.K.D., Sivakumar, D., Ananth, J.P., Kamalraj, S.: Identification of IoT device from network traffic using artificial intelligence based capsule networks. Wirel. Pers. Commun.. Pers. Commun. **123**(3), 2227–2243 (2022)

26. Tedeschi, S., Emmanouilidis, C., Mehnen, J., Roy, R.: A design approach to IoT endpoint security for production machinery monitoring. Sensors **19**(10), 2355 (2019)
27. DeMedeiros, K., Hendawi, A., Alvarez, M.: A survey of AI-based anomaly detection in IoT and sensor networks. Sensors **23**(3), 1352 (2023)
28. Elfrink, S., Garcia, M.A., Zhang, X., Li, Z., Cape Girardeau, M.O., Han, U.Q.: Teaching case unearth IoT security vulnerabilities by reverse engineering: a face-recognition IoT application-based lab exercises. In: Proceedings of the EDSIG Conference ISSN, vol. 2473, p. 4901 (2022)
29. Moustafa, N., Choo, K.K.R., Abu-Mahfouz, A.M.: Guest editorial: AI-enabled threat intelligence and hunting microservices for distributed industrial IoT system. IEEE Trans. Ind. Inf. **18**(3), 1892–1895 (2021)
30. Dunsin, D., Ghanem, M.C., Ouazzane, K., Vassilev, V.: A comprehensive analysis of the role of artificial intelligence and machine learning in modern digital forensics and incident response. Forensic Sci. Int. Digit. Invest. **48**, 301675 (2024)
31. Iqbal, S., Khan, T.M., Naqvi, S.S., Naveed, A., Usman, M., Khan, H.A., Razzak, I.: LDMRes-Net: a lightweight neural network for efficient medical image segmentation on IoT and edge devices. IEEE J. Biomed. Health Inform. **28**, 3860 (2023)
32. Li, S., Iqbal, M., Saxena, N.: Future industry Internet of Things with zero-trust security. Inf. Syst. Front., 1–14 (2022)
33. Basha, P.H., Prathyusha, G., Rao, D.N., Gopikrishna, V., Peddi, P., Saritha, V.: AI-driven multi-factor authentication and dynamic trust management for securing massive machine type communication in 6G networks. Int. J. Intell. Syst. Appl. Eng. **12**(1s), 361–374 (2024)
34. Trilles, S., Hammad, S.S., Iskandaryan, D.: Anomaly detection based on Artificial Intelligence of Things: a systematic literature mapping. Internet Things **25**, 101063 (2024)
35. Nazir, A., et al.: Collaborative threat intelligence: enhancing IoT security through blockchain and machine learning integration. J. King Saud Univ. Comput. Inf. Sci. **36**, 101939 (2024)
36. Nikbakht Bideh, P., Gehrmann, C.: RoSym: robust symmetric key based IoT software upgrade over-the-air. In: Proceedings of the 4th Workshop on CPS & IoT Security and Privacy, pp. 35–46 (2022)

Author Index

R. Silhavy and P. Silhavy (Eds.): CSOC 2024, LNNS 1120, pp. 705–707, 2024.
https://doi.org/10.1007/978-3-031-70518-2

The manufacturer's authorised representative in the EU is Springer Nature Customer Service Centre GmbH, Europaplatz 3, 69115 Heidelberg, Germany. If you have any concerns regarding our products, please contact ProductSafety@springernature.com

Printed and bound by CPI Group (UK) Ltd, Croydon, CR0 4YY

28/09/2026

02224057-0011